ENVIRONMENTAL BIOLOGY FOR ENGINEERS AND SCIENTISTS

ENVIRONMENTAL BIOLOGY FOR ENGINEERS AND SCIENTISTS

DAVID A. VACCARI
Stevens Institute of Technology

PETER F. STROM
Rutgers, The State University of New Jersey

JAMES E. ALLEMAN
Iowa State University

WILEY-INTERSCIENCE

A JOHN WILEY & SONS, INC., PUBLICATION

Library of Congress Cataloging-in-Publication Data:

Vaccari, David A., 1953–
    Environmental biology for engineers and scientists / David A. Vaccari, Peter F. Strom, James E. Alleman.
        p. cm.
    Includes bibliographical references.
    ISBN-13 978-0-471-72239-7 (cloth : alk. paper)
    ISBN-10 0-471-72239-1 (cloth : alk. paper)
    1. Biology–Textbooks.   I. Strom, Peter F.   II. Alleman, James E.
III. Title.
QH308.2.V33 2005
570–dc22   2005008313

Printed in the United States of America

10 9 8 7 6 5 4 3 2 1

5294076

At the conclusion of this project, my feelings are well expressed by the last lines of *Huckleberry Finn*:

... and so there ain't nothing more to write about, and I am rotten glad of it, because if I'd 'a' knowed what a trouble it was to make a book I wouldn't 'a' tackled it, and I ain't a-going to no more.

Dedicated to Liana and Carlo

D.A.V.

During the course of working on this book, and despite all the hours wrapped into its effort, my family learned a lot about the vital essence of life and the gifts we've all been given to share and treasure.

Dedicated to Carol, Amy, Matthew, Alison, Paul, and Elizabeth

J.E.A.

Dedicated to the environment, and to all those, past, present, and future, who make it worth caring about, especially to Daryl, Russell, Sue, Jean, and Arthur.

P.F.S.

CONTENTS

10 Microbial Groups   217

13   Microbial Transformations   387

14   Ecology: The Global View of Life   442

PREFACE

This book was originally developed to introduce environmental engineers to biology. However, we have realized that it will also fullfill a need for environmental scientists who specialize in nonbiological areas, such as chemists, physicists, geologists, and environmental planners. Much of what we say here about engineers applies also to these other specialists as well.

Those people coming from a biological science background might be surprised to discover that most engineers and many chemists and physicists do not have a single biology course in their bachelor's degree programs. Even environmental engineering students often receive only a brief exposure to sanitary microbiology, with a vast range of biological issues and concerns being neglected almost completely. Environmental chemists may study aquatic chemistry with little knowledge of biological activity in the aquatic system, and meteorologists studying global warming may have only a rudimentary understanding of the ecosystems that both affect and are affected by climate. However, the growth of the environmental sciences has greatly expanded the scope of biological disciplines with which engineers and scientists need to deal. With the possible exceptions of biomedical and biochemical engineering, environmental engineering is the engineering discipline that has the closest connection with biology. Certainly, it is the only engineering discipline that connects with such a wide range of biological fields.

The need to make engineers literate in biological concepts and terminology resulted in the development of a new graduate-level course designed to familiarize them with the concepts and terminology of a broad range of relevant biological disciplines. The first one-third of the semester introduced basic topics, covering each of the general biology topics in the first 10 chapters of the book. A college-level general biology text was used for this portion of the course, but no single text provided adequate coverage of the range of topics presented in the other chapters. This is the focus of and motivation for this book, which covers a much wider range of biology than has traditionally been taught to environmental engineers and scientists. Our intent in doing so is to strike off in a new direction with the approach to be used for training environmental professional in the future.

Specialists in every field have learned not to expect their colleagues trained in other areas to have certain basic knowledge in their own areas. This book aims to break one of these barriers of overspecialization. The objectives of a course based on this book will have been met if an engineer, chemist, or geologist who studied it is meeting with a biologist to discuss a situation of environmental concern, and the biologist at some point turns and says: "How did you know that?" It should not be a surprise that any well-educated person possesses some specialized knowledge outside his or her own profession.

The information herein is not limited to what environmental engineers or scientists "need to know" to do their jobs. The nonbiologist may occasionally need to read technical material written by biologists and should not be confused by the use of terminology standard to such material. Engineers and scientists who may eventually move to management positions of diversified organizations should be especially concerned about this.

A secondary need that this subject meets is the necessity for any technically literate person to be familiar with biology. Exposing nonbiologists to this field broadens their knowledge of the living world around them and of their own bodies. The biologically literate engineer or scientist will better understand and cope with the impact of technologically driven changes in the world. This understanding should encompass not only environmental issues such as pollution effects, ecosystem destruction, and species extinction, but also issues bearing on agriculture and medicine. Rapid progress in genetic engineering and medical technology makes it more essential to have such an understanding because it forces many societal and individual choices.

No single book can completely cover all biological topics relevant to environmental engineers and scientists. By design, this book has more information than could be covered in a single semester. Students should leave a course with a sense that there is more to know. It also gives students and instructors the choice of which topics to explore in more detail.

The first nine chapters are intended for use as a study guide and a summary of information that otherwise would have been learned in a course in general biology. Thus, they could be skipped in a course for, say, environmental science students who have already taken general biology. The rest of the chapters contain information that is specific to environmental applications. In broad terms, the important areas are traditional sanitary microbiology (health and biological treatment), ecology, and toxicology.

To play to the strengths of engineers, mathematical techniques are emphasized, as this was the initial focus of the book. Examples include population dynamics, microbial growth kinetics (focusing on batch systems, and stopping with the chemostat, short of treatment process models), pharmacokinetic models of toxicity, ecosystem modeling, statistical approaches to epidemiology, and probabilistic modeling of bioassay data. Other specialists, including biologists, could benefit from this treatment, as biology is becoming more and more quantitative. Nevertheless, the mathematical discussions can be skipped if time does not permit their development.

Familiarity with basic environmental concepts is assumed, such as the sources and types of pollutants, an understanding of acid–base relationships, oxygen demand, and other basic chemistry concepts.

There is sufficient information in this book for a two-semester course. We recognize that many programs have only a single semester to devote to this topic. Therefore, we offer the following as an outline on which to base such a course. The balance of the book will then be supplementary and reference material that instructors may draw from based on their special interests. The instructor may also consider assigning a research paper to be based on a topic from the book not included in the course.

Topic	Chapters/Sections
1. **Introduction**: the study of biology; complexity; ethics; biological hierarchies, evolution, taxonomy, interactions in biology	Chaps. 1 and 2
2. **Biochemistry**: organic structure and physicochemistry, carbohydrates, proteins, lipids, nucleotides **The cell**: structure and function, membranes	Chaps. 3 and 4
3. **Metabolism**: enzyme kinetics, glycolysis, fermentation, respiration	Secs. 5.1–5.3, 5.4.1–5.4.3
4. **Genetics**: heredity, DNA replication, protein synthesis, mutations and DNA repair, polymerase chain reaction	Secs. 6.1.1, 6.2, 6.3
5. **Plant and animal taxonomy**: including the fungi **Human physiology**: respiratory system, endocrine system, excretory system	Secs. 7.1, 8.1, 8.4.2, 9.1, 9.8, 9.11, 9.12
6. **Microbes**: stoichiometry, metabolism, classification, pathogenesis	Chaps. 10, 12
7. **Microbial growth kinetics**	Chap. 11
8. **Biogeochemical cycles**: nitrogen cycle reactions, etc.	Chap. 13
9. **Ecology**: energy pyramid, food web, biogeochemical cycles, population growth, diversity	Chap. 14
10. **Ecosystems**: forest, soil, aquatic, wetlands, microbial	Secs. 15.1–15.3, 15.5
11. **Biological pollution control**: activated sludge, anaerobic digestion	Chap. 16
12. **Toxicology**: mechanisms, effects, carcinogens, organ effects	Chap. 17
13. **Fate and transport**: uptake, absorption, distribution, biotransformation, excretion	Secs. 18.1–18.6
14. **Dose–response**: extrapolation, toxicity testing	Secs. 19.1–19.4, 20.1
15. **Toxicity**: effects of specific substances	Secs. 21.1–21.4

ACKNOWLEDGMENTS

Support for the lost productivity that resulted from writing this book came from two main sources: Stevens Institute of Technology, and my wife, Tien-Nye H. Vaccari. Without the support of both, this book would not have been finished. Moreover, without the encouragement of B. J. Clark, it would not have been started. B.J. was the first person other than the authors to recognize the need that this book would fulfill. Appreciation is also due to my relatives and friends in and around Castelgomberto, Italy, where the book got a good start under the most agreeable of situations. I also appreciate the help I received from Dina Coleman, Patrick Porcaro, James Russell, Alison Sleath, Zhaoyan Wang, and Sarath Chandra Jagupilla.

D. A. VACCARI

1

PERSPECTIVES ON BIOLOGY

Before immersing ourselves in the subject of environmental biology, in this chapter we consider factors to motivate, facilitate, and provide a context for that study. Thus, we start by discussing reasons, history, mindsets, and ethics that can guide our approach to the subject.

1.1 WHY ENVIRONMENTAL ENGINEERS AND SCIENTISTS SHOULD STUDY BIOLOGY

For an environmental scientist, the answer to the question posed in the title of this section is fairly evident. However, for environmental engineers it is worthwhile to consider this in more detail. For example, environmental engineers need to know a broader range of science than does any other kind of engineer. Physics has always been at the core of engineering, and remains so for environmental engineers concerned with advective transport (flow) in the fluid phases of our world. The involvement of environmental engineers with chemistry has increased. Formerly, it was limited to chemical precipitation and acid–base chemistry in water, and relatively simple kinetics. Now it is necessary also to consider the thermodynamics and kinetics of interphase multimedia transport of organics, the complex chain reaction kinetics of atmospheric pollutants or of ozone in water, and the organic reaction sequences of pollutant degradation in groundwater. In a similar way, the role of biology in environmental engineering has burgeoned.

Traditionally, the biology taught to environmental engineers has emphasized microbiology, because of its links to human health through communicable diseases and due to our ability to exploit microorganisms for treatment of pollutants. Often, there is a simple exposure to ecology. However, the ecology that is taught is sometimes limited

Environmental Biology for Engineers and Scientists, by David A. Vaccari, Peter F. Strom, and James E. Alleman
Copyright © 2006 John Wiley & Sons, Inc.

to nutrient cycles, which themselves are dominated by microorganisms. As occurred with chemistry, other subspecialties within biology have now become important to environmental engineering. Broadly speaking, there are three main areas: microbiology, ecology, and toxicology. The roles of microbiology are related to health, to biological pollution control, and to the fate of pollutants in the environment. Ecological effects of human activity center on the extinction of species either locally or globally, or to disturbances in the distribution and role of organisms in an ecosystem. Toxicology concerns the direct effect of chemical and physical pollutants on organisms, especially on humans themselves.

This book aims to help students develop their appreciation for, and awareness of, the science of biology as a whole. Admittedly, applied microbiology is often included in many environmental engineering texts, focusing on disease transmission, biodegradation, and related metabolic aspects. However, little if any material is provided on the broader realm of biology in relation to environmental control. Such an approach notably overlooks a considerable number of important matters, including genetics, biochemistry, ecology, epidemiology, toxicology, and risk assessment. This book places this broad range of topics between two covers, which has not been done previously.

There are other factors that should motivate a study of biology in addition to the practical needs of environmental engineers and scientists. The first is the need to understand the living world around us and, most important, our own bodies, so that we can make choices that are healthy for ourselves and for the environment. Another is that we have much to learn from nature. Engineers sometimes find that their best techniques have been anticipated by nature. Examples include the streamlined design of fish and the countercurrent mass-transfer operation of the kidney. An examination of strategies employed by nature has led to the discovery of new techniques that can be exploited in systems having nothing to do with biology. For instance, the mathematical pattern-recognition method called the *artificial neural network* was inspired by an understanding of brain function. New process control methods are being developed by reverse-engineering biological systems. Furthermore, there may be much that engineers can bring to the study of biological systems. For example, the polymerase chain reaction technique that has so revolutionized genetic engineering was developed by a biologist who was starting to learn about computer programming. He borrowed the concept of iteration to produce two DNA molecules repeatedly from one molecule. Twenty iterations quickly turn one molecule into a million.

Engineers can also bring their strengths to the study of biology. Biology once emphasized a qualitative approach called **descriptive biology**. Today it is very much a quantitative science, using mathematical methods everywhere, from genetics to ecology. Finally, it is hoped that for engineers the study of biology will be a source of fascination, opening a new perspective on the world that will complement other knowledge gained in an engineering education.

1.2   PRESENT PERSPECTIVES ON ENVIRONMENTAL ENGINEERS AND SCIENTISTS

Science is defined as "knowledge coordinated, arranged, and systematized" (Thatcher, 1980). The *McGraw-Hill Dictionary of Science and Engineering* (Parker, 1984) defines **engineering** as "the science by which the properties of matter and the sources of power in nature are made useful to humans in structures, machines, and products."

Thus, engineering is defined as one of the sciences. Yet in the professional world, those who classify themselves as scientists and those who call themselves engineers seem to distinguish themselves from each other. To be sure, engineers study the sciences relevant to their disciplines, although not as deeply as scientists might prefer (the subject of this book being an instance of this). On the other hand, wouldn't scientists benefit from a better understanding of how to apply engineering analysis to their fields? For example, aren't mass and energy balances or transport phenomena useful for analyzing the multiplicity of phenomena affecting a laboratory experiment?

The following comparison of engineers' and scientists' approaches were obtained from the authors' observations, plus informal discussions with science and engineering practitioners. They reflect the perception of differences between engineers and scientists, not necessarily reality. They certainly do not apply to all. A common complaint to be heard from a scientist about an engineer is that the latter "wants to reduce everything to a number" and tends not to look at the system holistically. This may be due to the basic function for which an engineer is trained—to **design**: that is, to create an arrangement of matter and energy to attain a goal or specification, using the minimum amount of resources. Ultimately, this is reduced to such things as how big, how much, or for how long an arrangement must be made. An engineer designing an in situ groundwater bioremediation project uses models and design equations (and judgment) to determine flow rates, well locations, nutrient dosages, duration of treatment, and finally, the financial resources needed to eliminate a subsurface contaminant.

However, nonquantitative factors may be just as important to the success of a project. Are microorganisms present that are capable of utilizing the contaminant? In fact, is the contaminant biodegradable at all? Are the by-products of the biodegradation process more or less toxic than the starting material? Although such considerations are taught in engineering courses, a quick look at homework and exams shows that the emphasis is on the "single-valued outcome," the bottom-line answer.

Paradoxically, there is also a sense in which an engineer's approach is *more* holistic than a scientist's. To make a problem tractable, the scientist may simplify a system, such as by considering it as a *batch* or *closed system*: one in which material cannot cross the boundary. Alternatively, a scientist may restrict systems to be either constant temperature or adiabatic (insulated): a reaction in a beaker, microbial growth in a petri dish, or a laboratory ecosystem known as a **microcosm**. An engineer deals more often with flow or open systems. He or she will literally turn on the pump, adding and removing material and energy. The underlying process is the same from a scientist's point of view. However, the study of open systems is dealt with much more often in engineering courses than in science courses. This study enables scientific principles to be applied more directly to real problems in industrial or environmental situations.

An engineer may be more likely to draw conclusions inductively from previous cases to find solutions to a problem. On the other hand, a scientist prefers to avoid assumptions and to base decisions on case-specific information only. For example, in examining a contaminated landfill, an engineer expects it to be similar to others in his or her experience, until information to the contrary appears. A scientist may be more deductive and may be reluctant to proceed until a thorough study has been conducted. The former approach is more economical, except when unanticipated problems arise. The latter approach is more rigorous but may sometimes be impractical because of cost. One must decide between the risk of overlooking unexpected problems vs. that of experiencing "paralysis by analysis."

A scientist is looking for knowledge; an engineer is looking for solutions. Thus, when an engineer is done with his or her work, a problem has been solved. When a scientist is done, there may be more questions than before.

Engineers are trained to consider the cost-effectiveness of their actions. Economic feasibility may be farther down the list of a scientist's priorities. Some engineers are more willing to make environmental trade-offs; an environmental scientist prefers to "draw the line" against any environmental costs. If wetlands would be destroyed by a project, an engineer may weigh it against the value of the project. If the project is important enough, the value of the wetlands may be compensated for by mitigation or by replacement methods. A scientist may accept the necessity of this, but may still consider the loss of the original wetlands a tragedy.

The tasks of both scientists and engineers are to explain (analyze), predict (forecast), and prescribe (design). A scientist needs to know about all of the individual phenomena that may affect a system and how they are related. An engineer often depends on having a mathematical model to represent the phenomena. If such a model can be created, engineers can produce quantitative results. However, science comes into play again when the limitations of the model are considered. These include extreme conditions where assumptions of the model may be violated and effects of phenomena (and every system will have such phenomena) that are not amenable to mathematical modeling.

For example, suppose that an industry that uses well water from a shallow aquifer is having problems with iron in the water. After use, the water is contaminated with biodegradable organics and ammonia and is treated in a lagoon system before discharge to surface water. In considering the pollutional inputs to the system and an analysis of the groundwater, a scientist, might identify the problem as being due to a recycling effect where some of the water seeps from the lagoon into the soil, whereupon microbially mediated nitrification consumes some of the alkalinity in the water, lowering the pH, dissolves iron from the soil, and then is again taken up in the well. The scientist might then prescribe a long-term solution involving adding alkalinity to the water after use. An engineering approach might be to model the system using acid–base chemistry to predict the effect of the nitrification on the effluent, followed by groundwater flow and water quality modeling to predict the water quality at the well. The model could be used to determine the alkalinity dosage required, leading to a design for the treatment process.

This process started with a scientific analysis and continued with engineering methods. Now, suppose further that after implementation, the lagoon effluent being discharged to the surface water is found toxic to fish. Again, a scientific analysis is called for: a toxicity identification evaluation (TIE). It may be found that at the higher pH, more of the ammonia in the effluent is in non-ionized form, which is more toxic to fish than is the ammonium ion. No one will doubt that science and engineering cannot be divorced from each other, but the better that engineers and scientists understand each other's disciplines, the better the outcome will be.

The engineering approach can be fruitful for scientific inquiry. Consider the problem of explaining why multicellular life-forms evolved. A biologist may think first of survival, fitness, and adaptation. An engineering approach might focus on considerations of energy efficiency and the effect of area/volume ratio on mass and energy transfer between the organism and the environment. Both approaches are valid and contribute to understanding.

Engineers and scientists have much to learn from one another, which they do by working together. In addition, both will profit from a direct study of each other's disciplines.

There are many instances in this book of mathematical or engineering analysis applied to biology. These should help convince engineers that biology really is a "hard" science. By learning and applying knowledge of biology, engineers can help convince scientists that engineering isn't always simplification and abstraction—that they can take into account the full complexity of a system. At the same time, engineers should be humble about their capability. The following statement by LaGrega et al. (2001) with respect to toxicology should be true of other biological disciplines as well: "What is the single, most important thing for an engineer to know about toxicology? The engineer cannot and should not practice toxicology." Still, the more the nonspecialist understands, the better his or her decision making will be.

1.3 PAST PERSPECTIVES ON ENVIRONMENTAL ENGINEERS AND SCIENTISTS

Environmental biology is not a unified discipline. Almost every chapter in this book could constitute a different field with its own history. What we can describe here is a bit of history of the development of the associations that have formed between biology and environmental engineers.

Barely a century ago, environmental and sanitary issues were largely the domain of scientists rather than engineers. In fact, long before the practice of environmental engineering had ever been conceived, physicists, chemists, and biologists were already hard at work investigating a range of pollution problems that they recognized as a serious threat.

These sorts of concerns about procuring clean waters and discarding wastes safely had actually been documented and addressed for at least two millennia. However, by the time that Charles Dickens had written his classic, *A Tale of Two Cities*, the "worst of times" had truly befallen most industrialized countries. Humankind's waste emissions had finally overtaxed nature's assimilatory capacity, and the telltale signs were readily evident. Skies were blanketed with coal smoke and soot, rivers were befouled with a sickening blend of filth, and waste piles offered frightful opportunities for the dissemination of disease.

In the latter half of the nineteenth century, the heightened level of pollution brought by the Industrial Revolution triggered a response from scientists, whose contributions continue to this day. For example, one of England's leading chemists, Edward Frankland, routinely monitored water quality changes in the Thames River, and his microbiologist son, Percy, tried to resolve the bacterial reactions found in sewage. John Tyndall, a renowned physicist, focused a considerable amount of his talent on air quality and contaminant analysis problems. Charles Darwin's eminent friend and staunch supporter, Thomas Huxley, played a major role in the cause of sanitary reformation as a practical extension to his expertise in biology.

The roots of our modern practice of environmental engineering sprang largely from the sciences. These investigators were not motivated by regulatory requirement, legal threat, or financial gain. Instead, their fledgling efforts were effectively compelled by personal concerns about an environment whose quality had already deteriorated seriously. These scientists had been trained to appreciate the balance of nature and were duly concerned about the stress imposed by a rapidly escalating range of pollution problems. It was the field of biology, however, which truly gave these yesteryear environmental efforts their highest motivation. Biologists working at the end of the nineteenth century demonstrated successive refinements in their sense of awareness and appreciation for the technical importance of the environmental problems that they faced.

The connection between biology and environmental engineering is best demonstrated by tracing the academic lineage extending beyond Thomas Huxley's seminal work with sanitary health. Although apparently self-educated, Huxley was a preeminent leader in the emerging field of biology late in the nineteenth century and gained worldwide recognition for his forceful backing of Charles Darwin's evolutionary theories. Having been invited to present the opening speech for Johns Hopkins University's inaugural ceremony in 1894, he used the opportunity to recommend one of his students H. Newell Martin, as chairman of the first biology department in the United States. Under Martin's leadership, students were imbued with an inherent sense of "environmental" concern. For William Sedgwick, this issue became a lifelong cause. After receiving his doctorate, Sedgwick joined the biology program at Harvard, where he introduced a radical new focus on sanitary matters. Together with his own student, George Whipple, he then cofounded public health programs at Harvard and the Massachusetts Institute of Technology (MIT), which in the coming years would lead formatively to distinct academic offerings in public health, sanitation, and industrial hygiene. At much the same time, Sedgwick and Whipple created a technical program at MIT that dealt with the applied aspects of sanitary science, thereafter known as the first-ever environmental engineering program.

1.4   AMBIGUITY AND COMPLEXITY IN BIOLOGY

Often in fields such as physics or engineering it is possible to identify all the variables that affect a process (in terms of a model). In fact, the number of variables can be reduced to a minimum by the use of dimensional analysis, producing a set of dimensionless numbers that describe a situation completely. In biology, it is more common for there to be unknown influences and obvious gaps in our knowledge. Consequently, many biological "facts" are conditional—answers often have to be prefaced with "it depends…" or important questions may be totally lacking an answer. Despite the recent explosion of knowledge in chemical genetics, we still cannot give a satisfactory explanation of how an embryonic cell "knows" to grow into part of a fingertip and not a hair follicle. We need to have a certain humility in our studies:

> In school we start each course at the beginning of a long book full of things that are known but that *we* do not yet know. We understand that beyond that book lies another book and that beyond that course lies another course. The frontier of knowledge, where it finally borders on the unknown, seems far away and irrelevant, separated from us by an apparently endless expanse of the known. We do not see that we may be proceeding down a narrow path of knowledge and that if we look slightly left or right we will be staring directly at the unknown. (Gomory, 1995)

Few appreciate the difficulty of making ironclad distinctions in the study of biology. Actually, the same is true for all fields, although it is perhaps more apparent in biology. In engineering, the mathematical constructions create an abstract ideal out of concepts. When we speak of the velocity of water in a pipe, the idea seems very clear and can be manipulated fairly unambiguously, such as to compute flow or pressure drop. However, if the turbulent nature of the flow is examined in detail, we find that we need to use probability distributions to describe the velocity, and even that is incomplete. We use the simplified abstraction so often, and with such success, that we often forget the underlying complexity.

In biology, the uncertainties behind the abstractions are often closer to the surface. For example, consider the concept of *species*. We would like to define species such that all living things belong unambiguously to a species. This turns out to be impossible. One definition is that a **species** is a group of organisms that interbreed with each other in nature and produce healthy and fertile offspring. Thus, the horse and the donkey are different species even though they can mate, because all their offspring are infertile. A number of problems occur with this definition. First, not all organisms reproduce by breeding (sexual reproduction). More important for this discussion is that there are populations we would like to define into separate species that can interbreed, such as the domestic dog and the African golden jackal. So perhaps we would alter the definition to include organisms that can *potentially* interbreed. However, there are cases in nature where organism A can breed with B, B with C, C with D, but D cannot breed with A.

Similar problems occur every time we make a classification. The euglena is a single-celled organism that can move at will through its aqueous environment. This motility, together with its lack of a cell wall, would indicate that it should be classified as an animal. However, it has the green pigment chlorophyll, which enables it, like a plant, to capture light energy. Biologists have created a separate category, the protists, in part to eliminate the problem of where to put the euglena. However, some protists, the algae, are very similar to plants; others, such as protozoans, are animal-like. Thus, the classifications lack an iron-clad quality. Textbook definitions sometimes include the word *mostly*, as in "animals are mostly multicellular."

Whether an organism is single-celled or multicellular is an important characteristic used in classification. However, some may either be at different stages in their life cycle or may simply change in response to environmental conditions. The slime mold is an unusual organism that grows on the forest floor and behaves at one stage as a mass of single-celled protozoans; at another stage the cells fuse into a single supercell with many nuclei; and at yet another stage it forms fruiting bodies on stalks resembling a plant.

Another idea that must be recognized as somewhat arbitrary is the notion of an *event*. This is another kind of useful fiction. In classical science (i.e., other than in quantum mechanics), there are no events, only processes. Consider the "moment of conception," when a sperm enters and fertilizes an egg. Examined more closely, we must realize that the event consists of a sequence of changes. If we say that fertilization occurs as soon as a sperm penetrates an egg's cell membrane, we must ask, "penetrates how far?" If, instead, we place the event at the moment that the chromosomes join into a single nucleus, we must ask how complete the joining must be. It is like asking when two asymptotic curves combine. The problem is not that science cannot say when the critical moment occurs. No such moment actually exists.

Another thing that is important to appreciate about biology is that a certain amount of caution is necessary when making predictions or judgments about the validity of reported observations. Living things often surprise and contradict. The following quote by the Dutch biologist C. J. Brejèr (1958) applies as well to biology as a whole: "The insect world is nature's most astonishing phenomenon. Nothing is impossible to it; the most improbable things commonly occur there. One who penetrates deeply into its mysteries is continually breathless with wonder. He knows that anything can happen, and that the completely impossible often does."

There is no perfect way around these difficulties; the complexity of nature sometimes resists our attempts to coordinate, arrange, and systematize it. All life-forms are unique, defying easy classification; so we create a working definition and go on. We must use the distinctions when they are useful and alter them when they're not.

Biology is also unique in the number of levels of scale that it is necessary to examine in its understanding. In Section 2.2 we describe how living systems can be examined at many levels of detail, from the chemical to the cell to the organism to the ecosystem. Within each level are numerous types of entities (e.g., cells or organisms) and myriad instances of each type. The number of potential interactions is staggering. Sometimes the reductionist approach is appropriate and an individual entity or interaction will be studied almost in isolation. Other times it is necessary to look holistically at the behavior of the group.

It is often the goal of scientific studies to "explain" behaviors observed at one level by looking at behaviors of its component parts. For example, the primary productivity (production of algal biomass) of a eutrophic lake can be predicted by measuring the productivity of individual species cultured in a lab under similar conditions. However, it is often the case that the aggregate behavior of numerous individuals cannot be predicted straightforwardly, even if the behavior of the individuals were well understood. For example, proteins are polymers of 20 different amino acids, in varying sequences. Although the individual amino acids do not function as catalysts, proteins do. A protein is not simply the "sum of its parts." "New" properties that arise from the interaction of numerous similar parts are called **emergent properties**. A mathematical field of study called **complexity theory** has arisen to study the relation between large numbers of interacting autonomous parts and resulting emergent properties.

A related source of complexity is chaotic behavior. **Chaos** is dynamic behavior characterized by "extreme sensitivity to initial conditions." Consider the sequence of real numbers, $(x_1, x_2, \ldots)$ between 0.0 and 1.0 generated by what is known as the **quadratic iterator**:

$$x_{i+1} = 4x_i(x_i - 1) \tag{1.1}$$

Figure 1.1 shows plots of two such series that start at nearly the same point, 0.9000 and 0.9001, plus a plot of the difference between the two series. Note that after a dozen iterations the two series become uncorrelated. The difference between the two series seems almost random. This demonstrates that although we know the rule generating the data, without perfect knowledge of the initial condition, we cannot predict very far into the future. This extreme sensitivity to initial conditions is also called the **butterfly effect**, after the analogy for chaos in weather systems which states that a butterfly flapping its wings in Beijing can cause a hurricane in New York City two weeks later.

A consequence of chaotic behavior is that extremely complex behaviors can result from very simple rules, as in the example just given. The paleontologist Steven Jay Gould proposed that small mutations can greatly alter body plans, producing great leaps in evolution. The benefit of this for the study of biology is that complex processes do not rule out the possibility of simple explanations. The difficulty is that it places a limit on the reductionist view. Having a high degree of understanding of the dynamics of nerve cells does little to explain how the human brain can so quickly recognize a face or decide on a chess move.

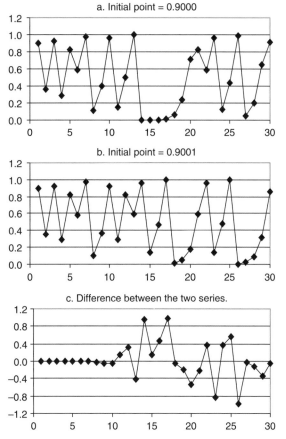

Figure 1.1 Quadratic iterator, illustrating chaotic behavior and extreme sensitivity to initial conditions.

1.5 CONSERVATION AND ENVIRONMENTAL ETHICS

The study of biology helps create an increased appreciation for the natural world. It causes us to question our collective and individual actions that may harm it. However, we seem to be constrained in our concern by our own need for survival. What is needed is an ethic to answer questions such as: How should we value the natural world? What is our place in it? What are our responsibilities toward it?

Why do we pollute, in effect spoiling our own nests? The problem is in a paradox of individual freedom that is called the **tragedy of the commons** (Hardin, 1968):

> The tragedy of the commons develops in this way. Picture a pasture open to all. It is to be expected that each herdsman will try to keep as many cattle as possible on the commons. Such an arrangement may work reasonably satisfactorily for centuries because tribal wars, poaching, and disease keep the numbers of both man and beast well below the carrying capacity of the land. Finally, however, comes the day of reckoning, that is, the day when the long-desired goal of social stability becomes a reality. At this point, the inherent logic of the commons remorselessly generates tragedy.

As a rational being, each herdsman seeks to maximize his gain. Explicitly or implicitly, more or less consciously, he asks, "What is the utility to me of adding one more animal to my herd?" This utility has one negative and one positive component.

1. The positive component is a function of the increment of one animal. Since the herdsman receives all the proceeds from the sale of the additional animal, the positive utility is nearly +1.
2. The negative component is a function of the additional overgrazing created by one more animal. Since, however, the effects of overgrazing are shared by all the herdsmen, the negative utility for any particular decisionmaking herdsman is only a fraction of −1.

Adding together the component partial utilities, the rational herdsman concludes that the only sensible course for him to pursue is to add another animal to his herd. And another. . . . But this is the conclusion reached by each and every rational herdsman sharing a commons. Therein is the tragedy. Each man is locked into a system that compels him to increase his herd without limit—in a world that is limited. Ruin is the destination toward which all men rush, each pursuing his own best interest in a society that believes in the freedom of the commons. Freedom in a commons brings ruin to all.

Morality is about when we should act against our own immediate self-interest. **Ethics** is the study of systems for deciding the morality, or rightness or wrongness, of our actions. **Environmental ethics** is the study of the morality of actions that affect the environment. An understanding of environmental ethics will help us to apply the knowledge and power we gain from our education.

It is a popular saying that one cannot teach someone to be ethical. This is not true. Public educational campaigns are often used with good effect to encourage "right" behavior, such as to institute recycling programs. Furthermore, an understanding of ethical principles can help counter specious or just plain wrong arguments that people sometimes use in support of self-serving actions. Finally, an understanding of ethics can help guide and reinforce those who want to do the right thing.

A **moral principle** is a rule or set of rules used to decide moral questions. Several moral principles have been proposed that are specific to our effect on the environment. A moral principle will not be an infallible guide to behavior. We will still be faced with dilemmas in which each alternative has its own moral cost.

What is the basis of moral principles? James Q. Wilson argues that we have inborn "moral senses," including the senses of sympathy, fairness, self-control, and duty. These senses then propel the development of moral principles. The Harvard biologist Edward O. Wilson postulates that concern for the environment stems from an innate affinity that people have for all living things, a principle he calls the **biophilia hypothesis**. This affinity becomes most apparent when contact with living things is limited. Antarctic researchers who are isolated over the winter must ration time spent in the plant growth chamber. The station's doctor prescribes such time to treat depression. NASA has conducted experiments in which crews are isolated in closed environmental systems for as long as 90 days to simulate space missions. The crew in one such experiment found that one of their greatest pleasures was growing their small lettuce crop. Planting and harvesting decisions were made only after considerable group discussion, and crew members often enjoyed opening the growth chamber just to look at the plants.

Two basic types of ethics are utilitarian ethics and rights-based ethics. **Utilitarianism** is the principle that rules or acts are moral if they produce the greatest amount of good for all concerned. A problem for utilitarianism is that some may suffer unfairly for the greater good of the majority. **Rights-based ethics** get around this problem by postulating moral rights that are universal (possessed by all), equal (no one has the right in any greater or lesser degree than another), inalienable (cannot be given up or taken away), and natural (not created by human acts, as are legal rights). A major problem with rights-based ethics is that different rights may conflict, and criteria need to be selected for choosing among them.

Both of the aforementioned types of ethics are focused on needs of individuals and thus are **humanistic**. Some propose a holistic ethic that places value on systems rather than individuals. This approach is used to raise to the level of moral principles ideas such as the diversity and integrity of ecosystems or the sustainability of economic systems.

The Judeo-Christian tradition forms the basis of much of Western thought. Various interpretations have been applied to its view of the relationship between humans and the rest of the natural world. A negative view has been blamed on the biblical injunction to "have dominion over the fish of the sea, and over the fowl of the air, and over the cattle, and over all the earth, and over every creeping thing that creepeth upon the earth. . . . Be fruitful, and multiply, and replenish the earth, and subdue it: and have dominion over the fish of the sea, and over the fowl of the air, and over every living thing that moveth upon the earth" Genesis 1:26–28.

However, other passages imply that all of creation has value. Genesis 1:31 states that "God saw every thing that he had made, and, behold, it was very good." The animals are also commanded to be fruitful and multiply. This leads to the **stewardship** concept, which states that humans have responsibility for the protection of creation. In any case, the Western tradition developed in which humans were viewed as the center and pinnacle of creation, a notion that is called **anthropocentrism**.

Some Asian religions, such as Zen Buddhism, Taoism, and Hinduism (especially the Buddhists), teach unity with nature, including compassion toward other humans as well as animals. For native Americans, unity means an interdependence and kinship between all animals, including humans, and natural systems. They believe that all animals have spirits that deserve respect. Animals can only be killed out of necessity, after which humans have to make apologies and atonement to the spirit of the killed animal. Many tribes also link their identity to prominent landscape features.

The Darwinian revolution dethroned humans from their special position in creation. Instead, they are part of a continuum with the animals, plants, and ultimately with the nonliving chemical world. The other parts of the living and nonliving world are seen as kin, which gives us an incentive to make our ethic include that which is good for them.

The wildlife biologist and amateur philosopher Aldo Leopold (1949) proposed such an ethical system in his book *A Sand County Almanac*. He calls it the **land ethic**. This book is considered the "gospel" of the conservation movement, much as Rachel Carson's (1962) *Silent Spring* sounded the alarm that stimulated the environmental movement. The idea of a land ethic is developed further by Callicott (1986).

Leopold describes how ethics developed from responsibilities toward other people. Eventually, these responsibilities were expanded to include the family and the clan, and then larger and larger groups, ultimately encompassing all of society. As the boundaries of the community expanded, the inner ones remained. These can be viewed as a concentric hierarchy of responsibilities to larger and larger communities (Figure 1.2): from self to nuclear family, extended family, clan, nation, and all of humanity. A person's responsibility toward the outer rings does not cancel the inner ones, but rather is layered

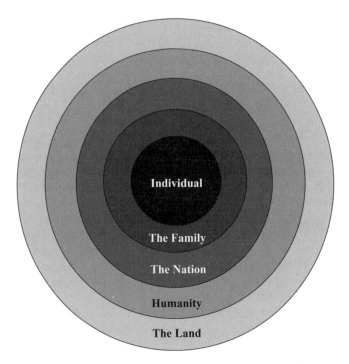

Figure 1.2   Hierarchy of responsibility in the land ethic.

over them. Leopold's contribution was to describe this hierarchy and to extend the idea of community to include the land. By the land he meant not just the soil but the pyramid of energy starting with the soil, to the plants it supports, the herbivores that live on the plants, the predators that depend on all the levels below it, and the organisms of decay that return the nutrients to the soil.

Furthermore, the particular responsibilities due each layer in the hierarchy are different. One does not owe all of humanity the same share of attention and resources that is owed to family members. So what is our responsibility to the land? According to Leopold, human use of the land should preserve and enhance the diversity, integrity, stability, and beauty of the biotic community. This does not prohibit our use of natural resources but requires that we do so in a sustainable way with a minimum of waste.

This leads us to one of the great moral challenges before our society. At the same time that the world's population continues to increase, the per capita use of natural resources is also increasing, and wilderness habitats are continually giving way to human uses. Optimists may expect that continuing technological developments will enable growth to continue. However, eventually we will have to achieve a steady-state condition in which resources are consumed at essentially the same rate at which they are regenerated. Humans have the capacity to plan, so society has the potential to bring about such a steady state without having it imposed by catastrophe. How will such a "sustainable society" look? What will be its population, how much will each person have to eat, and how much will remain of the wilderness and biodiversity that we have today? Although a sustainable society may take generations to become reality, it will never happen unless earlier generations begin to bring it about.

A final note of particular relevance to an engineer or scientist is **professional ethics**, often stated in a **code of ethics**. An example is the code of ethics adopted by the American Society of Civil Engineers, which states their responsibilities to the environment:

- Engineers uphold and advance the integrity, honor, and dignity of the engineering profession by using their knowledge and skill for the enhancement of human welfare.
- Engineers shall hold paramount the safety, health, and welfare of the public in the performance of their professional duties.
- Engineers should be committed to improving the environment to enhance the quality of life.

1.6   GUIDELINES FOR STUDY

In many of the fields of science the task of "coordinating, arranging, and systematizing" the knowledge can be difficult precisely because mathematics cannot easily be applied. Although mathematics has the reputation of being a difficult subject, it is tremendously efficient at compressing information. If a picture is worth a thousand words, an equation is worth a thousand pictures. Consider, for example, the effect of temperature, volume, and composition on the total pressure of a mixture of three solvents in equilibrium with its vapor, or the shape of the cone of depression around a well as it depends on soil permeability and water flow. We cannot easily visualize high-dimensional relationships, so they must be represented by a family of curves. Alternatively, the information could be contained concisely in a single mathematical representation. However, biological systems often are not described so compactly.

This leads us to discuss how to approach the study of biology. The method usually first thought of is to apply memorization. This is laborious and unproductive. It is extremely difficult to keep numerous unconnected facts in your head for any amount of time. The key, then, is to establish connections, to seek out relationships. This recovers some of the efficiency of the mathematical equation. We can think graphically about concepts, mentally plotting developmental trends, sequences, patterns, and networks of relationships.

Make studying an active process by keeping notes: Outline the reading and write lists, even lists of lists, under a unified topic. See if you can create an explanation of the concepts in each section of this book in your own words. If you have trouble with this at some point, you should formulate the difficulty as a question and seek the answer in the references or from your instructor. To learn a complex relationship from a figure such as the oxidation of glucose in Figure 5.5 or the nitrogen cycle in Figure 14.7, try copying it (possibly with less detail), and then try reproducing it again with the book closed.

Create what are known as **concept maps**. These are graphical representations of the relationships among information. For example, Figure 1.3 shows a concept map for science and engineering science topics in environmental engineering education, emphasizing the place of biology and leading to the twin roles of environmental engineers: design and prediction. To create a concept map, start with a list of related concepts. Then, state an **organizing principle**, which is a concept or idea used to arrange and connect items appropriately. The organizing principle behind Figure 1.3 is that connections lead from one topic to others that require their application. Another arrangement of

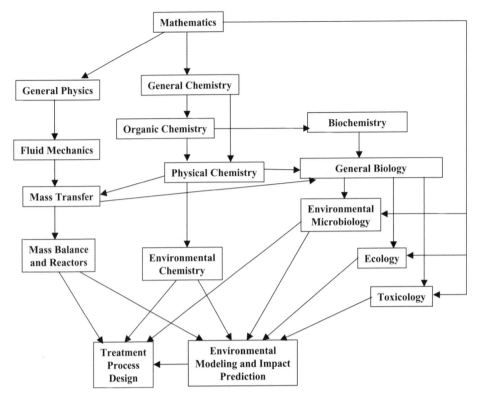

Figure 1.3 Concept map.

the items could have been created using a different organizing principle, for example as the sequence in which the subjects developed historically.

Such data structures organize and compress information, creating a mentally retrievable path to the information when called for. There is no right or wrong way to create structures. As long as it makes sense to you, it's right. The act of creating them results in learning. In addition, this technique is an excellent way to communicate any kind of technical data to others.

Besides the organizing techniques just described, the major activity for the study of biology is simply to read. Read beyond the required material. Seek out the references—these expand and reinforce the basic concepts. Multiple sources of information on the same topics can provide alternative viewpoints that highlight and clarify the key information.

PROBLEMS

1.1. Create a concept map using the chapters of this book as a starting list of items to be related. What organizing principle are you using?

1.2. How would responsibility to your employer or to your consulting client fit into Aldo Leopold's hierarchy? Where do government regulations come in?

REFERENCES

Briejèr, C. J., 1958. The growing resistance of insects to insecticides, *Atlantic Naturalist*, Vol. 13, No. 3, pp. 149–155.

Callicott, J. B., 1986. The search for an environmental ethic, in *Matters of Life and Death*, T. Regan (Ed.), Random House, New York.

Carson, Rachel, 1962. *Silent Spring*, Houghton Mifflin, Boston.

Gleick, J., 1987. *Chaos: Making a New Science*, Viking Penguin, New York.

Gomory, R. E., 1955. The known, the unknown and the unknowable, *Scientific American*, June.

Gunn, A. S., and P. A. Vesilind, 1988. *Environmental Ethics for Engineers*, Lewis Publishers, Chelses, MI.

Hardin, Garrett, 1968. The tragedy of the commons, *Science*, Vol. 162, pp. 1243–1248. See also http://dieoff.org/page95.htm.

Kellert, S. R., 1997. *Kinship to Mastery: Biophilia in Human Evolution and Development*, Island Press, Washington, DC.

LaGrega, M. D., P. L. Buckingham, and J. C. Evans, 2001. *Hazardous Waste Management*, McGraw-Hill, New York.

Leopold, Aldo, 1949. *A Sand County Almanac*, Oxford University Press, New York.

Martin, M. W., and Roland Schinzinger, 2005. *Ethics in Engineering*, McGraw-Hill, New York.

Parker, S. P. (Ed.), 1984. *McGraw-Hill Dictionary of Science and Engineering*, McGraw-Hill, New York.

Thatcher, V. S. (Ed.), 1980. *The New Webster Encyclopedic Dictionary of the English Language*, Consolidated Book Publishers, Chicago.

Waldrop, M. M., 1993. *Complexity: The Emerging Science at the Edge of Order and Chaos*, Touchstone Books, New York.

Wilson, J. Q., 1993. *The Moral Sense*, Free Press, New York.

2

BIOLOGY AS A WHOLE

In this chapter we provide an overview of the broad topic of biology. Much of biological science, as well as the chapters of this book, is *reductionist*. That is, systems are studied by focusing on their component parts. However, in this chapter we examine biology *holistically*, stepping back and looking at concepts that connect all of its disparate components.

The most basic concept, which forms the foundation of the entire structure of biology, is the definition of life. A "vertical" view of the structure generates a hierarchy of scale from biochemical reactions to the ecosystem. A "horizontal" view displays the classifications of living things, the pigeonholes that biologists use to group organisms into species, families, and so on. Connecting the parts of the structure is the theory of evolution, the grand organizing principle that gives the structure resilience, explaining why things are as they are. Finally, stepping back a bit, one can examine the two-way interactions between the structure and its surroundings, the environment.

2.1 WHAT IS LIFE?

First, we must define our subject. *Biology* is the study of living things. What, then, is the definition of *life*? For the most part, none of us would have trouble distinguishing living from nonliving things. Therefore, we will make our definition to include everything we consider to be alive and to exclude the nonliving things. This must be done carefully because we can sometimes find counterexamples to particular rules. Doesn't a crystal grow and possess order? Have you ever seen a polymer solution climbing out of a glass (motility)? Aren't there computer programs that reproduce and evolve? What about virus particles? Viruses are crystalline particles that must take over the cell of an

Environmental Biology for Engineers and Scientists, by David A. Vaccari, Peter F. Strom, and James E. Alleman
Copyright © 2006 John Wiley & Sons, Inc.

ordinary organism in order to reproduce. Should we make our definition include them? Even if we do, it turns out that there are infectious agents, viroids and prions, which are even simpler in structure, possibly even single molecules.

It should be obvious that this is one of those somewhat arbitrary questions that often arise in biology. Recognizing this, and also that there may be alternative definitions, let's go ahead and make our definition as best we can:

A living thing is a discrete entity, or **organism**, that has the following characteristics:

- It can **capture useful energy** from its surroundings.
- It can **extract materials** from its surroundings for growth and maintenance of its structure.
- It is **responsive**—it reacts to its surroundings, using energy and material to move or grow.
- It **reproduces**—it generates other organisms like itself, preserving its own characteristics by passing them on to succeeding generations.
- It **adapts** and **evolves**—successive generations can change as the environment changes; individuals exhibit **adaptations**, that is, they are specialized for a particular function in a particular environment.

Other characteristics sometimes mentioned are that life is **orderly** and **complex**; that it is capable of **motility**, or movement, which is a type of response; that it exhibits **development**, in which its complexity increases progressively; and that it has **heredity**, in which its characteristics are passed to its offspring as it reproduces.

So, is a virus alive? It is capable of evolution and reproduction, but it clearly fails in the first two of the criteria above. Strictly adhering to the criteria, the answer would be no. However, this might not be satisfying to some, who feel intuitively that they should be included. It may be that the best answer would be just to say that "viruses have several of the characteristics of living things" and leave it at that. To press the question further may be meaningless. Certainly, the topic of viruses belongs in a book on biology. On the other hand, the line may be drawn to definitively exclude viroids and prions. These may be considered to be chemical substances with infectious behavior.

Textbooks often do not mention at this point in the discussion another criterion for life that is commonly used by biologists:

- All living things are composed of cells.

A **cell** is the smallest entity completely surrounded by a **membrane** and containing a **nuclear region**, which stores the hereditary information used for growth and reproduction, and **cytoplasm**, which is a gel-like substance between the membrane and the nuclear region. With this additional criterion, viruses are clearly excluded, since they lack a membrane and cytoplasm.

Life may also be described in thermodynamic terms. The molecules in any isolated system (a system whose boundaries are sealed to both matter and energy) tend to rearrange themselves by random processes. For such systems of even moderate complexity, of all the possible random arrangements the "orderly" ones are extremely rare. Thus, such systems tend to rearrange into "disordered" systems. This behavior is described by the *second law of thermodymics*, which states that **entropy**, a measure of disorder, tends to increase *in isolated systems*.

But Earth is not an isolated system. It receives energy in the form of visible and ultra-violet light and emits the same amount of energy to space in the lower-grade form of infrared radiation. As this energy passes through the living world, part of it is used to do the work of countering the increase in entropy on Earth. In fact, living things use energy to cause a local decrease in entropy. The infrared emission to space results in an even greater increase in entropy outside the Earth, so that the total entropy of the universe still is found to increase.

Erwin Schrödinger (1956), developer of the quantum wave equation, described these ideas in a book entitled "What Is Life?":

> Thus a living organism continually increases its entropy—or, as you may say, produces positive entropy—and thus tends to approach the dangerous state of maximum entropy, which is death. It can only keep aloof from it, i.e. alive, by continually drawing from its environment negative entropy—which is something very positive as we shall immediately see. What an organism feeds on is negative entropy. Or, to put it less paradoxically, the essential thing in metabolism is that the organism succeeds in freeing itself from all the entropy it cannot help producing while alive.

Later, he continues:

> Hence the awkward expression "negative entropy" can be replaced by a better one: entropy, taken with the negative sign, is itself a measure of order. Thus the device by which an organism maintains itself stationary at a fairly high level of orderliness (= fairly low level of entropy) really consists in continually sucking orderliness from its environment. . ... After utilizing it they return it in a very much degraded form—not entirely degraded, however, for plants can still make use of it. (These, of course, have their most powerful supply of "negative entropy" in the sunlight.)

This "negative entropy" can be identified with the thermodynamic concept of Gibbs free energy, described in Section 5.1.1.

2.2 THE HIERARCHY OF LIFE

One way to examine living things is to look at them with different degrees of magnification, as it were, from the atomic level to cells, to organisms, to standing back and looking at Earth as a whole. The following lists the levels of detail in this hierarchy:

Metabolism   All of the chemical reactions within an organism that sustain life.

Organelle   Subcellular structures within cells that carry out specialized functions.

Cell   The basic unit of life, a structure completely surrounded by a membrane, containing a nuclear region and cytoplasm; the smallest structure capable of having all of the characteristics of life.

Tissue   A group of similar cells having the same function in a multicellular organism.

Organ   A single structure of two or more tissues that performs one or more functions in an organism.

Organ system   A group of organs that carry out related functions.

Organism	An individual entity that has all of the characteristics of life described above and possessing the same hereditary information in all its cells.
Population	A group of individuals of the same species living in the same environment and actively interbreeding.
Community	A group of interacting populations occupying the same environment.
Ecosystem	The combination of a community and its environment.

To understand the last several items, it is necessary in addition to define **environment**: the physical and chemical surroundings in which individuals live. Strictly speaking, the environment does not have boundaries. In practice, of course, it is common to restrict the discussion to a particular region, such as a single forest, a particular glen within that forest, or just the soil in a glen in a forest.

The first half of this book is organized roughly along the lines of the hierarchy. A more detailed description of the items in the hierarchy follows.

Metabolism   This is the view of life at the chemical level. The study of chemical reactions in organisms is called **biochemistry**. The reactions can be classified into (1) energy conversion reactions, (2) synthesis and breakdown of cell material, and (3) reactions associated with specific functions such as reproduction or motility.

The energy conversion reactions include those that capture energy from sunlight (photosynthesis) and convert it into food or that convert food energy into a simple form that is available for other reactions (respiration, fermentation). The synthesis reactions include production of the components of the major constituents of cells, such as proteins, carbohydrates, and lipids (which include fats), or minor ones (in terms of mass), such as nucleic acids, hormones, and vitamins. Reactions that break molecules down into simpler forms may facilitate the removal of waste products.

All of the metabolic reactions are connected to each other by a web of **metabolic pathways**, which consists of sequences of biochemical reactions. A particular compound can often participate in several reactions, thus becoming a link between those pathways.

Organelles   Some of the specialized functions carried out by these substructures include protein synthesis, photosynthesis (the conversion of light energy to chemical energy), and respiration (the release of chemical energy into a form usable by cells). The organelles that perform these functions are the **ribosomes**, the **chloroplasts**, and the **mitochondria**, respectively. There are many other organelles, especially in higher organisms. Not all cells have all types of organelles.

Cells   All living things are made of cells. Some are composed of a single cell and are called **unicellular**. Organisms made of more than one cell are called **multicellular**. A key structure of all cells is the **plasma membrane**. This separates the cell contents from its environment and forms the boundary of many of the cell's organelles. There are two primary types of cells. The **prokaryote** is the simpler of the two. It has no internal membrane structures, and the only organelle is the ribosome. Bacteria are typical prokaryotes. The other type is the **eukaryote**, which has internal membrane-bound structures such as mitochondria and chloroplasts. All the plants and animals are eukaryotic organisms, as are some single-celled organisms such as protozoans and algae.

Tissues   These are present only in multicellular organisms, of course. Examples include muscle tissue, bone tissue, or nerve tissue. Tissues should not be confused with organs. Although they may be composed predominantly of one type of tissue, organs may also be formed from a variety of tissues.

Organs   This is a complete unit such as an entire muscle, a bone, or the brain. Many organs serve a single function. Others organs have multiple functions, such as the pancreas and the adrenal gland. Both of these glands produce more than one hormone, which have unrelated functions.

Organ Systems   Some organs work in concert with each other and form a system. Examples are the nervous system, the circulatory system, the respiratory system, the skeletal system, and the endocrine system.

Organism   An organism is a living thing with a single **genome** (the set of hereditary information contained in all its cells' nuclear regions), which exists, for at least part of its life cycle, in the environment and separately from others of its kind. Some organisms, such as the algae and the fungi in lichens, must live together, but they have different genomes. Others live in colonies but exist separately at an earlier stage of their lives. An example of the latter is the jellyfish, which is really a colony of organisms that started out as free-swimming larvae. The individual cells of a jellyfish may come from different parents, yet it is difficult not to think of a jellyfish as a single organism. This is another example of the difficulty in making distinctions in biology.

Population   Examples of populations include all of the striped bass that breed in the Hudson River (although they spend most of their lives in the Atlantic Ocean), and the *Nitrosomonas europaea* bacteria in a biological wastewater treatment plant.

Sometimes the definition of population is extended to any group of organisms sharing a characteristic, such as the population of invertebrates (animals lacking an internal skeleton) in the soil, or the population of deer ticks that carry the Lyme disease bacteria. Thus, in this usage, a population may either incorporate more than one species or may be a subset of a species. Although such usage is common, it should be avoided. It would be more proper to label multiple-species groups as *populations* (note the plural) and to call the subset of a species a *subpopulation*.

Community   Strictly speaking, a community comprises all of the organisms in a region. Thus, an earthworm may be considered a member of a number of communities, depending on the region defined: It may be a member of the soil community, the forest community, or the community of organisms on a particular island in a lake.

The word *community* is sometimes used in a looser sense, similar to the second usage of the word *population* described above. For example, some may speak of the community of soil bacteria when they are concerned about the interactions among those members, even though there are other species in the same environment. Again, it would be more precise to use the term *populations*.

Ecosystem   The ecosystem is the most inclusive unit studied in biology. Again, the term may be restricted to local systems, such as a particular lake, or an island and nearby waters. Strictly speaking, since all of these systems are interconnected, there is really only one known ecosystem—the Earth's. This all-inclusive ecosystem is also known as the **biosphere**.

2.3   EVOLUTION

Evolution is the process of genetic change by which species adapt to their environment or develop new ways of coping with environmental stress. The theory of evolution is the most fundamental organizing principle in biology. It can be invoked to explain the origin of almost any feature of living things. By the early nineteenth century enough geological and fossil evidence had accumulated to challenge the prevailing idea that Earth and the life on it were unchanging. Jean-Baptiste Lamarck then proposed the ultimately rejected theory that species evolved by passing on new characteristics that were acquired by changes during their lifetime. For example, it was suggested incorrectly that early giraffes stretched their necks by reaching for leaves from high branches and that their offspring retained this change.

Two English naturalists then developed a better explanation for evolution. From 1831 to 1836, Charles Darwin served on the round-the-world voyage of HMS *Beagle*. His observations in South America, especially in the Galápagos Islands off Ecuador, led him to form the modern theory of evolution. Darwin wrote a summary of the theory in 1842 but did not publish it immediately. Independently, and several years later, Alfred Russell Wallace developed the same ideas, even using the same term, *natural selection*. After Wallace published several papers he sent one to Darwin, asking him to forward it to the Linnean Society for publication, which Darwin did. Darwin then rushed his own life's work into print in 1859 (giving proper credit to Wallace) in the form of the book *The Origin of Species*.

Our modern understanding of evolution is based on genetic theory. However, its original development preceded that knowledge. Here a detailed discussion of heredity and genetics is deferred to Chapter 6. Evolution can be described based on the understanding that individuals in a population vary in numerous traits (observable characteristics) that can be inherited by their offspring. The traits may be of various types, such as **morphological** (related to structure, such as height or shape of the eyes), biochemical (such as the ability to produce their own vitamin C), or even behavioral (e.g., aggressiveness in dogs).

The theory of evolution can be reduced to two mechanisms that act in combination:

1. **Random variation**. The sources of the variation in traits within a population are random mutations in the genetic code, and sorting and recombination of genetic material that occurs during *meiosis*, a type of cell reproduction. Only a minority of genetic changes may confer an advantage on an individual. In fact, most changes are probably fatal and are not passed on to future generations. Mutations are caused by errors in the biochemical processes of reproduction in which the genetic material is copied for progeny, or by damage from chemical or physical agents such as ionizing radiation.

2. **Natural selection**. Because organisms have an inherent reproductive growth rate that would cause the population to exceed the ability of the environment to support it, not all individuals survive to reproduce. Individuals with heritable traits that do confer an advantage tend to leave more offspring than those without such traits. Consequently, those traits become more common in succeeding generations.

These cause the traits held by a population to tend to change with time either because novel traits are developed randomly that confer an advantage in the current environment, or because different traits are selected for when the environment changes, such as by climate change, introduction of new competing species to the area, or various forms of human intervention. Favorable traits, which increase the fitness of a population to an environment, are called **adaptations**. The Galápagos Islands, which have become a field laboratory for evolution, furnish an example. In 1977 a drought wiped out 85% of a species of finch. Studies showed that the survivors were mostly birds with larger beaks. It was found that this was because during the drought there were fewer herbs and grasses that produced small seeds. The birds with small beaks were unable to eat the larger seeds that remained, and they did not survive to pass on their characteristics.

Prior to our modern understanding of genetic theory and molecular biology, the theory of evolution could be supported by three types of evidence: the fossil record, comparisons between the structure and function of different species, and by an analysis of the geographic distribution of existing species. The fossil record shows that (1) different organisms lived at different times, (2) different organisms lived in the past than are in existence today, (3) fossils found in adjacent sedimentary layers (and therefore relatively close to each other on a geological time scale) are similar, (4) intermediate forms of species are sometimes found, and (5) older rocks tend to have simpler forms.

Comparison of species falls into three categories: comparative anatomy, comparative embryology, and comparative biochemistry. **Comparative anatomy** shows that similar organisms have similar structures, but structures that serve different functions. For example, the same bones that a human has in the forearm are found in the flipper of a whale and the wing of a bat. It was easier for nature to modify existing structures of these mammals than to develop completely new, specialized structures. Sometimes a structure loses its function altogether, forming a **vestigial organ**. For example, whales and snakes retain the pelvis (hipbone) and femur (thighbone). **Comparative embryology** finds that similar organisms have similar **embryos** (the earliest multicellular form of an individual). For example, all vertebrate embryos, including humans, have gill slits, even if the adult does not. Evolution accounts for this by explaining that those features are retained from ancestral forms. In an example of **comparative biochemistry**, techniques of molecular biology have shown that similar species have similar genetic material. It is possible to compare species based on the degree of similarity between their DNA (the chemical in the nuclei of cells that stores the hereditary information). This has shown definitively that species that are similar on an evolutionary scale (based on other evidence) are also similar genetically. Furthermore, the code that converts DNA into proteins is the same in all living things from bacteria to humans (see Table 6.2). There is no fundamental reason that this should be so unless all these organisms developed from a common ancestor.

The third line of evidence is from **biogeography**, the study of geographic distribution of living things. This type of evidence was particularly striking to Darwin. He observed many unique species in the Galápagos, off South America, and in the Cape Verde Islands, off Africa. Although the two island groups have similar geology and climate, their species are more similar, although not identical, to those on the nearby mainland than to each other. This suggested that the islands were colonized from the nearby mainland by organisms swimming, flying, or rafting on floating vegetation, and that evolution continued through their subsequent isolation. At the same time, unrelated organisms of the two island chains had similar characteristics, suggesting that evolution formed similar structures in response to similar requirements.

The study of evolution has helped to understand changes in populations other than the formation of new species. Random changes in traits can occur in a population, resulting in what is called **genetic drift**. This is seen when a population becomes divided by some circumstance, such as the formation of an island from a peninsula by rising water levels. If two populations are isolated from each other long enough, they can diverge to form distinct species, even if both are in similar environments. This is called **divergent evolution**. The differences between Galápagos species and the mainland species with which they share ancestry is an example.

Another form of genetic drift is called the **bottleneck effect**. This occurs if some catastrophe destroys a large portion of a population. As a result, only those traits carried by the survivors are found in future generations. Many of the less common traits will disappear, and some rare ones may become common (see Figure 2.1). The bottleneck effect is also seen when a small number of a species are introduced to a new ecosystem and flourish there. The descendents share a few recent ancestors and limited genotypes.

Environmental stresses can cause changes in the genetic makeup of a population by favoring organisms with certain alleles more than others. This is, in fact, the normal way that populations can adapt rapidly to changes in their environment without mutations being required to produce new adaptations. It is also the reason why populations with genetic diversity are more likely to survive in the face of change. However, there is another side of this phenomenon related to human impacts on populations. Toxins added to the environment exert selection pressure for individuals that are more tolerant of the toxins. One negative impact of this is that it can reduce the genetic diversity of a population, making it vulnerable to further stresses. Another problem occurs if the organism is a pest and the toxicant is an agent such as a pesticide or antibiotic. As a result of the selection pressure, the population seems to develop tolerance or resistance to the agent, which then becomes less effective.

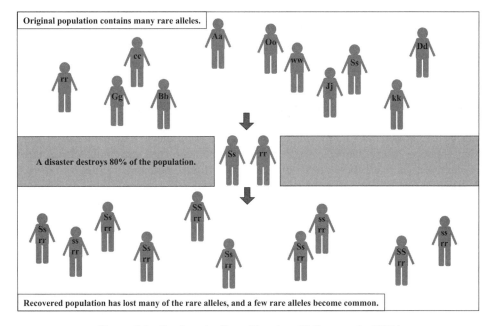

Figure 2.1   Bottleneck effect. (Based on Wallace et al., 1986.)

Sometimes very different species develop similar characteristics in response to similar environmental conditions. This is called **convergent evolution**. Because of this it is not always possible to consider species to be related evolutionarily because of superficial similarities. For example, most of Australia's original mammals are **marsupials**, which nurture their fetuses in pouches; whereas North America is dominated by **placental mammals**, whose fetuses grow internally in the uterus until ready to live independently. Despite their being very different groups, similar species have arisen on both continents, such as marsupial analogs to the wolf, mouse, and even the flying squirrel.

The Pace of Evolution   The traditional view of evolution has been that it proceeds by the accumulation of small increments of change. This view is called **gradualism**. Some fossil evidence supports this. For example, fossils have shown a series of species leading from a small, dog-sized animal, to the modern horse. However, the fossil record is more compatible with the view that species remained stable for millions of years before suddenly disappearing and being replaced by new ones. An alternative view, called **punctuated equilibrium**, was proposed in 1972 by Niles Eldredge and Stephen Jay Gould. This predicts that evolutionary changes occur rapidly over short periods, forming new species in small populations. These stay relatively unchanged for millions of years until they become extinct. Gould suggests that the rapid changes could be caused by small, yet influential genetic modifications. For example, radical changes in the body plan of an organisms could be mediated by a small number of mutations.

The theory of punctuated equilibrium answers the criticisms directed at evolution theory for the absence of "missing links" in the fossil record. However, gradualism also explains some of the features observed in the living world. The structure of the eye has been cited as being so complex as to defy explanation in terms of development from simpler forms. However, four different species of mollusk illustrate stages of a continuum in eye development (see Figure 2.2).

Extinction   Extinction is the elimination of a species from Earth. The term is also used to describe elimination of a species from a particular area or ecosystem. Today, there is serious concern because human activities are causing the extinction of numerous species each year. Activities that destroy ecosystems or even just reduce their size cause loss of species. Some biologists estimate that human population pressure on natural ecosystems could eliminate 20% of Earth's species over the next 25 or 30 years. The term **biodiversity** describes the taxonomic variation on Earth or in an ecosystem. The loss of species due to human activities eliminates adaptive information created by nature over eons. Many hope to protect against loss of biodiversity. Besides the moral motivation, there is also the utilitarian concern that lost species might have been useful, such as for drug development or to help control biological pests.

Extinctions also occur naturally, of course. A species may become extinct because of competition or increased predation from another species. This could occur if the other species is one that is newly evolved or that has invaded the ecosystem from other locations. Such invasions are caused by geological and/or climatic changes such as land bridges that form between major islands and continents due to fluctuation sea levels, or elimination of a climatic barrier between, say, ecosystems separated by mountains or desert. Human importation of exotic species, intentional or not, produces a similar effect.

The fossil record shows that a number of mass extinctions have occurred in the past. At the end of the Permian Period, some 250 million years ago, 96% of species, such as

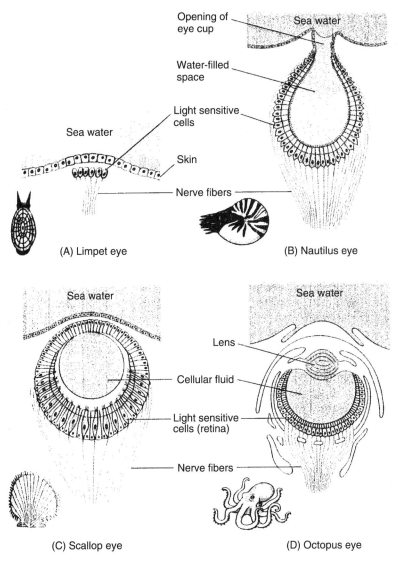

Opening of eye cup

Sea water

Water-filled space

Light sensitive cells

Sea water

Skin

Nerve fibers

(A) Limpet eye

(B) Nautilus eye

Sea water

Sea water

Lens

Cellular fluid

Light sensitive cells (retina)

Nerve fibers

(C) Scallop eye

(D) Octopus eye

Figure 2.2   Development of the eye in mollusks: limpet, nautilus, scallop, octopus. (From Postlethwait and Hopson, 1995.)

the trilobite, disappeared from the fossil record. Some 65 million years ago, at the end of the Cretaceous Period, known as the **K-T boundary**, as much as 76% of species disappear, including the dinosaurs. Recently, geologic evidence has established that this mass extinction was caused by a large meteorite or comet striking Earth at the Yucatan Peninsula in what today is Mexico. Shock waves traveling through Earth's crust seem to have focused on the opposite point of the globe, southwest of India, causing massive volcanic outpourings. The impact plus the volcanism are thought to have sent dust and smoke into the atmosphere, darkening the sun, directly wiping out many ecosystems and reducing the primary productivity of the Earth (Chapter 14), starving many species into extinction.

Mass extinctions seem to be followed by a period of accelerated evolution of new species. The elimination of dinosaurs, for example, paved the way for the further development of mammals. Human activities are currently being blamed for causing the extinction of great numbers of species, mostly by encroaching on their habitats.

2.4   TAXONOMY

There are an estimated 50 million species on Earth. Of the 1.5 million species that have been named, about 5% are single-celled organisms (prokaryotes and eukaryotes), and plants and fungi make up about 22%. About 70% of the known species are animals. Most of these are invertebrates, especially insects, of which there are about 1 million known species.The earliest and, until this century, the most dominant activity in biology has been the classification of species. Aristotle was concerned with it. Even Charles Darwin, who started developing his theory of evolution while on the famous round-the-world voyage of the *Beagle*, was occupied primarily with collecting and categorizing new species of organisms.

Why should environmental engineers and scientists study taxonomy? One reason is so they won't be at a loss to see how any particular organism fits in with others. Does that worm always live where it is, or will it someday metamorphose into an insect and fly away, perhaps carrying biological or chemical contamination with it? At a more fundamental level, classification is a prerequisite to the identification of patterns, which leads to generalizations or hypotheses, and ultimately to tests of hypotheses by experiment or further observations. Thus, classification can be seen as a basis for the scientific method. The generalizations came later to biology than they did to physics and chemistry, perhaps owing to the dazzling variety of life and the inherent complexity of the underlying mechanisms.

Taxonomy is the science of classification of organisms. Taxonomy (except for microorganisms) was a fairly settled area until recently. Now, genetic techniques are reopening old questions and revealing new things. The possibility that humans are causing a new mass extinction lends a new urgency to knowing the organisms that now exist on Earth. This loss in the diversity of life would have several potential consequences. From a utilitarian point of view, destruction of species results in a loss of genetic material that may include useful traits, such as production of chemicals that may have medical uses, or traits that may be useful for invigorating agricultural plants or animals. Ecological relationships may link the survival of one species to many others. Finally, it may be argued from several ethical viewpoints that we have a moral imperative to preserve Earth's biological heritage regardless of actual or potential utility.

Two ways to classify organisms are to group them (1) by structure and function or (2) by closeness of evolutionary descent. The two approaches often produce similar results, but not always. Convergent evolution can make dissimilar evolutionary branches form similar characteristics, whereas divergent evolution does the opposite. Barnacles and limpets both live in shells glued to rocks in the sea, but barnacles are arthropods, like crabs, whereas limpets are mollusks, like clams. The use of evolutionary descent was constrained in the past due to the incompleteness of the fossil record, resulting in ambiguity in classifying existing organisms. Recently, however, the development of genetic engineering techniques has led to quantitative measures of genetic similarity. These have settled old classification questions and even led to increased detail in classification by dividing old groups, forming new ones, and producing new species, phyla, and even king-doms (see below).

Biologists use the **Linnaean system** to classify organisms, which consists of a hierar-chy of groupings and a naming convention. The lowest level of the hierarchy is the *spe-cies*. Similar species are grouped into a **genus.** The naming convention, called **binomial nomenclature**, assigns to each species a two-word name. The first word, which must be capitalized, is the name of the genus, and the second, which is uncapitalized, is the name of the species. For example, modern humans are in the genus *Homo* and the species *sapiens*; thus, they're called *Homo sapiens*. Other species in our genus are extinct, such as *Homo habilis*. Genus and species names are either italicized or underlined when writ-ten, and the genus name may be abbreviated by a capitalized first initial once it has already been written out in full: *H. sapiens.*

In increasing degrees of generality, the other classifications are **family, order, class, phylum** or **division** (in plants and fungi), **kingdom**, and **domain**. Each of these may sometimes be subdivided: for example, subkingdom, subphylum, or subspecies. The fol-lowing phrase is a memory trick for the sequence from kingdom to species:

King	**Phillip**	**came**	**over**	**from**	**Greece**	**Saturday**.
Kingdom	**Phylum**	**Class**	**Order**	**Family**	**Genus**	**Species**

The highest-level category is the **domain**. It is based on the type of cell comprising the organism, of which there are three: the bacteria, the archaeans, and the eukarya. Each domain is subdivided into kingdoms. Some biologists consider the archeans and the bac-teria each to consist of only one kingdom. However, microbiologists note that molecular biology methods have shown that different groups of bacteria, for example, are more dis-tinct from each other genetically than plants are from animals. Therefore, archaea have been divided into three kingdoms, bacteria into 15, and eukarya into four. Although this method of classification has not yet become universal among biologists, in this book we adopt this approach.

One domain is that of the **bacteria**, also sometimes called **eubacteria**. Their cell type is the **prokaryote**, which is a simple type lacking any internal membrane structures. Their size is typically 0.2 to 2.0 micrometers (μm). Based on **morphology** (physical structure) and biochemical characteristics, the bacteria have classically been organized into 19 groups. Examples are *Pseudomonas*, a common soil bacterium that is also exploited in wastewater treatment processes; and the *Cyanobacteria*, or blue-green bacteria, important in the environment because they convert atmospheric nitrogen into a useful nutrient. Based on genetic techniques, new groups have been found and some of the older ones combined; one current classification system now has 15 high-level groupings.

Another prokaryotic domain is that of the **archaea**, which includes three kingdoms. Archaea includes the methanogenic bacteria that produce methane gas in a wastewater treatment plant's anaerobic digester. Archaeons were once considered to be bacteria. However, unsuccessful attempts to transfer genes from eubacteria to archaebacteria led to the discovery that their membranes, although structurally similar to the other domains, are chemically distinct. Many of the archaea are notable for the harsh chemical and phy-sical environments in which they thrive. Because of this it is thought that archeons may be relics of the earliest forms of life that still exist today.

The third domain is that of the **eukarya**, which comprises four kingdoms, including the familiar plants and animals. The cells of this domain are called **eukaryotic**. This is a more

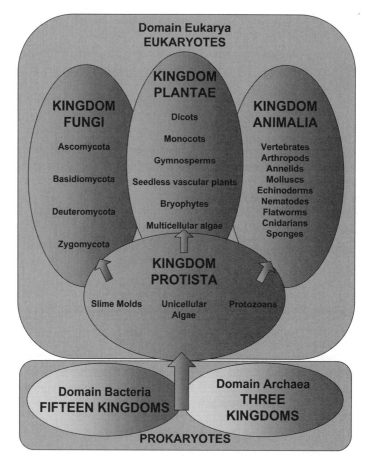

Figure 2.3   The six kingdoms, according to one taxonomic approach.

complex cell type that is characterized by internal membrane structures and membrane-based organelles. Eukaryotic cells are much larger than prokaryotes; animal cells are typically about 20 μm across, and plant cells are about 35 μm.

The next level of classification below kingdom is **phylum** (in animals) or **division** (in plants and fungi). Figure 2.3 summarizes the variety of life at the domain and kingdom level and shows some of the major phyla and divisions. Table 2.1 shows how several

TABLE 2.1   Classification of Several Familiar Eukaryotes

Kingdom	Animalia	Animalia	Animalia	**Kingdom**	Plantae
Phylum	Chordata	Chordata	Arthropoda	**Division**	Anthophyta
Subphylum	Vertebrata	Vertebrata	Uniramia	**Subdivision**	
Class	Mammalia	Mammalia	Insecta	**Class**	Dicotyledones
Order	Primates	Cetacea	Orthoptera	**Order**	Sapindales
Family	Pongidae	Mysticeti	Tettigoniidae	**Family**	Aceraceae
Genus	*Homo*	*Balenoptera*	*Scudderia*	**Genus**	*Acer*
Species	*H. sapiens*	*B. musculus*	*S. furcata*	**Species**	*A. rubrum*
Common name	Human	Blue whale	Katydid	**Common name**	Red maple

familiar organisms are classified according to this system. Biology does not have a central authority for determining how to organize animals into phyla. The classifications gain currency by general usage and adoption in textbooks.

Species may be further subdivided into **subspecies**, **strain** (especially in prokaryotes), or **cultivar** (in crop plants).

Formerly, the eukaryotic domain was considered to consist of only two kingdoms, plants and animals. Now it has been further subdivided to include fungi and protists. Several of these kingdoms are characterized based on their source of carbon. Autotrophs obtain their carbon from the inorganic form, in particular carbon dioxide. Heterotrophs are organisms that must obtain their carbon in the form of organic chemicals such as sugars, fats, or proteins.

The **fungi** kingdom is defined to include the plantlike, mostly multicellular heterotrophic organisms lacking the green pigment chlorophyll. Fungal cells are surrounded by a rigid capsule called a **cell wall**. Fungal cell walls are made of a polymer called chitin. Fungi have a major ecological role in breaking down dead organisms and making their nutrients available to other organisms. Four divisions have been recognized: the ascomycetes include yeast, mildew, and the prized edibles morels and truffles; basidiomycetes include most of the common mushrooms; deuteromycetes include the mold that produces penicillin; zygomycetes include common black bread mold.

Plants are multicellular photoautotrophs that reproduce sexually to form **embryos** (an early stage of multicellular development). Photoautotrophs are autotrophs that get their energy from light (as opposed to chemical sources of energy). Plant cells usually contain chloroplasts, which are organelles containing the chlorophyll. The cells of plants are surrounded by a cell wall composed largely of cellulose fibers. The major ecological role is the capture of light energy from the sun to fix atmospheric carbon dioxide and produce oxygen in the process called photosynthesis. This provides food for other organisms and oxygen for their respiration, the biochemical oxidation of organic matter to produce energy that is the opposite reaction to photosynthesis. The plant kingdom consists of 12 divisions: the *bryophytes* (mosses and relatives), four *seedless vascular plants* divisions (e.g., ferns), *gymnosperms* (including conifers), and *anthophyta* (also referred to as angiosperms, the flowering plants).

Like the fungi, animals, are multicellular heterotrophic eukaryotes. In addition, they reproduce primarily by sexual reproduction, are motile in some part of their life cycle, and their cells lack a cell wall. There are 33 animal phyla, including several marine organisms, such as sponges, corals, and jellyfish; several phyla containing wormlike organisms; mollusks; arthropods (which include crustaceans and insects); and the chordates. The latter include the vertebrates, which are the organisms with an internal skeleton. Thus, we humans are chordates. All of the other animal phyla, plus a small group within the chordates, are called invertebrates.

The remaining kingdom in domain eukarya is that of the protists. This kingdom somewhat arbitrarily combines single-celled eukaryotes that would otherwise be part of either the fungi, plant, or animal kingdoms. Thus, they are classified into three groups: animal-like protists, or protozoans, which have four phyla, including those with *paramecium* and the *amoeba*; plantlike protists, with six divisions, *euglena*, *dinoflagellates*, *diatoms*, *red algae*, *brown algae*, and *green algae*; and funguslike protists, with three divisions, including *slime molds*. Each of these groups function ecologically in ways that are similar to their analogous multicellular kingdom.

The six divisions of the plantlike protists are also called algae. Three of these divisions, the red algae, brown algae, and green algae, are dominated by multicellular forms. For example, kelp is a brown algae and sea lettuce is a green algae. For this reason, some textbooks place them in the plant kingdom. However, these divisions also include many unicellular forms. Many phytoplankton are green algae, including *Chlamydomonas* and a family called the desmids. In this book we adopt the convention of keeping them with the other algae in kingdom Protista, although in Chapter 10 their modern taxonomy is examined more closely. Plankton are very small aquatic or marine plants, animals or algae. Plant and algal plankton are referred to as phytoplankton, and animal plankton or animal-like protists are called zooplankton.

Viruses, viroids, and prions are infectious agents not considered to be members of a living kingdom. They do not have a cytoplasm or cell membrane and are not capable of any metabolic reactions within their own structure. They reproduce by infecting living cells and causing them to produce more of the infectious particles. The resulting damage is responsible for many diseases. They may be more properly considered to be chemical agents. Viruses are the largest and most complex, being a geometrical particle of DNA or RNA (the chemical agents of heredity), and surrounded by protein. They may be 1/1000 to 1/10,000 the size of prokaryotic cells. An important medical distinction between viruses and bacteria, which also cause disease in humans, is that since viruses do not metabolize, virus diseases cannot be treated by antibiotics, which act by killing or otherwise preventing the reproduction of microbes. However, viruses can often be prevented by vaccines, which stimulate the human immune system to destroy the virus particles within the body.

Viroids are even simpler than viruses, consisting of a single RNA molecule. They have been found to cause some crop diseases. **Prions** are similar in that they consist of a single molecule; however, in this case the molecule is a protein. They are poorly understood. Prions are suspected causes of several serious neurological diseases in humans and animals, including scrapie in sheep, bovine spongiform encephalopathy in cows (mad cow disease), and Creutzfeldt–Jakob disease and kuru in humans. They are remarkable in that they may be the only biological agent that reproduces yet does not have DNA or RNA, the chemical basis of heredity.

2.5   INTERACTION OF LIVING THINGS WITH THE ENVIRONMENT

Living things don't just live *in* the environment: To a great extent they create it. Photosynthetic plants and microorganisms are responsible for our atmosphere consisting of 21% oxygen and having only trace levels of carbon dioxide. Without them the atmosphere would more closely resemble that of the primitive Earth, which was composed of carbon dioxide, nitrogen, water vapor, hydrogen sulfide, and traces of methane and ammonia.

It is not only chemical changes that are wrought, but also physical ones, from the temperature of Earth to the form of our landscape. The temperature of Earth's surface is strongly affected by plants' ability to remove carbon dioxide, a *greenhouse gas* that prevents radiation of heat energy from Earth into space. Bacteria in the gut of termites have been found to be responsible for the release of methane gas, another greenhouse gas, into our atmosphere. Trees, grasses, and other plants reduce soil erosion, affecting the shape of the landscape.

The two-way relationship between the all the living things on Earth and the physical environment has been recognized in the **Gaia hypothesis**, put forward by the English scientist James Lovelock (1987). His statement of the hypothesis is: "The biosphere is a self-regulating entity with the capacity to keep our planet healthy by controlling the chemical and physical environment." Lovelock is also known for another important contribution to environmental protection. He is the inventor of the electron capture detector, used in gas chromatographs to analyze for organics containing halogens or oxygen with exceptional sensitivity. By revealing the fate and transport of minute chemical quantities such as pesticides, this device can take some credit for stimulating the environmental revolution.

Lovelock illustrated the Gaia hypothesis with a computer model that demonstrated how organisms could control the environment to their advantage. In a model he called Daisyworld, he postulates a planet like Earth receiving energy from the sun and covered with two species of daisy, one light-colored and the other dark (see Figure 2.4). Both species had an optimum temperature for growth, with the light daisy's optimum higher than that of the dark daisies. He then simulated the system and varied the amount of energy radiated by the sun. The simulation showed that when the energy input increased, raising the temperature, the population of light daisies would increase and the dark would decrease. This increases the planet's **albedo**, the fraction of incident energy that is reflected back into space. By reflecting more energy, the temperature of the surface is prevented from increasing as much as it would otherwise. At lower incident radiation, the cooler temperatures would favor dark daisies that decrease the albedo. Without the organisms, the surface temperature increases smoothly as radiation increases. With the daisies present, as radiation is increased, the temperature first increases, then forms a plateau within which the organisms stabilize the system.

As evidence for the Gaia hypothesis, Lovelock cites the fact that Earth's atmosphere is far from equilibrium. Without organisms to maintain it, he says that the nitrogen and oxygen would eventually combine to form nitrates, which would dissolve in the oceans, leaving an atmosphere consisting mostly of CO_2, like that of Mars.

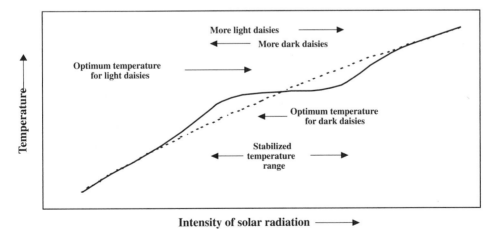

Figure 2.4 Temperature vs. radiation intensity for Daisyworld. The dashed line shows the global temperature in the absence of the two daisy populations. The solid line shows how temperature would vary with their influence included.

Individual organisms employ control mechanisms to regulate their internal environments, such as body temperature or blood chemistry in mammals. Because of this similarity, some have extended the Gaia concept to mean that Earth itself *is* a living organism, rather than saying that it *is like* one. This has created controversy for the hypothesis, but it is not part of Lovelock's conception. Although suggestive, Daisyworld is not proof of the Gaia hypothesis. Proof awaits stronger evidence that such feedback mechanisms actually operate on Earth.

Although the self-regulating nature of Earth's ecosystem as a whole remains to be proven, there are many examples of environmental conditions that are caused by biological activity, some of which involve self-regulation. Some examples include the cycling of minerals and elements in the environment, the formation of soil, and the control of pH in anaerobic digestion.

Biogeochemical cycles describe the various reservoirs for an element, and the reactions and transformations that convert the matter from one reservoir to another. For example, the major reservoirs for carbon are atmospheric CO_2, aquatic or marine CO_2, the biosphere (in the form of living or dead biological material), and the **lithosphere** (solid minerals, i.e., rocks or soil) as carbonate rocks or as fossil carbon (coal, oil, or natural gas). The cycles can be described quantitatively in terms of the amount in each reservoir and the rate of each reaction. Look ahead to Figure 14.5 to see that most of Earth's carbon is stored in carbonate rocks such as those that form the Rocky Mountains. It is interesting to realize that the Rockies and similar mountain chains worldwide are formed entirely from biological activity, by the gradual accumulation of the skeletons of microscopic plants in sediments on the ocean floor, followed by geological upheavals.

Biogeochemical cycles are a starting point for the analysis of the effect of humans on the environment. Since the heat-trapping effect of atmospheric CO_2 and methane are a serious concern, it is important to understand the rate of **anthropogenic** (human-caused) emissions of these gases and the rates at which they are naturally removed. Such an analysis has led to the conclusion that half of the CO_2 released by combustion of fossil fuel since the beginning of the Industrial Revolution remains in the atmosphere. It is thought that the ocean has absorbed the remainder. The details of biogeochemical cycles are discussed in Chapter 13 and Section 14.2.

Soils may be formed abiotically, but the result is very different when mediated by living things. First, the rate at which rocks are broken down into soil can be accelerated by the action of acids produced by **lichens** (an association of fungus and algae) and by mechanical effects of the roots of plants, which can widen cracks in rocks, exposing more surface area to weathering. Plants prevent the soil from eroding by absorbing moisture and by the action of their root structure. Dead plant matter is degraded by fungi and bacteria to form **humic substances**. These are complex polymeric phenolic compounds that resist further biodegradation. Their presence enhances soil water–holding capability and makes it more **friable** (easy to crumble), which makes it easier for roots to penetrate. Humic substances also enhance the ability of soils to adsorb organic pollutants. Insects and other invertebrates, especially earthworms, aerate and mix the upper soil. Charles Darwin's major preoccupation in the years after he published *The Origin of Species* was a detailed study of how earthworms affect the soil, especially by moving soil continually from the depths to the surface. Microorganisms in the soil and associated with the roots of plants convert atmospheric nitrogen to a form available to plants, effectively producing a fertilizer in place. These changes are both caused by organisms and benefit them.

Anaerobic digestion, an environmental engineering process used to biodegrade concentrated slurries of biomass, provides another good example of organism–environment interaction. Here, the pH is controlled by a delicate balance between two populations of microorganisms. The acidogens break down organics into simple volatile fatty acids such as acetic and propionic acids. This consumes alkalinity. The methanogens use the acids to make methane, restoring the alkalinity. The methanogens are more sensitive to environmental changes, including pH, than the eubacterial acidogens. Should some disturbance affect the methanogens adversely, the acid generation can cause a pH drop, further slowing acid destruction and causing a cessation of methane formation from which it is difficult for the system to recover. This is similar to Daisyworld, where the organisms control their environment until conditions are forced outside the range in which control can occur, and the system becomes unstable.

2.6   BRIEF HISTORY OF LIFE

With the overview of the types of living things and the mechanisms by which they evolve, as given above, let us take a brief look at the changes that have occurred in the biology of Earth. One of the biggest questions is: How did life originate from prebiotic conditions? After the formation of Earth and the cooling of its crust, the atmosphere consisted mostly of water vapor, carbon dioxide, carbon monoxide, and N_2. The dominant theory derives from the ideas of Haldane and Oparin, early twentieth-century biochemists. It holds that energy from sunlight, lightning, and volcanic action gradually formed simple organic molecules from these precursors. In the 1950s, Urey and Miller conducted famous experiments that proved that this could occur. They passed spark discharges through similar

TABLE 2.2   Timeline for Life on Earth

Million Years Ago	Event
4,600	Earth forms.
3,800	Life begins.
3,000	Prokaryotes appear.
1,000	Eukaryotes appear.
570	Algae and marine invertebrates (e.g., trilobites) dominate.
405	Terrestrial plants and vertebrates appear.
225	Dinosaurs appear.
135	Flowering plants appear.
65	Mass extinction, including dinosaurs and up to 70% of all animal species.
	Mammals begin to dominate.
40	Primates appear.
4	Hominids appear.
2	Genus *Homo* appears.
Years Ago	
30,000	*Homo sapiens neanderthalis* disappears.
	H. sapiens sapiens appears.
5,000	Development of agriculture.
146	Publication of *The Origin of Species* (1859).
52	Discovery of DNA structure (1953).
8	Cloning of mammals (1997).

primitive atmospheres in a sealed glass apparatus and produced aldehydes, carboxylic acids, and most important, amino acids, the building blocks of proteins. In the absense of life on early Earth, these compounds could accumulate until the seas consisted of **primordial soup**. These compounds could be further concentrated in drying coastal ponds or at hot springs. Experiments have shown that under realistic conditions, polymeric macromolecules could be formed. Other steps along the way to a viable self-replicating cell have been reproduced, but many critical steps have not. Overall, the process is still a mystery.

However the origin occurred, the fossil record shows that life developed in complexity continuously (Table 2.2). Humans are a very recent addition. Notice how long the dinosaurs were on Earth relative to our tenure here.

PROBLEMS

2.1. Consider a toxic pollutant such as gasoline, the dry-cleaning solvent perchloroethylene, the heavy metal mercury, or the pesticide DDT. Can you think of ways in which the pollutant might have some effect at each level of the biological hierarchy from metabolism to ecosystem?

2.2. How are ecosystem and environment distinguished from each other? Note that these two terms are often confused by the public.

2.3. How could anthropogenic effects change the course of evolution?

2.4. Can you think of some pollutants that have different effects on different biological kingdoms? How will these effects differ from each other?

2.5. All of the following have been used for biological waste treatment: bacteria, cyanobacter, green algae, fungi, green aquatic plants, earthworms. Try to think of how each could be used, and for what type of waste each would be applicable.

2.6. Think about how you would create a model like Daisyworld. What laws and relationships would you use?

2.7. Assuming that the Gaia hypothesis were true, which of the criteria for life does Earth's ecosystem as a whole satisfy, and which does it not satisfy?

REFERENCES

Doolittle, W. F., 2000. Uprooting the tree of life, *Scientific American*, Vol. 282, No. 2.

Lovelock, James, 1987. *Gaia: A New Look at Life on Earth*, Oxford University Press, New York.

Odum, Eugene, 1993. *Ecology and Our Endangered Life Support Systems*, Sinauer Associates, Sunderland, MA.

Postlethwait, J. H., and J. L. Hopson, 1995. *The Nature of Life*, McGraw-Hill, New York.

Schrödinger, Erwin, 1956. *What Is Life?*, Doubleday Anchor Books, Garden City, NY.

Wallace, R. A., J. L. King, and G. P. Sanders, 1986. *Biology: The Science of Life*, Scott, Foresman, Glenview, IL.

3

THE SUBSTANCES OF LIFE

A useful simplification of biological organisms sometimes made by environmental engineers and scientists is to view them as catalysts for chemical reactions, such as the oxidation of ammonia or ferrous iron, or production of methane and carbon dioxide from acetic
acid. Such a view hides the detailed mechanisms, including the sequence of chemical
intermediates and the specific chemical nature of the catalyst. Examining these details
will help us to understand more complex chemical interactions between organisms and
their environment, such as biodegradation of toxic organic chemicals or the effect of chemicals on the health of organisms and ecosystems. The details of biochemistry begin with
knowledge of the four most important types of chemical substances comprising living
things: carbohydrates, lipids, proteins, and nucleic acids. Later, in the chapters on toxicology, we consider the biochemical reactions involving **xenobiotic** compounds (those that
are, literally, "foreign to life," that is, substances not normally found in living things).

3.1 BASIC ORGANIC CHEMICAL STRUCTURE

A fairly simple summary of organic chemistry is sufficient for understanding much of the
structure and reactions of biochemicals. There are only two types of **chemical bonds** that
need to be considered at this point: ionic bonds and covalent bonds. The great majority of
biochemicals are composed of only six elements connected by those bonds: carbon,
hydrogen, nitrogen, oxygen, sulfur, and phosphorus. In addition, the attractions of dipole
forces such as hydrogen bonds and van der Waals forces act between molecules or
between different parts of the same molecule. The structure of most organic chemicals
can be described by combining the six elements with ionic and covalent chemical
bonds.

Environmental Biology for Engineers and Scientists, by David A. Vaccari, Peter F. Strom, and James E. Alleman
Copyright © 2006 John Wiley & Sons, Inc.

Physicochemical interactions are attractions or repulsions between molecules, or between different parts of the same molecule, which do not result in formation or breaking of ionic or covalent bonds. They are much weaker than chemical bonds and are much more sensitive to changes in temperature. These forces change the shape of molecules, causing them to bend, fold, or form liquids or crystals. This affects physical properties such as solubility or boiling point as well as the ability of molecules to enter into chemical reactions. Much of this physicochemical behavior can be related to a few types of atomic groups on the molecule that are called *functional groups*.

The main importance of this section is to arrive at an understanding of what controls the shape of biochemical compounds. Biochemicals are often large, polymeric molecules, whose function is intimately related to their shape. Thus, their behavior is not determined completely by their chemical formula, but also depends on how they are folded or coiled or by how they are arranged relative to neighboring molecules.

3.2   CHEMICAL BONDING

Atoms in a molecule are held together by covalent or ionic bonds. These bonds involve a sharing of electrons between pairs of atoms. Bonds can be classified in terms of how equally the electrons are shared or by the amount of energy required to break the bond. The electrons in **covalent bonds** are equally or almost equally shared, whereas those in **ionic bonds** are associated almost completely with one of the atoms in the bonded pair.

Electrons in bonds between identical atoms tend to be shared equally between the two. But if the two atoms are different, the electron cloud making up the bond may be displaced toward one of them. As a result, the center of charge of the electrons is different from the center of charge of the nuclei of the atoms, and the bond will produce an electrostatic field in its vicinity. A bond that produces an electrostatic field is said to be **polar**. Polarity produces attractions between molecules.

For example, Figure 3.1 shows an ethylene molecule. The bond between the two carbons will be nonpolar. A molecule of ethanol is also shown. The carbon and oxygen atoms

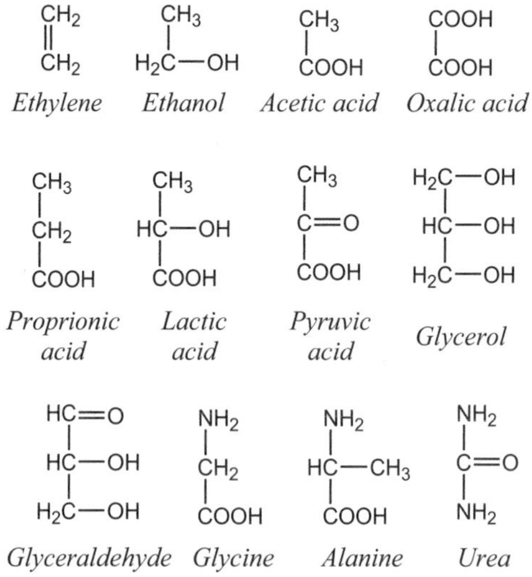

Figure 3.1   Structures of several small, biologically important organic molecules.

are very different in their tendency to attract electrons; thus, this molecule is polar. The tendency of atoms to attract electrons is called its **electronegativity**. Oxygen is more electronegative than carbon; therefore, the ethanol molecule will be polarized with a positive charge toward the carbon and negative charge toward the oxygen. Sometimes oxygen can attract electrons away from a neighboring hydrogen so completely that it leaves the hydrogen fairly free to dissociate from the molecule as a proton, producing an acid. This is seen most commonly in organic chemicals in the case of the carboxylic acid group:

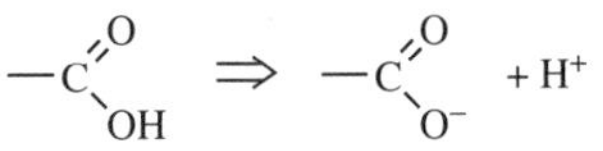

$$-C\overset{\displaystyle O}{\underset{\displaystyle OH}{}} \;\Rightarrow\; -C\overset{\displaystyle O}{\underset{\displaystyle O^-}{}} + H^+$$

also written —COOH. The two oxygen atoms act together to strip the electron from the hydrogen.

Fluorine is the most electronegative of all the elements. Figure 3.2 shows the electro-negativity of all the elements that have been shown to be essential in at least some living things. The electronegativity of the other elements (except the noble elements) tends to decrease with distance from fluorine in the periodic table. Thus, chlorine is somewhat less electronegative than fluorine. The scale decreases more rapidly by moving to the left; thus, oxygen, nitrogen, and carbon represent a sequence of decreasing affinities for electrons. The leftmost column contains the alkali elements that are so low in electron affinity that they tend to give up their electrons entirely in bonds formed with the halogens on the right. These form the most polar of the bonds, the ionic bond.

The symbol-and-stick diagram for atoms and bonds in Figure 3.1 shows features that can be used to recognize the covalent and ionic bonding arrangement for the vast majority of organic compounds. Atoms form bonds in order to give themselves enough electrons to fill an orbital shell, such as the noble elements have. In simpler terms, each element generally shares as many electrons (i.e., forms as many bonds) as corresponds to the number of columns that element is away from a noble element in the periodic table. For example, oxygen is two columns away from argon, so it forms two bonds and should be drawn as —O— or =O, the latter showing that two single bonds can be combined into a double bond. Carbon, nitrogen, and phosphorus can also form triple bonds, although the amount

H 2.1																
Li 1.0	Be 1.5											B 2.0	C 2.5	N 3.0	O 3.5	F 4.0
Na 0.9	Mg 1.2											Al 1.5	Si 1.8	P 2.1	S 2.5	Cl 3.0
K 0.8	Ca 1.0	Sc 1.3	Ti 1.5	V 1.6	Cr 1.6	Mn 1.5	Fe 1.8	Co 1.8	Ni 1.8	Cu 1.9	Zn 1.6	Ga 1.6	Ge 1.8	As 1.9	Se 2.4	Br 2.8
					Mo 1.8								Sn 1.8			I 2.5
					W 1.7											

Figure 3.2 Electronegativity of biologically important elements. This shows the portion of the periodic table containing many of the elements that are important to life. The numbers indicate the electronegativity of the corresponding elements. (From Pauling, 1970.)

TABLE 3.1 Number of Bonds Commonly Formed by the Six Principal Elements of Biochemical Compounds

Element	C	N	O	S	P	H
Number of bonds	4	3	2	2	5	1

of energy required makes these uncommon. Table 3.1 shows the number of covalent or ionic bonds formed by the six principal elements.

There are several important exceptions to this simple scenario for an element that forms the number of bonds corresponding to its valence state (see Table 13.1). One of these exceptions is the metals. For example, iron forms two bonds in the ferrous (II) state and three in the ferric (III) state. Sulfur typically forms two bonds in organic compounds but can have other numbers in inorganics. Even carbon has exceptions to the rule, as in the case of carbon monoxide. Also, phosphorus is almost always bound in the phosphate group (PO_4^{3-}), which is usually considered as a unit.

Ionic bonds are much weaker than covalent bonds. As a result, the atoms in ionic bonds can be separated by the relatively low-energy physicochemical forces. Although ionic and covalent bonds seem quite distinct, there is actually a continuum of bonds of varying polarity from ionic to covalent.

3.3 ACID–BASE REACTIONS

A special type of ionic bond is when one of the ions is either H^+ or OH^-, resulting in an acid or a base, respectively. Water serves as a source and sink for hydrogen ions that are generated or consumed by the dissociation of acids and bases. In fact, the very ability of other acids in solution to act as acids may depend on their being dissolved in water. Put another way, the presence of water shifts the acid dissociation equilibrium. The prototypical acid–base reaction is

$$HA \Leftrightarrow H^+ + A^- \tag{3.1}$$

where HA is the undissociated acid and A^- is its conjugate base. The equilibrium constant is called the *acid dissociation constant*, K_a (square brackets denote molar concentration):

$$K_a = \frac{[H^+][A^-]}{[HA]} \tag{3.2}$$

Taking the logarithm of equation (3.2), substituting the relations $pH = -\log[H^+]$ and $pK_a = -\log[K_a]$, and rearranging, we obtain an expression for the fraction of undissociated acid:

$$\frac{[HA]}{[HA] + [A^-]} = \frac{1}{1 + 10^{pH - pK_a}} \tag{3.3}$$

Thus, the relative proportion of HA and A^- depends on the pH. At low pH most of the acid is undissociated. At high pH most is dissociated. The pK_a is the pH at which the

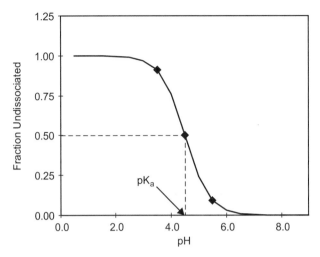

Figure 3.3 Fraction of undissociated acid for acetic acid ($pK_a = 4.7$). Points shown are for pH 3.7, 4.7, and 5.7.

equilibrium is positioned at the 50:50 point: half as HA and half as A^-. Figure 3.3 shows a plot of the fraction $[HA]/C_T$ for acetic acid, where $C_T = [HA] + [A^-]$ is the total acid concentration. Table 3.2 shows pK_a values for some important chemicals. Note that at 1 or 2 pH units below pK_a, the acid is almost completely undissociated, whereas at 1 or 2 units above pK_a, it is almost completely dissociated.

Example 3.1   What fraction, f, of acetic acid is undissociated at pH 7.0? The pK_a is 4.7; thus, the fraction undissociated is

$$f = \frac{1}{1 + 10^{7.0 - 4.7}} = \frac{1}{1 + 200} = 0.50\%$$

The pH strongly influences the shape of biochemicals by altering the proportion of dissociation of acidic groups on the molecule. Large biomolecules may have numerous acidic groups with various pK_a values. Varying the pH adds or removes hydrogen ions. This affects which portion of the molecule can form hydrogen bonds, or changes the polarity of a molecule, thereby significantly changing its shape and function.

TABLE 3.2 pK_a Values for Some Biologically Important Chemicals

Compound	pK_a	Compound	pK_a
Phosphoric acid (pK_1)	2.0	Carbonic acid (pK_1)	6.3
Citric acid (pK_1)	3.1	Citric acid (pK_3)	6.4
Formic acid	3.8	Phosphoric acid (pK_2)	6.7
Lactic acid	3.9	Boric acid	9.2
Benzoic acid	4.2	Ammonium	9.3
Acetic acid	4.7	Carbonic acid (pK_2)	10.4
Citric acid (pK_2)	4.7	Phosphoric acid (pK_3)	12.4

3.4 PHYSICOCHEMICAL INTERACTIONS

Atoms on a molecule can also be attracted to atoms on another molecule, producing **intermolecular forces**. These forces are primarily electrostatic. The strength of these forces range from that of hydrogen bonds, the strongest, to the van der Waals forces, the weakest. However, all of them are weaker than the forces involved in chemical bonding.

Hydrogen bonds are intermolecular attractions that occur between electronegative atoms on one molecule (e.g., O, N, or Cl) and with hydrogen atoms in another molecule. They will occur only if the hydrogen is bonded to other atoms more electronegative than carbon (e.g., O or N). The latter atoms pull the electrons away from the hydrogen, creating a strong local positive charge. This charge consequently has a strong attraction to electronegative atoms on the other molecules.

Even in the absence of polarity-producing asymmetry, a molecule can exhibit a temporary polarity as electrons shift from one side to another, or shift in response to the presence of the electron cloud of another molecule nearby. This produces a weaker electrostatic attraction, the **van der Waals force**.

Attractions due to hydrogen bonds, van der Waals forces, or polarity produce several important effects:

- The strength of these attractions allow chemicals to form liquids and solids. In more precise terms: They tend to raise the boiling point and melting point, increase the heat capacity and heat of vaporization, and decrease the vapor pressure.
- Dissimilar molecules that have sufficient attraction to each other can dissolve or mix, whereas those that do not attract tend to form separate phases.
- They contribute to the shape of large molecules because of attractions between different parts of the same molecule, modified by competition with other attractions, such as with solvent molecules or other dissolved species.
- They affect the rate of chemical reactions between like and unlike molecules by affecting the proximity and orientation of molecules to each other.

Water illustrates several physicochemical interactions and also plays a key role in biochemistry. Water is the most abundant chemical in living things. Living things are composed of between 50 and 90% water by weight. In humans, 99 of 100 molecules are water. The water molecule is bent; that is, rather than being directly on opposite sides of the oxygen, the two H—O bonds form a 104.5° angle with each other. The electronegativity of the oxygen pulls the electrons from the hydrogen. This leaves the molecule with a strong dipole, and each atom can form hydrogen bonds with other water molecules. These properties give it strong self-attraction. This results in the highest boiling point and heat of vaporization of any chemical of similar size or molar mass (i.e., molecular weight). Water's heat capacity and latent heat helps it regulate the temperature of living things, especially animals, which can generate heat at a rapid rate. Evaporation of water carries away a large amount of heat. If a human consuming 2000 Calories per day (note that 1 dietary Calorie = 1000 calories, distinguished in writing by their capitalization) were to release all that energy as heat, it could be consumed by evaporation of 3418 grams of water, less than 1 gallon. The actual evaporation is less because not all dietary calories are released as heat, and some heat is lost by convection.

Water's polarity also enables it to reduce the repulsive forces between charged particles such as ions, because it can align itself with the ion's electric field. The reduction in

TABLE 3.3   Properties of Various Chemicals Compared to Water

Substance	Molar Mass (g/mol)	Boiling Point (°C)	Specific Heat	Heat of Vaporization (cal/g)	Dielectric Strength
Water	18	100	1.0	585 (20°C)	80
Methanol	32	65	0.6	289 (0°C)	33
Acetone	58.1	56.2	0.51	125 (56°C)	21.4
Chloroform	119.5	61.7	0.24	59 (61°C)	5.1
Benzene	78	80.1	0.5	94 (80°C)	2.3

Source: Smith et al. (1983), Table 2.2.

electrostatic force relative to a vacuum is called the **dielectric strength**. This property is what makes water an excellent solvent for ions, since less work is required to place the ion into the water matrix. In addition, the polarity and hydrogen-bonding capability of water facilitates direct attraction to ions and other molecules that are polar or form hydrogen bonds. Polar or hydrogen-bonding compounds with appreciable solubility in water are termed **hydrophilic** (water-loving). Nonpolar compounds with negligible aqueous solubility are called **hydrophobic**, or **lipophilic**. Hydrophobic compounds are important for forming separate phases within cells. The resulting interfaces are an important location for many biochemical reactions and control the movement of hydrophobic toxicants into and out of cells. Table 3.3 summarizes some of the interesting physicochemical properties of water and compares them to other solvents.

The physicochemical forces associated with large biological molecules depend in a very complex way on the complete chemical structure of the molecule and on its chemical environment. However, much of their behavior can be described in terms of local chemical structures, called **functional groups**, some of the most important of which are listed in Table 3.4. Consider any hydrocarbon molecule, consisting only of a carbon backbone and hydrogens. Replace any of the hydrogens with one of the functional groups from Table 3.4 and you produce a compound with very different properties. A simple example is methane, CH_4. Replace a hydrogen with a hydroxyl group, and methanol results, which has a much higher boiling point and forms a liquid that is completely miscible with water. Replacing with carboxyl forms acetic acid. Taking the three-carbon hydrocarbon propane and replace one hydrogen from each carbon with a hydroxyl produces glycerol, a basic component of biological fats.

3.5   OPTICAL ISOMERS

One of the simplest of the sugars is *glyceraldehyde*, which has two isomeric forms:

d(+)-glyceraldehyde   l(−)-glyceraldehyde

These structures may seem practically identical, and in fact they have the same physicochemical properties, such as melting point and aqueous solubility. However, the difference between them is biochemically critical. One is an **optical isomer** of the other; that

TABLE 3.4   Functional Groups and Their Properties[a]

Hydroxyl (OH)	$R-OH$	Polar; increases solubility in water; forms hydrogen bonds; characteristic of alcohols
Carboxyl (COOH)	$R-C\begin{smallmatrix}O\\\\OH\end{smallmatrix}$	Polar; most important acidic group in biochemicals; pK_a depends on R (the more electronegative, the lower the pK_a)
Amine (NH$_2$)	$R-NH_2$	Weak base; electronegative nitrogen attracts an additional H^+ from solution to become positively charged; common in proteins
Aldehyde (CHO)	$R-C=O$ with H below	Polar and water soluble; common in sugars and as fermentation products
Keto (CO)	$R_1-C=R_2$ with O below	Polar and water soluble; common in sugars and as fermentation products
Methyl (CH$_3$)	$R-C-H$ with H above and below	Nonpolar; reduces water solubility
Phosphate (PO$_4$)	$R-P-O$ with OH above and below	Polar acidic group; important in energy metabolism; found in DNA, sugars; additional organic side chains can replace the two hydrogens; phosphorus is usually present only in the form of phosphate
Sulfhydryl	$R-S-H$	Polar; forms disulfide bonds to link molecules
Disulfide	$R_1-S-S-R_2$	Formed from two sulfhydryl bonds; important in protein folding

[a]R stands for the rest of the molecule.

is, they rotate polarized light in opposite directions due to their mirror-image asymmetry. To understand this, it is necessary first to recognize that the four bonds that a carbon atom can participate in are arranged to point to the vertices of a tetrahedron if each bond connects to identical structures. This can be seen in Figure 3.4, which shows the tetrahedral

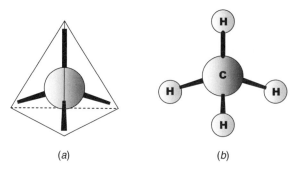

(a)   (b)

Figure 3.4   (a) Tetrahedral structure of carbon bonding and (b) a methane molecule. (Based on Gaudy and Gaudy, 1988.)

structure and a methane molecule having that form. If the functional groups on all four bonds are identical, the angle between any two bonds will be 109°.

The difference between the two types of glyceraldehyde is not obvious in the two-dimensional representation above. It seems that it should be possible to rotate the molecule around two of the bonds to change one form to the other. That it is not possible will be clearer if the molecule is viewed in its true three-dimensional form. The central carbon atom could be viewed as being at the center of a tetrahedron, with each of its four bonds pointing to a vertex. However, each of these bonds connects to a different group: an H, an OH, a CHO, and a CH_2OH. Such asymmetrical carbons are called **chiral centers**. Molecules with chiral centers can rotate polarized light either to the right, designated (+), or the left, designated (−). Biochemical compounds are often designated *d-* for *dextro*, meaning "right" or *l-* for *levo*, meaning "left," based on a relationship to the structure of (+)glyceraldehydes or (−)glyceraldehydes, respectively. More complex molecules may have multiple carbon atoms that can form centers for optical rotation. In such cases, the *d-* or *l-* notation indicates a relationship to the structure of glyceraldehydes, not whether the molecule actually rotates light to the right or left. Whether the molecule actually rotates polarized light to the right or left is designated by including (+) or (−) in the name, respectively.

Figure 3.5 shows three-dimensional views of *d-* and *l-*glyceraldehyde. In part (a) the two forms seem identical, but a close examination will show that bonds that point out of the plane of the page in one form point inward in the other. The difference between the two becomes more clear in part (b), which could be formed by taking hold of the hydrogen bonded to the central carbon of each compound and pointing it toward the viewer. The CH_2OH group is pointed to the left (the oxygen is hidden behind the carbon). Then it can clearly be seen that the two compounds cannot be superimposed on each other

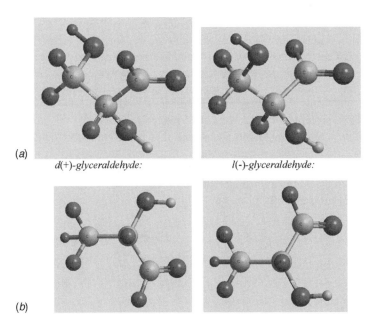

(a)

d(+)-glyceraldehyde: *l(-)-glyceraldehyde:*

(b)

Figure 3.5   Three-dimensional views of the glyceraldehyde structure: (*a*) mirror-image views; (*b*) views showing the orientation with hydrogen of the central carbon pointed toward the viewer.

because the three groups attached to the central carbon atom are arranged in opposite directions around the carbon.

The importance of chirality is that these isomers are biochemically different and will react differently from each other. Fundamentally, optical isomers have different shapes from each other. The shape of a molecule determines how it can form complexes with other molecules. As described below, the formation of complexes, especially with enzyme molecules, is a key step in a biochemical reaction. For example, *d*-glucose can be used for energy by the brain, but *l*-glucose cannot. Similar differences are true for many other stereoisomeric biochemical compounds. Another name for *d*-glucose is *dextrose*, the familiar ingredient in manufactured food items.

3.6   THE COMPOSITION OF LIVING THINGS

Four groups of compounds are of primary importance in living things: carbohydrates (including sugars, starches, cellulose, and glycogen), lipids (fats and oils), proteins, and nucleic acids (which form DNA and RNA). The first three of these form the majority of cell dry weight and are important for structural material, energy metabolism, and other metabolic functions. Nucleic acids are significant in reproduction and in energy metabolism. Finally, there are many compounds that do not fit neatly into these categories or may be hybrids of two or more.

3.6.1   Carbohydrates

Carbohydrates include sugars, starches, and structural materials such as cellulose. All have the empirical chemical formula $(CH_2O)_n$. For example, *glucose* is $C_6H_{12}O_6$, so *n* is 6. Glyceraldehyde is one of the simplest carbohydrates, with an *n* of 3. The large number of hydroxide groups on carbohydrates renders them hydrophilic. Carbohydrates are classified into several groups: **Monosaccharides** are the simplest and are building blocks for the others. They have relatively low molar masses, and *n* in the formula can range from 3 to 9. Monosaccharides can form chains, called **polymers**, producing **disaccharides**, which are formed from pairs of monosaccharides, or the long-chain **polysaccharides**, which can have molar masses as high as 1 million. Large molecules such as polysaccharides, proteins, or DNA are called **macromolecules**.

The structures of several monosaccharides that are commonly found in nature are shown below. *Ribose* is a five-carbon sugar; the others are six-carbon sugars. Glucose is a particularly important six-carbon sugar. It is the principal sugar formed by photosynthesis and is the main immediate source of energy for all cells. Nervous tissue in animals can only use glucose for energy.

H−C=O	H−C=O	H−C=O	CH₂OH
HO−C−H	H−C−OH	H−C−OH	C=O
H−C−OH	HO−C−H	H−C−OH	HO−C−H
H−C−OH	HO−C−H	H−C−OH	H−C−OH
H−C−OH	H−C−OH	CH₂OH	H−C−OH
CH₂OH	CH₂OH		CH₂OH
d(+)-glucose	*d*(+)-galactose	*d*(+)-ribose	*d*(−)- fructose

One end of monosaccharides such as glucose reacts spontaneously with either the other end or the adjacent carbon to form ring structures. For example, glucose can form a six-membered ring, and fructose, a five-membered ring.

d-glucose d-fructose

The rings are closed by an oxygen atom. The ring form and the open-chain form freely interconvert. However, for glucose, for example, the equilibrium highly favors the ring form.

A number of monosaccharide derivatives are important. Conversion of ends to $-COOH$ groups produces the **sugar acids**, such as glucuronic acid. Fermentation may result in the production of sugar acids as intermediates. **Amino sugars** are formed by replacing one of the hydroxyl groups on certain monosaccharides with a nitrogen-containing amino group (see below under proteins). Another derivative is the **deoxy sugar**, formed by replacing one of the hydroxy groups with a hydrogen; thus, one of the carbons will have two hydrogens. A very important deoxy sugar is deoxyribose, an important component of DNA.

Monosaccharides can also form a special type of bond with each other or with other molecules. This **glycosidic bond** forms between a hydroxide on the monosaccharide and a hydroxide on another molecule, which may be another monosaccharide, with the elimination of one molecule of water. Again, the link will be through an oxygen atom.

If the other molecule is another monosaccharide, the result is a disaccharide. Familiar table sugar is sucrose, a disaccharide formed from the monosaccharides glucose and fructose. Lactose, the sugar in milk, is formed from glucose and galactose. The same two monosaccharides can form disaccharides in several ways, depending on orientation and location of the connection. For example, *maltose* and *cellobiose* are both made from two glucose residues.

sucrose

Glycosidic bonds can form longer-chain carbohydrates called **oligosaccharides**, those with only a few monosaccharides, which include the disaccharides, and the much longer-chain polysaccharides. **Starch** is a type of polysaccharide produced by plants for energy storage. Humans obtain most of their dietary carbohydrate in the form of starch from grain. Starch is a chain of glucose residues bonded as in maltose. The chain may be straight, as in *amylose*, which makes up about 20% of potato starch; or it may be branched, as in *amylopectin*, which forms the other 80% of potato starch.

Animals store carbohydrates in a polysaccharide called **glycogen**. Glycogen is similar to starch except that the chain is much more highly branched. It is thought that by having more "ends" to the glycogen molecule, it is more available for rapid conversion to glucose for the sudden energy demands of animals. Depletion of glycogen in the muscle may cause the "wall" experienced by marathoners after several hours of running, which prevents them from continuing the race. With glycogen gone, the body switches to fat, which does not provide energy fast enough and produces other physiological stresses. In mammals, skeletal muscle contains about two-thirds of the body's glycogen, and the liver holds most of the rest. The liver uses the glycogen to control glucose levels in the blood. Figure 3.6*a* shows the basic structure common to both starch and glycogen.

Cellulose is an unbranched polysaccharide also composed exclusively of glucose residues, but with an important difference from starch. The glycosidic bond is reversed, as in the cellobiose disaccharide (see Figure 3.6*b*). This prevents the molecule from twisting, making it stiffer. As a result, it finds use as a structural material located in the cell walls of plants and forming the major component of wood. (It should be noted that the hardness of wood comes not from cellulose but from *lignin*, described below.) Another important fact about cellulose is that only a very few animals (such as garden snails) can digest it. Most cannot break it down to glucose to make the stored energy available. However, several animals, such as termites and cows, have developed associations with microorganisms that live in their digestive systems. The microorganisms accomplish the digestion for the animals.

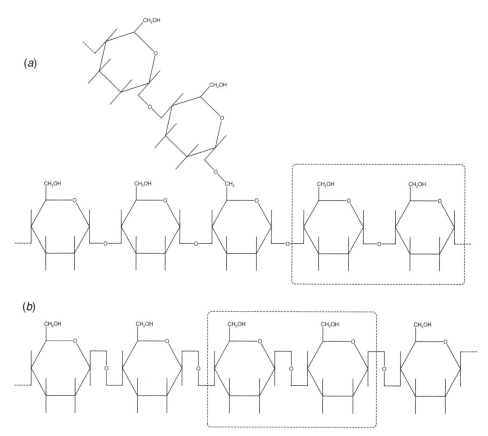

Figure 3.6   Polymers of glucose: (*a*) starch or glycogen showing a maltose repeating disaccharide unit; (*b*) cellulose with a cellobiose repeating unit.

Another structural polysaccharide is *chitin*, which forms the hard shell of arthropods such as insects and crabs, as well as the cell walls of most fungi. Chitin is a polymer of a sugar amine, *N*-acetyl glucosamine; crustaceans also include calcium carbonate in their shells. Chitin has been studied for use as an adsorption media for removing heavy metals from water. Other important structural polysaccharides include *agar* and *carrageenan*, which are extracted from seaweed. The former is used as a substrate for culturing bacteria, and the latter is used as a food thickener.

3.6.2 Lipids

Lipids refer to a loose category of compounds with the common property that they have fairly low solubility in water or are extracted from biological materials by solvents having polarity much less than water, such as ethanol or chloroform. There are five major types, of which the first four are described here and the fifth in a later section.

- Fatty acids, long-chain aliphatic carboxylic acids
- Fats, esters of fatty acids with glycerol
- Phospholipids, esters of phosphate and fatty acids with glycerol
- Lipids not containing glycerol, including waxes and steroids
- Hybrid lipids, such as those combined with carbohydrates or proteins

Fatty acids are simply straight-chain hydrocarbons with a carboxylic acid functional group at one end (Figure 3.7). They usually, but not always, have an even number of carbon atoms. It is the hydrocarbon chain that imparts hydrophobicity, since the carboxyclic acid group is water soluble. The larger the chain is, the more hydrophobic the molecule. The simplest fatty acid is *formic acid*, where a simple hydrogen makes up the variable R-group. The most familiar is *acetic acid*, formed with a methyl group. Vinegar is about 5% acetic acid. Both of these are quite water soluble. The melting point of fatty acids tends to increase with chain length.

Fatty acids usually do not accumulate in nature. Systems may be engineered to produce them, as is done in fermentation processes. An important environmental application is anaerobic digestion, in which fatty acids consisting mainly of acetic acid, but also

Basic structure:
$$R—COOH$$

Saturated fatty acids:

Formic	$HCOOH$
Acetic	CH_3COOH
Proprionic	CH_3CH_2COOH
n-Butyric	$CH_3(CH_2)_2COOH$
Caproic	$CH_3(CH_2)_4COOH$
Palmitic	$CH_3(CH_2)_{14}COOH$
Stearic	$CH_3(CH_2)_{16}COOH$

Unsaturated fatty acids

Oleic	$CH_3(CH_2)_7CH{=}CH(CH_2)_7COOH$
Linoleic	$CH_3(CH_2)_4CH{=}CHCH_2CH{=}CH(CH_2)_7COOH$
Arachidonic	$CH_3(CH_2)_4{-}(CH{=}CH{-}CH_2)_4{-}(CH_2)_2{-}COOH$

Figure 3.7   Structures of some of the more common fatty acids found in nature.

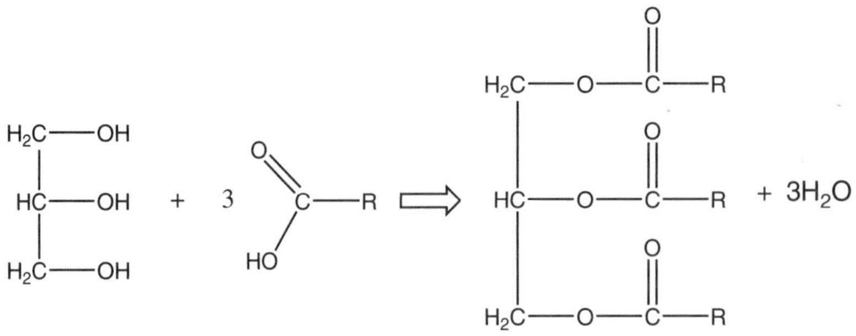

Figure 3.8   Formation of triglyceride from glycerol and fatty acids.

proprionic, butyric, and others, can accumulate to levels typically below 1%, as an intermediate in methane production.

Saturated fatty acids are those in which the maximum number of hydrogens have been bonded with the carbons in the hydrocarbon chain. If a hydrogen is removed at a point in the chain, the corresponding carbon has an extra bond to form, to fill its complement of four. It satisfies this requirement by forming a double-bond with its neighbor (which also loses a hydrogen). Fatty acids containing one or more double-bonded carbons are called **unsaturated fatty acids.** Unsaturation puts a kink in the chain, reducing the ability of the molecules to pack together. As a result, they are less likely to form solids (i.e., their melting points are lowered).

Except for formic acid (pK_a = 3.75), all of the saturated fatty acids have acid dissociation constants averaging around 4.85, similar to that of acetic acid (pK_a = 4.76).

Fats are formed from the covalent bonding of three fatty acids with the three hydroxides of a glycerol (Figure 3.8). For this reason they are often called **triglycerides.** The bonding of a hydroxyl of an alcohol with the hydroxyl of a carboxylic acid, with elimination of a water molecule, results in an **ester linkage**. This eliminates the hydroxide from the alcohol and the ionizable portion of the acid. With the loss of these functional groups, polarity is greatly reduced, ionization is eliminated, and water solubility is therefore decreased. Because the ester linkage is only slightly polar, the properties of fats are dominated by the properties of the hydrocarbon chains of the fatty acids that form them.

The reverse of the esterification reaction is **hydrolysis**, which means "splitting with water." Hydrolysis of ester bonds is catalyzed by H^+ or by enzymes called **lipases**. The ester bond can also be hydrolyzed in a strong basic solution, in a reaction called **saponification**. The salts of the fatty acids thus formed are soaps. In fact, soap was made in preindustrial households by reacting fats collected from meat with lye (sodium hydroxide) leached from ashes.

Oils are fats that are liquids at room temperature. Fats made from unsaturated fatty acids are **unsaturated fats**. As with the fatty acids, they will have lower melting points. This is why you may read that margarine or shortening contains "partially hydrogenated fats." Fats are hydrogenated by reacting unsaturated vegetable oils with hydrogen, reducing the number of double bonds, in order to decrease the melting point, allowing the oil to form a solid. In a similar way, living things select for the saturation of fats to obtain properties that are useful.

The major functions of fats in organisms are for energy storage and structural use. Fats yield twice as much chemical energy as carbohydrates or proteins; fats average about 9.3 Calories per gram vs. 4.1 Calories per gram for both proteins and carbohydrates. (*Note that the dietary Calorie, which must be capitalized, is equivalent to the kilocalorie*

TABLE 3.5   Percentage of Types of Fats in Some Dietary Fats and Oils

	Saturated	Mono-unsaturated	Poly-unsaturated	Other	Cholesterol (mg/tB)
Canola oil	6	62	31	1	0
Safflower oil	9	12	78	1	0
Olive oil	14	77	9	0	0
Beef fat	51	44	4	1	14
Butter	54	30	4	12	33
Coconut oil	77	6	2	15	0

commonly used in thermodynamics.) However, fat storage is long term. The energy contained in fats is not released as rapidly as with the polysaccharides starch or glycogen. Other functions of fats are for insulation for animals exposed to cold, and for flotation in marine animals. The blubber of sea mammals is an example of both of these functions.

The degree of fat saturation in the human diet is linked to human health. A small amount of unsaturated fat is required in the diet, because humans cannot form fatty acids with double bonds. One of these in particular, linoleic acid, can be converted into all the other fatty acids needed by humans. For this reason, it and several others are placed in a dietary group called the **essential fatty acids**.

A mammalian diet containing a high proportion of saturated fats is associated with cardiovascular disease, via its connection with cholesterol and blood lipoproteins. (Cholesterol is another type of lipid, described below. Lipoproteins are hybrid compounds, also described below.) Among the unsaturated fats, **monounsaturated fats**, those with only one double bond, seem to be even more healthful than **polyunsaturated fats**, those with more than one double bond. Table 3.5 shows how various fats compare in these dietary constituents. Olive oil has the highest monounsaturated fat content of all. This has lent support to the "Mediterranean diet," in which olive oil replaces butter and other animal fats in recipes, to the extent of placing a plate of oil on the table for dipping bread instead of a butter dish. Lest one should think that only animal fats are suspect, notice the high saturated fat content of coconut oil. High polyunsaturated fat content also seems to beneficially lower blood lipid concentration. Safflower oil is best in this regard, followed by corn, peanut, and cottonseed oils.

Phospholipids are particularly critical to life, as they form the basic structure of cell membranes. Glycerol can form ester linkages with inorganic as well as organic acids. Phospholipids consist of ester linkages of glycerol with two fatty acids and with phosphoric acid in the third position. Usually, another of the phosphate hydroxyl groups form, in turn, another ester bond with still another organic molecule, called a *variable group*. The variable group often contains nitrogen, adding to the polar character of the lipid.

phospholipid structure

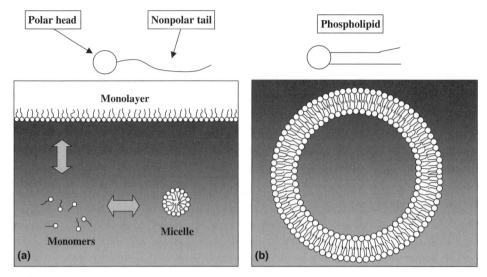

Figure 3.9 Surfactant structures formed in solution: (*a*) behavior of surfactants in solution; (*b*) phospholipid bilayer structure.

This structure results in a **surfactant**, a molecule that has a hydrophilic part and a hydrophobic part. Similar to soaps and detergents, if there are enough molecules in solution, they will form aggregates called **micelles**, a sort of "circling the wagons" in which the polar ends face the water and the hydrophobic ends form a separate phase in the interior of the micelle (Figure 3.9). One of the phospholipids is *lecithin*, which is important in metabolism of fats by the liver. Egg yolks are rich in lecithin, and its detergent nature helps maintain the emulsion between oil and vinegar in mayonnaise.

Biological lipids show an important difference in behavior from other surfactants. Under the right conditions, they form micelles with multilayered structures that can become **vesicles** (water-filled cavities) bounded by a **lipid bilayer**, as also shown in Figure 3.9. This important structure is the basis for the cell membrane, and therefore for the cell itself. The bilayer membrane forms a barrier to the uncontrolled passage of water-soluble constituents into and out of the cell. However, its lipophilicity makes it the site of action of many lipophilic pollutants, such as some industrial organic solvents.

Other lipids: Since the definition of lipids is based on physicochemical properties and not chemical structure, it is not surprising that the group is very diverse. A variety of other biological compounds fit the category. **Waxes** are long-chain fatty acids combined with long-chain alcohols other than glycerol. They are formed by plants to produce protective and water-conserving layers on their surfaces and by animals such as the honeybee, or for ear canal protection. **Terpenes** are long-chain hydrocarbons based on units similar to the compound isoprene. They include essential oils from plants, among other compounds.

One important group of compounds that is in this catch-all class of lipids is the steroids. **Steroids** consist of four fused rings, three with six carbons and one with five, and with a hydroxide at one end and a hydrophobic "tail" at the other. Steroids are important regulatory chemicals in plants and animals. They are able to move through cell membranes formed by lipid bilayers. The sex hormones estrogen and testosterone are steroids, as are the fat-soluble vitamins A, D, and E. The most abundant steroid in animals is *cholesterol*. It forms an essential part of the cell membrane, affecting its fluidity. However, factors related to diet and heredity can cause cholesterol to form obstructive deposits in blood vessels, causing strokes or heart attacks.

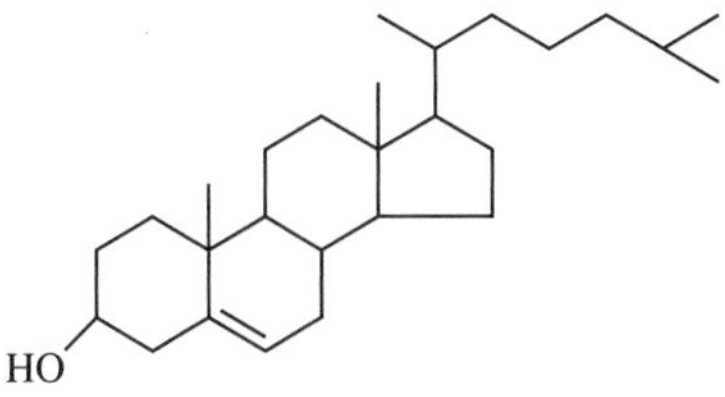

cholesterol

Another lipid that is important in biological wastewater treatment is *poly-β-hydroxybutyric acid* (PHB). This is used for energy storage by some bacteria.

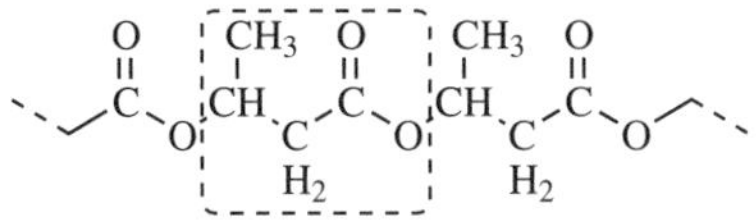

poly-β-hydroxybutyric acid

3.6.3 Proteins

Proteins have a central role in cell function. Like carbohydrates and lipids, they are involved in structure and in energy metabolism. More important is the function unique to proteins: Each of the thousands of enzymes in living things, each of which catalyzes a specific biochemical reaction, is a protein specialized for that task. In addition, they may act as biochemical regulators or hormones, such as insulin; transport chemicals such as hemoglobin, which transports oxygen in the blood; or they may be responsible for motility, as in the cilia and flagella of protists.

Proteins are composed of one or more chains of amino acid repeating units. Each individual chain is called a **polypeptide**. In turn, **amino acids** are compounds in which a central carbon is covalently bonded to three functional groups: an amine, a carboxylic acid, and a variable organic side group:

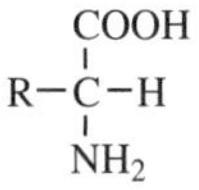

$$
\begin{array}{c}
\text{COOH} \\
| \\
\text{R}-\text{C}-\text{H} \\
| \\
\text{NH}_2
\end{array}
$$

Many amino acids are found in nature, but only 20 are commonly part of proteins. In contrast to polysaccharides, proteins are formed not from one or two repeating units, but from all 20, and in any sequence. This makes possible a virtually unlimited variety of structure and of corresponding function. A chain of n amino acids can form 20^n different proteins. For a length of only five acids, this gives 3.2 million combinations. Typical chain lengths are from 100 to several thousand. Of course, only certain combinations actually occur. The human body has about 100,000 different proteins.

Table 3.6 shows the 20 amino acids commonly found in proteins. Some have simple side groups (e.g., glycine and alanine), others are complex (e.g., tryptophan). Two contain sulfur (cysteine and methionine). Some have ionizable side groups, such as the carboxylic

TABLE 3.6   Amino Acids Commonly Found in Proteins and Their pI Values

Name	Symbol	Structure	pI
Amino Acids with Nonpolar R Groups			
Alanine	Ala		6.00
Isoleucine[a]	Ile		6.02
Leucine[a]	Leu		5.98
Methionine[a]	Met		5.74
Phenylalanine[a]	Phe		5.48
Proline	Pro		6.30
Tryptophan[a]	Trp		5.89
Valine[a]	Val		5.96
Amino Acids with Uncharged Polar R Groups			
Asparagine	Asn		5.41
Cysteine	Cys		5.07
Glutamine	Gln		5.65

TABLE 3.6   (*Continued*)

Name	Symbol	Structure	pI
Glycine	Gly	$H-\overset{\displaystyle H}{\underset{\displaystyle NH_3^+}{C}}-COO^-$	5.97
Serine	Ser	$HO-CH_2-\overset{\displaystyle H}{\underset{\displaystyle NH_3^+}{C}}-COO^-$	5.68
Threonine[a]	Thr	$H_3C-\underset{\displaystyle OH}{CH}-\overset{\displaystyle H}{\underset{\displaystyle NH_3^+}{C}}-COO^-$	5.60
Tyrosine	Tyr	$HO-\langle\bigcirc\rangle-CH_2-\overset{\displaystyle H}{\underset{\displaystyle NH_3^+}{C}}-COO^-$	5.66

Amino Acids with Acid R Groups (Negatively Charged at pH 6.0)

Name	Symbol	Structure	pI
Aspartic acid	Asp	$\overset{\displaystyle ^-O}{\underset{\displaystyle O}{\diagdown}}C-CH_2-\overset{\displaystyle H}{\underset{\displaystyle NH_3^+}{C}}-COO^-$	2.77
Glutamic acid	Glu	$\overset{\displaystyle ^-O}{\underset{\displaystyle O}{\diagdown}}C-CH_2-CH_2-\overset{\displaystyle H}{\underset{\displaystyle NH_3^+}{C}}-COO^-$	3.22

Amino Acids with Basic R Groups (Positively Charged at pH 6.0)

Name	Symbol	Structure	pI
Arginine	Arg	$^+H_3N-\underset{\displaystyle NH}{C}-NH-CH_2-CH_2-CH_2-\overset{\displaystyle H}{\underset{\displaystyle NH_3^+}{C}}-COO^-$	10.76
Histidine	His	$\underset{\underset{\displaystyle H}{\overset{\displaystyle ^+HN\diagdown\,\diagup NH}{C}}}{HC}=C-CH_2-\overset{\displaystyle H}{\underset{\displaystyle NH_3^+}{C}}-COO^-$	7.59
Lysine[a]	Lys	$^+H_3N-CH_2-CH_2-CH_2-CH_2-\overset{\displaystyle H}{\underset{\displaystyle NH_3^+}{C}}-COO^-$	9.74

[a]Essential amino acids.

acids in aspartic acid and glutamic acid, or the basic amines of lysine, histidine or arginine. Amines, including the one connected to the central carbon, ionize by accepting an additional hydrogen ion at pH levels below a characteristic pK value. Because all amino acids have both an acidic and a basic functional group, all will have an ionic charge at either sufficiently high or low pH values. For some, it is possible at certain pH values for both groups to be ionized:

$$
\begin{array}{ccc}
& \text{COOH} & & \text{COO}^- \\
& | & & | \\
\text{R}-\text{C}-\text{H} & \Longrightarrow & \text{R}-\text{C}-\text{H} \\
& | & & | \\
& \text{NH}_2 & & \text{NH}_3{}^+
\end{array}
$$

These behaviors are important in forming attractions between different parts of the protein molecule, affecting its shape. Table 3.6 also gives the **isoelectric point (pI)**, which is the pH at which the number of positive charges on the amino acid in solution equals the number of negative charges.

The fact that all amino acids contain nitrogen explains the importance of that element to all organisms. Nitrogen limitations can cause growth problems in biological wastewater treatment and in natural ecosystems. Humans also have a nutritional requirement for proteins. We can form many of the amino acids from other compounds. However, we cannot synthesize the following eight, which are called **essential amino acids**: *isoleucine, leucine, lysine, methionine, phenylalanine, threonine, tryptophan,* and *valine.* These must be provided in the diet daily, since free amino acids are not stored in the body. The absence of any one causes protein synthesis to stop. Most animal proteins contain all eight, but many plant proteins do not, or do not have them in balanced amounts. For this reason vegetarians need to be aware that they must eat vegetables in combinations that eliminate deficiencies. For example, rice is deficient in lysine but has sufficient methionine, whereas beans are low in methionine but contain adequate lysine. However, beans and rice together in a diet supply a complete protein supply.

Amino acids form polypeptides by covalently bonding the carbon from the carboxylic acid group to the nitrogen of the amine, with loss of a water molecule. The result is called a **peptide bond**:

$$
\text{R}_1-\overset{\overset{\text{H}}{|}}{\underset{\underset{\text{NH}_2}{|}}{\text{C}}}-\text{C}\overset{\nearrow\text{O}}{\underset{\searrow\text{OH}}{}} \ + \ \text{R}_2-\overset{\overset{\text{H}}{|}}{\underset{\underset{\text{NH}_2}{|}}{\text{C}}}-\text{C}\overset{\nearrow\text{O}}{\underset{\searrow\text{OH}}{}} \ \Rightarrow \ \text{R}_1-\overset{\overset{\text{H}}{|}}{\underset{\underset{\text{NH}_2}{|}}{\text{C}}}-\overset{\overset{\text{O}}{||}}{\text{C}}-\overset{}{\underset{\underset{\text{H}}{|}}{\text{N}}}-\overset{\overset{\text{COOH}}{|}}{\underset{\underset{\text{H}}{|}}{\text{C}}}-\text{R}_2 \ + \text{H}_2\text{O}
$$

peptide bond formation

Because of the amino acid functional groups, proteins also possess charges in solution, which vary with pH. Like individual amino acids, proteins have an isoelectric point. These can range from less than 1.0 for pepsin, the digestion enzyme that must act under acidic conditions in the stomach, to 10.6 for cytochrome *c*, which is involved in cellular respiration.

Unlike polysaccharides, which can have random lengths and random branching, each peptide is a single chain with a precise sequence of amino acids. Changing even a single amino acid can destroy the ability of the resulting protein to perform its function. Furthermore, the peptide chain must arrange itself into a complex shape, which is determined by the exact amino acid sequence, and often by the method by which the cell machinery

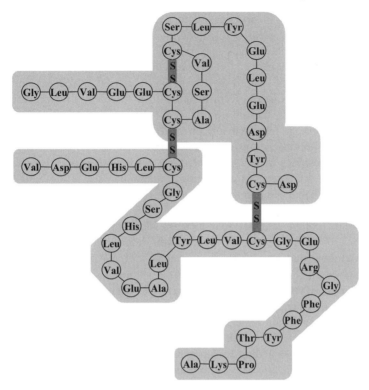

Figure 3.10   Tertiary and quarternary protein structure as shown in bovine insulin. This protein consists of two polypeptide chains joined by two disulfide bonds. Another disulfide bond within the smaller chain contributes to the molecule's shape. (Based on Bailey and Ollis, 1986.)

constructs the protein. The structure of a protein has three or four levels of organization. The **primary level** is the actual amino acid sequence. The **secondary level** refers to relatively local arrangements such as coiling into a helix or folding into a pleated sheet. The helix is held together by hydrogen bonds between the peptide bonds of every fourth amino acid. The **tertiary level** of organization is larger-scale folding and coiling, to give the overall shape to the molecule. Some proteins will exhibit the **quaternary level** of structure, in which several polypeptides are linked together by a variety of attractions, including hydrogen bonds, ionic attraction, or covalent disulfide linkage between cysteine amino acids on the two peptides. Hemoglobin, for example, consists of four polypeptide units. Figure 3.10 shows an example of protein structure. If the protein forms a compact, water-soluble state, which the majority of proteins do, they are called **globular proteins**. **Fibrous proteins** are elongated and often function in structural applications in connective tissue, contractile tissue, or as part of the hair or skin in mammals.

Proteins molecules often have other chemical compounds, called **prosthetic groups**, included in their structure, usually through noncovalent bonding. Often, they include metal ions. Hemoglobin contains four organic prosthetic groups, each containing an iron atom. Other proteins may contain chromium, copper, or zinc, for example. This is one of the reasons that humans and other organisms have a nutritional requirement for some heavy metals.

Since the higher levels of protein structure depend on relatively weak bonds such as hydrogen bonds, they are easily disrupted by increasing temperature or by changing pH or ionic strength. Such changes may result in conversion of the protein to a non-functional form, which is said to be **denatured**. These changes are often reversible. For example, hair can be curled by wrapping it around a rod and heating. This breaks hydrogen bonds, which re-form upon cooling, "freezing" the protein in the new shape. However, there is tension in the hair fibers, and with time the hydrogen bonds gradually rearrange into their former relationship, losing the curl. A "permanent" rearrangement can be made by using chemical treatment, which breaks disulfide bonds between cysteine residues in hair proteins, then re-forms them in the curled shape. A common example of irreversibly denaturing proteins by heat is the cooking of eggs. Heat disrupts the globular albumin proteins, which do not return to their native state upon cooling.

Enzymes are protein catalysts that increase biochemical reaction rates by factors ranging from 10^6 to 10^{12} over the uncatalyzed reactions. They often include non-amino acid portions that may be organic or consist of metallic ions. These are called **cofactors.**

Most enzymes are named with the suffix -*ase*. For example, *lipase* is an enzyme that digests lipids. Another enzyme is *lactase*, which catalyzes the breakdown of milk sugar, the disaccharide *lactose*, into monosaccharides glucose and galactose. Many adults, and almost all non-Caucasian adults, lose their ability to produce lactase after early childhood. However, some bacteria, including *Escherichia coli*, produce a different lactose-digesting enzyme. Adults lacking lactase who eat milk products have abdominal disturbances when the bacteria in the gut begin to produce gas using the lactose.

Enzymes are very specific; each catalyzes one or only a few different reactions, which is sensitively controlled by its shape. It is remarkable that contrary to reactions in aqueous media in the laboratory, enzyme-catalyzed reactions produce few side reactions. Equally remarkable is the fact that, with enzymes, a wide variety of reactions are promoted at mild conditions of temperature, pressure, and pH.

Each enzyme has at least one **active site**, the location on the molecule that binds with the **substrate(s)** (the reactants in the catalyzed reaction). The active site attracts the sub-strate(s) and holds it, usually by physicochemical forces. Two major mechanisms by which enzymes increase reaction rates are (1) by bringing the reactants close together, and (2) by holding them in an orientation that favors the reaction (Figure 3.11). It is also thought that enzymes can act by inducing strain in specific bonds of bound substrates, making certain reactions favorable.

Since the shape of a molecule is so sensitive to its environment, the cell can turn reactions on or off by changing conditions (e.g., pH) or by providing or withdrawing a cofactor or inhibitory compound. Figure 3.12 shows how a cofactor could promote binding of a single substrate with an enzyme. The cofactor binds first with the enzyme, changing the shape of the active site. This allows the substrate to bind, forming the complex. As with all proteins, denaturing stops the function of any enzyme.

Enzymes may also require **coenzymes**, which are molecules that function by accepting by-products of the main reaction, such as hydrogen. Coenzymes differ from cofactors and from enzymes themselves in that they are consumed by the reaction (although they may be regenerated in other reactions). Examples include NAD and FAD, discussed below. Some cofactors and coenzymes cannot be synthesized by mammals and must be included in their diet, making them what we call *vitamins*.

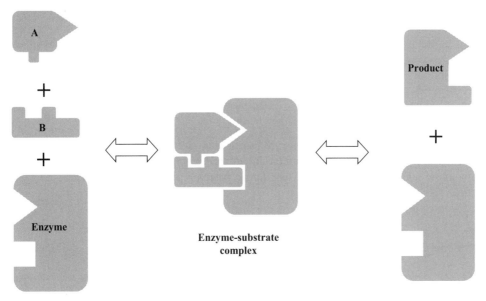

Figure 3.11   Enzyme control of proximity and orientation of substrates.

Another important protein function is their use as binding proteins. Hemoglobin is an example of a binding protein that transports oxygen in the blood. Other binding proteins are active in the immune system, which responds to foreign substances in animals. The cell membrane is studded with proteins that function in communicating substances

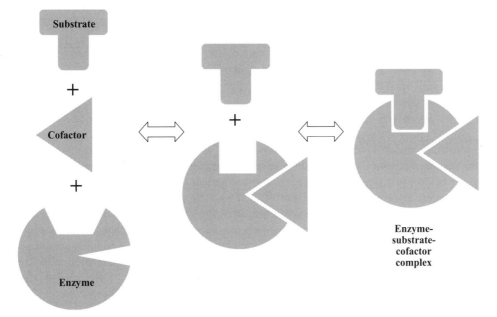

Figure 3.12   Hypothetical enzyme mechanism involving a cofactor.

and signals into and out of the cell. Cell membrane proteins are also a point of attack for infectious agents such as viruses, or may bind with drugs, leading to reactions that produce their characteristic effects.

3.6.4   Nucleic Acids

Nucleic acids do not form a large portion of the mass of living things, but make up for this in importance by being central to reproduction and control of cell function (DNA and RNA) and as the single most important compound in energy metabolism [adenosine triphosphate (ATP)]. DNA and RNA are linear polymers of nucleotides; ATP is a single nucleotide.

A **nucleotide** is a compound consisting of three parts:

pyrimidine or purine base + ribose or deoxyribose sugar + one or more phosphates

The **pyrimidines** are based on a six-membered ring containing two nitrogens. Only the following three pyrimidines are found in DNA and RNA: *thymine*, *cytosine*, and *uracil* (Figure 3.13). **Purines** have an additional five-membered ring fused to the pyrimidine. Only two purines are used in DNA and RNA, *adenine* and *guanine*. Nucleotides using these bases are labeled with their first letter: A, G, U, C, or T. The five-carbon sugars are bonded to the base, and the phosphate(s) are connected to the sugar via an ester.

A **nucleoside** is the same as a nucleotide, without the phosphate. The nucleoside formed from adenine and ribose is called **adenosine**. The nucleoside formed from thymine and deoxyribose is called **thymidine**.

Besides forming a chain of phosphates, the phosphate portion of the molecules can form ester bonds to two nucleotide sugars, forming a linear polymer with the phosphates and sugars as a "backbone" and the bases as branches (Figure 3.14).

Deoxyribonucleic acid (DNA) and **ribonucleic acid** (RNA) are polymers formed from nucleotides. In DNA the sugars are deoxyribose; in RNA the sugars are ribose. Another important difference between these polymers is that DNA does not contain uracil, while RNA never includes thymine. In other words, DNA includes only A, G, C, and T; RNA has only A, G, C, and U.

RNA is present only as a single chain. However, two strands of DNA form a fascinating structure called the **double helix**, discovered by Watson and Crick in 1953. It happens that thymine and adenine on two different strands can form two hydrogen bonds in just the right position relative to each other, and cytosine and guanine on two complementary strands form three such bonds (Figure 3.14). Thus, the sequence of bases on one strand determines the sequence on the other. The two complementary strands are held together by a large number of hydrogen bonds, making the pairing very stable. The resulting structure resembles a ladder, with each sugar–phosphate backbone making one side of the ladder, and the base pairs forming the rungs. A purine is always opposite a pyrimidine, to keep the lengths of the rungs all the same. In addition, the ladder has a right-hand twist to it, making a complete turn every 10 "rungs" or base pairs, producing the famous double helix structure.

Notice from Figure 3.14 that the point of attachment at one end of the chain of nucleotides is carbon 5 of the sugar, and at the other end it is carbon 3. This gives a direction to

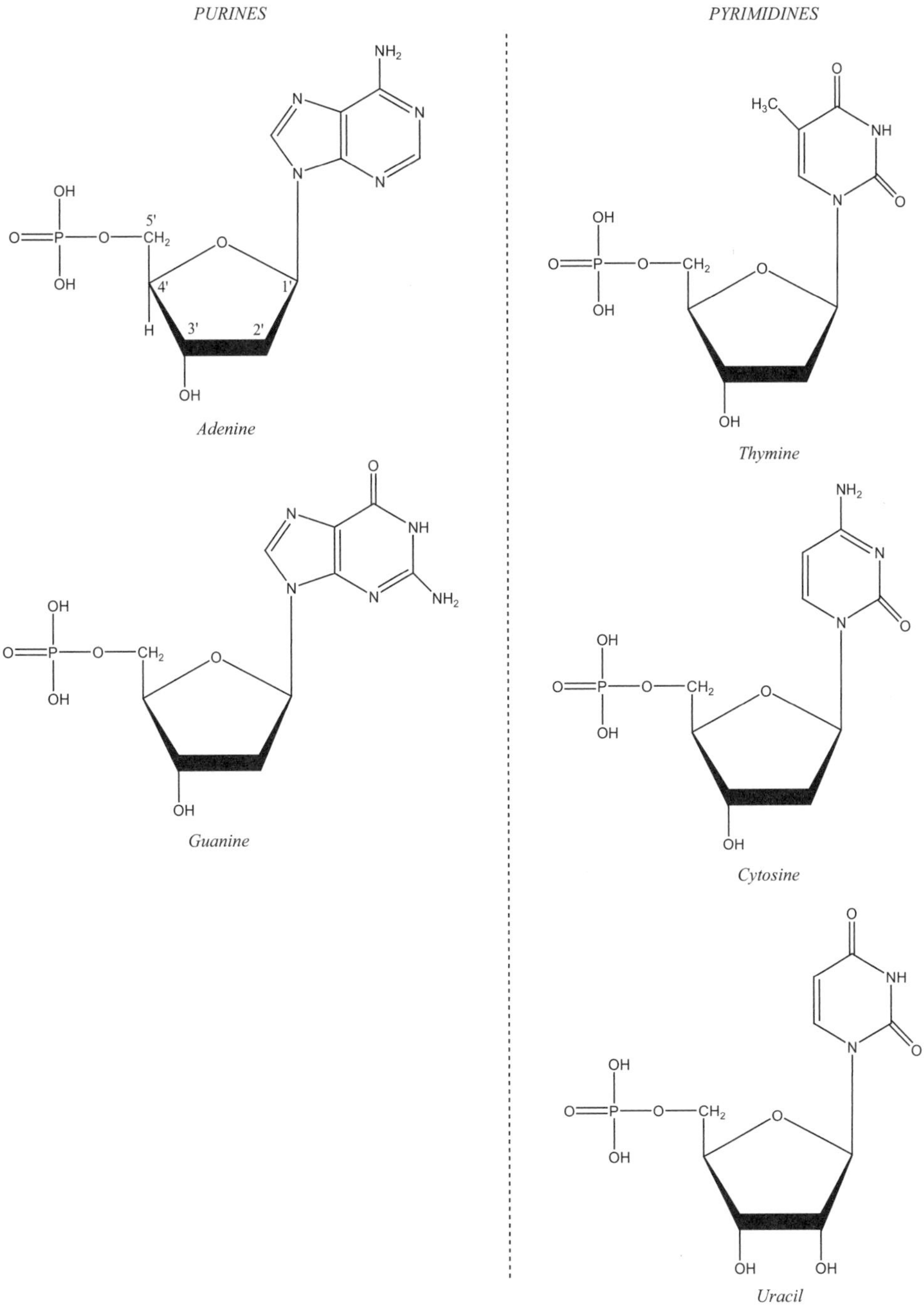

PURINES

PYRIMIDINES

Adenine

Thymine

Guanine

Cytosine

Uracil

Figure 3.13   Nucleotide structures.

the chain. The ends are labeled 3' and 5'. You can also see that the complementary DNA chains run in opposite directions.

The importance of DNA is that it carries all of the genetic information of all living things. The information is coded in the sequence of bases in the DNA molecule. The two strands, although different, are complementary and carry the same information. That

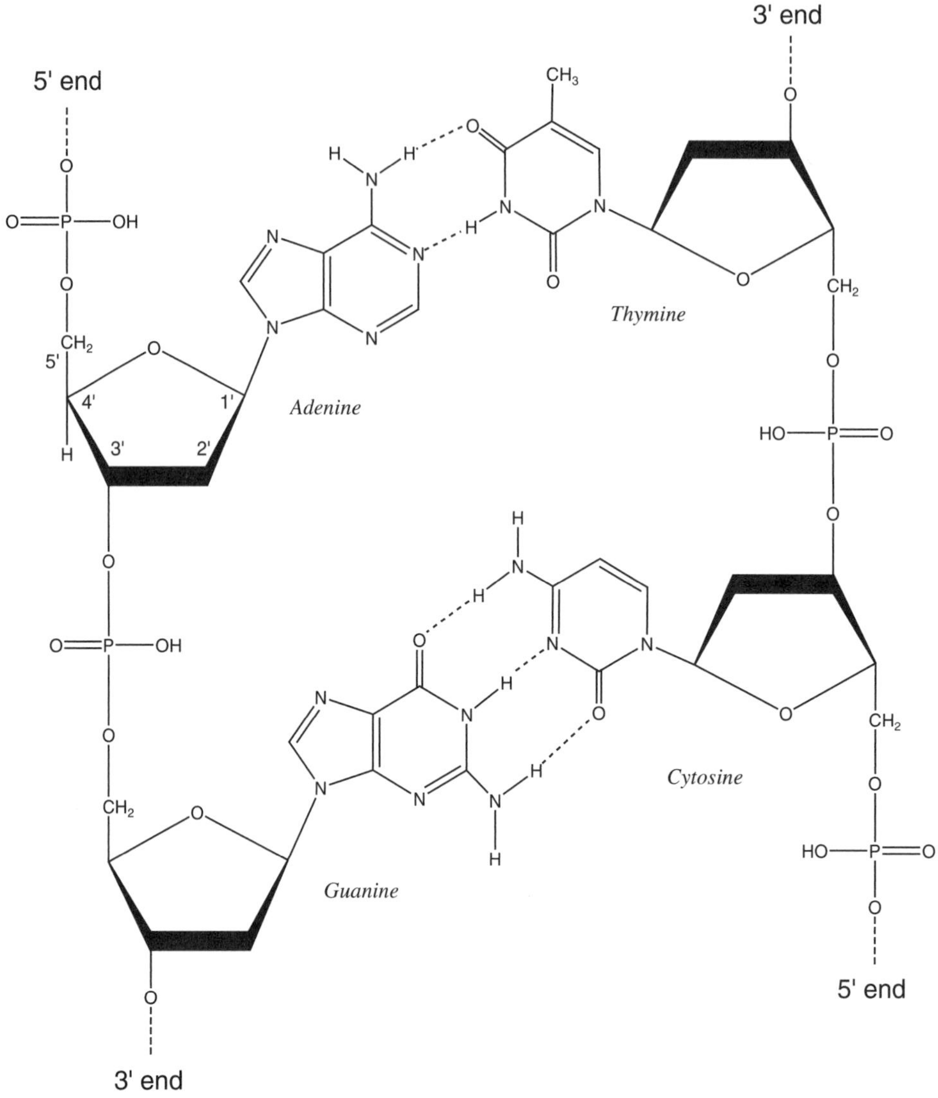

Figure 3.14   DNA molecule section showing the phosphate-sugar "backbone."

is, if one strand has the bases ATGCCACTA, the other strand must be TACGGTGAT, to form the following pairings:

$$3' \ldots \text{ATGCCACTA} \ldots 5'$$
$$5' \ldots \text{TACGGTGAT} \ldots 3'$$

What does DNA code for? Very simply, by specifying the sequence of amino acids, it contains instructions for the construction of all the proteins that the organism can make. Recall that proteins can have 20 different amino acids. So how can four nucleotide bases code for all 20? The answer is that the DNA bases code in groups of three.

The sequence for the bottom strand in the example above, and the corresponding amino acid sequence, is

$$
\begin{array}{llll}
\text{DNA}: & \text{TAC} & - \quad \text{GGT} & - \quad \text{GAT} \\
\text{AA}: & \text{methionine} & - \quad \text{proline} & - \quad \text{leucine}
\end{array}
$$

As mentioned above, RNA uses uracil in place of thymine and ribose sugar instead of deoxyribose. In addition, it is a single strand. The major function of RNA is to communicate the DNA code from the cell nucleus to the cytoplasm, where proteins are synthesized. More details on the mechanisms involved are provided in Section 6.2.1. RNA has another function, recently discovered. It can act as a catalyst, similar to protein enzymes. RNA with this capability are called **ribozymes**. One school of thought holds that because RNA can act as both a genetic template and as a catalyst, it may be that when life originated, it was based on RNA for both of those functions.

Several nucleotide monomers are important participants in biochemical reactions. More is said in Section 5.1.3 about *adenosine triphosphate* (ATP) and its central role in energy metabolism. The cell uses a number of other nucleotides. Cyclic adenosine monophosphate (cAMP) is involved in regulation of cell metabolism. Adenine is combined with other organic molecules to form a number of coenzymes, including:

flavin adenine dinucleotide (FAD)

nicotinamide adenine dinucleotide (NAD)

nicotinamide adenine dinucleotide phosphate (NADP)

guanosine triphosphate (GTP)

These compounds are important in the mechanisms for many biochemical processes, including photosynthesis and respiration, as discussed below.

3.6.5 Hybrid and Other Compounds

Hybrid compounds are those composed of a combination of two or more types of compounds, such as sugar combined with protein or with lipid. Some have already been discussed; examples are some proteins with their prosthetic groups; the nucleotides themselves, which contain sugars; and the nucleic acid coenzymes.

In addition, sugars commonly combine covalently with lipids and proteins. Many are important in cell membranes. For example, *peptidoglycans* have the interesting property that they form a two-dimensional polymer (covalently bonded in both the x and y directions) so that they encapsulate bacterial cells with a single huge macromolecule. Lipopolysaccharides include bacterial cell membrane components called **endotoxins**, which are responsible for powerful toxic effects in animals. Bacteria important in wastewater treatment secrete a coating of polysaccharides and lipopolysaccharides that enables them to flocculate into large aggregates or to form slime layers called **biofilms**. This improves their ability to capture particulate food matter.

Lipoproteins are noncovalently bound lipids and proteins. Since lipids are insoluble in water, they are transported in the blood by being associated with proteins in this way. In other words, proteins act as a "detergent" to solubilize lipids, including cholesterol. Despite cholesterol's bad reputation related to disease of the circulatory system, it is an essential component of animal cell membranes and a precursor for steroid hormones and bile acids (which aid in lipid digestion). However, people with high levels of cholesterol in the blood tend to have higher incidences of **arteriosclerosis**, the narrowing and blocking of arteries by cholesterol deposits. This can lead to strokes and heart attacks. Blood lipoproteins can be separated into fractions distinguished by density. One of the fractions is low-density lipoprotein (LDL). Blood cholesterol is concentrated in the LDL fraction, resulting in its being labeled as "bad" cholesterol. A high LDL level is associated with the intake of saturated fats and alleviated by the intake of monounsaturated fats.

Lignin is a polymer of various aromatic subunits, such as phenylpropane. It is produced by woody plants as a resin, bonding the cellulose fibers into a tough composite.

3.7   DETECTION AND PURIFICATION OF BIOCHEMICAL COMPOUNDS

The tremendous amount that is known about biochemistry may seem somewhat mysterious without an appreciation for the methods used to gain it. A brief description of some of the techniques used to purify and detect biochemical compounds may help with understanding how this body of knowledge came about.

Thin-layer chromatography is done with paper or a glass plate coated with silica powder or other adsorbent material. A solution containing a mixture of biochemical compounds, ranging from amino acids and nucleic acids to polypeptides, is placed in a spot near one corner. One edge is then placed in contact with a solvent, which carries the compounds along the plate as it is drawn up by capillary action. Compounds with various physicochemical properties move at different speeds. This spreads the compounds out in a line along one edge. That edge is then placed in a different solvent, which moves the compounds across the plate in the other direction. This distributes the individual compounds from the mixture across the two dimensions of the plate (Figure 3.15). The resulting spots can be made visible by chemical treatment or can be removed for further experiments.

If the molecules are electrically charged, such as amino acids and polypeptides at the appropriate pH, one of the solvent steps can be replaced with an electrostatic field to move the compounds. This is called **electrophoresis**.

Large molecules and subcellular particles can be separated by sedimentation. This is done in an **ultracentrifuge**, which creates accelerations up to $400,000g$ by spinning as fast as 75,000 rpm. Molar masses of large macromolecules can be determined from their sedimentation rate. Particle size is often given in terms of its settling velocity in a centrifuge, measured in **Svedberg units**.

Immunoassay uses antibodies to form a precipitate with specific compounds. *Antibodies* are special proteins produced by the body to bind with foreign substances so that they can be made harmless. Each antibody is highly specific, binding only to a single substance and binding extremely tightly in what is called a *lock-and-key relationship*. Molecular biology techniques have enabled the production of large quantities of antibodies of a specific type, called **monoclonal antibodies**. They are used for research purposes as well as to detect specific hormones in pregnancy tests and tests for prostate cancer. Immunoassay is a highly sensitive and selective detection method. Its use has been extended to organic pollutants and even to heavy metals.

The power of the methods described above is often increased by the use of **radioisotope labeling**. In this technique, radioactive compounds such as ^{3}H, ^{14}C, ^{32}P or ^{33}P, and ^{35}S are incorporated into substrates, making it easy to detect the products incorporating

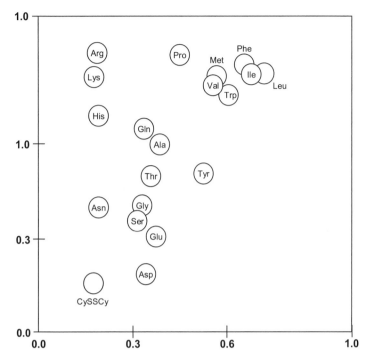

Figure 3.15 Two-dimensional thin-layer chromatograph separation of amino acids. (Based on White et. al., 1973.)

them. The radioactivity makes compounds easier to detect in small quantities. For example, exposing plants to carbon dioxide with carbon-14, and then analyzing plant matter periodically for radioactive compounds is a way to elucidate the steps in photosynthesis as the ^{13}C appears in one compound and then another. **Radioimmunoassay** can detect compounds such as hormones or drugs in blood plasma at quantities as low as several picograms (10^{-12} g).

Other analysis techniques, based on the methods of molecular biology, are discussed in Chapter 6.

PROBLEMS

3.1. Write symbol-and-stick diagrams for CO_2, NH_4, 1,1,1-TCA (1,1,1-trichloroethane), TCE (trichloroethylene), and formic acid (CH_2O_2).

3.2. Glucose, starch, and glycogen are very similar chemically. How will wastewater containing a high concentration of one or the other affect a biological wastewater treatment plant differently? How would the same wastewater affect a stream differently? Assume that the chemical oxygen demand (COD) of each wastewater is similar.

3.3. A rule of thumb for biological wastewater treatment plant nutrition is that for every 100 mg of oxygen demand there should be 5 mg of nitrogen. (**a**) Would a wastewater containing either only glycine or only leucine need a nitrogen supplement? (**b**) Which of the 20 amino acids is most balanced in this respect? [*Hint:* Answer by computing the theoretical oxygen demand for each substrate and compare to the mass concentration of nitrogen. The theoretical oxygen demand for $C_cH_hO_oN_n$ is $c + (h - 3n)/4 - o/2$ moles of oxygen per mole of substrate. Convert this to grams of oxygen per gram of substrate by the molar mass of O_2 (32 g/mol), dividing by the molar mass of the substrate.] Computing oxygen demand is discussed in detail in Section 13.1.3.

3.4. List all the elements that are more electronegative than carbon.

3.5. For each of the acids listed in Table 3.3, compute the fraction that would be undissociated at pH 7.4, the normal pH of human blood.

REFERENCES

Bailey, J. E., and D. F. Ollis, 1986. *Biochemical Engineering Fundamentals*, 2nd ed., McGraw-Hill, New York.

Fried, G. H., 1990. *Schaum's Outline: Theory and Problems of Biology*, McGraw-Hill, New York.

Kimball, J. W. http://users.rcn.com/jkimball.ma.ultranet/BiologyPages, and http://users.rcn.com/jkimball.ma.ultranet/BiologyPages/R/Radioimmunoassay.html.

Pauling, Linus, 1970. *General Chemistry*, 3rd ed., W.H. Freeman, San Francisco.

Smith, E. L., R. L. Hill, I. R. Lehman, R. J. Lefkowitz, P. Handler, and A. White, 1983. *Principles of Biochemistry*, 7th ed., McGraw-Hill, New York.

White, A., P. Handler, and E. L. Smith, 1973. *Principles of Biochemistry*, 5th ed., McGraw-Hill, New York.

4

THE CELL: THE COMMON DENOMINATOR OF LIVING THINGS

Cells were discovered by Robert Hooke of England in the mid-seventeenth century, and named by him after the small dormitory-style rooms inhabited by monks. He first saw cell wall remains in thin slices of cork, using a microscope. Around the same time, Anton van Leeuwenhoek of Holland advanced the art of building microscopes, achieving magnification up to 500 times. With these he was able to make detailed studies of living cells. Further study led to the development in the nineteenth century of **cell theory**:

- All living things are composed of one or more cells.
- Cells are the basic units of living things and are the site for the reactions of life.
- Under today's conditions, all cells come from preexisting cells.

The first tenet encompasses everything from single-celled bacteria to large animals and trees that can have trillions of cells. The second tenet recognizes that individual parts of cells are not by themselves viable. This tenet also excludes viruses from being classified as living things, since they do not metabolize. The third tenet leaves open the possibility of cells arising spontaneously under the conditions of the primitive Earth.

The light microscope opened a new world to examination, literally under our noses. Typical cell sizes are about $1 \mu m$ for bacteria to $10 \mu m$ for most human cells. The human eye can resolve down to only about $100 \mu m$ (0.1 mm). The light microscope extends resolution down to $0.2 \mu m$ [200 nanometers (nm)], which is half the wavelength of violet light. An advance similar in magnitude to Hooke and Leeuwenhoek's microscopes occurred in the 1960s with the development of electron microscopes. These can magnify by 30,000 to 100,000 times, yielding resolution down to 2 nm. This is enough to resolve some of the larger macromolecules such as proteins and nucleic acids. Another

Environmental Biology for Engineers and Scientists, by David A. Vaccari, Peter F. Strom, and James E. Alleman
Copyright © 2006 John Wiley & Sons, Inc.

leap was made in 1986, with the development of the atomic force microscope, which can resolve individual atoms and has been used to detect the shape of the DNA helix. These imaging tools, together with biochemical techniques, have led to continual advances in **cytology**, the study of cells.

4.1 PROKARYOTES AND EUKARYOTES

As mentioned in Section 2.4, the highest level of biological classification, the domain, is based on cell type. Bacteria are made of the simpler prokaryotic cell; protists, fungi, plants, and animals are eukaryotic. Prokaryotes are smaller than eukaryotic cells and lack any internal membrane-bound structures. Eukaryotes have organelles, which are specialized structures within the cell that are surrounded by their own membranes, almost like cells within a cell. Some organelles, such as mitochondria, even have their own DNA.

It is thought that prokaryotes are relatively primitive life-forms and that eukaryotes may have evolved from a symbiotic association in which an early form of prokaryotic cell incorporated other prokaryotes internally. For example, a large anaerobic prokaryote may have incorporated an aerobic bacterium. The latter eventually became mitochondria, the site of aerobic respiration in eukaryotes.

Table 4.1 shows some of the similarities and differences between prokaryotic and eukaryotic cell structures. Figure 4.1 illustrates a typical bacterial (prokaryotic) cell. The nuclear region is not surrounded by a membrane, as it is in eukaryotes. Photosynthetic bacteria include thylakoid membrane structures, an exception to the rule of not having internal membranes. Granules are structures that contain storage projects such as lipids or starches. Bacterial cells are typically about 1 μm in size.

Table 4.1 Contrast Between Prokaryotic and Eukaryotic Cells

Structure	Prokaryotes		Eukaryotes	
	Archaeans	Bacteria	Plants	Animals
Cell wall	No peptidoglycan; some have glycoprotein or protein walls	Peptidoglycan	Cellulose	None
Cell membrane	Lipid biolayer ether-linked branched hydrocarbon chains	Phospholipid bilayer composed of ester-linked straight hydrocarbon chains		
Motility	Flagella	Flagella	Flagella	Flagella or cilia
Genetic material	Single circular DNA	Single circular DNA	Several linear DNA molecules	
Ribosomes	Yes	Yes	Yes	Yes
Membranous organelles	Absent	Few: e.g., thylakoids for photosynthesis in cyanobacter	Nuclear envelope, endoplasmic reticulum, mitochondria, Golgi apparatus, lysosomes, centrioles	
			Chloroplasts and central vacuoles	

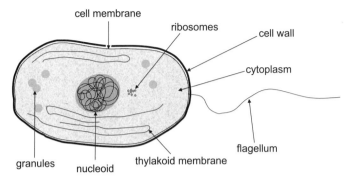

Figure 4.1   Structure of a prokaryotic cell.

Figure 4.2 shows typical plant and animal (eukaryotic) cells. These cells are much more complicated. The nuclear region (nucleus), endoplasmic reticulum, mitochondria, Golgi apparatus, chloroplasts in photosynthetic organisms, and a number of other structures are all surrounded by their own membranes. Eukaryotic cells are usually much larger than prokaryotic cells, on the order of 10 μm in size.

Functionally, prokaryotes are capable of a much wider range of basic metabolic processes, including many of environmental importance. These include many processes that catalyze key pathways of the biogeochemical cycles. All chemoautotrophs (organisms that obtain energy from inorganic chemicals) are prokaryotes. These include bacteria that oxidize minerals, such as Fe(II), NH_3, H_2S, and others. Other major processes limited to prokaryotes are nitrogen fixation and denitrification. Furthermore, with few exceptions, only prokaryotes are capable of using electron substitutes for oxygen, such as nitrate, sulfate, or carbonate, and can live their entire life cycles in the absence of oxygen. However, prokaryotes are specialized, and none can do all of these processes, and most do very few. What specialized advantage do eukaryotes have? One answer is that only eukaryotes form multicellular organisms.

4.2 THE BIOLOGICAL MEMBRANE

In many areas of science, it is at the interface where things get interesting, and difficult. Chemical reactions in homogeneous gas or liquid phases are complex enough, but when an interface is present, even the notion of chemical concentration is oversimplified. For example, the pH near the surface of a colloid can vary with distance from the surface. Surface-active agents will distribute themselves differently between the surface and the bulk fluid. Many surfaces act as catalysts. They can be gatekeepers, affecting the transport of substances between phases. The phospholipid bilayer membrane is the major interface formed by living things. It is the structure that separates "inside" from "outside." By increasing the complexity of the system, it also increases its possibilities.

The biological **membrane** is a flexible sheet forming a closed surface, whose basic structure is formed of a phospholipid bilayer (see Figure 4.3). The capability of phospholipids to form enclosed bilayer vesicles spontaneously was described in Section 3.7.2. The outer membrane of all cells is called the **plasma membrane** or **cell membrane**. Similar membranes also enclose cell organelles such as the mitochondria or the nucleus in eukaryotic cells.

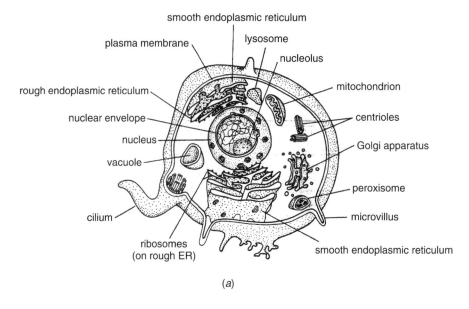

(a)

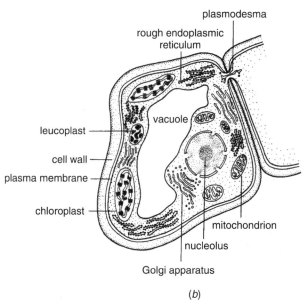

(b)

Figure 4.2 Animal (*a*) and plant (*b*) cells. (From Fried, 1990. © The McGraw-Hill Companies, Inc. Used with permission.)

Besides being flexible, the phospholipid molecules can move freely within the plane of the membrane, a behavior described as a *two-dimensional fluid*. Other molecules are intimately associated with the membrane. Eukaryotic membranes can contain large amounts of cholesterol, which increase the fluidity of the membrane. Fatty acids serve the same function in prokaryotes. Globular proteins are embedded in the membrane, somewhat like icebergs floating in the sea. Some penetrate both surfaces of the membrane and participate in the transport of substances across it. Others are embedded in one or the

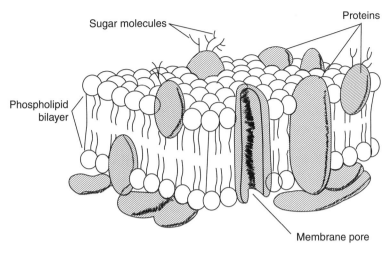

Figure 4.3 Plasma membrane structure. (From Van De Graaff and Rhees, 1997. © The McGraw-Hill Companies, Inc. Used with permission.)

other surface and function as chemical receptors or catalytic sites. Biological membranes are asymmetrical. The external surface of the plasma membrane and the internal surface of organelles have carbohydrates bonded to them, forming glycolipids and glycoproteins.

The plasma membrane of archaean cells is chemically distinct from eubacterial or eukaryotic cell membranes. Instead of being composed of lipids made from straight-chain fatty acids bonded to glycerol by ester bonds, archean membrane lipids are made of the branched hydrocarbon *isoprene* bonded to glycerol by ether bonds. This structure is thought to give archeans greater physical and chemical resistance to the relatively unfavorable environmental conditions in which they are often found.

Membranes are typically less than 10% carbohydrate by mass; the rest of the membrane mass is about equally divided between protein and lipid. In animals, about half the lipid is phospholipid and half is cholesterol.

The membrane controls the transport across it of both substances and information. Information is transported in the sense that substances, called **ligands**, can bind to **receptors** composed of transmembrane proteins (proteins that penetrate both sides) on one side of the membrane, producing a change in its conformation on the other side. The altered protein can then affect other reactions. This sends a signal across the membrane without a substance actually crossing over. Examples of this are the intercellular messengers called *hormones*. Insulin, for example, binds to a receptor and causes two separate effects. Primarily, it stimulates plasma membrane mechanisms for the transport of glucose, some ions, and amino acids. Second, it results in changes in intracellular metabolism that result in increased synthesis and storage of protein, glycogen, and lipid. Some toxic substances act by binding with receptors, either by stimulating an inappropriate response directly, or by competing with normal ligands.

4.3 MEMBRANE TRANSPORT

Membranes also control the transport of substances. This is one of the essential functions of life: the maintenance of different conditions between the interior and exterior of the cell, the distinction between the self and the rest of the world. Many toxic substances also must be transported across membranes to achieve their effect.

The simplest mechanism is **passive transport**, the movement of a chemical from an area of relatively high concentration, through a membrane, to an area of low concentration, by means of molecular diffusion. The transport occurs only when a concentration difference exists between the aqueous phases on the two sides of the membrane. To diffuse across a membrane, a chemical must first dissolve in it. Then, random molecular motion results in a net movement toward the side with the lower concentration. However, molecules that are polar, such as ions or sugars, have very low solubility in the nonpolar interior of the phospholipid bilayer. Thus, the transport of such compounds through the membrane is very slow by this mechanism.

Water also moves across membranes by passive transport. The movement is from low solute concentration to high solute concentration. Realize that the higher the concentrations of solutes (ions, proteins, sugars, etc.), the lower the concentration of solvent (water, in this case). From the point of view of the water, the movement is from high concentration (of the water) to low concentration. The passive transport of water due to its concentration gradient is called **osmosis**.

Consider a membrane that is permeable to neutral molecules such as water but impermeable to ions and that separates a solution of fresh water from salt water. The water will move by osmosis from the fresh water (which has a higher water concentration) to the salty side. The flow continues until equilibrium is achieved. If the pressures are equal, osmotic equilibrium means that the concentration of particles is the same on both sides. However, if the system is held at constant volume, the flow results in an increase in pressure on the salty side. The pressure difference across the membrane at equilibrium, π, is related to the difference in the molar concentration of total particles, Δc, by a version of the ideal gas law:

$$\pi = \Delta c\, RT \tag{4.1}$$

The effect of particle concentration is approximately independent of charge or size. For example, a $0.1\,M$ solution of sodium chloride has the same osmolarity as a $0.2\,M$ solution of glucose, because the sodium chloride dissociates to form 0.2 mol of ions per liter.

The concentration of particles expressed in moles per liter is called the **osmolarity** (Osmol/L). For example, blood plasma has a total particle concentration of about 325 mmol per liter (mOsmol/L). This produces an osmotic pressure of about 6000 mmHg (7.9 atm). (This is similar to the osmotic pressure of the ocean during the Precambrian Era, when animals with closed circulation evolved. Since then, the osmotic pressure of seawater has continued to increase about 3.5-fold.) If two solutions with different osmolarities are separated by a barrier to the solutes such as a plasma membrane, the lower-concentration solution is said to be **hypotonic** to the other. The higher-concentration solution is said to be **hypertonic**. If the two solutions have the same osmolarity, they are **isotonic**.

Because of osmotic pressure, the total concentration of particles inside a cell must be almost equal to that outside. Otherwise, water would cross the membrane, causing the cell to either shrink or to swell and burst. Plant and bacterial cells are surrounded by rigid cell walls that protect them from bursting, and can therefore exist in a hypotonic solution. A hypertonic solution, however, will cause them to shrink within their cell walls. Cells that have osmotic pressure are said to be **turgid**.

Plasma membranes have pores consisting of embedded proteins. In most cells, these allow molecules up to a molar mass of 100 to 200, including water itself, to pass by filtration. The membranes of capillary (blood vessel) cells have larger pores, which allow molecules up to molar mass 60,000 to pass.

Molecules too lipophobic for passive diffusion or too large for filtration may still enter the cell by specialized transport systems. These include four types of carrier-mediated transport: facilitated diffusion, active transport, co-transport, and countertransport. In **facilitated diffusion** the problem of lipophobicity is solved by forming a complex between the solute and a protein in the membrane (the *carrier*). Random molecular motion then transports the complex to the other face, where the solute is released and the protein is freed for reuse by other molecules. Facilitated diffusion has two important differences from passive diffusion: (1) Because there are a limited number of carriers, at high solute concentrations the carrier protein can become saturated, resulting in a maximum flux. Passive diffusion is limited only by the concentration gradient. (2) Facilitated diffusion is more selective than passive diffusion, since the solute must be able to form a complex with the carrier, and it can be inhibited by competition. An important compound that is carried across membranes by this mechanism is glucose. If it weren't for protein transporters, membranes would be impermeable to glucose. The hormone insulin affects the permeability of muscle and adipose tissue membranes by increasing the amount of transporter proteins in those membranes.

Passive transport, whether membrane or film theory, membrane filtration, or facilitated transport is a physicochemical process and requires no expenditure of energy by the cell. Instead, the cell relies on the free energy of the concentration gradient. Another form of carrier-mediated transport, **active transport**, involves the use of metabolic energy in the form of ATP. Active transport can occur when a membrane-spanning carrier absorbs the substrate from one side of the membrane and uses energy from ATP to pass the substrate through a channel and exude it at the other side. A second active transport mechanism can involve the consumption of a substrate on one side of the membrane, coupled with its production at the other side. The substrate is not really transported across the membrane, but the effect is as though it has been. An example of the latter mechanism is the "transport" of protons across the membrane of bacteria or mitochondria in the respiration process.

Active transport is an important mechanism for the transport of natural biochemical compounds in living things. It is the way they obtain nutrients and eliminate waste products. It is also a mechanism for the uptake and excretion of toxins. For example, lead is absorbed by active transport in the intestines. Active transport can move chemicals against a concentration gradient. Active transport has the following characteristics:

1. Chemicals can be moved against electrochemical gradients.
2. Like facilitated diffusion, active transport can be saturated and shows a maximum flux.
3. Also like facilitated diffusion, active transport is highly selective and exhibits competitive inhibition.
4. Active transport requires energy and so can be inhibited by metabolic poisons.

Active transport is exploited to transport water passively across membranes by osmosis. For example, the large intestine removes water from its lumen (interior) by using active transport to remove sodium. This decreases the osmolarity in the lumen, and water passively diffuses out to the higher-osmolarity tissues surrounding it. When solutes are transported across a membrane, water will follow if the membrane is permeable to water.

Cotransport and **countertransport** involve two substrates being brought across the membrane by the same carrier at the same time, either in the same direction or in opposite directions, respectively. At least one of the substrates must be transported downgradient and provides the energy to transport the second substrate. For example, a sodium gradient can drive the cotransport of glucose into a cell. A countertransport mechanism exchanges chloride for bicarbonate in the kidney. Since these mechanisms are carrier mediated, they can also be saturated.

Another type of specialized transport, limited to eukaryotes, is **endocytosis**, in which particles are engulfed by cells. The cell accomplishes this by surrounding the particle with part of the plasma membrane, which then pinches off inside the cell, forming a vesicle. If the particle is solid, the process is called **phagocytosis**; if liquid, it is called **pinocytosis**. The protozoan *amoeba* feeds by phagocytosis and the lungs use it to clear themselves of inhaled particulates, incidentally taking up any associated toxins.

More details on some membrane transport processes, including ways of describing them mathematically, are given in Section 18.2.

4.4   EUKARYOTIC CELL STRUCTURE AND FUNCTION

The emphasis here is on eukaryotic cells. More details on prokaryotic cells are given in Chapter 10.

Plants, some fungi, some protists and most prokaryotes form a **cell wall** outside their plasma membrane. The basic structure in plants is composed of cellulose fibers (cotton is a pure form of cellulose). Besides maintaining cell shape and preventing a cell from rupturing due to osmotic forces, a cell wall provides structural rigidity to multicellular plants. Many plants also lay down a secondary cell wall, which includes lignin, inside the primary wall. *Lignin* is a polymer of aromatic subunits whose presence increases the toughness of wood. Lignin biodegrades much more slowly in the environment than cellulose does. The degradation products of lignin are thought to be the major source of *humic substances*, which are the predominant forms of organic matter in soil. The humic substances, in turn, give topsoil its beneficial properties for plant growth and are responsible for the great capacity for soils to absorb nutrients as well as many types of pollutants.

Animal cells protect themselves against rupturing from osmotic forces partly by manipulating those forces. Multicellular animals can control the osmotic pressure of their intercellular fluids. In addition, they produce a network of collagen fibers. **Collagen** consists of polypeptide strands twisted into a ropelike structure. About 25% of the protein in mammals is collagen, and tendons are mostly collagen.

The basic matrix of the cell interior is the gel-like **cytoplasm**, in which the other cell materials and structures are suspended, including the organelles. A three-dimensional lattice of protein fibers called the **cytoskeleton** gives structure to the cytoplasm and helps orient organelles. Inside cells, the largest and most obvious structure is the **nucleus**, which contains most of a cell's DNA. In eukaryotes the nucleus is separated from the cytoplasm by a **nuclear membrane**, which consists of *two* phospholipid bilayer membranes, penetrated by numerous protein pores. (In prokaryotes, the DNA is not isolated from the cytoplasm by a membrane, and the structure is called a **nuclear region** rather than a nucleus.) Each of the DNA molecules in the nucleus is contained in a chromosome. A **chromosome** is a complex of a DNA molecule and associated proteins. Chromosomal DNA forms a template for protein synthesis.

Protein synthesis takes place outside the nucleus in the cytoplasm. The nucleus synthesizes RNA molecules, each having the code for production of a protein. The RNA then passes out through the nuclear pores to the cytoplasm. There, the RNA interacts with particles in the cytoplasm. These particles are the **ribosomes**, in which amino acids are linked up to form proteins, using the RNA templates as a guide. Ribosomes are made of several proteins and RNA molecules in two subparticles called 50S and 30S in prokaryotes, and 60S and 40S in eukaryotes. The designation refers to their size in Svedberg units as measured by the settling velocity in a centrifuge.

In eukaryotes, the ribosomes line a folded membrane structure called the **endoplasmic reticulum** (ER), which occupies a large portion of the cytoplasm. ERs serve as channels to transport newly synthesized substances within a cell. The part studded with ribosomes, called the **rough ER**, is responsible for the synthesis of proteins, including enzymes. Another portion of the ER, the **smooth ER**, lacks ribosomes. The smooth ER is associated with synthesis of lipids and the detoxification of lipid-soluble toxins. The liver has abundant smooth ER. It is the location of the enzyme complex *cytochrome P450 system*, which is responsible for much of the liver's detoxification activity as well as other biotransformation functions (see Section 18.5)

The endoplasmic reticulum pinches off small, self-enclosed sacs of synthesized compounds called **vesicles** for transport elsewhere in the cell. Often, the vesicles go to the **Golgi apparatus**, a structure consisting of flattened disks of membrane structures. The Golgi apparatus may perform finishing touches on the products, then pinch them off in another set of vesicles that go toward their final destination. For example, some digestive enzymes are activated in the Golgi apparatus, and migrate in vesicles to the plasma membrane, where their contents are discharged outside the cell.

Other vesicles, including lysosomes, remain in the cytoplasm. **Lysosomes** contain enzymes that can digest particles taken in by endocytosis or, interestingly, can be used by the cell to commit "suicide," when the cell is damaged or otherwise unneeded. Other vesicles contain enzymes that break down water-soluble toxins in the cytoplasm. **Peroxisomes** destroy peroxides in the cell, and peroxisomes in the liver and kidney perform about half of the work of ethanol detoxification in the body. **Microsomes**, vesicles originating in the smooth ER, contain enzymes for detoxification.

The **mitochondria** are the site where eukaryotes perform respiration, the oxidation of substances using inorganics such as oxygen as ultimate electron acceptors, for the production of energy. Thus, the mitochondria are often described as the cell's "powerhouse." They are the place where organics and oxygen are reacted to form carbon dioxide and water, with the capture of energy in the biologically useful form of ATP molecules. The structure and function of mitochondria are discussed in detail in Section 5.4.3 (see also Figure 5.6).

Several organelles are unique to plants and plantlike protists. Chief among these is the chloroplast. The **chloroplast** is the site where energy from light is captured in photosynthesis. Photosynthesis is the production of carbohydrates from CO_2 and H_2O using light energy; it is the reverse reaction to respiration and is the basis for the production of almost all organic matter by the biosphere. The chloroplast and the mitochondrion have many similarities in structure and function, although they are quite distinct. A detailed description is deferred to Section 5.4.5. The structure of the chloroplast is shown in Figure 5.9.

A **vacuole** is a membrane-bound storage vesicle within a cell. In plants they may be quite large, even filling the majority of the cell volume. The plant cell vacuole may serve to store nutrients, as in sap. It may also be a depository for waste products, since plants do not have the mechanisms that animals have for excreting wastes. Vacuoles in some protists, such as *Euglena* and *Paramecium*, are used to excrete water, which otherwise would cause the cells to burst.

Flagella and cilia are organelles of cell movement. **Flagella** are whiplike extensions of a cell, which may be many times longer than the main part of the cell. A cell can propel itself through a liquid by motion of a single flagellum. Examples of cells with a flagellum are the protist *Euglena* and the human sperm cell. Animal cells or some animal-like protists may instead have cilia. **Cilia** are short hairlike projections covering the surface. They propel the cell by a coordinated beating action, like oarsmen on an ancient warship. The protist *Paramecium* is an example of a cell that propels itself this way. Cilia may have another function: moving particles past stationary cells. This is their function in helping protists such as the "stalked ciliates" feed. Cilia also line the human respiratory tract, where they serve to expel inhaled particles. Damage to respiratory cilia from cigarette smoking impairs removal of harmful materials and causes "smokers' cough."

Another mechanism by which cells can move under their own power is by extending part of the cell membrane, "flowing" the cytoplasm into the extending part while withdrawing from the other side. The classic example of this is the protist *amoeba*. In humans, the white blood cells, or **leukocytes**, which protect the body against infection, have the same capability.

4.5   CELL REPRODUCTION

Growth and reproduction of organisms proceeds primarily by increasing the number of cells rather than by increasing cell size. The **cell cycle** is a sequence of growth, duplication, and division. Cell reproduction in prokaryotes is called **binary fission**. Cytoplasm is produced during growth and the cell enlarges. The single loop of DNA and other structures, such as the ribosomes, are duplicated and collect at opposite ends of the cell. A partition grows between the two ends, which then separate into two new cells, called **daughter cells**. Under optimum conditions of temperature, nutrients, and so on, bacteria can double as often as once every 10 minutes. For example, under ideal conditions, *E. coli* doubles in as little as 17 minutes, and *Bacillus sterothermophilus* can double in 10 minutes.

Cell division in eukaryotes is more complex. In eukaryotic cells, the chromosome is a complex structure consisting of a single linear DNA molecule wrapped up around a group of proteins. The eukaryotic chromosomes are large enough that at some stages of the cell cycle they can be made visible in a light microscope by staining. At this point it is worth reviewing the organization of DNA in a eukaryotic cell. DNA consists of two complementary strands of a linear polymer of nucleic acids, which forms a double helix. Each three units of the polymer codes for a single amino acid. A sequence of these units contains the code for a single protein. A gene is a sequence of DNA that codes for a single protein. One DNA molecule has many genes. Each DNA molecule is wrapped up with proteins in a chromosome structure. Most eukaryotic cells have their chromosomes in functional pairs. Thus, there will be two of each gene in the nucleus, both of which code for a protein

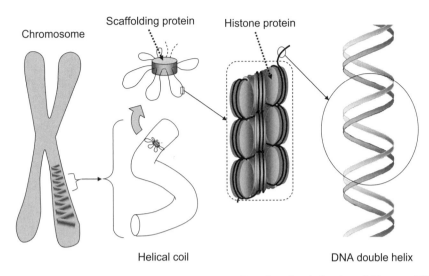

Chromosome   Scaffolding protein   Histone protein

Helical coil   DNA double helix

Figure 4.4   Chromosome structure in eukaryotes. (Based on Postlethwait and Hopson, 1995.)

having the same function. One chromosome of each pair comes from the organism's father, the other from its mother. It is important to note that the proteins are not necessarily identical, just that they have the same function. Humans, for example, have 23 pairs of chromosomes, 46 in all.

There are two main phases of eukaryotic cell reproduction. In **interphase,** the cell produces new protein, ribosomes, mitochondria, and so on, and this is followed by replication (copying) of the cell's chromosomes. Each chromosome is linked with its new copy at a point along its length, so they have the appearance of two sausagelike shapes tied together. This results in a number of X- or Y-shaped structures, depending on where they are linked (see Figure 4.4). The cell then enters the **division phase**. The division phase itself is composed of two stages. **Mitosis** is the separation of the replicated DNA into two new nuclei, and **cytokinesis** is the distribution of the cytoplasm components and a physical separation into daughter cells. The eukaryotic cell cycle may take much longer than prokaryotic binary fission. Red blood cells are produced by division of cells in the bone marrow, which divide every 18 hours. The events of mitosis are the most dramatic of the cell cycle, and can be readily observed under the microscope with suitable staining.

Figure 4.5 shows that mitosis is also separated into a number of stages of development. At the beginning of mitosis, the chromosomes are dispersed in the nucleus and not readily visible under the light microscope. As mitosis proceeds, the nuclear envelope disperses and a **spindle** forms, consisting of a web of microtubules. By this time the chromosomes can be made visible under a light microscope when stained, displaying the well-known X and Y shapes. (The X, for example, is actually two replicated chromosomes linked along their length, due to be separated into daughter cells. The Y shape is the same thing except that the chromosomes are joined near one end.) The chromosomes migrate into positions aligned along the center between the two poles of the spindle. The spindle then physically separates the chromosomes to the poles of the now-elongated cell. Finally, the spindle dissolves and the nucleus re-forms, completing mitosis. Cytokinesis continues, forming new plasma membranes to complete the cell division.

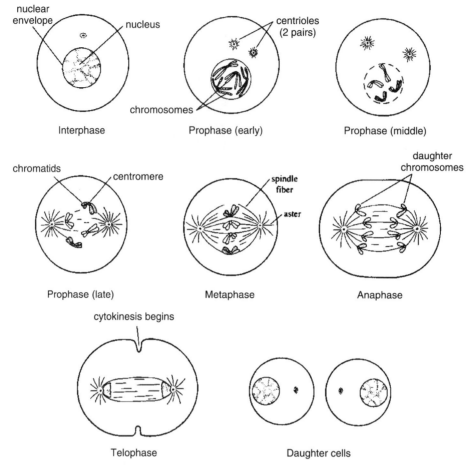

Figure 4.5 Stages of mitotic division. (From Fried, 1990. © The McGraw-Hill Companies, Inc. Used with permission.)

There are two salient facts regarding mitosis. First, mitosis results in two cells with identical genetic composition, barring errors in reproduction. Thus, each of our "body" cells are genetically identical. Skin cells have the same genes as pancreas cells, so skin cells have the genes to produce insulin, although they do not actually do so. The remarkable variety in cell structure and function despite their having the same control codes is one of the greatest mysteries in biology. How, during the formation of the body as well as later, does a cell "know" that it is a bone marrow cell and not a liver cell?

The second salient fact about mitosis is that it both starts and ends with cells having two copies of each chromosome. Such cells are called **diploid cells** (Figure 4.6). The "body" cells in most multicellular organisms are diploid and are called **somatic cells**. This is to distinguish them from cells that are destined to produce offspring by way of sexual reproduction, called **germ cells**, which are also diploid.

Germ cells produce **gametes**, which are reproductive cells such as the egg and sperm cells in animals, and the egg cells and pollen spores in flowering plants. Gametes have only one copy of each chromosome, a condition that is called **haploid**. In the process of **fertilization** the haploid germ cells from two parents fuse to create a new diploid

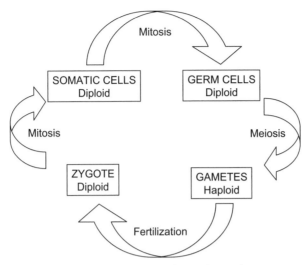

Figure 4.6   Life cycle of sexual organisms.

cell, the **zygote**, which then grows into a new organism. Thus, the life cycle of complex multicellular organisms such as animals and higher plants consists of growth by mitotic cell division, meiosis, and fertilization.

Meiosis is a form of cell division that converts a diploid cell into a haploid cell. Meiosis actually consists of two cell divisions, resulting in four cells that have chromosomes that are not only different from each other, but are different from those in the parent cell.

The process will be illustrated using the example of a cell with two sets of chromosomes (Figure 4.7). This diploid cell has two short chromosomes: one from the organism's father (shown in black) and one from the mother (white). Similarly, there are two longer chromosomes, one from each parent. As in mitotic division, the chromosomes replicate. But a critical difference is that the chromosome pairs come together and may intertwine themselves. At the points where the chromosomes cross each other, the DNA molecules can break and reconnect with the fragment from its complement. This results in new chromosomes that can have both maternal and paternal genes. This process is called **crossing over**, and the resulting exchange of genetic material between chromosomes is called **recombination**. The crossing-over point can occur at many locations on a single chromosome, although only one is shown in the figure. Thus, a huge number of possible combinations can result.

Recombination is an important source of genetic variability in organisms. In addition, it is important to science for several reasons. It is the basis of chromosome mapping, a technique that determines the location of genes on the chromosomes. Artificial control of recombination has become one of the most important procedures of genetic engineering (discussed further in Chapter 6). It has inspired a mathematical optimization method called appropriately the **genetic algorithm**, which has applications far beyond biology.

After crossing over, the replicated chromosomes line up and are separated by a spindle apparatus, just as in mitosis. However, note that this could occur in four different ways. The two daughter cells could have all paternal chromosomes in one cell and maternal chromosomes in the other; or one cell could have a long paternal and a short maternal, and the other would then have a long maternal and a short paternal. If there were more

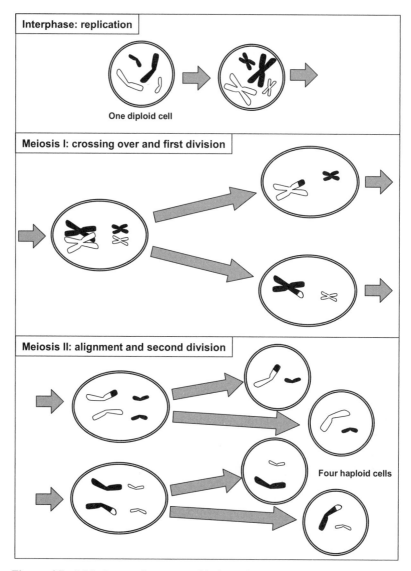

Figure 4.7   Meiosis: crossing over and independent assortment of chromosomes.

chromosomes, more combinations would be possible. In all, there would be 2^N, where N is the number of chromosome pairs. For humans with 23 pairs, this results in $2^{23} = 8.4 \times 10^6$ possible combinations. If crossing over did not occur, this is the number of genetically unique offspring that a single human couple could possible produce. The combinations occur randomly, depending on how the chromosomes align before being pulled apart by the spindle. The random distribution of chromosomes to the daughter cells in the first stage of meiosis is called **independent assortment**.

When random assortment is combined with crossing over, the number of possible daughter cells that can be formed is extremely large. This is an important source of the random variation that drives evolution. Even without mutations, crossing over can create new genes by combining pieces of two old ones. The great variation in offspring that results can help a species adapt rapidly to changing environmental conditions. This is one of the great evolutionary advantages of sexual reproduction.

The meiotic process just described is actually only the first step, called **meiosis I**. After that step, two daughter cells are produced that although already haploid, contain two of each type of chromosome that originated in the replication process. In **meiosis II**, these are separated in another division process, resulting in a total of four haploid cells, each of which probably has a unique genetic complement.

PROBLEMS

4.1 Name the kinds of physicochemical or chemical effects the following types of pollutants might have on plasma cell membranes: (**a**) dissolved hydrophobic solvents such as benzene or TCE; (**b**) surfactants such as detergents; (**c**) salts such as chloride or sulfate; (**d**) strong oxidizers such as chlorine or ozone.

4.2 Why do animal cells not need cell walls to maintain their structure?

4.3 Consider a solution of 500 mg of NaCl per liter as an analog for tap water. What is the particle concentration in mmol/L? What will be its osmotic pressure?

4.4 A fruit fly has four pairs of chromosomes. How many possible combinations can be produced in its gametes by random assortment (with no crossing over)? List the possibilities.

REFERENCES

Atlas, R. M., and Bartha Richard, 1981. *Microbial Ecology: Fundamentals and Applications*, Addison-Wesley, Reading, MA.

Bailey, J. E., and D. F. Ollis, 1986. *Biochemical Engineering Fundamentals*, McGraw-Hill, New York.

Fried, G. H., 1990. *Schaum's Outline: Theory and Problems of Biology*, McGraw-Hill, New York.

Gaudy, A. F., and E. T. Gaudy, 1988. *Elements of Bioenvironmental Engineering*, Engineering Press, San Jose, CA.

Grady, C. P., Jr., G. T. Daigger, and H. C. Lim, 1999. *Biological Wastewater Treatment*, Marcel Dekker, New York.

Henze, et al., 1986. *Activated Sludge Model No. 1*, International Association on Water, London.

Holland, J. H., 1992. Genetic algorithms, *Scientific American*, Vol. 267, No. 1, pp. 66–72.

Lim, Daniel, 1998. *Microbiology*, 2nd ed., WCB McGraw-Hill, New York.

Postlethwait, J. H., and J. L. Hopson, 1995. *The Nature of Life*, McGraw-Hill, New York.

Smith, E. L., R. L. Hill, I. R., Lehman, R. J. Lefkowitz, P. Handler, and White, 1983. *Principles of Biochemistry: General Aspects*, McGraw-Hill, New York.

Stryer, L., 1995. *Biochemistry*, 4th ed., W.H. Freeman, New York.

Van de Graaff, K. M., and R. W. Rhees, 1997. *Schaum's Outline: Theory and Problems of Human Anatomy and Physiology*, McGraw-Hill, New York.

Watson, J. D., 1968, *The Double Helix*, Atheneum, New York.

White, A., P. Handler, and E. L. Smith, 1973. *Principles of Biochemistry,* McGraw-Hill, New York.

5

ENERGY AND METABOLISM

All the biochemical reactions in a living organism, taken together, are the organism's **metabolism**. The metabolic processes of life can ultimately be described as a complex set of chemical reactions distributed in space and time, and coupled by shared intermediate compounds and transport mechanisms. In this chapter we focus on the chemical reactions themselves, but the reader should also pay attention to the *where* and *when* of the chemical choreography of the cell.

Our first concern will be whether or not a particular biochemical reaction can proceed. This question is answered by thermodynamics. In general, a reaction is feasible if the products are collectively more stable (in a lower-energy state) than the reactants. Thermodynamics also describes how energy is produced or consumed by a reaction. Although a reaction may be thermodynamically feasible, there may be an energy barrier between products and reactants that limits the rate at which it occurs. We describe the rate at which reactions occur using chemical kinetics. Finally, we examine several of the most important metabolic reactions found in living things.

5.1 BIOENERGETICS

Thermodynamic relationships govern whether a reaction *can* occur, although not whether it *will*. A reaction may be thermodynamically feasible but may occur too slowly for practical consideration. We will also be very concerned with whether a reaction provides energy for doing biochemical work or requires such energy from other sources.

5.1.1 Some Basic Thermodynamics

The second law of thermodynamics states that processes in an **isolated system**, including chemical reactions, tend to proceed from a less likely to a more likely state: that is, to

Environmental Biology for Engineers and Scientists, by David A. Vaccari, Peter F. Strom, and James E. Alleman
Copyright © 2006 John Wiley & Sons, Inc.

increase **entropy** (S), or randomness. However, organisms are not isolated systems. They are better approximated as a **closed system**, in which energy, but not matter, can cross their boundaries. In particular, organisms approximate closed systems at constant temperature and pressure. The appropriate measure for energy added to a closed system at constant pressure is the change in **enthalpy**, ΔH,

$$\Delta H = \Delta U + P \Delta V \tag{5.1}$$

where ΔU is the resulting change in the internal energy of the system, P is its total pressure, and ΔV is the change in volume of the system.

In closed systems a decrease in entropy can be compensated for by adding energy from outside the system. This causes the entropy outside the system to increase enough so that the total entropy change is positive. For a closed system at constant temperature and pressure, the relevant quantity to determine if a process is feasible is the change in the **Gibbs free energy**, ΔG, which includes the effects of both entropy and enthalpy:

$$\Delta G = \Delta H - T \Delta S \tag{5.2}$$

To summarize, in isolated systems changes in energy are described in terms of the internal energy, U, and the system will always change in a way that increases its entropy, S. However, for closed systems at constant pressure, it is more convenient to describe both of these processes in terms of enthalpy, H, and Gibbs free energy, G, respectively.

Thus, a reaction is feasible in closed systems only if it causes a *decrease* in Gibbs free energy. For an individual chemical, the Gibbs free energy increases with concentration logarithmically. For example, the Gibbs free energy of compound A would be

$$G_A = G_A^\circ + RT \ln[A] \tag{5.3}$$

where G_A° is the **standard Gibbs free energy** of A, corresponding to the Gibbs free energy at a molar concentration of A, [A], equal to 1.0 M in solution or 1.0 atm partial pressure for gases. The standard Gibbs free energies of formation for several biochemical compounds are give in Table 5.1. (Note that *chemical activity* should be used instead of concentration, to be strict. However, we will use concentration, which approximates activity, in order to make the development easier to follow.)

Consider the reaction

$$a\mathrm{A} + b\mathrm{B} \Leftrightarrow c\,\mathrm{C} + d\,\mathrm{D} \tag{5.4}$$

in which a moles of A and b moles of B react to form c moles of C plus d moles of D. As a result, the Gibbs free energy of the mixture will be decreased by $aG_A + bG_B$ and increased by $cG_C + dG_D$, producing a net change in Gibbs free energy for the reaction, ΔG:

$$\begin{aligned}
\Delta G &= cG_C + dG_D - aG_A - bG_B \\
&= c(G_C^\circ + RT \ln[C]) + d(G_D^\circ + RT \ln[D]) \\
&\quad - a(G_A^\circ + RT \ln[A]) - b(G_B^\circ + RT \ln[B])
\end{aligned} \tag{5.5}$$

TABLE 5.1   Standard Gibbs Free Energies of Formation for Several Biochemical Compounds[a]

Compound	Formula	$G°$ (kcal/mol)
Oxalic acid (1)	$(COOH)_2$	−166.8
Proprionic acid (liq) (2)	$C_3H_5O_6$	−91.65
Pyruvic acid (liq) (2)	$C_3H_4O_3$	−110.75
Acetic acid*	$C_2H_4O_2$	−93.8
Lactic acid (cr) (2)	$C_3H_6O_3$	−124.98
Glucose (3)	$C_6H_{12}O_6$	−216.22
Glycerol (3)	$C_3H_8O_3$	−116.76
Ethanol (3)	C_2H_6O	−43.39
Ethylene (gas) (1)	C_2H_4	16.282
Methanol (liq) (1)	CH_4O	−57.02
Methane (gas) (1)	CH_4	−12.14
Alanine (3)	$C_3H_6O_2N$	−88.75
Glycine (2)	$C_2H_5O_2N$	−88.62
Urea (aq) (1)	CH_4ON_2	−48.72
Ammonia (gas) (1)	NH_3	−4.0
Water (liq) (3)	H_2O	−56.69
CO_2 (gas) (3)	CO_2	−94.45
Nitric acid (1)	HNO_3	−19.1

[a]Values for 1 M aqueous solutions at pH 7.0 and 25°C, except as noted; cr, crystal form.
Source: (1) *CRC Handbook of Chemistry and Physics*, 60th ed.; (2) *Lange's Handbook of Chemistry*; (3) Lehninger, *Biochemistry*.

Some rearrangement gives the following expression for the change in Gibbs free energy of the reaction at any concentration of reactants:

$$\Delta G = \Delta G° + RT \ln \frac{[C]^c [D]^d}{[A]^a [B]^b} \tag{5.6}$$

where

$$\Delta G° = cG_C° + dG_D° - aG_A° - bG_B° \tag{5.7}$$

is the *standard* **Gibbs free-energy change** for the reaction.

   If one starts with a mixture of A and B, they would begin to convert to C and D because if the latter have a low enough concentration, their Gibbs free energy will be less than that of the reactants. As the reaction proceeds, the combined Gibbs free energy of A and B will decrease, and the contribution from C and D will increase. Eventually, a point is reached where any change in the Gibbs free energy of the products exactly balances changes in the Gibbs free energy of the reactants. Any further reaction would cause an increase in the total Gibbs free energy. The total Gibbs free energy is then at a minimum, and if the reaction proceeded infinitesimally, the change in Gibbs free energy given by equation (5.6) would be zero. Thus, substituting equilibrium concentrations into (5.6) and setting ΔG

to zero, we have

$$\Delta G^\circ = -RT \ln \frac{[C]_{eq}^c [D]_{eq}^d}{[A]_{eq}^a [B]_{eq}^b} = -RT \ln K_{eq} \qquad (5.8)$$

where K_{eq} is defined as the equilibrium coefficient:

$$\frac{[C]_{eq}^c [D]_{eq}^d}{[A]_{eq}^a [B]_{eq}^b} = K_{eq} \qquad (5.9)$$

From equation (5.8) we also have that

$$K_{eq} = \exp\left(-\frac{\Delta G^\circ}{RT}\right) \qquad (5.10)$$

The equilibrium constant is a unique thermodynamic constant for each reaction. We have already met one example of an equilibrium constant in the K_a for acid–base reactions as defined in equation (3.2).

By convention, when biochemists compute K_{eq} and ΔG° for reactions that involve water or hydrogen ions, they assume that their concentrations are held constant at 1.0 and 10^{-7} M (pH 7.0), respectively.

Example 5.1 Compute ΔG° and K_{eq} for the complete oxidation of oxalic acid.
    Answer First write the balanced equation for the oxidation:

$$C_2H_2O_4 + \frac{1}{2}O_2 \rightarrow 2\,CO_2 + H_2O$$

Then use Table 5.1 and equation (5.7):

$$\Delta G^\circ = cG_C^\circ + dG_D^\circ - aG_A^\circ - bG_B^\circ$$
$$= 2(-94.45) + 1(-56.69) - (-166.8) - 0.0 = -78.79\,\text{kcal/mol}$$

Note that ΔG° for elements (oxygen, in this case) equals zero. Next, use equation (5.10):

$$K_{eq} = \exp\left(-\frac{\Delta G^\circ}{RT}\right) = \exp\left[\frac{-78.79}{(1.987)(298.15)}\right] = 1.14$$

Keep in mind that equation (5.8) holds only at equilibrium, whereas equation (5.6) is valid at any combination of reactant and product concentrations. If equation (5.6) evaluates to a negative value, the reaction will tend to proceed in the forward direction as written. If it is positive, the reverse reaction will tend to occur. This also means that the actual amount of Gibbs free energy released by a biochemical reaction will depend on the concentrations in the cell. As reaction products build up in relation to reactants, the Gibbs free energy yield decreases. If product concentrations are held low enough, any reaction can be made to produce a large amount of Gibbs free energy.

Put another way, it is important to distinguish between ΔG and $\Delta G°$. The latter is the change in Gibbs free energy *only when the reactants are all at concentrations of* 1 *M* (if a reactant is a gas, its partial pressure is used instead of concentration, and the partial pressure is assumed to be 1 atm). A positive value for $\Delta G°$ does not mean that the reaction cannot proceed. It only means that the equilibrium is tilted toward the reactants. The reaction can be made to proceed by keeping the concentration of one or more of the products low. On the other hand, ΔG depends on the actual concentration of reactants and products. When the numerator of the quotient on the right-hand side of equation (5.6) is less than the denominator, the logarithm will be negative. If it is negative enough so that its absolute value exceeds $\Delta G°$, ΔG will be negative and the reaction will tend to proceed spontaneously toward equilibrium. If, on the other hand, the numerator is more than the denominator, ΔG will be positive, and the reaction will tend to proceed in the reverse direction.

As an analogy, $\Delta G°$ is like the amount of potential energy that would be released by a fluid falling through a standard elevation change, say, 1 meter. This is a property of the fluid (a function of its specific gravity). On the other hand, ΔG is like the potential energy released by an actual change in elevation. This can be made arbitrarily large, or reversed in sign, by changing the beginning and ending elevations for the flow.

Equation (5.8) embodies what is known as **Le Châtelier's principle**. This states that if a reaction at equilibrium is disturbed, such as by adding one of the reactants or products to the solution, the reaction will proceed in a direction so as to partially eliminate the disturbance. For example, cells convert glucose-6-phosphate to fructose-6-phosphate as part of glycolysis. If the fructose-6-phosphate were not subsequently consumed, allowing it to accumulate, the reaction could stop or even reverse itself, so as to maintain the proper ratio between product and reactant. This is an important mechanism for the control of biochemical reactions.

Also remember that the thermodynamic relationships say nothing about how fast a reaction will proceed. The sucrose crystals in a sugar bowl are unstable in contact with air in terms of equation (5.6), but do not react at a measurable rate. However, in the presence of microbial enzymes and other requirements of microbes, such as moisture and nutrients, the sugar and oxygen are soon converted to carbon dioxide and water.

The enthalpy, ΔH, is the actual amount of energy released by a reaction in a closed system under constant pressure and temperature. So why do we use ΔG, which is the enthalpy reduced by a factor involving the change in entropy $T\,\Delta S$ [equation (5.2)] when we discuss the energy provided by a reaction? The answer is that the Gibbs free energy is the *energy available to do work*. This work may include the driving of other reactions "uphill" in a thermodynamic sense. For example, the Gibbs free energy released in oxidizing sucrose can be used to synthesize amino acids. The ΔH value for the oxidation of glucose to carbon dioxide to water is 680 kcal/mol, whereas the $\Delta G°$ value for this reaction is -686 kcal/mol, which gives $K_{eq} = 3.046$.

The Calorie counts for fats, proteins, and carbohydrates represent values for ΔH for the complete oxidation (except for proteins, because the nitrogen is excreted mostly as urea, not as nitrate). However, in humans the processes of digestion and absorption also require energy, amounting to about 6%, 4%, and 30% of the energy in fats, carbohydrates, and proteins, respectively. This energy is wasted as heat. This accounts for why people often want to eat less in hot weather. It also suggests that they would be less uncomfortable if they substituted pasta for meat on warm days.

5.1.2   Oxidation–Reduction

Among the energy-intensive reactions are oxidation–reduction, or *redox, reactions*. Oxidation is the loss of an electron, reduction is the gain. Consider how easy it is for some molecules to lose a proton. The acids readily give up a proton (hydrogen ion) in aqueous solution, but the hydrogen's electron is left behind. Subsequent removal of the electron constitutes oxidation and liberates energy. As a result, removal of a hydrogen atom (with its electron) often is the same as oxidation in biochemical systems. This is referred to as **dehydrogenation**. For example, methanol can lose two hydrogens in a reaction with oxygen to make formaldehyde:

$$2\,HC\!\!\begin{array}{c} H \\ | \\ | \\ H \end{array}\!\!OH + O_2 \Leftrightarrow 2\,C\!\!\begin{array}{c} H \\ | \\ \| \\ H \end{array}\!\!=O + H_2O$$

Sometimes it is less obvious if a reaction is an oxidation. Reactions of covalently bonded carbon atoms rarely involve complete transfer of electrons. Whether an organic compound was oxidized or reduced can be determined if there is an increase or decrease in the **oxidation state** or **oxidation number** of its atoms, particularly of its carbon atoms. The oxidation state is determined for each carbon in the molecule by assuming that all of the electrons that participate in bonding with the carbon are assigned to the more electronegative of the bonded pair. The oxidation state will then be the electrical charge that remains on the carbon.

For example, because carbon is more electronegative than hydrogen, a -1 is added to its oxidation state for each hydrogen bonded to it. Since oxygen is more electronegative than carbon and a pair of electrons are shared in a carbonyl bond ($-C{=}O$), the carbonyl bond contributes $+2$ to the carbon's oxidation state. An $-OH$ connected to a carbon contributes $+1$. Although the oxidation state really refers to a single atom, the oxidation state of an organic compound can be considered to be the sum of the oxidation state of all its carbon atoms.

Thus, the carbon in methane has an oxidation state of -4, and carbon dioxide is $+4$. The carbons in sugars have an oxidation state of 0, whereas the alcohol methanol is -2. Reaction by addition of a water molecule (**hydrolysis**) does not change the oxidation state of the carbons in an organic compound.

The oxidation state of different compounds can be compared using the **mean oxidation state of carbon** (MOC). The MOC is the sum of the oxidation states of all the carbon atoms in a molecule, divided by the number of carbon atoms. The concept can be extended to complex mixtures including suspensions such as wastewaters and sludges, leading to the MOC. The MOC can be estimated using the **total organic carbon** (TOC) and the **theoretical oxygen demand** (ThOD):

$$MOC = 4 - 1.5\left(\frac{ThOD}{TOC}\right) \tag{5.11}$$

The TOC can be computed from the chemical formula as the mass of the carbon per mole of the compound divided by its molar mass. ThOD is the mass of oxygen required stoichiometrically to oxidize a material completely. The ThOD can be computed from the balanced equation for the oxidation. The calculation of TOC, ThOD, and MOC is shown in Example 5.2. ThOD can also be estimated in the laboratory by calculating the **chemical oxygen demand** (COD). A sample is oxidized using potassium dichromate with strong acid and a catalyst under boiling conditions and the amount of oxygen equivalent to the dichromate is computed.

Example 5.2   Compute the oxidation number of oxalic acid by counting electrons shared across bonds using equation (5.11).

Answer   Each carbon in the oxalate molecule is bonded to two oxygens, one by a double bond. Since oxygen is more electronegative than carbon, one electron from the single bond and two from the double bond are assigned to the oxygens, leaving each carbon with a charge of $+3$. The carbon–carbon bond does not redistribute an electron. Therefore, the average oxidation number of the two compounds is 3.0.

According to the balanced equation for the oxidation of oxalic acid in Example 5.1, $\frac{1}{2}$ mol of O_2 is required to oxidize each mole of oxalic acid. The molar mass of oxygen is 32 g/mol, and for oxalate it is 90 g/mol. Therefore, the ThoD is $(0.5)(32/90) = 0.178$ g O_2/g oxalate. The molecular formula of oxalate is $C_2H_2O_4$, so the TOC $= (2)$ $(12/90) = 0.267$ g C/g oxalate. Using these values with equation (5.11) yields MOC $= 3.0$, which agrees with the value computed directly.

Equation (5.11) assumes that there are not significant amounts of nitro, azo, or halogenated compounds present. These can result in MOC values that are too low. It is also assumed that inorganic interferences with the COD measurement are absent. These can include cyanide, sulfide, cyanate, thiocyanate, sulfite, nitrite, thiosulfate, Fe^{2+}, Mn^{2+}, or H_2O_2. The MOC can be used as an indication of the extent to which an organic waste has been treated by oxidation processes.

In oxidation, electrons are never released directly, but are taken up by other compounds, called **electron acceptors**, which thus become reduced. Hydrogens (and their electrons) are usually picked up in pairs. Thus, oxygen accepts a pair of electrons and becomes H_2O, and nitrate becomes ammonia. Organic molecules can also accept electrons, as in fermentation reactions, for example. The oxidized forms of some metals [e.g., Fe(III) and Mn(IV)], can also act as electron acceptors.

A cell will often need to oxidize or reduce compounds. Since every oxidation must be accompanied by a reduction (one compound loses an electron, another gains), the cell could do this by coupling the oxidation of one compound to another that needs to be reduced. However, the great variety of individual oxidations and reductions that may be needed would result in a huge variety of combined reactions, and it would be necessary to have an enzyme system for each. A simpler approach is to have a smaller number of intermediary compounds that accept electrons in oxidations and then give them up in separate reduction reactions. This is called reaction **coupling**.

Several of the important compounds involved in coupling redox reactions are nucleotides that have been mentioned previously, including NAD, NADP, and FAD. In their reduced forms they are referred to as $NADH_2$, $NADPH_2$, and $FADH_2$ (often written without the subscript "2"), respectively. NAD serves as an oxidizing agent in many biochemical reactions, extracting a hydrogen with its electron. Conversely, $NADH_2$ is a reducing agent. These acronyms are a shortened form. NAD is actually NAD^+, and $NADH_2$ is actually $NADH + H^+$, and similarly for NADP and FAD. In this book we use the simpler notation.

5.1.3   Phosphate Compounds and ATP

Another form of reaction coupling involves the liberation of energy in reactions, because the energy liberated by one reaction could be used to drive another. As in redox coupling, however, there is a need for an intermediary to serve as a sort of broker in the transfer of energy from one set of reactions to another. This function is served mostly by a single compound, adenosine triphosphate (ATP).

Energy is stored in a high-energy bond between phosphate and organic compounds. Such bonds have standard Gibbs free energies of hydrolysis in the range -2 to -13 kcal/mol. The reaction that bonds phosphate to an organic is called **phosphorylation**. The reverse reaction is hydrolysis. Phosphorylation of an organic is often a step in its oxidation. For example, the first step in oxidation of glucose is conversion to glucose-6-phosphate.

A number of compounds have energies in the higher end of the range given above, including adenosine triphosphate (ATP). ATP is used to provide energy to almost all biochemical reactions that require energy. The standard Gibbs free energy for the hydrolysis of ATP to adenoside diphosphate (ADP) is -7.3 kcal/mol:

$$\text{ATP} \Leftrightarrow \text{ADP} + \text{P}_i \qquad \Delta G^\circ = -7.3 \, \text{kcal/mol} \tag{5.12}$$

However, since other reactions remove phosphate continually, this reaction is not at equilibrium in the cell. The actual ΔG can be as negative as -11 or -12 kcal/mol or more. ATP is not stored in significant quantities by cells. If production stopped, a muscle cell would deplete its supply in a few minutes. However, the turnover rate is quite high. Humans produce and use about 40 kg in a day.

adenosine triphosphate (ATP)

5.1.4 Reaction Coupling

To show how reaction coupling works, consider the following hypothetical reaction:

$$\text{A} \Leftrightarrow \text{B} \qquad \Delta G^\circ = +4.0 \, \text{kcal/mol} \tag{5.13}$$

The equilibrium ratio for this reaction will have the following value at 310 K (based on equation (5.7):

$$K_{A/B} = \frac{[\text{B}]_{eq}}{[\text{A}]_{eq}} = \exp\left(-\frac{4.0}{RT}\right) = 1.5 \times 10^{-3} \tag{5.14}$$

Thus, at equilibrium only a small portion of A will be converted to B. However, if by some mechanism, reaction (5.13) could be combined with reaction (5.12), the net reaction and standard Gibbs free-energy change would be

$$A + ATP \Leftrightarrow B + ADP + P_i \qquad \Delta G^\circ = 4.0 - 7.3 = -3.3 \, \text{kcal/mol} \qquad (5.15)$$

The equilibrium constant for reaction (5.15) is

$$K'_{A/B} = \frac{[B]_{eq} \, [ADP]_{eq} [P_i]_{eq}}{[A]_{eq} \quad [ATP]_{eq}} = \exp\left(+\frac{3.3}{RT}\right) = 216.0 \qquad (5.16)$$

However, ATP in cells is never in equilibrium with ADP and P_i. Under physiological conditions the relationship is approximated by

$$\frac{[ATP]}{[ADP][P_i]} \simeq 500 \qquad (5.17)$$

Now we can see that the equilibrium ratio of B to A in the coupled reaction is

$$K_{A/B} = \frac{[B]_{eq}}{[A]_{eq}} = K'_{A/B} \frac{[ATP]}{[ADP][P_i]} = (216)(500) = 1.1 \times 10^5 \qquad (5.18)$$

Thus, the equilibrium ratio of B to A would be increased by a factor of more than 10^7 by coupling the reaction with the conversion of ATP to ADP. Note that for this coupling to work, there must be some actual connection between the two reactions. In this example, A may first react with ATP, transferring a phosphate to A. The phosphorylated A might then be an unstable compound that converts into phosphorylated B. The final stop might then be the dissociation of the phosphate from B, leaving the free product. Thus, there is no step where A is converted directly to B. If all the steps just described each can proceed to a reasonable extent, this mechanism would circumvent the activation energy barrier. The net effect of this coupling is that as long as ATP is present, the equilibrium between A and B will be displaced toward formation of the product. This kind of coupling is exploited in numerous biochemical reactions.

5.2   ELEMENTARY KINETICS

One last preliminary subject is needed before we look at specific metabolic reaction systems. If the thermodynamics is favorable, the reaction may proceed. However, thermodynamics tells us nothing about how fast the reaction will go. As noted above, a solution of sugar in water in contact with air is thermodynamically unstable. However, the reaction proceeds immeasurably slowly. How fast the reaction occurs is part of the study of **chemical kinetics**. We start with some basic kinetics, since the same approach is useful to model other kinds of biological changes, including microbial growth, growth of plant or animal populations, and transport of pollutants in the environment or within organisms.

Recall reaction (5.4). In a constant-volume batch system the **specific reaction rate**, r, is defined in terms of any of the reaction participants as

$$r = \frac{1}{a}\frac{d[A]}{dt} = \frac{1}{b}\frac{d[B]}{dt} = -\frac{1}{c}\frac{d[C]}{dt} = -\frac{1}{d}\frac{d[D]}{dt} \qquad (5.19)$$

The specific reaction rate, r, depends on the concentrations of the reacting species as well as factors such as temperature, pressure, and ionic strength. We hold those last three constant and examine the dependence on reactant concentration alone. We will also examine only the forward reaction, r_f, the elimination of A and B by conversion into products, neglecting any formation of A and B from the reverse reaction of C and D. The equation relating the reaction rate to the concentrations of reactants is called the **rate law**.

At this point we invoke an assumption that the reaction proceeds exactly as written, without the formation of intermediate compounds. Such a reaction is called an **elementary reaction**. Although not strictly applicable, elementary kinetics are a good approximation for many complex biochemical reactions, such as biodegradation of organics by microorganisms.

The rate *of an elementary reaction* is proportional to the product of the reactants raised to their stoichiometric coefficients. This is called **the law of mass action**. Referring to the reaction rate for A ($r_A = ar$) instead of the specific reaction rate, the law of mass action gives the following rate law:

$$r_A = \frac{d[A]}{dt} = k[A]^a[B]^b \tag{5.20}$$

Thus, if we know the stoichiometry of the reaction and can assume it to be an elementary reaction, we can write down the rate law. The **order** of the reaction is defined as the sum of the exponents of the concentrations in the rate law. In this case, the order is $a + b$. The coefficient k is called the **rate coefficient**.

For example, the air pollutant nitrogen pentoxide (N_2O_5) decomposes spontaneously in air according to the following rate law:

$$\frac{d[N_2O_5]}{dt} = -k[N_2O_5] \tag{5.21}$$

This is an example of a **first-order rate law**. The rate coefficient for this reaction at 45°C has a value of 0.0299 min^{-1}. First-order rate laws are used very commonly. For example, the bacterial biodegradation of organic compounds in lakes and rivers is usually approximated as a first-order decay process, even though it is a complex metabolic process.

Equation (5.21) is easily integrated to yield concentration, c, as a function of time for a closed system (such as a reaction in a beaker) given an initial concentration c_0:

$$\frac{c}{c_0} = \exp(-kt) \tag{5.22}$$

Equation (5.22) shows that for first-order reactions the concentration changes by fixed ratios in fixed time periods. One such ratio leads to the concept of **half-life**, $t_{1/2}$. Setting $c/c_0 = 0.5$, and rearranging gives the relationship between the half-life and the rate coefficient:

$$t_{1/2} = \frac{-\ln 0.5}{k} = \frac{0.693}{k} \tag{5.23}$$

Bear in mind that this relationship, and indeed the very idea of a half-life, is valid only for first-order reactions. In the nitrogen pentoxide example, the half-life is 23.2 min. Thus a vessel containing 100 ppmv initially would have 50 ppmv after 23.2 min and 25 ppmv after 46.4 min.

Example 5.3   The oxidation of ferrous iron(II) to ferric(III) in water follows first-order kinetics if oxygen is not limiting. A solution of 20 mg/L ferrous iron at pH 8.0 is aerated for 60 min. The ferric iron that is formed precipitates and can be removed by filtration. The remaining iron concentration is 4.0 mg/L. What are the rate coefficient and the half-life?

Answer   Rearrange equation (5.22) gives $k = -\ln(c/c_0)/t = 0.027 \, \text{min}^{-1}$. Then use equation (5.23), we have $t_{1/2} = 0.693/0.027 = 25.8 \, \text{min}$.

Sometimes a higher-order reaction will behave as if it were first order. For example, if the reaction is $A + B \Rightarrow C$, the rate law is $d[A]/dt = -k[A][B]$. Now suppose that B is present far in excess of A. Then as A is depleted, B will decrease only a small amount proportionally. For a specific example, suppose that the initial concentrations $[A]_0 = 2.0$ and $[B]_0 = 100.0$. Now as A becomes depleted by 50%, B will only decrease by 1%. Thus, B can be assumed to have a constant concentration. The concentration of B can be combined to form a new rate coefficient, $k' = k[B]$, and the new rate law becomes first order: $d[A]/dt = -k'[A]$. This assumption is called a **pseudo-first-order reaction**. This effect is one reason why a great many reactions can be treated as being first order even if the underlying reaction mechanism is very complicated.

5.3   ENZYME KINETICS

Enzymes function not only to catalyze reactions but to enable control of those reactions via sensitivity to environmental conditions and to the presence of cofactors, which can increase or decrease rates. A detailed treatment of the kinetics of enzyme reactions will be beneficial for two reasons. First, it will lead to a clearer understanding of the action of enzymes. Second, the kinetic equations themselves can be useful in modeling biochemical processes in environmental systems. This can be true even when the process involved is not a simple enzyme reaction. For example, the first system we describe will be Michaelis–Menten kinetics. The resulting model is derived rigorously. However, an empirical equation of the same form, the *Monod equation*, is used effectively to model substrate utilization by systems consisting not only of many different enzymes and of substrates, but of mixed populations of microbial organisms.

Recall from general chemistry that a catalyst is a substance that affects the rate of a reaction but is not changed by it, and that it acts by lowering the **activation energy**, E_a. The activation energy refers to the height of a barrier between reactants and products of a reaction. The molecules in a given system have a distribution of energies. If the activation energy is high, fewer molecules possess enough to surmount the barrier. Lowering the height of the barrier means that a higher proportion of the molecules has enough energy to get over it, and the rate at which molecules pass is increased. The effect of activation energy and temperature on the rate constant of a first-order reaction is given by the *Arrhenius equation*:

$$\ln k = B - \frac{E_a}{RT} \tag{5.24}$$

where k is the rate constant for the reaction, B is an arbitrary constant, R is the gas law constant, and T is the temperature in kelvin. As an example, the hydrolysis of sucrose is catalyzed by hydrogen ions, with an activation energy of 25,600 cal/mol. The yeast enzyme invertase lowers the activation energy barrier to about 8000 cal/mol. By writing equation (5.24) first for the hydrogen catalysis, then for the enzyme, assuming that B is the same for both reactions, and subtracting the two, one gets the following expression for the ratio of the two rate constants:

$$\log \frac{k_{enzyme}}{k_{hydrogen}} = \frac{25,600 - 8000}{2.3\,RT} = 12.4 \tag{5.25}$$

at the human body temperature of 310 K. Thus, by reducing the activation energy 69%, the enzyme enhances the rate of this reaction by a factor of 2.5×10^{12}!

5.3.1 Single-Substrate Kinetics

The simplest enzyme system we can study is one with a single substrate, S, and single product, P: that is, the reaction

$$S + E \rightarrow P + E \tag{5.26}$$

The reaction rate, r, is the mass of product produced or substrate consumed per unit time per unit volume. This is expressed as

$$r = \frac{-d[S]}{dt} = \frac{d[P]}{dt} \tag{5.27}$$

Experiments with such systems have been conducted at varying concentrations of total enzyme $[E]_t$ and substrate $[S]$. Two observations are commonly made based on experimental observations:

- The rate of the reaction at any particular substrate concentration is proportional to the total concentration of enzyme if the substrate concentration is held constant (Figure 5.1a).

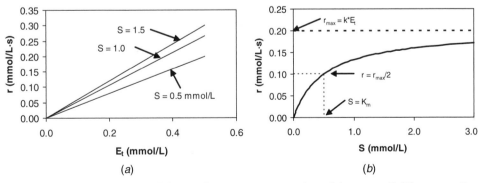

Figure 5.1   Typical behavior of single substrate–enzyme reactions: (a) constant S; (b) constant E_t.

- If the total enzyme concentration, E_t, is held constant, the rate of reaction shows a saturation effect of substrate concentration. At low concentrations of substrate, the rate is proportional to substrate concentration. However as substrate concentration becomes large, the rate levels off, until it is independent of concentration (Figure 5.1b).

An empirical expression was proposed in 1902 by Henri that satisfied these observations. Michaelis and Menten developed a theoretical derivation, later improved by Briggs and Haldane, resulting in what is now called the *Michaelis–Menten equation*. The derivation starts by assuming the following reaction mechanism:

$$S + E \overset{k_1}{\underset{k_{-1}}{\rightleftharpoons}} ES \qquad (5.28)$$

$$ES \overset{k_2}{\rightarrow} P + E \qquad (5.29)$$

The species ES represents a combined enzyme–substrate complex in which the substrate is bound to the active site of the enzyme. The first reaction, in which the complex is formed, is assumed to be reversible and in equilibrium. The second part, decomposition of the complex into product with recovery of the free enzyme, is assumed to be irreversible. The total enzyme concentration is given by

$$[E]_t = [E] + [ES] \qquad (5.30)$$

Experimentally, $[E]_t$ is known since it is the amount of enzyme present initially. The species E and ES usually cannot be measured directly. Now the rate equations for the reaction rate and the rate of change of [ES] can be written assuming elementary reaction kinetics:

$$\frac{d[S]}{dt} = -k_1[S][E] + k_{-1}[ES] \qquad (5.31)$$

$$\frac{d[ES]}{dt} = k_1[E][S] - (k_{-1} + k_2)[ES] \qquad (5.32)$$

At this point another key assumption is introduced that simplifies the result. Equations (5.30) to (5.32) do not have an analytical solution but can be solved numerically. Examination of numerical solutions shows that after a short startup period, the concentration of the enzyme–substrate complex is fairly constant. Thus, equation (5.32) can be set to zero. This is the **quasisteady-state assumption**. With this assumption, and using equation (5.30) to eliminate [E], equation (5.32) can be rearranged into

$$\frac{[S]([E]_t - [ES])}{[ES]} = \frac{k_{-1} + k_2}{k_1} = K_m \qquad (5.33)$$

where K_m is known as the **Michaelis constant**. Solving equation (5.33) for [ES] gives

$$[ES] = \frac{[E]_t[S]}{K_m + [S]} \qquad (5.34)$$

Finally, it is noted the rate of the reaction in equation (5.29) must be equal to the overall rate of reaction, r (with the units $[M \cdot t^{-1}]$. Therefore, from elementary reaction kinetics we can say:

$$r = k_2[ES] \tag{5.35}$$

Using equation (5.35) to eliminate [ES] from equation (5.34), we obtain

$$r = \frac{k[E]_t[S]}{K_m + [S]} \tag{5.36}$$

This is the final form of the Michaelis–Menten equation, which satisfies the observations made above. Note that $k = k_2$. We have dropped the subscript 2 from the rate constant for simplicity. The two coefficients of equation (5.36) are determined by fitting to experimental data. Usually, the equation is linearly transformed so that linear regression methods can be used. However, modern statistical software makes nonlinear regression easy and eliminates accuracy problems with the linear transformation methods.

The Michaelis–Menton equation approaches its maximum value of $k[E]_t$ as the substrate concentration increases without limit. When the substrate concentration is equal to K_m, the reaction rate will be one-half its maximum value at that same enzyme concentration.

Example 5.4   The reaction rate for an enzyme reaction is 0.1 mol/L·min a substrate concentration 0.01 M. Doubling the substrate concentration increases the rate by 50%. What is K_m for this reaction? What is the maximum rate for this reaction?

Answer   Taking equation (5.36) and writing the first rate r_1 at concentration S_1, and the second r_2 at S_2, we can cancel out $k[E]_t$ by taking the ratio r_1/r_2 and then solve for K_m:

$$\frac{r_1}{r_2} = \frac{K_m + S_2}{K_m + S_1} \frac{S_1}{S_2}$$

$$K_m = \frac{S_1 S_2(r_2 - r_1)}{r_1 S_2 - r_2 S_1} = \frac{(0.01)(0.02)(0.15 - 0.10)}{(0.10)(0.02) - (0.15)(0.01)} = 0.02\,M$$

The maximum rate is equal to $k[E]_t$, which from equation (5.36) is

$$k[E]_t = r_1 \frac{K_m + S_1}{S_1} = (0.10)\left(\frac{0.02 + 0.01}{0.01}\right) = 0.30\,mol/L \cdot min$$

The Michaelis–Menten equation is well known and often used. However, it is instructive to consider conditions under which the equation is not accurate. One is if the ratio of $[E]_0/[S]$ is large (see Problem 5.2). Also, the pseudosteady-state assumption does not apply at the beginning of an experiment, when enzymes are just mixed with substrate and no complex has been formed. Of course, there are many situations in which the basic mechanism may be more complicated, such as in the case of multiple substrates or the use of cofactors.

What is the effect of assuming that reaction (5.29) is irreversible? If it is assumed reversible, with an equilibrium constant equal to K_P, and renaming K_m to K_S, and

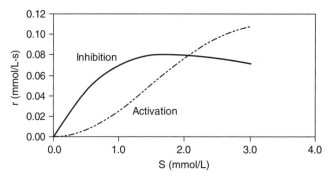

Figure 5.2   Substrate activation and inhibition.

again invoking the quasisteady-state assumption, we obtain

$$r = [E]_t \frac{(k_2/K_S)[S] - (k_{-1}/K_P)[P]}{1 + [S]/K_S + [P]/K_P}$$
(5.37)

One important thing to note about this case is that the rate depends on the product concentration as well as substrate concentration, and that the rate decreases with increasing product concentration.

Two other single-substrate concentration effects are worth noting. **Substrate activation** describes the situation observed in Figure 5.2, in which the effect of substrate concentration on reaction rate shows a sigmoidal shape. It can be modeled by adding a reaction to (5.28) and (5.29) in which the free enzyme is in equilibrium with an inactive form. Activation is different from *enzyme induction* (discussed in Section 6.2.2). In induction the presence of substrate actually stimulates the organism to produce the enzyme; that is, the enzyme is absent when not needed.

Substrate inhibition is the case when adding substrate beyond an optimum amount causes a reduction in the reaction rate (Figure 5.2). It can be modeled by assuming that a second substrate molecule complexes reversibly with the enzyme but that this new complex does not produce product directly. The resulting rate equation is

$$r = \frac{k[E]_t}{1 + K_m/[S] + [S]/K_I}$$
(5.38)

where K_m and K_I are constants. This expression is known as the **Haldane equation**, and is equivalent in form to the Andrews equation (Section 11.7.7), often used to model microbial biodegradation of toxic substances. Equation (5.38) has a maximum at the substrate concentration:

$$[S]_{max} = \sqrt{K_m K_I}$$
(5.39)

Many industrial organic chemicals are both biodegradable and toxic to microorganisms, and thus their biodegradation may be modeled by the Haldane equation. Examples of such compounds include benzene and phenol.

Inhibition can be important in the normal control of metabolism. For example, the amino acid isoleucine inhibits one of the enzymes involved in its formation, preventing an oversupply from being produced. This is an example of **feedback inhibition.**

5.3.2   Multiple Substrates

In the case where two substrates, S_1 and S_2, bind reversibly to a enzyme to make a single product, we may assume the following reaction mechanism:

$$S_1 + E \underset{k_{-1}}{\overset{k_1}{\rightleftharpoons}} ES_1$$

$$S_2 + E \underset{k_{-2}}{\overset{k_2}{\rightleftharpoons}} ES_2$$

$$S_2 + ES_1 \underset{k_{-12}}{\overset{k_{12}}{\rightleftharpoons}} ES_1S_2 \qquad (5.40)$$

$$S_1 + ES_2 \underset{k_{-21}}{\overset{k_{21}}{\rightleftharpoons}} ES_1S_2$$

$$ES_1S_2 \overset{k}{\rightarrow} P + E$$

where the first four reactions are in equilibrium with dissociation constants of K_1, K_2, K_{12}, and K_{21}, respectively. The resulting equation for the rate is

$$r = k[E]_t \frac{[S_1]}{K_1^* + [S_1]} \frac{[S_2]}{K_2 + [S_2]} \qquad (5.41)$$

where

$$K_1^* = \frac{K_{21}[S_2] + K_1 K_{12}}{[S]_2 + K_{12}} \qquad (5.42)$$

It is interesting that an empirical equation of the same form as (5.41) is sometimes used to model dual substrate microbial kinetics, except that K_1^* is treated as a constant. For example, such a model has been used where the two substrates are chemical oxygen demand (COD) and oxygen, or COD and nitrate.

Besides the substrate inhibition described above, inhibition can come from compounds other than the substrate. Some inhibitors will react irreversibly with the enzyme, removing it from availability permanently. These are poisons. An example is cyanide, which deactivates an enzyme in the respiration process. Others act by competing with substrate in reversible reactions with the enzyme. Their effect can be categorized in terms of their effect on the coefficients of the Michaelis–Menten equation and in terms of how they bind to the enzymes.

Competitive inhibitors bind to the same active site as the substrate. They do not change the maximum reaction rate, but do change K_m. Their effect can be overcome by increasing substrate concentration. For example, TCE is cometabolized with substrates such as methanol. The presence of TCE reduces the biodegradation rate of the methanol but not its maximum biodegradation rate. **Noncompetitive inhibitors** bind at a different site than the substrate. They act by changing the shape of the enzyme, and therefore its activity. The maximum rate is changed but not K_m. **Uncompetitive inhibitors** bind to the enzyme–substrate complex, preventing the formation of products. Both maximum rate and K_m are reduced.

5.3.3 Effect of pH

Like all proteins, enzymes have a variety of acidic and basic groups, each with its own ionization constant. For the enzyme to have optimum activity, it must exist in a particular ionization state. If the pH changes enough to change the ionization significantly on any one group, the enzyme's activity will decrease. This can be modeled by assuming that the enzyme is in equilibrium with two versions of itself, one containing one additional proton over the optimum, E^+, and another with one proton less than optimum, E^-, with equilibrium constants associated with the conversion to each of those states:

$$E^+ \overset{K_1}{\Leftrightarrow} E \overset{K_2}{\Leftrightarrow} E^- \tag{5.43}$$

With this mechanism it is possible to derive the following equation for the fraction of the total enzyme in the optimum state as a function of the hydrogen ion concentration:

$$\frac{[E]}{[E]_t} = \frac{1}{1 + [H^+]/K_1 + K_2/[H^+]} \tag{5.44}$$

Notice that this has the same form as the Haldane equation for substrate inhibition as in equation (5.38). In a similar fashion, the optimum pH is found to be

$$pH_{optimum} = \frac{pK_1 + pK_2}{2} \tag{5.45}$$

Figure 5.3 shows the effect of pH on enzyme activity according to equation (5.44). All curves have an optimum pH of 7.0. The outer curve has pK_1 and pK_2 values of 4 and 10, respectively; inner curves are 5 and 9, 6 and 8, and 6.75 and 7.25, respectively. Note that the farther apart the pK_a values, the broader the optimum. Furthermore, if the pK_a values are close to each other, the maximum activity falls substantially below 100%. This means

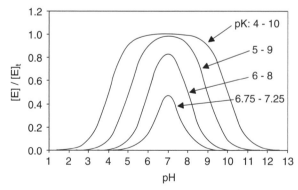

Figure 5.3   Effect of pH on enzyme activity for various ranges of pK_a.

that even at the optimum pH, much of the enzyme is in an inactive form because one or the other of the acid–base groups is in the wrong state of association.

5.3.4   Effect of Temperature

At first it would seem that the Arrhenius equation (5.24) would be all that is needed to describe the temperature effect. In fact, it does hold, but only in the lower range of temperatures associated with life. At higher temperatures another reaction occurs: the denaturation of the enzyme. One way to handle this is to treat the denaturation reaction as a simultaneous equilibrium. The dissociation equilibrium constant, K_d, is related to temperature by

$$K_d = \exp\left(\frac{-\Delta G_d}{RT}\right) = \exp\left(\frac{-\Delta H_d}{RT}\right)\exp\left(\frac{-\Delta S_d}{R}\right) \qquad (5.46)$$

where ΔG_d, ΔH_d, and ΔS_d are Gibbs free energy, enthalpy, and entropy of deactivation, respectively. For example, the enthalpy and entropy of dissociation for trypsin are 68 kcal/mol and 213 cal/mol·K, respectively, and the Gibbs free energy for the reaction is 1.97 kcal/mol. With this and a similar relationship for the rate constant, k, in equation (5.36) the following expression for maximum rate in the Michaelis–Menten equation can be derived:

$$r_{\max} = k[\mathrm{E}]_t = \frac{\beta T \exp(-E/RT)}{1 + \exp(\Delta S_d/R)\exp(-\Delta H_d/RT)} \qquad (5.47)$$

where β is a kinetic rate coefficient. This is curve (a) in Figure 5.4.

Figure 5.4 shows this relationship. Note that the side of this curve below the temperature optimum is slightly concave upward. This portion can be approximated empirically as a simple exponential, as shown in Figure 5.4b:

$$r = r_{20}\theta^{T-20} \qquad (5.48)$$

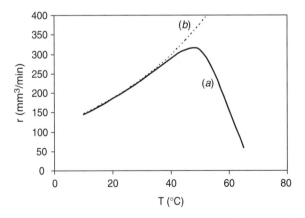

Figure 5.4   Effect of temperature on hydrogen peroxide decomposition by catalase; (a) is equation (5.47) with E = 3.5 kcal/mol, ΔH_d = 55.5 kcal/mol, ΔS_d = 168 kcal/mol · K, β = 258 mm^3/min; (b) is equation (5.48) fitted to equation (5.47) at 20°C and 25°C; r_{20} = 185.8 mm^3/min, θ = 1.024. (Based on Bailey and Ollis, 1986.)

where T is the temperature in degrees Celsius, r_{20} the reaction rate at $20°C$, and θ an empirical coefficient. This expression is commonly used to describe the effect of temperature on biological growth rates in biological waste treatment processes. In the example of Figure 5.4, the denominator of equation (5.47) is very close to 1.0 up to a temperature of $40°C$. Thus, the numerator carries most of the effect of temperature and the exponential form of (5.48) holds.

5.3.5   Other Considerations

Keep in mind that the deactivation due to temperature described above is reversible. Irreversible denaturation can be modeled in a number of ways, depending on the mechanism. In the simplest case, first-order decay may suffice. Irreversible denaturation by heat may be one of the principal mechanisms of heat sterilization.

It is also important to know that many enzymes function when associated with membranes or other biological structures and may not function at all when extracted into solution. Some, such as pancreatic lipase, function only when absorbed at the interface between a lipid droplet and an aqueous solution.

5.4   BIOCHEMICAL PATHWAYS

With the preceding discussion of the types of compounds that form living things (Chapter 4) and of the enzymes that facilitate and control their chemical transformations and of thermodynamic relationships governing individual reactions, we can now examine some of the most important of these transformations. We are usually interested not in individual reactions but in a sequence of reactions, called a **pathway**, leading from initial reactants to final products. For example, the oxidation of the carbon in glucose to carbon dioxide involves some 21 reactions, usually divided into two pathways: glycolysis, which forms an intermediate called *pyruvate*, and the Krebs cycle, which produces the CO_2.

The detailed reactions of each pathway are not given here. Instead, we examine the overall reactions for several major pathways, with a few details about what occurs within the pathway. The rate of an overall reaction pathway is often governed by a single limiting step in the sequence of reactions. In the case of biodegradation of xenobiotic compounds, it is often the first step that limits the overall rate. Biochemical pathways can be classified in several ways. One of the simpler ways is to divide them into the following: **Catabolic pathways** are those that break down organic compounds, usually to provide energy; **anabolic pathways** are those that synthesize complex organics, such as to form new cell material, from simpler precursors. However, many metabolic pathways do not fit easily into these two categories.

5.4.1   Glycolysis

First we consider ways of extracting energy from organics without external oxidizers such as oxygen. Essentially, this means producing ATP from ADP (**phosphorylation**) for short-term use (on the order of seconds or minutes). Phosphorylation of ADP by reaction with organic molecules, and without an electron acceptor such as oxygen, is called **substrate-level phosphorylation**. For longer-term energy supply, cells store glucose (as starch in plants or as glycogen in animals) or lipids. Thus, our starting point is glycolysis, which provides a ready source of ATP by partial oxidation of glucose. It incorporates substrate-level phosphorylation. Its products feed into respiration pathways, for more energy extraction, or to fermentation pathways for waste elimination.

Glycolysis is the conversion of glucose to pyruvic acid, an important metabolic intermediate. It is the first of several pathways for the production of ATP from glucose (Figure 5.5). Glycolysis consists of nine reactions involving nine intermediate compounds. The six-carbon glucose first enters into several reactions involving ATP, forming a fructose with two phosphate groups. Since this consumes ATP, it represents an investment by the cell to get the reaction going, forming an activated intermediate compound. This splits into two three-carbon sugars, each with one phosphate, which is oxidized by an NAD to form $NADH_2$. Both three-carbon sugars then go through a series of steps that generate two ATPs each, for a total of four. Thus, the net reaction of glycolysis is

$$glucose + 2ADP + 2P_i + 2NAD \Rightarrow 2\,pyruvic\ acid + 2ATP + 2NADH_2 \qquad (5.49)$$

Since each ATP has a standard Gibbs free energy of 7.3 kcal/mol and the complete oxidation of glucose has 686 kcal/mol, this reaction has an efficiency of only 2.1%. Additional energy is contained in the pyruvic acid and the $NADH_2$, but these are unavailable without an oxidizing agent. This is why anaerobic bacteria yield less cell mass than aerobes do, since they do not harvest as much energy for synthesis. In the aerobic process of respiration, described below, both the pyruvate and the $NADH_2$ can be made to yield more ATP.

Pyruvate is a link to numerous other biochemical pathways, besides feeding into the respiration pathway. It can be made into alanine, leading to synthesis of amino acids. Pyruvate is also the starting point for many fermentation products described below. Some of the other glycolysis intermediates are also used to synthesize other sugars or amino acids. There also are other less important pathways for glucose catabolism. The cell shifts to them to control the level of ATP in the cell or to form other compounds.

5.4.2   Fermentation

Besides ATP and pyruvate, glycolysis produces the reducing agent $NADH_2$. Unless this were reoxidized to NAD, the cell's supply of NAD would be depleted. Also, pyruvate is retained by the cell. A buildup of pyruvate would decrease the rate of the glycolysis by the law of mass action. Many cells use the pyruvate and $NADH_2$ in respiration with oxygen (described in the next section) to produce more ATP. In the absence of oxygen, cells use a different pathway, in which the $NADH_2$ is used to reduce the pyruvate to various products that can leak out of the cell as waste products. This process, which forms such partially oxidized by-products to regenerate NAD, is called *fermentation*. More generally, **fermentation** is an anaerobic biochemical process in which organic compounds serve as both electron donors and acceptors. Thus, some organics become reduced (e.g., NAD) and others are oxidized (e.g., ethanol to acetic acid), and no inorganic electron acceptor (such as oxygen) is needed.

In animals, the rapid generation of ATP needed for muscle power can only be generated by glycolysis. Animal cells eliminate pyruvate and regenerate NAD by a single-step reaction that forms lactic acid. However, lactic acid is a "blind alley" in animals and cannot be used further. But unlike pyruvate, it can diffuse out of the cell, where the blood transports it to the liver. There it is converted back to pyruvate to enter other pathways.

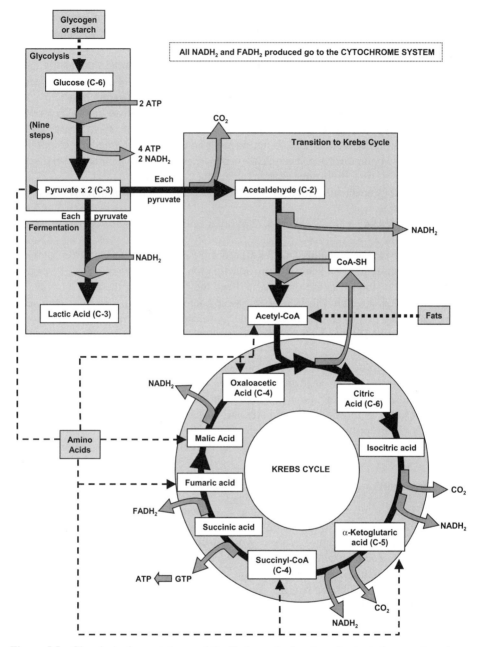

Figure 5.5   Glycolysis, fermentation, and the Krebs cycle. Inputs and outputs from each cycle are shown, as well as connections with the metabolism of other biochemical compounds.

The accumulation of lactic acid in the muscles causes the pain that results from vigorous exercise. Rest allows time for the elimination and conversion of the lactic acid.

Bacteria and yeast can produce other products. The most important is ethanol, which is formed by yeast in two steps, with acetaldehyde as an intermediate. Specific organism and cultivation conditions yield specific end products. Yeast can produce glycerol in addition to ethanol, although it is derived from intermediates of glycolysis and not from pyruvate. *Clostridium* produces acetone, isopropanol, butyrate, and butanol. Proprionic acid bacteria produce proprionate. Coliforms produce formic acid, acetic acid, hydrogen, and CO_2. *Enterobacter* produces ethanol, 2,3-butanediol, formic acid, and lactic acid.

Fermentation produces other compounds of commercial importance, some of which may originate with pyruvate, such as penicillin. However, although they are produced during fermentation, they are not end products used by the cell. Thus, these are properly referred to as *secondary metabolites*, not fermentation products.

5.4.3 Respiration

Respiration is a process in which organic compounds or reduced inorganics (such as hydrogen or ferrous iron) are oxidized by inorganic electron acceptors for the production of energy. Eukaryotes can only use oxygen as a final electron acceptor, in what is called **aerobic respiration**. Microorganisms can also use nitrate, sulfate, some metals, and even carbon dioxide, in what is generally called **anaerobic respiration**. In environmental applications the term **anoxic respiration** is generally used for nitrate reduction, and anaerobic respiration is limited to the other forms.

Strictly speaking, glycolysis is not a part of respiration, although it is the first step leading to respiration for glucose. Nevertheless, when discussing respiration, many scientists and engineers are referring to the overall conversion of glucose or other carbohydrates to carbon dioxide and water:

$$C_6H_{12}O_6 + 6O_2 \Rightarrow 6CO_2 + 6H_2O \tag{5.50}$$

Respiration occurs in two phases. The first is the **Krebs cycle** [also called the **citric acid cycle** or **tricarboxylic acid (TCA) cycle**]. The Krebs cycle completes the job of oxidizing the carbon that originated with glucose, forming CO_2 and ATP. However, much of the energy is left in the reducing power of $NADH_2$ or $FADH_2$. These are converted to ATP in the second phase, called the **electron transport system** or **cytochrome system**. The cytochrome system is also responsible for reducing oxygen to water, the other product of the overall reaction for the oxidation of glucose. Both the Krebs cycle and the cytochrome system are cyclic because intermediates involved in each reaction are regenerated by other reactions.

Both of these process are mediated by enzymes bound to membranes and require the presence of the membranes in order to function. In eukaryotes this occurs within the mitochondria. Pyruvate must diffuse into the mitochondria to enter the process. Prokaryotes do not have internal membrane structures. Their respiratory enzymes are bound to their cell membrane. The discussion here focuses on eukaryotes, but the process is similar in prokaryotes. Mitochondria consist of an inner and an outer membrane, forming an inner and an outer compartment. The inner membrane is folded extensively (Figure 5.6). The Krebs cycle occurs within the inner compartment. The cytochrome system is integral to the inner membrane and involves reactants in both compartments.

Glycolysis forms two three-carbon pyruvate molecules for each glucose. When oxygen or another suitable electron acceptor is available, the pyruvate enters into respiration instead of fermentation. Respiration begins when the pyruvate diffuses into the inner compartment of the mitochondria. There, each pyruvate becomes covalently bonded through a sulfhydryl bond with a coenzyme, called **coenzyme A CoA**, which itself is a derivative of

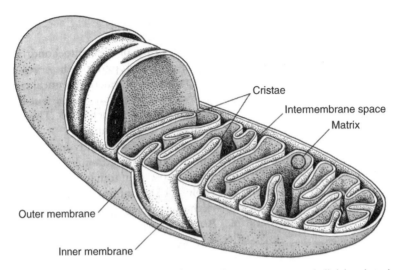

Figure 5.6 Mitochondrion structure, showing membrane structure and division into inner and outer compartments. (From Smith et al., 1983.)

ADP. The process is an oxidation and requires an NAD. It results in a compound called **acetyl-CoA**:

$$
\begin{array}{l}
CH_3 \\
| \\
C=O + CoA-SH + NAD \implies \\
| \\
COOH
\end{array}
\begin{array}{l}
CH_3 \\
| \\
C=O + CO_2 + NADH_2 \\
| \\
S\text{-}CoA
\end{array}
\qquad (5.51)
$$

pyruvate   *acetyl-CoA*

The removal of one CO_2 for each pyruvate accounts for two of the six carbons originating in the glucose. The $NADH_2$ carries much of the energy of oxidation and will be used to generate ATP in the electron transport system.

Acetyl-CoA is another important metabolic intermediate. It is a point of connection between a number of pathways, including the breakdown of lipids and amino acids for energy. But for purposes of the present discussion, a pair of acetyl-CoAs serve the purpose of carrying four carbons from the original glucose molecule into the Krebs cycle.

Krebs Cycle   The Krebs cycle oxidizes the two carbons remaining from each pyruvate to CO_2. One ATP is formed as a result, and the remaining four pairs of hydrogens (with their energetic electrons) reduce the cofactors NAD or FAD. The overall reaction is

$$
\begin{aligned}
&\text{acetyl-CoA} + 3\text{NAD} + \text{FAD} + \text{ADP} \\
&\Rightarrow 2CO_2 + \text{CoA} + 3\text{NADH}_2 + \text{FADH}_2 + \text{ATP}
\end{aligned}
\qquad (5.52)
$$

Since two such reactions occur for each glucose, all of the original carbon atoms are accounted for. Only two additional ATPs are formed, for a total of four when combined with glycolysis. Thus, the efficiency so far, in terms of standard Gibbs free energy, is four times 7.3 kcal/mol for the ATPs, divided by 686 kcal/mol glucose, or 4.3%. However, there are a total of 10 pairs of energetic hydrogen electrons that have been captured by cofactors.

Some details of the cyclical reactions of the Krebs cycle may help clarify the picture. The cycle starts when the C_4 oxaloacetate combines with two carbons of the acetyl-CoA that originated with pyruvate. The products of this first reaction regenerate the CoA and form the C_6 tricarboxylic acid citric acid:

$$
\begin{array}{ccccc}
 & & & \text{COOH} & \\
 & & & | & \\
\text{COOH} & & & \text{CH}_2 & \\
| & & & | & \\
\text{CH}_2 & + \quad \text{acetyl-CoA} & \Longrightarrow & \text{HO}-\text{C}-\text{COOH} & + \quad \text{CoA} \\
| & & & | & \\
\text{C}=\text{O} & & & \text{CH}_2 & \\
| & & & | & \\
\text{COOH} & & & \text{COOH} & \\
\textit{oxaloacetate} & & & \textit{citric acid} &
\end{array}
\qquad (5.53)
$$

A series of eight reactions complete the cycle, in which the two carbons added by the first step are effectively removed, along with the four pairs of hydrogen. (Some of the hydrogen originates in H_2O, which is incorporated at several steps.) Some of the intermediates may be used to synthesize other compounds needed by the cell. The reducing power of the $NADH_2$ may be used in synthesis reactions of other pathways. Both of these reduce the energy yield of the Krebs cycle; however, this is not necessarily inefficient since useful products are made. However, the cell must maintain a supply of oxaloacetate so that the cycle can continue. If intermediates are drawn off, oxaloacetate can be formed from pyruvate and CO_2.

The Krebs cycle is normally controlled by feedback inhibition by the products of the cycle. High levels of the ratios $NADH_2/NAD$, ATP/ADP, or acetyl-CoA/CoA indicate that the cell has ample energy, and slow the cycle.

The Cytochrome System   The six carbons from the glucose have been disposed of, but the substantial reducing power in the form of $NADH_2$ and $FADH_2$ has yet to be converted to energy in a useful form: namely, as ATP. In eukaryotes this conversion occurs with a series of enzymes and other compounds bound to the inner mitochondrial membrane, which comprise the **cytochrome system** (Figure 5.7). $NADH_2$ reduces the first of these compounds, *flavin mononucleotide* (FMN), which spans the membrane from one side to the other. FMN takes the two hydrogens, passes their electrons to the next membrane compound (at a lower energy), and releases the protons to the outside of the membrane. It may be useful to think of the electrons as dropping to a lower voltage, to use an electronic analogy, each time they pass from one electron carrier to the next. Thus, the net effect is to use some of the energy from the electrons to move a pair of protons to the outer compartment of the mitochondrion.

The compound that receives the electrons does a similar thing; it passes them on at a yet lower energy, which may or may not result in moving protons to the outer compartment. In all, six protons are transported. Consider the result: Energy is used to transport protons, creating a pH gradient. This pumping of protons across a membrane into an area of increasing proton concentration stores energy, just as to pumping air into a tank produces a pressure difference that also stores energy. But there isn't only a pH gradient. Since the protons are transported without their electrons or complementary anions, an electric charge gradient, an actual voltage, is also accumulated. The combined chemical and electrical potential across the inner membrane of the mitochondrion is called the **chemiosmotic potential**.

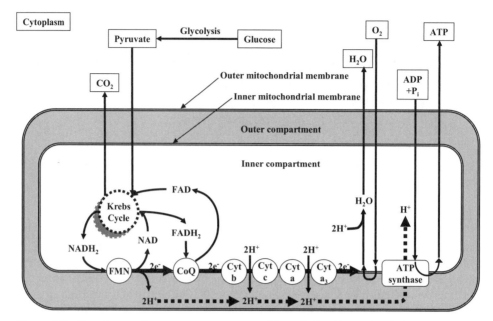

Figure 5.7   Cytochrome electron transport system, with a relationship to glycolysis and the Krebs cycle.

The energy of the chemiosmotic potential is used when the protons flow back across the membrane to the inner compartment through a complex of transmembrane proteins called **ATP synthase**, producing ATP from ADP. Think of air that has been pumped into a pressure vessel then flowing back out through a turbine to produce electricity. Altogether, 3 mol of ATP is produced for each mole of $NADH_2$. The $FADH_2$ acts similarly, except that its electrons have a lower energy to start with, so it enters the chain farther along. It provides only enough energy to produce 2 mol of ATP per mole of $FADH_2$.

Finally, what becomes of the energy-depleted electrons in the transport chain? The final components of the transport chain are a series of membrane-bound compounds called **cytochromes**. The final cytochrome performs one last reduction, that of oxygen. Each $\frac{1}{2}$ mol of O_2 receives 2 mol of electrons and 2 mol of H^+, forming H_2O and completing the process of respiration. Because of the use of an electron acceptor, ATP production in this system is called **oxidative phosphorylation**.

Assuming that none of the intermediate compounds or $NADH_2$ have been shunted off for other cellular purposes, the final tally for ATP is a maximum of 36 mol per mole of glucose oxidized. Now the efficiency based on standard Gibbs free energy is 38%, a big improvement over glycolysis or the Krebs cycle. Since organisms will use every opportunity for growth, a more typical number of ATPs actually produced by glycolysis and respiration is 20 to 25.

In prokaryotes, the electron transport chain is located in the cell membrane. Bacteria perform the function of the electron transport system without mitochondria by pumping protons outside the cell, depleting it within to create the chemiosmotic potential.

Alternative Electron Acceptors   Some bacteria can switch to other electron acceptors, such as nitrate, when oxygen is absent. The nitrate becomes reduced ultimately to nitrogen by the process of **denitrification**. It is thought that this occurs in a series of steps as follows:

$$NO_3^- \Rightarrow NO_2^- \Rightarrow NO \Rightarrow N_2O \Rightarrow N_2$$

However, the nitrate requires a higher-energy electron for its reduction. It gets it at an earlier point in the electron transport system, so that only 2 mol of ATP is formed per mole of $NADH_2$. This is why organisms that can will always use oxygen when it is present and switch to nitrate only in the absence of oxygen.

Sulfate and carbon dioxide can also function as electron acceptors for certain microorganisms. However, the organisms that do so cannot also use oxygen. When sulfate is reduced, the product is hydrogen sulfide, a poisonous gas. Carbon dioxide is reduced to methane by molecular hydrogen by a special group of organisms within the archaea. Oxidized metal ions such as ferric (III) iron can also be an electron acceptor. This occurs, for instance, in the sediments of wetlands, where oxygen is limiting. The iron becomes reduced to ferrous (II) iron. If an environment contains oxygen, nitrate, sulfate, and carbon dioxide, the oxygen will tend to be used up first, producing H_2O. Then the nitrate, then sulfate, and then CO_2 will be used. In homogeneous environments, one electron acceptor will not be utilized until the energetically more favorable one is depleted. However, in some situations microenvironments may form, allowing several of these reactions to proceed simultaneously. For example, in biological slime layers in aquatic systems, microorganisms near the slime layer surface may have access to oxygen while organisms below the surface may be depleted of oxygen and will use nitrate.

5.4.4 Oxidation of Fats and Amino Acids

Obviously, glucose is not the only fuel used by living things. Our foods contain other sugars, such as lactose in milk and fructose in the disaccharide sucrose (table sugar). The success of dieters hinges on the body's ability to use fat as fuel; of course, this is why the body stores fat in the first place. Under starvation conditions, the body obtains its energy for basic cell function by cannibalizing itself, by oxidizing its proteins.

The sugars are converted fairly easily into either glucose or another intermediate in the glycolysis pathway. In the case of sucrose and glycogen, these polysaccharides are split into simple sugars by phosphorolysis (splitting by phosphate) instead of hydrolysis. This results in glucose-6-phosphate, the intermediate in glycolysis that just follows the point where an ATP is reacted with glucose to get things going. Thus, the extra ATP is not needed, and glycolysis yields one more ATP than for glucose itself.

The human body maintains only enough stored glycogen to last about a day. Then it must switch to fats as a fuel. Most fats are stored in the body as triglycerides. Because they are not soluble in water, they must be transported in the blood as lipoproteins. After a fatty meal the concentration in the blood may be sufficient to give it a milky opalescence. Lipase enzymes in the blood or in fatty tissues hydrolyze the lipoproteins to glycerol and to fatty acids bound to blood proteins. The glycerol enters glycolysis after a few steps involving ATP. The fatty acids enter the cell and react with CoA similar to the way pyruvate does at the start of the Krebs cycle. The fatty acid-CoA compounds are then transported into the inner compartment of the mitochondria, where a series of reactions split off the last two carbons of the fatty acid, with the CoA, forming acetyl-CoA and a shorter fatty acid-CoA. The process, called **beta oxidation**, is repeated until the fatty acid has been consumed. The acetyl-CoAs produced then enter the Krebs cycle, where they are further oxidized. In addition, for every acetyl-CoA formed, one $NADH_2$ and one $FADH_2$ are formed, feeding the electron transport system production of ATP. This accounts for the high-energy yield of fats in comparison to carbohydrates.

Proteins are first hydrolyzed into their component amino acids, followed by **deamination**, removal of the amino group. Finally, each of the 20 amino acids is converted to either pyruvate, acetyl-CoA, or one of the other intermediates in the Krebs cycle, for further oxidation. Free amino acids are not stored in the body. Excess proteins in the diet thus must be eliminated by the mechanism just described. Deamination releases ammonia to the blood, which can be toxic and must be rapidly removed. This can be accomplished by incorporation into new amino acids or by excretion either directly as ammonia (fish), uric acid (birds and reptiles), or urea (mammals).

5.4.5   Photosynthesis

Virtually all the organic carbon in our environment, from the carbon in a person's fingernails to the carbon in a plastic pen, was formed by plants from CO_2 in the air. The energy for this conversion comes entirely from sunlight. Among the few known exceptions in nature are ecosystems found on the ocean floor and hot springs that obtain their energy from oxidation of reduced inorganic compounds issuing from deep below the ocean floor in hot-water vents. Most human-made sources of energy, such as fossil fuels and weather-driven electric plants (wind and hydroelectric), ultimately come from the sun. Only nuclear, geothermal, and tidal electrical facilities, plus a portion of wind energy driven by Earth's rotation, do not derive their energy from the sun.

Use of the sun's energy to synthesize carbohydrates is called **photosynthesis**. Only certain bacteria, algal protists, and green plants are capable of photosynthesis. Organisms that can synthesize their own carbohydrates from inorganic precursors are called **autotrophs**. Those that use sunlight to provide the energy for this are called **photoautotrophs.** Some bacteria can use inorganic energy sources, such as H_2, H_2S, NH_3, or reduced metallic salts such as manganese or ferrous iron, to form carbohydrates from CO_2. These are called **chemoautotrophs** or **lithoautotrophs**. All other organisms, including all animals and fungi, and most bacteria, depend ultimately on the autotrophs for organic carbon and energy. Organisms such as animals, fungi, and many bacteria that must obtain their organic carbon ultimately from autotrophs are called **heterotrophs**.

Photosynthesis is somewhat more complicated to describe than respiration. However, once we have understood respiration, there are enough similarities to respiration in reverse to describe it in those terms. Recall that in respiration, there is a separation between the oxidation of organic carbon to CO_2 (glycolysis and the Krebs cycle), which produces reducing power as $NADH_2$, and the electron transport system, which consumes the reducing power and reduces oxygen to water.

In photosynthesis, the CO_2 gets reduced, producing glucose. The reducing power comes from photons of light. The overall net reaction of photosynthesis is

$$6\,CO_2 + 6\,H_2O \Rightarrow C_6H_{12}O_6 + 6O_2 \qquad (5.54)$$

The two major parts of photosynthesis are the light reactions, which are analogous to electron transport in respiration, and the dark reactions, which can be compared to the reverse of the Krebs cycle and glycolysis. Several basic experimental facts support the division of photosynthesis into light and dark reactions: (1) Plants give off oxygen only in the light; and (2) if a suspension of algae is illuminated for some time in the absence of CO_2, then placed in the dark with CO_2, the CO_2 is incorporated into carbohydrate for a brief time. Additional work with tracer elements further supports the idea.

In the **light reactions**, energy from light is used to oxidize H_2O to O_2 and produce ATP and/or $NADPH_2$. As in respiration, electron transport is involved and even uses cytochromes. In fact, one type of organism, the purple nonsulfur bacteria, use the same electron transport chain for both respiration and photosynthesis. In algae and plants there are actually two electron transport systems just for photosynthesis, which act in concert. The light reactions require a membrane structure, the chloroplast (except in cyanobacter), which is similar to the mitochondrion. Furthermore, ATPs are formed as a result of the generation of a chemiosmotic potential by the electron transport system. In the **dark reactions**, the ATP and $NADPH_2$ are used in a cyclic reaction to form sugars from CO_2. Some of the reactions involved are the reverse of parts of glycolysis.

Chloroplasts are the cellular organelles in plants and algae where all of the light reactions and some of the dark reactions are located. Bacteria perform the function of chloroplasts using their cytoplasmic membrane, as they do for the respiratory electron transport system. Like mitochondria, they contain an internal membrane structure that divides the interior into stacks of flattened hollow disks, **thylakoids** (Figure 5.8). The inner compartment of the thylakoid is called the **lumen**, the outer compartment is called the **stroma**.

The membranes of the thylakoids are studded with three groups of proteins and other compounds (Figure 5.9). Two of these groups are **photosystem I** and **photosystem II**, which perform the principal task of capturing the light energy and transporting the electrons. Each photosystem consists of pigments, proteins, and electron transport compounds, such as cytochromes. The third group is actually a single complex, called the **CF1 particle**. Like the *ATP synthase* particle in the mitochondrion, The CF1 particle uses the energy stored in the chemiosmotic potential created by the electron transport systems (in this case of the photosystems) to generate ATP.

The photosystems consist of an association of membrane-bound particles, one of which is called the **antenna**. The antenna is a complex of numerous chlorophyll molecules, most bound to proteins. **Chlorophyll** is the green pigment of plants, which absorbs much of the light energy for photosynthesis. The antenna then transfers the energy to a "special pair"

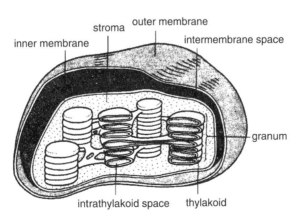

Figure 5.8   Chloroplast structure. (From Fried, 1990. © The McGraw-Hill Companies, Inc. Used with permission.)

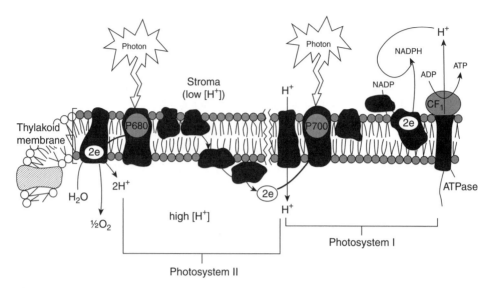

Figure 5.9   Photosynthetic apparatus. (From Fried, 1990. © The McGraw-Hill Companies, Inc. Used with permission.)

of chlorophyll molecules, called the **reaction center**, which uses the energy to promote an electron to a higher energy level and transfer it to the electron transport system. The reactions centers of photosystem I and photosystem II are known as P700 and P680, respectively, for the optimum wavelengths at which they operate.

The chlorophyll molecule consists of a hydrocarbon tail connected to a ring system with a nonionic magnesium atom at its center (Figure 5.10). The ring system has the alternating single and double bonds that often characterize compounds that absorb visible light. There are two main types of chlorophyll: a and b, but chlorophyll a is most common in plants. Chlorophyll a absorbs most strongly in the blue (400 to 450 nm) and red (640 to 680 nm) regions of the spectrum, leaving green to please our eyes (Figure 5.11). However, photosystem antennas contain other pigments, such as carotenoids, which capture green and yellow light energy and transfer it to the chlorophyll. This is particularly important for plants that live in deep waters, where the longer wavelengths do not penetrate. Diatoms, the brown algae, and dinoflagellates (including the *red tide* organism) contain pigments that absorb strongly in the range 400 to 550 nm. The carotenoids become visible when the

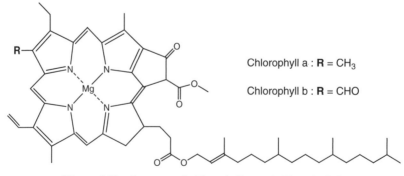

Chlorophyll a : **R** = CH_3

Chlorophyll b : **R** = CHO

Figure 5.10   Structure of chlorophyll a and chlorophyll b.

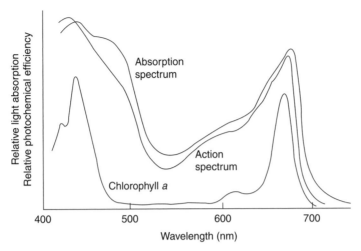

Figure 5.11   Absorption spectrum of chlorophyll and of a green leaf, and the action spectrum for the rate of photosynthesis vs. wavelength. (From Fried, 1990. © The McGraw-Hill Companies, Inc. Used with permission.)

chlorophyll degrades in deciduous tree leaves in certain areas in the fall, creating the spectacular fall color displays.

Before examining photosynthesis in plants, it may be useful to consider bacterial photosynthesis first, because it is simpler and because its mechanism is retained in algae and plants. Bacteria have only one photosystem. Absorption of a photon by the photosystem starts the process of **cyclic photophosphorylation**, which results in the production of a single ATP. The photon excites an electron in the chlorophyll, which transfers to the electron transport chain. As the electron is passed down the chain, two protons are secreted outside the cell using energy from the electron. At the end of the chain, the low-energy electron returns to the chlorophyll, ready to start the cycle again. The proton secretion creates a chemiosmotic potential across the membrane, which produces ATP as it passes back into the cell through an enzyme complex. To form carbohydrates, the bacteria need ATP and $NADPH_2$. Bacteria require an external chemical reducing agent to form $NADPH_2$, such as H_2, H_2S, or organic matter. The ATP and $NADPH_2$ then feed the dark reactions to produce carbohydrates as described below. No oxygen is produced as it is by plants. Indeed, oxygen is toxic to these bacteria. An exception is the prokaryotic nitrogen-fixing *cyanobacter*, which does produce oxygen.

Light Reactions   Plants and algae retain the cyclic photophosphorylation pathway but have developed **noncyclic photophosphorylation**, which has the advantage of not requiring an external reducing agent to replace the electron given up by the chlorophyll (Figure 5.12). Instead, in photosystem II, water is hydrolyzed to hydrogen and oxygen; the hydrogens provide the needed electrons, and the remaining protons enter the thylakoid lumen to augment the chemiosmotic gradient (ultimately for ATP production). The oxygen formed by hydrolysis of the water is released. After passing through the electron transport chain of photosystem II, using their energy to pump more protons, the low-energy electrons can then replace electrons promoted by another photon absorbed by photosystem I. The reenergized electrons, now in photosystem I, pass through another electron transport chain and ultimately are used to reduce NADP to $NADPH_2$. (Note

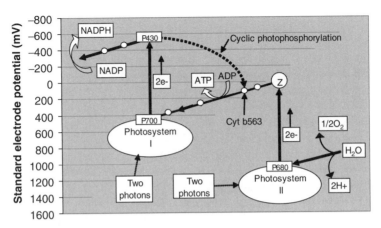

Figure 5.12   Noncyclic photophosphorylation.

the use of NADP instead of NAD as in respiration. NADP is used preferentially when the reducing power is to be used for synthesizing compounds, not just for transporting electrons.) The sequence of events takes about 5 ms. The overall reaction for noncyclic photophosphorylation is

$$12\,H_2O + 48\,h\nu + 12\,NADP + 12\,ADP + 12\,P_i \Rightarrow 6\,O_2 + 12\,NADPH_2 + 12\,ATP$$
$$(5.55)$$

Equation (5.55), as written, involves two electrons, each of which requires two photons. The protons produced by photolysis of water, plus additional protons pumped by the electron transport system (not shown), wind up in the thylakoid lumen. The chemiosmotic gradient across the thylakoid membrane can be as much as 3 pH units, with an electric potential of up to 100 mV. About one ATP is produced per $NADPH_2$. Overall, then, noncyclic photophosphorylation requires four photons to produce one ATP and one $NADPH_2$.

Noncyclic photophosphorylation produces $NADPH_2$ and ATP in about equal amounts; but glucose production requires additional ATP, and the cell has other uses for ATP as well. To get more ATP, plants are able to exploit cyclic photophosphorylation, similar to bacteria, which does not produce oxygen or $NADPH_2$. It does this by a sort of "short circuit" that shunts electrons excited by photosystem I back to the electron transport chain of photosystem II (see the dotted line in Figure 5.12). The process is controlled by $NADPH_2$ levels; high levels hinder the flow of electrons out of photosystem I, forcing them into the other pathway. Thus, the cell can control somewhat independently the relative amounts of $NADPH_2$ and ATP that it produces.

The energy of a mole of photons, E, is related to the frequency, ν (or wavelength, λ) by

$$E = h\nu = \frac{hc}{\lambda} \qquad (5.56)$$

where h is Planck's constant and c is the speed of light. The two peak wavelengths shown in Figure 5.11 are 430 and 650 nm. The corresponding energy levels are 67 and 43.5 kcal/mol, respectively. Four moles of photons of light energy must be absorbed by each of the two photosystems (eight photons in all) to produce one molecule of O_2, two of $NADPH_2$, and two of ATP. To produce a six-carbon sugar by the dark reactions, the cell will need 12 $NADPH_2$ and 18 ATP. This can be satisfied most efficiently by 48 photons in noncyclic photophosphorylation (making $12NADPH_2$ and 12ATP) and 12 photons in cyclic photophosphorylation (six more ATP), for a total of 60 photons. Assuming that a conservative 40 kcal/mol of photons gives a total of 2400 kcal/mol glucose formed since the standard Gibbs free energy of formation of glucose is 686 kcal/mol, the potential efficiency is 29% (based on photons *absorbed*). The actual efficiency is less, due largely to a wasteful side reaction called photorespiration, described below.

Dark Reactions The conversion of CO_2 to glucose is called **CO_2 fixation**. It is accomplished by two pathways. The first is a series of 12 reactions called the **Calvin cycle** (Figure 5.13). The cycle starts in the stroma of the chloroplast when CO_2 combines with a pentose (five-carbon sugar) to form two three-carbon acids:

$$\text{ribulose-1,5 biphosphate} + CO_2 \Rightarrow \text{2,3-phosphoglycerate} \qquad (5.57)$$

Note that this step may be considered to be the actual point of carbon fixation. Yet it does not require energy from light or even ATP. But ATP and the reducing agent $NADPH_2$ are needed to ensure a supply of the ribulose to keep this reaction going. Several

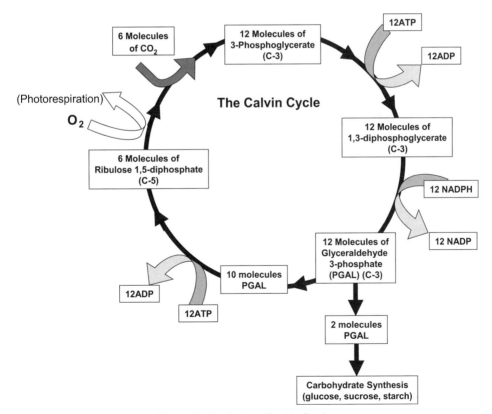

Figure 5.13 Carbon dioxide fixation.

subsequent reactions use two $NADPH_2$ and two ATP to produce *phosphoglyceraldehyde* (PGAL, or glyceraldehyde-3-phosphate). Most of the PGAL goes on to be converted back into the ribulose, to keep the cycle going. This requires another ATP.

Two of every 12 PGALs formed are converted to another triose, which leaves the chloroplast for the cytoplasm. There they can be converted to glucose or other carbohydrates in the second pathway. The enzymes for forming glucose are the same as the ones in glycolysis, acting in reverse. In fact, PGAL is one of the key intermediates in glycolysis. PGAL can also be converted into glycerol and fatty acids for fats, or into amino acids for proteins. Overall, since each turn of the cycle incorporates a single CO_2, it takes six cycles to produce one molecule of glucose. The overall stoichiometry for the Calvin cycle is

$$6\,CO_2 + 12\,NADPH_2 + 18\,ATP \Rightarrow C_6H_{12}O_6 + 12\,NADP + 18\,ADP + 18\,P_i + 6\,H_2O$$
(5.58)

Strangely, most plants have a wasteful side reaction, called **photorespiration**, which uses the same enzyme as reaction (5.57). The reaction occurs during hot, dry conditions when leaf pores close to conserve water. CO_2 is depleted inside the leaf and O_2 builds up. The oxygen competes with the CO_2 for the enzyme, producing a side reaction with the ribulose-1,5 biphosphate. Waste products are formed instead of carbohydrates, and ATP is consumed. There seems to be no benefit to the plant. This is a major drain on biological productivity for the world's plants and can reduce the net efficiency of photosynthesis below 1%.

A minority of plants (about 100 species are known) have developed a mechanism to tilt the balance back toward normal CO_2 fixation. This has occurred in plants that grow in hot, arid regions. Closing the stomates to limit water loss also limits CO_2 entry to the leaf. In these plants, instead of forming the three-carbon phosphoglycerate in reaction (5.57), cells near the outside of the leaf use a different reaction that forms the four-carbon *oxaloacetate*, which is then reduced by $NADPH_2$ to the four-carbon *malate*. The malate diffuses to other cells in the leaf interior, where it reacts to form pyruvate, CO_2, and $NADPH_2$. This has the net effect of transporting CO_2 and $NADPH_2$ to the inner cells at a higher concentration, where they enter the Calvin cycle. Twelve additional ATPs are used per glucose, but the net efficiency increases because photorespiration is relatively low. Plants that do this include the important food crops sorghum, corn, and sugarcane. Because they form a four-carbon product with CO_2, they are called **C₄ plants**. The more common plants without this ability are called **C₃ plants**.

Actual maximum photosynthetic efficiencies of the C_4 crops sugarcane (*Saccharum officinale*), sorghum (*Sorghum vulgare*), and corn (*Zea mays*) have been measured to range from 2.5 to 3.2%; for the C_3 crops alfalfa, sugar beet, and the alga *Chlorella*, the range is 1.4 to 1.9%. C_3 plants can fix 15 to 40 mg of CO_2 per dm^2 of leaf surface per hour, while C_4 plants can fix 40 to 80 mg/dm^2 per hour. C_4 plants lose much less water by transpiration than do C_3 plants. The optimum daytime temperature for growth of C_4 plants is higher: 30° to 35°C vs. 20° to 25°C for C_3 plants. Examples of C_3 crop grasses are wheat (*Triticum aestivum*), rye (*Secale cereale*), oats (*Avena sativa*), and rice (*Oryza sativa*).

A third strategy for photosynthesis is carried out by succulents such as the jade plant and some cacti. These are called **CAM** (for "crassulacean acid metabolism") **plants**. CAM plants fix carbon at night with the stomata (pores in the leaves) open. During the heat of the day the stomata close, and the Calvin cycle obtains its carbon from the nighttime storage. Note that C_4 and CAM plants have all the photosynthetic machinery of C_3 plants: namely, the photosystems and the Calvin cycle. However, they have additional mechanisms that enable them to utilize CO_2 more efficiently.

The difference between C_3 and C_4 plants has caused concern about the effect of the global CO_2 increase that is under way. Fossil fuel consumption has caused CO_2 in the atmosphere to increase from 280 ppmv in preindustrial times to 356 ppmv in 1993. C_3 plants are more sensitive to low CO_2 levels. Experimental results have verified that they can benefit more from an increase than C_4 plants. This has the potential to cause ecological changes as the competitive balance between plant species changes. It is also feared that some of the C_4 food plants could face increased competition from C_3 weeds. On the other hand, the lawn grasses Kentucky Bluegrass (*Poa pratensis*) and creeping bent (*Agrostis tenuis*) are C_3 plants, whereas crabgrass (*Digitaria sanguinalis*) is a C_4 plant, ordinarily giving it an advantage during hot, dry summers.

5.4.6 Biosynthesis

Many of the smaller molecules of the cell, whether those used directly or those used as building blocks of macromolecules, are synthesized from precursors found among the intermediates of respiration. These anabolic pathways connect respiratory intermediates with other sugars, amino acids, fatty acids and fats, and nucleic acids. Virtually all the pathways in a cell are interconnected in a network of reactions.

Nitrogen is assimilated by cells by forming glutamate and glutamine from ammonia and ketoglutaric acid. In nitrogen-poor environments, energy must be expended. Rooted plants and algae can use ammonia, but primarily take nitrogen in as nitrate. Some microorganisms can take in nitrate or fix atmospheric nitrogen, but first convert these to ammonia. Other amino acids are then formed using the glutamate, possibly with other precursors, many of which are intermediates in respiration, such as pyruvate or oxaloacetate. Formation of proteins from amino acids is discussed in Chapter 6.

Nucleotide synthesis begins with ribose for purines and glutamine for pyrimidines. The nitrogens are obtained from the amino acids glycine, glutamine, and aspartate. Fatty acids are formed using acetyl-CoA, similar to a reverse of the fatty acid oxidation described above. The process occurs in the cytoplasm and requires $NADPH_2$, CO_2, and Mn^{2+}. Mammals can synthesize the saturated and monounsaturated fatty acids.

The biopolymers (starch, glycogen, protein, and nucleic acids) are produced by successive addition to the ends of the molecules. Often, ATP is converted to another nucleoside triphosphate for the reaction, such as uridine triphosphate, cytosine triphosphate, guanidine triphosphate, or thymine triphosphate. The reaction usually splits the second phosphate bond, producing a monophosphate instead of a diphosphate. For example, if a monomer M is joined to a polymer of n units,

$$ATP + M + [M]_n \Rightarrow AMP + [M]_{n+1} + PP_i$$

Then the resulting pyrophosphate (PP_i) is hydrolyzed to orthophosphate. By eliminating the pyrophosphate reaction product, the first reaction is pushed toward completion by the mass action law.

Sulfur in the form of sulfide (S^{2-}) is needed for the amino acid cysteine, which forms the disulfide bonds that are so important to protein folding. However, our aerobic environment provides sulfur mostly in the form of sulfate, SO_4^{2-}. The sulfide can be formed anaerobically by microorganisms that use the sulfate as a terminal electron acceptor in the electron transport chain. Sulfate can also be reduced by a series of steps starting with ATP, followed by use of one $NADPH_2$ to form sulfite (SO_3^{2-}) and then three more $NADPH_2$ to reduce the sulfite to sulfide.

PROBLEMS

5.1. Compute ΔG° and K_{eq} for the complete oxidation of ethanol.

5.2. Compute ΔG° and K_{eq} for methanogenesis $C_6H_{12}O_6 \Leftrightarrow 3CO_2 + 3CH_4$.

5.3. Compute the oxidation states of glucose, pyruvate, and lactic acid. Is the conversion of glucose into two pyruvates an oxidation? How about the conversion of glucose into lactate?

5.4. For pyruvic acid, glucose, ethanol, and glycine: Using the molecular formula, compute the oxidation state, the theoretical oxygen demand, and the total organic carbon content (TOC). Use equation (5.9) to compute the mean oxidation state of carbon. Compare with the value computed directly from the formula. Discuss the results.

5.5. Use the Arrhenius equation to compare the temperature increase that doubles the reaction rate coefficient starting from 25°C for the hydrolysis of glucose example given in the text, for cases with and without enzyme catalysis.

5.6. In Example 5.4, at what concentration will the reaction rate be double the first value measured?

5.7. (**a**) Write a computer program or develop a spreadsheet to solve Michaelis–Menten equations (5.30) to (5.32) numerically using Euler's rule with $\Delta t = 0.001$ min. Start by writing a differential equation based on elementary kinetics for $d[P]/dt$ based on equation (5.29). Use values for all rate coefficients equal to 1.0 (units are min^{-1} or $min \cdot L/mg$), initial values are $[S]_0 = 100$, and $[E]_0 = 50.0$, and $[ES] = [P]_0 = 10$. Plot the values of $[S]$, $[ES]$, and $[P]$ from 0.0 to 0.5 min. How accurate is the quasisteady-state assumption?
(**b**) Repeat part (a) using $[E]_0 = 100$. How is the rate of conversion of S to P affected?

5.8. Using the data of Figure 5.4, equation (5.47) gives reaction rates of 145.2, 185.8, and 234.2 mm^3/min for temperatures 10, 20, and 30°C, respectively. Use the data for 20 and 30°C in equation (5.48) and solve for θ. Then use (5.48) with your value of θ to estimate the reaction rate at 10°C. How does this compare to the value from equation (5.47)?

5.9. What is the difference between Le Châtelier's principle and the law of mass action?

5.10. If a first-order reaction has a half-life of 1 h, what is the rate coefficient for that reaction? How long will it take for the reactant to fall to 1% of its initial concentration?

5.11. If K_m in the Michaelis–Menten equation is equal to 0.02 M, at what concentration will the reaction proceed at 90% of the maximum rate? 99%?

5.12. A sample of water containing photosynthetic algae (and negligible amounts of heterotrophs) is placed in a sealed bottle in sunlight. After a time it is observed that the dissolved oxygen level in the water has increased and dissolved carbon dioxide decreases. How will the rate of change of these two substances change immediately after the bottle is placed in the dark? (Don't forget about the dark reactions of photosynthesis.) What about several hours after being in the dark?

REFERENCES

Bailey, J. E., and D. F. Ollis, 1986. *Biochemical Engineering Fundamentals*, McGraw-Hill, New York.

Fried, G. H., 1990. *Schaum's Outline: Theory and Problems of Biology*, McGraw-Hill, New York.

Grady, C. P., Jr., G. T. Daigger, and H. C. Lim, 1999. *Biological Wastewater Treatment*, Marcel Dekker, New York.

Henze, et al., 1986. *Activated Sludge Model No. 1*, International 1 Association on Water, London.

Lange's Handbook of Chemistry, 1987. McGraw-Hill, New York.

Lehninger Principles of Biochemistry, 3rd ed., W.H. Freeman, New York.

Smith, E. L., R. L. Hill, I. R. Lehman, R. J. Lefkowitz, P. Handler, and A. White, 1983. *Principles of Biochemistry: General Aspects*, McGraw-Hill, New York.

Stryer, L., 1995. *Biochemistry*, 4th ed., W.H. Freeman, New York.

Vogel, F., J. Harf, A. Hug, and P. R. von Rohr, 2000. The mean oxidation number of carbon (MOC): a useful concept for describing oxidation processes, *Water Research*, Vol. 34, No. 10, pp. 2689–2702.

Watson, J. D., 1968. *The Double Helix*, Atheneum, New York.

Weast, R. C. (Ed.), 1979. *CRC Handbook of Chemistry and Physics*, 60th ed., CRC Press, Boca Raton, FL.

6

GENETICS

Genetics is one of the most fascinating areas of biology. It has effects at all scales from the molecule to populations. Its study involves a wide variety of tools, from biochemical tests to microscopy to breeding experiments. It relies on sophisticated mathematical analysis, especially from probability and statistics. It is one of nature's best examples of amplification, in which a small change in a single molecule can spell the difference between disaster and delight at the birth of a child.

Genetics is the science of heredity. **Heredity** is the transmission of traits from one generation to another. By **traits**, we mean a distinguishing observable characteristic of an organism, such as color or shape of one of its parts, or its ability to metabolize a particular substance. Each trait is controlled by one or more genes.

6.1 HEREDITY

Gregor Mendel was one of the first biologists to use mathematical analysis to prove hypotheses. He used elementary probability theory in his work. He was also inspired by the atomic theory of matter (an appropriate example of cross-fertilization in science). He wondered whether hereditary traits were also passed as particles to future generations. Previously, it was thought that traits from parents blended to form a new mix in offspring. Mendel proved his hypothesis and published his results starting in 1865. However, his contribution was not recognized until the principles were rediscovered independently by others some 35 years later.

Mendel deduced several of the basic principles of heredity. Here we summarize all of the principles as known today. Then several of Mendel's and others' results will be described as examples. The principles of heredity can be summarized as eleven rules:

Environmental Biology for Engineers and Scientists, by David A. Vaccari, Peter F. Strom, and James E. Alleman
Copyright © 2006 John Wiley & Sons, Inc.

1. A **gene** is a sequence of nucleotides that codes for a specific polypeptide. [Recall that a protein can consist of several polypeptides, and possibly other factors such as metal atoms (e.g., hemoglobin, which has four polypeptides and iron atoms).] Not all genes in a particular cell actually produce their polypeptide. Cells that do produce the polypeptide associated with a gene are said to **express** the gene.

2. The genes are contained in the chromosomes.

3. A gene may come in a variety of mutated forms, called **alleles**. For example, the gene for eye color may have an allele for blue eyes and another allele for brown eyes. Although all the organisms of a single species will have the same genes, each individual can have a unique combination of alleles.

4. The particular combination of alleles that an individual has is called its **genotype**. The particular set of genetically controlled traits possessed by an individual is called its **phenotype**. That is, the phenotype is the result of expression of the genotype.

5. Most higher organisms, in particular most plants and animals, are diploid, having chromosomes in pairs. Thus, they have two copies of each gene.

6. The two copies can have the same or different alleles. If both chromosomes have the same allele, the organism is said to be **homozygous** for that gene. If the alleles are different, the organism is **heterozygous**. For example, a human can have a blue-eyed allele on one chromosome and a brown-eyed allele on the other. That individual would be heterozygous for eye color. Alternatively, both alleles could be for blue eyes, and the person would be homozygous for eye color.

7. When the two alleles for a gene are different, several situations can occur:

   a. **Complete dominance**: one allele is expressed; the other is not. An allele that is expressed at the expense of another is said to be **dominant**. The allele that is not expressed in a heterozygous genotype is said to be **recessive**. For example, brown eye color is dominant and blue eye color is recessive. Blue eye color is expressed only if both alleles are for blue eyes. When a gene is homozygous for a recessive allele, it is said to be **double recessive**, and the recessive allele can be expressed.

   b. **Incomplete, or partial, dominance**: In this case, a mixed phenotype results in a heterozygous individual. For example, if snapdragons with red flowers (homo-zygous for red) are crossed with those with white flowers (homozygous for white), all the offspring will be heterozygous and will have pink flowers. However, if these are then crossbred with each other, about one-fourth of the next generation will be red, one-fourth white, and one-half pink.

   c. **Codominance** is when both traits are *fully* expressed in the heterozygous organism. This is usually seen only in biochemical traits. An example is human blood type. If a heterozygous individual has allele A and allele B, their blood cells will have both type A and type B antigens.

8. Meiosis separates the two alleles of a diploid organism. Thus, gametes have only one allele. This is called **segregation**.

9. Random assortment in meiosis distributes the chromosomes to the gametes independently. That is, the chromosomes of an individual's gametes can have any combination of its maternal and paternal chromosomes. In genetics, this is

called **independent assortment**. For example, the fruit fly *Drosophila* has four pairs of chromosomes, one of each pair coming from each parent. However, its egg cells, which are haploid, may contain chromosome 2 from the father and chromosomes 1, 3, and 4 from the mother, or any other combination.

10. All of the genes on a chromosome tend to segregate together. Therefore, genes on the same chromosome are not assorted independently. Their traits tend to be passed on to the next generation together. The tendency for genes on the same chromosome to be inherited together is called **gene linkage**.

11. **Recombination** caused by **crossing-over** in meiosis acts to weaken linkage between genes on the same chromosome. Because of crossing-over, a maternal chromosome in a gamete may include pieces of the paternal chromosome. The closer together two genes are located, the less likely that crossing-over will separate them and the stronger will be their linkage. Two genes that are at extreme ends of the same chromosome will exhibit weak linkage and in fact may show independent assortment, behaving as if they were on different chromosomes.

6.1.1 Mendel's Experiments

Mendel studied the inheritance of seven traits in the pea plant. For example, one trait was seed shape. Mendel had a variety that produced either round or wrinkled seed. He started with two sets of true-breeding plants, one that produced only round seeds and another that produced only wrinkled seeds. When he crossbred them with each other, all of the offspring had round seeds (see Figure 6.1). It would seem that the ability to produce wrinkled seeds had been lost. However, if a number of plants produced in this way were allowed to self-fertilize (self-pollination is the normal mode of fertilization in the pea plant), one-fourth of the offspring had wrinkled seeds!

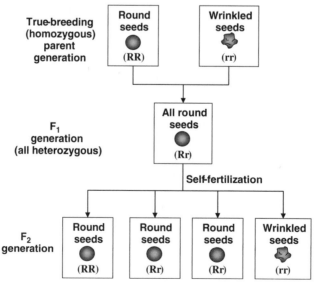

Figure 6.1 Cross-breeding between two pea plants homozygous for seed shape. (Based on Postlethwait and Hopson, 1995.)

The explanation for this result is as follows. *True-breeding plants* are those that have been self-bred for many generations. As Problem 6.1 shows, this results in plants that are homozygous in all their genes. If we represent the gene for seed shape by **R** for round seed and **r** for wrinkled seed, the true-breeding round-seed plant would have its two alleles represented by **RR**, and the homozygous wrinkled-seed plant would have genotype **rr**. The round-seed plant could produce only one type of gamete: **R**. Similarly, the wrinkled-seed plant could only produce **r** gametes. Thus, the offspring produced by mating these two plants could only be heterozygous: **Rr**. The fact that all of the offspring of this generation had round seeds demonstrates that **R** is the dominant allele and **r** is the recessive.

The offspring, being all heterozygous, produce gametes half of which are **R** and half **r**. Randomly mating these would produce genotypes of about 25% **RR**, 50% **Rr**, and 25% **rr**. Since **R** is dominant, both **RR** and **Rr**, or about 75% of the offspring, would have round seeds. The remaining 25% are homozygous recessive and would express the gene for wrinkled seeds. The reappearance of the wrinkled phenotype and the proportions involved proved Mendel's hypothesis of the particulate transmission of hereditary traits, the principles of paired alleles, the existence of dominant and recessive alleles, and the existence of segregation of alleles in gametes.

Mendel demonstrated the principle of independent assortment by examining the phenotypes for two genes at a time in similar experiments. Another phenotype of Mendel's pea plants had to do with seed color. Yellow seeds were dominant over green seeds in his variety. True-breeding plants with yellow round seeds would be homozygous for both genes (**YYRR**). True breeding plants with green wrinkled seeds would also be homozygous, but in this case, they would be double recessive for both traits (**yyrr**). A cross between these two plants could only be **YyRr**, heterozygous for both traits, and therefore expressing the dominant yellow, smooth seed phenotype.

However, the gametes of these offspring could be of four types: **YR**, **Yr**, **yR**, or **yr**. When allowed to self-fertilize, the possible crosses can be found by constructing a *Punnett square* (Figure 6.2). The possible gametes are listed along the top and left. Each type has equal probability of occurring. Thus, each cross represented by a box within the square occurs in the same proportion of the population. Just as expected from the single- trait experiment, one-fourth of the population will have wrinkled seeds

Gamete Genotype	YR	Yr	yR	yr
YR	*YYRR* Yellow, Smooth	*YYRr* Yellow Smooth	*YyRR* Yellow, Smooth	*YyRr* Yellow, Smooth
Yr	*YYRr* Yellow, Smooth	*YYrr* Yellow, Wrinkled	*YyRr* Yellow, Smooth	*Yyrr* Yellow, Wrinkled
yR	*YyRR* Yellow, Smooth	*YyRr* Yellow, Smooth	*yyRR* Green, Smooth	*yyRr* Green, Smooth
yr	*YyRr* Yellow, Smooth	*Yyrr* Yellow, Wrinkled	*yyRr* Green, Smooth	*yyrr* Green, Wrinkled

Figure 6.2   Dihybrid cross. The gametes are formed by plants that are heterozygous for both traits.

and one-fourth will be green. Assuming that two events are independent, elementary probability theory gives the chance of them occurring simultaneously as the product of the two probabilities: $\frac{1}{4} \times \frac{1}{4} = \frac{1}{16}$. This is the proportion shown in the Punnett square to be both green and wrinkled.

If the seed color and shape genes were linked by being on the same chromosome, the gametes listed across the top of Figure 6.2 would not occur with equal probability. For example, suppose that the two genes were so close together that meiotic recombination practically never separates them. Then one chromosome comes from the **YYRR** plant and the other from the **yyrr**. Only **YR** and **yr** gametes would be produced, and one-fourth of the next generation would have green, wrinkled seeds. Of course, the strength of the linkage can vary, producing a range of distributions.

By looking at the proportions of different phenotypes in breeding experiments, geneticists can determine if genes are linked, establishing that they are on the same chromosome. By examining the strength of the linkage, a relative distance can be determined between genes. This led to the development of **genetic mapping**, the determination of the relative position of genes on a chromosome. Genetic mapping was invented by an undergraduate student, Alfred H. Sturtevant, while he was working with Thomas H. Morgan, the scientist who first explained linkage.

A gene may have more than two types of alleles associated with it. For example, the basic human blood types A, B, AB, and O are caused by the combination of three alleles: A, B, and O. Allele A results in a sugar called *type A* to occur on the surface of a person's red blood cells. Allele B results in a different sugar called *type B*. Allele O does not result in the formation of any sugar. Alleles A and B are codominant; that is, both are expressed. A person who is heterozygous AB produces cells with both sugars and has blood type AB. A person with genotype either AA or AO will have only type A sugar and will have blood type A. Genotypes BB or BO result in type B blood. Allele O is recessive, so to have type O blood, a person must have homozygous genotype OO. In this case, a single gene with three alleles results in six possible genotypes but only four phenotypes.

Some factors, such as height or stature, are controlled by two or more pairs of genes, giving rise to a range of phenotypes. Traits resulting from two or more genes are called **polygenic**. In addition, some alleles can be associated with more than one phenotypic effect. Other genes can have their expression modified by environmental effects.

6.1.2   Sex Chromosomes

Recall that the somatic cells in animals contain chromosomes in pairs that are similar. After replication, each chromosome looks like a pair of sausages tied together. The pairs may have varying lengths and they may be bound at different points. If bound near the middle, they resemble an X. If bound near one end, they look like a Y. However, in most animals one pair can be dissimilar. The chromosomes of one of the 23 pairs in humans are called **sex chromosomes**. It contains the genes that determine a person's gender, among other things. In females, both chromosomes of this pair are X's; in males one of the pair is much smaller and has the Y form. The Y chromosome contains a gene for a factor that stimulates the formation of male sexual structures, and thence the male sex hormones. The absence of this gene results in the formation of a female. Thus, each cell in a female contains an XX pair of sex chromosomes, and each cell in a male contains an XY pair. Chromosomes other than the sex chromosomes are called **autosomes**.

Because the chromosome pairs are separated by meiosis in the formation of gametes, the gametes from females (egg cells) can have only X chromosomes. Half of the gametes from males (sperm cells) have X and half have Y chromosomes. Thus, the sperm determines whether the individual resulting from fertilization will be male or female.

6.1.3   Genetic Disease

Many diseases are caused by genetic defects. The corresponding alleles may be dominant or recessive and may be found on autosomes or on sex chromosomes. Some genetic diseases are associated with extra chromosomes or missing pieces. An example of a dominant autosomal genetic disease is Huntington's disease. It results in neurological degradation leading to death. Because the symptoms do not appear until a person is about 40 years old, the victim may already have produced a family, potentially passing the trait on. The children of the victim are left knowing that they have a 50% chance of having the disease themselves. Since the 1980s, it has been possible to test these offspring for the gene, which, however, results in several types of ethical dilemmas (see Problem 6.2).

Many genetic diseases are caused by recessive genes on autosomal chromosomes. Albinos have a recessive defect on chromosome 11, which prevents formation of pigments in skin, hair, and eyes. *Cystic fibrosis* is caused by a chromosome 7 gene, which results in a defect in a membrane protein. Affected individuals produce excessive mucus, which harms respiratory and digestive function. One in every 11,000 persons born in this country inherits a defect in chromosome 12, resulting in *phenylketonuria*. Those affected lack an enzyme to metabolize the amino acid phenylalanine. The condition exhibits partial dominance. Heterozygous individuals have about 30% of the enzyme activity as normal people. Unchecked, it can cause retardation. However, routine screening at birth makes it possible to avoid disease symptoms by careful dietary control. In addition, affected people must avoid foods containing the artificial sweetener Aspartame, which is an amino acid dimer of phenylalanine and aspartic acid.

Several conditions are caused by recessive alleles on the X chromosomes. Since males have only one copy of X, they can inherit these conditions only from their mother. Examples of these conditions are color blindness for green, inability to form the enzyme glucose-6-phosphate dehydrogenase, Duchenne muscular dystrophy, and two forms of hemophilia. Since many X-linked genetic diseases are fatal when homozygous, only heterozygous females survive gestation. They themselves do not exhibit the disease and are called **carriers**. They pass the gene for the disease to (on average) half their daughters, who also become carriers, and to half their sons, who exhibit symptoms of the disease.

Another type of genetic disease is that related to chromosomal abnormalities such as extra or missing chromosomes, chromosome breakage, or **translocation** (the exchange of genetic material between chromosomes). Errors during meiosis can cause extra or missing chromosomes. A person with a single X chromosome and no Y becomes a sterile female. Some people have one or more extra sex chromosomes. One in 1000 males is XXY. These individuals are sterile, have tall stature and diminished verbal skills, yet normal intelligence. One in 1200 females is XXX. Many are normal, but some exhibit sterility and mental retardation. XYY males tend to be above average in height and below normal in intelligence. Prison inmates have been found to have a higher incidence of this disorder, although XYY males in the general population do not show any unusual tendencies.

Missing autosomes are usually fatal before birth, and extra autosomes are fatal in infancy, with the exception of the smallest autosome, number 21. Having three chromosomes is called **trisomy**. Trisomy 21 is the cause of 95% of *Down syndrome*. Affected persons show numerous characteristics, including an eyefold similar to those of oriental people, short stature, and small heads. They are retarded and susceptible to various diseases, including respiratory and heart problems, and to Alzheimer's disease in older persons. The rate of Down births increases rapidly with the age of the mother, from 1 in 1000 at age 30, to 1 in 100 at age 40, and 1 in 50 at age 45.

Cancer is a result of a series of mutations (see Section 6.2.3). If some of these are inherited, a person is more susceptible to cancer, since fewer mutations remain to occur. For example, a gene associated with breast cancer has been discovered on chromosome 17 that is responsible for about 10% of all cases. Women with the mutation have an 80 to 90% chance of contracting the disease.

6.2 MOLECULAR BIOLOGY

In recent times the most dramatic advances in biology are coming from the field of molecular biology. Although this title could describe any area of biochemistry, it is usually taken to represent the study of processes involving genetic material that controls the activity and destiny of every individual cell. Genetic processes determine how cells differentiate into brain cells and stomach cells. Genetic control ensures that the brain cells do not secrete hydrochloric acid, as the stomach cells do.

It may seem incredible that the 23 DNA molecules in the haploid human genome can contain enough information to describe a human being completely. A simple comparison can be made to memory in a binary computer. Each of the 3.2 billion positions in the human genome can be occupied by one of four nucleotides. The number of possible combinations is $4^{3.2 \text{ billion}}$, which is $2^{6.4 \text{ billion}}$. Thus, the human genome has the information capacity of a 6.4-gigabyte binary memory. Think about how much memory would be required to specify the design of a human. The brain alone consists of about 10^{12} nerve cells, each with an average of 10^3 connections! A way around this paradoxical situation is to consider that the genes might only specify a set of rules that generate the structure rather than the detailed design of the structure. The new mathematical fields of chaos and complexity theory described in Section 2.5 show that exquisitely detailed structures can result from exceedingly simple rules.

Even more surprising is to discover that the human genome codes for only about 25,000 polypeptides (Table 6.1). Details of the remarkable mechanism by which the

TABLE 6.1 Amounts of Genetic Material in Several Organisms

Organism	Nucleotides	Genes	Chromosomes
Homo sapiens	3.2 billion	20,000–30,000	23
Drosophila melanogaster (fruit fly)	120 million	12,500	4
Saccharomyces cerevisciae (yeast)	15 million	6,000	
E. coli	4 million	4,000	1
Bacteriophage λ	45,000 (est.)	35–40	1
HIV (AIDS virus)		9	

Source: Based on Klug and Cummings (1997).

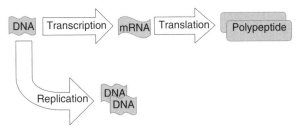

Figure 6.3   Cell processes involving genetic material.

DNA code is transformed into those proteins are summarized below. Several of these processes have environmental significance. Others are of interest simply because they explain the working of the machine of life, or for the utilitarian reason that scientists are learning to manipulate (i.e., "engineer") the processes to our own purposes, and we will have to make societal and individual choices about their use. We are interested primarily in three processes (Figure 6.3), which we discuss in turn.

Replication is the production of two DNA molecules from one. As described in Section 4.6, this occurs during the interphase of mitosis as the cell prepares to divide. The basic mechanism is similar for prokaryotes and eukaryotes. The DNA of eukaryotes exists in the nucleus wrapped around a complex of special proteins called **histones**. The complex must be dissociated for replication to occur. Furthermore, eukaryotes typically have on the order of 10,000 times as much DNA as prokaryotes, and it is present in a number of linear DNA molecules rather than a single ring as in prokaryotes. It would be worthwhile at this point to review the description of nucleic acids in Section 3.6.4.

The replication process starts with an uncoiling of the DNA double helix. Then the double strands are separated at a special origin, forming a **replication bubble** (see Figure 6.4). Replication then proceeds along the DNA molecule, expanding the bubble

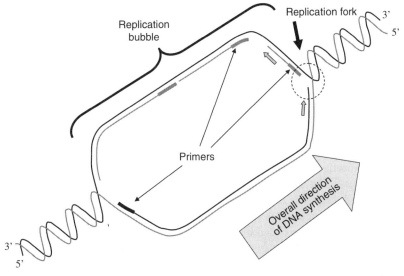

Figure 6.4   DNA replication, showing how the replication fork proceeds, opening up the replication bubble and leaving two complete double strands behind it.

in one direction only. The point where the DNA separates at the end of the replication bubble in the direction of synthesis is called the **replication fork**. Prokaryotes have only one replication fork, whereas eukaryotic DNA has tens of thousands. Thus, eukaryotic DNA replication can occur simultaneously at numerous points on the molecule, increasing the overall speed of the process.

A short piece of RNA known as a **primer** is attached to the DNA at the origin. This provides a free nucleotide end for the DNA chain to start building on. Free nucleotides diffuse into the replication fork and pair up with the exposed nucleotides on the single strands of DNA. Recall, as shown in Figure 3.14, that the two strands of the DNA run in opposite directions, and the two ends can be designated by the terminal bonding carbon number on the base: namely, either 3 or 5. DNA synthesis can proceed only in a direction toward the 5' end of the molecule. Since the two strands run in opposite directions, the "wrong" strand has to add new primers repeatedly as the fork moves along, and synthesis proceeds backward in short segments. The primers are removed periodically and the gaps filled.

The final step is the formation of the ester bonds between the phosphate groups and the sugars, linking the adjacent nucleotides together. This is accomplished by a group of enzymes known as **DNA polymerase**. A DNA polymerase molecule moves along the pair of separated DNA strands just behind the replication fork. The replication fork and the DNA polymerase travel along the DNA strand, leaving two complete double helix strands in their wake. Each cell has several types of DNA polymerase. Because prokaryotes have only a single DNA molecule with a single origin of replication, they have few molecules of the main form of DNA polymerase (15, in the case of *Escherichia coli*). Eukaryotes, on the other hand, have about 50,000, so as to be able to copy the large genome rapidly. As we shall see, DNA polymerase is also an important tool in genetic engineering.

Not infrequently, the wrong nucleotide will have moved into position for polymerization. DNA polymerase can detect these mistakes, back up, remove the offending nucleotide, and then continue forward again. This is called **proofreading**, and it greatly increases the accuracy of the replication. The overall error rate is about 10^{-9}. Thus, a human egg or sperm with its 3.2 billion base pairs will have an average of three errors each. Some of these errors will have minor effects; others will render the zygote nonviable, resulting in a spontaneous abortion. Some errors may result in hereditable defects, and a very few may actually produce hereditable advantages to offspring. Advantageous errors contribute to the natural variation that drives evolution.

6.2.1   Protein Synthesis

Each gene contains the instructions that determine the sequence of amino acids in a single polypeptide. (Recall that a protein consists of one or more polypeptides). As described in Section 3.6.4, each sequence of three nucleotides, which is called a **codon**, codes for one amino acid. The four nucleotides can form codons in $4^3 = 64$ ways. Since there are only 20 amino acids, most can be coded by more than one codon. Thus, it is said that the code is **degenerate**. In addition, there is one codon to indicate START, the beginning of a gene, and several to indicate STOP at the end. The START codon also indicates the amino acid methionine, making it the only codon with two purposes. Table 6.2 is a complete list of codons with their corresponding amino acids.

TABLE 6.2 mRNA Coding Dictionary

First Position	Second Position								Third Position
	U		C		A		G		
U	UUU	phe	UCU	ser	UAU	tyr	UGU	cys	U
	UUC		UCC		UAC		UGC		C
	UUA	leu	UCA		UAA	STOP	UGA	STOP	A
	UUG		UCG		UAG	STOP	UGG	trp	G
C	CUU	leu	CCU	pro	CAU	his	CGU	arg	U
	CUC		CCC		CAC		CGC		C
	CUA		CCA		CAA	gln	CGA		A
	CUG		CCG		CAG		CGG		G
A	AUU	ile	ACU	thr	AAU	asn	AGU	ser	U
	AUC		ACC		AAC		AGC		C
	AUA		ACA		AAA	lys	AGA	arg	A
	AUG	met START	ACG		AAG		AGG		G
G	GUU	val	GCU	ala	GAU	asp	GGU	gly	U
	GUC		GCC		GAC		GGC		C
	GUA		GCA		GAA	glu	GGA		A
	GUG		GCG		GAG		GGG		G

Remarkably, the coding scheme is nearly universal in all organisms, from bacteria to humans. This underscores the unity of life and makes it seem likely that all living things on Earth arose from the same primitive organisms. Of practical significance is the fact that a gene from a human, when inserted into bacteria by genetic engineering techniques described below, can be expressed by the bacteria. In other words, bacteria can be made to produce human proteins.

Now, let's examine the mechanism by which the DNA code results in protein synthesis. The two fundamental steps are called transcription and translation. **Transcription** converts the DNA code into an intermediate molecule called **messenger RNA** (mRNA). The mRNA leaves the nucleus for the cytoplasm, where it combines with ribosomes in the process of **translation**, wherein the mRNA sequence is translated into an amino acid sequence, producing a polypeptide. Translation occurs on the ribosomes, using the mRNA as a template. Another type of RNA, called **transfer RNA** (tRNA) acts as a "shuttle," bringing individual amino acids to the mRNA–ribosome complex, where they are polymerized into the protein. With this outline of the process, let us now examine it in more detail.

The DNA never leaves the nucleus, so transcription occurs in the nucleus. Transcription, or mRNA synthesis, is catalyzed by the enzyme **RNA polymerase**. The RNA polymerase particle is thought to "patrol" the DNA molecule until it comes upon one of several specific sequences of nucleotides called **promoters**. The promoter is a signal to start forming mRNA from the DNA template. It is located a fixed number of nucleotides before the actual beginning of the gene (usually, 10 to 35 bases upstream). Unlike DNA replication, no primer is needed, just the AUG start codon. The RNA polymerase simply begins to link nucleotides together, complementary to the template strand of DNA. Wherever the DNA has an adenosine residue, the mRNA is given a uracil, T produces

A, C produces G, and G produces C. For example, if a section of the DNA template has the bases TACGGTGAT, the resulting mRNA will be AUGCCACUA, with the following resulting amino acid (AA) sequence:

DNA :	TAC	–	GGT	–	GAT
RNA :	AUG	–	CCA	–	CUA
AA :	methionine	–	proline	–	leucine

Note that the other complementary DNA strand is not directly involved. Although it could be used to form mRNA by another RNA polymerase working in the opposite direction, in most cases this is not a functional gene. Furthermore, in eukaryotes only, most genes have sections within their sequence that do not ultimately code for amino acids. These sections are called **introns**. The sections that do code for amino acids are called **exons**. Initially, the eukaryotic mRNA contains both. The introns are excised and the remaining exons spliced together before the mRNA goes on to participate in protein synthesis. Whether introns perform any function is unknown.

In prokaryotes, the mRNA can enter the translation process to produce proteins even before its synthesis is completed. As one end of the RNA is being synthesized, proteins can be produced at several points along the already completed length. In eukaryotes, however, the mRNA must first diffuse out of the nucleus into the cytoplasm. In either case, the mRNA then associates with a ribosome particle to begin the translation process (Figure 6.5).

The transfer RNA (tRNA) is relatively small, consisting of only 75 to 90 nucleotides (Figure 6.6). One form exists for each amino acid. The tRNA are folded and looped to form a "cloverleaf" structure. The amino acid binds to a section near the paired ends of the molecule. One of the loops has three chemically modified nucleotides that are complementary to the mRNA codons corresponding to its amino acid.

The ribosome is a particle consisting of several RNA strands and several dozen proteins. Recall that it is associated with the endoplasmic reticulum. Ribosomes are the cellular machine that brings all the players together for protein synthesis. These include the mRNA, which forms the blueprint, and the tRNA with associated amino acids. First, the mRNA is bound to the ribosome. One of the codons is presented, and a corresponding tRNA binds to it. The next codon of the mRNA is presented and attracts another tRNA. The two amino acids are then joined in a peptide bond, the mRNA moves over by one codon, and the assembly is ready for another tRNA to bring in another amino acid. Repeated cycles elongate the polypeptide chain until a STOP codon is encountered on the mRNA and the new polypeptide is released.

The flow of information in the process just described is from DNA to mRNA to protein. However, certain viruses, called **retroviruses**, can form DNA from an RNA template. They use a special enzyme called **reverse transcriptase**. Human immunodeficiency virus (HIV), which causes the disease AIDS, is a retrovirus.

6.2.2 Gene Regulation

Cells do not continually produce all the proteins coded for by their genes. This would be wasteful or even disruptive. Therefore, there must be some means to turn genes on or off, or to control their rate of expression. For example: Prokaryotes respond to the presence of substrates to produce enzymes to metabolize a particular substrate only when needed.

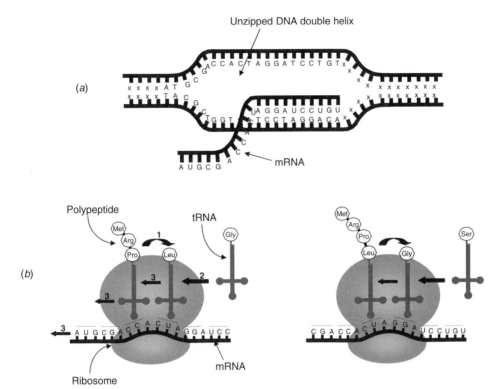

Figure 6.5 (*a*) Transcription and (*b*) translation. In transcription, messenger RNA is formed from the DNA template at the chromosome. Subsequently, the mRNA is transported to the cytoplasm, where it encounters ribosomes and participates in translation. In translation, mRNA is used as a template to produce polypeptides. The numbered arrows indicate the approximate sequence of events.

This has significance in the biodegradation of chemicals. Multicellular eukaryotes can also turn genes permanently on or off, to reflect the specialized roles of particular tissues.

Some enzymes are produced continuously regardless of the environmental conditions. Others are produced only when the substrate is present. The latter is called **enzyme induction**. For example, the cytochrome P450 enzymes responsible for biotransformation of xenobiotic compounds are induced by many of these substances, including components of tobacco smoke and polynuclear aromatic hydrocarbons.

The production of some enzymes is inhibited by the presence of some molecule in the environment. This is called **enzyme repression**. Both of these types of regulation are **transcription-level control**, controlling protein synthesis by affecting the production of mRNA.

Studies of *E. coli* led the French geneticists François Jacob and Jacques Monod to propose the operon model of genetic control in 1961 to explain both induction and repression. In the **operon model**, a biochemical function is associated with a cluster of genes, including **structural genes**. Structural genes are genes that actually code for enzymes. Other genes control the expression of the structural genes. The control genes include a repressor

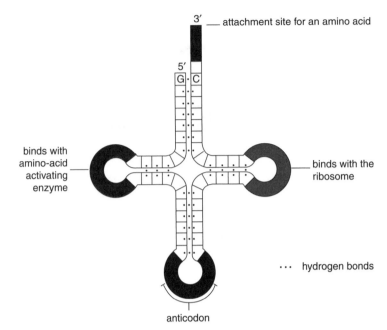

3′ — attachment site for an amino acid

5′
G·C

binds with amino-acid activating enzyme

binds with the ribosome

··· hydrogen bonds

anticodon

Figure 6.6 Structure of the transfer RNA molecule. (From Fried, 1990. © The McGraw-Hill Companies, Inc. Used with permission.)

gene and an operator region. The **operator region** is the location where RNA polymerase binds to the DNA to begin the transcription process. The **repressor gene** produces a protein that binds to the same location in the absence of substrate, preventing gene expression. When the substrate is present, it binds to the repressor protein, causing it to release from the operator region, and allowing transcription to proceed.

For example, *E. coli* has three enzymes that play a role in cleaving the disaccharide lactose into glucose and galactose. When there is no lactose in the medium, *E. coli* does not produce the enzymes because the repressor protein is bound to the operator region of the operon containing the three genes. When lactose is present, it binds to the repressor protein, changing its shape and causing it to release from the operator region. The enzymes are then produced by the transcription and translation process.

In enzyme repression, the repressor protein normally cannot bind to the operator region. However, a corepressor molecule can bind to the repressor, changing its shape to one that does bind the operator region, stopping expression of the structural genes in the operon. For example, *E. coli* can produce enzymes to synthesize the amino acid tryptophan. However, when sufficient tryptophan is present in the growth medium, there is no need for the enzyme, and indeed, it represses its own production.

Eukaryotes also have transcription-level control. However, they also exert control in posttranscription stages, such as by controlling mRNA splicing, transport, and stability, by controlling translation, or by modifying the activity of the protein product.

Similar to operator regions, eukaryotes have **promoter** and **enhancer** nucleotide sequences that interact with proteins called **transcription factors** to activate transcription (in contrast to prokaryotic repressors). Steroid hormones act through this mechanism. They bind to protein transcription factors, and this complex then binds to DNA and activates transcription.

Another mechanism for preventing expression of a gene is by changing the DNA chemically. Specifically, a methyl group is added to some of the cytosine nucleotides, preventing transcription to mRNA.

Certain white blood cells in vertebrates produce antibodies in the blood. Antibodies complex with and inactivate foreign substances that may be toxic or infectious. The antibodies are highly specific proteins, each designed to react with a different foreign substance. The body produces millions of different kinds—many more than the number of genes in the chromosomes! To achieve this type of variety, the cell can actually rearrange the DNA sequences for antibody production. Rapidly growing tissues, such as egg-producing tissues or cancerous tumors, can increase the rate of gene expression by making many copies of the gene or by stockpiling mRNA for later use.

A remarkable and poorly understood feature of gene regulation in multicellular organisms is their ability to choreograph gene expression from embryonic development through specialization in the adult. Somehow, the ability of a cell to produce insulin must be turned off in the zygote, then turned on later, but only in certain cells in the pancreas. Similarly, each tissue must have specialized genes that determine its function, distinct from all others in the organism. The ability to "clone" animals, even mammals, from a single somatic cell taken from an adult is proof that each somatic cell contains the entire genome.

6.2.3 Mutations

A **mutation** is a change in the hereditary information contained in an organism. With most mutations, if the genotypic change is expressed phenotypically, cell death results. In a few cases survival is possible, with altered function. In rare cases, the mutation may actually aid the cell. Usually, it does not. In somatic cells, some mutations may be an initiating step in the formation of cancer. In germ (reproductive) cells, mutations may be passed on to the offspring, resulting in either teratogenesis, genetic disorders, or in some cases producing cancer that develops later in the offspring. Mutations can be caused by chemical or physical agents such as natural or industrial chemicals or radiation.

Mutation results from one or more changes in the sequence of nucleic acid bases in the DNA molecule. The effect can be illustrated by looking at what happens when a single change is made in the base sequence, in what is called a **point mutation**. There are three types of point mutation: base-pair substitution, insertion, or deletion. In the first, **base-pair substitution**, only a single amino acid in the final protein is affected (Figure 6.7). This type of mutation has a reasonable probability of not having a large effect. For example, if the protein being coded for is an enzyme, and the affected amino acid is not at the active site nor greatly affects the shape of the enzyme, the enzyme may still function normally, or almost normally. Also, note that because of the degeneracy of the genetic code, there is a chance that a base-pair substitution will still result in coding for the same amino acid.

The other two types of mutations have a much greater potential for disastrous consequences: **base-pair insertion** and **base-pair deletion** mutations. These are collectively called **frame-shift mutations**, because by changing the number of base pairs, they affect the reading of every codon that follows. Unless the mutation occurs near the end of the protein, a large number of amino acids will be changed, probably destroying its function. The sequence in Figure 6.7 is read from left to right, with location of mutations and altered amino acids marked by underscores. Shown are the sequence for the DNA strand

NORMAL SEQUENCE								
Compl-DNA	ATG	TCC	TGT	CCA	TGG	GGA	CGA	
DNA	TAC	AGG	ACA	GGT	ACC	CCT	GCT	
mRNA	AUG	UCC	UGU	CCA	UGG	GGA	CGA	
Amino acid	Met	Ser	Cys	Pro	Trp	Gly	Arg	
BASE-PAIR SUBSTITUTION								
Compl-DNA	ATG	TCC	T<u>A</u>T	CCA	TGG	GGA	CGA	
DNA	TAC	AGG	A<u>T</u>A	GGT	ACC	CCT	GCT	
mRNA	AUG	UCC	U<u>A</u>U	CCA	UGG	GGA	CGA	
Amino acid	Met	Ser	<u>Tyr</u>	Pro	Trp	Gly	Arg	
BASE-PAIR INSERTION								
Compl-DNA	ATG	TCC	T<u>G</u>G	TCC	ATG	GGG	ACG	A
DNA	TAC	AGG	A<u>C</u>C	AGG	TAC	CCC	TGC	T
mRNA	AUG	UCC	U<u>G</u>G	UCC	AUG	GGG	ACG	A
Amino acid	Met	Ser	<u>Trp</u>	<u>Ser</u>	<u>Met</u>	<u>Gly</u>	<u>Thr</u>	
BASE-PAIR DELETION								
Compl-DNA	ATG	TCC	TT_C	CAT	GGG	GAC	GA	
DNA	TAC	AGG	AA_G	GTA	CCC	CTG	CT	
mRNA	AUG	UCC	UU_C	CAU	GGG	GAC	GA	
Amino acid	Met	Ser	<u>Phe</u>	<u>His</u>	<u>Gly</u>	<u>Asp</u>		

Figure 6.7 Types of point mutations with examples.

that is being expressed, its complementary DNA strand, the messenger RNA formed from the sequence, and the resulting amino acid sequence.

If the mutation does not leave both strands of the DNA molecule with the proper complementary pair at a point, special repair enzymes can attempt to undo the damage. Of course, the mechanism is not perfect. Sometimes there is so much damage that information is lost. The repair enzymes excise or replace damaged parts of the DNA molecule, but consequently, may leave point mutations.

Mutations have the potential for causing heritable genetic disorders. However, this is less of a concern than the potential for causing cancer. Most heritable genetic disorders are due to genes already in the human gene pool. Some even confer advantages in special situations. For example, sickle-cell anemia, although disabling for those who are double recessive, may give partial immunity from malaria for those who are single recessive. Similarly, it is thought that single-recessive carriers of the cystic fibrosis gene may be relatively resistant to cholera. Nevertheless, exposure to mutagens can increase the frequency of genetic disorders such as those in Table 6.3.

6.2.4 DNA Repair

Cells have a variety of processes for repairing damage to DNA. First, the damage must be of a form that can be detected by a repair mechanism. Deletion of one or more base pairs results in a chemically correct DNA molecule and cannot be fixed by enzymes. However, removal of one base from a pair, replacement of one base with a noncomplementary base, or other chemical changes to the nucleotides produce recognizable distortions in the DNA molecule.

TABLE 6.3   Partial List of Human Genetic Disorders

Chromosome abnormalities
    Cri-du-chat syndrome (partial deletion of chromosome 5)
    Down syndrome (triplication of chromosome 21)
    Klinefelter syndrome (XXY sex chromosome constitution; 47 chromosomes)
    Turner syndrome (X0 sex chromosome constitution; 45 chromosomes)

Dominant mutations
    Chondrodystrophy
    Hepatic porphyria
    Huntington's chorea
    Retinoblastoma

Recessive mutations
    Albinism
    Cystic fibrosis
    Diabetes mellitus
    Fanconi syndrome
    Hemophilia
    Xeroderma pigmentosum

Source: Based on Williams and Burson (1985).

For example, a particular type of damage caused by UV radiation is the formation of covalent bonds between the pyrimidine residues of two adjacent thymidine bases. (Normally, the bases are joined only through the sugar–phosphate groups.) This **thymine dimer** produces strain in the DNA molecule. The bonds can be removed by a special enzyme called **photoreactivation enzyme**. Interestingly, this enzyme requires a photon of blue light to complete the repair. It is present in humans as well as prokaryotes.

Another type of repair involves several enzymes. **Excision repair** uses a variety of enzymes that "patrol" the DNA molecule, remove bases that they recognize as damage, and patch the molecule using DNA polymerase. Any damage that distorts the double helix structure can stimulate this mechanism.

Some humans have an inherited autosomal recessive condition called *xeroderma pigmentosum*. These persons are exceedingly sensitive to getting skin cancer from exposure to sunlight. The condition can be diagnosed in infancy, enabling people to lead fairly normal lives by protecting themselves from exposure to the sun. The condition seems to be related to a number of different mutations, including about seven that affect excision repair, and to reduced activity of the enzyme that uses blue light to repair thymidine dimers.

A third repair mechanism, called **recombinational repair**, comes into play when damage is missed by the other mechanisms, but comes to interfere with the replication process.

6.3   GENETIC ENGINEERING

Knowledge of the chemistry of DNA has made it possible to manipulate this material in useful ways. This leads to a category of techniques variously called *molecular biology*, *biotechnology*, or *genetic engineering*. Most of the applications involve the ability either to place genes from one organism into the chromosomes of another or just to be able to analyze the DNA at various levels of detail. The former application makes it possible, for example, to produce human proteins such as hormones in bacterial cultures or to place genes for hereditable human diseases in laboratory mice to facilitate the experimental study of therapies. In the emerging area of **gene therapy**, humans with genetic diseases such as cystic fibrosis can potentially be "infected" with viruses carrying the normal version of the genes, replacing the function of the defective genes.

There are other ways to exploit our knowledge of molecular biology for medical uses. The drug zidovudine (AZT) is similar to normal thymidine but lacks an OH group at a point where it is linked to other nucleotides to form DNA. AZT competes with the thymidine, preventing the completion of DNA synthesis. Fortunately, this effect is more pronounced in DNA synthesis caused by the AIDs virus than the cell's normal DNA synthesis.

Another therapy on the horizon is **antisense drugs**, which are synthetic nucleotide polymers that bind selectively to mRNA from a particular gene, interfering with synthesis of the corresponding protein. If the mRNA structure for a protein is known, an antisense drug can be designed to be complementary to 17 to 25 of its bases. This gives the drug high specificity, and since cells may make thousands of protein molecules from each mRNA, the antisense drugs could be effective in very low doses. Antisense drugs are being designed to block expression of cancer and virus genes, as well as normal human genes in special cases. However, delivery of these drugs to the cell interior is a problem.

DNA analysis includes *DNA fingerprinting*. This technique can be used forensically, such as to identify the person that a blood stain or hair root came from. It can be used to determine paternity conclusively. It can also be used to determine another type of relationship—that between species, in the development of taxonomic family trees. Another emerging area with direct relevance to the environmental field is the use of DNA analysis to make positive identification of disease-causing organisms, including waterborne parasites.

6.3.1 DNA Analysis and Probes

DNA sequencing is the complete determination of the sequence of nucleotide bases in a DNA molecule. Several methods have been developed to do this, involving radioactive tracers or fluorescent dyes and electrophoresis separation. The technique using dyes has been automated, making the process faster and relatively inexpensive. Sequencing has made it possible to identify the nucleotide sequences associated with origins of replication, regulatory regions adjacent to genes, and of course the genes themselves. For example, in eukaryotes, the location of a gene is marked by a promoter sequence, followed a fixed number of nucleotides later by a START codon, and ending with a STOP codon. When the sequence is known, the amino acid sequence of the corresponding protein can be determined. Although the gene may include introns, comparison with known proteins can often find the corresponding protein.

Another powerful technique is the ability to cut DNA molecules at specific locations. **Restriction enzymes** have the property of being able to make a precise cut in a double-stranded DNA molecule at or near a specific nucleotide sequence. About 200 different ones are known, each of which recognizes a different sequence. For example, one called *Eco*RI comes from *E. coli* and catalyzes the following reaction:

$$\cdots\text{GAATTC}\cdots \quad \Rightarrow \quad \cdots\text{G} \qquad + \quad \text{AATTC}\cdots$$
$$\cdots\text{CTTAAG}\cdots \quad Eco\text{RI} \quad \cdots\text{CTTAA} \qquad \text{G}\cdots$$

Restriction enzymes have a number of important uses. One is the production of recombinant DNA, which is described below. Other uses have to do with DNA analysis. Treatment with a restriction enzyme breaks a DNA molecule up into fragments of various sizes. Gel electrophoresis separates these according to size and provides a measure of the number of base pairs in each fragment. By examining the fragment sizes produced when DNA is treated by a variety of restriction enzymes, singly and in combination, it is possible to locate the positions of all the cutting sites on the molecule. This is called **restriction mapping**. By combining restriction map information with genetic maps based on linkage, individual genes can be located on the chromosome. This is a first step in the isolation of individual genes.

In **DNA fingerprinting**, the pattern of DNA segments produced by treatment with restriction enzymes is used for forensic purposes such as identification of the person from which a tissue sample, such as semen or hair follicles, originated. It is also used to establish paternity. In most cases, determinations can be made with a very high degree of confidence.

Individual genes can be marked using molecular probes. A **probe**, in molecular biology, is a cloned section of DNA that is complementary to part of a gene or a section of RNA. For example, suppose it is desired to locate the gene that produces a particular protein. If the protein could be purified, its amino acid sequence could be determined. Using the genetic code, one can determine nucleotide sequences that could produce a segment of the protein. Since the genetic code is degenerate, there are many such sequences. Then, using chemical methods, segments of single-stranded DNA containing radioactive nucleotides are synthesized that are complementary to all the possible gene sections (Figure 6.8).

Amino Acid	...	met	his	gly	...
Eight candidate oligonucleotides	...	AUG	CAU	GGA	...
	...	AUG	CAU	GGT	...
	...	AUG	CAU	GGC	...
	...	AUG	CAU	GGG	...
	...	AUG	CAC	GGA	...
	...	AUG	CAC	GGT	...
	...	AUG	CAC	GGC	...
	...	AUG	CAC	GGG	...
Selected oligonucleotide	...	AUG	CAC	GGA	...
Complementary gene	...	TAC	GTG	CCT	...

Figure 6.8 Development of DNA probes complementary to part of a gene for a short amino acid sequence. Methionine has only one possible codon, but histidine has two and glycine has four. Thus, there are eight possible oligonucleotide combinations for this sequence of amino acids. A mixture of all eight is prepared, and the one that is complementary to the gene will bind to it selectively.

These are cloned (copied) and placed in contact with restriction enzyme-cut DNA segments from a gel electrophoresis plate. The probes that are actually complementary to parts of the DNA will be bound by the same hydrogen-bonding mechanism that holds the double helix together. The rest are then rinsed away, and the restriction fragments that retain radioactivity are detected using x-ray film. All the fragments containing DNA complementary to one of the probes will be marked by this procedure, which is called the **Southern blot**. The Southern blot is also one form of DNA fingerprinting, used to identify the individual source of biological samples.

Due to normal genetic variation, two individuals of the same species can have different fragment patterns. Differences in locations of restriction enzyme cuts are called **restriction fragment length polymorphisms** (RFLPs). RFLPs are inherited according to Mendel's laws. Several thousand have been identified at locations randomly distributed throughout the human genome. Since they are inherited like genes, they can be mapped by linkage analysis, and from knowledge of their nucleotide sequence, their location can be further specified by restriction mapping and the use of probes and the Southern blot. RFLPs have greatly accelerated the mapping of the human genome. By studying their linkage, they help to locate specific genes, such as those associated with disease.

These techniques and other described below have made it possible to map the DNA of an organism completely. This could be done at several levels of detail: from RFLPs, to individual genes, to a complete nucleotide sequence. These **genome projects** have been conducted on hundreds of organisms from bacteria to humans. The **human genome project** cost at least $3 billion and required the efforts of many laboratories around the world. The project produced a complete nucleotide sequence for a particular human cell culture commonly used in laboratory experiments, which originates from a single person. A surprise result of the human genome project is that there are only 20,000 to 30,000 human genes, much fewer than the 100,000 or more thought to be present based on earlier evidence. Genome projects have already yielded new knowledge. For example, geneticists were surprised to learn that the traditional mapping techniques discovered only about half of the genes on a segment of *E. coli* that was sequenced. Several new modes of mutations leading to human genetic diseases have been discovered. Ultimately, it is hoped that these projects will produce a windfall for science, in terms of understanding the basis of life, and for medicine, in the development of treatments for genetic disease and cancer.

Probes may also be produced that are covalently linked to fluorescent molecules, producing what is known as the **fluorescence in situ hybridization** (FISH) **probe**. This makes it possible to visually identify cells containing specific DNA or RNA sequences using fluorescent microscopy. For example, it is possible to determine where a strain of nitrifying organisms is located within an activated sludge floc using this method. To do this, a sample of activated sludge is placed on a microscope slide and fixed, such as by heating. The slide is then treated with a solution containing a fluorescent probe that is complementary to the ribosomal RNA of the bacterial strain of interest. The probe can penetrate the membrane of the fixed cells and attach to (hybridize with) the ribosomal RNA. The probes thus become concentrated in the bacterial cells. When illuminated by the appropriate wavelength of light under a microscope, the cells of the selected strain "light up," making them visually distinguishable from other cells or cellular debris.

FISH probes typically have about 20 nucleotide base units. They are designed to be complementary to sections of the 16S rRNA of the microorganism. Probes are available commercially that select for individual species of bacteria, particular bacterial groups or genera, or even all bacteria.

6.3.2   Cloning and Recombinant DNA

A **clone** is a genetically identical copy of a DNA molecule, a cell, or an organism. Clones of DNA are produced in order to obtain sufficient genetic material for applications, such as DNA analysis or for producing recombinant DNA. **Recombinant DNA** is DNA formed by joining segments of DNA from different organisms.

Recall that recombination occurs naturally between chromosomes in the same cell during meiosis and results in segments of DNA from one chromosome being included in another. Recombinant DNA can also be produced artificially. The procedure uses restriction enzymes that leave uneven lengths of DNA on the two strands, such as *Eco*RI. For example, if it is desired to join a segment of human DNA to a bacterial DNA, both are first treated with the *Eco*RI. The uneven ends will have nucleotide sequences of either TTAA or AATT. These are termed **sticky ends** because they are complementary and easily form hydrogen bonds. In a mixture, some of the human strands will bond with some of the bacterial strands. The enzyme **DNA ligase** is then added, which repairs the strands at the cuts. DNA ligase is different from DNA polymerase in that the latter forms the chain of nucleotides by adding nucleic acids one by one, whereas the former links two strands of DNA together end to end.

The recombinant DNA can be cloned by insertion into prokaryotic or eukaryotic cells, which then reproduce, also reproducing the extra DNA. After the cell multiplies, the recombinant DNA can be extracted and purified, or the cells themselves may be used to express the inserted gene, producing the corresponding protein or expressing a trait in the organism. The process of inserting recombinant DNA into a cell is called **transfection**. Transfection is accomplished in various ways. One way is by direct insertion using a very fine needle under microscopic examination. Another way to accomplish transfection is by forming the recombinant DNA from a special DNA molecule called a vector. A **vector** is a DNA molecule that can reproduce itself independently inside cells. There are several main types: plasmids, bacteriophages, and yeast artificial chromosomes, plus some others that are hybrids or forms of these.

Recall that bacteria possess a single circular chromosome. In addition, they, as well as some eukaryotes (e.g., yeast) and some plants, can also contain an additional piece of circular double-stranded DNA called a **plasmid**. In bacteria, the plasmid often contains genes for antibiotic resistance. Bacteria can exchange plasmids, and plasmids can exist outside the cell. Their possession of antibiotic genes makes it easy to select for bacteria that have been transfected by recombinant plasmid DNA.

A **bacteriophage** is a virus that infects bacteria. A virus is itself an infectious particle consisting solely of a protein coating with DNA or RNA genetic material. A virus infects a cell by injecting its genetic material, which then directs the cell to produce new virus particles. This behavior makes it a natural means of cloning recombinant DNA in bacteria.

A **yeast artificial chromosome** (YAC) is a linear DNA molecule with yeast telomeres (the chromosome ends), an origin of replication, and a yeast centromere. These features allow the YAC to be replicated when inserted into eukaryotic cells. The YAC has the further advantage in that larger genes (up to 10^6 base pairs) can be inserted, unlike most plasmids or phages.

Whichever vector is used, it is transfected into an appropriate host cell using chemical and/or physical techniques. It is somewhat of a random process; only a fraction of the cells in a culture will be transfected successfully. Antibiotic resistance or other marker genes are included in the transfection to help select the desired cells. These are then cultured for further application.

Cloning for the purpose of getting an organism to express a gene from a different species has been applied in a number of ways. Animal hormones, including from humans, used to have to be purified from large quantities of tissue samples. Now some, such as growth hormones for humans and cows, are produced in bacterial cultures.

Sometimes bacterial gene expression is not sufficient, and eukaryotes must be used. Plants or animals containing a foreign gene are called **transgenic**. A human blood clot–dissolving substance called tPA can save the lives of heart attack victims. However, tPA was still too expensive when produced by bacteria. The solution was to insert the gene for tPA into a sequence of goat DNA involved in milk production. One such goat produced milk containing 3 g of tPA per liter of milk, worth about $200,000 per day at previous production rates. Transgenic crop plants have been developed with increased resistance to viral disease or with resistance to agricultural herbicides.

Sometimes, DNA is inserted into the genome of an organism to disrupt a particular gene, as a way to study the effect of the gene or to mimic human genetic disorders. This has been done with mice, and the resulting animals are called **knockout mice**.

Gene therapy is the transfer of normal human genes into human somatic cells to treat genetic disease. There has been some success at the experimental level, such as in treating severe combined immunodeficiency (SCID), a hereditary inability to fight infections. The therapy works as follows: A viral vector is prepared with the nondefective SCID gene inserted, and several viral genes responsible for its replication are removed. This makes the virus particle capable of infecting but not destroying the cell. Next, tissue is obtained from the patient and a type of leucocyte, or white blood cell, is separated. The leucocytes are infected with the viral vector in vitro (literally, "in glass," or in the test tube). The leucocytes are then injected back into the patient, where they function normally in fighting infections. The procedure has to be repeated periodically to maintain a population of normal cells.

6.3.3 Polymerase Chain Reaction

Although one of the uses of cloning is to produce quantities of identical DNA, a chemical procedure has largely taken the place of this function. The **polymerase chain reaction** (PCR) is a technique that repeatedly replicates DNA in vitro. Each cycle takes a matter of minutes and forms two DNA molecules from one. Twenty of these doublings produce more than 1 million molecules starting from a single strand. Variations of the basic procedure are used for detection of mutations and genetic diseases, identification of viruses and bacteria, and recovery of DNA from tiny samples for analysis.

PCR was developed in 1983 by Kary Mullis. He was working in a biotechnology laboratory producing DNA probes for Cetus Corporation in Emeryville, California. Recent automation techniques had left him "creatively underemployed," with lots of time to think. The personal computer revolution was also under way, and Mullis had learned about the power of iteration in computer programs. Driving late at night on his way to a mountain cabin for the weekend, with his girlfriend asleep in the passenger seat,

Mullis had the insight that would revolutionize molecular biology and win him a Nobel prize.

The procedure works as follows:

1. The sample containing the DNA to be copied is heated to separate the two strands of the double helix, and then cooled.
2. Two primers are added, one for each strand. The primer is a short piece of RNA similar to a probe but that can be recognized by DNA polymerase as a place to start DNA replication. The primers are chosen to be complementary to each strand close to its one end, since the next step extends the molecule toward the other end.
3. DNA polymerase and a solution of nucleotide bases are added. The DNA polymerase extends the primer using nucleotides complementary to the original strand until it completes the replication process.
4. Repeat.

The process is easily performed today using a small automated tabletop instrument. It turned out that all the steps needed for PCR had been well known for a decade. Many biologists had to wonder "why didn't I think of that?"

6.3.4   Genetic Engineering and Society

Along with the benefits of genetic engineering described above come numerous actual and potential dangers in its application. Experience with introduction of "natural" nonnative species into an ecosystem has shown the potential for disaster. Often introduced deliberately, they have often been found to occupy unexpected niches in the ecosystem, displacing other organisms. Many examples exist. The kudzu vine was introduced to the southern United States to control erosion, but proliferated beyond control and is destroying some forests in the region. The starling was introduced to the United States from England by groups who wanted to import a representative of each species mentioned by Shakespeare. Starlings soon displaced native songbirds such as the bluebird from the northeast.

Suppose that a new variety of rice were developed by genetic engineering techniques that could be cultured in salt water, increasing arable land availability. Would it become a weed that will choke salt marshes around the world? If a cancer gene were transfected into a common infectious virus, could it spread a new epidemic? If a bacterium were engineered to biodegrade xenobiotic toxic pollutants or oil spills, would it spread to destroy chemicals in useful applications, or infect petroleum stocks?

Some of these risks can readily be discounted, although others remain. The possibility of bacterially transmitted cancer was taken seriously enough by biologists so that in the mid-1970s they agreed to a moratorium on recombinant DNA research until tests showed that transmission did not occur. Infection of oil wells would be limited not by the ability of bacteria to biodegrade oil, which they already are capable of, but rather by the limitations in that environment of water, oxygen, and nutrients such as nitrogen.

Other risks exist in the uses of biotechnology. From the late nineteenth century until World War II, a school of thought called **eugenics** suggested that the methods of genetics should be turned to improving the human gene pool. This idea led to forced sterilization: first, of various criminal populations, and eventually, of alcoholics and epileptics. The policies were used to restrict immigration of certain Asian and European populations that were termed genetically inferior. Eugenics had its ultimate expression when it provided the "scientific" basis for the racial policies of the Nazis before and during World War II. Where the capability exists, so will the temptation. Will parents seek to amplify the gene for human growth hormone in their offspring so that their children could become heftier football linemen or taller basketball players? The ability to select the gender of one's offspring by amniocentesis and abortion is already causing problems in some cultures.

A more immediate set of risks are the problems of confidentiality and discrimination associated with the newly developed capability to test for genetic predisposition to certain diseases. As mentioned previously, a gene has been discovered that predisposes women to breast cancer. Women who test positive for this gene can protect themselves by aggressive monitoring or by preemptive surgery. However, insurance companies may refuse to issue life or health insurance to women who possess this gene although not all will actually contract the disease. As a result, many women whose family history suggests that they might have the gene refuse to be tested.

Extensive genetic screening of African Americans for the genetic disorder **sickle-cell anemia** was conducted in the 1970s. However, it has now been practically halted because affected persons were not adequately counseled, some suffered job discrimination as a result, and suspicions arose between the African Americans and the predominantly white medical establishment that administered the program. Three lessons were learned from this:

1. A person must be able to realize a direct benefit from participation in genetic screening, such as by its leading to therapy or prevention.
2. Results must be held in strict confidence.
3. Advice must be provided about the meaning of the outcome for that person, and for his or her relatives.

Controversy exists over whether we should allow new life-forms to be patented. Without patent protection, companies would be reluctant to develop useful new organisms. The U.S. Patent and Trademark Office has issued patents for oil spill–eating bacteria and pesticide-resistant crops. Some have sought to patent portions of the humane genome that were sequenced without even knowing their function. However, these patent claims have been denied.

6.4 GENETIC VARIATION

Genetic variation drives adaptation and evolution. The early eugenicists and modern proponents of "ethnic purity" ignore the fact that heterozygous organisms tend to be more vigorous. Many alleles are responsible for lethality, sterility, or other defects when they are homozygous. Thus, inbreeding in populations with diverse genotypes usually results in a loss of **fitness**, defined as the probability that an organism will produce offspring. In one species of fruit fly, almost 70% of chromosomes 2 and 3 contain alleles that cause male sterility when homozygous.

Many domesticated plants and animals have been inbred for centuries, with nonvigorous offspring eliminated through artificial selection. Nevertheless, the vigor of some of these can be increased by crossing them to form **hybrids**. Hybrid crops are often more productive. However, the vigor commonly declines with subsequent generations in hybrids, so the hybrids must be regenerated from the original breeding stocks for each planting.

In natural populations, inbreeding is more often deleterious. About 6.7% of human genes are heterozygous. Inbreeding in human populations increases the rates of spontaneous abortions, birth defects, and genetic disease. Many endangered species have very low heterozygosity. The cheetah is only 0.07% heterozygous, giving it a low capacity to adapt to changes such as to challenges from new disease organisms. Within humans, the variation between racial subgroups is accounted for by about 10% of the genome. The difference in skin color between Europeans and Africans is controlled by only two to five genes.

Genetic variation between species can be studied by looking at the number of differences in (1) the amino acid sequence of one of their proteins, or (2) the nucleotide sequence of a particular gene. At the genetic level, the distinction between similar species can be very small. Two species of fruit fly, for example, differ in DNA sequence by only about 0.55% located at about 15 to 19 genes. Yet their morphology and breeding behaviors are drastically different.

The amount of time that has elapsed since two species diverged from a common ancestor is sometimes known from paleontological data. By examining the number of protein or nucleotide sequence differences that separate two such species, it is possible to calibrate a **molecular clock**, to measure the time since other species diverged by evolutionary change. These data can be used to construct genetic "family trees," called **phylogenetic trees**, to show the degree of relationships. For example, the phylogenetic tree in Figure 6.9 was developed on the basis of a measure of DNA hybridization. This value was calibrated by comparison with fossil evidence to determine the rate at which mutations accumulate in the genome. The figure shows that chimpanzees and humans are more closely related to each other than either is to the gorilla. This type of analysis has led to revision of many taxonomic classifications formerly based on measures that are more ambiguous.

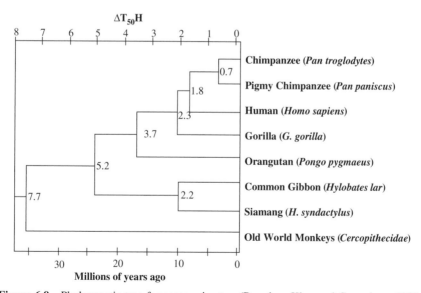

Figure 6.9 Phylogenetic tree for some primates. (Based on Klug and Cummings, 1997.)

Phylogenetic trees can also be constructed using differences among individuals in a single species. The mitochondrion has its own DNA, separate from the chromosomes of the nucleus. This DNA is inherited only from the mother. A molecular clock was constructed using RFLPs of human mitochondrial DNA. The results indicate that all humans have a common female ancestor who lived 200,000 years ago.

6.5   SEXUAL REPRODUCTION

Simpler organisms, such as unicellular ones, reproduce by **asexual reproduction**, in which offspring are produced by a single individual. The offspring have the same genotype as the parent. Examples of asexual reproduction range from binary fission in bacteria to vegetative propagation of plants by cuttings or the budding of the *hydra*, a jellyfishlike animal.

In **sexual reproduction,** offspring are produced by combining genetic material from two distinct individuals. The offspring are genetically different from either parent by virtue of having a combination of the genes from each. Sexual reproduction evolved as a strategy to greatly increase variation within a population. Recombination during meiosis can combine pieces of two genes from alleles from different parents to create novel genes. This facilitates adaptation to changing conditions. When environmental conditions change, it is likely that some subset of the population will have a genotype that is more appropriate than the majority of the population to the new conditions. This minority will tend to flourish, establishing a new majority that is better adapted. To put it succinctly: *Sex produces variation without mutation.* This is the advantage that sexual reproduction confers on a species.

In diploid organisms that reproduce sexually, the germ cell line alternates between haploid and diploid with each generation. The transformations from one to the other are performed by the processes of meiosis and fertilization. Meiosis converts diploid cells into haploid cells, which become the gametes. In fertilization, two gametes unite to create the diploid **zygote** cell. Around this basic plan, there are three variations (Figure 6.10).

In **zygotic meiosis**, the haploid cells produced by meiosis are called *spores*. **Spores** are haploid cells that can undergo mitosis, producing a multicellular haploid organism. [In contrast, **gametes** are haploid cells that must fuse (in the process of fertilization) to form diploid cells, which in turn divide mitotically to form a multicellular diploid organism.] Spores reproduce to form a haploid multicellular individual (or many unicellular ones). These individuals eventually produce gametes, which through fertilization produce a diploid zygote. The zygote then immediately undergoes meiosis. The zygote is the only diploid form in organisms with this type of life cycle. Organisms with this life cycle include the fungi and some algae, such as *Chlamydomonas*.

Gametic meiosis is a life cycle wherein meiosis produces haploid gametes almost directly. Subsequent fusion of two gametes in fertilization results in a diploid zygote, as in zygotic meiosis. However, in gametic meiosis the zygote now forms a multicellular individual. Thus, the diploid form is the only multicellular form, and it dominates the life cycle. This life cycle is characteristic of most animals, some protists, and at least one alga. In animals, the female gametophyte is the **egg**, and the male gametophyte is the **sperm**.

In **sporic meiosis**, multicellular forms appear after both meiosis and fertilization. The haploid organism is called a **gametophyte**; the diploid form is a **sporophyte**. This plan is typical of all plants and many algae. Sometimes the gametophyte looks like the

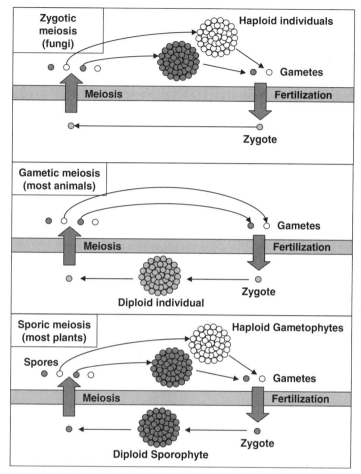

Figure 6.10 Types of life cycles in sexual reproduction. (Based on Raven et al., 1992.)

sporophyte (e.g., certain brown algae). In most plants, they are distinct. In simpler plants, such as mosses, the gametophyte is larger than the sporophyte and is self-sufficient nutritionally. In the higher vascular plants the sporophyte is dominant and the gametophyte depends on the sporophyte for nutrition. In the most extreme example, in flowering plants the female gametophyte has only seven cells and the male gametophyte (pollen) has only three.

PROBLEMS

6.1. Consider what would happen in Mendel's experiment with a single trait if another generation were produced by self-fertilization. What proportion of the total population would be heterozygous? What about after yet another generation?

  You should see that the proportion of heterozygous organisms becomes smaller and smaller. If repeated many times, every organism will ultimately have a genotype consisting of a mix of double recessive and double dominant genes. This is how a "true-breeding" population is produced. In many species, especially of animals, this can be harmful, because although many recessive traits may be helpful to the organism as a hybrid, they can be harmful as a double recessive. The gene for sickle-cell anemia is an example. When heterozygous it is thought to confer resistance to malaria, but as a double recessive it causes a crippling sensitivity to oxygen deprivation. The risks from true breeding provides a biological justification for social taboos against incest and marriage between close relatives.

6.2. What are some of the ethical problems that could be created by the existence of a test for the Huntington's disease gene?

6.3. Use your calculator to compute the number of copies produced by 20 doublings, which would occur in that many cycles of the polymerase chain reaction.

REFERENCES

Fried, G. H., 1990. *Schaum's Outline: Theory and Problems of Biology*, McGraw-Hill, New York.

Glieck, James, 1987. *Chaos: Making a New Science*, Viking Penguin, New York.

Klug, W. S., and M. R. Cummings, 1997. *Concepts of Genetics*, Prentice Hall, Upper Saddle River, NJ.

Mullis, Kary, The polymerase chain reaction, *Scientific American*.

Oerther, D. B., S. Jeyenayagam, and J. Husband, 2002. Fishing for fingerprints in BNR systems, *Water Environment and Technology*, Vol. 14, No. 1, pp. 22–27.

Postlethwait, J. H., and J. L. Hopson, 1995. *The Nature of Life*, McGraw-Hill, New York.

Raven, P. H., R. F. Evert, and S. E. Eichhorn, 1992. *Biology of Plants*, Worth Publishers, New York.

Rawls, R. L., 1997. Optimistic about antisense, *Chemical and Engineering News*, June 2.

Williams, P. L., and J. L. Burson (Eds.), 1985. *Industrial Toxicology: Safety and Health Applications in the Workplace*, Van Nostrand Reinhold, New York.

7

THE PLANTS

Plants and fungi were once classified into a single kingdom (together with the plantlike protists). They are now distinguished into separate kingdoms. They have in common the facts that they are mostly multicellular, mostly nonmotile, and their cells possess a cell wall (although of different compositions). Fungi are heterotrophic, eukaryotic organisms. They are discussed in Chapter 10 with the microorganisms, even though they include multicellular macroscopic forms.

Plants are multicellular, photosynthetic, eukaryotic organisms. Their environmental importance begins with their being the basis of almost all ecosystems, fixing atmospheric CO_2 into organic matter that provides food for other organisms, and producing oxygen. Other environmental concerns involving plants (both positive and negative) that are more directly affected by human activities include:

- Protection of soil from erosion
- Sequestration of carbon dioxide produced by fossil-fuel combustion
- Forest damage from acid deposition
- Damage to plants from urban air pollution

In addition, plants are now being used increasingly for pollution control, in a process called **phytoremediation**. For example, certain grasses have been found to concentrate heavy metals, including radionuclides, from contaminated soil. Poplar trees are being investigated for the remediation of groundwater contaminated with volatile organic compounds.

Plant cells are characterized by cellulosic cell walls, they store carbohydrates as starch, and they contain both chlorophyll *a* and *b*. The major trend in the evolutionary development of plants is increased dominance of the sporophyte. As you move to the higher

Environmental Biology for Engineers and Scientists, by David A. Vaccari, Peter F. Strom, and James E. Alleman
Copyright © 2006 John Wiley & Sons, Inc.

divisions, the gametophyte phase becomes less and less conspicuous, and increasingly dependent on the sporophyte for survival. Another structure that distinguishes more complex plant divisons from the simpler ones is the presence of **vascular bundles**. These are bundles of cells arranged in columns spanning the length of the plant. They conduct water and minerals from the roots upward and the products of photosynthesis from the leaves to the rest of the plant. The development of vascularization enabled plants to grow to great heights. We first describe the taxonomy and then examine the physiology of the dominant plant division, the flowering plants.

7.1   PLANT DIVISIONS

There are 12 existing divisions of plants, arranged in four groups (Table 7.1). The **bryophytes** comprise the mosses and related divisions. They lack specialized vascular tissues for transport of nutrients within the plant, which limits them to a height of only several centimeters. Instead of roots, they have **rhizoids**, which are filamentous cells that act primarily as anchors, not for absorption. They require moist conditions for their motile sperm cells to reproduce. (Some plants that are called *mosses* are not bryophytes. For example, reindeer moss is a lichen, and the Spanish moss that hangs from branches of trees in the southeastern United States is actually an angiosperm.) The genus *Sphagnum* includes 350 species of peat mosses. The gametophyte dominates, forming masses of green plants. They grow in temperate and cold bogs, where they acidify their environment to a pH as low as 4.

The second group is the **seedless vascular plants**, which includes the ferns. They developed several adaptations suited to drier conditions. These include vascular systems

TABLE 7.1   The Divisions of the Plant Kingdom, Organized into Groups

Bryophytes	The nonvascular plants
Division Bryophyta	Mosses (9500 species)
Division Hepatophyta	Liverworts (6000 species)
Division Anthocerophyta	Hornworts
Seedless vascular plants	Vascular plants that reproduce by spores
	Four divisions are extinct, four remain
Division Psilotophyta	Includes *Psilotum*, a tropical plant and greenhouse weed that lacks both roots and leaves
Division Lycophyta	The club mosses, 1000 living species; during the Carboniferous age some grew to 40 m in height
Division Sphenophyta	A single genus: *Equisetum* (horsetails), 15 species
Division Pterophyta	The ferns (11,000 species, mostly in tropical climates)
Gymnosperms	"Naked seed" plants: provide protection and nutrition
Division Cycadophyta	Cycads: palm-like plants (palms are monocot angiosperms)
Division Ginkgophyta	Only one surviving species: *Ginkgo biloba*
Division Gnetophyta	Gnetophytes: 3 genera with 70 species
Division Coniferophyta	Conifers: 50 genera, 550 species
Angiosperms	The flowering plants
Division Anthophyta	235,000 species
	Monocots (65,000 spp.)
	Dicots (170,000 spp.)

for the transport of water and nutrients to upper plant parts, specialized roots for absorption, a waxy cutin layer on leaf surfaces to reduce evaporation, and the production of lignin to provide structural strength for taller plants. Before the Cretaceous age, 144 million years ago, in the time of the dinosaurs, ferns dominated in tropical climates, forming large trees. Although the living plants are now gone, their carbonized remains are the basis of today's coal deposits. The familiar fern plant is a sporophyte that produces spores on the underside of the fronds. When the spores fall on a suitable surface, they produce a small gametophyte. This, in turn, produces eggs and motile sperm cells that fertilize and grow into a mature sporophyte. Ferns may also reproduce asexually from horizontal stems called **rhizomes**.

The third group is the **gymnosperms**, or "naked seed" plants. The seed is formed after fertilization. It contains the **embryo**, which is a young sporophyte. It also contains a quantity of starch for use by the embryo during germination. An outer **seed coat** provides protection. In conifers, gnetophytes, and angiosperms, seeds are formed when the **ovule**, which contains the egg, is fertilized by the **pollen grain**, the male gametophyte. This arrangement eliminates the requirement that there be free water present in order for fertilization to happen, as is needed by all seedless plants.

Of the four gymnosperm divisions, by far the most familiar is that of the conifers. They include 50 genera with 550 species. One is the redwood (*Sequoia sempervirens*), the tallest terrestrial plant, at up to 11 m in diameter and 117 m high. Others are the pines, firs, spruces, hemlocks, cypresses, junipers, and yews. Most bear their seeds in cones, except for the yew, which contains them in a fleshy fruitlike structure. The bark of the Pacific yew (*Taxus brevifolius*) has been found to contain a substance, called *taxol*, that is used as a treatment for breast and ovarian cancer. Most conifers keep their leaves for several years and do not lose them all at once. A few conifers, however, are **deciduous**; that is, they lose their leaves at the end of the growing season. These include the European larch (*Larix decidua*) and the bald cypress (*Taxodium distichum*).

The **cycads** form tall trees that resemble palms, with the difference that the trunk is covered with the bases of leaves that have been shed as the tree grew taller. New leaves are grown only at the top of the trunk. (Palm trees are monocot angiosperms.) Cycads produce compounds that are neurotoxic and carcinogenic. They harbor cyanobacteria and thus contribute to nitrogen fixation. **Gnetophytes** include many unusual plants, such as *Welwitschia*, which grows mostly below the soil, as well as trees, vines, and shrubs. Some are very similar to dicotyledonous angiosperms. **Ginkgo** includes a single species of tree, *Ginkgo biloba*, that was known to the West from fossils but was thought to be extinct. It was subsequently found cultivated on temple grounds in China and Japan, but apparently was not found anywhere in the wild. It is now cultivated throughout the world and is easily identified by its unique fan-shaped leaves. It is exceptionally resistant to air pollution and thus has found use in urban parks and along roadsides.

The final group is that of the **angiosperms**, the flowering plants. Angiosperms have dominated the land for 100 million years. Their rise paralleled that of the mammals, both of which were relatively insignificant during the tenure of the dinosaurs. The single division of the angiosperms is called **anthophyta**, making these two terms synonymous. They are distinguished by the presence of flowers, fruit, and a distinctive life cycle. Most familiar plants are angiosperms, from the millimeter-sized duckweed (*Lemmaceae*) to 100-m-tall *Eucalyptus* trees; from the aquatic water lily (*Nymphaea odorata*) to the Saguaro cactus (*Carnegiea gigantea*). Although mostly autotrophic, it includes parasitic and saprophytic species. For example, the Indian pipe (*Monotropa uniflora*) lacks

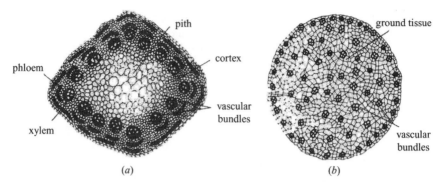

Figure 7.1   Microscopic cross sections through dicot (*a*) and monocot (*b*) stems. (From Fried, 1990. © The McGraw-Hill Companies, Inc. Used with permission.)

chlorophyll but obtains nutrients from the roots of other plants through an association with fungi.

The angiosperms are divided into two classes: **monocotyledones** and **dicotyledones** (**monocots** and **dicots**, for short). The names refer to the embryonic leaves of the seed. Monocots have a single leaf (called a **cotyledon**), dicots have two. They can be distinguished by looking at the starchy food-storage part of the seed, called the **endosperm**. Rice and corn are monocots and have a single endosperm. Dicots, such as peanut or bean, have two. Even without seeing the seeds, the two are easily distinguished. Monocot leaves have parallel veins, as in grasses. Dicots have a network venation, such as in oak or maple leaves. In monocots the vascular bundles (described in the next section) are usually arranged throughout the cross section of the stem, as in celery. The vascular bundles of dicots are arranged in a ring (Figure 7.1).

7.2   STRUCTURE AND PHYSIOLOGY OF ANGIOSPERMS

Since angiosperms are the dominant terrestrial plants and are of such economic importance, we describe them in more detail than the other divisions. Agriculture results in several types of environmental impacts, including extensive monoculture and contamination from pesticides and fertilizers. Therefore, understanding plants would be a useful background for managing these sources of pollution.

7.2.1   Water and Nutrient Transport

The vascular bundles consist of two types of conducting tissues, xylem and phloem. **Xylem** is the main water- and mineral-conducting tissue. It consists of columns of cells, some with perforations through the ends of each cell, through which materials move. At maturity the xylem cells lose their protoplasm, forming nonliving hollow tubes. **Phloem** is the food-conducting tissue that transports substances to and from the roots and leaves.

The vascular bundles consist of xylem and phloem, as well as other tissues. In dicot stems there is a layer of growing cells called the **cambium** layer between the xylem and phloem in each bundle. The cambium extends from one bundle to the next, forming a ring all the way around the stem. This narrow layer just below the bark of trees is the only

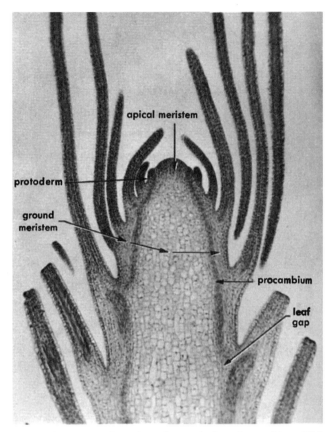

Figure 7.2   Microscopic cross sections through an apical meristem. (From Weier et al., 1970.)

place where growth occurs in the stem. Xylem is formed toward the inside of the cambium, and phloem toward the outside. Each season produces a new layer of xylem, visible as the "tree rings" of a stump. Wood consists of dead xylem cells, most of which still conducts water and minerals. A tree can be killed without cutting it down simply by destroying the cambium all the way around the tree, a practice known as **girdling**. Lengthening of a plant occurs mostly by growth at the tip of the shoot, called the **terminal bud** or the **apical meristem** (Figure 7.2). Other buds produce branches or leaves.

   The **root** is the underground portion of the sporophyte. Their main function is anchorage and absorption, but they may also be used in storage (as in carrots and potatoes). Monocots form a shallow fibrous root system. Gymnosperms and most dicots form a main root called a **taproot** that grows straight down. Roots of some trees have been found to penetrate 30 to 50 m into the soil. Most of the tree roots involved in absorption are in the top 15 cm and extend out beyond the crown of the tree. Roots of the corn plant (*Zea mays*) penetrate up to 1.5 m, and spread horizontally 1 m around the plant. The surface area of roots is greatly increased by the formation of **root hairs**, which grow from cells at the surface of the roots (Figure 7.3). As the plant grows, it maintains a balance between the leaf surface area and root surface area, so that water, minerals, and carbohydrates are formed in the proper proportion for growth. The roots absorb minerals by active

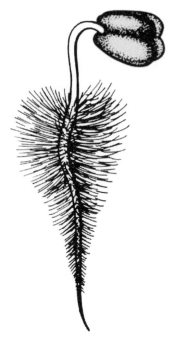

Figure 7.3   Microscopic view of root with root hairs. (From Simpson and Orgazaly, 1995.)

transport, which requires energy. Water transport, on the other hand, is mostly passive, powered by evaporation in the leaves.

The **leaf** is the main photosynthetic organ of the plant. Its large surface area is designed to capture sunlight. A cross section through the leaf (Figure 7.4) shows that the upper and lower surfaces of the leaf are formed of a layer of cells called the **epidermis**, which is covered by a waxy cuticle to minimize water loss. The cells between these two layers form the **mesophyll**, which has two parts. The **palisade parenchyma** form vertical columns under the upper surface. They are the main photosynthetic tissues of the leaf. The rest of the mesophyll consists of **spongy parenchyma**, which has a about 15 to 40% void volume filled with air and a large surface area for exchange of carbon dioxide, oxygen, and water vapor.

The epidermis is punctuated by special pores called **stomata** (singular, **stoma** or **stomate**). There can be as many as 12,000 stomata per square centimeter, mostly on the lower surface of the leaf, where they occupy about 1% of the leaf surface. Each stoma is made up of a pair of banana-shaped **guard cells**. The guard cells have chloroplasts, whereas the other epidermal cells do not. The stomata control the passage of gases into and out of the leaf, thus affecting photosynthesis and water use. They also affect the transport of pollutants into the leaf. When the guard cells accumulate water, they swell, causing them to bow outward, opening the pore. Conversely, loss of water causes the pore to close. This occurs in response to several conditions. First, and most obviously, if water is scarce, a general wilting will cause the stomates to close, limiting water loss. This gives the plant control over **transpiration**, the evaporation of water from the plant. This "loss" is actually necessary for the plant's survival, giving it the means to transport minerals

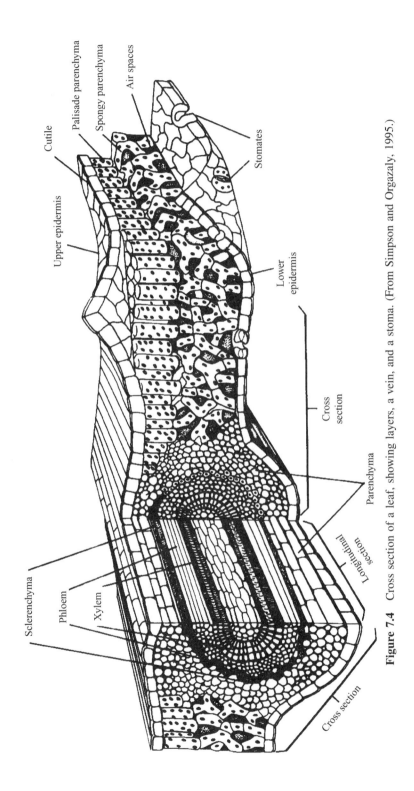

Figure 7.4 Cross section of a leaf, showing layers, a vein, and a stoma. (From Simpson and Orgazaly, 1995.)

Cutile

Palisade parenchyma

Spongy parenchyma

Air spaces

Upper epidermis

Stomates

Lower epidermis

Cross section

Parenchyma

Longitudinal section

Cross section

Sclerenchyma

Phloem

Xylem

from the roots. A field of corn in Kansas transpires the equivalent of 28 cm of rain per year. A stand of red maple transpires an estimated 72 cm. These figures are far in excess of the water actually used by the plant in its tissues or for photosynthesis.

When water is not limiting, the plant needs the stomata to open to admit CO_2 for photosynthesis. When light is present, the plant actively transports salts into the guard cells. This increases their osmotic potential, drawing in water and opening the pore. Photosynthesis in the guard cells also contributes to their osmotic potential by producing glucose in the cytoplasm. Temperatures above 30 to 35°C or high CO_2 levels can cause stoma closing. Because of all these effects, a plant may close its stomata at night, open them in the morning, then close them again in the heat of the midday, finally to open them again when the afternoon cools. The arid-climate CAM plants are different: They open their stomata at night and close them all day to conserve water. CO_2 is stored during the night for use in photosynthesis during the day.

An interesting problem for tall plants is how they are able to move water from the soil to great heights, as high as 111 m. Either the water is pushed by the roots, in which case the water column would be under pressure, or it is drawn from the top, which would place the water in the xylem under vacuum. Experiments show that when the xylem is punctured, air is drawn in. This eliminates the pressure hypothesis. However, it is a simple hydrodynamic fact that even a perfect vacuum pump cannot raise water much more than about 10 m. The accepted explanation is called the **cohesion-tension theory**. The suction is provided by evaporation of water from small pores in the leaves. The water column is prevented from breaking because of adhesion to the walls of the xylem. Experiments have shown that water in a fine capillary can withstand a tension of up to −26.4 megapascal (MPa). But only −2.0 MPa (about −20 atm) is needed to lift water to the top of a tall redwood. (Actually, the roots of many plants do produce some pressure. They do so by active transport of ions into the xylem, producing osmotic pressure. Root pressures as high as 0.3 to 0.5 MPa have been measured.)

Transport of sugars via the phloem is described by the **pressure-flow hypothesis**. First, the sugars (mostly sucrose) are moved into the phloem from leaves or storage by active transport. This increases the osmotic potential, causing water to be absorbed into the phloem, which generates hydrostatic pressure. The pressure causes the fluid to flow toward tissues that are removing sugars. The removal of sugar at those sites causes water to leave the phloem osmotically at that point. Most of the water then returns to the xylem to complete the circulation. Thus, plants, like animals, can be thought of as having a complete circulatory system, although it is not a single closed system.

7.2.2   Plant Growth and Control

Plants produce a number of chemical messengers, called **hormones**, which affect the way the plant grows. There are five main hormones or groups of hormones:

1. **Gibberellins** promote growth, especially the lengthening of stems and formation of fruit and leaves.
2. **Auxins** promote growth by stimulating cells to elongate. They control the bending of plants toward light. They are used commercially to stimulate plant cuttings to form roots. Auxins are produced by the growing tip of the plant and diffuse downward through the plant. Pruning encourages the formation of lateral buds by

removing the auxin-forming tip of the plant. Developing seeds produce auxin, which stimulates the formation of the fruit. There is only one naturally occurring auxin: indole-3-acetic acid, which is chemically similar to the amino acid tryptophan. However, many synthetic analogs have been developed, including the herbicides 2,4-D and 2,4,5-T.

3. **Cytokinins** promote growth by stimulating cell division. They have an effect on differentiation of plant cells. Leaves that turn yellow after being picked can be kept green longer by treatment with cytokinins. Chemically, they are derivatives of adenine.

4. **Abscisic acid** has the opposite effect of some of the other hormones: It inhibits growth and development. It is produced during water stress, causing the stomata to close and thereby inhibiting photosynthesis. It is involved in causing the seed embryo to become dormant so that it does not germinate prematurely. It also stimulates. A some processes, such as protein storage in seeds.

5. **Ethylene** is the simple hydrocarbon $H_2C{=}CH_2$. It stimulates maturation, promoting the ripening of fruit and the dropping of fruit, leaves, and flowers. Being a gas, it is released to the air by plants and ripening fruit. This explains the adage "one bad apple spoils the bunch," as ethylene from one fruit hastens ripening in the others. Plants also release it when injured. Ethylene is used commercially to stimulate ripening of fruit such as the tomato that were picked green so that they could be transported to market while firm and less liable to damage. Ethylene is active at air concentrations as low as 0.06 ppmv (parts per million by volume).

Plants also change their growth patterns in response to environmental stimuli, such as by light and gravity. The response to gravity is called **geotropism**. If a potted plant is placed on its side, the cells on the lower side of the stem will elongate, causing the stem to bend upward. Auxin seems to be involved. The turning of plants toward the light is called **phototropism**. Experiments have established that light coming from the side, especially blue light, causes auxin to migrate to the shadow side of a stem. This causes elongation on the shaded side, bending the plant toward the light.

Plants exhibit another behavior, familiar to those who raise houseplants, called **photo-periodism**, in which the length of the night controls when the plant flowers. Plants exhibit one of three photoperiodism behaviors: **Short-day plants** set flower only when the length of the night exceeds a critical period, which varies from plant to plant. Examples include some chrysanthemums, poinsettias, and strawberries. Short-day plants flower in the early spring or fall. Ragweed, for example, blooms in the fall and needs at least 9.5 hours of continuous darkness. Interestingly, a single flash of light in the middle of the night can fool a short-day plant and prevent its flowering. **Long-day plants** require a period of darkness less than some critical value and tend to flower in the summer. Spinach, lettuce, and some varieties of potato and wheat are long-day plants. Spinach will bloom only if there is less than 10 hours of darkness. In contrast to short-day plants, a flash of light during a long night can fool a long-day plant into flowering. Finally, there are **day-neutral plants**, such as cucumber, sunflower, rice, and corn, which are not controlled by photoperiod. Photoperiodism is controlled by a membrane-bound protein complex called **phytochrome** that acts as a detector for light. Plants also use phytochrome to detect if light is totally absent, such as if the plant is shaded by other plants or by a fallen log. In response, plants do not produce chlorophyll but instead, devote their energy to growing longer, seeking

TABLE 7.2   Elemental Plant Nutrients

Element	Form in Which Absorbed	Typical Plant Concentration (% or ppm dry weight)	Functions
Macronutrients			
Carbon	CO_2	~44%	Component of organic compounds
Oxygen	H_2O or O_2	~44%	Component of organic compounds
Hydrogen	H_2O	~6%	Component of organic compounds
Nitrogen	NO_3^- or NH_4^+	1–4%	Component of amino acids, proteins, nucleotides, chlorophylls, and coenzymes
Potassium	K^+	0.5–6%	Involved in osmosis and ionic balance, stomata opening and closing, and coenzymes
Calcium	Ca^{2+}	0.2–3.5%	Cell wall component; enzyme cofactor; involved in membrane permeability
Phosphorus	$H_2PO_4^-$ or HPO_4^{2-}	0.1–0.8%	Component of energy-storing compounds (ATP), nucleic acids, several essential coenzymes, phospholipids
Magnesium	Mg^{2+}	0.1–0.8%	Part of the chlorophyll molecule; activator of many enzymes
Sulfur	SO_4^{2-}	0.05–1%	Component of some amino acids and coenzyme A
Micronutrients			
Iron	Fe^{2+} or Fe^{3+}	25–300 ppm	Required for chlorophyll synthesis; component of cytochromes and nitrogenase
Chlorine	Cl^-	100–10,000 ppm	Involved in osmosis and ionic balance; probably essential in photosynthetic reactions that produce oxygen
Copper	Cu^{2+}	4–30 ppm	Activator or component of some enzymes
Manganese	Mn^{2+}	15–800 ppm	Activator of some enzymes; required for integrity of chloroplast membrane and for oxygen release in photosynthesis
Zinc	Zn^{2+}	15–100 ppm	Activator or component of many enzymes
Molybdenum	MoO_4^{2-}	0.1–5.0 ppm	Required for nitrogen fixation and nitrate reduction
Boron	$B(OH)_3$ or $B(OH)_4^-$	5–75 ppm	Influences Ca^{2+} utilization, nucleic acid synthesis, and membrane integrity
Essential to some plants			
Sodium	Na^+	Trace	Involved in osmotic and ionic balance; probably not essential for many plants; required by some desert and salt marsh species and may be required by all plants that utilize the C_4 photosynthesis pathway
Cobalt	Co^{2+}	Trace	Required by nitrogen-fixing microorganisms

Source: Raven et al. (1992), Table 27.2.

152

the sunlight. The result can be seen in the spindly, yellow growth of grass that has been covered by a board for some days. Some seeds require total darkness in order to germinate; others, such as lettuce, grow only if exposed to light. The difference relates to whether the seeds need to be buried or on the surface of the soil to germinate. This behavior is also controlled by phytochromes.

Plants also show 24-hour cycles in movements of their leaves that persist even if light cues to day length are removed. Some plants open their flowers in the morning; others fold their leaves at night. In the absence of light cues, the timing is not precise. It ranges from 21 to 27 hours. These cycles are called **circadian rhythms** and have been referred to as **biological clocks**. Circadian rhythms are now known to occur in all eukaryotes, including animals. However, they are absent in prokaryotes.

The herbicides *2,4-dichlorophenoxyaceticacid* (2,4-D) and *2,4,5-trichlorophenoxyacetic* (2,4,5-T) were active ingredients in *Agent Orange*, the dioxin-contaminated herbicide used in the Vietnam war to deprive the enemy of their forest cover. The use and manufacture of 2,4,5-T is now banned in the United States. 2,4-D is still used in household broadleaf weedkiller, because dicots (such as dandelion) are much more sensitive to it. They act by causing the plant to grow out of control, plugging the phloem. Another important herbicide is *glyphosate*, which is sold by Monsanto under the name Roundup. Glyphosate effectively kills all plants, although it is nontoxic to animals. It acts by blocking the action of a single enzyme involved in the production of aromatic amino acids.

7.2.3 Plant Nutrition

Plants, being autotrophs, do not require organic food sources. In fact, they do not need any organic growth factors. They can synthesize all of the amino acids and vitamins they require. Thus, their only known nutritional requirements are the inorganic nutrients listed in Table 7.2. Besides roles as components of organic compounds, the elements are used as enzyme cofactors, as intermediates in electron-transfer reactions, and for regulating osmotic pressure and membrane transport.

Nitrogen is absorbed by plants as either nitrate or ammonium. If it is in the form of nitrate, it is reduced to ammonium by the plant. This requires energy. The ammonium is then transferred to organic compounds such as amino acids, in a process known as **amination**. These processes occur mostly in the roots. Forms of nitrogen rising toward the shoot in the xylem are almost all organic, mostly amino acids.

Plants can also take up toxic minerals such as lead or some radionuclides from the soil. This is being exploited in the phytoremediation of contaminated land (see Section 16.2.4).

PROBLEMS

1.1. By looking at the Table 7.2, can you find any minerals that may be soil pollutants and that might be removed from soil by plants? Are there any minerals not in the table that are more toxic, yet have similar chemical behavior?

1.2. Sometimes industrial sites can be seen to have large areas of soil with few or no plants. Other than the presence of toxic substances, what could account for this?

REFERENCES

Fried, G., 1990. *Schaum's Outline: Theory and Problems of Biology*, McGraw-Hill, New York.

Raven, P. H., R. F. Evert, and S. E. Eichhorn, 1992. *Biology of Plants*, Worth Publishers, New York.

Simpson, B. B., and M. C. Ogorzaly, 1995. *Economic Botany: Plants in Our World*, McGraw-Hill, New York.

Weier, T. E., C. R. Stocking, and M. G. Barbour, 1970. *Botany: An Introduction to Plant Biology*, Wiley, New York.

8

THE ANIMALS

Animals are multicellular heterotrophs whose cells lack cell walls and that have a motile stage at some part of their life cycle. They are diploid and reproduce primarily by sexual reproduction. Some of the interactions between animals and the environmental effects of human activities are:

- Loss of wildlife habitat may cause extinction.
- Overharvesting may lead to depletion or extinction of species.
- Animals may be reservoirs for human diseases.
- Control of pests may cause pollution.

The animal kingdom is divided into 33 phyla. A few of them include familiar groups or organisms. However, quite a few are relatively unfamiliar, and most of these are small marine organisms. About 1.5 million animal species have been named, but these are probably only a fraction of those that exist. Table 8.1 summarizes the animal phyla, grouped according to an informal subdivision system.

8.1 REPRODUCTIVE STRATEGIES

Animals have a great variety of means of reproduction. Only a few of the simplest animals reproduce asexually. Cnidarians, such as hydra, reproduce by **budding**, the unequal division of an organism. Many invertebrates, such as sea stars, can reproduce by **fragmentation**, in which the organism can simply be broken in two, with each part then regenerating the missing portion. Asexual reproduction produces *clones*, individuals

TABLE 8.1 Summary of Animal Phyla

Phylum	Description
Mesozoa	Small worms 0.5 to 7 mm long, composed of only 20 to 30 cells; exclusively marine parasites.
Porifera	Sponges, ~5000 species (150 freshwater) made of gelatinous matrix with calcium carbonate or silicate skeleton. Flagellated cells pump as much as 1500 L of water per day to capture food.
Placozoa	Contains a single species, *Trichoplax adhaerens*, a small (2 to 3 mm diameter) plate like organism, it glides over food, secreting digestive enzymes, absorbing the products.

Coelenterates (the Only Animals with Radial Symmetry)

Phylum	Description
Cnidaria	Mostly marine, includes jellyfish, corals, anemones, and freshwater hydras. First phylum with specialized tissues (e.g., nerve cells).
Ctenophora	Common name *comb jellies*. Fewer than 100 species known. Only **biradial** phylum.

Flatworms, etc.: More Complex Organ Specialization

Phylum	Description
Platyhelminthes	Flatworms and flukes, turbellarians (e.g., *Planaria*), trematodes (e.g., *Schistosomona*), tapeworms.
Gnathostomulida	Jaw worms: found in sediments, where they tolerate low oxygen levels; feed on fungi and bacteria scraped from rocks.
Nemertea	Ribbon worms: simplest organisms with a circulatory system.

Vermiform (Worm-Shaped)

Phylum	Description
Rotifera	Rings of beating cilia draw food to the mouth, giving impression of rotating wheels.
Nematoda	Roundworms; includes organism that causes trichinosis.
Gastrotricha	These resemble rotifers, but without "wheel" of cilia. Together with rotifers
Kinorhynca	and nematodes, form a group called **aschelminthes**, and are common
Loricifera	aquatic organisms.
Nematomorpha	Horsehair worms, include 250 species that are arthropod parasites.
Acanthocephala	Spiny-headed worms; all are internal parasites that cause painful intestinal injury.
Entoprocta	Sessile organisms that superficially resemble hydra, but with ciliated tentacles.
Priapulida	15 species of cold-water marine worm.

Development of a "True" Body Cavity

Phylum	Description
Mollusca	Bivalves (e.g., clams), octopus and squid, snails and slugs.
Annelida	Segmented worms: polychaetes (marine worms), oligochaetes (e.g., earthworms, tubifex), leeches.
Arthropoda	Insects, crustaceans, spiders.

Lesser Protostomes: Minor Ecological Significance

Phylum	Description
Echiurida	
Sipunculida	
Tardigrada	Water bears, common in aquatic habitats.
Pentastomida	
Onychophora	
Pogonophora	Includes tube worms from deep-sea vent communities.

TABLE 8.1   (*Continued*)

Phylum	Description
	Animal Weeds: Sessile Wormlike Marine Animals
Phoronida	Only 10 known species.
Ectoprocta	Moss animals.
Brachiopoda	The "lamp shells", resemble bivalves.
	Deuterostomes
Chaetognatha	Arrowworms: wormlike planktonic predators, 2.5 to 10 cm long.
Hemichordata	Wormlike organisms.
Echinodermata	Sea stars, sea urchins, etc.
	Higher Animals
Chordata	Tunicates, lancelets, vertebrates.

that are genetically identical to the "parent." It has the advantage that it can produce new individuals rapidly. However, it reduces genetic variablity and the ability of a population to adjust to drastically changed environmental conditions.

Sexual reproduction is the formation of a new individual from the union of two potentially genetically distinct gametes. The larger, nonmotile gamete is called the **egg**, produced by meiotic division in female sex organs. In animals the smaller motile gamete is the **sperm**, which is produced by meiosis in male sex organs. The haploid gametes combine in the process of **fertilization** to form a diploid single cell called the **zygote**, which will develop into a new individual.

The most familiar form of sexual reproduction is **bisexual reproduction**, in which the male and female organs that produce the gametes are on separate individuals. Species that use this strategy are called **dioecious**. Almost all vertebrates, and many invertebrates, are dioecious. Mating requires time, energy, and coordination and results in offspring that do not share all the genes of either of its parents. Nevertheless, sexual reproduction confers on a species a great ability to adjust to environmental change.

Some animals reproduce by **parthenogenesis** (virgin origin). No fertilization occurs, although the diploid condition is restored by chromosome doubling. Some flatworms, rotifers, insects, and even fish reproduce parthenogenetically. Among the bees, a queen may allow only some of her eggs to be fertilized. The fertilized eggs develop into diploid female worker bees; the haploid males are drones. Certain strains of turkeys have been found capable of parthenogenetic reproduction. Even in mammals, such as mice, eggs will on rare occasions start to develop into embryos and fetuses without fertilization. However, they have never been found to survive to full term.

Some animals are **hermaphrodites**, having both male and female sex organs on the same individual. They are also referred to as **monoecious**. This is a reproduction strategy that is appropriate for organisms that are sessile, such as barnacles, or live in burrows or parasitically within other organisms, such as the tapeworm. The advantage of this is that it makes it possible for every individual to bear young, not just half the population. Some hermaphroditic fish actually start out as a member of one sex, then change to the other during their life span.

8.2 INVERTEBRATE PHYLA OTHER THAN ARTHROPODS

Coral (cnidaria) form colonies, building huge structures as their calcium carbonate skeletons add to others below. This accretion eventually builds into reefs and atolls. Coral reefs have bright colors due to symbiotic algae that live inside the coral cells. In recent years many reefs have been losing their algal symbionts, leading to the death of the reef. The cause of this **coral bleaching** is unknown, but it is feared that pollution and global warming are involved.

The three phyla in the group called flatworms show more organ specialization, including a primitive nervous system. Some have a gut with only a single opening rather than a separate mouth and anus. Others lack a gut and absorb food directly through their surface.

The phylum **Platyhelminthes** comprises the flatworms and flukes. There are four classes, of which we will describe three. The **turbellarians** include the small aquatic flatworm *Planaria* (Figure 8.1), found on the bottom of rocks in streams. The **trematodes** include the blood fluke *Schistosomona* (also called *Bilharzia*), which causes the important waterborne parasitic infection **schistosomiasis** (or **bilharziasis**). This disease is common in warm climates. The schistosomes are discharged from the intestines of infected people in their feces. If the fecal contamination reaches a stream, the fluke can infect a specific kind of snail, where it develops. Released back to the water, the schistosomes can reinfect humans through the bare skin. It then travels via the bloodstream to infection sites in the liver and digestive system. Symptoms in humans include extreme diarrhea and bloody stool or urine. It can be controlled by proper sanitation and by control of conditions favorable to the snail host.

Tapeworms are in another class, the **cestodes**. Tapeworms are long (up to 10 m), very flat worms that infect vertebrate digestive tracts. Humans can contract them from improperly cooked meats. When ingested, they attach inside the intestines and grow to great

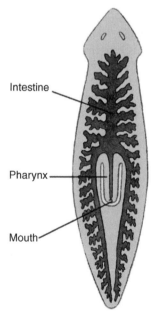

Intestine

Pharynx

Mouth

Figure 8.1 Planarian, showing internal structure. (From Hickman et al., 1997. © The McGraw-Hill Companies. Used with permission.)

lengths. They have no digestive system, but absorb food from the host gut through tiny projections similar to the villi of the intestines themselves. They shed reproductive organs from their end, which leave with the feces or crawl out the anus, spreading larvae for subsequent infections. Pork tapeworm can migrate to the eye or brain of humans, causing blindness or death. About 1% of American cattle are infected, and 20% of slaughtered cattle are not federally inspected. Furthermore, inspections miss 25% of infections. These facts point up the importance of cooking meat thoroughly, which kills tapeworms.

The nine phyla of the **vermiform** (worm-shaped) group have a complete mouth-to-anus digestive system. Phylum **rotifera** (the rotifers, Figure 8.2) have one or more rings of cilia at the head to draw food toward the mouth. When the cilia are beating, they give the impression of a rotating wheel. The foot has one to four toes that it uses for attachment. There are about 1800 species, all of which are dioecious. Most live in benthic aquatic habitats. They can withstand desiccation for long periods.

The phylum **nematoda** contains the nematodes, or roundworms. Nematodes are among the most abundant animals on Earth. There can be 3 billion per cubic meter of soil, where they help to recycle organic matter. They are common parasites in pets and other animals. *Ascaris lumbricoides* is a large roundworm that commonly infects humans. A female may lay 200,000 eggs a day, which pass in the feces. *Trichinella spiralis* causes the disease **trichinosis**. In the past this disease was contracted by eating poorly cooked pork, but it has been virtually eliminated in the United States. *T. spiralis* is actually an intracellular parasite that can control the gene expression of its host cell. Eight species of **filarial**

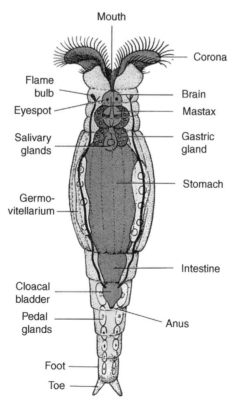

Figure 8.2   The rotifer *Philodina*, showing internal structure. (From Hickman et al., 1997. © The McGraw-Hill Companies. Used with permission.)

worms infect humans, including one that causes the disease called *river blindness*, which is carried by black flies and is common in hot climates.

8.3 MOLLUSKS, SEGMENTED WORMS, ARTHROPODS

This group is distinguished as having the simplest phyla with a "true coelum," a body cavity completely lined with mesoderm tissue.

8.3.1 Mollusks

The mollusks are soft-bodied animals whose unique features include a muscular foot for motility and a **radula** (except in bivalves), a rasplike tongue for scraping food off substrates. A skin covering called the **mantle** protects the body and, in most species, secretes the shell. Mollusks are the first phylum with a specialized respiratory system, consisting of lungs or gills. Most have an open circulatory system, in which blood flows not only in vessels but also outside the vessels, where it can bathe the tissues directly. There are about 50,000 known living mollusk species.

There are eight classes of mollusks, of which we discuss three. The largest and most diverse class is the **gastropods**, which includes snails, slugs, conches, and limpets. If there is a shell, it is a **univalve** (one piece). Gastropods show a wide range of feeding behaviors. Most are herbivores that rasp algae off surfaces. Others are scavengers or carnivores. The oyster borer *urosalpinx cinerea* uses its *radula* to drill through the shell of the oyster. Members of the genus *Conus* can eat fish, harpooning them with their radula at the tip of a proboscis and using a poison to stun the fish.

The class **bivalvia** contains all the mollusks with two-part shells, including mussels, clams, scallops, and oysters. Most bivalves are marine and most are filter feeders. Their gills are specialized for both oxygen transfer and feeding. Cilia are used to draw water through pores in the gills. The pores trap the particles and ensnare them in mucus secretions. The mucus is then transported to the mouth by ciliary action. This feeding behavior causes them to concentrate pathogens from wastewater pollution, making them a possible source of infection for humans who feed on them.

A third class of mollusks is composed the **cephalopods**. This class contains marine predators, including octopuses, squid, nautiluses, devilfish, and cuttlefish. The foot, as the name implies, has become part of the head and is modified for expelling water from the mantle to provide locomotion. The edge of the head forms a circle of arms or tentacles. Squids and cuttlefish have a small internal shell, and octopuses do not have shells at all. Cephalopods have the largest brains of any invertebrates. Octopuses can even be trained. Cephalopods have well-developed eyes. The eyes of the giant squid can be up to 25 cm in diameter, the largest in the animal kingdom.

8.3.2 Annelids

The **annelids** are the segmented worms, which include the common earthworm. Segmentation facilitates the formation of larger organisms and the motility of those organisms. Annelids have a completely closed circulatory system, which allows higher blood pressures and the consequent control over blood flow. They lack specialized respiratory organs. Their digestive system is more developed than that of other phyla we have discussed, containing two specialized areas: The **crop** stores food; the **gizzard** grinds it. Annelids are usually divided into three classes: polychaetes, oligochaetes, and hirudinea.

Polychaetes (many bristles) include 10,000 species, mostly of marine worms.They are usually 5 to 10 cm long, and many are brightly colored. They are distinguished from the other annelids by the presence of a pair of appendages on each segment. They are commonly found burrowed in the sediments of estuaries. They may achieve densities of thousands per square meter in mud flats. Others move about, and these are usually predatory.

Oligochaetes (few bristles) are the class that includes the common earthworm *Lumbricus terrestris*. Most are scavengers, feeding on decayed plant and animal matter. Earthworms are hermaphroditic but must copulate with another individual. Eggs and sperm are then deposited in a mucus and chitin pouch, or **cocoon**. Earthworms play an important role in the ecology of the soil. Their burrows improve soil drainage and aeration. They mix the soil by ingesting it at depth and depositing it at the surface, and they bring organic matter from the surface down into the burrows, where the nutrients released by their decay can become available to plant roots. Freshwater oligochaetes are smaller but more mobile than their terrestrial cousins. They can be an important source of food for fish. One is *Tubifex*, a small reddish worm that forms tubes in the sediment in which it lives head down, with tails waving in the water. Tubifex can form carpets on the bottoms of heavily polluted streams.

The class **hirudinea** are the leeches, most of which live by sucking blood from other organisms. They have specialized sucker organs to attach themselves to prey. Some aquatic leeches prey on fish, cattle, horses, and humans. Some terrestrial leeches prey on insect larvae, earthworms, and slugs, or may climb bushes and trees to reach birds.

8.3.3 Arthropods

Arthropods (jointed legs) include the spiders, centipedes, millipedes, insects, and crustaceans. They represent an evolutionary advance that allowed them to form the greatest diversity and number of species of all the phyla, about 1 million species known. The advance was the development of a jointed exoskeleton made of chitin, a nitrogenous polysaccharide, bound with protein. In crustaceans the exoskeleton also contains calcium salts for added strength. Other innovations with this phylum are increased specialization of the body segments; locomotion via muscles in external appendages; improved sensory capabilities; more efficient respiratory organs, which enables high activity rates and even flight; and even the development of social organization.

Because the rigid exoskeleton places limits on the organism's growth, it must shed its covering periodically in a process called **molting**. It then secretes a new cuticle, which hardens into a new shell. Each stage in the life of an arthropod between molts is called an **instar**.

Besides taking on virtually all the kinds of ecological roles there are, arthropods are essential to the survival of many other species, from the many flowering plants that depend on insects to pollinate them, to the many larger animals that depend on them for food. In aquatic ecosystems, for example, microscopic crustaceans form much of the **zooplankton**, which forms the first animal level of the food chain, and insect larvae are the food source for many fish species.

There are numerous arthropod classes (see Table 8.2). We focus on several of the most familiar and ecologically important. The diversity of this group is so great that we will

TABLE 8.2   Classification of Familiar Arthropods

Phylum	Subphylum	Class	Order
Arthropoda			
	Trilobites (extinct)		
	Chelicerata		
		Merostoma [includes horseshoe crabs (*Limulus*)]	
		Pycnogonida (sea spiders)	
		Arachnida	
			Araneae (spiders)
			Scorpionida (scorpions)
			Opiliones (harvestmen, or "daddy longlegs")
			Acari (ticks and mites)
	Uniramia (two classes not shown)		
		Chilopoda (centipedes)	
		Diplopoda (millipedes)	
		Insecta (the terrestrial mandibulates; 16 of 27 orders shown)	
			Anoplura (lice)
			Coleoptera (beetles, fireflies, weevils)
			Dermaptera (earwig)
			Diptera (true flies, mosquitoes)
			Ephemeroptera (mayfly)
			Isoptera (termites)
			Lepidoptera (moths, butterflies)
			Hemiptera (true bugs, bedbugs, water striders)
			Homoptera (aphids, cicadas)
			Hymenoptera (bees, wasps, ants)
			Neuroptera (lacewing, dobsonflies)
			Odonata (dragonfly)
			Orthoptera (roaches, grasshoppers, locusts, crickets, mantids)
			Plecoptera (stoneflies)
			Siphonaptera (fleas)
			Trichoptera (caddis flies)
	Crustaceans (the aquatic mandibulates; two minor classes not shown)		
		Branchiopods (several other classes not shown)	
			Anostraca (brine shrimp)
			Cladocera (water fleas, or *Daphnia*)
		Maxillopoda	
		Subclass Ostracoda	
		Subclass Copepoda (e.g., *Cyclops*)	
		Subclass Cirripedia (includes barnacles)	
		Malacostraca (14 orders, not all shown)	
			Isopoda (pill bugs or sow bugs)
			Amphipoda (scuds, sideswimmers)
			Euphausiacea (krill)
			Decapoda (crayfish, lobsters, crabs, "true" shrimp)

find numerous familiar species to be in surprising relationships. For example, horseshoe crabs are more closely related to spiders than to crustaceans, and the "daddy longlegs" is not really a spider. The "pill bugs" often found upon overturning a rock in the garden are not insects, but one of the few terrestrial crustaceans.

Arthropods share the segmentation of the annelids; however, the segments are fused into functional parts. For example, the insects have three such parts: the **head**, **thorax**, and **abdomen**. In spiders and crustaceans the head and thorax are further fused into a **cephalothorax**.

Arachnids include the spiders and several other orders. There are about 35,000 species of spiders. All are venomous predators. However, few are dangerous to humans, not even the tarantula, *Rhechostica hentzi*. Among the few that are dangerous are the black widow spider, *Latrodectus mactans*, and the brown recluse, *Loxosceles reclusa*. These can be recognized by a red "hourglass" shape on the bottom of the abdomen and a "violin-shaped" mark on the cephalothorax, respectively. Even most scorpions are not very dangerous, except, again, for certain species in Africa and Mexico. Arachnids have six appendages, but only four are used as legs. The other two are modified into poison fangs. The protinaceous web is spun from an organ on the abdomen, producing silk that is stronger than steel of the same thickness.

Ticks and mites are the most abundant arachids and the most important from an economic and medical viewpoint. They attach to plants and animals, puncturing their surface to suck out fluids or blood. Besides the direct harm this causes, they are the most important insect disease vector after mosquitoes. They spread Lyme disease and Rocky Mountain spotted fever in humans, as well as many cattle diseases.

Crustaceans and insects make up some 80% of all the named species of animals. Whereas insects dominate the land, crustaceans rule the waters. In fact, it may be that the most abundant animals in the world belong to the copepod genus *Calanus*, a zooplankter.

Crustaceans are distinguished from other arthropods in that they always have *two* pairs of antennas, and their chitin exoskeleton is hardened with calcium salts. The head and thorax is fused into a cephalothorax, as with spiders. Each of the typically 16 to 20 body segments has a pair of appendages, adapted for various uses, such as grasping, walking, and swimming. They have gills for respiration. An example is the lobster, a decapod (Figure 8.3).

Hold a glass of pond water up to the light and you will see "specks" swimming or darting about. Most likely, you will be seeing one of the planktonic crustaceans: cladocerans, ostracods, or copepods (see Figure 15.10). The **cladoceran** *Daphnia* spp. are also called water fleas. They feed on algae. *Daphnia*, along with the brine shrimp, are important in the laboratory for their use in toxicity testing. **Ostracods** resemble the cladocerans, except that they are enclosed by a two-piece carapace that makes them look like a very small clam. As mentioned above, **copepods** are extremely common. In the oceans the copepod *Calanus* and the euphasid krill form the major food source for herring, menhaden, sardines, and some whales and sharks. The copepod *Cyclops* is very common in freshwater systems.

Barnacles start their lives as planktonic forms resembling copepods. Eventually, most attach to a solid surface and form the familiar shell of calcareous plates that we see on rocks exposed by low tide. When covered with water, they open the slitlike apertures and wave their fanlike legs to filter food particles from the water.

The class **Malacostraca** is the most diverse. It contains the most familiar crustaceans of all, the decapods, which include crabs, lobsters, and shrimp. As the name implies,

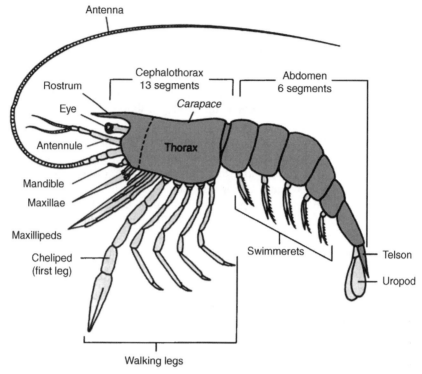

Figure 8.3   Typical body plan of the lobster, order Decapoda, class Malacostraca, subphylum crustacean. (From Hickman et al., 1997. © The McGraw-Hill Companies, Inc. Used with permission.)

decapods have five pairs of walking legs, the first of which may be modified into a claw. The **euphasids** have only about 90 species but form the important marine food source called *krill*. Most of them are bioluminescent. The **amphipods** have a laterally flattened form with immobile compound eyes and two types of legs. For example, one set of legs may be for swimming, the other for jumping. An example is *Orchestia*, the beach hopper. **Isopods** are the only crustacean group to include terrestrial species, such as the familiar pill bug. They have a flattened form and compound eyes. Although adapted for land, they are not efficient at conserving water and require moist conditions to survive. Some are parasites of fish or other crustaceans.

Around 1 million species of insects have been named, and this is thought to be only a fraction of the number that exist. There are more species of insects than of all other animals combined. They dominate the land and fresh water, although few marine insects exist. Their success is attributed to the arthropod characteristics, plus their small size and their development of flight. Table 8.2 shows some of the variety in the insect class. In all there are 27 orders. The word "bug" is popularly used to denote any insect, or sometimes any small organism. Environmental engineers often refer to the organisms populating biological wastewater treatment plants, including bacteria and protozoans, as bugs. To an **entomologist** (a scientist who studies insects), however, a "bug" is an insect of a single order only, the Hemiptera.

An **insect** is an arthropod with three pairs of legs and (usually) two pairs of wings on the thorax. Some insects have only one pair of wings or none at all. For example, the true flies (order Diptera) have only one pair, and female ants and termites have wings only at certain times. Lice and fleas have no wings. The head of an insect usually has two large compound eyes and one pair of antennas. Most insects eat plants, although parasitic insects are common. Some are predators on other insects or animals. Fleas feed on the blood of mammals. Many wasps lay their eggs in or on spiders, centipedes, or other insects, on which the larvae feed after hatching. Many predatory insects benefit humans by attacking crop pests. Many larvae and beetles eat dead animals.

To accommodate the high oxygen consumption rate necessary for flight, insects have evolved an efficient **tracheal system**, consisting of tubes to transport air directly to the tissues. Insects, along with arachnids, have evolved an efficient water-conserving excretion mechanism that eliminates nitrogen waste in the form of insoluble uric acid.

Insects are dioecious with internal fertilization. Most female insects lay a large number of eggs. The eggs must be laid in highly specific settings. For example, the monarch butterfly must lay its eggs on milkweed plants, which its caterpillar stage needs to eat. Some insects lay their eggs on water, and their juvenile stages are aquatic.

About 88% of insects go through **complete metamorphosis**, in which the juvenile form is radically different from the adult insect. The hatchling of an insect that undergoes this process is wormlike and is called a **larva**, also known as a caterpillar, maggot, grub, and so on, depending on the species. The larva grows through a series of molts and eventually surrounds itself with a cocoon, transforming into a transitional nonfeeding stage called a **pupa** or **chrysalis**. When it molts out of this stage, the result is an adult, a form that no longer molts. Other insects undergo **gradual metamorphosis**, in which the juvenile somewhat resembles the adult and becomes an adult by gradual change. The juvenile forms are called **nymphs**. Examples of species that have nymphs are grasshoppers, cicadas, mantids, and terrestrial bugs. Both larvae and nymphs pass through several instars before going on to the next stage.

Besides those that attack crops, many insects harm humans directly. Mosquitoes transmit malaria, encephalitis, yellow fever, and filariasis. Fleas transmit plague; the housefly, typhoid; and the louse, typhus fever. Our battle with insects has had signal success—only for us to have found that in some cases, the cure is worse than the disease. Unrestricted use of pesticides starting after World War II had become a crisis by the 1960s, giving rise to the modern environmental movement. Elimination of the more persistent pesticides and more careful use of others has greatly reduced environmental and health risks today. However, alternative approaches are being developed. Parasitic wasps are sometimes cultured to fight crop pests. Viral, bacterial, and fungal insect pathogens are being recruited. For example, the bacteria *Bacillus thuringiensis* has been found to be effective against lepidopterans (moths and caterpillars). A strategy called **integrated pest management** (IPM) combines plant culturing techniques such as crop rotation and selection of resistant varieties with minimum timed use of pesticides to control insects without the heavier pesticide dose that would otherwise be needed.

8.3.4 LESSER PROTOSTOMES

The **lesser protostomes** are a group of phyla whose species have mostly minor ecological significance today. Most of them are worms or wormlike. One phylum that may be familiar is the **tardigrades**, the water bears, which are common in aquatic habitats. Phylum **pogonophora** live mostly below 200 m in the ocean in tubes made of chitin. This phylum includes one of the most interesting of the recently discovered animals: the tube worms of the deep-sea vent community (see Section 15.4.3).

8.4   DEUTEROSTOMES (STARFISH, VERTEBRATES, ETC.)

The deuterostomes constitute a group of phyla distinguished by the apparently arcane distinction that the gut forms from the blastula, starting at the anus rather than the mouth, as in protostomes. There are seven phyla in this group. The six that are exclusively invertebrate can be clustered into a group called the *lesser deuterostomes*. Only one of these, the echinoderms, includes familiar organisms, the sea stars and the anemones.

8.4.1   Echinoderms

The phylum **echinodermata** includes sea stars, sea urchins, brittle stars, sea lilies, and sea cucumbers. They are among the most fascinating of marine organisms for the beachcomber. The name means "spiny skinned," a reference to the spikes, bumps, or other projections all echinoderms have to protect them from predators. Curiously, although they are among the higher phyla, they have primitive features, such as radial or biradial symmetry. This characteristic is usually associated with sessile animals, but echinoderms are motile. Their larvae are, however, bilateral. They also lack a brain and excretory and respiratory systems. However, they do have one evolutionary advance: an **endoskeleton** consisting of calcareous plates. Echinoderms cannot osmoregulate and thus are not found in brackish waters, where the salinity may vary with tidal changes. Movement, excretion, and respiration are performed by means of a unique **water-vascular system**, which uses canals connected to the environment by pores. By pressurizing parts of the canal system, the echinoderm can slowly move part of its body.

Sea stars, or starfish, are mostly predators feeding on other invertebrates and small fish. Strong suckers on the legs enable movement and gripping prey. Some can force open a clamshell, and then discharge their stomach through their mouth to digest the clam. Sea urchins have their mouths on the bottom, where they graze algae and detritus from the substrate.

8.4.2   Chordates, Including the Vertebrates

Finally, we come to the phylum that includes ourselves. The chordates incorporated evolutionary innovations that made it possible for them to grow to great size, forming the largest animals on land (dinosaurs and elephants) and in the water (whales). The four unique characteristics of the phylum **Chordata** are:

1. They have the presence of a **notocord** at some point in their development. The notocord is a flexible skeletal rod that runs the length of the organism. It remains in adult lampreys and hagfish, but in other chordates it is mostly replaced by vertebrae.
2. They have a single, hollow tubular **nerve cord**, located dorsal to the notocord, usually enlarged at the anterior end to form a brain.
3. They have paired openings in the anterior digestive tract called **gill slits** at some point in their development. In humans these are present in the embryonic stage.

One pair remains after this stage and forms the **Eustachian tubes**, which connect the throat to the middle ear. It is the Eustachian tube that allows you to equalize the pressure in your ear by swallowing.

4. The notocord, nerve cord, and segmented muscles continue into a **tail** that extends beyond the anus. In humans the tail is present only briefly in the embryo.

Other important, although less unique characteristics are: They have **segmented muscles** in an otherwise unsegmented body. In humans these muscles lie along the spinal column. Chordates have a closed blood system with a ventrally located heart, and they have a complete digestive system.

Not all chordates are easily recognizable as being closer relatives to us than, say, echinoderms. Two subphyla within the Chordates are marine invertebrates (Table 8.3). The subphylum **Urochordata** constitutes the *tunicates*. These include about 2000 species of filter feeders. Many, called *sea squirts*, are sessile. One group, called the *salps*, form meters-long colonies of transparent lemon-shaped bodies. Salps are brightly bioluminescent. The phylum **Cephalochordata** includes the 25 or 30 species of lancelets. Also called *amphioxus*, they resemble a small, translucent fish without fins or mouth or eyes to mark the head. They are common in sandy coastal waters around the world. Four species are found off the coast of North America.

If you ask someone to name any animal that comes to mind, they will probably think of a member of the third chordate subphylum: the **vertebrates**. The unique characteristic of the subphylum **Vertebrata** (also called **Craniata**) is that they possess a cartilaginous or boney endoskeleton with a **cranium**, or brain case. Bone is a living tissue, not just calcareous deposits. As the name implies, almost all vertebrates have a spine made of a series of bones called **vertebrae**. The segmented muscles attach to the vertebrae. The only exception is the cartilaginous fishes (such as sharks). The subphylum Vertebrata includes eight classes that are placed in five groups, each of which contributes new adaptations:

1. The **fishes** comprise four classes with about 30,000 species. They are the oldest vertebrates and represent the development of the skull, bones, jaws, fins, and vertebrae.

2. The class of **amphibians** has about 3000 species. They innovated with legs, fully functioning lungs, a chambered heart, and separate circulation systems for the lungs and the rest of the body.

3. The class of **reptiles** has 6500 species. Their innovations include dry, scaly skin, an expandable rib cage, and the transfer of legs to beneath the body.

4. The **birds** (class **Aves**) include 9600 species. They represent the development of warm-bloodedness, although many now believe this first occurred among the dinosaurs. They also uniquely have feathers, air sacs, and hard-shelled eggs.

5. The class of **mammals** has about 5000 members. They independently developed warm blood, plus their unique characteristics of body hair, a placenta for nurturing embryos (in most orders), and mammary glands with milk production for nurturing juveniles.

Hagfish and lampreys are in two classes that together make a superclass called the **jawless fish**. Hagfish are blind scavengers that find prey by smell and attach themselves by the mouth. They are the only vertebrates that keep their osmotic equilibrium with the

TABLE 8.3 Classification of Living Members of Phylum Chordates

Subphylum	Class	Order	Family	Genus	Species

Urochordata (tunicates; sea squirts and salps)
Cephalochordata (lancelets)
Vertebrata

    Myxini (hagfish (43 spp.)
    Cephalaspidomorphi (jawless fish, e.g., lamprey)
    Chondrichthyes (cartilagenous fish: sharks and rays, 850 spp.)
    Bony fish (45 living orders, about 30,000 spp.)
    Amphibia (3900 spp.)
        Caudata (salamanders, 360 spp.)
        Anura (frogs and toads, 21 families, 3450 spp.)
    Reptilia (13 orders, including the extinct dinosaurs)
        Testudines (turtles)
        Squamata (lizards and snakes)
        Crocodilia (crocodiles and alligators)
    Aves (birds) (27 living orders)
    Mammalia
        Nonplacental mammals:
        Monotremes (only 3 spp., including duck-billed platypus);
        Marsupials (kangaroos, opossums, koalas, 260 spp.)
        Placental mammals:
        Insectivora (shrews, hedgehogs, moles, 390 spp.)
        Macroscelidea (elephant shrews, 15 spp.)
        Dermoptera (flying lemurs, 2 spp.)
        Chiroptera (bats, 986 spp.)
        Scandentia (tree shrews, 16 spp.)
        Xenarthra (anteaters, armadillos, sloths, 30 spp.)
        Pholodota (pangolins, 7 spp.)
        Lagomorpha (rabbits, hares, pikas, 69 spp.)
        Rodentia (gnawing mammals: squirrels, rats, 1814 spp.)
        Cetacea (whales, dolphins, porpoises, 79 spp.)
        Carnivora (dogs, wolves, cats, bears, weasels, 240 spp.)
        Pinnipedia (sea lions, seals, walruses, 34 spp.)
        Tubulidentata (aardvark, 1 spp.)
        Proboscidea (elephants, 2 spp.)
        Hyracoidea (coneys, 7 spp.)
        Sirenia (sea cows and manatees, 4 spp.)
        Perissodactyla (odd-toed hoofed mammals:
          horses, asses, zebras, tapirs, rhinoceroses, 17 spp.)
        Artiodactyla (even-toed hoofed mammals: swine, camels, deer,
          hippopotamuses, antelopes, cattle, sheep, goats, 211 spp.)
        (Perissodactyla and Artiodactyla are called ungulates)
        Primates (233 spp.):
          prosimians (lemurs, tarsiers), monkeys, apes, humans
              Great Apes
                  Homo
                    Homo sapiens
                    Homo habilis (extinct)
                 Pan (chimpanzees)
                 Gorilla
              Lesser Apes

surrounding seawater instead of controlling their internal condition. The lamprey is a destructive eel-like parasite. It attaches itself to live fish by a suckerlike mouth and uses sharp teeth to rasp at the flesh of the victim. Accidental introduction of the sea lamprey *Petromyzon marinus* to the Great Lakes in the nineteenth century led to devastation of one commercial species of fish after another. Larvicides are applied to streams where they breed. This, combined with fish stocking programs, are producing a recovery of the fisheries.

The **cartilaginous fish**, which include sharks and rays, surprisingly evolved from boney ancestors. They track prey by smell and use an interesting **lateral line system** of sensors along the side that detect vibrations and bioelectrical fields produced by all animals. The electric ray can produce a low-voltage but high-current field from large organs alongside its head. The power output can be several kilowatts and can stun prey or repel predators.

The **boney fish** include most of the familiar fresh- and saltwater fish. They have developed ray fins and a **swim bladder**, a gas-filled chamber below the spine that the fish uses to control buoyancy. The importance of this can be seen by considering the sharks, which lack a swim bladder and therefore must swim continuously to keep from sinking to the bottom of the sea.

Among amphibians, the life cycle of frogs comes close to proving a famous biological principle proposed in the nineteenth century which states that **ontogeny** (the development of an individual from embryo to adult) recapitulates (repeats) **phylogeny** (the evolutionary development of a species). This principle, termed **recapitulation**, is invoked to explain embryonic features that disappear in the adult, such as gill slits in human embryos. As new features evolve, they are added to the embryonic development. But the principle turns out to have too many exceptions to be generally valid. Evolution sometimes deletes characteristics, as well as adding them, so the ontogeny is not an accurate record of the organism'sevolution. What we observe in frogs is the development of an aquatic species into a terrestrial species. For example, all frogs hatch from the egg resembling fish, having gills, a tail, and no limbs. This early stage is called the **tadpole**. Gradually, the tadpoles sprout limbs, grow lungs and eyelids, and the tail shortens, until the adult frog is finished.

The most abundant frog genus is probably *Rana*. All frogs, and most adult amphibians, are carnivores. They feed on anything that moves and that is small enough to swallow whole. Some amphibians, such as some salamanders, remain aquatic; others do not have an aquatic stage. However, all have thin water-permeable skins that require moist conditions. This may make them among the terrestrial animals that are most vulnerable to environmental problems. There is considerable anecdotal evidence today that frog populations in many areas are in serious decline. The cause is a mystery, but some of the factors that have been blamed include habitat destruction, acid rain, pesticides, and damage from increased ultraviolet radiation due to destruction of stratospheric ozone.

Reptiles show numerous evolutionary developments. Many of these developments freed reptiles from dependence on moist conditions. Their dry skin limits evaporative loss. Internal fertilization and shelled, waterproof eggs eliminates the need for water for reproduction. In turtles and crocodilians the temperature of incubation of the eggs determines the sex ratio. More efficient respiration allows greater size, up to the 115-kg Komodo dragon. Reptile jaws are capable of applying crushing force. They have higher blood pressure and more efficient circulation. Their lungs are more efficient. Whereas amphibians force air into their lungs with mouth muscles, reptiles developed the ability to suck air in by expanding the rib cage. They excrete nitrogen as insoluble uric acid, further conserving water. The nervous system of reptiles is much more complex than that of amphibians. Crocodilians developed a true cerebral cortex. Placement of the legs below the body instead of at the sides allows better support and motion.

Birds (class Aves) have many distinguishing characteristics other than the obvious ones of having feathers and forelimbs developed into wings. Their necks are disproportionately long. Their skeleton contains air cavities, making them strong, yet light. Their beaks lack teeth. They have a four-chambered heart and are warm-blooded. They excrete nitrogenous wastes as uric acid. Fertilization is internal and they produce eggs with a large amount of yolk. Some birds feed on insects; other invertebrates, such as worms, mollusks, and crustaceans; and vertebrates. About one-fifth feed on nectar. Many eat seeds. The beak is specialized for the type of feeding behavior. Bird behavior is highly developed, including complex activities such as nest building and social activity. The farther from the equator one gets, the higher the percentage of birds that migrate for the winter. The Arctic tern breeds above the Arctic Circle and then migrates to the Antarctic region! Day length seems to be the factor that stimulates migration. Birds can navigate using the sun and stars as cues. Amazingly, it has been proven that they can also detect Earth's magnetic field.

Birds are very sensitive to the use of pesticides in the environment. Some pesticides tend to biomagnify, increasing in concentration as they are passed up the food chain. This has had the greatest impact on birds of prey, such as eagles and osprey. Other contaminants that affect birds include lead shot discharged by hunters in wetlands. Waterfowl ingest them, resulting in lead poisoning. The loss of wintering lands in tropical regions may be a cause of the reduced songbird populations that have occurred in the last 40 years.

The classification of birds and reptiles may soon change. This is because the common ancestor of all reptiles is also the ancestor of the birds. Thus, they should be classified together, or perhaps distinguished as feathered or featherless reptiles. Birds developed from certain dinosaurs. Their closest relatives today are the crocodilians. For the present, however, the traditional classification is retained.

The main distinguishing characteristic of mammals is the presence of hair or fur and the production of milk. Hair or fur provides insulation, giving greater advantage to the warm-blooded metabolism. It has other functions as well, such as camouflage, waterproofing, and sensing (as by whiskers). Hair is produced from dead cells that leave a fibrous protein called **keratin**, the same substance that forms nails, claws, hooves, and feathers. Mammalian skin contains a variety of glands, including sweat, scent, and mammary glands. The latter produce the milk to nurture the young. Herbivorous animals harbor anaerobic bacteria and protozoans in a stomach chamber to ferment cellulose into sugars, fatty acids, and starches. No vertebrates are able to use cellulose on their own. Carnivores feed mostly on herbivores. The need for hunting behavior has selected for intelligence in these animals. The herbivores, on the other hand, have developed keen senses as protection against predators.

Most mammals have mating seasons, timed to produce young at a favorable season for rearing. Mating is limited by the female, who is receptive to mating only during a brief period in the mating season known as **estrus**. Old World monkeys and humans have a different cycle, the menstrual cycle. Mammals exhibit three ways of giving birth. Monotremes, such as the duck-billed platypus, is a mammal that lays eggs. Animals that lay eggs are called **oviparous**. The platypus still nurses its young with milk, however. Marsupials are **viviparous**; that is, they give live birth. However, the newborns are essentially still embryos. For example, the red kangaroo spends about a month in the uterus. When it is born, it climbs by itself through the mother's fur to a pouch, where it attaches itself to a teat to suckle for 235 days until it is capable of living independently.

The third way to give birth is that of the **placental mammals**, which comprise 94% of all mammals. Gestation in utero is prolonged. The embryo is nourished by a **placenta**, a membrane structure produced by and surrounding the embryo. The placenta grows thousands of tiny fingerlike projections called **villi** into the lining of the mother's uterus to absorb nutrients and oxygen from the maternal blood supply without there being an actual exchange of maternal and fetal blood. The fetus is connected to the placenta by the **umbilical cord**. Once born, the mammal may be more or less dependent on parents for a time.

Humans are members of the primate order. The first primate fossils date back about 40 million years. About 8 million years ago, forests in eastern Africa gave way to savannas, giving an advantage to apes that could stand upright. But it was 4 million years ago that the first fossils that could be called **hominids** (humanlike) appeared. This species has been named *Australopithecus afarensis*. One of the best such fossils, discovered in 1974, is of a female that has been named "Lucy." It had a brain the size of a chimpanzee's, about 450 mL. The first species of our genus was *Homo habilis*, which lived from 2 million years ago until about 1.5 million years ago. This species used stone and bone tools and had a brain size of about 640 mL. This was followed by *Homo erectus*, which had a brain size about 1000 mL and lived in social groups. *Homo erectus* disappeared about 300,000 years ago and was replaced by our species, *Homo sapiens*. About 130,000 years ago a subspecies of *Homo sapiens* appeared, called the **Neanderthals**. They had a brain size within the range of modern humans, about 1100 to 1700 mL.

About 30,000 years ago Neanderthal man was replaced by a new subspecies, *Homo sapiens sapiens*—us! These ancestors of ours were distinct from the Neanderthals in a number of ways. Although Neanderthals made stone blades, those of *sapiens* were of much higher quality. *Sapiens* buried their dead with ritual objects, suggesting a belief in an afterlife. Finally, their vocal apparatus and related portions of the brain show increased development. This suggests that early humans may have been capable of language. Language capability confers the ability to do abstract reasoning. Thus, language is the basis of intellect, the one thing that most decisively distinguishes us from the other animals. It is interesting to realize that modern humans coexisted with a distinct subspecies not long before the beginning of recorded history. Many suspect that Neanderthal man was wiped out in conflicts with our ancestors. Others think that the two groups merged by interbreeding.

Some interesting information on our more recent origins has been obtained from studies of mitochondrial DNA. Although we inherit the DNA in our cells' nuclei from both parents, there is also DNA in our mitochondria that we inherit from our mother alone. Mitochondrial DNA can be used to trace genetic heritage because it is not confounded by recombination and the mixing due to sexual reproduction. The diversity in mitochondrial DNA from different ethnic populations has been compared with the divergence that would have occurred due to the rate of mutation from natural effects. From this it has been possible to establish that if the assumptions used in the analysis are reasonable, all humans in the world today are descendents of a single individual that lived in Africa 200,000 years ago.

PROBLEMS

8.1. Which phyla consist of organisms with no specialized excretory organs and therefore would be most susceptible to toxic water pollutants? What other mechanisms would these organisms have for excreting toxins?

8.2. The most common type of organism on land is virtually absent from the sea, and vice versa. What are these two groups, and what do they have in common?

8.3. What class of vertebrates, other than the fish, are most susceptible to water pollution, and why?

REFERENCE

Hickman, C. P., Jr., L. S. Roberts, and Allan Larson, 1997. *Integrated Principles of Zoology*, Wm. C. Brown, Dubuque, IA.

9

THE HUMAN ANIMAL

Now we begin a detailed examination of the physiology and anatomy of the human animal. *Homo sapiens* has adapted itself for life under the widest conditions of any animal species, from the equator to the poles, from the bottom of the ocean to the top of the atmosphere and beyond. The phenomenal success of this animal has not been without negative effects on other organisms. Its activities have resulted in extinction of other species and loss of habitat for many more. It may be the only species that by itself is responsible for global changes in the environment. It produces chemicals that deplete stratospheric ozone, with effects that are yet to be well understood. Its activities produce CO_2 fast enough to result in a doubling of the atmospheric concentration by the mid-twenty-first century, and which is expected to alter Earth's climate. *Homo sapiens* has turned a good portion of the world's biological productivity toward its own use, but it is at risk of depleting its own resources. It has the greatest ability to plan of any species by far, including the capability of controlling its own resource consumption to prevent the risk of depletion. However, it remains to be seen whether it will actually do so to prevent catastrophe in the future.

We will study the biological functioning of the human body, stopping short of human behavior, which is beyond the scope of this book. We focus in turn on each of the 11 major organ systems. An understanding of the structure and function of the human body will help in understanding the effect of toxins.

The cells in the body are differentiated into about 200 different kinds. These are classed into four basic types, which also correspond to the four basic types of tissues:

1. **Muscle** cells and tissues generate mechanical force and movement. They are derived from the mesoderm. There are only three types of muscle tissue: smooth, skeletal, and cardiac.

Environmental Biology for Engineers and Scientists, by David A. Vaccari, Peter F. Strom, and James E. Alleman
Copyright © 2006 John Wiley & Sons, Inc.

2. **Nerve** cells and tissues produce and conduct electrical signals over large distances within the organism and at great speed. They can stimulate response from other cells, such as in muscles or glands. They are derived from the ectoderm.

3. **Epithelial** cells and tissues, or **epithelium**, form a lining over the organs, line body cavities such as the inside of the intestines or the lungs, and cover the organism as a whole, as the outer part of the skin. Thus, almost all absorption of pollutants must first pass through the epithelium. In keeping with this role, its cells are specialized to be selective in their absorption and secretion of ions and organic compounds, enabling them to act as a barrier to the passage of chemicals into and within the body. Epithelium is formed from ectoderm, mesoderm, and endoderm. Epithelial tissue never has its own blood supply; instead, it receives nutrients by diffusion from the underlying connective tissue. Cells that secrete substances such as hormones or digestive enzymes are epithelial cells called **gland cells**.

4. **Connective** cells and tissues connect and support parts of the body. Examples include bones, cartilage, fat storage cells, blood cells, and lymph fluid. Connective tissue includes a number of cells that respond to tissue injury. These include a variety of **leukocytes**, or white blood cells, which protect against infection, and **mast cells**, which release the compounds *histamine* and *heparin*, which are responsible for the inflammation response. Connective tissue underlies most epithelial layers. Connective cells are often embedded in a large amount of extracellular material. For example, bone cells are surrounded by a crystalline matrix. Tendons and ligaments consist almost completely of **collagen**, a tough, ropelike fibrous protein that resists stretching. Some ligaments have a stretchy fiber called **elastic fiber**, which is made up of the protein **elastin**.

The organs themselves are made of combinations of different kinds of the four basic types of tissues. In a typical example, blood vessels have an inner lining of epithelium, surrounded by a layer of connective tissue, and then by another layer of muscle tissue. The muscle layer controls the movement of blood by constricting or dilating. For example, when the body is cold, the muscle layer surrounding the blood vessels near the skin constricts, reducing the transport of heat to the skin surface. Blushing occurs when the opposite happens: The capillary muscles of facial skin **dilate** (open wider), bringing more blood to the surface. **Adipose** tissues are the fatty deposits below the skin. They can be an important storage site for toxins.

9.1   SKIN

The **integumentary system** consists of the skin and associated structures, such as hair, nails, and glands. The skin is one of the body's main protections against toxins in the environment. Looking at this from a different point of view, the skin is also an important route of exposure for many toxins. The skin consists of two main layers, the dermis and the epidermis (Figure 9.1). The **dermis**, the inner layer, is made up primarily of connective tissue, plus sweat glands, hair follicles, nerve endings, blood vessels, and the small muscles that make your hair "stand up on end." The dermis contains a large amount of intracellular collagen fibers and elastic fibers. The elastic fibers give the skin the ability to stretch. The collagen fibers limit the stretching and give skin strength.

The **epidermis**, the outer layer, is about 1 mm thick. The epidermis does not have its own blood supply; it is supplied from capillaries in the dermis. The innermost part of the epidermis is the **basal cell layer**, a single layer of cells that replicate to replace lost

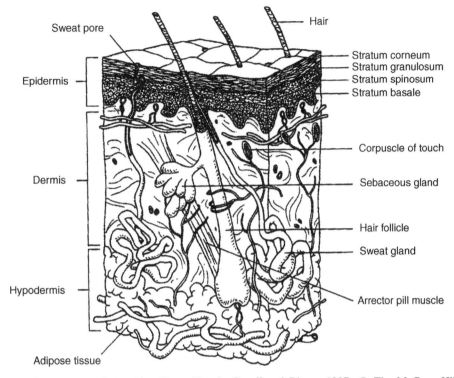

Sweat pore

Hair

Stratum corneum
Stratum granulosum
Stratum spinosum
Stratum basale

Epidermis

Corpuscle of touch

Dermis

Sebaceous gland

Hair follicle

Sweat gland

Hypodermis

Arrector pill muscle

Adipose tissue

Figure 9.1 Layers of the skin. (From Van de Graaff and Rhees, 1997. © The McGraw-Hill Companies, Inc. Used with permission.)

epidermal cells. These new cells push up as other new cells are produced below. Once they are several cells away from the basal cell layer, they start producing large amounts of **keratin**, a tough fibrous protein, and the cells gradually die off and dehydrate. This process produces the outermost layer of the epidermis, called the **stratum corneum**. This layer consists of dead epidermis cells filled with keratin and sandwiched between phospholipid membranes. These cells are sloughed off continuously and regenerated by the epidermis. It takes about 15 to 30 days for a cell to go from the basal layer to the stratum corneum, and another 14 before it is shed.

The stratum corneum is resistant to water and other substances, although pores in the skin provide a route for substances such as drugs or toxins to enter. The solvent *dimethyl sulfoxide* (DMSO) easily crosses the skin and enters the circulation. Substances that otherwise do not easily cross the skin may do so if dissolved in DMSO. **Melanin** is the dark pigment in the skin of black people and of Caucasians with substantial exposure to sunlight (a tan). It is produced by cells in the basal layer called **melanocytes**. The melanin protects skin by absorbing ultraviolet light from the sun. Ultraviolet radiation from sunlight is a common cause of skin cancer.

9.2 SKELETAL SYSTEM

The **skeletal system** consists of the bones, cartilage, and joints. Its functions are to provide support, protection, movement, mineral storage, and the production of red and white blood cells. Of environmental interest, the bones may be a significant storage site for toxic divalent heavy metals such as lead, radium, or strontium.

Bone is a living tissue, although most of its mass consists of calcium salt crystals and collagen fibers. Almost two-thirds of the mass of the bone is *hydroxyapatite*, $Ca_{10}(PO_4)_6(OH)_2$, plus sodium, magnesium, and fluoride salts. About one-third of bone mass is made up of collagen fibers. Living cells make up only 2% of the bone's mass. The cells include blood vessels and bone cells, called **osteocytes**. The combination of crystalline hardness and collagen toughness give bone structural properties comparable to high-quality steel-reinforced concrete. Bone structure often consists of a solid **compact bone** exterior, with a strong, yet light **spongy bone** layer inside. Many bones have an internal cavity containing **marrow**, which is the bone tissue that produces red and white blood cells and that stores fat reserves.

Bones are dynamic tissues. Some osteocytes, called **osteoblasts**, produce new bone; others, called **osteoclasts**, dissolve bone material. For example, when a long bone grows in diameter, the osteoblasts produce bone on the outside while osteoclasts dissolve bone on the intererior to enlarge the marrow cavity. Physical stress, such as from exercise, causes bones to grow at the point of attachment of the muscle tendons. (Exception: In children, excessive strenuous activity can damage sites of bone lengthening and result in shortened stature.) Baseball pitchers and tennis players have 35 to 50% more bone in their playing arm than in the other arm. Inactivity, on the other hand, can cause a loss of bone mass. Weightlessness causes astronauts to lose significant bone mass even if they exercise regularly. Older women are vulnerable to bone loss due to hormonal changes, a condition called **osteoporosis**. However, this problem can be at least partially forestalled by physical activity and/or hormonal therapy.

Humans have about 206 bones. The exact number varies because a variable number fuse together during development. The head contains 29 bones: 22 facial bones, 8 cranial bones, 6 bones in the ears, plus the *hyoid bone*. The latter is the only bone in the body not connected to the skeleton. It is located just under and behind the tongue, and is involved in swallowing. The spine consists of 26 bones: 24 are vertebrae, divided into the cervical, thoracic, and lumbar regions. At the bottom of the spine are the *sacrum*, which connects to the pelvis and is formed from five fused bones, and the *coccyx*, a vestigial tail formed from three to five fused bones. The rib cage plus the sternum makes another 25 bones. The upper and lower extremities plus the pelvic and pectoral girdles makes up another 126.

Cartilage is a clear, glassy material composed of cartilage cells surrounded by a mesh of collagen fibers. It lines the moving surfaces of the skeletal joints. Cartilage has few blood vessels, a reason that injury to it is slow to heal. **Tendons** are bands of dense tissues containing parallel fibers of collagen that connect muscles to bones. **Ligaments** are like tendons, but form connections from one bone to another. Some ligaments, such as those in the spine, contain more elastic fibers than collagen, making them springy.

9.3 MUSCULAR SYSTEM

Muscles are organs containing cells, called **muscle fibers**, which are specialized to contract. There are three types of muscle tissue. **Cardiac muscle** is present only in the heart. **Smooth muscle** is responsible for involuntary actions, such as the contractions that move food through the digestive system. It is also the type of muscle that surrounds the blood vessels, controlling their contraction and dilation. Both of these types of muscle make up about 3% of the body. However, the great majority of the body's muscle is of the third type: **skeletal muscle**, which makes up about 40% of the body's weight. We focus on this type of muscle.

Each skeletal muscle is made up of **muscle fibers**, each one a single cell that can extend the length of the muscle (up to 20 cm) and 10 to 100 μm in diameter. Each of these giant cells has hundreds of nuclei. Within each muscle fiber are many parallel bundles called **myofibrils**, which run the length of the cell. Finally, within each myofibril are numerous **thick filaments** composed of bundles of the protein **myosin**, and **thin filaments** made of **actin** and other proteins. Muscle contraction is caused by the **sliding filament mechanism**, in which the myosin filaments pull on the actin filaments like a person pulling on a rope.

The ATP for muscle contractions can come from four sources: aerobic oxidation, or glycolysis with fermentation to lactic acid, creatine phosphorylation, and free ATP. Under resting conditions, muscle cells produce ATP from the oxidation of fatty acids via aerobic respiration. Because it requires oxygen, this mechanism can provide energy only as fast as the required amount of oxygen can be provided to the muscle by the bloodstream. Although this is most efficient, aerobic respiration does not provide ATP fast enough for continuous contractions such as those during athletic performances. It can, however, support moderate continuous activity such as hiking or, literally, "aerobic" exercise.

Recall from Section 5.4.1 that glycolysis, which occurs in the cytoplasm, can produce a few ATPs from a partial oxidation of glucose to pyruvate. A fermentation step then converts the pyruvate to lactic acid. The glucose itself is stored in the muscle in the form of glycogen. Glycolysis provides enough energy for vigorous activity lasting 1 to 3 minutes, such as an 800-m run. It has the advantage that it does not require oxygen. However, the lactic acid accumulates in the muscle fiber, reducing the pH and disrupting the cell's activity. This is the "burn" felt after heavy exercise that forces us to rest to recover from the exertion. Muscle cells cannot oxidize the lactic acid. They must release it to the bloodstream, where it will reach the liver. Only the liver can eliminate lactic acid. The liver will gradually oxidize about 30% via respiration and use the resulting ATP to convert the rest of the lactic acid back to glucose. Thus, the rate of oxygen utilization is elevated for a time after the exertion stops, to repay the accumulated **oxygen debt**.

For very brief, explosive activity, such as a single leap or a throw, the muscle cell has two immediate sources of ATP. One is the small amount of free ATP always present in cells. This can support about 2 seconds of contractions. The second immediate source is a high-energy compound called *creatine phosphate* (CP). The muscle fibers store about five times as much CP as they do ATP.

An analogy for these sources of energy is in sources of money. Free ATP is like change in the pocket. It's ready in an instant, but it doesn't last long. If you take a moment more, you can fish for paper money in your wallet or purse. This is like the creatine phosphate. Glycolysis is like using a credit card. It lasts longer but produces uncomfortable side effects (financial debt, in this case), which must be paid back. Finally, the reactions of respiration are like getting a job to obtain money. It comes in slow and steady, although not fast enough to sustain a marathon shopping spree. To extend the analogy, fats can be compared to ownership of stocks. Although they can be a significant reserve, they must be converted to cash (liquidated, in the financial parlance) in order to be useful.

The muscles contain sensors that provide information to the central nervous system about the state of contraction of the muscles. This gives us a sense of body position and helps provide **muscle tone**, which is the contraction of some muscle fibers even when resting. Muscle tone helps keeps bones and joints in position and protects against sudden shocks.

9.4   NERVOUS SYSTEM

The **nervous system** is one of the two main control mechanisms of the body, the other being the endocrine system. Nervous tissue is among the most sensitive in the body to toxins. Because nervous tissue has a very high metabolic rate, its role is so critical that even small amounts of damage can have significant effects, and some toxins (i.e., the insecticides) are actually designed to be neurotoxic in order to be effective against insects at low dosages.

The primary functional components of the nervous system are the nerve cells, or **neurons**. They can have many shapes; Figure 9.2 shows a common type. The main parts are the cell body, or soma, the highly branched dendrites, and the axon. The **soma** contains the nucleus and cytoplasm, with its typical cell machinery, such as mitochondria and ribosomes. The neuron lacks a centriole, which makes it impossible for it to divide. Thus, neurons cannot regenerate once damaged. This also means that neurons cannot become cancerous. Brain cancer in adults occurs in other nervous system cells, called **glial cells**. The **dendrites** provide most of the sites for the reception of nerve signals. The **axon** is a long extension of the cell that serves to transmit the nerve signal over a great distance. It may have branches, and it is often sheathed in a lipid coating called **myelin** that serves as an insulator for the signal transmission. Neurons transmit signals from one cell to another at a specialized point of contact called a **synapse**. A **nerve** is a bundle of neuron axons.

9.4.1   Nerve Signal Transmission

Nerve signals are transmitted along the axon as a wave of electrochemical charges caused by movement of potassium and sodium ions across its membrane. A membrane would be

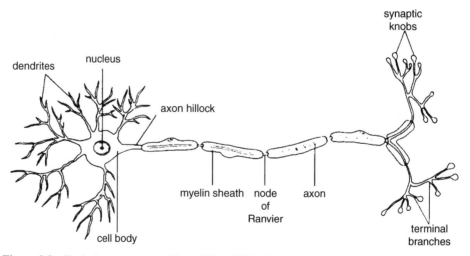

Figure 9.2   Typical motor neuron. (From Fried, 1990. © The McGraw-Hill Companies, Inc. Used with permission.)

fairly impermeable to ions except for the presence of special pores called **membrane channels** that provide a path for the ions. There are special membrane channels for potassium and others for sodium. One type of membrane channel allows a continuous but slow leak of ions across the membrane. Another type opens and closes in response to voltage across the membrane. In addition, there is a **sodium–potassium exchange pump** that uses ATP to exchange three intracellular sodium ions with two extracellular potassium ions. In other words, it pumps two potassium ions into the cell for every three sodium ions pumped out. When the nerve cell is resting, the exchange pump maintains an electrostatic potential, or voltage, of -70 mV across the membrane. That is, the inside of the cell is negatively charged, or has a deficiency of positive anions. This voltage is called the **resting potential** of the cell. The pump also ensures that under resting conditions, the inside of the cell has far more potassium than sodium, and the reverse is true outside the cell.

Nerve signal propogation along the axon begins when a sodium channel is opened at some point, such as by a chemical signal from a synapse. This causes the voltage to increase toward zero, a process called **depolarization**. Depolarization triggers a sequence of sodium and potassium pore openings and closings. First one, then the other ion floods across the membrane, causing the voltage to increase and then decrease back to the resting potential. This sequence moves along the axon in a wave, transmitting the signal. The entire sequence at one point may occur in 1 ms.

The largest axon fibers, from 4 to 20 μm in diameter, are well insulated by myelin and can propagate signals at speed up to 140 m/s. Others are unmyelinated and are less than 2 μm in diameter; their transmission speed is only about 1 m/s. The faster fibers are used to transmit the senses of balance, body position, and delicate touch. Slower neurons take less space and are used to transmit information on temperature and pain as well as information for organs and glands. About one-third of an adult's neurons are myelinated. Children lack myelination until early adolescence, which partly accounts for their reduced coordination ability. **Multiple sclerosis** is a disease involving the loss of myelination of axons, which results in muscle paralysis and loss of sensations.

9.4.2 Synaptic Transmission

At the end of the axon, the neuron must transmit the signal to another neuron or to an effector. This is done through the synapse. Most synapses involve the use of a chemical, called a **neurotransmitter**, to communicate the signal across a gap from one cell to another. One of the most widespread neurotransmitters is *acetylcholine* (ACh). This neurotransmitter is a target of many insecticides, in particular the organophosphorus and carbamate pesticides.

The axon ends in a **synaptic knob** (Figure 9.3), which contains as many as 1 million vesicles containing ACh. The membrane of the synaptic knob is separated from the membrane of a target neuron by a gap of about 20 nm width. When the depolarization wave from an axon reaches the synaptic knob, ACh is released and diffuses across the gap to the target neuron. There it stimulates pores to open, depolarizing the target neuron membrane and initiating a nerve signal transmission. An enzyme, *acetylcholinesterase*, rapidly decomposes the ACh into *choline* and *acetate*. The choline is reabsorbed by the synaptic knob and recycled into more ACh. It is the enzyme acetylcholinesterase that is affected by organophosphorus and carbamate pesticides (see Section 17.4.7).

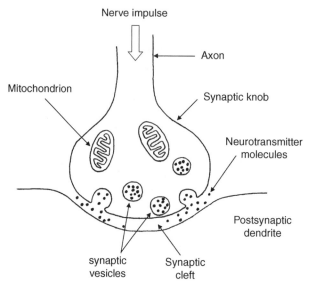

Figure 9.3   Neural synapse. (Based on Fried, 1990.)

There are many other neurotransmitters. Many hormones serve this function, including *epinephrin* (*adrenaline*), ADH, *oxytocin, insulin,* and *glucagon.* The amino acids glycine, glutamine, and aspartic acid are neurotransmitters, as are the gases carbon monoxide and nitric oxide. A group of compounds called *endorphins* modifies the effect of neurotransmitters and may be involved in mood and pain reduction. They are similar in structure to morphine. It is thought that exercise produces a natural release of endorphins. *Dopamine* is a central nervous system neurotransmitter that can be inhibitory or excitatory, depending on the receptor. A decline in dopamine production produces **Parkinson's disease**, in which the inhibitory action of dopamine is missing. As a result, the neurons that control muscle tone become overstimulated. All movement requires overcoming the tension of the opposing muscle. Dopamine cannot cross the blood–brain barrier, but the drug L-*dopa* can, and it is converted to dopamine in the brain, providing relief from symptoms.

9.4.3   Nervous System Organization

The nervous system may be the best example of "the whole is more than the sum of its parts." Even accounting for the fact that the behavior of individual neurons is much more complex than described above, it is difficult to explain our higher behaviors, such as language, abstract reasoning, and self-consciousness, in terms of them. That is a far greater task than explaining the functioning of a computer in terms of the action of individual transistors. Those higher behaviors depend on neuronal activity, but in ways far from well understood and beyond our scope here, in any case. Here we can only summarize the basic organization of the nervous system.

The nervous system can be divided into a **central nervous system** (CNS) and a **peripheral nervous system** (PNS) (Figure 9.4). The CNS consists of the brain and spinal cord. The CNS performs integration of information and coordination of actions. The actual source of information and distributor of commands to the body is the PNS, which includes all of the other neurons in the body.

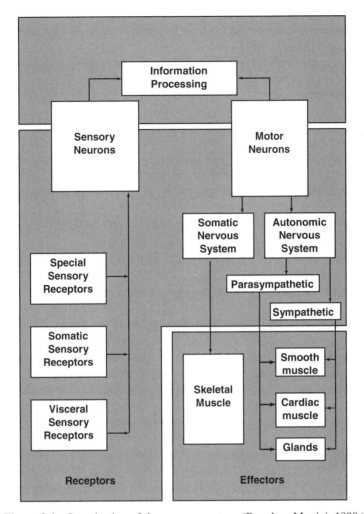

Figure 9.4   Organization of the nervous system. (Based on Martini, 1998.)

The **brain** is where all the higher-level activity occurs. The human brain weighs about 1.4 kg and has a volume averaging 1200 cm^3. Brain size in humans correlates with body size and not with intelligence. It contains about 10^{12} neurons, each with an average of about 1000 connections. The brain has three main parts: the cerebrum, the cerebellum, and the brainstem. The **brainstem** is the connection between the brain and the spinal cord. It regulates some of the functions most basic to our survival. Within the brainstem, the **medulla oblongata** controls basic functions such as breathing, heartbeat, and blood pressure. Above the medulla are the **pons** and **midbrain**, which act as relay centers for sensory information. The **hypothalamus** is above this. It regulates the pituitary gland, forming a connection between the nervous system and the endocrine system. It also regulates many functions of homeostasis: body temperature, salt and water balance, hunger, and the digestive system. It also contains areas associated with pleasure and ecstasy. The **thalamus** is above the hypothalamus (as the name implies) and also serves as a relay station. From the thalamus down to the spinal cord runs the **reticular formation**, which is connected with sleeping and awareness and the ability of a student to concentrate during long, boring lectures.

The **cerebellum** is located dorsal to the medulla and controls fine motor skills. Although the decision to make some motion is made in the cerebrum, the cerebellum provides feedback control to accomplish the task with precision. It is most developed in birds and mammals, since they have much higher dexterity than that of the other animals.

The **cerebrum** is the part of the brain responsible for voluntary control of skeletal muscles, speech, vision, memory, thoughts, and consciousness. Most of its activity relates to sensory and motor functions. It is the largest part of the brain, forming the familiar highly folded structure divided into two parts, a right and a left hemisphere. The outer layer of the cerebrum is the **cerebral cortex**. It is only 3 mm thick but accounts for 90% of the brain's cell bodies. Many specific capabilities can be localized to very small areas of the cerebral cortex. For example, the ability to recognize your mother's face can be eliminated by the destruction of a very small area of the cortex, such as by a stroke. Another area controls learned eye movements. People with damage to this area can understand writing but cannot read because they cannot follow lines of type on a page with their eyes.

The endothelial cells that line the blood vessels in the brain are joined to each other very tightly. Thus, for any chemical to pass from the blood to the brain, it must pass into and out of the epithelial cells rather than through spaces between them as is possible elsewhere in the body. This is called the **blood–brain barrier**, and it limits the passage of hydrophilic (water-soluble) compounds, thus protecting the brain against many types of toxins and drugs. The brain uses glucose for energy almost exclusively and does not store glycogen; thus, it is dependent on a continuous supply in the blood. It also uses oxygen at a high rate. If the oxygen is cut off for more than 4 or 5 minutes, or if glucose is cut off for more than about 15 minutes, brain damage will occur.

The peripheral nervous system has two parts, corresponding to the two functions of obtaining information and distributing commands. Information is provided by the nerves associated with the senses through the neurons of the **sensory system**. Action is stimulated by the **motor system**, which connect to the muscles and initiate their contractions. Both of these systems branch out from the brain and spinal cord. Some sensory neurons, such as some of those connected with pain sensors in the skin, are connected to motor neurons in the spinal column. This forms a **reflex arc**, which enables rapid response to danger without taking the time to transmit signals all the way to the brain for additional CNS signal processing. For example, the sudden pulling of a finger away from a hot object is accomplished by the reflex arc.

The motor system is itself divided into two parts: the somatic nervous system and the autonomic nervous system. The **somatic nervous system** transmits signals for voluntary control of the skeletal muscles. Their neurons, called **motor neurons**, have large-diameter myelinated axons and tend to pass directly from the brain or spinal cord to the target muscle without intermediate synapses. All the motor neurons stimulate muscle contractions; none are inhibitory. The only neurotransmitter used is ACh.

The **autonomic nervous system** controls involuntary activities, the body functions that occur without our conscious awareness. The **effectors** (the target organs that act upon receiving a nerve signal) include smooth muscle, the heart, and glands. Most of the neurons of the autonomic nervous system do not pass directly from the CNS to the effector. Instead, each path consists of two neurons, joined by a synapse in a structure called a **ganglion**, most of which are located alongside the spinal column. Autonomic system neurons may be stimulatory or inhibitory. Most autonomic neurons use ACh, but other neurotransmitters are also involved. One "ganglion" is actually a gland, the **adrenal medulla**, in which the neurotransmitters epinephrine and norepinephrine are released to the bloodstream instead of to another neuron. Thus they act as hormones in this situation. This activity is described in Section 9.5.

The autonomic nervous system is divided into the sympathetic and parasympathetic divisions. These are both anatomically and functionally distinct. The **parasympathetic division** neurons originate in the brainstem and in the sacral region at the base of the spine. The parasympathetic division stimulates activity of the visceral organs, which occurs while in a relaxed state. Its effects can be summarized as follows: (1) metabolic rate is decreased; (2) heart rate and blood pressure are decreased; (3) secretion by salivary and digestive glands increases; (4) muscle contractions in the digestive tract increase; and (5) urination and defecation are stimulated.

The **sympathetic division** branches out from the spinal column in the thoracic and lumbar regions. Most sympathetic paths release norepinephrine at the effector, although some release ACh or nitric oxide. It is the sympathetic division that stimulates the release of epinephrine (adrenaline) and norepinephrine by the adrenal gland. In fact, the overall effect of sympathetic division stimulation can be summarized as producing the **fight or flight response**, which prepares the body for an emergency that might require intense physical activity. It is stimulated by emotions such as fear or stress. In summary, the effects of the sympathetic division are (1) increased mental alertness, (2) increased metabolism, (3) inhibited digestive and urinary function, (4) activation of energy reserves, (5) increased respiration, (6) increased heart rate and blood pressure, and (7) stimulation of sweat glands.

Both the sympathetic and parasympathetic divisions innervate most of the same organs and other effectors, but they usually have opposite effects: One will stimulate while the other inhibits. This gives finer control over activity, like having both a brake and an accelerator on a car. The heart, for example, is stimulated to decrease its output by ACh from the parasympathetic division and to increase output by norepinephrine from the sympathetic division. Some of the neurotransmittor receptors in the autonomic system are stimulated by *nicotine*, the active ingredient in tobacco. As a result, nicotine poisoning produces increased heart rate and blood pressure, plus vomiting and diarrhea.

Information is provided by **sensory receptors**, which are neurons that are specialized to produce signals in response to physical or chemical stimulus. The receptors can be divided into three divisions. The **special sensory receptors** are associated with complex sensory organs; the special senses are vision, hearing, taste, smell, and balance. The **somatic sensory receptors** include the senses of touch, pain, temperature, pressure, vibration, and proprioception. **Proprioception** is the sense of position of the skeletal muscles and joints. The **visceral sensory receptors** monitor the internal organ systems, including the cardiac, digestive, respiratory, urinary, and reproductive systems. Since an action potential is always of the same strength, a receptor signals the strength of a sensation by varying the frequency of action potentials. That is, a stronger sensation produces a more rapidly repeated action potential. Ultimately, sensory signals are interpreted by the central nervous system in what is called **perception**. A receptor can be as simple as an ordinary dendrite of a neuron. Pain receptors may be of this form. They respond to many different types of stimulus. Other receptors are enclosed in complex structures that admit only a highly selective type of stimulus. The special sensory receptors are of this type.

9.5   ENDOCRINE SYSTEM AND HOMEOSTASIS

The nervous system is not the only way the body controls bodily functions. It also uses chemical messengers. Some cells communicate directly with their contact neighbors through special junctions. This is usually to coordinate local activity such as ciliary movement or muscle contractions. Others release chemicals into the intercellular spaces that primarily affect cells in the same tissue. An example is the *prostaglandins*, a powerful fatty acid with many functions. Prostaglandins are released by damaged tissues and stimulate inflammation and the sensation of pain. *Aspirin* and other analgesics act by inhibiting the formation of prostaglandins and similar compounds. Many tissues issue chemicals that inhibit cell division locally. This prevents uncontrolled growth such as occurs in cancer tumors. Compounds such as these are called **local hormones** or **paracrine factors**. Some specialized cells produce chemicals that are excreted through ducts onto epithelial surfaces, such as inside the intestines or onto the skin. These cells are in structures called **exocrine glands**. Saliva and sweat are secretions of exocrine glands.

The focus in this section is on chemicals that are secreted by glands into the blood supply for the regulation of bodily function. These glands form the **endocrine system**, which consists of **endocrine organs** that produce hormones. The chemical messengers produced by the endocrine system are called **hormones**, which are chemical messengers that influence the response of cells and tissues at locations remote from the hormone-producing cells. The endocrine system is just as vital for proper functioning of the body as the nervous system is. Although its response time is not as fast, its effects can be long lasting. Of interest from an environmental point of view is the idea that some pollutants, including 2,3,7,8-TCDD (dioxin), mimic the female sex hormone estrogen. Such toxins are called **xenoestrogens** or **endocrine disrupters**.

Figure 9.5 shows some of the glands of the endocrine system. The pituitary gland has two main parts. The *anterior pituitary* produces a number of important hormones, the release of all of which is controlled by other hormones produced by the hypothalamus. The *posterior pituitary* does not produce its own hormones. However, it stores several hormones produced by the hypothalamus and releases them upon receiving a neural command from the hypothalamus.

The adrenal glands are located atop each kidney and also have two main parts. The outer part, or **cortex**, produces steroid hormones; the inner part, called the **medulla**, produces epinephrine and norepinephrine. Ninety-nine percent of the *pancreas* serves an exocrine function, producing digestive enzymes. The other 1% performs a critical endocrine function: controlling blood glucose.

9.5.1   Homeostasis

The nervous system and the endocrine system share major responsibility for maintaining **homeostasis**, a stable internal environment, in the face of external changes. Homeostasis requires that a number of vital factors be controlled within a proper range. Some of these vital factors include body temperature, blood glucose concentration, the concentration of individual electrolytes (sodium, potassium, calcium) in the blood, and blood pressure and volume.

Homeostatic control is accomplished mostly by the use of **negative feedback**, in which the movement of a condition outside the vital range stimulates an effect that tends to move it in the opposite direction. For example, most adults have a resting body temperature between 36.7 and 37.2°C (98.1 to 99.0°F). The "normal" value for a person's vital factor is termed the **set point**. If the temperature rises about 0.2°C above the set-point

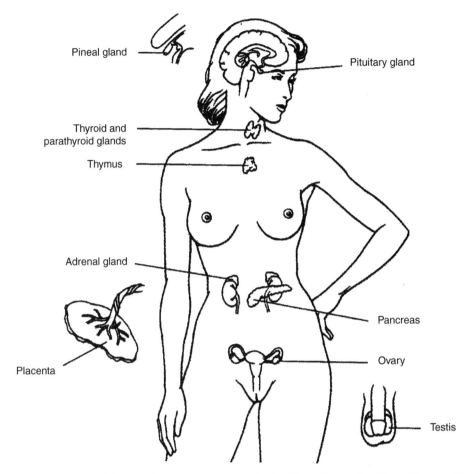

Figure 9.5   Some of the most important endocrine glands. (From Van de Graaff and Rhees, 1997. © The McGraw-Hill Companies, Inc. Used with permission.)

temperature, the hypothalamus stimulates blood vessels in the skin to dilate and sweat glands to increase their secretion. Both of these tend to increase the loss of heat from the body, limiting or reversing the temperature increase. A body temperature decrease causes the reverse effects, plus other activities, such as shivering.

Body temperature also involves **feedforward control**, in which events that would alter a vital factor are detected and the body responds before the alteration actually occurs. The hypothalamus also receives signals from temperature sensors in the skin. These stimulate the temperature control effects before the sensors within the hypothalamus actually detect an internal change.

A few mechanisms involve **positive feedback**, in which a stimulus produces an effect that increases the stimulus. This usually involves processes that need to be completed quickly, such as blood clotting. When childbirth begins, stretch receptors in the wall of the uterus stimulate the brain to cause the pituitary gland to release stored *oxytocin*, a hormone that stimulates uterine contractions. This increases the rate of contractions, expelling the baby faster. Once the baby has left the birth canal, the uterine receptors relax, leading to a drop in oxytocin levels, breaking the positive feedback loop.

Factors that cause deviation from homeostasis are termed **stresses**. Stress can be physical, emotional, environmental, or metabolic. When stress initiates, the body enters what is called the **alarm phase** and responds with the fight or flight response described in Section 9.4.3. If the stress continues for the long term, the body switches to the **resistance phase**, in which glucocorticoid hormones are released, especially cortisol, plus lesser amounts of epinephrine, growth hormones, and thyroid hormones. These maintain the rate of energy supply at elevated levels by mobilizing lipid and protein reserves and saving glucose for nervous tissue. If the stress is starvation, the resistance phase ends when reserves run out. Otherwise, it may end due to the side effects of the hormones, such as the slowing of wound healing due to the anti-inflammatory effects of glucocorticoids, elevated blood pressure and volume, and altered mineral balance (especially loss of potassium in the blood) due to aldosterone and ADH, or exhaustion of the ability of the adrenal cortex to continue glucocorticoid hormone production, destroying the ability of the body to maintain blood glucose levels. If any of these occur, the result is the **exhaustion phase**, in which homeostasis breaks down. One or more organs may malfunction. For example, excessive potassium loss can cause heart failure.

9.5.2 Hormones

Hormones are classified into three groups based on chemical structure:

- *Amino acid derivatives*: derivatives of either tyrosine (e.g., catecholamines and thyroid hormones) or tryptophan (e.g., melatonin).
- *Peptide hormones*: glycoproteins, short peptides, or small proteins (under 200 amino acids). Examples include growth hormone and insulin.
- *Lipid derivatives*: either *eicosanoids* (derivatives of arachidonic acid), such as prostaglandins, or *steroids* (derivatives of cholesterol), such as hydrocortisone.

Tables 9.1, 9.2, and 9.3 summarize most of the hormones and their functions. Although others are known, some are still poorly understood.

TABLE 9.1 Peptide Hormones

Gland	Hormone	Target and Function
Hypothalamus	*Antidiuretic hormone, or vasopressin (ADH)*	Stored in posterior pituitary; when released, it increases water resorption by the kidneys; it is inhibited by alcohol.
	Oxytocin	Stored in posterior pituitary; when released, stimulates milk ejection and uterine contractions in females, ductus deferens and prostate in males; secreted by uterus and fetus; released in orgasm in both sexes.
	Releasing and inhibiting hormones	Controls the release of hormones of the anterior pituitary, a different one for each.
Anterior pituitary	*Thyroid-stimulating hormone (TSH)*	Stimulates synthesis and secretion of thyroid hormones; stimulated by hypothalamic thyrotropin-releasing hormone (TRH).
	Follicle-stimulating hormone (FSH)	Females: stimulates follicle maturation and estrogen synthesis; males: stimulates production of sperm; stimulated by hypothalamic gonadotropin-releasing hormone (GnRH).

TABLE 9.1   (*Continued*)

Gland	Hormone	Target and Function
	Leuteinizing hormone (LH)	Females: stimulates ovulation, formation of corpus luteum, and synthesis of estrogen and progesterone; males: stimulates synthesis of testosterone; stimulated by hypothalamic GnRH.
	Prolactin (PL)	Stimulates growth of mammary gland and synthesis of milk.
	Growth hormone (somatotropin, GH)	Stimulates growth of all cells, but the cells of the bone and cartilage are especially sensitive.
	Adrenocorticotropic hormone (ACTH)	Stimulates production of steroids in adrenal cortex, part of the response to stress.
Intermediate pituitary	*Melanocyte-stimulating hormone (MSH)*	Increases melanin synthesis in epidermis (not active in adults except pregnant women).
Parathyroid	*Parathyroid hormone (PTH)*	Increases calcium in circulation by stimulating bones to release it, enhances digestive uptake, and inhibits kidney removal.
C cells of the thyroid	*Calcitonin (CT)*	Opposite effect of PTH, decreases calcium in blood.
Heart	*Atrial natriuretic peptide (ANP)*	Release stimulated by stretch receptors in cardiac muscle; it promotes loss of sodium and water in kidney, supresses thirst and water-conserving hormones.
Thymus	*Thymosins*	A blend of hormones that induce T-cell differentiation in the immune system.
Pancreas	*Insulin*	Stimulates the uptake and use of glucose by cells throughout the body, the production of glycogen by the liver and skeletal muscles, and the formation of fats by adipose tissue. The effect of insulin on glucose uptake by cells is indirect; it stimulates an increase in the production of membrane proteins that transport glucose through the membrane by facilitated diffusion.
	Glucagon	Released when blood glucose is low: stimulates cells throughout the body to release glucose, the liver to break down glycogen into glucose, and the breakdown of fats by adipose tissue.
Digestive tract	*Gastrin*	Stimulates hydrochloric acid secretion by the stomach.
	Cholecystokinin	Stimulates the exocrine cells of the pancreas to secrete digestive enzymes.
Kidneys	*Erythropoietin (EPO)*	Produced in response to low oxygen in the kidney; stimulates red blood cell production by bone marrow.
	Renin (actually an enzyme)	Converts angiotensinogen to angiotensin I in the blood.
Liver	*Angiotensinogen (a plasma protein)*	After conversion by renin in the blood and other enzymes in the lungs, it becomes angiotensin II, which stimulates thirst and production of aldosterone and ADH, causing sodium and water retention. This is part of the response to low blood volume.

TABLE 9.2   Amino–Acid–Derived Hormones

Gland	Hormone	Target and Function
Pineal gland	*Melatonin*	Pineal gland receives stimuli from visual neurons that reduce melatonin production during the day and increase it at night.- Thought to be involved in circadian (daily) rhythms of the body. Increases during dark winters thought to cause seasonal affective disorder (SAD), a disorder that affects mood, sleeping, and eating.
Thyroid	*Thyroxine (T_4) and triiodothyronine (T_3)*	Synthesis of thyroxine and triiodothyronine requires iodide. Synthesis and release is stimulated by hypothalamic regulatory enzymes. Produces a rapid, short-term increase in metabolic rate by binding to mitochondria, and by activated genes that code for glycolysis enzymes.
	Catecholamines	
Adrenal medulla	*Epinephrine (adrenaline) and norepinephrine*	Epinephrine makes up about 75% of the release of the adrenal medulla; the rest is norepinephrine. Release is stimulated by the sympathetic division of the autonomic nervous system as part of the fight or flight response. Causes skeletal muscles to break down glycogen to glucose as a ready source of energy, adipose tissue to break fats down to fatty acids for other tissues to use, the liver to break down glycogen for use by the neurons that cannot use fatty acids.
Hypothalamus	*Dopamine*	Controls release of prolactin.

The molecules that the hormones interact with are called **receptors**. The effect of a hormone depends not only on how much hormone is in circulation but how many receptors are present and on which cells. This is an important aspect of hormone control. Different tissues may vary in their numbers of receptors for a particular hormone. This enables the action of hormones to be selective. Change in receptors is also a mechanism for the control of hormone response. For example, thyroid hormone stimulates adipose cells to produce receptors for epinephrine, increasing their sensitivity to epinephrine's causing the release of fatty acids. The interaction of a hormone with a cell-surface receptor ultimately results in the activation or inhibition of an enzyme within the cell. Many drugs are designed to either stimulate or interfere with receptors.

The lipid hormones enter the cell and complex with the DNA in the nucleus. They then either stimulate or inhibit gene expression, and ultimately, the cell function. For example, testosterone stimulates production of structural proteins in skeletal muscles, increasing muscle size and strength. Thyroid hormones, in addition to complexing with DNA, may bind to mitochondria, causing an increase in ATP production.

Hormones are also controlled by changes in the composition of intercellular fluids, by other hormones, or by neural stimulation. The hypothalamus and the adrenal medulla are

TABLE 9.3   Lipid-Derived Hormones

Gland	Hormone	Target and Function
Eicosanoids		
Many tissues	*Leukotrienes*	Released by white blood cells, they coordinate tissue responses to injury and disease.
	Prostaglandins	Generated by most tissues of the body, converted by platelets in the blood to thromboxanes.
	Thromboxanes	Causes platelets to aggregate and the smooth muscles of the blood vessel walls to contract.
Steroids		
Reproductive organs	*Androgens (testosterone and others)*	
	Testosterone	Produced by testes, it promotes sperm growth, stimulates overall growth, especially of the skeletal muscles, and produces aggressive behavior.
Follicles of ovaries	*Estrogens (e.g., estradiol)*	Stimulates maturation of egg cells and growth of the lining of the uterus.
Corpus luteum of ovaries	*Progestins (e.g., progesterone)*	Prepares uterus for embryo, stimulates movement of egg or embryo, stimulates enlargement of mammary glands.
Adrenal cortex	*Several dozen hormones, e.g.:*	
	Aldosterone	In response to drop in blood sodium, volume, or pressure, stimulates retention of sodium by kidneys, sweat glands, salivary glands, and pancreas; increases sensitivity of salt-sensing taste buds in tongue; also stimulated by angiotensin II.
	Glucocorticoids (e.g., cortisol, hydrocortisone, corticosterone)	Accelerates glucose synthesis and glycogen formation, especially in the liver; stimulates tissues to use fatty acids for energy instead of glucose, reduce inflammation by inhibiting white blood cells and mast cells. Cortisol inhibits its own production by a negative feedback loop involving the hypothalamus and ACTH.
	Androgens	The "sex" hormones, including testosterone. Some are converted to estrogen.
Kidneys	*Calcitriol*	One of the D vitamins, released in response to PTH. Stimulates absorption of calcium and phosphate by the digestive tract, stimulates differentiation and behavior of bone cells, and inhibits PTH production.

stimulated directly by nervous activity. The hypothalamus controls the release of many hormones from the anterior pituitary, which in turn affects numerous other hormones and tissues. It also controls the release of vasopressin and oxytocin, which it produces, but which are stored in the posterior pituitary. This arrangement makes it possible to have a rapid response for urgent needs such as blood pressure control.

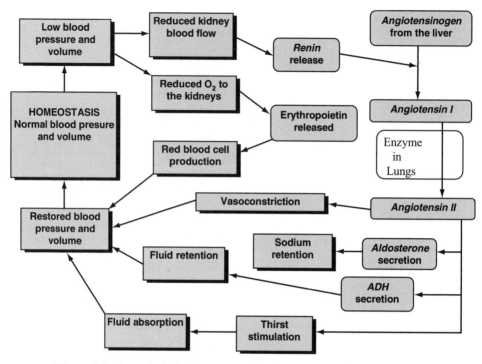

Figure 9.6   Control of blood pressure and volume. (Based on Martini, 1998.)

Hormones, endocrine organs, and the nervous system interact in complex ways. Figure 9.6 shows one case that illustrates this complexity schematically. That this system is much more complex than portrayed in the figure is apparent when one considers that each box represents a complex stimulus–response process in itself.

One of the most common diseases of the endocrine system is *diabetes mellitus*, which results in unstable blood glucose levels. Diabetes is the result of either inadequate insulin production, production of abnormal insulin, or production of defective receptors. These, in turn, may be determined genetically. Without effective levels of insulin, cells cannot remove glucose from the bloodstream. As a result, the liver, brain, and other tissues "starve" despite the ready supply. The kidneys work to remove the excess glucose, resulting in the loss of large amounts of water. This produces the symptoms of copious production of sweet urine, thirst, and dehydration.

Hydrocortisone creams are used in the treatment of rashes or excessive itching. These symptoms are overreactions of the immune or nervous systems. However, these medicines should never be applied to wounds, because they inhibit the inflammation response that would otherwise speed healing.

9.6   CARDIOVASCULAR SYSTEM

The **cardiovascular system** includes the blood, heart, and blood vessels. The heart and blood vessels serve to transport the blood throughout the body. The blood, in turn, has the following functions: (1) transport of gases, nutrients, hormones, and wastes; (2) regulation of pH and electrolyte concentration of intercellular fluids; (3) control of heat transport;

(4) removal of pathogens and cell debris; and (5) restriction of fluid loss from injury, by clotting. The blood is the mechanism by which systemic toxins are carried around the body to the tissues they injure. Each of its parts may also be attacked by various toxins.

Blood makes up about 7% of human body weight. In females this averages about 4.5 L of blood; males about 5.5 L. When a tube of blood is centrifuged, the upper 55% of the volume will be a pale yellow liquid called plasma. **Plasma** is 90% water, with 10% solutes including proteins and electrolytes; nutrients such as glucose, amino acids, and lipids; and waste products such as urea and *bilirubin*, a waste product that gives plasma, as well as urine and feces, their color. About 96% of the plasma proteins consist of **albumins** and **globulins**. These have overlapping functions, including control of osmolarity and the transport of insoluble lipids, vitamins, metals, hormones, and so on. Globulins are an important component of the immune system, discussed in Section 9.7. The rest of the protein is **fibrinogen**, involved in blood clotting. The dominant electrolytes are sodium and chloride. Intercellular fluid is similar in composition to plasma.

The other 45% of blood volume consists of the blood cells and platelets (Figure 9.7). The great majority of these, about 5,000,000 per cubic millimeter, are the **red blood cells**, or **erythrocytes**. These are produced in the bone marrow under stimulation by erythropoetin from the kidney. When mature, they lose their ribosomes, mitochondria, nucleus, and so on, giving them the appearance of a small (about 7 μm) disk indented in the center. Without their organelles, erythrocytes cannot long maintain their structure. They are replaced continuously, at the rate of a little less than 1% per day. Old red blood cells are destroyed and recycled by the spleen and liver.

The function of erythrocytes is to carry oxygen from the lungs to the other tissues and to carry carbon dioxide on the reverse path. These compounds are carried by the red pigment **hemoglobin** (Figure 9.8), which constitutes one-third of the mass of the erythrocyte. Hemoglobin is composed of four polypeptide chains complexed with four *heme* groups, each containing a ferrous iron ion at its center. Note the similarity between

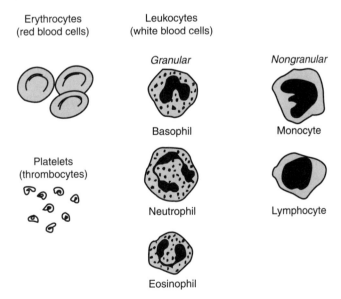

Figure 9.7 The formed elements of the blood. (From Van De Graaff and Rhees, 1997. © The McGraw-Hill Companies, Inc. Used with permission.)

Figure 9.8   Hemoglobin. (From Van De Graaff and Rhees, 1997. © The McGraw-Hill Companies, Inc. Used with permission.)

the heme molecule and chlorophyll shown in Figure 5.10. Each of the four ferrous ions can complex with one molecule of O_2. This gives blood an oxygen-carrying capacity of about 74 mg/L, about 10 times the solubility of oxygen in water at physiological temperature.

heme subunit

Just 0.1% of the blood volume is composed of **white blood cells**, or **leukocytes**, and platelets. All are descendants of the same cells in the bone marrow that produce erythrocytes, although some leukocytes go through additional processing in the thymus and organs of the lymphatic system. There are four types of leukocytes, of which 60 to 70% are **neutrophils** and 20 to 25% are **lymphocytes**. Functionally, leukocytes are part of the immune system and are discussed in Section 9.7. Although transported by blood, they usually leave the blood vessels for sites of infection or damage to do their work.

Blood **platelets** are cytoplasmic cell fragments that are much smaller than erythrocytes. They are involved in the formation of blood clots to prevent blood loss and promote tissue healing after a physical injury. Such injury disrupts the endothelial lining of the blood vessels, exposing underlying connective tissue. Platelets adhere to collagen, and a plug of platelets is formed rapidly. Simultaneously, a complex series of reactions, some of which involve platelets, results in the conversion of fibrinogen into a form that aggregates to form a mesh of interlocking strands called **fibrin** that forms the clot. Several factors limit clotting to the area of the wound, including prostaglandins produced by normal tissue. *Hemophilia* is a gender-linked genetic disease in which the victim cannot produce one of the factors involved in fibrin formation. It has been mentioned that aspirin inhibits prostaglandins. However, at low doses it inhibits thromboxane more. Thus, it has been found that men with no previous history of heart disease who take one aspirin every other day reduce their chance of heart attack (caused by clotting of the coronary arteries) by 50%.

The **blood vessels** are the tubes that conduct blood around the body. They are lined with a type of epithelial cells, which in this case are called **endothelium**. **Arteries** carry blood away from the heart toward the tissues. They contain muscles that help maintain blood pressure. Arteries branch successively into **arterioles**, which by their contraction and dilation control blood flow to various tissues. These divide further until they form a weblike network of tiny thin-walled vessels called **capillaries**. These are about 5 to 8 μm in diameter, just wide enough for erythrocytes to pass in single file. Capillaries permeate almost all tissues; few cells are more than 1 mm from a capillary. The capillaries are responsible for the transfer of heat and substances to and from tissues. **Veins** carry blood back to the heart. (Note that the definition does not depend on whether or not the blood is oxygenated. Specifically, the pulmonary arteries are deoxygenated but carry blood from the heart to the lungs, and the pulmonary veins are oxygenated but carry blood back to the heart.) Veins are less muscular and under much lower pressure than arteries. Thus, movement is aided by contraction of skeletal muscles surrounding them and by the presence of one-way check valves to prevent flow reversal.

The **heart** is actually two connected pumps. The right side pumps deoxygenated blood through the lungs, the left distributes it through the body. The heart of a resting adult pumps about 5.0 L/min in 72 beats/min, or about 70 mL/beat. When exercising, a non-athlete may pump 20 L/min at 192 beats/min. A well-trained athlete may pump up to 35 L/min. Blood is delivered to the **right atrium** of the heart by the **superior and inferior vena cava**. This collects a volume of blood to fill the more muscular **right ventricle**, which then provides the pressure to force the blood out through the pulmonary arteries to the lungs. After CO_2 is exchanged for O_2, the blood flows back into the **left atrium**. The left atrium fills the **left ventricle**, the most muscular part of the heart, which must develop enough pressure to circulate the blood throughout the body. The left ventricle discharges through the main artery, called the **aorta**. A pair of **coronary arteries** branch out from the aorta to provide blood for the heart's own use.

Although innervated by the autonomic nervous system, the heart can beat independently. Nervous stimulation can serve to increase or decrease heart rate as needed. The contraction phase of the cardiac cycle is called **systole**; the relaxation phase is called **diastole**. Pressure peaks at a normal 120 mmHg (over 2 m of hydraulic head) during systole. During diastole it drops to about 80 mmHg. These measurements are usually expressed in the form 120/80. Blood pressure above 140/90 is abnormally high. Surprisingly, the heart also produces a hormone to regulate blood volume in concert with the kidneys.

9.7   IMMUNITY AND THE LYMPHATIC SYSTEM

The **immune system** consists of defenses against foreign matter that gains entry to the body. It consists of the lymphatic system plus components of numerous other systems of the body. Many toxic pollutants either stimulate or suppress the immune system.

The **lymphatic system** consists of lymph, lymphatic vessels, lymphocytes, and lymphoid tissues and organs. **Lymph** is a fluid similar to plasma. **Lymphatic vessels** are similar in structure to veins. They conduct lymph from peripheral tissues to the veins. Capillaries deliver more liquid to tissues than they carry away. The rest forms intercellular fluid that collects as lymph. The most important **lymphoid organs** are the lymph nodes, the spleen, and the thymus. The **lymph nodes** contain immune system cells that remove pathogens from the lymph before they reach the bloodstream. The *tonsils* are lymph nodes positioned to respond to infections arriving by way of the mouth or nose. The **thymus** produces T cells (described below). The **spleen** performs functions for the blood similar to those the lymph nodes perform for the lymph, such as removal of cellular debris and responding to pathogens in the blood. It also stores iron from recycled erythrocytes.

The defenses of the body are classified into two forms: specific and nonspecific. **Nonspecific defenses** protect without discriminating the exact type of threat, and act relatively rapidly. There are seven kinds of nonspecific defenses:

1. **Barrier defenses** include the skin, mucous membranes, and hair, which act to prevent physical access to the interior by disease agents.

2. **Phagocytes** are cells that engulf pathogens (the process of phagocytosis) and debris. Several of the white blood cells serve this function. In addition, the white blood cells known as *monocytes* are converted into phagocytes called **macrophages**, which perform this function in tissues and the lymphatic system.

3. **Immunological surveillance** involves a kind of lymphocyte called **natural killer (NK) cells** that can recognize and kill virus-infected and cancerous cells. They are also recruited by the specific defense immune system.

4. **Interferons** are small proteins released by lymphocytes and macrophages that stimulate virus-infected cells to produce antiviral proteins.

5. **Complement** consists of 11 special proteins in plasma that work either alone, forming pores in the membrane of foreign cells, killing them, or in conjunction with the specific defenses to have the same effect. They can also attack virus structure, attract phagocytes, and stimulate inflammation.

6. **Inflammation** results when damage stimulates the release of histamines, heparin, prostaglandins, potassium, and other substances which produce dilation and increased permeability of capillaries. These produce pain, swelling, warmth, and redness of the injured area. Clots isolate the area, and macrophages are attracted. The increased temperature can reduce pathogen growth.

7. **Fever** is the increase of body temperature above 37.2°C (99°F). Fever is caused by proteins in the blood called **pyrogens**, Some pyrogens are produced by macrophages, which affect the temperature set point. High body temperature may be a strategy to inhibit growth of pathogens while enhancing the body's own metabolism. Metabolism increases about 10% for each 1°C increase in temperature.

Specific defenses are highly selective: for example, recognizing a particular strain of bacteria but ignoring all others. However, they take longer to get into action. The protection of specific defenses is called **immunity**. Immune responses are stimulated by foreign substances called **antigens**. Specific defenses refer to immune responses that recognize particular antigens. Immunity is mediated by cells that differentiate from lymphocytes that have left the bloodstream. These can form three basic types: NK cells, mentioned above; B cells; and T cells. Both B and T cells are produced in the lymphoid organs.

There are several basic forms of T cells. The most important are the **cytotoxic T cells** (**T$_C$ cells**, also called **killer T cells**). T$_C$ cells recognize antigens bound in the membranes of other cells. These other cells can be any cell in the body. All our cells constantly take protein fragments from the cytoplasm and place them on their membranes complexed with a glycoprotein called the **major histocompatibility complex** (MHC). If a cell is damaged or contains infectious material such as proteins, some of the antigens presented on the surface will be abnormal. The deranged cell will also produce other membrane proteins that signal, in effect, "I am deranged." A T$_C$ cell can bind to the MHC and the second signal protein. The presence of both binding sites stimulates the T$_C$ cell to kill the deranged cell either by producing toxins or by stimulating the cell to kill itself (a process called **apoptosis**). The now-activated T$_C$ cell can then reproduce many times, generating an army of cells to attack other cells carrying the same antigen. The reproduction process will also generate **memory T$_C$ cells**, which remain in reserve in case of future reinfections by the same antigen-bearing pathogen.

B cells act to eliminate the source of antigens directly instead of killing infected cells. The lymph nodes store millions of different populations of B cells, each capable of reacting to a different antigen. Almost any biological substance, whether from a fungus or a transplanted organ, can find a B cell that can react to it. Each population of B cells has a specific type of globulin protein studding its surface. These are Y-shaped molecules called **antibodies**, or **immunoglobulins** (Figure 9.9). Antibodies are produced by B cells. Once an antigen complexes with the antibodies on the surface of a B cell, the B cell becomes *sensitized*. It then reacts with a helper T cell and becomes *activated*. The activated B cell then starts reproducing rapidly. Some of the offspring become **plasma cells**, which produce antibodies that circulate in the blood. Others become **memory B cells**, which remain in storage in case of a repeat infection.

The attack is now carried on by the circulating antibodies. The top of the Y can complex with antigens. When macrophages encounter a substance covered with antibodies, it is stimulated to phagocytize it, destroying it. Complement also reacts with the antibodies, and if they are attached to a cell or a virus, the complement will destroy them.

Several varieties of polysaccharides on the surface of red blood cells differ between persons to give them their characteristic *blood type*. These sugars are natural antigens that act when blood from one person is given to another in a transfusion. The genetic

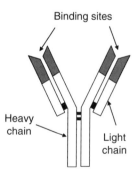

Figure 9.9 Basic structure of an immunoglobulin protein. (Based on Van de Graaff and Rhees, 1997.)

TABLE 9.4   Human ABO Blood Types and Their Compatibilities

Blood Type	Percent of U.S. Population	Antigen on RBC	Genetic Makeup	Antibodies in Blood	Compatible Blood Types
A	42	A	AA or AO	Anti-B	A or O
B	10	B	BB or BO	Anti-A	B or O
AB	3	A and B	AB	None	Any
O	45	None	OO	Anti-A and anti-B	A, B, or AB

basis of blood type was described in Section 6.1.1. The gene for blood type has three alleles: A, B, or O. Alleles A and B code for the production of a corresponding surface sugar. The allele for O does not produce a surface sugar. Thus, a person with alleles AA or AO have only type A sugars on the surface of their RBCs, and those people have type A blood. Unlike standard antibody response, people with type A blood always have anti-B antibodies even if they have never been exposed to the type B antigen. Thus, if blood of types B or AB is transfused into type A persons, their antibodies will produce a reaction that involves **hemolysis** (RBC rupture) or clumping of RBCs. Table 9.4 shows the similar relationships for other blood types. Note that type AB persons can accept transfusions of any blood type. Thus, they are called *universal recipients*. Type O persons can donate blood to anyone and are called *universal donors*.

Another type of T cell, the **helper T (T_H) cell**, is produced by a mechanism similar to the production of cytotoxic T cells. The T_H cell produces cytokines that stimulate the maturation of cytotoxic T cells and the B cells. The *human immunodeficiency virus* (HIV), which causes the disease known as *acquired immune deficiency syndrome* (AIDS) infects and kills a type of T_H cell. These T cells slowly disappear over a period of years in infected individual, eventually disabling the entire specific defense system.

Vaccines consist of inactivated antigens, such as killed infectious bacteria or virus. When injected into a human, the person develops antibodies for that bacteria or virus just as if it were faced with a real infection. If a real infection occurs subsequently, circulating antibodies and memory T cells and B cells will be ready to respond more quickly to stop the infection.

There are several types of diseases of the immune system. AIDS is a type of **immunodeficiency disease**. Besides the infectious type, this problem can be caused by radiation or congenital problems. **Autoimmune disorders** occur when the immune system treats part of the body as if it were a foreign substance. Examples are rheumatoid arthritis, lupus erythematosis, psoriasis, and diabetes mellitus. **Allergies** are excessive or inappropriate immune responses. A first exposure to an antigen stimulates the production of antibodies, which mediate an exaggerated response on subsequent exposures.

As described earlier, stress produces the release of glucocorticoids, which inhibit inflammation responses and ultimately, the healing of wounds. This makes the stressed person more susceptible to disease.

9.8   RESPIRATORY SYSTEM

The **respiratory system** has the primary function of exchanging the gases O_2 and CO_2 between the blood and the atmosphere. Along with the integument and the digestive tract, it is a major area of contact between the body and the environment. In fact, it is the major area of exposure, in terms of the volume of environmental material contacted. Besides infectious illness, the respiratory system is susceptible to gaseous and particulate pollutants, ranging from irritants to carcinogens.

The **upper respiratory system** includes the nose and nasal cavity, the nasal sinuses, and the pharynx. The **sinuses** are open chambers in the face, connected to the nasal cavity by passages. Their function is to lighten the head, add timbre to the voice, and to produce mucus to moisten and lubricate the surfaces of the nasal cavity. Air passing through the nasal cavity is warmed, humidified, and filtered of large particles. The **pharynx** is the part of the respiratory system shared with the digestive system, extending from the back of the mouth down to the larynx.

The **lower respiratory system** includes the larynx (voice box), trachea (windpipe), and the bronchi, bronchioles, and alveoli of the lungs. The **glottis** is a narrow passageway between the pharynx and the trachea. It can open and close much like a pair of sliding doors. In speech, air passing through the glottis vibrates ligaments at the inner edge of the glottis, the **vocal cords**. The **larynx** is a group of cartilaginous structures that surround and protect the glottis. At the top is a spoon-shaped cartilage called the **epiglottis**. When you swallow, the larynx is pushed up, forcing the epiglottis to fold down and seal off the trachea. At the same time, the glottis closes. If particles get past the epiglottis, they fall on the glottis and stimulate the **coughing reflex**. In a cough, the glottis is held closed against air pressure from the lungs and then is opened abruptly to eject any material.

Air is then conducted down through the **trachea**, which then branches into two **bronchi** and enters each of the lungs (Figure 9.10). The trachea and each bronchus has a similar tubelike structure surrounded by C-shaped cartilage rings. These branch successively into smaller and smaller bronchi. At the smallest level they form **bronchioles**, with an inside diameter of 0.3 to 0.5 mm. Bronchioles lack cartilage. Like arterioles, they can expand and contract under the influence of the autonomic nervous system to control flow.

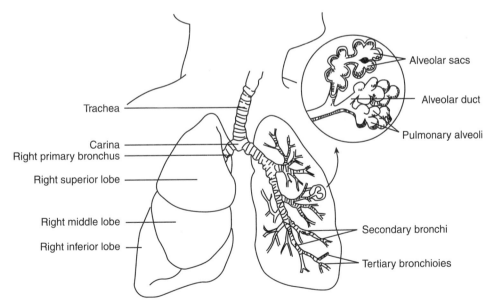

Figure 9.10   Trachea, bronchi, and lungs. (From Van De Graaff and Rhees, 1997. © The McGraw-Hill Companies, Inc. Used with permission.)

Asthma is a spasm of the bronchioles that greatly restricts the air-exchange capability of the lungs. The bronchioles lead to groups of sacs arranged like clusters of grapes. These are dead-end chambers called **alveoli**, which are the site of actual mass transfer of oxygen and CO_2. Each alveolus is surrounded by capillaries. The gases need to diffuse as little as 0.1 μm to cross between the lumen of the alveoli and the blood. Both oxygen and CO_2 are lipid soluble, so diffusion is relatively easy.

Most of the epithelium of the respiratory tract, from the nasal passages to the bronchi, secretes sticky, viscous mucus onto the surface. The epithelial cells are lined with cilia that sweep continuously the mucus toward the pharynx, where it can be swallowed, along with trapped particles. This mechanism, called the **mucociliary escalator**, traps particles larger than 10 μm. The mechanism is damaged by tobacco smoking, as evidenced by "smokers' cough," which is needed to replace the mechanism's throat-clearing effect. The mucociliary escalator does not cover bronchioles or alveoli. However, special macrophages called **dust cells** roam about and consume particles of size 1 to 5 μm that may be deposited in the alveoli. Particles smaller than about 0.5 μm remain suspended and pass out of the lung with expiration. In the genetic disease *cystic fibrosis* the mucus is unusually thick and cannot be removed by the cilia, resulting in lung blockage and frequent infections.

The atmosphere, with its total pressure of 760 mmHg, is 20.9% oxygen (a partial pressure of 159 mmHg). It has 400 ppmv CO_2 (0.04%, or 0.3 mmHg) and about 2 to 20 mmHg of water vapor, depending on the humidity. The air we breathe out is saturated with water at physiological temperature: 6.2% or 47 mmHg. Oxygen has dropped to about 15.3% (116 mmHg) and CO_2 increased to 3.7% (28 mmHg). Oxygen-depleted blood arriving in the lungs absorbs oxygen due to the mass transfer driving force between the air and the blood. For the same reason, it releases carbon dioxide.

Another mechanism enhances this transfer. The capacity of hemoglobin for oxygen depends on the pH of the blood. This, in turn, is affected by how much CO_2 is being carried by the blood. The hemoglobin carries some CO_2, but most is dissolved in the blood plasma. Carbon dioxide reacts with water to form *carbonic acid*, H_2CO_3, which dissociates at a normal blood pH range of 7.2 to 7.6. Since CO_2 is being produced in the tissues, its concentration increases there. As blood passes through, its pH drops, from the formation of carbonic acid. As a result, the capacity of hemoglobin for oxygen also drops, forcing the hemoglobin to release oxygen. The reverse situation occurs in the lungs.

The force for inspiration (inhalation) comes from the **diaphragm**, a sheet of muscle that spans the chest under the lungs, plus other muscles that lie over the ribs. Expiration may occur passively by elastic rebound or by contraction of muscles under the ribs and of the abdomen. A resting adult breathes at a *frequency*, f, of about 12 to 18 min^{-1}. The volume inhaled or exhaled with each breath under resting conditions is called the **tidal volume**, V_T. The respiratory flow, Q, is the product $f \times V_T$. For example, a typical tidal volume for an adult is 0.5 L. At a frequency of 12 breaths/min, the flow would be 6.0 L/min. The U.S. Environmental Protection Agency uses a respiratory flow rate of 20 m^3/day (about 13.9 L/min) to estimate exposure to atmospheric toxins for risk assessment purposes.

Physicians use an instrument called a **spirometer** to measure tidal volume and other respiratory air volumes, as shown in Figure 9.11. The **vital capacity** is measured by having the patient draw in a very deep breath and then blow out as completely as possible into a spirometer. This measurement is used to detect harm in populations exposed to air pollutants.

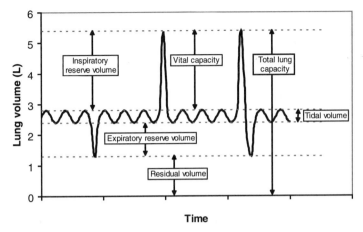

Figure 9.11   Spirogram, showing respiratory air volumes. (Based on Van de Graaff and Rhees, 1997.)

Breathing is controlled by centers in the medulla oblongata of the brain. Two groups of nerves fire alternately to stimulate inspiration and expiration. Of course, the brain can control these centers voluntarily as well as involuntarily. The brain is very poor at detecting low oxygen concentration and will not respond to oxygen deprivation by increasing the breathing rate. However, it strongly senses CO_2 accumulation in the blood and will increase the breathing rate in response to CO_2 buildup.

9.9 DIGESTION

The digestive system has as its main function the absorption of nutrients. However, it also has an important role in excretion of waste products. Furthermore, the digestive tract is one of the three main routes of exposure to toxins from the environment. **Digestion** is the physical and chemical breakdown of food into components that can be absorbed and used. It is accomplished by the action of mechanical mixing, acid conditions (in the stomach), and digestive enzymes. Digestive enzymes may include:

- Protein-digesting **proteases**, which produce peptides and amino acids
- Fat-digesting **lipases**, which produce fatty acids
- Complex carbohydrate–digesting **amylases**, which produce simple sugars or small oligosaccharides
- Nucleic acid–digesting **nucleases**, which produce nucleotides

Some digestive enzymes are secreted in an inactive form called a **zymogen** or **proenzyme**, designated by the suffix "-ogen" or the prefix "pro-." The zymogen is then converted, or activated, in the lumen of the digestive tract. For example, the pancreas produces the zymogen *trypsinogen*, which is converted in the intestines to the protease *trypsin*.

The **digestive system** consists of the digestive tract and accessory organs. The **digestive tract**, or **gut**, is basically a muscular tube including the oral cavity, pharynx, esophagus, stomach, small intestine, and large intestine. The **accessory organs** include the teeth, tongue, and several exocrine glandular organs, such as the salivary glands, liver, and pancreas. The digestive tract is lined with an epithelial tissue called the **mucosa**, or **mucous membrane**, underlain by loose connective tissue. The mucus protects the epithelium from digestive acids and enzymes.

Food is crushed by teeth in the oral cavity and mixed with **saliva**, which contains lubricating glycoproteins and amylase. Amylase begins the breakdown of polysaccharides into their component sugars. After swallowing, the ball of food, which is now called a **bolus**, is propelled down the esophagus to the stomach by a wavelike muscle contraction called **peristalsis**.

The **stomach** serves four main functions: (1) storage, (2) mechanical breakdown of food, (3) digestion by acids and enzymes, and (4) production of *intrinsic factor*, a glycoprotein essential for the absorption of vitamin B_{12} in the intestines. When full, the stomach can contain 1.0 to 1.5 L of material. The stomach produces about 2 L/day of hydrochloric acid, to maintain the pH at 1.5 to 2.0. The acidic conditions disinfect the food by killing microorganisms, denature enzymes and other proteins in the food, and break down plant cell walls and animal connective tissue in the food.

Hydrochloric acid is secreted by *parietal cells* by an interesting mechanism that avoids the production of the acid within the cells themselves (Figure 9.12). CO_2 and water form carbonic acid, which dissociates into bicarbonate ions and hydrogen ions. The bicarbonate passes into the bloodstream by a passive membrane transport mechanism that is coupled with the transport of chloride ions out of the blood. (Thus, the alkalinity of the blood increases.) This maintains electroneutrality across the membrane. The chloride accumulation in the cell is secreted toward the stomach through calcium channels. At

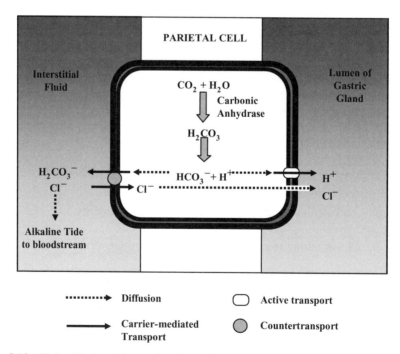

Figure 9.12 Hydrochloric acid secretion by parietal cells of the stomach. (Based on Martini, 1998.)

the same time, active transport uses ATP to move the hydrogen ions from the carbonic acid toward the stomach.

The zymogen *pepsinogen* is secreted into the stomach and converted by the low pH into the protease *pepsin*. By secreting an inactive form, the cells of the stomach themselves are protected from being digested. The stomach also secretes hormones, including *gastrin*, which stimulates contraction of the stomach for mixing and the secretion of hydrochloric acid, intrinsic factor, and pepsinogen. Gastrin secretion is stimulated by protein in the food. It is also stimulated by coffee (both caffeinated and decaffeinated). Food stays in the stomach several hours. Then the mixture, now called **chyme**, is squirted into the beginning of the small intestine by repeated waves of stomach contractions. Proteins and complex carbohydrates in chyme are partially digested, and fats are undigested.

The small intestine averages 6 m in length and has three sections. In the first 25 cm of the small intestine, called the **duodenum**, chyme is mixed with digestive secretions from the liver and pancreas. The next 2.5 m is the **jejunum**, where most of the digestion and absorption occurs. The last 3.5 m is the **ileum**, which performs additional absorption. It is lined with lymph nodes that protect the small intestine against bacteria from the large intestine. The inside of the small intestine, especially the jejunum, has numerous folds. The surface, especially of the duodenum and jejunum, is covered with tiny finger-like projections called **villi**. Furthermore, each epithelial cell on the villi has fingerlike projections of its plasma membrane called **microvilli**. Taken together, these projections greatly increase the intestinal surface area for absorption. This means, of course, not only adsorption of food, but potentially of toxins as well.

The **pancreas** produces about 1 L/day of digestive juices containing enzymes and zymogens. It also secretes bicarbonate alkalinity to neutralize the acid pH of the stomach by a mechanism similar to the stomach secretion of HCl in reverse. The enzymes produced by the pancreas or its secretions include the proteases *trypsin*, *chymotrypsin*, *carboxypeptidase*, and *elastase*. Each attacks peptide bonds between specific amino acids and leave the others, producing a mixture of dipeptides, tripeptides, and amino acids. It also produces amylase, lipase, and nucleases.

The **liver**, the largest visceral organ, is remarkable for the number of vital roles it plays in the body. Over 200 different functions have been identified. Besides its digestive function, it has major control of the following metabolic and blood functions:

Carbohydrate metabolism. Controls blood sugar by forming and breaking down glycogen reserves, or synthesizing glucose from fats, proteins, or other compounds.

Lipid metabolism. Maintains blood levels of triglycerides, fatty acids, and cholesterol.

Amino acid metabolism. Removes excess amino acids from circulation. Some may be destroyed by removal of the amine group, producing ammonia. This is then converted to urea, which is removed by the kidneys.

Vitamin storage. The fat-soluble vitamins A, D, E, K, and B_{12} are stored in reserve.

Mineral storage. Iron from the breakdown of red blood cells is stored in the liver.

Detoxification. Drugs and other toxins are transformed into easier-to-excrete forms.

Erythrocyte recycling. Red blood cells are broken down.

Production of plasma proteins. The liver produces albumins, various transport proteins, clotting proteins, and the complement proteins of the immune system.

Removal of hormones. The liver removes and recycles hormones such as epinephrine, norepinephrine, thyroid hormones, and steroid hormones.

Storage or excretion of toxins. Lipophilic toxins such as DDT are removed from circulation and stored in the liver. Others are excreted in the bile.

Production of bile. The last function returns us to the role of the liver in digestion. **Bile** is a digestive secretion of the liver. It is stored in the gallbladder, which releases it to the duodenum upon stimulation by cholecystokinin. Bile contains six major components: (1) bile salts (synthesized from cholesterol), (2) cholesterol, (3) lecithin (a phospholipid), (4) bicarbonate ions and other inorganic salts, (5) bile pigments such as bilirubin, and (6) trace amounts of metals. The first three act as an emulsifier for lipids in the food. The bicarbonate contributes to neutralizing stomach acid. The bile pigments and metals are excretions of the liver destined for elimination with the feces.

The blood supply of the liver is unusual. Besides the usual oxygenated blood, it receives all the deoxygenated blood from the intestines, and these are mixed in the liver. The liver then processes nutrients and may attack toxins. By receiving all the blood from the intestines, the liver serves as a first line of defense against toxins, before substances absorbed in the intestines are spread throughout the body. After they have done their work, most of the bile salts are absorbed in the ileum and recycled in the liver. The circulation of bile salts from liver to intestines and back to the liver via the blood is called the **enterohepatic circulation**. Enterohepatic circulation may also result in the recycling of lipophilic toxins.

It takes an average of 5 hours for food to move through the small intestine. Contractions gradually move the remaining unabsorbed food to the large intestine. The **large intestine** connects the small intestine to the anus. It is about 1.5 m long and 7.5 cm in diameter and consists of the cecum, the colon, and the rectum. The **cecum** is a pouch at the entrance attached to which is a small worm-shaped structure about 9 cm long called the **appendix**. The appendix is part of the lymphatic system and contains lymph nodes. The second, and major part of the large intestine is the **colon**, which absorbs water, vitamins, and minerals from the chyme. Bacteria are a natural component of the colon. They include *Escherichia coli*, *Lactobacillus*, and *Streptococcus*, collectively called the **intestinal flora**. These bacteria, especially *E. coli*, are used as indicators to detect fecal contamination of water. The bacteria in our gut benefit us by displacing harmful bacteria and by producing vitamins, including thiamine, riboflavin, B_{12}, and K. Some materials in our food are poorly digested. For example, beans contain large amounts of indigestible polysaccharides. These are used by the intestinal flora, producing gas called **flatus**, consisting mostly of CO_2 and nitrogen, plus smaller amounts of hydrogen, methane, and hydrogen sulfide.

Material takes 12 to 36 hours to move through the colon. The resulting product is about 150 g per day of **feces**, which is about two-thirds water. The 50 g of dry solids in feces is more than 60% bacteria, plus undigested polysaccharides (including dietary fiber), bile pigments, cholesterol, and salts such as potassium. The last 15 cm of the large intestine is the **rectum**, which stores feces until discharge through the **anus** (the exterior opening).

The colon absorbs water by first absorbing sodium, and the water then follows by osmosis. Some diseases, notably *cholera*, disable the main sodium transport mechanism, preventing water absorption. Cholera is transmitted by contaminated drinking water due to poor sanitary conditions. The result is **diarrhea**, the loss of large amounts of water and minerals with the feces. Diarrhea is the major cause of death in children in the developing world, killing about 4 million under the age of 5 each year. This makes the lack of proper sanitation the most important environmental problem in the world today.

Cholera by itself is not deadly. If the loss of water and minerals is compensated for, the prognosis is quite good. However, in the past this has meant intravenous injection, which was not widely available in poorer parts of the world. Over the last several decades a simpler treatment has been developed, called **oral rehydration therapy** (ORT). ORT is based on the observation that there are other mechanisms by which the intestines transport sodium that are not affected by cholera. Specifically, there is a mechanism involving the cotransport of glucose and sodium that can be stimulated by drinking a dilute solution containing both solutes, plus potassium chloride and trisodium citrate. This kind of solution has reduced cholera mortality from 50 or 60% to less than 1%.

Popular soft drinks have concentrations of sugar and other solutes that make them hypertonic. Thus, by osmosis, they increase the flow of water into the gut, water which then has to be removed. To avoid this problem, "sports drinks" are formulated to be isotonic.

The digestive system is controlled both by hormones (at least four, and perhaps up to 12) and neural control. As noted above, gastrin, is a hormone that stimulates stomach activity. The endocrine hormone *secretin* is secreted by the duodenum and stimulates the pancreas to start producing enzymes and bicarbonate alkalinity. Secretin was the first hormone to be discovered. The intestinal mucosa secrete the hormone *cholecystokinin*, which further stimulates the pancreatic secretions.

9.10   NUTRITION

Males between the ages of 19 and 75 require between 2000 and 3300 Cal in their diet. Females of the same age range need between 1400 and 2500 Cal. The body requires about 50 substances that are either not produced by it or not produced as fast as they are used. These substances, called **essential nutrients**, must be provided in the diet.

Adult females need 46 g of protein per day, males about 55 g. About 170 g (6 ounces) of lean roast beef is all that is needed to satisfy this requirement. The eight **essential amino acids:** *isoleucine, leucine, lysine, methionine, phenylalanine, threonine, tryptophan,* and *valine*, must be obtained in the diet. Furthermore, these amino acids must be present in proper proportions and taken together at the same meal. The body does not store amino acids. Individual plants may be deficient in some essential amino acids and should be consumed in combination with other plant protein that makes up for the lack. If protein is consumed in excess of requirements for maintenance and growth, the body deaminates the amino acids to produce carbohydrates or fats. The nitrogen goes into urea, which must be excreted in the urine.

Two **essential fatty acids** contain double bonds and cannot be synthesized: *linoleic acid* and *linolenic acid*. However, deficiency of these is rare. Linoleic acid can be synthesized from *arachidonic acid*. If the diet lipids are mostly saturated fats (such as from meat), the blood lipid levels may be twice as high as if polyunsaturated fats dominate the diet.

There are no essential carbohydrates. However, if carbohydrates are omitted from the diet, fats must make up the caloric requirement, since fairly small amounts of protein produce a feeling of satiety that causes people to limit their intake.

Several other compounds are important in cell metabolism and regulation. Examples are the sugar *inositol* and the precursor molecule *choline*, which are widely distributed in foods. A deficiency is generally found only in laboratory animal experiments.

The body requires about 1500 g of water per day. Carbon, hydrogen, oxygen, and nitrogen make up 99.3% of the atoms in the body. The next most important elements, called the **macronutrients**, make up about 0.7%. These are *calcium, phosphorus, potassium, sulfur, sodium, chlorine*, and *magnesium*.

Another 13 **trace elements** or **micronutrients** make up less than 0.01% of the body's atoms. Deficiency of these elements is rare. Deficiency of zinc has been observed in communities in the Middle East that use unleavened breads and eat grains high in *phytic acid*, which prevents zinc absorption. Yeast has enzymes that break down phytic acid. In addition to the elements in Table 9.5, other elements that may be essential are *cobalt, nickel, tin, vanadium*, and possibly even *arsenic*!

Vitamins are organic molecules required in very small amounts, whose absence readily produces deficiency diseases. There are 13 in all (14 if you include choline, as some people do). They are classified as either **water-soluble vitamins** (Table 9.6) or **fat-soluble**

TABLE 9.5   Mineral Total Body Reserves and Adult Requirements

Mineral	Function	Total Reserves	Recommended Daily Intake
Macronutrients			*More than 300 mg*
Sodium	Essential for membrane function, mostly in intercellular fluid and plasma	110 g	0.5–1.0 g (up to 3.3 is safe)
Potassium	Essential for membrane function, mostly in cytoplasm	140 g	1.9–5.6 g
Chloride	Major anion in body fluids	89 g	0.7–1.4 g
Calcium	Essential for muscle and neuron function and bone structure	1.36 kg, mostly in skeleton	0.8–1.2 g
Phosphorus	Part of nucleic acids, ATP, and bone structure	744 g, mostly in skeleton	0.8–1.2 g
Magnesium	Enzyme cofactor and membrane functions	29 g (17 g in skeleton)	0.3–0.4 g
Micronutrients			*Less than 20 mg*
Iron	Required for hemoglobin, myoglobins, and cytochromes	3.9 g (incl. 1.6 g in storage)	10–18 mg
Zinc	Enzyme cofactor	2 g	15 mg
Copper	Cofactor for hemoglobin synthesis	127 mg	2–3 mg
Manganese	Enzyme cofactor	11 mg	2.5–5 mg
Iodine	Thyroid hormones		150 µg
Fluoride	Essential for dental health		1.5–4.0 mg
Chromium	Affects sensitivity of tissues to insulin		0.05–0.2 mg
Selenium	Component of enzymes		60 µg
Silicon	Present in large amounts in arterial walls		
Cobalt	Component of vitamin B_{12}		
Molybdenum			0.15–0.5 mg

Source: Martini (1998).

TABLE 9.6   Water-Soluble Vitamins

Vitamin	Function	Sources	Daily Req't
B_1 (thiamine)	Coenzyme in decarboxylation of pyruvate and other compounds. Deficiency causes *beriberi*, which exhibits muscle weakness, anxiety and confusion, acute cardiac problems. Deficiency often found in alcoholics. Excess causes low blood pressure.	Outer layer of seeds, meat (esp. pork)	1.9 mg
B_2 (riboflavin)	Used to synthesize the flavins FAD and FMN. Deficiency not a prime factor in human disease, but found with other deficiencies.	Liver, yeast, wheat germ, milk, meat	1.2–1.8 mg
Niacin (nicotinic acid)	Forms coenzyme NAD and NADP. Deficiency of this and other B vitamins and protein causes *pellagra*, which exhibits dermatitis in sun-exposed skin, gastrointestinal tract deterioration, diarrhea, dementia.	Can be made from tryptophan; meat, liver, whole grain	12–20 mg
B_5 (pantothenic acid)	Part of acetyl-CoA and ultimately about 70 enzymes. Deficiency not observed in humans.	Yeast, liver, eggs, meat, milk	5–10 mg
B_6 (pyridoxine)	Three equivalent forms. Involved in amino acid conversions. Deficiency can cause convulsions.	Germ of grain, egg yolk, yeast, liver, kidney	Varies with amount of protein in diet
Folic acid (folacin)	Involved in synthesis of amino acids and nucleotides. Deficiency causes growth failure and anemia.	Widely distributed in plants and animals	400 µg; 800 µg in pregnancy
B_{12} (cobalamin)	Formed of porphyrin rings such as hemoglobin, but with cobalt in center instead of iron. Only microorganisms, including those in the gut, produce it.	Meat	4.5 µg
Biotin	Forms prosthetic group in enzymes (e.g., acetyl-CoA carboxylase). Deficiency has been caused in lab animals by feeding large amounts of raw egg whites, which has a protein that prevents biotin absorption. Symptoms of deficiency include dermatitis, fatigue, nausea.	Widely distributed; esp. in beef liver, yeast, peanuts, chocolate, eggs	10 µg, although bacteria in gut may supply needs
C (ascorbic acid)	Biochemical reducing agent (hydrogenator), antioxidant. Deficiency causes *scurvy*, deterioration of epithelial and mucosal tissues, sore gums, loose teeth, subcutaneous hemorrhage, joint pain, anemia.	Fresh fruits and vegetables, leafy greens, citrus (heat labile, i.e., destroyed by cooking)	60 mg

TABLE 9.7   **Fat-Soluble Vitamins**

Vitamin	Function	Sources	Daily Req't
A (retinol)	Precursor of vision pigments, necessary for epithelial tissues. Available in several active forms, including the *carotenes*. First symptom of deficiency is night blindness, also cessation of bone growth in children. Excess dosage causes bone fragility and other problems.	Yellow and leafy green vegetables, fish liver oils	1 mg [1000 retinol equivalents (REs)]
D	Several forms available, including *cholecalciferol*, which is formed from sterols in the skin upon irradiation by UV light. Mediates absorption of calcium and mobilization in bone. Deficiency causes *rickets* when bone mineral formation is slowed, causing curved bones to form. Ten times the normal level can cause bone fragility.	Fish liver oils, fortified milk	5 µg, or equivalent exposure to sunlight
E (tocopherols)	Protects vitamin A against oxidation, protects red blood cells from oxidation by peroxides, may protect plasma membranes. In lab animals deficiency causes infertility, hepatic necrosis, RBC lysis, vitamin A deficiency, and muscular dystrophy. In humans, deficiency is associated with poor lipid absorption.	Plant oils (esp. wheat germ)	12 mg
K	Required for liver production of several blood clotting proenzymes. Deficiency may result in bleeding or hemorrhage. Newborns may be deficient due to intestinal flora not yet being established.	Intestinal flora, vegetables, meat	0.7–0.14 mg

vitamins (Table 9.7). The water-soluble vitamins are mostly components of coenzymes. The fat-soluble vitamins have a variety of specialized functions. They are produced by plants and obtained by eating either plants or animals that eat plants.

Despite the beneficial reputation that vitamins have, an excess of vitamin intake, particularly of the fat-soluble vitamins, can produce toxic symptoms collectively called **hypervitaminosis**. Vitamin A is the most often overdosed vitamin. Children who received 500,000 RE per day showed tender swelling over the long bones. Higher dosages cause severe headache, nosebleed, anorexia, nausea, weakness, and dermatitis. Chronic excess vitamin D may leave bones prone to fracture.

Other growth factors may await discovery. Some deficiencies are difficult to develop, either because intestinal flora satisfy the normal requirement, or because a long-term supply is stored in the body or even transmitted from mother to child in utero.

9.11   EXCRETORY SYSTEM

Excretion is the elimination of waste products from the body. We excrete substances mostly in the urine or the feces, but also by sweat, milk, other body fluids, and even in our hair. Excretion is important from an environmental viewpoint for two reasons: (1) It is a means by which the body eliminates toxic substances; and (2) excretory organs may themselves be susceptible to the action of toxic substances, damaging their ability to maintain homeostasis.

In this section we focus on urinary excretion, that is, on kidney function. The kidneys are one of the most fascinating organs from an engineering point of view. They are at the center of several important control mechanisms, and their excretory function illustrates several important principles of mass transfer. The kidneys have a wide variety of functions, like the liver, but less so. Only the first four of these functions have to do with excretion:

- Regulation of water and electrolyte balance, including blood pH.
- Removal of metabolic waste products, such as nitrogen wastes, from the blood.
- Removal of foreign chemicals from the blood.
- Control of blood pressure and volume through the secretion of *renin*, which ultimately affects water and sodium excretion (Section 9.5.2).
- Control of red blood cell formation through the secretion of *erythropoietin* (Section 9.6).
- Control of calcium levels by formation of the active form of vitamin D.
- Conversion of amino acids into glucose during prolonged fasting.

Wastes from the metabolism of carbohydrates and fats produces CO_2 and metabolic water, which do not need to be excreted by the kidney. However, the breakdown of nitrogenous compounds such as amino acids produces ammonia, which would be toxic if accumulated in the blood. The liver converts the ammonia to urea, which is relatively nontoxic. We excrete about 21 g of urea per day, plus small amounts of ammonia and uric acid. Some animals excrete nitrogen mostly as uric acid, which is insoluble. This further improves water conservation. The white material in bird droppings is mostly uric acid. We also produce about 1.8 g of *creatinine* per day for excretion, a breakdown product of creatine phosphate. Sodium, potassium, and chloride are lost with the urine and must be replaced in the diet. To conserve water, the kidneys concentrate these solutes to a total concentration over four times that of blood plasma. The kidney must remove these substances from the blood without losing vital solutes such as sugars and amino acids. The kidney concentrates solutes from the osmolarity of blood plasma, about 300 mOsmol/L, by more than four times, to 1200 to 1400 mOsmol/L.

The kidney has three major regions, arranged in layers (Figure 9.13). The outer layer is the **renal cortex**, the middle layer is the **renal medulla**, and the innermost part is a cavity the **renal pelvis**. The fundamental unit of the kidney's mass transfer and urine production is a tiny tubule called a **nephron**, which is surrounded by blood vessels. The nephron starts in the cortex, passes into the medulla, and then back out to the cortex, where it connects to a collecting duct that conducts the urine into the pelvis. Each of the two kidneys is connected by a **ureter** to the **bladder**, which stores urine until a person is ready to release it by **urination**. The tube that drains the bladder to the outside of the body is called the **urethra**.

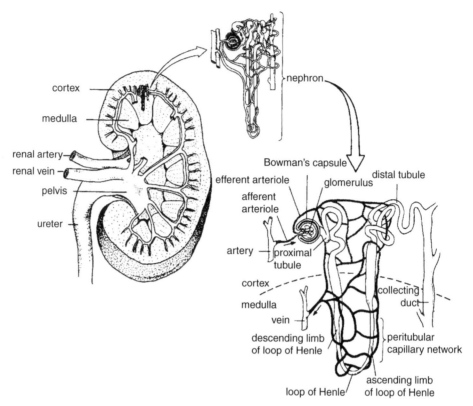

Figure 9.13 Human kidney with details of the nephron. (From Fried, 1990. © The McGraw-Hill Companies, Inc. Used with permission.)

The nephron is about 50 μm long, 15 to 60 μm in diameter, and has four main parts. The ball-shaped **glomerular capsule** in the cortex contains a tuft of about 50 capillaries called the **glomerulus**. This connects to the **proximal tubule**, which, in turn, is connected to the **loop of Henle**, with a descending limb that passes down into the renal medulla and an ascending limb that goes back into the cortex. Finally, the **distal tubule** passes through more of the cortex before emptying into a **collecting duct**. Each of these four sections of the nephron play a role in forming urine using three distinct processes: (1) glomerular filtration, (2) tubular resorption, and (3) tubular secretion.

The capillaries of the glomerulus have pores with a diameter between 50 and 100 nm. In **glomerular filtration**, blood pressure forces plasma through these pores into the nephron. Blood cells and most blood proteins are retained, but the liquid, salts, and small organic molecules pass through to form a filtrate. The kidneys receive about 20 to 25% of the blood flow from the left ventricle of the heart. Of this, some 10%, about 125 mL/min, passes into the nephrons. This filtrate has a composition similar to plasma, except without the large proteins. The filtration rate is maintained at a fairly constant rate within the kidney but can be modified by hormones or by the autonomic nervous system. A disorder called **glomerulonephritis** can follow a severe infection of bacteria or viruses. The infection can produce a high concentration of antibody–antigen complex to circulate in the blood. The complexes plug the pores of the glomerulus, reducing the filtration rate and producing an inflammation of the renal cortex.

As the filtrate passes along the nephron, various materials are removed or added by diffusion, osmosis, or carrier-mediated transport such as facilitated transport active transport, cotransport, or countercurrent transport. Carrier-mediated transport has a limited capacity, and if the blood concentration of the transported compound exceeds a threshold, the excess will be lost in the urine. This is the fate of much of the water-soluble vitamins taken in high concentrations in pill form. After a high-sugar meal, your blood sugar may briefly exceed the threshold of 180 mg/dL, and some sugar will be lost. This is a chronic problem with diabetics.

Using facilitated transport and cotransport, the proximal tubule removes 99% of organic nutrients such as sugars, amino acids, and vitamins, as well as drugs and toxins, from the urine. It is an important site for the removal of peptide hormones, such as insulin, from the blood. Many ions, such as sodium, potassium, magnesium, bicarbonate, phosphate, and sulfate, are actively transported out of the urine. Water follows the solutes by osmosis. Overall, the proximal tubule reduces the volume of the filtrate by 60 to 70%. Some secretion also occurs, as described below. The proximal tubule and the loop of Henle remove most of the calcium filtered by the glomerulus.

The osmolarity of the urine is still at the level of plasma, about 300 mOsmol/L. The urine now passes to the loop of Henle, where half the remaining water and two-thirds of the sodium and chloride ions are removed. The descending loop runs parallel to the ascending loop through the medulla. The thin descending loop is permeable to water but not to solutes. The thick ascending section is impermeable to both, but has active transport mechanisms that pump sodium and chloride out of the tubule and into the medulla. The collecting duct also passes through the medulla, and urea diffuses from the urine in the duct into the medulla. The sodium chloride and the urea increase the osmolarity of the medulla near the turn in the loop to about 1200 mOsmol/L. As a result, water passively diffuses out of the descending loop by osmosis. As the urine passes up the ascending loop, the removal of sodium chloride reduces the osmolarity to levels below that of plasma, as low as 100 mOsmol/L. This mechanism for the removal of water and salts involving opposite flow directions of the descending and ascending loops is called the **countercurrent multiplier effect** (Figure 9.14).

The total flow is now about 15 to 20% of the original filtrate, and continues on to the distal tubule. Both the proximal and distal tubules are active in tubular secretion. Drugs such as penicillin and phenolbarbitol are removed from the blood in this way. The urinary drug testing of athletes is possible because of tubular secretion. Sodium and chloride is removed from urine by active transport, but at the expense of two potassium ions for each three sodium ions. This is stimulated by the hormone *aldosterone*, which, as was noted above, conserves water in response to stress, but produces potassium loss. The distal tubule also secretes hydrogen ions and exchanges them for bicarbonate. This gives the kidneys some control over blood pH. Both the proximal and distal tubules produce and secrete ammonia as a way to remove hydrogen ions from the blood without decreasing urine pH excessively.

Contrary to what the name implies, the collecting ducts do more than just act as pipelines. They are critical for the final processing of urine and in the kidney's role in controlling blood pressure and volume. The walls of the ducts are permeable to water. As the duct passes into the high-salinity medulla, water is removed by osmosis until it is in equilibrium with the medulla at an osmolarity approaching 1200 mOsmol/L. At this point, its volume has been reduced to about 1% of the amount filtered in the glomerulus. The permeability of the distal tubule and the collecting ducts is controlled by the hormone ADH

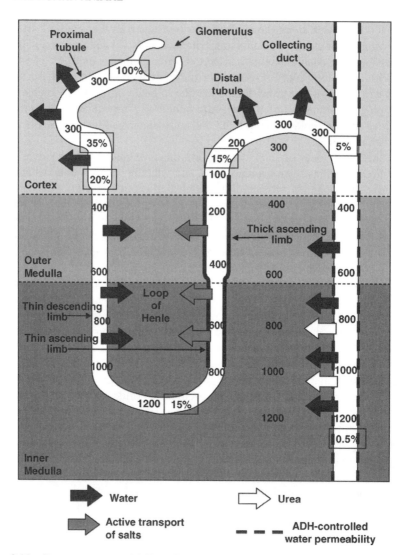

Figure 9.14 Countercurrent multiplier effect in the nephron. Percentages refer to fraction of glomerular flow remaining. The other numbers inside the nephron are the milliosmolarity. (Based on Smith et al., 1983.)

(vasopressin). In the absense of the hormone, the ducts become impermeable to water. No water is absorbed from the distal tubule on, and the person secretes large amounts of dilute urine. This is what occurs in the disease *diabetes insipidus* (not to be confused with the insulin production disorder diabetes mellitus). Normal persons secrete ADH continuously to closely control water recovery. ADH is opposed by the hormone ANP that is produced by the heart (Section 9.5.2). ANP increases glomerular filtration, suppresses sodium absorption by the distal tubule, blocks release of ADH and aldosterone, and inhibits the response of the distal tubule and collecting ducts to ADH and aldosterone.

The water and solutes removed from the filtrate reenter blood vessels that are intimately associated with the nephron. These blood vessels, not shown in Figure 9.14, form a loop parallel to the loop of Henle after passing through the glomerulus. The blood picks up solutes as it passes down into the medulla, increasing its osmolarity to about 1200 mOsmol/L. It then loops back up to the cortex, absorbing water as it goes, until its osmolarity returns to a normal 300 mOsmol/L.

The pH of urine is normally between 5.5 and 6.5, but may range from 4.6 to 8.0. The value is influenced by diet. Drinking milk can produce acidic urine with a pH about 6.0. A diet high in fruits and vegetables produces alkaline urine. The concentration of urea varies directly with nitrogen in the diet, particularly protein. Creatinine is excreted in proportion to a person's muscle mass.

Some of the normal constituents of urine may precipitate in the ureter or urethra, forming **kidney stones**. One-third of these are associated with alkaline urine or calcium problems and include $Ca_3(PO_4)_2$, $MgHN_4PO_4$, $CaCO_3$, or a mixture of these. About half of all kidney stones are calcium oxalate, caused by eating large amounts of spinach or rhubarb, which have high levels of oxalic acid. Stones may also be formed from organics such as uric acid.

9.12   REPRODUCTION AND DEVELOPMENT

The reproductive system is the only system that is not essential to a person's survival. It is, however, essential to the survival of the species. Because cells replicate continually, they are subject to errors of replication, either due to the inborn rate of error or because of environmental agents such as chemical pollutants or radiation. This makes the reproductive organs susceptible to diseases associated with genetic damage. These range from cancer, which affects the individual, to birth defects and mutations, which affect a person's offspring. Some pollutants are thought to mimic the sex hormones, affecting reproduction and development.

The main reproductive organs are the **gonads**: the **testes** in the male and the **ovaries** in the female. The gonads have two functions: production of haploid gametes by meiosis and the production of the steroidal sex hormones. The male gametes are **sperm** and the female gametes are **ova** (singular, *ovum*). In addition, there are a number of **accessory reproductive organs**. These include the ducts that transport sperm and ova, and associated glands.

Each ejaculation releases 2 to 5 mL of **semen**, the fluid containing the sperm. Semen normally contains between 20 and 100 million sperm per milliliter. The sperm consists of a 5-μm head containing the nucleus, a midpiece containing mitochondria, and a 55-μm-long flagellum, or tail, which can propel the sperm at a speed of 1 to 4 mm/min. The **seminal vesicles** and the **prostate gland** produce much of the liquid that makes up semen, which includes fructose to provide energy for the sperm, and alkalinity to counter the acidity of the vagina. When the male becomes sexually excited, nerves of the autonomic nervous system produce dilation of arterioles in the penis, causing blood to enter that organ faster than it can leave. This causes the **erectile tissue** to become engorged, causing an **erection**. The male then inserts the penis into the female's **vagina** in the act of **copulation**, injecting the semen by rhythmic contractions called **ejaculations**.

The ovaries contain about 400,000 ova arrested in prophase of meiosis I since birth. Some 400 of these will eventually develop completely, one per menstrual period. Cilia along the **fallopian tubes** transport the ova to the **uterus**. Fertilization may occur in either of these organs, and development takes place mostly in the latter.

The main sex hormone in the male is **testosterone**. It is produced mainly by the testes. It is part of a group of hormones produced by both males and females in the adrenal cortex that have a masculinizing effect. Thus, they are collectively called **androgens** ("man-maker"). The androgens other than testosterone are relatively low in potency. The main sex hormones in the female are **progesterone** and a group of hormones called the **estrogens**. The most important estrogen is called **estradiol**.

Production of sex hormones is controlled by a series of hormones originating with the brain stimulating the hypothalamus to release **gonadotropin-releasing hormone** (GnRH). GnRH is released in pulses and stimulates the anterior pituitary to release **follicle-stimulating hormone** (FSH) and **luteinizing hormone** (LH). Although these names are based on their functions in females, they are produced in both sexes.

In males, FSH stimulates the testes to produce sperm and another hormone called **inhibin**. The inhibin inhibits FSH production, thus forming a negative feedback loop to control sperm production. LH stimulates the testes to produce testosterone. The testosterone inhibits GnRH and LH secretion, forming a negative feedback control loop. Testosterone has many functions, including (1) combining with FSH to stimulate sperm production, (2) affecting aggressive behavior and sexual desire through the CNS function, (3) stimulation of protein synthesis and muscle growth, (4) stimulation of secondary sex characteristics such as facial hair, (5) maintaining accessory organs, and (6) in fetuses it stimulates the formation of male reproductive system. In contrast to females, GnRH production in males is relatively constant from hour to hour or day to day, thus keeping its effects also at a constant level.

The hormone levels in females, on the other hand, varies in a cycle with a period of 22 to 35 (average 28) days. This cycle, together with associated physiological changes, is called the **menstrual cycle**. The menstrual cycle is unique to humans, monkeys, and apes. Females who menstruate may be receptive to males at any part of the cycle, although they can only conceive at certain points. Other mammals follow a simpler **estrous cycle**, in which the female will only copulate at certain times, called **estrus** or **heat**. Some animals, such as dogs and foxes, enter estrus only once per breeding season. Others, especially the mammals in tropical regions, have repeated estrus during the breeding season. The estrous cycle provides control over the season in which young are born.

The menstrual cycle consists of two ovarian phases or three uterine phases. In the **follicular phase** the structure containing the ovum develops into a follicle, and the ovum continues its meiosis to produce a mature ovum, or egg cell. The follicular phase ends with **ovulation**, when the ovum erupts from the surface of the ovary and enters the fallopian tube. This event marks the beginning of the second, **luteal phase** of menstruation, in which the remains of the follicle, now called the **corpus luteum** (yellow body), produces hormones that prepare the uterus for pregnancy. If pregnancy does not occur, the corpus luteum degenerates after about 12 days and the luteal phase ends.

Simultaneous with the ovarian phases, the uterus goes through its own cycle (Figure 9.15). The uterus is a muscular organ that supports and nourishes the fetus. It is lined with a glandular mucosa called the **endometrium**, which performs the nourishing function. The first phase of the uterine cycle is **menses**, in which the thickened endometrium from the previous cycle deteriorates. Over a period of 3 to 5 days it is sloughed off and discharged along with 35 to 50 mL of blood. This endometrial sloughing is called **menstruation**. Menses is followed by the **proliferative phase**, in which the endometrium regenerates inside the uterus and produces a glycogen-rich mucus. Menses and the proliferative phase coincide with the follicular phase of the ovarian cycle. After ovulation, the uterus enters the **secretory phase** (simultaneous with the ovarian luteal phase), in which

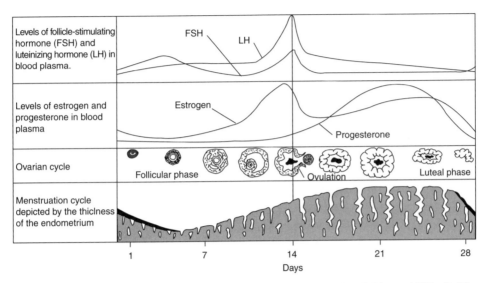

Figure 9.15 Menstrual and ovarian cycle. (From Van De Graaff and Rhees, 1997. © The McGraw-Hill Companies, Inc. Used with permission.)

further secretions occur and the endometrium is ready to receive the conceptus. (The **conceptus** is a name for the products of conception, such as zygote, embryo, fetus, and so on, including the placenta, at any stage of pregnancy.) If pregnancy does not occur, the cycle ends and menses begins.

The menstrual cycle is controlled by an intricately choreographed sequence of hormonal secretions. In males the director of the choreography is GnRH. GnRH is secreted in pulses from the hypothalamus and stimulates the anterior pituitary to produce FSH and LH. Unlike males, in females these secretions vary considerably from day to day, and the timing is controlled by feedback mechanisms involving ovarian hormones. At the beginning of the menstrual cycle an increase in GnRH produces a broad peak a few days later in FSH and a gradual increase in LH production. As its name implies, FSH stimulates development of the follicles. The follicles secrete inhibin, which exerts a negative feedback effect on FSH production, and estrogen, which exerts positive feedback on GnRH pulses and negative feedback on LH. Estrogen also stimulates the uterine endometrium to thicken. When estrogen levels exceed a threshold level for more than about 36 hours, estrogen's effect on LH switches from negative to positive feedback. This causes a massive release of LH by the pituitary, triggering the completion of meiosis I by the ovum and ovulation.

LH also stimulates formation of the corpeus luteum, which produces estrogen plus large amounts of progesterone, which inhibits GnRH pulse production and stimulates the endometrium further. If pregnancy does not occur, the corpus luteum degenerates, and estrogen and progesterone levels fall sharply. Without the inhibition from the progesterone, GnRH pulses increase again, to initiate another menstrual cycle. The drop in estrogen and progesterone also results in the degeneration of the endometrium in the uterus, leading to menses.

During the follicular phase the **basal body temperature** (the resting body temperature measured upon waking in the morning) is about 0.3°C below the temperature during the luteal phase. In addition, the temperature dips measurably at the time of ovulation. Thus, the sudden increase in temperature can be used to determine when ovulation occurs. This is used for birth control (either to prevent or increase the likelihood of conception), since fertilization is most likely to occur within 1 day of ovulation. Oral contraceptives (the birth control pill) come in several forms. The *combination pill* contains both progesterone and estrogen. They are taken starting about 5 days after the start of menses and continued for 3 weeks. They act by inhibiting GnRH production, and therefore FSH release, and ultimately preventing follicle development and ovulation.

9.12.1   Prenatal Development

Development begins when the haploid sperm and ovum unite in the process of **fertilization** (or **conception**), to form a diploid zygote. After ovulation, a layer of cells surrounds the ovum. At least 100 sperm are needed to release an enzyme that creates gaps in this layer. Then one of the sperm may penetrate the cytoplasm of the ovum. This triggers reactions that render the ovum impermeable to additional sperm. In addition, the ovum completes meiosis II. The zygote then divides by mitotic division without an increase in the size of the conceptus. This division is called **cleavage**. After one or more cleavages, or even in the blastocyst stage, the cells may separate and develop independently, resulting in genetically *identical twins*. *Fraternal twins* result when two ovulations occur and are fertilized by different sperm. In the United States about one pregnancy in 90 produces twins, and 70% of twins born are fraternal.

From fertilization until all of the organs are formed, the developing individual is called an **embryo**. In humans this stage lasts two months. For the rest of the **prenatal** (before birth) development the individual is called a **fetus**. Table 9.8 lists the events of this development.

The **amnion** is a membrane that grows to completely enclose the fetus, which floats in the **amniotic fluid** to protect it from shock. Surrounding the amnion is another membrane called the **chorion**. Eventually, as the fetus and amnion grow, the amnion fuses with the chorion, forming the single membrane whose rupture discharges the amniotic fluid, signaling the start of labor. The portion of the chorion that is in contact with the endometrium is called the **placenta**. The fetus is connected to the placenta by the **umbilical cord**. The placenta has fingerlike projections into the endometrium called **chorionic villi**. These provide surface area for the exchange of nutrients and waste products between the fetus and the mother.

The placenta also has important endocrine functions. Soon after implantation, **chorionic gonadotropin** (CG) is produced. This maintains the corpus luteum so that progesterone production is continued. Progesterone maintains the thick endometrium. After month three or four of the pregnancy, CG drops sharply and the placenta produces progesterone itself; the corpus luteum is allowed to degrade. The placenta also produces large amounts of estrogens, especially estriol, from testosterone produced by the fetal adrenal glands. The high levels of progesterone and estrogen are thought to contribute to the "morning sickness" experienced by some pregnant women. The hormone **human placental lactogen**, placental **prolactin**, plus maternal hormones prolactin and thyroid hormones prepare the mammary glands for milk production and has other effects similar to growth hormone. The peptide hormone **relaxin** prepares for birth by causing dilation of the cervix and suppressing oxytocin production by the hypothalamus.

TABLE 9.8   Events of First-Trimester Development in the Human

Time Since Fertilization	Events
1 day	First cleavage produces two-cell stage.
4 days	The embryo has developed into a solid ball of cells, called a morula.
5 days	The ball becomes hollow, and is called a blastocyst. The outer layer is the trophoblast, the smaller inner cells will form the embryo.
7–10 days	The embryo implants itself into the uterine endometrium.
10–16 days	By process of gastrulation the embryo forms first two, then three distinct germ layers: ectoderm ultimately gives rise to the skin, epithelium of mouth, nose, and anus, the nervous system, pituitary gland, and adrenal medullae; mesoderm, produces muscle, cardiovascular system, lymphatic system, kidneys, adrenal cortex, gonads and ducts, bone, linings of body cavities, all connective tissues; endoderm forms the lungs, thymus, thyroid, pancreas, most of the digestive tract including the liver, and stem cells that produce gametes.
Week 2–3	Formation of extraembryonic membranes from germ layers and the trophoblast: yolk sac, amnion, allantois, and chorion. The part of the chorion that connects with the endometrium is called the placenta.
Week 3	Neural tube forms, distinguishing dorsal, ventral, right and left sides.
Week 6	Fetal heart begins to pump.
Week 13	End of first trimester, in which organogenesis (organ formation) is begun. External genitalia become distinctive of their sex.

By the end of the first trimester, all the major organ systems have become initiated. It is during this time that the fetus is most sensitive to agents, such as drugs or other toxins, that can cause birth defects (teratogens). Development of sex organs is quite interesting. At the end of the first trimester the embryo possesses gonads and external genitals, but the genitals have not differentiated into male or female. Furthermore, the embryo possesses two sets of ducts. One is destined to become the vas deferens of males, the other the uterus, etc., of females. Which one developes is determined by a single gene on the Y chromosome that stimulates formation of the testes. The testes then produce testosterone, which stimulates development of the male ducts and external genitalia. The testes also produce a hormone that blocks development of the female ducts. The absense of testosterone results in development of female characteristics. Occasionally, hormonal disorders result in individuals who are genetically of one sex developing external genitalia of the other sex. These individuals may grow, apparently normal, to adulthood without knowing that they are genetically of the opposite sex, until they find out that they are unable to have children.

In the second trimester, development continues. In the third trimester most organ systems become ready to perform their normal functions. A number of changes occur in the mother to support the pregnancy: Respiration rate and tidal volume increase, blood volume increases, nutrient and vitamin requirement increases 10 to 30%, causing increased hunger, glomerular filtration in the kidney increases 50% along with urine production; the uterus increases in size; and mammary glands increase in size and begin secretory activity.

After 280 days (40 weeks) the fetus is said to be **full term** and is ready for birth. Medical technology makes it possible to deliver premature babies safely as early as the twenty-seventh week. The entire process of birth is called **parturition**.

Estrogen levels increase throughout the pregnancy, making the uterine muscle increasingly sensitive and likely to start contractions. Progesterone has an inhibiting effect, but estrogen production increases faster just before birth. Estrogen also increases the sensitivity of uterine muscle to **oxytocin**. Oxytocin is produced by the hypothalamus but released by the posterior pituitary. Its release is stimulated by increasing estrogen levels and by distortion of the uterine cervix by the weight of the fetus. Estrogen and oxytocin stimulate production of prostaglandins in the endometrium that also stimulates muscle contractions. Once these factors reach a critical level, parturition begins and is maintained by a positive feedback loop. Parturition occurs in three stages. (1) In the **dilation stage** the cervix dilates completely, contractions occur at increasing frequency, and the amniochorionic membrane ruptures, releasing the amniotic fluid (the "water break"). This stage usually lasts 8 hours or more. (2) In the **expulsion stage**, which lasts less than 2 hours, contractions occur at maximum intensity, and the newborn infant, or **neonate**, is delivered through the vagina. (3) In the **placental stage** the uterus contracts to a much smaller size, tearing the placenta from the endometrium. Within an hour of delivery the placenta is delivered in what is called the **afterbirth**.

PROBLEMS

9.1. Which organ systems are most susceptible to oxygen deprivation, and why?

9.2. Nerve transmission is often described as an electrical transmission. In what sense is this true? Is there any transport of electrons, as in electrical current in wires?

9.3. Which will cross the blood–brain barrier more easily: ethanol or chloroform? Which class of hormones will cross most easily?

9.4. How would hydrophobic toxins such as chloroform be carried by the blood?

9.5. Name two organs that both remove and are affected by toxins.

REFERENCES

Fried, G., 1990. *Schaum's Outline: Theory and Problems of Biology*, McGraw-Hill, New York.

Hickman, C. P., Jr., L. S. Roberts, and Allan Larson, 1996. *Integrated Principles of Zoology*, Wm. C. Brown, Dubuque, IA.

Hirschhorn, N., and W. B. Greenough III, 1991. Progress in oral rehydration therapy, *Scientific American*, Vol. 264, No. 2, pp. 50–56.

Martini, F. H., 1998. *Fundamentals of Anatomy and Physiology*, Prentice Hall, Upper Saddle River, NJ.

Needham, J. G., and P. R. Needham, 1962. *A Guide to the Study of Freshwater Biology*, Holden-Day, San Francisco.

Smith, E. L., et al., 1983. *Principles of Biochemistry: Mammalian Biochemistry*, 7th ed., McGraw-Hill, New York.

Van de Graaff, K. M., and R. W. Rhees, 1997. *Schaum's Outline: Theory and Problems of Human Anatomy and Physiology*, McGraw-Hill, New York.

Vander, A. J., J. H. Sherman, and D. S. Luciano, 1994. *Human Physiology: The Mechanisms of Body Function*, McGraw-Hill, New York.

10

MICROBIAL GROUPS

Some organisms are too small to be seen with the naked eye. Such organisms are referred to as **microorganisms**, or **microbes**, from the Greek word *mikros* (small), and their study is referred to as **microbiology**. This is not a true taxonomic grouping since not only is there tremendous diversity among microbes (including all three domains), but some species may be microscopic whereas their close relatives are readily visible (macroscopic).

Although individual microorganisms are small, their collective impact is not. Their activities led to the development of Earth's present aerobic atmosphere, are a major factor in biogeochemical cycling of elements, and have a substantial impact on every ecosystem. Their numbers are staggering, typically ranging from a million to a billion (10^6 to 10^9) per gram of soil, biofilm, or sludge. They have also been widely used by humans to make products ranging from foods and medicines to industrial chemicals. On the other hand, they may cause food to spoil and materials to corrode or degrade.

To the environmental engineer and scientist, microbial activity can be both a blessing and a curse. On the positive side, microbes often afford the most efficient and cost-effective means of treating many of society's wastes. Thus, they are used routinely in engineered waste treatment systems such as sewage treatment plants. They are also of critical importance in the recovery processes of natural environments degraded by human activities, such as in the self-purification of streams receiving sewage and runoff, and the natural attenuation of industrial contaminants leaked or spilled onto soil.

On the other hand, microorganisms have the potential to create substantial environmental problems. For example, they may deplete oxygen, generate unpleasant tastes and odors, clog equipment, and corrode pipes. Perhaps of greatest concern, however, is that some microorganisms are capable of producing disease in plants and animals, including, of course, humans.

In this chapter we consider the prokaryotic groups, Bacteria (including blue-green algae) and Archaea. We also examine the eukaryotic groups containing single-celled organisms: protozoans, algae, fungi, and slime molds, even though they (except protozoans) also include many multicellular, macroscopic species. The **metazoans**, microscopic

Environmental Biology for Engineers and Scientists, by David A. Vaccari, Peter F. Strom, and James E. Alleman
Copyright © 2006 John Wiley & Sons, Inc.

animals such as rotifers and nematodes, were covered in their respective phyla in Chapter 8. We also consider noncellular biological forms, which are actually submicroscopic (for the light microscope): viruses, viroids, and prions.

10.1   EVOLUTION OF MICROBIAL LIFE

It is believed that Earth is 4.6 billion years old. The oldest known rocks are almost 4 billion years old, and because some of these are sedimentary rocks, it is believed that liquid water has been present for at least that long. Fossilized evidence of microbial life dates back at least 3.5 billion years, and undoubtedly developed even earlier.

The atmosphere of the early Earth was much different from that of today. Although elemental nitrogen (N_2) was present, oxygen was not. Rather, carbon dioxide, hydrogen, and ammonia were abundant. Under such reducing conditions, it has been shown that many of the organic molecules characteristic of life can be formed if sufficient energy—such as from ultraviolet light, lightning, and volcanic activity—is available.

The best present theory of the evolution of early life is that it started with self-replicating RNA molecules. These might then have been incorporated in lipoprotein vesicles. Later, enzymatic proteins were incorporated as better catalysts than the original RNA, which was retained for its coding function. Eventually DNA, because of its greater stability, replaced RNA as the carrier of genetic information.

This organization, utilizing DNA, RNA, and proteins, has been so successful that it apparently has been the basis of all cells on Earth for billions of years. Because of their fundamental similarity at the cellular level, it is believed that all three domains—Bacteria, Archaea, and Eukarya—developed from a "universal ancestor" among these early forms. Table 10.1 shows some of the highlights of the evolution of microorganisms from these early beginnings. Since free oxygen was not present, early cells were necessarily anaerobic. Although it is believed that the three domains separated fairly early, cells similar to those of modern eukaryotes did not evolve until much later.

About 3 billion years ago, cyanobacteria (previously called blue-green "algae") evolved from photosynthetic anaerobes. This had major consequences for life on Earth because they produced oxygen as a waste product. Over a period of hundreds of millions

TABLE 10.1   Key Evolutionary Steps for Microbial Life

Time Frame (billion years before present)	Duration (billion years)	Geological and Biological Activity	Geologic Time (%)
~4.6–3.9	0.7	Earth formed; no life; chemical evolution	~15
~3.9–2.9	1.0	Origin of life; anaerobic environment	~22
~2.9–2.0	0.9	Oxygen production by cyanobacteria; emergence of aerobic bacterial life	~20
~2.0–0.8	1.2	Shift to aerobic atmosphere; emergence of more complex eukaryotic cells	~26
~0.8–0	0.8	Development of more advanced life	~17

of years, this oxygen accumulated to the point where it became a major atmospheric constituent. This meant that anaerobes, for which oxygen is often toxic, became limited to environments in which oxygen did not penetrate. However, it also made aerobic life possible. The first aerobes were probably bacteria, but because of its much more favorable energetics, aerobic metabolism also meant that the evolution of higher life, especially eukaryotes, could occur.

It is now generally believed that most modern eukaryotic cells actually contain what originally were bacterial **endosymbionts**. Mitochondria are considered to have originated as aerobic bacteria that entered and grew symbiotically within the eukaryote's cytoplasm. The chloroplasts of algae and plants similarly evolved from cyanobacteria.

10.2 DISCOVERY OF MICROBIAL LIFE

10.2.1 Antonie van Leeuwenhoek

Antonie van Leeuwenhoek (Figure 10.1) was the first to use magnifying lenses for the study of microbial life, which he referred to as *animalcules* (Figure 10.2). Leeuwenhoek

Figure 10.1   Antonie van Leeuwenhoek. (Portrait by Jan Verkolje, 1686.)

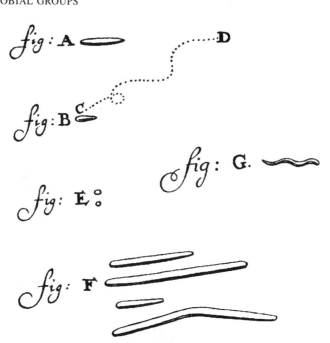

Figure 10.2   Some of Leeuwenhoek's *animalcules* from the "Scurf of the Teeth," drawings of bacterial shapes. A and B appear to show rods, with C and D showing the movement of B; E shows cocci; F, rods or filaments; and G, a spiral. From Leeuwenhoek's letter of 1683.

was a Dutch textile merchant, who probably used his magnifying lenses initially to study the quality of textile weaves. Although he lacked formal academic training, his letters (which carefully documented his findings) in the 1670s and 1680s to England's newly formed Royal Society captured an immediate, and undoubtedly astonished audience with many of the world's most prominent scientists. Leeuwenhoek's microscope (Figure 10.3) was deceptively simple, with a single, nearly spherical lens (his records indicated that he ground more than 400 such lenses in his lifetime) affixed to a back plate; the distance from an opposing mounting pin was coarsely adjusted with focusing screws.

This device provided magnifications ranging from 50 to 300 power. The realm newly revealed beneath this tool offered an astounding array and abundance of life, leading him to comment that "there are more animals living in the scum of the teeth in a man's mouth, than there are men in an entire kingdom." Indeed, the world depicted within Leeuwenhoek's extremely accurate drawings covered a wide range of *animalcules*, which are now recognized as bacteria, protozoans, algae, and fungi.

10.2.2   Spontaneous Generation and the Beginnings of Microbiology

With the surprising existence of these microbes discovered, considerable debate as to their origin developed over the next several centuries. Since ancient times, and extending well into the Renaissance, the concept of **spontaneous generation** had been widely accepted with some forms of life, including weeds and vermin. The fundamental premise of this theory was that nonliving matter developed into viable life-forms through some

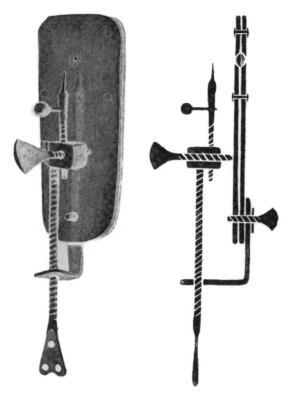

Figure 10.3   Leeuwenhoek's microscope.

form of unknown, spontaneous transformation. Leeuwenhoek offered a competing suggestion that these microbes were formed from "seeds" or "germs" released by his *animalcules*.

An Italian physician, Francesco Redi, had offered support for such a view with his studies (circa 1665) of the growth of maggots on putrefying meat. As opposed to spontaneous growth, his findings showed that these maggots actually represented the larval stages of flies that laid eggs on the unprotected surfaces. Meats held either in a closed vessel or covered with fine gauze (protected from direct fly contact) exhibited no such growth. Italian naturalist Lazzaro Spallanzini, roughly five decades later, used heat to prevent the appearance of *animalcules* in sealed infusions. Spurred by a prize of 12,000 francs established by Napoleon Bonaparte to find a means of preserving fresh foods for his soldiers, French inventor Francois Appert used Spallanzini's heating strategy in 1795 to develop a commercial *appertization* process for preparing canned foods. In 1856 the eminent French chemist Louis Pasteur (Figure 10.4) similarly applied this knowledge about the use of controlled heating, developing his **pasteurization** process as a means of carefully protecting wines.

The success Pasteur enjoyed with pasteurization prompted his subsequent research focus on microbial metabolism, by which he finally laid the spontaneous generation premise to rest. Using swan-necked flasks (Figure 10.5), he showed that sterile solutions would not yield any "germ" growth unless reexposed to the particles within contaminated air (proving that it was not the air itself that "generated" new life).

Figure 10.4   Louis Pasteur. (Portrait courtesy of Robert A. Thom. Reproduced with permission of Pfizer Inc. All rights reserved.)

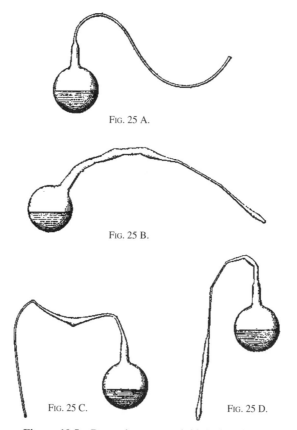

FIG. 25 A.

FIG. 25 B.

FIG. 25 C.   FIG. 25 D.

Figure 10.5   Pasteur's swan-neck biological flasks.

During the era of inquiry regarding spontaneous generation, British naturalist John Tuberville Needham had conducted another sterilization study (1740), whose results at the time were, considered inconsistent and inconclusive. Since his heated broths still generated new cells, some suspected that his methods had been flawed. However, several years after the spontaneous generation concept had effectively been laid to rest, English physicist John Tyndall (1870) discovered that hay infusions subjected to prolonged boiling (and seemingly sterilized) could, in fact, still germinate new viable cells. What he found, though, was that these cultures had not developed through spontaneous means or improper handling, but rather, from the growth of heat-resistant agents. Tyndall's work was confirmed by German botanist Ferdinand Cohn, who visually confirmed the presence of these heat-resistant bodies, known today as **endospores**. Tyndall's subsequent work with these sporulating cells also led to his development of a sterilization procedure (*tyndallization*) by which discontinuous heating effectively kills all cells, even those able to develop such spores.

Inspired by Pasteur's efforts, Cohn published a three-volume treatise on bacteria in 1872 that many consider to be the true origin of the field of bacteriology. This book offered several classic insights, including the first attempt at bacterial classification and the first description of bacterial spores.

10.2.3 Virus Discovery

A new means of filtering solutions had been developed during the 1880s at the Pasteur Institute that would effectively remove minute bacterial cells from suspensions. However, Russian scientist Dimitri Ivanowsky in 1892 found that tobacco plants were still susceptible to infectious attack by a filtered suspension. This observation led to a conclusion that even smaller and hitherto unknown viral particles were present. Martinus W. Beijerinck produced a similar set of findings independently.

10.2.4 Discovery of Archaea

Relatively recently (1970s), largely as a result of the work of Carl Woese and his group on 16S ribosomal RNA analysis (Section 10.4.4), a startling discovery was made. A number of the microorganisms that had long been called bacteria were actually examples of an entirely different form of life. This new group was at first called Archaebacteria, and the true bacteria were then called Eubacteria. However, in view of how different the two groups truly are (more different than plants are from animals, for example), microbiologists now refer to the two as Archaea and Bacteria.

There was considerable opposition at first to the idea that these were such distinct groups. Even today, when it is widely accepted, many people still refer to members of Archaea as "bacteria." However, since both are prokaryotes, this term can, when needed, be used to refer to the two together.

10.3 DIVERSITY OF MICROBIAL ACTIVITIES

When viewed on the basis of elemental or biochemical composition, the fundamental ingredients of life are surprisingly similar among the three domains and widely divergent species. However, there is a wide range of diversity within the world of microorganisms in terms of *survival strategy*: where they find energy, how they grow, and what environments they prefer. This section provides a brief overview of these alternatives.

10.3.1   Energy Sources

The two major sources of energy are chemical oxidation and photosynthesis. Cells that draw their energy from the conversion of reduced chemical substrates (including both organic and inorganic materials) are referred to as **chemotrophs**; those that use light energy are known as **phototrophs** (Table 10.2). Phototrophy is common to algae, cyanobacteria, and some other bacteria, but does not occur (with few exceptions) in other microbial forms. Furthermore, not all phototrophs are strictly dependent on light energy; many (including green and purple nonsulfur bacteria and some cyanobacteria; see Table 10.4) are able to chemotrophically oxidize chemical substrates in the dark.

Chemotrophs can be further divided between **organotrophs**, which oxidize organic compounds, and **lithotrophs** (from the Greek *lithos*, meaning "rock"), which utilize a variety of inorganics (such as hydrogen or reduced inorganic nitrogen, sulfur, or iron compounds) as their energy source.

10.3.2   Carbon Source

Since it is a major constituent of cell materials, all organisms need a source of carbon. **Heterotrophs** (including fungi, protozoans, and most bacteria) require organic carbon, whereas **autotrophs** (algae and some bacteria) consume inorganic carbon (carbon dioxide or bicarbonate). These terms can be combined with those above to refer, for example, to chemolithoautotrophs (Table 10.2). The term *heterotrophy* is also used more loosely to refer to chemoorganotrophy, since the two categories usually coincide (microbes using organic carbon as an energy source also use it for a carbon source). However, some lithoautotrophs are able to utilize amino acids, hence getting some of their carbon from organic sources, and a few lithotrophs, such as most species of the sulfur-oxidizer

TABLE 10.2   Categorization of Organisms by Energy and Carbon Source, with Examples

		Energy Source Used		
		Phototroph	Chemotroph	
			Organotroph (Organic Carbon)	Lithotroph (Inorganic Compounds)
Carbon source used	Autotroph (CO_2, HCO_3^-)	Photoautotroph: plants, algae, cyanobacteria	Chemoorganoautotroph or organoautotroph: methanotrophs	Chemolithoautotroph or lithoautotroph: nitrifying bacteria, *Thiobacillus*
	Heterotroph (organics)	Photoheterotroph *Chloroflexus*, some cyanobacteria	Chemoorganoheterotroph or "heterotroph": animals, protozoans, fungi, *Pseudomonas*, *Escherichia*	Chemolithoheterotroph or mixotroph:[a] most *Beggiatoa*

[a]Some primarily chemolithoautotrophic microbes can also utilize amino acids.

Beggiatoa, get their carbon from small organic molecules. Such organisms are often referred to as **mixotrophs**.

10.3.3   Environmental Preferences

Microbial cells are also commonly classified on the basis of the environments they prefer. Several factors are generally considered, including the presence of oxygen, temperature, salt tolerance, and pH.

Oxygen   Energy-yielding metabolism removes electrons from (oxidizes) a substrate (electron donor); these electrons must eventually be "discarded" (see Chapter 5). **Aerobic** respiration utilizes molecular oxygen (O_2) as the terminal electron acceptor, reducing it to water (H_2O). **Strict** aerobes require oxygen; cells able to grow at very low oxygen levels may be referred to as **microaerophilic**.

Microorganisms that can grow in the absence of oxygen are **anaerobic**. **Strict** or **obligate** anaerobes cannot grow in the presence of oxygen, and may actually be killed by exposure to air if they are not **aerotolerant**. **Facultative** anaerobes, on the other hand, can grow with or without oxygen. Anaerobic metabolism may be respiratory (using a variety of inorganic terminal electron acceptors such as nitrate, nitrite, ferric iron, sulfate, or carbon dioxide) or fermentative (using an organic terminal electron acceptor).

Strictly speaking, **anoxic** means in the absence of oxygen, and thus is equivalent to anaerobic. In environmental engineering and science, however, *anoxic* is frequently used to refer to conditions in which oxygen is absent, but nitrate and/or nitrite are present. The ability to utilize nitrate and/or nitrite as alternative terminal electron acceptors (**denitrification**), in the absence of oxygen, is widespread among many diverse groups of aerobic bacteria. Historically, this distinction was of particular importance because unpleasant "anaerobic" odors such as hydrogen sulfide ("rotten eggs") are not produced under "anoxic" conditions.

Temperature   Microorganisms have preferred temperature ranges. **Psychrophilic** microbes thrive under cold temperature conditions, ranging from below 0°C to the mid-teens. Bacteria adapted to living in polar ocean waters (or modern refrigerators), for instance, would be classified as psychrophiles. Organisms that prefer moderate temperatures are referred to as **mesophilic**. In this case, their temperature preferences range from the mid-teens to the high-30s or low-40s °C. The vast majority of microbial life would fall within this category. A relatively few organisms, mainly bacteria, archaea, and fungi, prefer elevated temperatures above 45 to 50°C and are called **thermophilic**. Some prokaryotic **extremophiles** (tolerate or prefer an extreme environmental condition) are **hyperthermophilic** (temperature optimum above 80°C), a few even growing at temperatures above 100°C. Organisms that can tolerate high temperatures despite having mesophilic temperature optima may be referred to as **thermotolerant**.

Salinity   Microbial cells typically maintain a different ionic strength (salt level) within their cytoplasm than that found outside the cell. Water tends to migrate across the cell membrane toward the higher salt zone by osmosis, thereby "attempting" to dilute it and eventually equilibrate the inner and outer salt levels. Microbial cells can resist this fluid movement (and the harm it may cause by excessive shrinking or swelling), and the resulting osmotic pressure, to varying degrees. **Halophilic** (salt-loving) microbes require NaCl; extreme halophiles may require concentrations above 15% and tolerate up to 32% NaCl (saturation). (Seawater contains approximately 3% NaCl.) At the other extreme, there are also many nonhalophilic cells that favor (or require) far lower levels of external salt, such as those found in fresh water. Cells that can tolerate but do not require elevated salt concentrations are referred to as **halotolerant**.

pH   Most microorganisms have a pH preference that falls within the range 5 to 9, and thus would be labeled **neutrophiles**. Outside this range, the metabolic activity of a neutrophile can be expected to decline rapidly, particularly as extreme pH values start to affect the ionic (i.e., electrically charged) properties of their constitutive functional groups. Some neutrophiles, such as many algae, nitrifying bacteria, and methanogens (Archaea) prefer slightly alkaline conditions. This is true of many marine organisms as well, since seawater typically has a pH of 8.3.

However, there are many microorganisms that are able to tolerate, or that even prefer or require, pH levels outside the neutral range (either acidic or alkaline). Fungi, as a group, tend to favor acidic environments (often with optima at pH 4.5 to 5). There are also a number of pH extremophiles among the bacteria and archaea.

Organisms growing at low pH are referred to as **acidophiles**, and some tolerate surprisingly low pH. *Ferrobacillus ferrooxidans* in acid-mine drainage waters and *Sulfolobus acidocaldarius* growing in acidic hot spring waters, for example, will readily proliferate at a pH of 1 to 2.

At the other extreme, **alkaliphiles** prefer pH levels above 9. For example, water bodies carrying high salinities (e.g., the Dead Sea), as well as soils high in carbonates, support bacteria and archaea able to tolerate pH levels between 9 and 11. These microorganisms, such as *Natronobacterium* and *Natronococcus*, consequently tend to be both halophilic and alkaliphilic (i.e., both salt- and alkali-loving).

It should be noted that these extreme pH values refer to the hydrogen ion concentration outside the cell. These organisms survive through mechanisms that allow them to maintain a more neutral pH within the cell.

10.4   MICROBIAL TAXONOMY

10.4.1   Basis of Identification

Traditionally, the classification, or taxonomy, of higher organisms has been based on their **morphology**: their form and visible structure. However, this has been difficult for microorganisms because of their small size. Typically, prokaryotes were classified based mainly on their biochemical activities (including usable carbon substrates, nitrogen and other nutrient sources, electron acceptors, metabolic products, and resistance to inhibitory compounds), and on staining, which gave an indication of some aspects of their composition, in addition to such morphological features as size, shape, pigmentation, sporulation, intracellular inclusions (including those visible after staining), and the presence and location of flagella or the presence of another form of motility. Together, the observable characteristics of an organism can be referred to as its **phenotype**.

Habitat (and niche) might also be used for classification, including preferences with regard to such environmental factors as temperature and pH. Gross DNA composition (%G + C content) and immunological properties (e.g., serotyping) have also been used for many years. More recently, detailed analysis of the composition of the cell wall and of other cell constituents (e.g., fatty acid analysis) have been added as other useful tools in distinguishing among microorganisms.

Precise characterization of DNA and RNA sequences is now being used extensively to determine relationships among organisms. Such analysis of an organism's genetic makeup reveals its **genotype**. Genotype obviously affects phenotype, but not all genetic capabilities are necessarily expressed in an organism at a particular time. As might be expected, knowledge of genotypes has greatly increased and modified our understanding of the relationships among microorganisms. In fact, the separate domain of Archaea was not recognized until these newer tools were applied; previously, they were simply considered types of (and may still often be referred to out of habit as) bacteria.

One aspect of taxonomy is the grouping of organisms in progressively higher **taxa** (singular, *taxon*), such as families, orders, classes, and phyla. Modern taxonomists base these groupings on **phylogeny**, or natural evolutionary lines inferred from genetic similarities.

10.4.2 Prokaryotic "Species"

Another problem besides size complicates the taxonomy of prokaryotes: What exactly is a species? In the long-used definition developed for plants and animals, a species is considered a grouping of similar organisms that under natural conditions is capable of mating to produce fertile offspring. However, this definition is not very useful for organisms that reproduce asexually. Perhaps an even greater difficulty is that the genetic exchange that does occur among prokaryotes can be between organisms that are obviously not closely related to each other. This can lead to cross-linkings of otherwise unrelated species in taxonomic trees: organisms that, in a sense, are not close relatives of one of their biological "parents"! (The human genome has been found to contain considerable bacterial DNA, for example.)

Yet the idea of a species is still useful for prokaryotes. Some individual cells clearly are very similar, sharing numerous traits, and having a name to describe this grouping is desirable. *Salmonella typhi* causes typhoid fever, *Pseudomonas putida* does not; *Bacillus stearothermophilus* is a thermophilic sporeformer, whereas *Nitrosomonas europaea* is a mesophilic autotroph. Some organisms exist that appear to be intermediate between closely related species, and arguments arise among taxonomists between "splitters," who think that even minor differences are sufficient to define a new species, and "lumpers," who feel that various strains can be included in a single species unless there are numerous substantial differences. A **strain** is a group of cells that have all derived from a single cell in the not-too-distant past and are thus genetically identical except for any random mutations that might have occurred; molecular biologists call a strain recently obtained from a single cell a **clone**.

One view taken by some microbiologists is that nearly unlimited genetic variation can occur among microorganisms, but that some of the potential combinations of characters are more successful and stable than others. This can be pictured (Figure 10.6) as a surface in three-dimensional space, with hills and valleys. A marble dropped on the surface potentially could end up on a hillside or even the peak of a hill, but most will end up in the valleys. Incorporating every potentially useful trait is not necessarily the best survival strategy for a microorganism, in the same way that the camper who takes every potentially useful item is not likely to be the one who does the best on a backpacking trip.

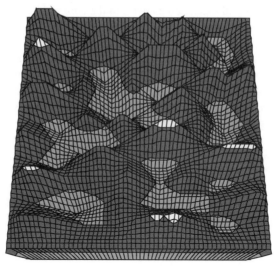

Figure 10.6 Three-dimensional surface: visualizing prokaryotic species as more stable (valleys), and hence more likely, combinations of characteristics, although intermediates may exist.

The new genetic tools (Section 10.4.4) have provided some quantitative means of determining whether individual strains of prokaryotes belong to the same species. One approach compares the degree of DNA **homology** (similarity in sequence of DNA base pairs) among microorganisms. Generally, if the DNA homology of two organisms is more than 70%, they will be considered the same species. If homology exceeds 20%, they may be considered to belong to the same genus. Using another tool, similarities of ribosomal RNA sequences of greater than 97% has led to considering two organisms to be the same species, while similarities of greater than 93 to 95% usually indicates the same genus. [The higher similarity for the rRNA analysis stems from the more highly conserved (less variable) nature of these sequences.]

The use of such techniques has led to the renaming of a number of microorganisms, as well as the "discovery" of the Archaea (Section 10.2.5). For example, the well-established species *Pseudomonas cepacia* was renamed *Burkholderia cepacia* when it was realized that it belonged to a different group of Proteobacteria than (and thus was not a close (relative of) other *Pseudomonas* species (Section 10.5.6).

10.4.3 Naming of Microorganisms

When microorganisms were first discovered, there were only two kingdoms, Plant and Animal. Thus, protozoans were considered "single-celled animals" studied by zoologists, and fungi, algae, and bacteria were "plants," studied by botanists. As a result, the latter organisms are still sometimes referred to as "flora" (and protozoans as "fauna"). However, based on our improved knowledge of their taxonomy, a better term today is **biota**.

The formal assignment of the Latin and Greek names (**nomenclature**) used in microbial taxonomy is now closely controlled. The official journal for publication of new prokaryote species is the *International Journal of Systematic Bacteriology* (note the holdover from the time when all prokaryotes were considered bacteria). If the new species description is first published elsewhere, it must be forwarded to IJSB to be formally accepted and included in the next approved list of names. The official representative, or **type**, culture of a newly labeled microorganism is held in an approved culture collection such as that maintained in the United States by the American Type Culture Collection (ATCC) or in Germany by the Deutsche Sammlung von Mikroorganismen und Zellkulturen (DSMZ). Rediscovered species (original type culture lost or never saved) are deposited as **neotype** strains. Species of some microorganisms have been subdivided further as specific cell strains, subspecies, or types (e.g., *Escherichia coli* type 0157:H7).

When a new bacterial or archaeal strain is isolated and characterized, it is compared to the existing information on described species, and/or directly to the type species. Traditional (mainly phenotypic) comparisons are made using keys (Figure 10.7) and standard references, particularly *Bergey's Manual of Determinative Bacteriology*. Genotypic information is included in *Bergey's Manual*, but large databases are also now available online (e.g., the Ribosomal Database Project maintained by the Center for Microbial Ecology at Michigan State University, for ribosomal RNA sequences, http://www.cme.msu.edu/RDP/).

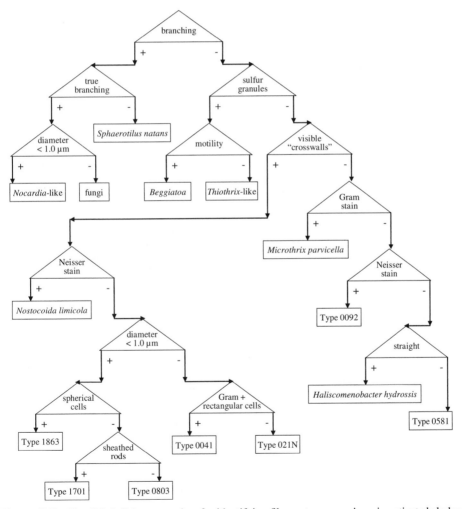

Figure 10.7 Simplified dichotomous key for identifying filamentous organisms in activated sludge.

Phylogenetic relationships (as opposed to identification) are the focus of *Bergey's Manual of Systematic Bacteriology.*

The designation of genus and species names can be based on a variety of factors. The following examples demonstrate some of the ways in which these names might be chosen:

- The **environment** in which they grow (e.g., *aquaticus, marina, coli*)
- Their usual **host** (*bovis, avium*)
- The environmental **conditions** they favor (*thermophilus, halophila, acidophilus*)
- Their **shape** (*ovalis, longum, sphaericus*)
- Their **color** [*aureus* (golden), *niger* (black)]
- Their **substrates** (*denitrificans, ferrooxidans*)
- Their **products** [*Methanobacterium, cerevisiae* (beer)]
- The **disease** with which they are associated (*typhi, botulinum, pneumoniae*)
- After a **person**: the discoverer or as a tribute (*winogradskii, Beijerinckia*)

In the following sections we provide an overview of many of the important groups of microorganisms. However, it may be useful to keep in mind that our view of microbial taxonomy (particularly for prokaryotes) is continuing to evolve, and that the names and groupings of organisms may be in dispute and may change in the future.

10.4.4   Characterization of Prokaryotes

Before looking at the various groups of Bacteria and Archaea, it is useful to describe briefly some of the characteristics commonly used to differentiate among them. In many cases it is necessary first to **isolate** the organism and then to grow it in **pure culture** (only one species present) before testing.

Shape   Prokaryotic cells show a remarkable diversity of shapes (Figure 10.2 showed a few). Most common are cylindrical **rods** (Figure 10.8), also called **bacilli** (singular,

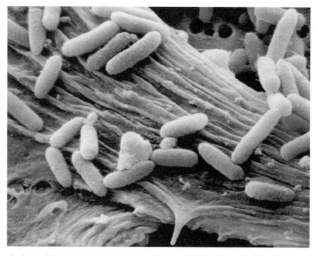

Figure 10.8   Rod-shaped bacteria: *Pseudomonas.* (SEM image courtesy of the University of Iowa Central Microscopy Research Facility.)

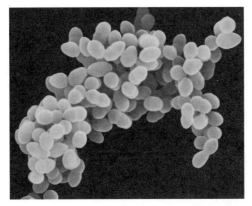

Figure 10.9   Cocci: *Staphylococcus*. (SEM copyright Dennis Kunkel Microscopy, Inc.)

bacillus), and spherical cells (Figure 10.9), called **cocci** (singular, **coccus**). Variations include short rods (**coccobacilli**), bent or comma-shaped rods (**vibrios**), and greatly elongated rods. There are also a variety of **spiral** cells (Figure 10.10), cells with specialized appendages, and those with irregular, indefinite, or special shapes.

The shape of a cell can also change during its life cycle. For example, *Arthrobacter* (Section 10.5.7) cells routinely alternate between cocci and rods.

Size   Typical rods may be 0.5 to 1.0 µm in diameter and 2 to 4 µm long, but size can vary tremendously, from diameters of 0.1 µm or less up to lengths of 200 µm or more. Most cocci have diameters of 0.5 to 2.0 µm, but again there is a considerable range.

Growth Form   Many microorganisms grow as individual, single cells. However, some grow in chains, or **filaments**, composed of a single species (Figure 10.11). Others may grow in clusters, or be found in clumps (**floc**; also in Figure 10.11) or other associations (such as biofilms or floating mats), sometimes with characteristic shapes, and often with

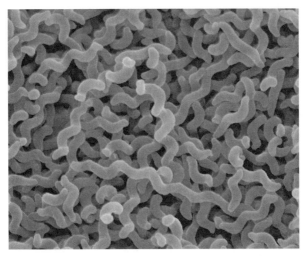

Figure 10.10   Spiral-shaped bacteria: *Campylobacter*. (SEM copyright Dennis Kunkel Microscopy, Inc.)

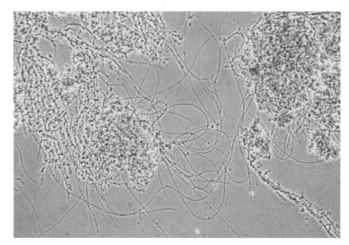

Figure 10.11 A filamentous bacteria growing with floc in an activated sludge wastewater treatment plant.

multiple species. Some go through a life cycle in which their form or type of association changes in a predictable way.

Prokaryotic cells usually reproduce by **binary fission**, or splitting into two roughly equal "daughter" cells. However, some bacteria reproduce by **budding**, in which a smaller new cell separates from the original one, and some bacteria produce special structures for releasing new cells for dispersal.

Stalked bacteria (Figure 10.12) produce an appendage that usually serves for attachment to a surface. If the stalk consists of an extrusion of cytoplasm surrounded by the cell membrane and wall, it is called a **prostheca** (plural, *prosthecae*). Some cells may attach directly to a surface with a small structure known as a **holdfast**.

Staining   The way in which their cells respond to certain colored chemicals (**stains**) is used to differentiate some species. The most widely used staining technique for bacteria is the *Gram stain* (Figure 10.13), developed in 1884 by Dutch bacteriologist Hans Christian

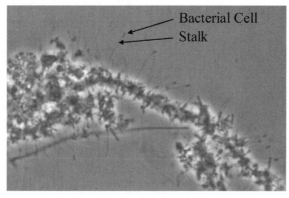

Figure 10.12   Stalked bacteria growing on a filament in activated sludge.

<u>Solutions</u>

Solution 1: Hucker's crystal violet
    Component A: 2 g crystal violet in 20 mL 95% ethanol;
    Component B: 0.8 g ammonium oxalate in 80 mL distilled water;
    Mix A and B (lasts 3 months).

Solution 2: modified Lugol's solution
    Mix 1 g iodine and 2 g potassium iodide in 300 mL distilled water;
    Store in dark (lasts 3 months).

Solution 3: decolorizing agent
    95% ethanol

Solution 4: counterstain
    Safranin O solution: 2.5 g in 100 mL 95% ethanol;
    Mix 10 mL in 100 mL distilled water.

<u>Procedure</u>

Make thin smear of culture on glass slide.
Allow to thoroughly air dry.
Cover with Solution 1 for 1 minute.
Gently rinse with tap or distilled water.
Cover with Solution 2 for 1 minute.
Hold slide at angle.
Decolorize drop by drop with Solution 3 until no more color runs off.
Rinse with water.
Counterstain with Solution 4 for 1 minute.
Rinse with water, then blot dry.
Put drop of immersion oil directly on slide.
Observe under microscope at 1000 X.

<u>Results</u>

Blue/violet = Gram positive;
Red = Gram negative.

Figure 10.13   Gram stain technique. This modification (one of many) is recommended for staining of activated sludge mixed liquor.

Joachim Gram. Although the method was developed empirically and seemed to show important differences between cells, it was not until much later that it was learned that a fundamental difference in cell wall composition was being highlighted (Figure 10.14). Gram-positive cells stain blue-violet because they retain the crystal violet, gram-negative cells are red because they are decolorized by ethanol and then take up the safranin counterstain.

Other stains also may be of great utility in differentiating among organisms or in visualizing cell structures. Figure 10.15 shows several types of staining approaches. Various staining procedures were particularly necessary prior to development of phase contrast and electron microscopy, and still may be highly useful.

Motility   Many prokaryotes have **motility**, the ability to move, by means of one or more **flagella** (singular, **flagellum**), long, thin, hairlike organelles that protrude from the cell. Flagella at the end of a rod-shaped cell are referred to as **polar**; those on the sides are **peritrichous**. The flagella of prokaryotes are too small to be seen with a light microscope

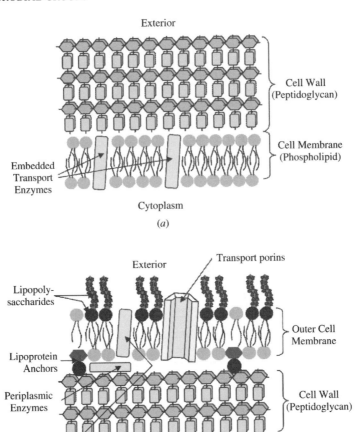

Figure 10.14 Bacterial cell wall and membrane: (*a*) gram positive; (*b*) gram negative.

but can be made visible through staining or observed with an electron microscope. Also, in some cases there may be visible tufts, consisting of numerous flagella. Both bacteria and archaea may be flagellated (as are some algae and protozoa). Rates of movement can exceed 50 μm/s.

Some cells are motile even without flagella. Several species, particularly of filamentous bacteria, are able to "push" against a surface or other object, resulting in **gliding** motility, or can flex the filament, leading to **twitching** motility. However, speeds typically are much slower (<10 μm/s) than for flagellated organisms.

Organism movement may be in response to a gradient. **Chemotaxis**, for example, is movement toward (or occasionally away from) a higher concentration of a chemical; **phototaxis** is in response to light. **Magnetotaxis**, response to a magnetic field, has also been observed in a few specialized species, such as *Magnetospirillum* (Section 10.5.6).

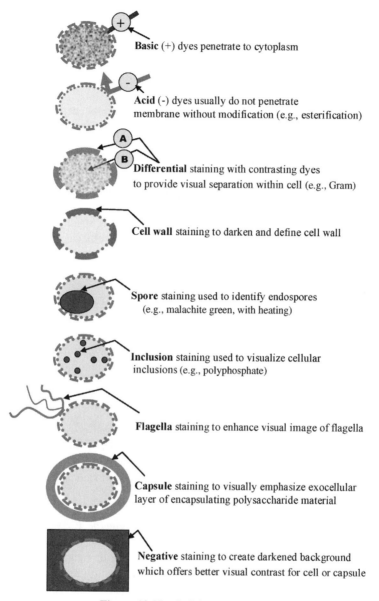

Basic (+) dyes penetrate to cytoplasm

Acid (-) dyes usually do not penetrate
membrane without modification (e.g., esterification)

Differential staining with contrasting dyes
to provide visual separation within cell (e.g., Gram)

Cell wall staining to darken and define cell wall

Spore staining used to identify endospores
(e.g., malachite green, with heating)

Inclusion staining used to visualize cellular
inclusions (e.g., polyphosphate)

Flagella staining to enhance visual image of flagella

Capsule staining to visually emphasize exocellular
layer of encapsulating polysaccharide material

Negative staining to create darkened background
which offers better visual contrast for cell or capsule

Figure 10.15   Staining approaches.

Pigmentation   Most prokaryotes appear colorless under the microscope, and colorless or whitish in liquid culture and on agar plates. However, a number of groups are photosynthetic and have appropriate pigments for this purpose. There are also a number of other pigments that might be formed by particular strains, so that cells and cultures may have a variety of colors (including red, pink, orange, yellow, and purple). Usually, any pigments formed will be within the cell, but in some cases they are released and may color the growth medium.

Sporulation   A number of prokaryotes produce resting stages, known as **spores**, that are resistant to hostile environmental conditions such as desiccation. Some gram-positive bacteria produce an especially heat-resistant structure within the cell, called an **endospore**. Some spores may be seen directly under a light or phase-contrast microscope, whereas others may be made visible by staining.

Intracellular Inclusions   A wide variety of materials may be deposited within cells, often as a means of energy storage. Some are readily visible under a microscope, whereas others require staining before they become apparent. Examples include poly-β-hydroxybutyrate (PHB), glycogen, polyphosphate (volutin), and elemental sulfur.

Nitrogen and Sulfur Sources   Many microorganisms are able to use inorganic nitrogen in the form of ammonium and/or nitrate as their nitrogen source. Others require organic forms of nitrogen, especially certain amino acids. A relatively specialized ability is utilization, or **fixation**, of elemental nitrogen. Similarly, many cells can utilize sulfur in the form of sulfate, or perhaps sulfide, but others may require organic sulfur. Some organisms are able to use inorganic nitrogen or sulfur compounds as energy sources or as electron acceptors.

Carbon and Energy Sources; Terminal Electron Acceptors and Relationships to Oxygen   As discussed above (Sections 10.3.1 and 10.3.2), organisms can be phototrophic or chemotrophic, organotrophic or lithotrophic (if chemotrophs), and autotrophic or heterotrophic. The various relationships to oxygen (e.g., aerobic, anaerobic; Section 10.3.3) and the terminal electron acceptors they are able to use are also important for characterization.

Habitat Preferences/Tolerances   In addition to oxygen, other environmental conditions preferred or tolerated can be useful in characterization. These include such factors as temperature, pH, and salinity, as mentioned above, but also include association with more specific habitats, such as the intestines of warm-blooded animals or plant root nodules.

Range of Substrates; Reaction Products; Presence of Specific Enzymes   The range of substrates metabolizable by a microorganism, and the reaction products formed, constitute a major part of the traditional identification process. Wide ranges of substrates might be tested in what is potentially a very laborious process. Ability to grow, the presence of specific enzymes, the general nature of the products formed (e.g., acid and/or gas production), and specific chemical products might be looked for. Production of **catalase**, which decomposes potentially toxic hydrogen peroxide to oxygen and water, and **oxidase**, which mediates the oxidation of some cytochromes, are two common tests for enzymes; both are related to the ability to grow aerobically. Modern techniques include some commercial tests, such as Biolog (Figure 10.16; Biolog, Inc., Hayward, California, www.biolog.com), capable of screening many compounds at once. Other systems, such as Enterotube (Becton, Dickinson and Company, Franklin Lakes, New Jersey, www.bd.com) and API strips (bioMérieux, Inc., Durham, North Carolina, www.biomerieux-inc.com), have been developed for identifying species within a particular family, such as Enterobacteriaceae (the enteric Proteobacteria; see Section 10.5.6).

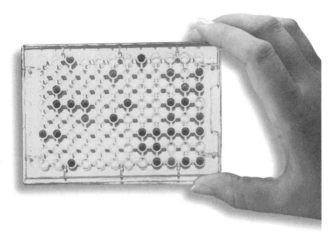

Figure 10.16 Biolog test; 95 test compounds and a control well are included in each plate. The plate shown was used to identify a gram-negative bacterium as *Leminorella grimontii*, based on comparing the pattern of positive (dark) and negative tests to results in a database. (Photo courtesy of Biolog, Inc.)

Pathogenicity   Although most prokaryotes are free-living **saprobes** or **saprophytes** (organisms that feed on dead organic material), some bacteria are associated with diseases of humans, other animals, or plants (Chapter 12). Often, these **pathogens** (disease-causing organisms) infect a single species, although some have a relatively wide range of hosts. Bacteria are also part of the normal biota of higher organisms (e.g., on the skin or in the intestinal tract of humans) without causing disease. Occasionally, **opportunistic** pathogens, which normally are free-living, may attack a **compromised** (weakened) host.

Immunological Properties   Antibodies may be used to detect specific antigens, which are typically cell surface proteins. Such testing is used widely in clinical microbiology for identification of pathogens. It can be highly specific, capable of differentiating among the sometimes numerous strains, or **serotypes**, of a single species.

Inhibition/Resistance Patterns   Inhibition of growth or activity by certain chemicals may be a helpful diagnostic tool. The pattern of resistance to specific antibiotics, with their different modes of action, can be particularly useful.

Analysis of Fatty Acids   Different species have different fatty acid compositions of their cell membrane lipids. Composition will also change based on growth conditions, but if these are standardized and the fatty acids extracted and analyzed, the resulting pattern can be highly useful in identification. The most widely used procedure involves forming the methyl ester of the fatty acid to make it readily analyzable by gas chromatography. **FAME** (fatty acid methyl ester) profiles (Figure 10.17) can be compared to a database for rapid identification of many isolates.

% G + C Composition   Recall from Chapter 3 that guanine is always paired with cytosine in DNA, and adenine with thymine. Thus, the base-pair composition of an organism's

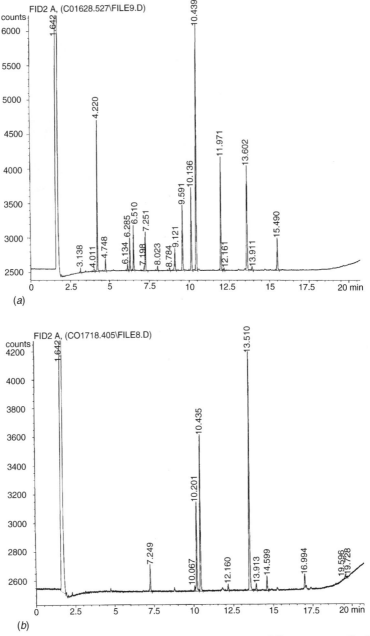

Figure 10.17   Fatty acid methyl ester (FAME) profiles showing different patterns for (a) *Serratia marcescens* and (b) *Tsukamurella paurometabolum*. (Courtesy of Michael Fleming.)

DNA can be specified either by any single nucleotide or by the sum of either A + T or G + C. By convention, % G + C is used as an indication of genetic relatedness. Organisms with very different % G + C compositions cannot be closely related. On the other hand, % G + C does not indicate the specific sequences of the bases, so that organisms with similar % G + C values are not necessarily related.

DNA Hybridization   One means of genotyping cells is by comparing the similarity of sequences of bases in their DNA. To compare two organisms, the DNA of one is first labeled (often with radioactive ^{32}P). The DNA extracted from both organisms is then broken mechanically into small fragments and heated to separate the two complementary strands (**denatured**). The labeled DNA and an excess of the unlabeled DNA are next mixed together and cooled to allow them to **reanneal** (re-form a double strand). The extent to which the two different DNAs are able to combine with each other, or **hybridize**, is a measure of how similar the base sequences are. This can be determined by the amount of radioactivity present in the reannealed DNA.

Molecular Probes   A large number of molecular probing techniques are now available or under development. One group of methods of particular interest is referred to as **fluorescent in situ hybridization** (FISH). In a generalized FISH approach, an **oligonucleotide** (short chain of nucleotides, typically 15 to 25 bases for FISH) containing a sequence characteristic of a particular taxon of interest is labeled with a fluorescent dye. The labeled oligonucleotide is then allowed to hybridize with the complementary sequence in the organism in situ. With fluorescence microscopy, it is then possible to see the numbers and location of the brightly colored target organism within a natural community. Depending on the degree of specificity of the sequence chosen, this approach potentially can be used to identify organisms belonging to any taxonomic level ranging from domain (e.g., Bacteria) to a particular species or even strain.

Another type of probe is used in **ribotyping**. Restriction enzymes break DNA at points where specific sequences of nucleotides occur. If DNA is treated with one or more such enzymes, fragments of DNA are produced. These fragments are then separated by size on a gel and probed with 16S rRNA genes. For treatment with a particular set of restriction enzymes, the resulting pattern of DNA fragment sizes will be characteristic of the organism.

Denaturing gradient gel electrophoresis (**DGGE**) analysis offers a characterization of the GC composition of DNA segments coding for rRNA or other genes. Following tagging and amplification using PCR (Section 6.3.3), these fragments are drawn electrophoretically (i.e., attracted by their electrical charge) through gel tracks laden with a gradient matrix of a denaturing agent (e.g., urea ranging from 30 to 55%). Since the G–C triple bond is stronger than the A–T double bond, it does not denature until farther along in the gradient. Fully denatured fragments stop migrating, resulting in a linear separation of the fragments into a banding pattern that is characteristic of the organism (Figure 10.18). A similar approach is used in thermal gradient gel electrophoresis (**TGGE**), except that a temperature gradient is used for denaturing the DNA.

16S rRNA Analysis   The primary tool used today in determining phylogeny is 16S rRNA analysis, developed largely by Carl Woese beginning in the early 1970s. Prokaryotic ribosomal RNA (rRNA) is composed of three molecules, referred to by their sedimentation coefficients measured in Svedberg units (S) as 5S (containing $\sim$120 bases), 16S ($\sim$1500), and 23S ($\sim$2900). Woese initially worked with 5S fragments but found they were too small to yield sufficient information. The emphasis then shifted to 16S rRNA (and the comparable 18S rRNA found in eukaryotes) based on its larger, yet still manageable number of nucleotides.

One important advantage of using the nucleotide sequences of 16S (or 18S) rRNA for determining relationships is that these molecules occur in all organisms. Also, because of

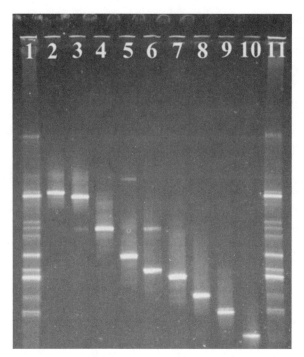

Figure 10.18 Denaturing gradient gel extraction track profiles.

their central role in the process of gene expression, there are strong selective pressures, making it unlikely that they will undergo rapid changes. This leads to the presence of extended regions of the molecule that are highly **conserved** (little changed) and which therefore are useful for establishing distant phylogenetic relationships. On the other hand, they also have an adequate number of variable regions that can be used to examine closer relationships. Thus, 16S rRNA can be used as an evolutionary clock, or chronometer, to measure the phylogenetic distance between taxa at a variety of levels based on the number of nucleotide differences—representing stable mutations—that have occurred over time.

One method to sequence 16S rRNA begins with extraction of a cell's RNA. A small DNA oligonucleotide primer (15 to 20 nucleotides in length) that is complementary in its base sequence to a conserved region of the 16S rRNA molecule is added. Reverse transcriptase can then be used to generate complementary DNA (cDNA), which in turn is amplified using PCR. The nucleotide sequences of these cDNA are then determined, and the original sequence of the rRNA is deduced from them. Another common approach is to extract and sequence the DNA of the gene that codes for the 16S rRNA rather than the RNA itself.

The 16S rRNA sequences of thousands of species have now been determined. Using advanced computer algorithms, short (6 to 10 nucleotides) **signature sequences** have been found that are highly consistent within groups of organisms. Also, **phylogenetic trees** can be generated in which the evolutionary distance between two groups is indicated by the length of the connecting lines (Figure 10.19).

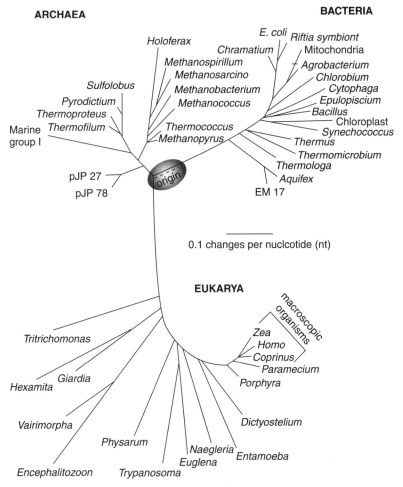

Figure 10.19 Phylogenetic tree indicating evolutionary branching and distance between groups based on on rRNA analysis. Fungi are represented by *Coprinus* (a mushroom), plants by *Zea* (corn), and animals by *Homo* (humans). (From Atlas, 1997; all rights reserved.)

10.5   BACTERIA

As a group, the domain Bacteria is extremely diverse, including phototrophs and chemotrophs, organotrophs and lithotrophs, heterotrophs and autotrophs, aerobes and anaerobes, psychrophiles and mesophiles and thermophiles and hyperthermophiles, halophiles and nonhalophiles, acidophiles and neutrophiles and alkaliphiles, saprophytes and parasites. They are able to utilize a vast array of organic compounds (i.e., all naturally occurring and almost all synthetic organics) as carbon and energy sources, many reduced inorganics as electron donors (for energy), and many oxidized inorganics as electron acceptors.

One way of discussing microorganisms is by function, grouping together all of those with a particular ability or characteristic (e.g., phototrophs). However, sometimes organisms that have similar functions are actually quite different from each other. (This is referred to as *convergent evolution*, since they have approached a similar form from different starting points.) Thus, microbiologists generally prefer, where possible, to group bacteria based on phylogeny. That is the approach that is taken here, for the most part, although it should be recognized that with our growing knowledge, these groupings continue to change.

Figure 10.19 showed the phylogeny of Bacteria using an approach based mainly on 16S rRNA analysis. The domain was broken into about eight known major groupings (although other 16S rRNA sequences have been found in environmental samples, indicating that additional "unknown" groups also exist). What should these major groupings of Bacteria (and Archaea) be called?

Previously, they may have been referred to as phyla, or divisions (reflecting bacteriology's roots in botany), or commonly, groups (reflecting phenotypic similarities rather than phylogenetic relationships). However, in looking at phylogenetic trees, such as that in Figure 10.19, some microbiologists have come to believe that if the differences between plants and animals (and fungi) warrant assignment to groupings at the level of kingdoms, the much greater differences among bacteria should also.

Of course, not all microbiologists agree, and this has not yet been widely accepted among other biologists, who traditionally have focused on Eukarya, and particularly, plants and animals. One argument is that 16S (or 18S for eukaryotes) rRNA analysis is not necessarily the "one true" measure of phylogeny and degree of evolutionary divergence. Still, as more is learned about microbial diversity using a variety of tools, it seems quite possible that some categorization scheme like this eventually will be accepted. Thus for Chapters 10 through 13, where microbiology is emphasized, these groupings will be referred to as "Kingdoms," even though elsewhere we do not make this distinction.

Table 10.3 shows 14 major groupings of Bacteria and indicates some further subdivisions into classes and orders. The newest (second) edition of *Bergey's Manual of Systematic Bacteriology* decided not to refer to the major groups as kingdoms, and instead, listed 23 phyla. (The term *domain* for Bacteria was retained; there had been a proposal to use "empire" instead.) The table also includes a number of representative, interesting, and environmentally important genera. Comparison with the phylogenetic tree of Figure 10.19 shows many similarities, but also some differences reflecting the changing taxonomy of bacteria.

Some of the kingdoms have few genera, perhaps none of which are of known major importance in environmental engineering and science. Other kingdoms have numerous genera of great relevance to humans. The three largest are Proteobacteria, Firmicutes (gram positives), and Cyanobacteria. It should also be noted that a number of bacteria (generally not shown in the table) do not appear to fit in any of the kingdoms and are thus of "uncertain" affiliation. (One extreme example is the genus *Chrysiogenes*, containing only one recognized species, which may merit its own kingdom!) Remember, these divisions are based on similarities in the genetic code, not on any attempt to distribute species equally among groups. One way to look at this unevenness (which also occurs among plants and animals) is to realize that some approaches to survival lend themselves to greater diversification because of the range of niches available.

It should also be recognized that our knowledge of bacterial groups is uneven. Naturally, because more effort has gone into their study, we tend to know more about groups that appear to be of greater importance to us. This includes bacteria used commercially, as well as pathogens of humans, domestic animals, and crop plants. On the other hand, it is not unusual to find that an organism isolated from the natural environment—or even a

TABLE 10.3   The Bacteria: A Proposed Phylogeny Including Some Representative, Interesting, and Environmentally Important Genera[a]

Kingdom 1. *Aquaficae*
   Class 1. *Aquaficae*
        Aquifex
   Class 2. *Thermotogae*
        Thermotoga
   Class 3. *Thermodesulfobacteria*
        Thermodesulfobacterium
Kingdom 2. *Xenobacteria*
   Class 1. *Deinococci*
        Deinococcus
   Class 2. *Thermi*
        Thermus
        Nitrospira (*or perhaps a separate Kingdom*)
        Leptospirillum
        Magnetobacterium
Kingdom 3. *Thermomicrobia*
   Class 1. *Chloroflexi*—green nonsulfur bacteria
   Order 1. *Chloroflexales*
        Chloroflexus
   Order 2. *Herpetosiphonales*
        Herpetosiphon
   Class 2. *Thermomicrobia*
        Thermomicrobium
Kingdom 4. *Cyanobacteria*—blue-green bacteria
                (formerly, blue-green algae)
   Class 1. *Prochlorophyta*
   Order 1. *Chroococcales*
        Chroococcus
        Prochloron
        Synechococcus
   Order 2. *Pleurocapsales*
        Pleurocapsa
   Order 3. *Oscillatoriales*
        Lyngbya
        Oscillatoria
        Spirulina
   Order 4. *Nostocales*
        Anabaena
        Calothrix
        Nostoc
   Order 5. *Stigonematales*
        Stigonema
Kingdom 5. *Chlorobia*—green sulfur bacteria
        Chlorobium
Kingdom 6. *Proteobacteria*[b]
   Class 1. α-*Proteobacteria* (*Rhodospirilli*)
   Class 2. β-*Proteobacteria* (*Neisseriae*)

(Continued)

TABLE 10.3   (*Continued*)

Class 3. γ-*Proteobacteria* (*Zymobacteria*)
Class 4. δ-*Proteobacteria* (*Predibacteria*)
Class 5. ε-*Proteobacteria* (*Campylobacteres*)
Kingdom 7. *Firmicutes*—Gram positives[b]
Class 1. *Clostridia*
Class 2. *Mollicutes*
Class 3. *Bacilli*
Class 4. *Actinobacteria*
Kingdom 8. *Planctomycetacia*
Class 1. *Planctomycetacia*
Order 1. *Planctomycetales*
Planctomyces
Order 2. *Chlamydiales*
Chlamydia
Kingdom 9. *Spirochetes*
Borrelia
Leptospira
Spirochaeta
Treponema
Kingdom 10. *Fibrobacteres*
Fibrobacter
Kingdom 11. *Bacteroidetes*
Bacteroides
Kingdom 12. *Flavobacteria*
Flavobacterium
Kingdom 13. *Sphingobacteria*
Cytophaga
Flexibacter
Haliscomenobacter
Saprospira
Kingdom 14. *Fusobacteria*
Fusobacterium
Streptobacillus
Kingdom 15. *Verrucomicrobia*
Prosthecobacter

[a]All classes listed if more than one. All orders listed if more than one except for the two phyla *Proteobacteria* and *Firmicutes*. Families not shown. (Based in part on outline available from *Bergey's Manual Trust*.)
[b]Large important group; additional breakdown and genera given in Tables 10.5 (*Proteobacteria*) and 10.7 (*Firmicutes*).

sewage treatment plant—is not a previously known, or at least not well described, species. Also, our knowledge of bacteria has been limited by our ability to **culture** (grow) them in the laboratory; those that are not readily grown on laboratory media have been difficult to study. Interestingly, we do know a fair amount about thermophilic bacteria. This is at least in part because of the early efforts of Thomas D. Brock, who began studying hot springs with the idea that the extreme conditions limited the microbial diversity, making microbial ecology studies simpler!

Phylogenetic grouping, as in Table 10.3 (and Figure 10.19), does not always lend itself to neat phenotypic groupings of organisms with similar characteristics. Still, we mainly use this organization below for a brief description of the Bacteria. No attempt is made to cover every subgroup or genus, and many details are omitted.

10.5.1 Aquaficae

This kingdom, which is considered to be deeply rooted (close to the original bacterial forms), is so diverse that there are proposals to break it into three (or alternatively, to lump it with the next). The known species are thermophilic or hyperthermophilic. *Aquifex* is the most thermophilic known true bacteria (optimum 85°C, maximum 95°C) and is a chemolithotrophic autotroph (growing on H_2, S^0, or $S_2O_3^{2-}$). It requires oxygen or nitrate as an electron acceptor but cannot tolerate high O_2 concentrations. *Thermotoga* is also a hyperthermophile but is a fermentative chemoorganotroph. *Thermodesulfobacterium* is a thermophilic sulfate reducer using fermentation products such as lactate and pyruvate as its carbon and energy source.

Although they are undoubtedly of significance in the environments in which they are found (mainly hot springs and marine thermal vents), these organisms are generally not of concern to environmental engineers and scientists. However, their unique features make them of special interest to microbiologists studying the evolution of life or biochemical diversity.

10.5.2 Xenobacteria

This kingdom contains two classes, Deinococci and Thermi. *Deinococcus radiodurans* was first isolated from food that had been irradiated to sterilize it. It is extraordinarily radiation resistant, able to withstand more than 1.5 million rad of gamma irradiation (3000 times the lethal dose to humans). Many deinococci, which are frequently reddish in color due to caretenoid pigments, also are resistant to ultraviolet radiation and other mutagens and are able to tolerate drying. Their resistance to genetic damage appears to result in large part from an exceptional ability to repair damaged DNA rapidly. It is hoped that these bacteria may prove useful in the future for certain types of hazardous waste site cleanups.

Thermus aquaticus, first found in hot springs in Yellowstone National Park by a group led by Thomas Brock, was one of the first extreme thermophilic organisms isolated. It has also been found in hot-water heaters. Its heat-stable *Taq* DNA polymerase is the enzyme commonly used in PCR (Section 6.3.3).

Nitrospira is an autotroph that derives energy from the oxidation of nitrite to nitrate. Thus, it is often grouped with the other nitrifying bacteria (Proteobacteria; see Section 10.5.6). However, it also has been proposed to include it in the class Thermi, or perhaps in its own kingdom, based on phylogeny.

10.5.3 Thermomicrobia (Including Green Nonsulfur Bacteria)

Although *Thermomicrobium* is a thermophilic (75°C optimum) chemoorganotroph, most of the known members of this small kingdom are phototrophic green nonsulfur bacteria. *Chloroflexus* is an **anoxygenic** (non-oxygen producing) phototroph that can grow autotrophically but grows better as a photoheterotroph (using organic carbon sources). It also can grow aerobically as a chemoorganotroph. It is thought to be descended from the earliest phototrophs. Table 10.4 compares several characteristics of the various groups of phototrophic bacteria.

10.5.4 Cyanobacteria: Blue-Green Bacteria (Formerly, Blue-Green Algae)

The Cyanobacteria is a large, diverse, and environmentally important bacterial group. As discussed in Section 10.1, their early forms probably were responsible for producing the oxygen of Earth's atmosphere, allowing the development of advanced eukaryotic life forms. Their present relevance to environmental engineers and scientists includes their role in the eutrophication of lakes and the production of tastes and odors in drinking waters drawn from surface supplies.

Because they are oxygen-producing photoautotrophs, cyanobacteria were long considered to be algae (Cyanophyta). However, once it was realized that they were prokaryotes, they were renamed and reclassified. It is now believed that they probably shared a common ancestor with chloroplasts. They are compared with the other phototrophic bacteria in Table 10.4. Their green color comes from chlorophyll a, while the blue color comes from other pigments called phycocyanins.

Most Cyanobacteria are strictly photoautotrophic, but a few can utilize simple organics as carbon sources (photoheterotrophs) or even grow in the dark as chemoorganotrophs. Many (e.g., *Oscillatoria*) have gliding motility. Although they are usually thought of as aquatic freshwater organisms, they are also found in the ocean and in soils. A number can withstand extreme environments, such as hot springs with temperatures of 70°C, the surface of desert soils, bare rock, saline lakes, and shallow tidal seas (where they may form large mats). Different species grow as individual cells, filaments, or various types of aggregates.

Many Cyanobacteria have the unusual ability to be able to **fix nitrogen** (convert N_2 to a combined form, usually as ammonium or an amine compound). Interestingly, **nitrogenase**, the enzyme responsible for nitrogen fixation, is sensitive to oxygen; this means that nitrogen fixation cannot readily occur in the light (when oxygen is being produced) in most species. However, some filamentous Cyanobacteria (e.g., *Anabaena*, Figure 10.20) form differentiated cells, known as **heterocysts**, for nitrogen fixation. These special structures do not photosynthesize and have thickened walls to minimize oxygen diffusion from the outside. They receive nutrition from neighboring cells in the filament, and in return make fixed nitrogen available to them.

The five major groups of Cyanobacteria of Table 10.3 correspond to general morphological forms: Chroococcales, single cells or small aggregates; Pleurocapsales, small spherical cells formed through fission (for reproduction); Oscillatoriales, filaments; Nostocales, filaments with heterocysts; Stigonematales, branched filaments. Except for the Chroococcales, which are very diverse, these groupings also hold up fairly well to phylogenetic analysis.

10.5.5 Chlorobia: Green Sulfur Bacteria

Another distinct group of phototrophs included in Table 10.4 is the green sulfur bacteria. Like the green nonsulfur bacteria, they are anoxygenic, and can use hydrogen sulfide (H_2S) as their electron donor. The sulfide is oxidized first to elemental sulfur, which produces granules outside the cell, and then to sulfate (SO_4^{2-}). Also, they contain unique

TABLE 10.4 Characteristics of Phototrophic Bacteria

	Green Sulfur	Green Nonsulfur	Purple Sulfur	Purple Nonsulfur	Blue-Green	Heliobacteria
Kingdom	Chlorobia	Thermomicrobia	γ-Proteobacteria	α- or β-Proteobacteria	Cyanobacteria	Firmicutes
Light metabolism	Obligately anaerobic, photo autotroph (some also photo heterotophic)	Usually anaerobic, photoheterotroph or photoautotroph	Obligately anaerobic, photoautotroph	Usually anaerobic, photo-heterotroph or photoautotroph	Aerobic, photoautotroph	Obligately anaerobic, photoheterotroph
Electron donor for lithotrophy	H_2, H_2S, S	H_2, H_2S	H_2, H_2S, S	H_2, H_2S	H_2O	—
Dark metabolism	None	Chemoorgano-heterotroph	None	Chemoorgano-heterotroph	None (some are chemoorgano-heterotrophs)	Chemoorgano-heterotroph
Carbon source	CO_2	CO_2 or organic carbon	CO_2	CO_2 or organic carbon	CO_2 (some can use organic carbon)	Organic carbon
Sulfur deposition	Outside cell	None	Within cell	None	None	None
N_2 Fixation	Yes	No	Yes	Yes	Yes (many)	Yes
Endospores	No	No	No	No	No	Yes
Chlorophyll	Bact.-a, and c, d, or e	Bact.-c	Bact.-a, b	Bact.-a, b	Chlor.-a	Bact.-g

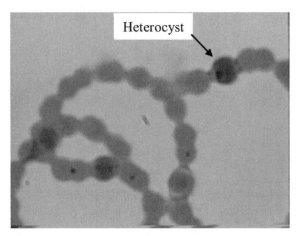

Figure 10.20   *Anabaena*, a filamentous cyanobacteria; with heterocyst.

structures called **chlorosomes**, in which much of their bacteriochlorophyll is concentrated. They are mostly autotrophic, but some photoheterotrophy has also been observed. *Chlorobium* is the best known genus.

10.5.6   Proteobacteria

The Proteobacteria is a vast kingdom, including many of the gram-negative species and many of the metabolic activities known among the bacteria. Based on 16S rRNA, it is broken into five classes, but these are more often referred to as the α, β, γ, δ, and ϵ subdivisions. Interestingly, the 16S rRNA of mitochondria (in eukaryotes) places them in this group also (Figure 10.19), suggesting that this organelle evolved from endosymbiotic early Proteobacteria. Table 10.5 lists some of the important and interesting genera. Because the kingdom is so large, and because organisms with similar activities may belong to different classes, we discuss them based mainly on phenotypic characteristics (i.e., based more on observable traits rather than on genetic similarity).

Phototrophs: Purple Sulfur and Nonsulfur Bacteria   The purple sulfur bacteria, such as *Chromatium*, and the purple nonsulfur bacteria, such as *Rhodospirillum*, are both anoxygenic phototrophic Proteobacteria (Table 10.4). Both groups show considerable morphological variability. They include species that are rods, spheres, and spirals, and may be flagellated or not. The purple nonsulfur group also includes some budding and appendage forming bacteria. All of the known purple sulfur bacteria, which deposit sulfur granules internally (unlike the external deposits of green sulfur bacteria), are γ = Proteobacteria. Purple nonsulfur bacteria are found in both the α and β groups.

How do the different phototrophs all survive—or where are their places (i.e., niches; see Chapter 14)? The Cyanobacteria occupy the upper, oxygenated regions of a lake, or other oxic environments, perhaps competing with algae and plants. The purple sulfur bacteria are normally found in the anoxic depths of lakes, where hydrogen sulfide has been released from the underlying sediments. The green sulfur bacteria, which typically can tolerate higher concentrations of hydrogen sulfide, and which also can survive with lower levels of light, would be found in even deeper zones. Both might also occupy

**TABLE 10.5   Kingdom Proteobacteria: Many of the
Gram-Negative Bacteria**[a]

Class 1. Alphaproteobacteria	*Azotobacter*
Acetobacter	*Beggiatoa*
Agrobacterium	*Chromatium*
Azospirillum	*Francisella*
Beijerinckia	*Haemophilus*
Brucella	*Legionella*
Caulobacter	*Methylomonas*
Hyphomicrobium	*Moraxella*
Magnetospirillum	*Nitrococcus*
Methylocystis	*Nitrosococcus*
Nitrobacter	*Pasteurella*
Paracoccus	*Photobacterium*
Rhodospirillum	*Pseudomonas*
Rickettsia	*Thiothrix*
Rhizobium	*Vibrio*
Sphingomonas	*Xanthomonas*
Xanthobacter	Family Enterobacteriaceae
Class 2. Betaproteobacteria	*Citrobacter*
Achromobacter	*Enterobacter*
Alcaligenes	*Escherichia*
Bordetella	*Klebsiella*
Burkholderia	*Leminorella*
Gallionella	*Proteus*
Leptothrix	*Salmonella*
Methylovorus	*Serratia*
Neisseria	*Shigella*
Nitrosomonas	*Yersinia*
Nitrosospira	Class 4. Deltaproteobacteria
Ralstonia	*Bdellovibrio*
Sphaerotilus	*Desulfovibrio*
Spirillum	*Myxococcus*
Thiobacillus	*Nitrospina*
Zoogloea	*Polyangium*
Class 3. Gammaproteobacteria	Class 5. Epsilonproteobacteria
Acinetobacter	*Campylobacter*
Aeromonas	*Helicobacter*

[a]All five classes (subdivisions α to ε) are listed, but not orders or families (except
Enterobacteriaceae). Only some representative, important, and interesting genera
are included.

other anaerobic environments where light and reduced sulfur are present, such as some
hot springs. Purple nonsulfur bacteria have an advantage in lighted anaerobic environments in which organic matter is present, since they are able to grow well heterotrophically. Green nonsulfur bacteria also grow heterotrophically and in low light, particularly
in thick microbial mats found in hot springs and shallow marine systems. Because of differences in the chlorophylls and other pigments each contain, the groups also have light
absorption profiles with optima at different wavelengths (Figure 10.21).

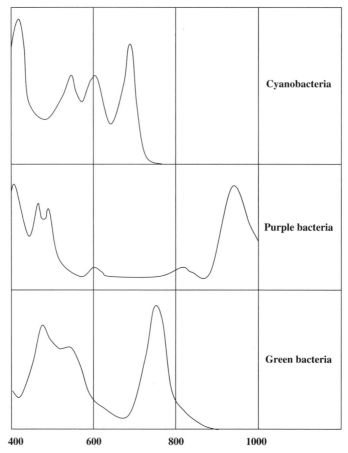

Figure 10.21 Schematic of light absorption characteristics (absorption vs. wavelength in nm) of phototrophic bacteria. (Based on Prescott et al., 1999.)

Chemolithotrophs The Proteobacteria include a number (but not all) of the organisms able to obtain energy from the oxidation of reduced inorganic compounds, including hydrogen and forms of nitrogen, sulfur, and iron (Table 10.6). Most of these organisms are also autotrophic (getting their carbon from carbon dioxide), and aerobic, although some are also able to use nitrate as an electron acceptor. They also tend to be slow growing, probably because the amount of energy available from oxidizing most of these compounds is relatively small.

The ability to utilize hydrogen is actually quite widely distributed. Examples of **hydrogen-oxidizing** Proteobacteria include some species of *Pseudomonas* and *Alcaligenes*. Most such bacteria are able to utilize organic substrates as well (unlike most other chemolithotrophs). Some (e.g., *Xanthobacter*) can fix nitrogen gas. Since hydrogen is typically produced under anaerobic conditions, and because the enzyme (hydrogenase) involved in its oxidation is oxygen sensitive, most hydrogen oxidizers prefer lower oxygen conditions (i.e., 5 to 10% O_2 rather than the 21% of air). Some hydrogen oxidizers are also able to grow on carbon monoxide, and such **carboxydotrophic** bacteria probably play a major role in elimination of this toxic gas.

TABLE 10.6   Examples of Chemolithotrophic Proteobacteria

Group	Representative Genera[a]	Reduced Substrate	Electron Acceptor	Oxidized Product
Hydrogen-oxidizing	*Alcaligenes* *Pseudomonas* *Ralstonia*	H_2 (hydrogen)	O_2 (or NO_3^-) (nitrate)	H_2O
Carboxydo-trophic	*Xanthobacter* *Alcaligenes* *Pseudomonas*	CO (carbon monoxide)	O_2	CO_2
Nitrifying Step 1	*Nitrosomonas* *Nitrosospira*	NH_3/NH_4^+ (ammonia/ammonium)	O_2	NO_2^-
Step 2	*Nitrobacter* *Nitrococcus* *Nitrospina*	NO_2^- (nitrite)	O_2	NO_3^-
Sulfur-oxidizing	*Beggiatoa* *Thiobacillus* *Thiothrix*	H_2S (hydrogen sulfide), S^0 (elemental sulfur), and/or $S_2O_3^{2-}$ (thiosulfate)	O_2 (or NO_3^-)	S^0 and SO_4^{2-} (sulfate)
Iron-oxidizing	*Gallionella* *Leptothrix* *Thiobacillus*	Fe^{2+} (ferrous)	O_2	Fe^{3+} (ferric)
Manganese-oxidizing	*Leptothrix*	Mn^{2+} (manganous)	O_2	Mn^{4+} (manganic)

[a]Not necessarily all strains can carry out the indicated reaction.

Nitrifying bacteria are aerobic autotrophs that oxidize reduced nitrogen in two separate steps. **Ammonium oxidizers** such as *Nitrosomonas*, *Nitrosospira*, and *Nitrosococcus* convert ammonium to nitrite. (The first two are β-Proteobacteria, the third a γ-Protoebacteria.) **Nitrite oxidizers** convert nitrite to nitrate; examples are *Nitrobacter* (α-Proteobacteria) and *Nitrococcus* (γ-Proteobacteria). (Another nitrite oxidizer, *Nitrospira*, is a member of the Xenobacteria.) There appear to be relatively few types of nitrifying bacteria, but they are widespread (soil, freshwater, and marine systems), common, and play a crucial role in the nitrogen cycle. They are relied on in most wastewater treatment systems that control ammonia, may exert a substantial oxygen demand on streams receiving ammonia in effluents, and play an important role in nitrate contamination of groundwater. All known nitrifiers prefer slightly alkaline, mesophilic conditions; they are inhibited by even moderately acidic pH, and none appear able to grow at temperatures much above 40°C.

Nonphototrophic prokaryotes able to oxidize various forms of reduced sulfur can be found among the Archaea and several bacterial kingdoms. Hydrogen sulfide is the most commonly used form, but elemental sulfur, metal sulfides, and thiosulfate also are used by many species. Some **sulfur-oxidizing** Proteobacteria, including members of the genus *Thiobacillus*, produce so much sulfate (sulfuric acid) from the oxidation of pyritic (metal sulfide) minerals that the pH may drop to below 2. This is a major cause of the environmental problem known as acid mine drainage (Section 13.4.3) and can also occur in clay soils containing pyrites if they are exposed to the air. Some hot springs and marine thermal vents also are sources of sulfides.

Other sulfur oxidizers prefer or require neutral pH. Many of these rely on the production of sulfide by the anaerobic sulfate-reducing bacteria (see later) growing on organic material (rather than release from minerals). Since they need oxygen (or in some cases nitrate), such bacteria tend to grow at the interface between aerobic and anaerobic environments (e.g., the surface sediments in salt marshes). One of these is the filamentous form *Thiothrix*, which may occasionally cause settling problems in activated sludge wastewater treatment plants if sufficient sulfide is present in the influent or generated in the process. Another, the gliding bacteria *Beggiatoa* (Figure 10.22), is the most common filamentous organism observed in rotating biological contactor (RBC) treatment plants, sometimes producing such heavy growths that the steel shafts collapse. Deeper layers of the biofilm become anaerobic in many such plants, so that hydrogen sulfide is then produced and released to the overlying aerobic portion of the film. It is thought that *Beggiatoa*'s ability to glide helps it in remaining at the interface of the aerobic and anaerobic zones. Unlike many other lithotrophs, *Thiothrix* and *Beggiatoa* are also able to use small organic molecules as their carbon source and are thus referred to as **mixotrophs**.

Some bacteria are able to utilize the oxidation of ferrous (II) iron as a source of energy. For example, some of the acidophilic *Thiobacillus* (especially *T. ferrooxidans*) that grow on iron pyrites also are **iron oxidizers**. Ferrous iron is stable at low pH, allowing time for biochemical activity, but at neutral pH it is chemically oxidized to the ferric (III) form fairly rapidly. Interestingly, some bacteria, such as *Gallionella* and *Leptothrix*, can grow at the interface between oxic and anoxic zones and oxidize the iron before the chemical reactions occur. This is sometimes a problem when anaerobic well waters containing soluble ferrous iron are brought to the surface, as the growths (containing large amounts of insoluble ferric iron) can cause clogging. Similarly, some **manganese-oxidizing** bacteria can oxidize the manganous (II) to the manganic (IV) form.

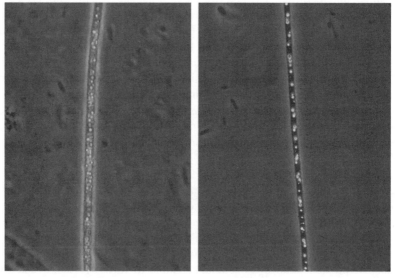

Figure 10.22   Two species of *Beggiatoa* in samples from RBC wastewater treatment plants; a gliding filamentous sulfur-oxidizing proteobacteria. Note internally deposited sulfur granules.

Methanotrophs   Methane (CH_4) is a major product of the anaerobic degradation of organic material, especially in the absence of large amounts of sulfate. **Methanotrophs** (e.g., *Methylomonas* and *Methylocystis*) are a relatively specialized but environmentally widespread group of obligately aerobic Proteobacteria (mainly α and γ groups) able to oxidize this methane. They are autotrophic, and some are also able to fix nitrogen. Most species produce a resting stage resistant to desiccation, either a **cyst** or an **exospore** (which is also quite heat resistant) produced by budding.

Most methanotrophs are also able to utilize methanol (CH_3OH) and many can utilize at least some other one-carbon compounds, such as formaldehyde (CH_2O), formic acid (CHOOH), methyl amine (CH_3NH_2), or carbon monoxide (CO), or other compounds without carbon–carbon bonds such as dimethylamine [$(CH_3)_2NH$], dimethyl sulfide [$(CH_3)_2S$], or dimethyl ether [$(CH_3)_2O$]. The ability to utilize such compounds is referred to as **methylotrophy** and is more widespread (e.g., including some *Bacillus*, a gram-positive bacteria) than growth on methane.

There has been recent interest in the use of methanotrophs for soil bioremediation systems. Their activity can be promoted by pumping methane into the subsurface environment, leading to increased generation of the enzyme methane monooxygenase (MMO). This enzyme can cometabolically convert a number of industrial contaminants, such as trichloroethylene, to less hazardous or harmless products (Section 16.7.2).

Nitrogen-Fixing Proteobacteria   **Nitrogen fixation**, the conversion of elemental nitrogen (N_2) to a more readily utilizable form, is an ability that is found scattered among many bacterial kingdoms (e.g., many of the Cyanobacteria and the gram-positives *Clostridium* and *Frankia*). Among the Proteobacteria this includes some methanotrophs, and most strains of the enteric bacteria *Klebsiella. Rhizobium*, and some similar species are of special importance because of the **mutualistic** (beneficial to both organisms) relationship they have with **legumes** (plants such as beans, peas, soybeans, alfalfa, clover, vetch, and mimosa). They are able to infect the plant roots to form special **nodules** in which they grow. Thereafter, they fix nitrogen (which is often in limited supply in soils), to the benefit of the plant, while they utilize organic substrates produced by the plant. Other important nitrogen fixers in soils (and water) are free-living, such as *Azotobacter*. Both the **symbiotic** (having a close relationship with another organism) and the free-living nitrogen fixers are aerobic, even though the required **nitrogenase** enzyme is sensitive to oxygen.

Proteobacteria with Special Morphologies: Appendaged, Sheathed, and Spiral Forms   A number of the Proteobacteria are known for their specialized morphologies, although these are not always indicative of phylogenetic relationships. *Hyphomicrobium* is a common aerobic soil and aquatic Proteobacteria whose prosthecae take the form of short hyphae from which buds are produced. *Caulobacter* is another common aerobic prosthecate bacteria; it uses its stalk for attachment. The chemolithotrophic iron oxidizer *Gallionella* also produces a stalk, but it consists of twisted exocellular fibrils coated with ferric hydroxide.

Sphaerotilus is a filamentous bacteria that produces a sheath (Figure 10.23). As part of its life cycle, single cells with polar flagella are formed that swim away to start new filaments. It is aerobic, but can still grow well at low dissolved oxygen concentrations (e.g., 0.5 mg/L). As a result, it is a common inhabitant of polluted streams and biological wastewater treatment systems. Large masses, commonly referred to as *sewage fungus* (although they are not fungi), sometimes form on rocks or other aquatic surfaces in such systems. Heavy growths in activated sludge wastewater treatment systems are one cause of the settling problem known as **bulking**. *Leptothrix*, mentioned above as an iron- and manganese-oxidizing organism, is a related species.

Spirillum is an example of a spiral Proteobacteria (not related to the spirochetes, Section 10.5.9). It is motile by tufts of flagella at both ends and is aerobic or microaerophilic.

Myxobacteria   The fruiting myxobacteria do not appear to be particularly important in environmental engineering and science but are of great interest to some microbiologists because of their life cycle, the most complex of any known prokaryote. The single, usually rod-shaped vegetative cells have gliding motility and often leave a slime trail as they move about on solid surfaces. When nutrients become limiting, the cells "swarm," coming together to form a visible, often brightly colored fruiting body. Cells within the fruiting body develop into myxospores, which are resistant to drying and some other environmental stresses. These spores can later germinate to form new vegetative cells.

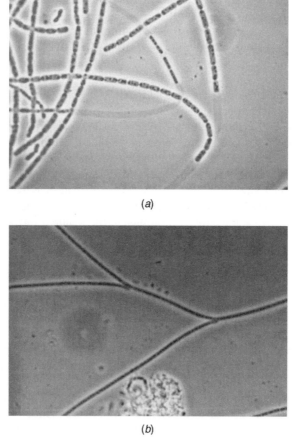

(a)

(b)

Figure 10.23 *Sphaerotilus natans*: (a) pure culture showing sheath and PHB granules; (b) branching filament in activated sludge sample.

Myxobacteria are aerobic chemoorganotrophs found in soil. Many, such as *Myxococcus*, are **bacteriolytic** (they **lyse**, or rupture, bacteria to feed on their cellular constituents—particularly proteins), and commonly occur on animal dung (which includes a high percentage of bacteria). However, a number (including some species of *Polyangium*) are instead **cellulolytic** (digest cellulose), and grow on decaying vegetation.

Pseudomonads   The pseudomonads are a large group of aerobic (although a few can utilize nitrate or nitrite when oxygen is unavailable), nonfermentative, chemoorganotrophic (except that a few can grow on hydrogen or carbon monoxide), heterotrophic, non-spore-forming gram-negative rods with polar flagella. *Pseudomonas* (Figure 10.8) is the major genus; however, when it was realized that bacteria called *Pseudomonas* were members of the α- and β- as well as the γ-proteobacteria (based on 16S rRNA), it was split into several genera, including *Sphingomonas*, *Burkholderia*, and *Ralstonia*. Still, there is much diversity in this group and disagreement among characterization methods as to where the natural groupings lie.

Pseudomonads are common soil and water bacteria, and because of their metabolic diversity, many are important in biodegradation of a very wide variety of natural and human-made organic compounds (including for bioremediation applications). A few, such as *Pseudomonas aeruginosa*, are opportunistic pathogens of humans. Some *Pseudomonas* and most *Xanthomonas* are plant pathogens. Another pseudomonad, *Zoogloea ramigera*, is found in some activated sludge wastewater treatment systems, where it forms a special type of zoogloeal **floc** (randomly organized aggregations) that can lead to poor settling conditions (Figure 10.24).

Other Aerobic Proteobacteria   Other aerobic proteobacteria include the nonflagellated species *Neisseria*, *Moraxella*, and *Acinetobacter* and the flagellated *Acetobacter*. Several *Neisseria* (including *N. gonorrhoeae*, the cause of gonorrhea; Section 12.6.2) and *Moraxella* (e.g., Section 12.7.4) are pathogenic to humans or other animals. *Neisseria* are cocci, whereas *Moraxella* are plump rods. *Moraxella*, which gives negative results for many classic biochemical tests, is of interest for another reason as well: Bacteria

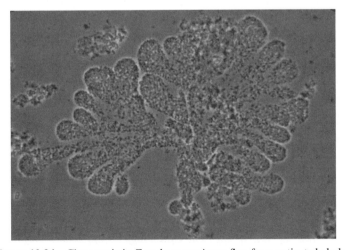

Figure 10.24   Characteristic *Zoogloea ramigera* floc from activated sludge.

isolated from the environment are also often negative for these tests, and hence have been misidentified as *Moraxella*. *Acinetobacter* is unusual in that it is an aerobe that is oxidase negative. These rods are common in soil and water. It sometimes shows twitching motility, as do some *Moraxella*.

Some *Acinetobacter* strains have been reported as phosphate accumulating organisms (PAOs) in activated sludge systems that achieve biological phosphorus removal (BPR). (This identification has since been challenged, and other aerobic proteobacteria using the same mechanism are now believed to be the major PAOs in these systems.) They contain **metachromatic (volutin)** granules composed of polyphosphate, which is used as a method of energy storage. This type of BPR is induced by having an anaerobic zone followed by an aerobic zone in the reactor. In the anaerobic zone, the PAOs utilize the stored polyphosphate for energy (releasing phosphate) to accumulate and store fermentation products. Subsequently, in the aerobic zone, they utilize the stored organics and remove phosphorus to very low levels (storing it again as polyphosphate). The low-P water is then separated and discharged, and the PAOs are recycled to the anaerobic zone to repeat the reaction. This is a fascinating application in which favoring a strictly aerobic organism in an ecosystem is accomplished by including a periodic anaerobic condition!

Acetic acid bacteria such as *Acetobacter*, although obligate aerobes, produce acetic acid from ethanol. This can lead to spoilage of alcoholic beverages but is also used commercially for vinegar production. Some *Acetobacter* are also able to produce a very pure form of cellulose.

The Family Enterobacteriaceae   The Enterobacteriaceae includes many bacteria of importance to humans and hence has been studied extensively. They also form a relatively stable taxonomic grouping; although originally based on testing by traditional (phenotypic) methods, there has been little change after genotype testing was employed. These members of the γ-proteobacteria are gram-negative, nonsporeforming rods that are typically facultatively anaerobic, capable of fermenting sugars. They either have peritrichous flagella or are nonmotile. Many live within the intestinal tract of warm-blooded animals (where they may play an important role in digestion) and hence are referred to as **enteric** bacteria. Some are important pathogens; others live in soil or water.

Escherichia coli (Figure 10.25) is probably the best known of all bacteria. *E. coli* is present in large numbers in the human intestines and is one of the **coliforms** used as indicator organisms to monitor fecal pollution of water (Section 12.9.2). A few strains are pathogenic (Section 12.3.2).

Important pathogenic members of this family include *Salmonella*, the cause of typhoid fever (*S. typhi*) and salmonellosis (but also used for testing chemical mutagenicity in the Ames test; Section 20.1.13); *Shigella*, which causes bacterial dysentery; and *Yersinia* (*Y. pestis* is the cause of plague). Other common species that may be intestinal or environmental (soil and water) include *Enterobacter* (formerly called *Aerobacter*), *Citrobacter*, and *Klebsiella*. *K. pneumoniae*, which is able to fix nitrogen, can also occasionally cause pneumonia. *Proteus* is well known to microbiologists because of the characteristic swarming of cells of some strains on the surface of agar plates. *Serratia* is also widely recognized in the lab because of the red colonies it often produces on plates.

Other Facultatively Anaerobic Proteobacteria   Other facultative bacteria include *Vibrio*, *Photobacterium*, *Aeromonas*, and *Chromobacterium*. These species differ from the Enterobacteriaceae in that they are polarly flagellated (*Chromobacterium* may have

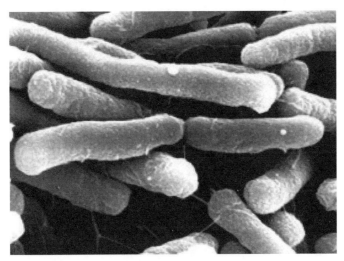

Figure 10.25 *Escherichia coli.* (SEM image courtesy of the University of Iowa Central Microscopy Research Facility.)

other flagella as well) and usually oxidase positive. They differ from *Pseudomonas* in that they are capable of fermentative metabolism. They are common soil and/or water organisms but include a few pathogens, such as *Vibrio cholerae* (cause of cholera) and *V. parahaemolyticus* (a marine species that can cause enteritis, usually from eating raw fish). *V. fischeri* and *Photobacterium* are marine species that are **luminescent**: They emit blue-green light (at ~490 nm) through reactions similar to those of the firefly. Some live in the specialized light-emitting organs of certain fish; others live freely and give some sea-waters their luminescence. They are also utilized in the Microtox (Azur Environmental, Strategic Diagnostics Inc., Newark, Delaware, www.azurenv.com) toxicity test. *Aeromonas* sometimes gives false-positive results in the coliform test. It is an aquatic organism sometimes associated with diseases of fish and frogs. *Chromobacterium* produces a violet pigment (violacein), so that colonies on agar plates are very noticeable.

Sulfur-Reducing Proteobacteria   Some strictly anaerobic proteobacteria, such as *Desulfovibrio* (Figure 10.26), are able to utilize oxidized forms of sulfur, especially sulfate and elemental sulfur, as the terminal electron acceptor for respiratory metabolism. As their energy source they use fermentation products (such as hydrogen, lactate, and pyruvate) produced by other bacteria in the same ecosystem. Sulfate reducers are very common in habitats containing organic material and sulfate, such as salt marsh sediments and animal intestinal tracts. They may also be active in anaerobic digesters used for sewage sludge treatment, and in the deeper layers of the biofilms in wastewater treatment processes such as trickling filters and rotating biological contactors. The hydrogen sulfide produced may be released, giving the characteristic rotten egg odor, and/or may combine with iron or some other metals to give the black color characteristic of such anaerobic systems. In confined spaces, hydrogen sulfide can build up to toxic levels. In sewers with inadequate flushing, the hydrogen sulfide released can cause odor problems and also dissolve in the moist films that form on the aerobic, unsubmerged surfaces; there the sulfide may undergo biological oxidation by chemolithotrophic bacteria, resulting in production of sulfuric acid and crown corrosion of the sewer.

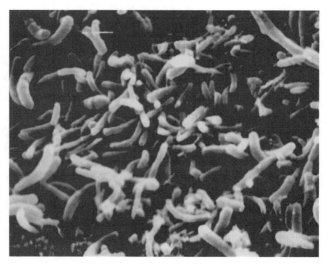

Figure 10.26   *Desulfovibrio*. Note the bent rods (vibrios).

Rickettsias   Rickettsias are obligate intracellular parasites—they grow within the cells of their hosts. *Rickettsia* species are responsible for typhus (Section 12.5.4) and Rocky Mountain spotted fever (Section 12.5.5). They are typically spread by fleas, lice, and ticks.

ε-*Proteobacteria*   *Campylobacter* (Figure 10.10) and *Helicobacter* are microaerophilic, motile, spiral, and usually, pathogenic. *Campylobacter* can cause serious cases of enteritis and has been responsible for some waterborne outbreaks (Section 12.2.5). *H. pylori* has been found to be responsible for many cases of stomach ulcers.

Other Proteobacteria   In addition to the Proteobacteria described briefly above, there are many other interesting and important species. For example, several bacteria, such as the marine *Magnetospirillum*, contain organelles with small magnetic mineral deposits that enable them to detect magnetic fields. *Bdellovibrio* hunts and feeds on other bacteria, growing inside the cell wall of its prey. *Legionella* grows in relatively clean, warm aquatic systems, including cooling towers and commercial hot-water and air-conditioning systems. If *L. pneumophila* becomes airborne and is inhaled, it can cause a severe form of pneumonia known as Legionnaires' Disease (Section 12.2.3).

10.5.7   Firmicutes: Gram Positives

This large kingdom is often broken into two broad groups based on G + C content (low G + C and high G + C) and then further subdivided (Table 10.7). It contains many bacteria of interest to the environmental engineer and scientist, including common soil organisms and normal inhabitants of our skin, mucous membranes, and digestive tract, as well as important pathogens. Members of the Firmicutes are also utilized in producing many foods, in industrial processes, and as sources of antibiotics. They are chemoorganic heterotrophs, including both aerobes and anaerobes.

Some of the Clostridia and Bacilli form **endospores**, specialized resting structures that are very resistant to heat, drying, and other environmental stresses. Many actinomycetes (high G + C) also produce spores, but these are not formed in the same way, nor are they

TABLE 10.7   Kingdom Firmicutes: The Gram-Positive Bacteria[a]

Low G + C	High G + C
Class 1. Clostridia	Class 4. Actinobacteria
Clostridium	Order Actinomycetales
Desulfotomaculum	*Actinomyces*
Epulopiscium	*Arthrobacter*
Eubacterium	*Micrococcus*
Heliobacterium	*Brevibacterium*
Sarcina	*Cellulomonas*
Class 2. Mollicutes	*Microbacterium*
Mycoplasma	*Corynebacterium*
Class 3. Bacilli	*Gordonia*
Order Bacillales	*Skermania*
Bacillus	*Mycobacterium*
Listeria	*Nocardia*
Staphylococcus	*Rhodococcus*
Order Lactobacillales	*Tsukamurella*
Enterococcus	*Actinoplanes*
Lactobacillus	*Micromonospora*
Streptococcus	*Propionibacterium*
	Streptomyces
	Microbispora
	Streptosporangium
	Thermomonospora
	Frankia
	Order Bifidobacteriales
	Bifidobacterium

[a] All classes are listed, but not all orders. Only some representative, important, and interesting genera are included.

as resistant. The endospore develops within the **vegetative** cell (Figure 10.27) and is then released through cell lysis. Under suitable conditions, the endospore can later **germinate** to produce a new vegetative cell. This ability is of obvious survival value in the soil, where conditions suitable for growth may alternate with unsuitable conditions. The degree of resistance shown by endospores is quite amazing, including survival in boiling water and autoclaving (steam heating at 121°C for 20 minutes) of soil. Probably the most incredible, however, was the recent finding of viable *Bacillus* endospores in the gut of a bee preserved in amber for at least 25 million years!

Low G + C   The genus *Clostridium* contains anaerobic endosporeforming rods. A variety of fermentation products (some have been used commercially) may be produced from carbohydrates by different species, including carbon dioxide, hydrogen, acetate, acetone, butanol, butyric acid, ethanol, isopropanol, lactic acid, propionic acid, and succinic acid. Cellulose-degrading species are probably the major decomposers of cellulose in anaerobic soils. Similarly, nitrogen fixation by some species probably represents the major source of new fixed nitrogen in anaerobic soils. Several species can ferment amino acids, producing the highly unpleasant odors characteristic of putrefying protein by forming compounds such as putrescine and cadaverine. Some species are thermophilic.

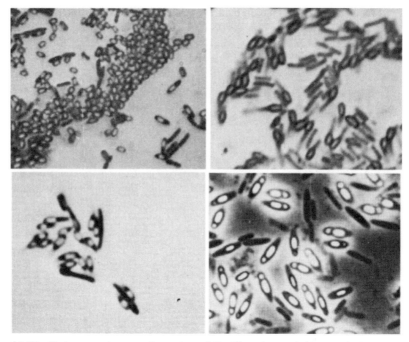

Figure 10.27 Endospores in several species of *Bacillus*: (upper left) *B. subtilis*, (upper right) *B. circulans*, (lower left) *B. stearothermophilus*, (lower right) *B. popilliae* (cause of milky spore disease in Japanese beetles). (From Gordon et al., 1973.)

Several common and normally free-living soil species of *Clostridium* produce toxins that may be fatal to humans, (Sections 12.3.1, 12.7.1, and 12.7.2). Botulism results from production (usually in food) of an exotoxin (a toxin released outside the cells) by *C. botulinum*. If infected food is eaten, the fatality rate is close to 100%. Also, the reason that infants should not be fed honey is the possibility of infection of the immature intestinal tract with *C. botulinum*, leading to toxin production. *C. tetani* can grow in deep, anaerobic wounds, releasing the exotoxin that causes tetanus, or "lockjaw." *C. perfringens* can be a normal inhabitant of the intestines (and has even been used as an indicator organism for sewage sludge). However, it can also cause gastroenteritis and is one of the clostridia responsible for gas gangrene.

Heliobacterium, a close relative of *Clostridium*, is of interest in that it is an endospore-forming phototroph (Table 10.4). It is also an important nitrogen-fixing soil bacteria, especially in paddies (rice fields). *Desulfotomaculum* is an endospore-forming sulfate-reducing bacteria. *Epulopiscium fishelsoni*, a rod-shaped symbiont in surgeonfish, is of interest because of its incredible cell size: 50 μm in diameter and up to 600 μm long!

Eubacterium is an anaerobic, non-spore-forming, rod-shaped bacteria that produces a mixture of organic acids as fermentation products. It is found in soil but is also one of the predominant intestinal bacteria. Some species are pathogens.

The Mollicutes lack a cell wall and hence are **pleomorphic** (of variable shape). They include the smallest (0.2 μm) known free-living cells. This small size and lack of rigidity means that they are sometimes able to pass through filtration systems intended to sterilize water or media. There are both facultative and obligate anaerobes. A number, including many *Mycoplasma*, are pathogenic.

Bacillus is a large, heterogeneous group of aerobic and facultative, rod-shaped endo-spore-formers. They are very common in soil, degrading a wide variety of compounds, and some are among the major species active during thermophilic composting (optimum temperature for *B. stearothermophilus* is 65°C). Some produce antibiotics (e.g., bacitra-cin), others are used commercially for industrial enzyme production. Several species attack insects, and in fact the insecticide Bt, effective against a number of pests, is actually *B. thuringiensis*. *B. anthracis* causes anthrax (Section 12.7.5), which is usually a disease of animals such as sheep. However, because it is so easy to culture and store (since it forms endospores), and because as an aerosol it can be very infective and nearly always fatal in humans, it is also of concern as a biological warfare agent.

The lactic acid bacteria, which include *Lactobacillus*, *Streptococcus*, and *Enterococ-cus*, grow mainly on sugars and produce lactic acid as their major fermentation product. Although they are obligate anaerobes, they tend to be aerotolerant. They require complex organic media, as they are unable to produce many basic cell constituents themselves from only simple compounds. All are cocci, except for *Lactobacillus* (rods), and tend to grow in chains. *Lactobacillus* is used to produce a number of foods, including some cheeses, yogurt, sourdough, sauerkraut, pickles, and acidophilus milk (for lactose-intoler-ant people), as well as being of great importance (because of the acids it produces) in the agricultural crop preservation technique known as **ensilage**. It can still grow well at pH 5 and below. It is also a common inhabitant of the mouth, intestines, and vagina of humans and other warm-blooded animals, but is not pathogenic.

Streptococcus species also are involved in ensilage, as well as production of some foods, such as buttermilk. Others are important pathogens, causing strep throat, scarlet fever, rheumatic fever, and pneumococcal pneumonia (Sections 12.4.1 and 12.4.2). Some species are normal inhabitants of the respiratory tract, intestines, or mouth, where they can be involved in dental caries (cavities). *Enterococcus* species (formerly, fecal streptococci) are common intestinal bacteria of warm-blooded animals; in fact, they are typically more common than *Escherichia coli* and other coliforms, except in humans.

Staphylococcus (Figure 10.9) are aerobic and facultatively anaerobic, typically grow in clusters, and are halotolerant, usually able to grow in 15% NaCl. They are common inha-bitants of the skin and mucous membranes. *S. aureus*, which is yellow, can be pathogenic, causing pimples, abscesses, boils, and impetigo of the skin, pneumonia, meningitis, and toxic shock syndrome. It also produces **enterotoxins** (exotoxins that affect the intestines), which makes it the most common cause of food poisoning (Section 12.3.1).

Listeria is widespread in nature, being found in soil, vegetation, and fecal material, and as an animal pathogen. One species causes listeriosis, which is most commonly a food-borne illness (Section 12.3.2). The short rods are aerobic or microaerophilic and are able to grow at refrigerator temperatures (4°C).

High G + C   The Actinobacteria include the actinomycetes and a number of related organisms. Most are soil bacteria (in fact, the earthy odor of soil comes from the **geos-mins** produced by many species of Actinobacteria and some cyanobacteria), but some are aquatic, and a number are pathogens. Actinomycetes develop filamentous masses referred to as **mycelia** (singular, *mycelium*), which superficially resemble fungal growths. The individual filaments are called **hyphae** (singular, *hypha*). Special sporeforming structures, called **sporangia** (singular, *sporangium*), may be produced, or filaments may fragment into spores in some species, but endospores are not formed. Many actinomycetes produce antibiotics, including such important ones (from *Streptomyces*) as streptomycin, tetracycline, chloramphenicol, and neomycin.

Among the Actinobacteria that are not actinomycetes are the Corynebacteria, which are aerobic or facultative, and include *Corynebacterium*, *Cellulomonas*, and *Arthrobacter*. *Corynebacterium* are irregular or club-shaped rods, often in V-shaped groups. It includes both free-living and pathogenic species, including *C. diphtheriae*, which causes diphtheria. *Cellulomonas* are very similar, but degrade cellulose and are nonpathogenic. *Arthrobacter*, strict aerobes unable to utilize cellulose, are among the most common soil bacteria. They begin as small cocci, which then elongate into irregular rods under growth conditions, forming cocci again as the medium is exhausted.

Propionic acid bacteria, such as *Propionibacterium*, are anaerobes that produce propionic acid as a main fermentation product. They help give Swiss cheese its characteristic taste, as well as its holes (from carbon dioxide production).

Mycobacterium are rod-shaped, sometimes forming filaments and even branches; however, they do not produce true mycelia. Included are free-living forms from soil and water, as well as several important pathogens. A key characteristic of this group is that they are **acid fast**; when stained with basic fuchsin and phenol, they are not decolorized by an acid–alcohol mixture. This property has been found to be due to the presence of mycolic acids on the cell surface, which occur only in this group. *Mycobacterium* tend to be slow growing, but also resistant to many environmental stresses. *M. tuberculosis* is the cause of tuberculosis (Section 12.4.3), a once-devastating disease that is still a major problem worldwide and is increasing again in the United States. *M. leprae* causes leprosy (Section 12.7.6).

Nocardioform bacteria (Figure 10.28) are usually considered the simplest of the actinomycetes, as well as being similar in some ways to the Corynebacteria and Mycobacteria. Mycelia are typically formed, but as the culture ages they tend to fragment into small sporelike elements. Almost all are strictly aerobic (a few are facultatively anaerobic), and most grow relatively slowly. *Nocardia* are very common in soils, where they seem to play a role in degradation of some recalcitrant compounds. A few are known to be opportunistic pathogens. They also have been found to be involved in foaming problems in activated sludge wastewater treatment plants, as have *Rhodococcus*, *Gordonia*, *Skermania*, and *Tsukamurella* (Section 16.1.3).

Actinomyces are anaerobic or facultative actinomycetes which produce easily fragmenting hyphae and no extensive mycelium. *Frankia* is microaerophilic, forms true mycelia, and is able to fix nitrogen. *Streptomyces* are very common, with the genus containing over 500 species; they are aerobic, with extensive mycelia that produce spores. A number of actinomycetes are thermophilic, including, for example, *Thermomonospora*, and some *Microbispora* and *Streptomyces*, and several species may be important in composting.

10.5.8 Planctomycetacia

Some authors break this kingdom into two, one for the Planctomycetales (containing *Planctomyces* and a few other genera), the other for the Chlamydiales (containing only

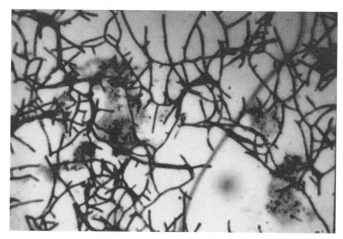

Figure 10.28 *Nocardia*-like filamentous bacteria in activated sludge foam (gram-stained preparation).

the genus *Chlamydia*). *Planctomyces* is a stalked, budding bacteria, but it is unrelated to the more common appendaged and budding bacteria such as *Caulobacter* (see earlier discussion of proteobacteria; Section 10.5.6), with which it was previously grouped. The stalk is made of protein (rather than cell wall and cytoplasm), and in fact the cell lacks the peptidoglycan that in other bacteria gives the cell wall much of its strength. Planctomycetes are heterotrophic chemoorganotrophs found mainly at the surface of freshwater lakes but also occur in marine and other aquatic environments. Recent genetic probe studies suggest that they are present in activated sludge wastewater treatment systems.

Chlamydia are obligate intracellular parasites, usually of birds or mammals, and so were originally grouped with the Rickettsias (α-proteobacteria). Trachoma, caused by *C. trachomatis*, is the leading worldwide cause of human blindness. Strains of this species also cause chlamydial infections of the genitourinary tract, which is probably the most widespread sexually transmitted disease, and apparently can be involved in heart disease through their role in the buildup of deposits on the interior walls of arteries. Other species may cause pneumonia in humans, and the disease psittacosis in birds and sometimes mammals, including humans. The bacteria have an infective form that invades a host cell, then grows inside it.

10.5.9 Spirochetes

The spirochetes are heterotrophic chemoorganotrophs with distinctive, long, thin, tightly coiled spiral cells. They have a "wriggling" type of motility that is now known to result from an unusual arrangement of **endoflagella**. Instead of sticking out from the cell, spirochete flagella bend back and run along the cell, and together with it are enclosed in a flexible sheath.

Members of the genus *Spirochaeta* are free-living obligate or facultative anaerobes. They are probably the spirochetes occasionally seen in activated sludge wastewater treatment systems (Figure 10.29), where they are indicative of anaerobic conditions. In some species the length may exceed 200 μm.

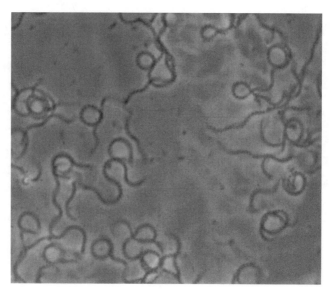

Figure 10.29 Spirochetes in activated sludge.

Several other spirochetes are important pathogens of humans and other animals. One species of *Treponema* (*T. pallidum*), which is microaerophilic or anaerobic, is the cause of the important venereal disease syphilis (Section 12.6.1). Other *Treponema* species cause yaws and other diseases in humans and animals, and many live commensally in the mouth or digestive tract, including in the **rumen** (the cellulose-digesting forestomach of ruminant animals, such as cattle, sheep, and deer). Microaerophilic *Borrelia* is the cause of Lyme disease (Section 12.5.6) and relapsing fever. *Leptospira* is aerobic and can be free living or pathogenic, causing leptospirosis (Section 12.2.3).

10.5.10 Fibrobacter

This kingdom contains only a few known species. *Fibrobacter* are gram-negative, rod-shaped (usually), non-spore-forming anaerobes that ferment cellulose and a few other compounds. They are found in the intestines or rumen.

10.5.11 Bacteroids

Bacteroids are obligately anaerobic chemoorganotrophic heterotrophs. Members of the genus *Bacteroides* are the most common bacteria in the human lower digestive tract, with concentrations in feces usually exceeding 10^{10} per gram. Bacteroids are often an important part of anaerobic environments (including the rumen) in which organics are being degraded to simpler compounds that can then be utilized by methanogens (archaea that generate methane; Section 10.6.3).

10.5.12 Flavobacteria

Flavobacteria also are chemoorganotrophic heterotrophs, but they are aerobic or occasionally facultatively anaerobic and are widely distributed in aquatic, marine, and soil systems. Colonies of the common genus *Flavobacterium* are often yellowish.

10.5.13 Sphingobacteria

The Sphingobacteria are aerobic or facultatively anaerobic chemoorganotrophic heterotrophs. *Cytophaga* are common in water and soil, have gliding motility, and can produce small spherical resting stages called **microcysts**. *Cytophaga* are able to digest cellulose and chitin, making them very important in the cycling of carbon in aerobic environments. They can also digest the agar that is used routinely for solidifying media in microbiology labs. *Haliscomenobacter* is a common small filament in activated sludge.

10.5.14 Fusobacteria

The Fusobacteria are gram-negative, anaerobic or facultative, non-spore-forming rods. *Fusobacterium* is typically found in the mouth or intestines and is sometimes pathogenic. *Streptobacillus* is usually found in the mouth of rats and can cause one type of rat bite fever.

10.5.15 Verrucomicrobia

There are only two known genera of Verrucomicrobia, both unicellular nonmotile chemoorganotrophic heterotrophs. *Prosthecobacter* has a single prostheca, is strictly aerobic, and is **oligotrophic** (grows at low levels of available nutrients). It is found in aquatic systems, soils, and wastewater treatment plants. *Verrucomicrobium* has numerous prosthecae, is facultatively anaerobic, and has been found in soils and a eutrophic (nutrient-rich) lake.

10.6 ARCHAEA

Although a number of Archaea species have been known for many years, until recently they were considered bacteria and not recognized as belonging to a separate major domain. Three kingdoms of Archaea (Table 10.8) are now recognized, and with the exception of the methane producers (methanogens), most of the known species are extremophiles (high temperature, high or low pH, and/or high salinity). They include both aerobes and anaerobes, chemoorganotrophs and chemolithotrophs, heterotrophs and autotrophs. Some are flagellated, others are nonmotile. None are known to be pathogenic. However, it is still too early in the investigation of these organisms to tell how diverse and widespread the group may truly be.

10.6.1 Korarchaeota

Members of the Korarchaeota have been observed in hot springs but have only recently been cultured. For this reason, little is yet known about them other than that the ones observed are hyperthermophiles. They are of particular interest to microbiologists because they may be the closest living relatives to the earliest life-forms to develop on Earth.

10.6.2 Crenarchaeota

Most of the known Crenarchaeota are hyperthermophiles, including *Pyrolobus*, the organism with the highest known growth temperatures (minimum, 90°C; optimum, 106°C; maximum, 113°C). Such organisms growing at temperatures above 100°C are found in

**TABLE 10.8   The Archaea: A Proposed Phylogeny Including Some
Representative, Interesting, and Environmentally Important Genera**[a]

Kingdom 1. Korarchaeota	Order 2. Methanomicrobiales
Kingdom 2. Crenarchaeota	*Methanomicrobium*
Class 1. Thermoprotei	*Methanospirillum*
Order 1. Thermoproteales	Order 3. Methanosarcinales
Pyrobaculum	*Methanosarcina*
Order 2. Desulfurococcales	Class 3. Halobacteria
Pyrodictium	*Halobacterium*
Pyrolobus	*Natronococcus*
Order 3. Sulfolobales	Class 4. Thermoplasmata
Sulfolobus	*Picrophilus*
Kingdom 3. Euryarchaeota	*Thermoplasma*
Class 1. Methanobacteria	Class 5. Thermococci
Methanobacterium	*Pyrococcus*
Class 2. Methanococci	*Thermococcus*
Order 1. Methanococcales	Class 6. Archaeoglobi
Methanococcus	*Archaeoglobus*
	Ferroglobus
	Class 7. Methanopyri
	Methanopyrus

[a]All classes and orders are listed if more than one.

marine hydrothermal vents, areas of volcanic activity on the ocean floor (where high pressures raise the boiling point of water). *Pyrolobus* is a chemolithotrophic autotroph that grows on hydrogen, utilizing nitrate (which is reduced to ammonium), thiosulfate, or oxygen as an electron acceptor. *Pyrodictium* can grow at 110°C (optimum, 105°C), utilizing hydrogen or organic material, with elemental sulfur as the electron acceptor. Other species (such as *Pyrobaculum*) can also utilize ferric iron (Fe^{3+}) as an electron acceptor.

Other Crenarchaeota grow in hot springs. *Sulfolobus*, for example, grows aerobically, oxidizing hydrogen sulfide to sulfuric acid and ferrous iron (Fe^{2+}) to ferric form. Its maximum temperature is "only" 87°C, but it is also an acidophile.

Not all Crenarchaeota are thermophilic. They have now been found to be widespread in the marine environment, including in Antarctic waters, where temperatures are typically below 0°C.

10.6.3   Euryarchaeota (Including Methanogens)

The Euryarchaeota include thermophiles and hyperthermophiles, hyperhalophiles, and methanogens. Among the thermophiles are *Thermoplasma* (optimum, 55 to 60°C), an acidophile (optimum, pH 1 to 2) found in hot springs and in coal refuse piles (large mounds of waste soil and rock from coal mining operations). Interestingly, *Thermoplasma* (like the mycoplasmas) lacks a cell wall, despite the harsh environments in which it thrives. *Picrophilus* (which does have a cell wall) can grow at a pH below zero.

Coal refuse piles represent an interesting ecosystem. They contain residual coal, iron pyrite (FeS_2), and other organic and inorganic compounds. Aerobic chemolithotrophs oxidize the pyrites, leading to highly acidic conditions (see acid mine drainage, Section 13.4.3), while oxidation of organics leads to self-heating (see composting, Section 16.2.3) and further depletion of oxygen. The elevated temperatures are believed to effect a partial chemical breakdown of high-molecular-weight organics present in the coal into smaller, more readily biodegradable compounds. This sets the stage for *Thermoplasma* to utilize the organics aerobically or through sulfur respiration.

Hyperthermophilic Euryarchaeota include *Thermococcus*, which grows on organic matter using elemental sulfur as its electron acceptor, and *Archaeoglobus*, a sulfate reducer. *Ferroglobus* can oxidize ferrous iron to ferric form utilizing nitrate.

The hyperhalophiles, such as *Halobacterium*, are found in very salty waters, such as the Dead Sea (Israel/Jordan) and Great Salt Lake (Utah), as well as in salt drying ponds and on salted fish. Others, including *Natronococcus*, thrive in highly alkaline soda lakes, such as those of the African Rift Valley. Of particular interest to some biochemists is that several of these organisms contain pigments that allow them to obtain energy (produce ATP) from light through a nonphotosynthetic pathway.

Methanogens are strict anaerobes that produce methane. Most commonly this is done by the reduction of carbon dioxide used as an electron acceptor during growth on hydrogen, but some methanogens can reduce methanol or other methyl compounds, cleave acetate (to methane and carbon dioxide), or carry out a very small number of related reactions. Methanogens are found in a variety of natural habitats, such as freshwater sediments and flooded soils (producing "swamp gas") and the digestive tracts of animals ranging from termites to humans. One genus found at hydrothermal vents, *Methanopyrus*, is hyperthermophilic (optimum 100°C, maximum 110°C).

The methanogens are the Archaea of greatest interest to environmental engineers and scientists. In addition to their critical role in the carbon cycle in anaerobic environments, their activities are widely utilized in such anaerobic organic waste treatment processes as anaerobic digestion of sewage sludge. Without the final conversion of anaerobic metabolic products to methane (and the potential for energy recovery), such anaerobic processes would be of very limited usefulness. Methanogens are also responsible for the production of methane at sanitary landfills.

10.7   EUKARYA

As discussed earlier, the third domain of life is Eukarya. Early biologists recognized only two kingdoms, Plants and Animals. Fungi and algae (and later bacteria) were considered plants, and protozoans, once they were discovered, were placed with the animals. (Interestingly, recent evidence suggests that fungi may actually be more closely related to animals than to plants!) Later, the eukaryotes were broken into four kingdoms (with the prokaryotes considered a fifth): Animals; Plants; Fungi; and **Protista**, which included protozoans, algae, and slime molds. However, as with the prokaryotes, genetic testing (in this case, 18S rRNA) has shown that if the differences between plants and animals warrants assignment to separate kingdoms, the protists can be considered to consist of a fairly large number of "kingdoms," including several each of protozoans and algae, two of slime molds, and one that was formerly considered a fungus. Here though, for convenience, we will still use the three groupings within the "protists." Also, although many fungi and algae and the slime molds may be large, multicellular, macroscopic organisms, they will still be discussed below with their microscopic "cousins".

10.7.1 Protozoans

Protozoans are chemoorganotrophic unicellular heterotrophic eukaryotes. They may absorb dissolved nutrients, but most feed mainly by ingestion of small particles (such as bacteria, algae, bits of organic matter, or macromolecules) through one of three methods. In **pinocytosis** water droplets are drawn into a channel formed by the cell membrane, while in **phagocytosis** solid particles are engulfed and enclosed by the membrane. In many ciliates, feeding is by a mouth and **gullet** through which particles are pushed—like swallowing in animals. In each case the ingested particle is enclosed in a membrane-bound **food vacuole** into which digestive enzymes are secreted. (This is in contrast to the fungi, which excrete enzymes to digest food particles externally.) Since they do not have a cell wall, to maintain osmotic pressure most freshwater protozoans also have a **contractile vacuole** that is used to expel excess water. Both asexual reproduction (often by fission) and sexual reproduction occur in many species. Most are aerobic, but a few contain a special structure, the **hydrogenosome**, instead of mitochondria, and are obligate anaerobes. Most protozoans are aquatic or marine, but a large number are parasitic or symbiotic, and others are important members of soil ecosystems. Many free-living types are seen in wastewater treatment systems, where they are thought to aid in purification.

Protozoans are usually motile by one (or more) of four means, at least in one part of their life cycle, and this has led to their being broken into the four major groups described below (Table 10.9). However, it is now recognized that protozoans are a highly diverse group, and include several different phyla and probably even distinct kingdoms. Although there are some colonial forms, no protozoans are multicellular (unlike some algae and most fungi). Still, the single cell may be highly complex, with many specialized organelles, especially among the ciliates. Although some of the simpler flagellates may be only 5 to 10 μm in size, many of the ciliates are 30 to 500 μm, and some Sarcodina exceed 1 mm (although most are much smaller).

Mastigophora (Flagellates)   The Mastigophora are motile by means of one or more flagella. This group actually includes several distinct lineages, including the Diplomonads [e.g., *Giardia* (Figure 10.30), a waterborne human parasite that causes giardiasis; Section 12.2.5], which lack mitochondria and are the most phylogenetically ancient known Eukarya. Another relatively ancient lineage, the Trichomonads, includes the human parasite *Trichomonas vaginalis,* which can cause vaginal and urinary tract infections (Section 12.6.6). Other flagellates include the Trypanosomes, such as *Trypanosoma gambiense*, the cause of African sleeping sickness (Section 12.5.2). *Bodo* is a common free-living flagellate (Figure 10.31) frequently seen in biological wastewater treatment. Other flagellates live in a small individual case called a **lorica**, often attached to the surface of a rock or another organism. Another group of flagellates, the Euglenophyta, contains chloroplasts and are discussed with the algae. (Often, the flagellates without chlorophyll have been referred to as "Zoomastigophora," to distinguish them from these "Phytomastigophora.") Sponges may have evolved from some of the flagellates. Some flagellates produce resistant **cysts**, helping them to withstand unfavorable conditions such as drying or exposure to toxic compounds (including disinfection).

Anaerobic flagellates include symbiotic inhabitants of the hindgut of termites and wood-eating cockroaches. In fact, it is the flagellates (or actually, probably the endosymbiotic bacteria within the flagellates) that produce the enzymes to digest the wood.

TABLE 10.9   The Protozoans

Group	Motility[a]	Feeding[a]	Comments	Examples
Flagellates	Flagella	Pinocytosis	Several distinct groups	*Bodo*
				Giardia
				Leishmania
				Monas
				Trichomonas
				Trypanosoma
Sarcodina	Pseudopodia	Phagocytosis	Some have tests	*Amoeba*
				Arcella
				Chaos
				Difflugia
				Entamoeba
Ciliates	Cilia	Gullet	Micro- and macro-nuclei	*Aspidisca*
				Colpidium
				Epistylis
				Euplotes
				Paramecium
				Podophrya
				Tetrahymena
				Vorticella
Sporozoans	None; flexing	Absorption	Obligate parasites	*Cryptosporidium*
				Plasmodium
				Toxoplasma

[a] Primary or characteristic method.

If "cured" of its flagellates, the insect starves to death no matter how much wood it eats.

Sarcodina (Amoebas)   Sarcodina, such as amoeba (Figure 10.32), typically move by **pseudopodia** ("false feet") formed by cytoplasmic streaming, and feed by phagocytosis. Some, such as *Arcella* (Figure 10.33; common in activated sludge), form shells, or **tests**. A few are pathogens, such as *Entamoeba histolytica,* the cause of amoebic dysentery (Section 12.2.4). Resistant cyst formation occurs in many species. Foraminiferans are often considered a separate phylum of marine testate organisms; deposits of their shells helped formed the famed white cliffs of Dover in England. Actinopods, which have numerous long projections protruding from their tests, also are now being considered a separate phylum.

Ciliophora (Ciliates)   The ciliates appear to be a relatively recent lineage. Unlike other protozoans, they contain two different nuclei; a **macronucleus**, which carries out most normal nuclear functions, and a **micronucleus**, which is involved in sexual reproduction. A typical member of this phylum, such as *Paramecium* (Figure 10.34a), has numerous short hairlike projections, called **cilia**, which can be moved in a coordinated way for locomotion. In *Aspidisca* and *Euplotes* (Figure 10.34b and c) additional cilia fused together form organelles that act like legs for walking over a surface, and perhaps as spines for protection. Food particles are usually ingested by means of a gullet. As opposed to these **free-swimming ciliates**, some attach to a surface by a stalk; by coordinating the

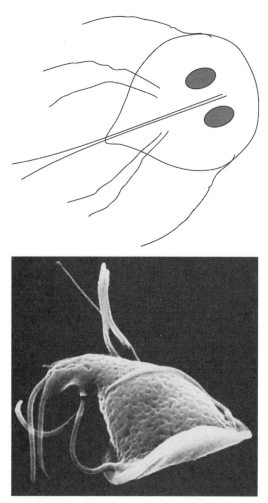

Figure 10.30 *Giardia*: sketch and electron micrograph. (SEM courtesy of the US National Park Service.)

movement of its cilia, such a **stalked ciliate** can create water currents (like small whirl-pools) that pull particles into its mouth. Other than the common *Vorticella* (Figure 10.35), which is solitary, most stalked ciliates (e.g., *Epistylis*) are colonial, with many organisms on a single branched stalk. One group of ciliates, the suctoreans (e.g., *Podophrya*, Figure 10.36), are stalked predators, waiting for a free-swimming ciliate to swim by; upon contact, a suctorean holds its prey by means of tentacles, and sucks out its cyto-plasm. Some ciliates are anaerobic and are common members of the rumen ecosystem.

Apicomplexa (Sporozoans)   The sporozoans are all obligate parasites. They feed by absorption of soluble nutrients from their host. The adult phases lack flagella or cilia but can move by flexing. They usually have a complex life cycle and form sporelike spor-ozoites to infect new hosts. Malaria (Section 12.5.1), one of the most important diseases worldwide, is caused by *Plasmodium*. *Cryptosporidium* (Figure 10.37), which has become a major concern in drinking water systems because of its extreme resistance to disinfection

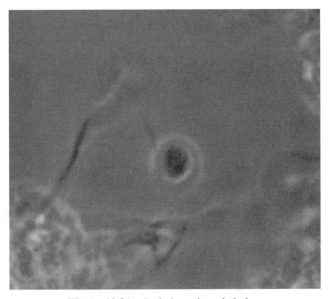

Figure 10.31 *Bodo* in activated sludge.

by chlorination, is also a member of this group. In 1993, over 400,000 people were sickened in Milwaukee, Wisconsin—the largest known outbreak of any waterborne disease in the United States (Section 12.2.5). Toxoplasmosis (Section 12.3.3), caused by *Toxoplasma*, is of particular concern for pregnant women who eat undercooked meat or come in contact with cat feces.

10.7.2   Algae

Algae are photosynthetic, oxygenic autotrophs. Most are unicellular, but many are colonial, and some are multicellular. Unlike plants, they do not have fully differentiated roots,

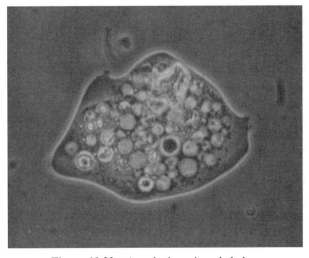

Figure 10.32 *Amoeba* in activated sludge.

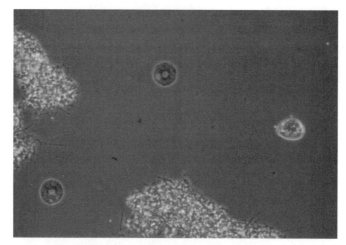

Figure 10.33   *Arcella, a* testate amoeba.

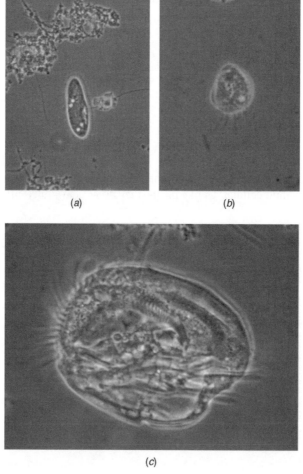

(a)

(b)

(c)

Figure 10.34   Free-swimming ciliates: (*a*) *Paramecium*; (*b*) *Aspidisca*; (*c*) *Euplotes*.

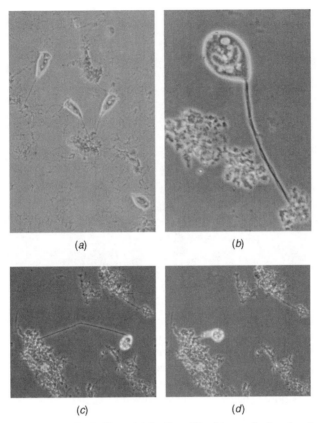

(a)   (b)

(c)   (d)

Figure 10.35   *Vorticella*, a stalked ciliate: (*a*) feeding; (*b*) with mouth closed and myoneme visible (dark line in stalk); (*c*) stalk extended; (*d*) seconds later, myoneme contracted to form a corkscrew-shaped stalk.

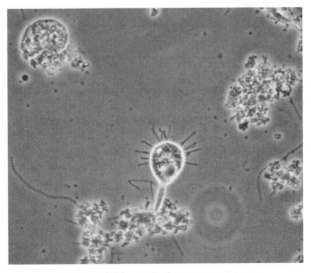

Figure 10.36   *Podophrya*, a suctorean.

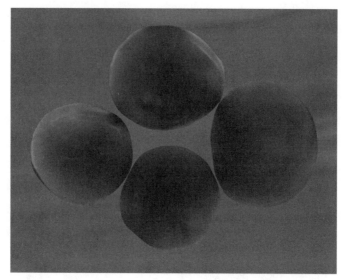

Figure 10.37   *Cryptosporidium* oocysts. (SEM copyright Dennis Kunkel Microscopy, Inc.)

stems, and leaves, although some of the brown algae, especially, have superficially similar structures. Six phyla (Table 10.10) are generally recognized by **phycologists** (scientists who study algae). The phyla are distinguished mainly on pigments, food storage materials, cell walls, and flagella. For the most part, the phyla are not considered to be closely related; in fact, it is now believed that they evolved separately from different nonphotosynthetic ancestors.

In Eukarya, photosynthesis occurs within membrane-bound organelles called **chloroplasts**. Since the **photosynthetically active range** (PAR) of light absorbed by algae is roughly from wavelengths of 400 to 700 nm, several different types of pigments are involved. These include various chlorophylls (including *a*, which is found in all algae, and *b*, *c*, or *d*), caretenoids, and phycobilins, which are associated with proteins to form light-harvesting complexes. Some of these pigments may also provide protection against harmful ultraviolet radiation. It is the combination of photosynthetic pigments present within algal cells that give them their distinctive coloration (e.g., green, red, brown).

Algae are the major primary producers of organic material in most aquatic and marine ecosystems, and hence many other organisms are directly or indirectly dependent on them. Some algae may also be found in or on soil and on terrestrial surfaces (e.g., trees, rocks). On a global level, they are of critical importance because of their production of oxygen and absorption of carbon dioxide.

From the perspective of environmental engineering and science, algae also can be problematic. If waters receive increased amounts of nutrients from wastewater discharges or runoff, excessive growth of algae, cyanobacteria, and plants can occur. This can lead to a number of problems, ranging from physical interference with recreation (or even navigation) to fish kills from oxygen depletion or production of toxins. In fact, this problem, referred to as *eutrophication* (Section 15.2.6), can even lead to the rapid filling in of a lake or pond, converting it eventually to a wetland and then dry land. On the other hand, algae are also utilized in some wastewater treatment processes to produce oxygen or remove nutrients. *Standard Methods* (Clesceri et al., 1998), a reference book available

TABLE 10.10 The Algae

Phylum	Euglenophyta	Phaeophyta	Chrysophyta	Pyrrophyta	Rhodophyta	Chlorophyta
Common name	Euglenoids	Brown algae	Golden algae/ diatoms	Dinoflagellates	Red algae	Green algae
Pigments[a]	chl.-b	chl.-c, xanthophylls	chl.-c	chl.-c	chl.-d, phycobilins	chl.-b
Cell wall	None	Cellulose	Silica (diatoms)	Cellulose plates	Cellulose	Cellulose
Main storage product[b]	β-1,2-glucan	β-1,3-glucan	Lipids	α-1,4-glucan	α-1,4- and α-1, 6-glucan	α-1,4-glucan
Motility	Two flagella (one very short)	Two flagella (reproductive cells)	Gliding (some)[c]	Two flagella	None	Zero, two, or four flagella
Major habitats	Mostly freshwater	Mostly marine	Freshwater, marine, soil	Mostly marine	Mostly marine	Mostly freshwater, soil
Usual form	Single cell	Multicellular	Single cell, colonial	Single cell	Multicellular	Single cell, colonial, multicellular
Example genera	*Euglena* *Phacus*	*Fucus* *Laminaria* *Sargassum*	Golden: *Tribonema* Diatoms: *Asterionella* *Diatoma* *Fragilaria* *Navicula* *Nitzschia* *Synedra* *Tabellaria*	*Ceratium* *Gonyaulax* *Gymnodinium* *Peridinium* *Pfisteria*	*Polysiphonia* *Porphyra*	*Ankistrodesmus* *Chlorella* *Chlamydomonas* *Cladophora* *Scenedesmus* *Selenastrum* *Spirogyra* *Ulothrix* *Volvox*

[a] In addition to chlorophyll-a (chl.-a) and carotenoids.

[b] Starch is α-1,4-glucan; β-1,2-glucan is called paramylon, and β-1,3-glucan is laminarin.

[c] Some chrysophytes have one or two flagella.

to most environmental engineers and scientists, includes very useful (and beautiful!) classic color drawings (by C. M. Palmer) of many important freshwater algae.

Euglenophyta   The Euglenophyta are considered to be the first of the algae to evolve, probably from unpigmented flagellates. In fact, they are considered protozoans (Phytomastigophora) as well as algae, mainly because they usually have two flagella (one so short it may be internal) and have no cell wall. Most, such as *Euglena* (Figure 10.38), are able to live chemoorganotrophically in the dark and can also survive in this way if their chloroplasts are removed. Their pigments are similar to those of the green algae and plants, but they are not felt to be closely related to these groups.

Phaeophyta: The Brown Algae   The brown algae are multicellular and mainly marine. They include most seaweeds. Giant kelp can exceed 60 m in length. The Sargasso Sea in the North Atlantic Ocean is known for its extensive beds of floating *Sargassum*. Brown algae are thought to have evolved from early pigmented flagellates and still produce flagellated asexual spores and sexual reproductive cells. Many produce leaflike blades and a holdfast, but not true leaves, stems, or roots.

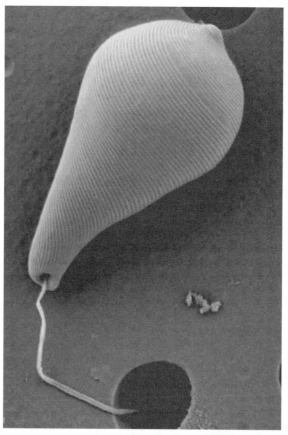

Figure 10.38   *Euglena gracilis.* (SEM image courtesy of Brian Leander and Michael A. Farmer, Center for Ultrastructural Research, University of Georgia, Athens, GA.)

Chrysophyta: Chrysophytes and Diatoms   The Chrysophyta, named from the Greek *chrysos* (gold), includes three major classes: **yellow-green**, **golden-brown**, and **diatoms**. They may be either unicellular or colonial. The yellow-green and golden-brown algae (**chrysophytes**) are mainly found in fresh water, although there are also a few marine species. Diatoms are common in both fresh and salt water, as well as in moist soils. Diatoms are typically single celled, although there also are filamentous species. They do not have flagella, but some have gliding motility.

Diatoms form silica-rich **frustules** as their cell walls; the two halves fitting together like a petri dish, with one side overlapping the other. These shell-like structures, such as those shown in Figure 10.39, give many diatoms a distinctive, often beautiful appearance, which frequently can be used for classification purposes. Geologic deposits of diatoms are mined to produce the diatomaceous earth used as a common filtration medium. During asexual reproduction, each half of the shell becomes the larger half of one of the two new frustules. Thus, one of the daughters is smaller than the parent. This continues until very small cells are formed, at which point sexual reproduction will occur.

Pyrrophyta: The Dinoflagellates   Most **dinoflagellates**, such as *Peridinium* (Figure 10.40), are single-celled, mainly marine algae with a cell wall made of tough cellulose–silica plates.

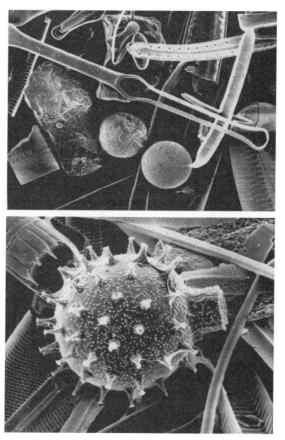

Figure 10.39   SEM images of various diatom frustules. (Courtesy of the University of Iowa Central Microscopy Research Facility.)

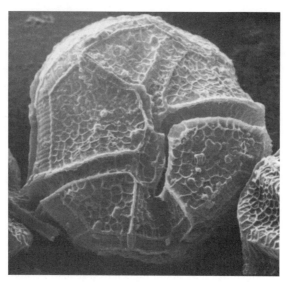

Figure 10.40 *Peridinium.* (SEM image courtesy of C. J. O'Kelly, D. H. Hopcroft, and R. J. Bennett.)

They have two flagella, one wrapped beltlike around the middle in a transverse groove, the other extending backward in a longitudinal groove. This gives them a toplike motion, from which they get their name (from the Greek for "whirl"). Although most are free-living, some are symbionts growing in reef-forming corals, and some are parasitic. The Pyrrophyta are believed to be close relatives of the ciliated protozoans.

Blooms of dinoflagellates in coastal or estuarine waters may sometimes become sufficiently dense to cause **red tides** (Section 15.2.2). Further, a few dinoflagellate species, such as *Gymnodinium, Gonyaulax,* and *Pfisteria,* are able to release a neurotoxin into the water, which may lead to fish kills or to human poisoning due to the consumption of contaminated shellfish (Section 12.3.3). In recent years, *Pfisteria* has received considerable attention in the Chesapeake Bay, apparently finding a selective advantage by killing fish to secure degradable substrates!

Rhodophyta: The Red Algae   The red algae, such as *Porphyra,* are mostly multicellular and marine, although there are also a few unicellular forms, and some that are freshwater or soil inhabitants. They have pigments similar to those of cyanobacteria, from which their chloroplasts may be derived. They produce no flagellated cells, further suggesting they are not derived from flagellates, as most other algae appear to be. Several red algae produce mucilaginous polymers that are used commercially, including the widely used food additive, carrageenan. Microbiologists also owe them a debt of gratitude, as they are the source of the **agar** widely used to solidify laboratory media for plating.

Chlorophyta: The Green Algae   The green algae are believed to be the group from which higher plants evolved, and in fact they have the same pigments, cell wall composition, and storage products. This is an extremely varied and ecologically widespread group, with habitats that not only include soils and waters (both fresh and salt) but also the surfaces and internal regions of other organisms. Many are unicellular, but there are

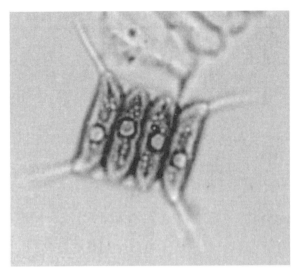

Figure 10.41 *Scenedesmus.*

also colonial (e.g., *Scenedesmus* Figure 10.41) and multicellular (but undifferentiated) forms, and a few that are **coenocytic** (filaments in which the cross-walls between cells deteriorate or do not form, so that they become single cells with many nuclei), like fungi.

10.7.3 Slime Molds

The slime molds are aerobic, mainly terrestrial (a few are aquatic) chemoorganotrophs. Although they appear to have no major role in environmental engineering or science, they will be discussed briefly for completeness and because of their interesting life cycle. There are two separate phyla (Table 10.11), both with similarities to some protozoa.

Myxomycota   Myxomycetes, or **plasmodial slime molds**, often form colorful visible growths on decaying wood or vegetation. An example is the genus *Physarum*, which is bright yellow. The vegetative form is composed of a plasmodium, which is a macroscopic multinucleated mass of cytoplasm with a single plasma membrane enclosing it (hence, their other common name, **acellular slime molds**). Other genera, such as *Echinostelium*, however, produce only microscopic plasmodia. The plasmodium crawls over surfaces with amoeboid motion (cytoplasmic streaming), engulfing food particles—mainly bacteria, but also yeasts, spores, and small pieces of organic material—through phagocytosis. Thus, it behaves much like a giant amoeba. However, when there is insufficient food, or the surface dries, the plasmodium forms a stalked sporangium and, after meiosis, releases haploid spores. Under proper conditions a spore germinates to produce a swarm cell (biflagellated) if there is sufficient water for swimming, or a myxamoeba (amoeboid cell) if the surface is only moist enough for crawling. Two such cells fuse to produce a diploid cell, which then undergoes mitosis (without cell division) to produce a new plasmodium.

Acrasiomycota   **Cellular slime molds**, such as *Dictyostelium* and *Polysphondylium*, spend most of their active life as free amoeboid cells. Under unfavorable conditions

TABLE 10.11   The Fungi, Oomycetes (Water Molds), and Slime Molds

Group	Phyla	Common Name	Motility	Feeding	Comments	Examples
Fungi	Zygo-mycota	Lower fungi	None	Absorption	Most mycorrhyzae	*Entomphthora* *Mucor* *Pilobolus* *Rhizopus*
	Asco-mycota	Sac fungi	None	Absorption	Most yeasts and lichens; truffles and morels	*Ceratocystis* *Claviceps* *Endothia* *Morchella* *Neurospora* *Tuber*
	Basidio-mycota	Club fungi	None	Absorption	Most mushrooms	*Agaricus* *Amanita* *Coprinus* *Phanerochaete* *Psilocybe*
	Deutero-mycota	Fungi imperfecti	None	Absorption	Sexual stage unknown (most are probably Ascomycetes)	*Arthrobotrys* *Aspergillus* *Candida* *Coccidioides* *Dactylella* *Epidermophyton* *Fusarium* *Geotrichum* *Histoplasma* *Monilia* *Penicillium* *Rhizoctonia* *Saccharomyces* *Trichophyton* *Verticillium*
Water molds	Oo-mycota	Water molds	Flagella	Absorption	Mycelia produce flagellated zoospores	*Allomyces* *Phytophthora* *Plasmopara* *Pythium* *Saprolegnia*
Slime molds	Myxo-mycota	Plasmodial or acellular slime molds	Plasmodium/ amoeboid/ flagella	Phago-cytosis	Like giant amoeba; forms sporangium	*Echinostelium* *Physarum*
	Acrasio-mycota	Cellular slime molds	Amoeboid/ slug	Phago-cytosis	Amoebas swarm, produce "slug"	*Dictyostelium* *Polysphondylium*

(inadequate food or water), these individuals aggregate to form a macroscopic **pseudoplasmodium** (because the cells remain individuals with separate membranes), or "**slug**" (unrelated to true slugs, which are mollusks), which crawls to an exposed location, forms a fruiting body, and releases spores. (In this way they resemble superficially the prokaryotic myxobacteria.) The spores can later germinate to produce new amoeboid cells. In addition to this asexual process, cellulose encased macrocysts may occasionally form in which two amoeba fuse.

10.7.4   Fungi

Fungi are chemoorganotrophic heterotrophs. Most are saprobic, but some are parasites (the majority of plant pathogens are fungi, as are some animal pathogens) or symbionts. Although some are found in aquatic or marine environments, most are terrestrial. In addition to not being photosynthetic, they also differ from plants and algae in having cell walls composed primarily of chitin rather than cellulose. Fungi are nonmotile, unlike protozoans and slime molds (and some algae). Mitosis in fungi is different from that in plants and animals.

Like some bacteria, fungi excrete enzymes to digest food externally, then absorb the smaller organic molecules produced. This makes them especially suited for degradation of organic solids and for their important ecological role as decomposers of dead biomass (such as leaves and wood) as well as high-molecular-weight anthropogenic compounds. They play a major role in the global carbon cycle, especially because of the ability of many to degrade cellulose and of some to attack lignin. Fungi store energy either as glycogen or in lipids.

Fungi usually have one of two growth forms, although a few can take both. Both may reproduce sexually through spores. **Yeasts** are unicellular and reproduce asexually by budding (see below). **Molds**, which are the much more common form, are multicellular, filamentous organisms and usually reproduce asexually with spores. As with the bacterial actinomycetes (Section 10.5.7), an individual filament is referred to as a **hypha** (plural, *hyphae*), and the intertwined, matted growths they form are known as **mycelia** (singular, *mycelium*). However, unlike with bacterial filaments, which are essentially colonies of single-celled organisms, molds are multicellular. Typically, there is considerable interchange of materials among cells within the filament. In fact, often the cross-walls separating cells have holes in them or disappear completely. This can lead to the presence of several nuclei within a single "cell," a growth form referred to as **coenocytic**. Hyphae are usually haploid and produce microscopic spores, usually at the tips of specialized filaments called **conidia** (singular, *conidium*; Figure 10.42) or in sacs called **sporangia** (singular, *sporangium*). After dispersal, the spores may germinate and grow into new hyphae if deposited on a suitable substrate.

The fungal mycelium typically permeates the soil or substrate on which the organism is growing. **Mushrooms** are the familiar, large, fruiting bodies of some fungi and do not represent the major portion of the organism, which consists of the mycelia below the surface.

Both bacteria and fungi degrade organic material, but bacteria typically grow faster. However, fungi are less inhibited by dry conditions than bacteria, and so often make up a majority of the living biomass in drier soils and at the dry edges of compost piles. Fungi are also able to withstand higher concentrations of salt and sugar, allowing them to cause spoilage in foods that most bacteria cannot grow on. Some thermophilic fungi can grow at temperatures up to 62°C. As a group, fungi also can tolerate lower pH than many common organotrophic bacteria, allowing them to grow competitively, for example, on fermented fruit. In activated sludge plants for wastewater treatment, if the pH drops below 5.5 to 6, fungi may grow excessively and interfere with the settling process (fungal bulking).

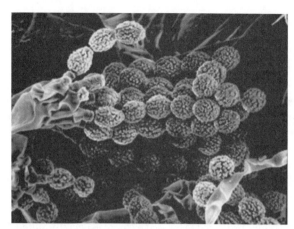

Figure 10.42   *Aspergillus* with conidia. (SEM image courtesy of the University of Iowa Central Microscopy Research Facility.)

The fungi are now classified into four phyla or divisions: Zygomycetes, Ascomycetes, Basidiomycetes, and Deuteromycetes (Table 10.11). Previously, another phylum, the Oomycetes, was included, and is discussed here, but they are now considered a separate protist group.

Zygomycota   The Zygomycetes, or **lower fungi**, include about 800 known species, including *Rhizopus stolonifer*, the common black bread mold. Zygomycete sporangia consist of small stalks that produce spores. In *Pilobolus*, the "shotgun fungus," which grows on horse and cow dung, spore release is dramatic—the sporangium can be shot to a height of 6 feet! Sexual reproduction occurs through contact of two hyphae (which are haploid) of opposite mating types (called + and −) and fusing of their nuclei, leading to the production of a diploid zygospore. This resistant form later undergoes meiosis, germination, mitosis, production of a sporangium, and release of haploid spores. Most zygomycetes, such as the common *Mucor*, are decomposers in soils and decaying plant material, with some causing food spoilage. One group forms mycorrhyzae (see below), and a few are parasitic on plants or animals. *Entomophthora*, for example, parasitizes houseflies, eventually killing them. (Look for a white powdery halo of spores around the next dead fly you find on a windowsill.)

Ascomycota   The Ascomycetes, or **sac fungi**, is the largest fungal phylum, with 30,000 species known, including most yeasts, a few mushrooms (such as truffles, morels, and cup fungi), and many molds. The blue-green mold that grows on citrus fruit and the pink bread mold *Neurospora* are Ascomycetes. Many plant diseases are caused by ascomycetes, including powdery mildew, apple scab, and several rots. Chestnut blight, caused by *Endothia parasitica*, has virtually eliminated the American chestnut, once a major forest tree in the northeastern United States. *Ceratocystis ulmi*, the cause of Dutch elm disease, has had a similar effect on the American elm, another major forest and shade tree. *Claviceps purpurea* infects rye and other cereals, leading to production of a structure called an *ergot* in the seed head. Ingestion by cattle or humans can lead to severe nervous system disorders, including delusions, spasms, and often, death. [One of the chemicals present is lysurgic acid, a precursor of the hallucinogen, lysurgic acid diethylamide (LSD).] In

Europe in the Middle Ages, whole villages would sometimes appear to go insane from eating bread made with infected rye.

Ascomycete molds (yeasts are described below) reproduce asexually by pinching off conidiospores from special hyphae called *conidia*. Sexual reproduction occurs through fusing of two hyphae to form a sac called an **ascus** (plural, *asci*), in which (after meiosis and mitosis) ascospores develop.

Basidiomycota   The Basidiomycetes, or **club fungi**, consist of over 16,000 species, including most of the familiar mushrooms, toadstools, stinkhorns, puffballs, and jelly and shelf fungi. In fact, Basidiomycetes are thought to form the major part of living biomass in most soil. Although some mushrooms (e.g., the commercially produced *Agaricus campestris bisporus*) are edible, others are poisonous (e.g., *Amanita verna*, the destroying angel) or hallucinogenic (e.g., *Psilocybe*). Mushrooms are the fruiting bodies that result from sexual reproduction, serving to release the haploid spores. Some important crop parasites, such as the rusts and smuts, also are basidiomycetes.

White rot fungi (e.g., *Phanerochaete*) are basidiomycetes that cause decay of wood by digesting lignin and cellulose. Because the lignin-degrading enzymes are also effective against many xenobiotic compounds, *P. chrysosporium* is being studied for use in engineered processes to degrade industrial pollutants. Brown rot fungi attack cellulose but not lignin.

Deuteromycota   The Deuteromycetes are also referred to as **fungi imperfecti**, reflecting the fact that they have no known sexual reproduction phase. This is not a true phylum but rather a form-phylum, containing more than 11,000 "form-species." Once the sexual phase is discovered, species in this group are reclassified into one of the other phyla. Almost all appear to be Ascomycetes, but a few are Basidiomycetes. Included are molds of the genus *Penicillium*, well known as the source of the antibiotic penicillin and also the source of the flavor and color of Roquefort and other "blue" cheeses. *Aspergillus* (Figure 10.42) is a common soil saprobe, but during growth on peanuts or other stored foods, some make **aflatoxins**, among the most toxic natural compounds known. *Aspergillus* also can be an **aeroallergen** (cause allergic reactions) when the spores become airborne in agriculture and at composting sites, and occasionally can be responsible for lung disease in immunocompromised hosts. Other species of *Aspergillus* are used in fermentation of soy sauce and saki. *Verticillium* cause pink rot of apples and also a wilt that is affecting many Norway maples planted as municipal street trees in the northeastern United States. *Fusarium* causes wilts of several plants, and *Rhizoctonia* (a Basidiomycete) can cause damping off and root rot.

On the other hand, some fungi have been found to be helpful in controlling nematode damage of plants. *Arthrobotrys* and *Dactylella*, for example, trap and feed on these small animals—serving as the Venus flytraps of the microbial world! These fungi also are occasionally observed in wastewater treatment systems (Figure 10.43). *Geotrichum* is another fungus that is sometimes found in activated sludge plants, where its excessive growth (at low pH) leads to a problem referred to as *fungal bulking*.

Among the Fungi Imperfecti that are human pathogens are several species (e.g., *Epidermophyton* and *Trichophyton*) that cause the skin infections known as athlete's foot and ringworm (Section 12.7.7). Histoplasmosis, which is a potentially fatal respiratory disease (Section 12.4.7), is caused by *Histoplasma capsulatum*. *Candida* is a common source of vaginal yeast infections (Section 12.6.7). Both *Histoplasma* and

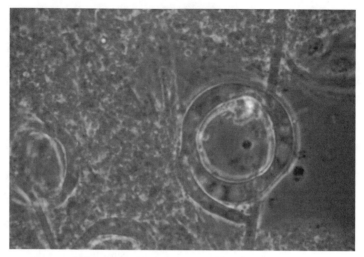

Figure 10.43   Nematode trapping fungus.

Candida are soil organisms and opportunistic pathogens and can invade a variety of tissues in compromised hosts. They also grow in both yeast and mold form.

Oomycota   Because they are nonphotosynthetic, have a cell wall, and produce a mycelium, the Oomycetes, or **water molds**, were considered fungi until fairly recently. However, unlike fungi, they produce motile spores, and whereas the cell wall contains chitin in some species, it is made of cellulose in others. Analysis of their 18S rRNA has now confirmed that they are a separate group. Asexual reproduction is by release of biflagellated zoospores from a sporangium that forms at a hyphal tip. Sexual reproduction involves contact of male and female hyphae, leading to production of oospores. Most oomycetes are aquatic, with a few, such as *Saprolegnia parasitica*, parasitic on fish. A few are terrestrial, including some very important plant pathogens. Late potato blight, which may have led to the starvation of as many as 1 million Irish in the 1840s (and prompted emigration of millions more), is caused by *Phytophthora infestans*. *Pythium* causes damping-off disease of seedlings, and *Plasmopara* and other Oomycetes cause downy mildew.

Yeasts   **Yeasts** are single-celled fungi. There are 60 genera with 500 known species. Most are ascomycetes, but some are basidiomycetes, zygomycetes, or deuteromycetes. They reproduce asexually by **budding**, a mitotic division producing a large and a small cell (the bud). Yeasts can respire using oxygen, but in the absence of oxygen, they are limited to glycolysis, which feeds into a fermentation pathway to produce ethanol and CO_2. The ascomycete *Saccharomyces cerevisiae* has been "domesticated" and developed into both baker's and brewer's yeast.

Mycorrhyzae   Some molds, called **mycorrhyzae** ("fungus roots"), grow either within or surrounding the roots of many plants as a symbiotic infection. Their mycelia extend into the soil, increasing the surface area available for absorption. This increases a plant's ability to extract nutrients. In return, the fungus shares in the primary productivity of the

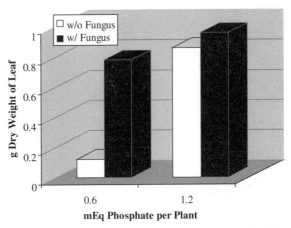

Figure 10.44   Combined effect of mycorrhyzal fungus and phosphate fertilizer on tomato growth. (Based on Ricklefs, 1993.)

plant. The effect is particularly significant when minerals are scarce. For example, in one study (Figure 10.44) tomato plants grown in sterile soil deficient in phosphate grew 6.5 times as much leaf mass when a mycorrhyzal fungus was inoculated into the soil.

Mycorrhyzal fungi are associated with some 90% of terrestrial plants. Most are zygomycetes that grow within the root, with mycelia extending out. Certain trees and shrubs have mycorrhyzal fungi that grow only on the outside of the root. Most of these are basidiomycetes, but some are ascomycetes such as truffles. Mycorrhyzae seem to help plants grow under adverse conditions such as high altitudes or acidic soils. Plants of the heather family (*Ericaceae*) show increased resistance to toxic heavy metals when mycorrhyzae are present. Orchid seeds will not germinate if an associated fungus is not present.

Lichens   Lichens are symbiotic associations between ascomycetes and either algae or cyanobacteria. The fungus receives carbohydrates and other nutrients from the algae, and in return, it provides moisture and protection. Lichens are familiar to hikers as the often-colorful flat growths adhering to rocks. About 17,000 lichen combinations are known. One of their ecological roles is to hasten the breakdown of the rock substrate, forming soil. They are very sensitive to sulfur dioxide, and thus can be used as an indicator of air pollution.

10.8   NONCELLULAR INFECTIVE AGENTS: VIRUSES, VIROIDS, AND PRIONS

Viruses, viroids, and prions are submicroscopic particles that are not composed of cells. They can carry out no metabolic activities on their own, and therefore are not considered "alive." However, they can infect living host cells and cause them to produce new copies of the infective agent. Thus, many of them produce disease. The viruses (from the Latin *virus*, meaning "poison") are the major type of such agent, but we also discuss the others briefly below.

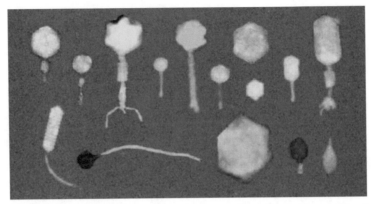

Figure 10.45 Tableau of viruses (mostly bacteriophage) from saline wetland ponds in Saskatchewan, Canada. (Photo by David Bird; courtesy of D. Bird and R. Robarts.)

10.8.1   Viruses

A virus particle, or **virion**, is composed of a nucleic acid core and a protein coat, or **capsid** (Figure 10.45). Some also have an outer envelope composed mainly of lipid and protein. They are too small—typically 20 to 300 nm (0.02 to 0.3 μm)—to be visible with a light microscope (or to be removed by normal filtration), but they can be "seen" using electron microscopy. Although they are strictly intracellular parasites, they also are characterized by having an extracellular form in which they can be transmitted to infect other host cells. Although individual virus types often are very specific as to the organisms (often a single species, or even a single strain) they can attack, there are varieties of viruses for almost (if not) all hosts, including animals, plants, fungi, and algae. Viruses that attack bacteria are called **bacteriophage**, or simply **phage**.

The nucleic acid in a virus genome is either DNA or RNA (but not both) and is either single or double stranded. It encodes for replication of the virion by the host cell. Some RNA viruses first produce a strand of DNA during replication (the reverse of the normal production of an RNA strand from DNA); such viruses are called **retroviruses**. The human immunodeficiency virus (HIV), which causes the disease acquired immune deficiency syndrome (AIDS; Section 12.6.4), is a retrovirus. Other RNA viruses are able to replicate their RNA directly without the intermediate DNA.

The protein coat is composed of subunits called **capsomeres**, which self-assemble after production by the host to form the capsid. The two basic forms are helical and the 20-sided icosahedron (which looks almost spherical), but some capsids are combinations of the two (e.g., an icosahedron head with a helical tail).

The classification of viruses usually depends on the nature of the genetic material, the kingdom of the host, and the form of the capsid. The Baltimore classification system (Table 10.12) has established six different major groups with a variety of subgroups on this basis. A "negative" RNA strand must be reversed (by transcription) to the complementary RNA strand before it can be used as a template for the synthesis of proteins.

There are typically several stages to a virus infection. First the virus will attach to the new host cell. It must then penetrate the cell surface and inject its genome into the host. With **lytic** viruses, replication of the nucleic acid and synthesis of the capsid protein then occurs, followed by assembly of the new virions. Finally, the host cell lyses (ruptures), releasing the new viruses.

TABLE 10.12   Baltimore Classification Groups for Animal, Plant, and Bacteria Viruses

Class	Form	Host	Group	Architecture[a]	Representative
I	Double-stranded DNA	Bacteria	Myoviridae	C	T2, T4
			Syphoviridae	C	λ
			Podoviridae	C	T7
		Animal	Papovaviridae	I	Papillomavirus, polyomavirus
			Adenoviridae	I	Adenovirus
			Herpesviridae	I	Herpes simplex, mononucleosis
			Poxviridae	C	Smallpox
			Hepadnaviridae	I	Hepatitis B
		Plant	Caulimoviruses	I	Cauliflower mosaic
II	Single-stranded DNA	Bacteria	Microviridae	I	φX174
		Animal	Parvoviridae	I	Parvovirus
		Plant	Geminiviruses	I	Maize streak
III	Double-stranded RNA	Bacteria			φ6
		Animal	Reoviridae		Rotavirus
		Plant			
IV	Positive-strand RNA	Bacteria	Leviviridae	I	MS2, F2
		Animal	Picornaviridae	I	Poliovirus, rhinovirus, hepatitis A, coxsackievirus
			Togaviridiae	I	Rubella
			Flaviviridae	I	Yellow fever, dengue, West Nile
			Coronaviridae	H	Murine hepatitis
			Caliciviridae	I	Norwalk virus
		Plant	Potyvirus	H	Potato Y
			Tymovirus	I	Turnip yellow mosaic
			Tobamovirus	H	Tobacco mosaic
			Comovirus	I	Cowpea mosaic
V	Negative-strand RNA	Bacteria			
		Animal	Rhabdoviridae	H	Rabies
			Filoviridae	H	Ebola
			Paramyxoviridae	H	Mumps, measles, distemper
			Orthomyxoviridae	H	Influenza
			Bunyaviridae	H	Hantavirus
			Arenaviridae	H	Lassa
		Plant			
VI	Retrovirus	Bacteria			
		Animal	Retroviridae	I	HIV, oncoviruses
		Plant			

[a]C, complex; H, helical; I, icosahedral.
Source: Based in part on Voyles (2002).

Lytic infection kills the host cell. With **temperate** viruses, after entering the host cell the genome inserts itself into a host chromosome. In that way, it is replicated as the host cell grows and divides. Eventually, it may direct the formation of new virions and cause lysis of the cell, but this may be only after many generations.

Viruses are the cause of many important plant and animal diseases. They may be spread through contact, food, water, and/or air. Among humans, they are responsible for the common cold, influenza, polio, herpes, hepatitis, measles, smallpox, and AIDS, among many others. Viruses also have been implicated as factors in a number of types of cancer.

10.8.2   Viroids and Prions

Viroids consist of a single short, circular strand of RNA, with no protein coat. Further, it appears that the RNA does not even code for any proteins. Thus, they are not considered viruses. They are usually found within the nucleus of the infected cell, where the host enzymes apparently can replicate them. They cause several important plant diseases and also infect some animals.

Prions are small proteinlike particles that contain no nucleic acid. However, they are still able to infect host cells, where they are then replicated. They are now known to be the cause of several fatal brain diseases in mammals, including mad cow disease (bovine spongiform encephalopathy), scrapie (in sheep and goats), and Creutzfeldt–Jakob disease in humans (formerly thought to be caused by a "slow" virus).

PROBLEMS

10.1. What are some of the difficulties associated with classification of bacteria as species?

10.2. How do phylogenetic and phenotypic classification schemes differ?

10.3. What are some of the inorganic chemicals that can serve as energy sources for prokaryotes?

10.4. A particular prokaryotic microorganism is photosynthetic. Develop a simple dichotomous key based on Table 10.4 that would allow you to place it in one of those six major groupings.

10.5. Describe the range of environmental conditions under which microorganisms live.

10.6. Is a virus "alive"? Why or why not?

10.7. Recent evidence strongly suggests that Mars at one time had liquid water on its surface. Since Mars probably went through a stage at which it was similar to Earth at the time microbial life evolved here, it has been suggested that microbial life once existed there also. Speculate on whether microbial life exists on Mars today. What factors make such life more or less likely?

REFERENCES

American Society for Microbiology. www.asm.org.

Atlas, R. M., 1997. *Principles of Microbiology*, 2nd ed., Wm. C. Brown, Dubuque, IA.

Barns, S. M., C. F. Delwiche, J. D. Palmer, and N. R. Pace, 1996. Perspectives on archaeal diversity, thermophily and monophyly from environmental rRNA sequences, *Proceedings of the National Academy Science USA*, Vol. 93, pp. 9188–9193.

Bergey's Manual Trust, www.bergeys.org.

Bitton, G., 1999. *Wastewater Microbiology*, 2nd ed., Wiley-Liss, New York.

Brock, T. D. (Ed.), 1975. *Milestones in Microbiology*, American Society for Microbiology, Washington, DC.

Center for Microbial Ecology, Michigan State University, Ribosomal Database Project, http://www.cme.msu.edu/RDP/.

Clesceri, L. S., A. E. Greenberg, and A. D. Eaton, 1998. *Standard Methods for the Examination of Water and Wastewater*, 20th ed., American Public Health Association, Washington, DC. Updated regularly, and available at www.wef.org.

Garrity, G. M. (Ed.-in-Chief), 2001–2007 (expected). *Bergey's Manual of Systematic Bacteriology*, 2nd ed., Vols. 1 to 5, Springer, New York.

Gordon, R. E., W. C. Haines, and C. H.-N. Pang, 1973. The Genus *Bacillus*, Agricultural Handbook No. 427, Agricultural Research Service, U.S. Department of Agriculture, Washington, DC.

Holt, J. G. (Ed.-in-Chief), 1984–1989. *Bergey's Manual of Systematic Bacteriology*, Vols. 1 to 4, Williams & Wilkins, Baltimore.

Holt, J. G., N. R. Krieg, P. H. A. Sneath, J. T. Staley, and S. T. Williams (Eds.), 1994. *Bergey's Manual of Determinative Bacteriology*, 9th ed., Williams & Wilkins, Baltimore.

Jahn, T. L., and F. F. Jahn, 1949. *How to Know the Protozoa*, Wm. C. Brown, Dubuque, IA.

Lee, J. J., S. H. Hutner, and E. C. Bovee (Eds.), 1985. *Illustrated Guide to the Protozoa*, Society of Protozoologists, Lawrence, KS.

Madigan, M. T., J. M. Martinko, and J. Parker, 2003. *Brock Biology of Microorganisms*, 10th ed., Prentice Hall, Upper Saddle River, NJ.

Pace, N. R., 1997. A molecular view of microbial diversity and the biosphere, *Science*, Vol. 276, pp. 734–740.

Patterson, D. J., 2003. *Free Living Freshwater Protozoa*, ASM Press, Washington, DC.

Prescott, G. W., 1970. *How to Know the Freshwater Algae*, 2nd ed., Wm. C. Brown, Dubuque, IA.

Prescott, L. M., J. P. Harley, and D. A. Klein, 1999. *Microbiology*, 4th ed, WCB/McGraw-Hill, New York.

Ricklefs, Robert E., 1993. *The Economy of Nature: A Textbook in Basic Ecology*, 3rd ed., W. H. Freeman, New York

Voyles, B. A., 2002. *The Biology of Viruses*, 2nd ed., McGraw-Hill, New York.

Woese, C. R., 1987. Bacterial evolution, *Microbiological Reviews*, Vol. 51, pp. 221–271.

11

QUANTIFYING MICROORGANISMS AND THEIR ACTIVITY

In Chapter 10 we discussed the great variety of microorganisms and their abilities in a mainly qualitative way. In many cases in environmental engineering and science, it is desirable or necessary to deal with microorganisms in a quantitative fashion. What is their density in a system: How many are there, or what is their mass? How rapidly are they growing, transforming substrates, or producing products? First, we look at the elemental composition of microorganisms. Next, we consider the microscope, an important tool of the microbiologist. Then, after some preliminary consideration of sampling and preparation, we look at a number of ways that the biomass of microorganisms can be determined, their numbers estimated, and their effect on a system quantified. Finally, we examine some of the simple models that can be used to help understand or even predict microbial activities. The focus is on general approaches rather than detailed specific applications.

11.1 MICROBIAL COMPOSITION AND STOICHIOMETRY

11.1.1 Elemental Makeup

Most microorganisms are 70 to 90% water on a mass basis. The remaining dry weight is typically about 15% **ash** (minerals that remain upon combustion) and 85% **volatile** (mainly organic) material. The elemental composition of the dry matter of typical bacteria such as *Escherichia coli* is shown in Table 11.1. Of course, these values will vary among different strains and will also depend on the physiological state of the cell.

Environmental Biology for Engineers and Scientists, by David A. Vaccari, Peter F. Strom, and James E. Alleman
Copyright © 2006 John Wiley & Sons, Inc.

TABLE 11.1   Elemental Composition of a Microbial Cell

Element	Symbol	Atomic Weight	Cell Dry Weight (%)[a]	Element Ratio[b]	Formula[c]	Weight (%)
Carbon	C	12.01	50	4.2	5	53.1
Hydrogen	H	1.00	8	8.0	7	6.2
Oxygen	O	16.00	20	1.3	2	28.3
Nitrogen	N	14.01	14	1.0	1	12.4
Phosphorus	P	30.97	3	0.097		
Sulfur	S	32.07	1	0.031		
Potassium	K	39.10	1	0.026		
Calcium	Ca	40.08	0.5	0.012		
Magnesium	Mg	24.30	0.5	0.021		
Iron	Fe	55.85	0.2	0.0036		
Other			~1.8			

[a]Based on E. coli.
[b]Apparent stoichiometric formula of E. coli based on cell dry weight.
[c]Useful stoichiometric ratio often used to write the components of a cell as a chemical compound formula.

It is sometimes useful to write an apparent chemical formula for microorganisms. Commonly, $C_5H_7O_2N$ has been used for this purpose, and these values have also been included in Table 11.1. Note that this formula gives reasonably good agreement with the values from E. coli for C, H, O, and N, which make up 92% of the total dry mass, but totally ignores the other elements. The elemental composition of some important cell constituents, including some storage materials, is shown in Table 11.2. Thus, although $C_5H_7O_2N$ is a useful simplification, it is not a true chemical formula nor an exact stoichiometric expression.

The cell's requirements for C, O, and H are typically supplied by some combination of organic material, carbon dioxide, elemental oxygen, and water (or occasionally, hydrogen sulfide or methane). The other requirements can be loosely categorized as **macronutrients**, **micronutrients**, and **trace elements**, although the boundaries between these groups are not applied uniformly.

TABLE 11.2   Elemental Composition (Mass %) of Some Important Microbial Cell Components

	Stoichiometric Formula	C	H	O	N	P	S
Glucose	$C_6H_{12}O_6$	40	6.7	53.3			
Cellulose, starch, glycogen	$(C_6H_{10}O_5)_n$	44.4	6.2	49.3			
Chitin	$(C_8H_{13}O_5N)_n$	47.3	6.5	39.4	6.9		
Protein[a]	$(C_{5.35}H_{7.85}O_{1.45}N_{1.45}S_{0.1})_n$	54.0	6.7	19.5	17.1		2.7
DNA[b]	$(C_{9.75}H_{12}O_6N_{3.5}P_1)_n$	38.3	4.1	31.4	16.1	10.1	
PHB	$(C_4H_6O_2)_n$	55.8	7	37.2			
Palmitic acid[c]	$C_{16}H_{32}O_2$	74.9	12.6	12.5			

[a]Assuming about equal prevalence of all the amino acids.
[b]Assuming 50% G + C content (equal prevalence of all four bases).
[c]A common fatty acid.

For microorganisms, N and P are typically considered macronutrients (for plants, K would be added). These are needed in a mass ratio of about $5:1$. The required C/N/P ratio, as a rule of thumb, is commonly said to be $100:5:1$. However, in this case, much of the carbon is used as an energy source rather than to make cell constituents. As can be seen in Table 11.1, the **C/N ratio** of a typical cell itself is around 3.6 (or 4.3 in the formula $C_5H_7O_2N$) rather than 20.

The term *micronutrients* usually includes S and Fe, and probably K, Ca, and Mg. Trace nutrients would include the many other elements, such as cobalt (Co), nickel (Ni), copper (Cu), and zinc (Zn), needed in only very small amounts, usually for specific enzymes. Roles that a number of elements play in cell metabolism are summarized in Table 11.3.

Some organisms may need fairly high concentrations of another element for a special purpose. Diatoms, for example, need substantial amounts of silicon (Si) to construct their silica shells, and many testate amoeba need calcium for theirs. Other organisms may

TABLE 11.3   Roles of Various Elements within Microorganisms

Element	Symbol	Important Cellular Roles
Carbon, hydrogen, oxygen	C, H, O	Major constituents of organic matter
Nitrogen	N	Proteins; nucleic acids; peptidoglycan
Phosphorus	P	Nucleic acids; membrane phospholipids; coenzymes; energy utilization (phosphorylation and ATP); present as phosphate (PO_4^{3-})
Sulfur	S	Amino acids cysteine and methionine, which give proteins much of their three-dimensional structure; coenzymes (including CoA)
Iron	Fe	Cytochromes and other heme and nonheme proteins; enzyme cofactor
Potassium	K	Major inorganic ion (K^+) in all cells; enzyme cofactor
Calcium	Ca	Major divalent ion (Ca^{2+}); enzyme cofactor; endospores; some amoeba tests (shells)
Magnesium	Mg	Major divalent ion (Mg^{2+}); enzyme cofactor; active in substrate binding; chlorophyll
Cobalt	Co	Coenzyme (vitamin) B_{12}
Copper	Cu	Specialized enzymes, including cytochrome oxidase and oxygenases
Manganese	Mn	Specialized enzymes, including superoxide dismutase; enzyme cofactor
Molybdenum	Mo	Specialized nitrogen enzymes (nitrate reductase, nitrogenase, nitrite oxidase) and some dehydrogenases
Nickel	Ni	Urease; required for autotrophic growth of hydrogen oxidizers
Selenium	Se	Specialized enzymes, including glycine reductase and formate dehydrogenase
Tungsten	W	Some formate dehydrogenases
Vanadium	V	Some nitrogenase enzymes
Zinc	Zn	Specialized enzymes, including RNA and DNA polymerases; enzyme cofactor
Silica	Si	Cell walls of diatoms (algae)
Sodium, chlorine	Na, Cl	Transport processes; osmoregulation; required by halophilic bacteria

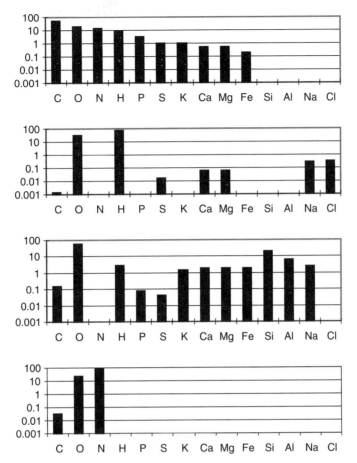

Figure 11.1 Elemental distribution of microbial biomass (top) relative to (bottom) the hydrosphere, lithosphere, and atmosphere (consecutively).

require small amounts of other elements. Molybdenum (Mo), for example, is needed in trace amounts by the nitrifying bacterium *Nitrobacter* to oxidize nitrite.

Other elements, such as sodium and chloride, may be present in fairly high concentrations within a cell. However, even if they are required for survival (e.g., for osmoregulation), they usually are not referred to as nutrients.

Figure 11.1 provides a comparison of a typical microbial cell's major and minor elements to their presence within the water, land, and air of our planet. In general, there are large reservoirs in water or soil for most of the cell's requirements. One exception is nitrogen, most of which is present in the atmosphere as N_2, a form that is available only to nitrogen-fixing organisms (Section 13.2.1).

11.1.2 Growth Factors

Many microorganisms are able to grow in a system with a single organic carbon and energy source, such as a sugar, and inorganic forms of all other nutrients (e.g., N as ammonium). This means that they are able to synthesize all of the other organic molecules that they require. Photo- and chemoautotrophs, in fact, may be able to grow without the need for any organic substances at all, since they get their carbon from CO_2.

However, other microbes may require a few or many specific essential organic molecules that they are unable to synthesize. Referred to as **growth factors**, they usually fall into one of three categories:

1. Vitamins, which are typically components of certain coenzymes
2. Amino acids, the building blocks of proteins
3. Purines and pyrimidines, the nitrogen-containing bases of nucleic acids

In some cases the growth factor required may depend on the other compounds present. The filamentous bacterium *Sphaerotilus natans*, for example, can grow with ammonium as the only nitrogen source if vitamin B_{12} (cyanocobalamin) is present, but otherwise requires methionine (a sulfur-containing amino acid).

In growing microorganisms in the laboratory, the growth **medium** (plural, *media*) is considered **defined** if its exact chemical composition is known. Growth factors can be added individually as part of a defined medium when required, but often a preparation such as yeast extract, made from natural products and containing many compounds of unknown composition, will be used instead. Such "undefined" media are referred to as **complex**.

11.1.3   Molecular Makeup

Most of the dry mass of cells is composed of macromolecules. As indicated in Table 11.4, proteins typically account for over half of the total. Nucleic acids, polysaccharides, and lipids are also major components.

11.2   MICROSCOPY

The fact that microbes are extremely small and relatively transparent imposes substantial constraints on "seeing" them. Thus, a variety of microscopes and microscopic techniques have been developed to magnify them.

TABLE 11.4   Typical Molecular Composition of Bacteria

	Cell Fraction (% dry weight)	Typical (Approx.) Molecular Weight (g/mol)	Cellular Role
Proteins	52	10^5	Structure and enzymes
RNA	16	10^5–10^6	Genetic
DNA	3	10^9	Genetic
Polysaccharides	17	10^3–10^6	Structure, genetic, storage
Lipids	9	10^3	Structure, storage
Inorganics and small organics	3	10^2	Enzyme cofactors, osmoregulation

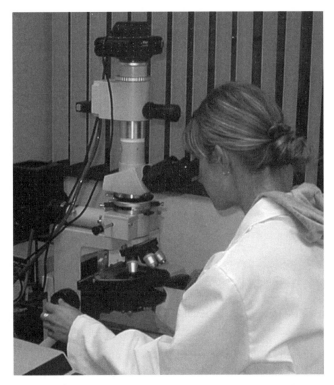

Figure 11.2   Modern compound light microscope. This model includes a fluorescence attachment and a camera.

11.2.1   Light Microscopes

Rather than the simple single lens instruments of van Leeuwenhoek (Section 10.2.1), today the **compound light microscope** is the conventional tool used (Figure 11.2). It routinely gives magnifications of up to 1000 power ($\times$) by combining a 10$\times$ **ocular** (eyepiece) lens with an **objective** of up to 100$\times$. The total magnification is determined by multiplying the two together. Objectives of 4, 10, 20, and 40$\times$ are also common, and oculars of 15$\times$ are available.

It is **resolution**, however, rather than magnification itself, that is the primary limitation of observation with a light microscope. For the average human eye, as two points move closer together they would become one point—could no longer be resolved as separate—at a distance of about 75 µm. (Most prokaryotes, remember, are about 1 to 2 µm in size.) At 1000$\times$ this would occur at a distance of 0.075 µm. However, visible light has wavelengths of about 450 to 720 nm (0.45 to 0.72 µm), which effectively limits resolution to about 0.2 µm. Any additional magnification does not lead to further resolution if visible light is used; the object simply looks grainy or fuzzy.

When the compound microscope is used for standard **bright-field microscopy** (Figure 11.3), the **specimen**, typically in water, is placed on a glass **slide** and a small glass **coverslip** is placed over it. The sample is then seen against a bright background. This is particularly useful in determining the color of cells, such as cyanobacteria and algae, and observing the yellow of sulfur inclusions. However, for most colorless

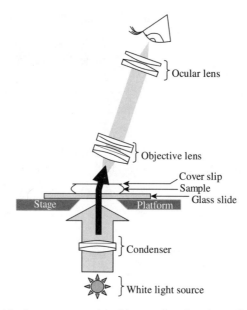

Figure 11.3   Bright-field microscopy: an original image viewed against a light or bright background.

microorganisms, there is little visible contrast between them and the water, and thus little detail is observable. This is a primary reason that so much effort was put into development of staining procedures (Section 10.4.4 and Figure 10.15) by early microscopists.

A disadvantage of traditional staining techniques is that they normally involve killing the cells. Thus, observations of motility and any other activity of the microorganisms is lost. One variation that is sometimes used is **dark-field microscopy** (Figure 11.4), in which objects are illuminated from the side. This makes them appear bright against a dark background and can improve visibility of some transparent or fine structures, such as the flagella of eukaryotes.

A major advance for observing live cultures was the development of **phase-contrast microscopy**. This specialized modification of the compound microscope makes small changes in refractive index visible, allowing improved observation of both the cell surface and internal structures (Figure 11.5). Frederik Zernike of the Netherlands was awarded a Nobel Prize in Physics in 1953 for this invention.

Another modification is the **fluorescence microscope** (Figure 11.6). Fluorescent compounds absorb light at one wavelength to become "excited" or "activated" and then emit the energy at another wavelength. A variety of fluorescent molecules have been developed that can penetrate or be taken up by cells. The specimen is then exposed to the specific activation wavelength of light, and through filters, the specific emission wavelength is observed.

The size of microorganisms can be measured using an **ocular micrometer**. This is a small "ruler" placed in one of the eyepieces, so that it appears superimposed on the magnified image. This scale is then calibrated at each magnification using a **stage micrometer**, a special glass slide with a precisely etched 1-mm scale broken down into tenths and hundredths of a millimeter (Figure 11.7). Usually, the ocular micrometer is designed to have markings at 10-μm intervals at 100× magnification, and at 1 μm for 1000×, but this may depend on the specific microscope.

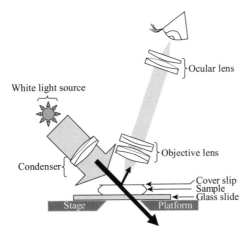

Figure 11.4 Dark-field microscopy: a bright reflected image is viewed against a darkened or black background.

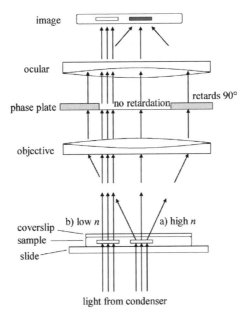

Figure 11.5 Phase-contrast microscopy. Object (*a*), with high refractive index (*n*), bends and retards the light more; after further retardation at the edges of the phase plate, the recombined light (retarded and unretarded) creates destructive interference, leading to a darker appearance. Object (*b*), with a low *n*, does not bend or retard the light, creating no interference, and thus appears bright. Thus, contrast was created between the two nearly transparent objects that was not otherwise visible. A special condenser (not shown) with a phase ring is also required.

Example 11.1   At $100\times$ ($10\times$ ocular and $10\times$ objective) a ciliated protozoan is observed to have a length of 9.3 units on the ocular micrometer scale. If from a previous calibration it is known that at $100\times$ each micrometer unit is 10 µm, what is the length of the organism? What is the field diameter if it is observed to correspond to 161 units?

Answer   10 μm/unit × 9.3 units = 93-μm-longciliate.   10 μm/unit × 161 units = 1610 μm = 1.61-mm field diameter.

11.2.2   Electron Microscopes

Electron microscopes utilize electron beams instead of light and thus are able to achieve greater resolution (Figure 11.8). The two major types are the transmission electron microscope (TEM) and the scanning electron microscope (SEM).

The TEM is analogous to a light microscope, with a beam of electrons being transmitted through the sample rather than light. Instead of glass lenses, the TEM uses electromagnets to control and focus the beam. The result is the potential for much higher magnification, with levels of resolution down to a few tenths of a nanometer (nearly 1000 times better than what can be achieved with a light microscope).

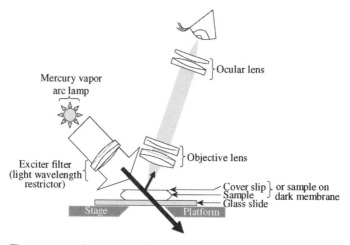

Figure 11.6   Fluorescence microscopy: a fluorescent image is viewed against a darkened or black background.

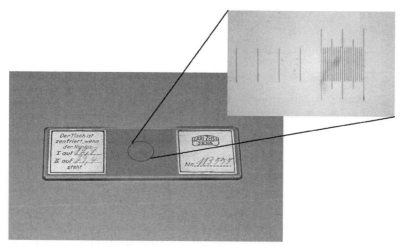

Figure 11.7   Stage micrometer used to calibrate dimensions under microscope.

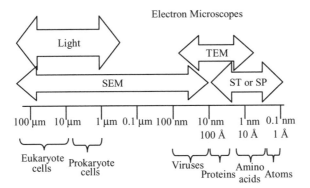

Figure 11.8   Resolution with various microscopes. Electron microscopes include scanning (SEM), transmission (TEM), and scanning tunneling (ST) or scanning probe (SP).

The disadvantages of TEM microscopy include the much higher cost and complexity, and the need for sophisticated preparation of samples, such as cutting **thin sections** of the cells to be observed. Of course, these techniques kill the cells. A main use of TEM is looking at internal cell structures.

SEM offers somewhat less magnification, with resolutions down to around 10 nm. It is based on the detection of backscattered rather than directly transmitted electrons. Typically, the specimen's surface is treated with an electron reflective coating, such as gold. As the applied electron beam is scanned back and forth across this sample, the scattered electrons provide a three-dimensional image of the coated surface. However, care must be taken that the coating procedure itself does not produce **artifacts** (false observations).

11.3   SAMPLING, STORAGE, AND PREPARATION

11.3.1   Sampling

Sampling is a critical, although often insufficiently considered, part of most analyses. The best that any analytical procedure can hope to do is to determine the analyte in the sample accurately. Thus, the quality of the information received about the system of interest can only be good as the samples on which it is based.

Sampling of microorganisms can provide some additional concerns because of their special nature. Extra care must be taken to avoid contamination, since even a small number of inadvertently added organisms could grow quickly and produce a large error. On the other hand, toxic agents present in the sample could continue to kill organisms, leading to underestimates of their numbers. The chlorine residual left after disinfection of water or wastewater is a common example of this. Also, in many systems microorganisms will adhere to surfaces. This might make special sampling techniques necessary.

11.3.2   Storage

Many types of microbial analysis must be done almost immediately. If not, the microbes may continue to grow or die, changing their total numbers and also the relative ratio of various species. Refrigeration will help in many cases, but storage times are still typically limited to only a few hours. This means, among other things, that **composite** samples

(samples made by mixing several individual, or **grab**, samples) taken over a 24-hour period to try to determine average conditions are not acceptable for many microbial tests. On the other hand, some methods (e.g., some DNA analyses) allow rapid freezing or other preservation techniques for long-term storage.

11.3.3 Preparation

For some samples, the numbers of microorganisms may be too low for the desired analysis, and a **concentration** technique will first be employed. This might involve centrifuging a large sample to collect the organisms in a smaller, more concentrated volume. Alternatively, filtration might be used to collect the organisms from a large volume on the filter surface. In another type of filtration, **planktonic** (floating) eukaryotes may be concentrated during collection by using a plankton net that is dragged through the water. For viruses (and a few cellular organisms), adsorption in a column of packed solid material, followed by elution into a small volume, may be used.

In other samples, concentrations of microorganisms may be too high for counting, resulting in the need for dilution. This might be with filtered water from the same source, or more commonly with a standard phosphate buffer or saline solution. In most cases, distilled or other high-purity water cannot be used, as it would cause an osmotic shock to the organisms (and possibly their death). A common approach is to make **serial dilutions**. Often, this consists of making a 10-fold dilution, then using that to make another 10-fold dilution, and repeating until the desired concentration is reached.

If microorganisms are adhering to a surface, it may be necessary to try to remove them. If they are adhering to each other, or to small particles of soil or other material, a dispersion technique such as blending is sometimes used in an attempt to separate them. Dispersing adhered or aggregated microorganisms remains an important problem in the analysis of many samples.

Removing extraneous or interfering materials from a sample also may be useful. Some chemicals can be removed by adding a reagent to react with them (e.g., sodium thiosulfate, $Na_2S_2O_3$, can be added to eliminate residual chlorine). Soluble interfering substances might be separated from the microbes by filtration or centrifugation. Removing particulates can be more difficult, since microorganisms may adhere to them.

11.4 DETERMINING MICROBIAL BIOMASS

11.4.1 Measurements of Total Mass

Prokaryotic cells are usually incredibly minute, with a typical individual mass on the order of only 10^{-12} g [1 picogram (pg)]. Since an average adult human typically has a mass of 50 to 100 kg, we are more than 10^{16} times larger. This number is so huge that it does not even have a commonly known name (a trillion is "only" 10^{12}). To put things in perspective, this is about the same ratio as the few seconds you spend reading this sentence compared to the entire 4.6 billion year existence of our planet.

Thus, the mass of these cells cannot normally be measured individually. However, in a number of cases it may be important to know the collective **biomass** of microorganisms present in a system. In laboratory cultures grown on soluble media, this can be measured by filtration followed by washing (to remove salts), drying, and direct weighing of the material collected, giving a **dry weight**. This is also done for suspensions in systems such as activated sludge, even though most of the material included as "biomass" is not actually living cells. (It is believed that more than 90% is commonly dead cells or other nonliving organic particulate material.) Similarly, biofilms and growths on the surface of a laboratory agar plate can be scraped off, suspended in buffer, and measured in this way, although for biofilms, again, much of the "biomass" will be film polymers and other nonliving materials. Biomass in the air (or other gases) might be estimated in a similar way after collection by filtration, although in most cases the nonliving material would again be a major interference.

One correction that is sometimes used for these methods based on **suspended solids** (SS) is to consider only the **volatile suspended solids** (VSS) fraction. The SS sample is **ashed** (or burned off by heating to $\sim500°$C) to drive and burn off the volatile materials (which are usually mainly organic), and the change in mass determined. This can be used to correct for the presence of inorganic materials, such as salts and metal oxides, but does not differentiate between living and nonliving organic matter. Also, $\sim15\%$ of a microorganism's dry mass typically will be ash (Section 11.1.1).

In some cases, centrifugation can be substituted for filtration. The compacted solids, referred to as the **pellet**, may be dried and weighed, usually after **washing** (resuspending in liquid and centrifuging again). In an application used at some sewage treatment plants, operators centrifuge a sample of the sludge and compare the pellet volume with a calibration chart to provide a rough estimate of the mass. This test can be rapidly done (10 minutes) for process monitoring using only an inexpensive small centrifuge, compared to the more time-consuming and expensive steps of filtration, drying (requiring an oven), and precise weighing (requiring an analytical balance).

These types of measurements are not practical for many other types of samples, such as soils, solid wastes, and most sludges. In those cases, as well as often for those described above, counts are performed instead of mass measurements. If desired, mass might then be estimated from the number of microorganisms (Section 11.5.3). However, counting has its own limitations in such samples (see below).

One other approach that has been used with varying success in soils is referred to as a **fumigation technique**. The microorganisms in a sample are killed by exposure to a toxic gas such as chloroform. The gas is then removed and the sample is **inoculated** with a small number of newly added bacteria. Since the killed microorganisms then serve as food for new growth, the amount of carbon dioxide that is produced by the microbial activity can be related (using proper controls and calibration) to the amount of biomass originally present.

11.4.2 Measurements of Cell Constituents

Rather than try to measure the total mass of organisms present, sometimes it is more practical or desirable to measure the mass of a particular cell constituent. For example, it may be possible to treat a sample to extract the microbial **protein** or **DNA** that is present and to quantify it by weight or by a chemical means. The amount of such fundamental cell constituents present may then be correlated to the biomass present (see Table 11.4, for example). In some cases these measures may be even more useful then total biomass estimates, since the amounts per cell vary less than for some other constituents, such as exocellular polymer or food storage products.

Another cell constituent of interest for estimating biomass is **ATP**. It is present in all organisms, and since it disappears quickly upon death of a cell, it can serve as a useful indicator of **viable** (living) biomass. However, the amount of ATP present in a cell can vary considerably with its physiological state.

Planktonic algae concentrations can often be estimated by measurement of **chlorophyll a**. This method also can be used for cyanobacteria, and in fact includes them in the estimates obtained. The sample is prepared by grinding the cells and extracting the pigment with acetone. The chlorophyll is then quantified by measuring absorbance at 664 nm with a spectrophotometer, or fluorescence with a fluorometer, or by high-performance liquid chromatography (HPLC). Corrections may be necessary due to the presence of other pigments.

Some methods have been developed to extract and identify cell constituents that are characteristic of a particular group of microorganisms, including cell wall constituents, fatty acids, and DNA or RNA sequences that are specific for particular taxa. So far these are used primarily to help identify what types of microorganisms are present, but quantitative methods are also being developed that can be used to estimate biomass.

11.5 COUNTS OF MICROORGANISM NUMBERS

Counts of microorganisms may be made by direct microscopic techniques or through indirect methods such as culturing.

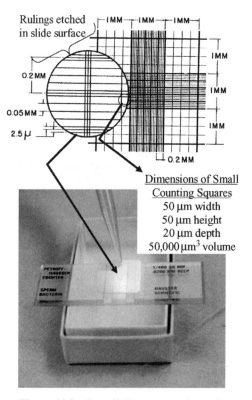

Figure 11.9   Petroff–Hauser counting cell.

11.5.1 Direct Counts

In some samples, microorganisms can simply be counted under the microscope. Counting chambers—specialized microscope slides with wells holding a fixed volume—are available for this purpose. Some, such as the Petroff–Hauser cell (Figure 11.9), also have a grid marked on them. An example of a chamber with a deeper well used for larger organisms or associations such as filaments or floc is the Sedgwick–Rafter cell (Figure 11.10); it has an area of 1000 mm^2 and a depth of 1 mm, thus holding a volume of 1.0 mL. Fresh live samples are used when possible, especially for protozoa and metazoa, so that activity such as motility or feeding can be observed.

Alternatively, a fixed volume (usually 1 drop, or about 0.05 mL = 50 μL) can be added to a regular slide, and a coverslip (usually square, 22 to 25 mm on a side) placed over it (spreading the sample under the coverslip). The number of cells seen in one **field** (the visible area seen through the eyepiece, usually about 1.6 mm in diameter at 100× and 0.16 mm at 1000×) under the microscope is then multiplied by the ratio of the area of the coverslip to the field area to get the concentration (count per volume).

Example 11.2   An average of 3.2 ciliated protozoans were seen per field when 10 fields were counted at 100×. If the sample size was ∼0.05 mL, the microscope field at that magnification is 1.6 mm in diameter, and a 22 × 22 mm coverslip was used, what is the concentration of ciliates present?

Answer   Area of field $A_f = \pi r^2 = \pi (1.6/2 \, \text{mm})^2 = 2.01 \, \text{mm}^2/\text{field}$

Area under coverslip $A_c = 22 \, \text{mm} \times 22 \, \text{mm} = 484 \, \text{mm}^2/\text{coverslip}$

$$\text{Conc. of ciliates} = 3.2/\text{field} \times \frac{484 \, \text{mm}^2/\text{coverslip}}{2.01 \, \text{mm}^2/\text{field}} \div 0.05 \, \text{mL}/\text{coverslip}$$

$$= 15{,}000/\text{mL}$$

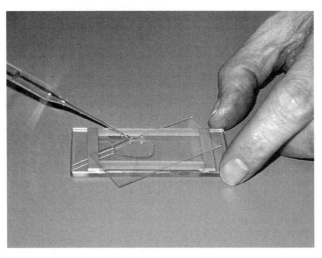

Figure 11.10   Filling a Sedgewick–Rafter cell.

For observation of highly motile protozoa, it is sometimes desirable to slow down their activity. This can be done by adding a compound such as methylcellulose to increase viscosity, or by using an inhibitory compound such as nickel sulfate.

Although small metazoa such as rotifers and nematodes can often be enumerated along with the protozoa, larger forms may require specialized techniques. Some, in fact, may actively avoid being drawn into a small sampling or subsampling device, such as a pipette. Others, as well as larger associations of prokaryotes such as large floc, may simply not fit through the opening of fine tip pipettes.

Because of their silica shells, a special technique can be used for diatoms. After washing of the sample with distilled water, organic matter is destroyed by heat or an acid-oxidation step, leaving the shells for counting.

Note that using approximations from the above example, 5000 cells/mL (500 mm^2/ 2 mm^2 divided by 0.05 mL) would be needed in a sample for there to be one cell on average per microscope field at 100×. For prokaryotic cells, which are usually too small to count at low magnification, concentrations of above $10^6 mL^{-1}$ (2 per field at 1000×) are usually necessary for direct counting. If there are too many cells to count ($>10^8 mL^{-1}$), the sample might be diluted first. If the sample contains too few microorganisms, filtration might then be used to concentrate them on a filter surface for viewing.

However, even with dilution or concentration, this simple counting approach is limited to samples with little extraneous material that would prevent clear viewing and that contain individual, dispersed cells rather than cells in flocs or biofilms. To allow direct counting in samples such as activated sludge (flocculated), biofilms, and soils, a variety of stains and probes have been developed, including many that are fluorescent. These allow differentiation between inert particles and cells, between living and dead cells, or even among specific strains of organisms.

One such approach uses a diacetate ester of **fluorescein**, which is able to pass into the cytoplasm of cells. Once inside, the ester is rapidly hydrolyzed by nonspecific esterase enzymes. The free fluorescein (which, as its name suggests, is fluorescent) that is released is trapped in the cell, thereby making it readily visible with a fluorescence microscope.

The **acridine orange** direct count (AODC) uses another fluorescent dye that passes into the cytoplasm and then binds to nucleic acids, giving an orange or green color. This does not require the activity of enzymes within the cell and thus may give a higher, **total count**, including both viable and nonviable (dead) cells.

One problem with the AODC method is that clay particles also may appear orangish, potentially interfering with cell counts. This has contributed to the increased popularity of a blue fluorescent dye, 4′,6-diamido-2-phenylindole (**DAPI**), which binds more specifically with DNA.

If antibodies to the surface of a specific organism can be produced, the **fluorescent antibody** technique can be used. In this method the antibody is conjugated with (chemically attached to) a fluorescent dye. When this preparation is added to the sample, it attaches to the cell so that the surface of the target organisms fluoresce.

Genetic probes can also be used to stain specific groups of organisms for counting. The fluorescent in situ hybridization (**FISH**) technique (Section 10.4.4) is one popular example.

Another alternative is the use of **metabolic stains**. These typically utilize a dye that is reduced by (accepts electrons from) a cell's metabolic activities to form a colored product. A number of **tetrazolium** dyes such as 2-(*p*-iodophenyl)-3-(*p*-nitrophenyl)-5-phenyl tetrazolium chloride (**INT**), for example, are capable of passing through the cell membrane and then accepting electrons by way of dehydrogenase enzymes. The result is conversion to dark red to deep-purple **formazan** crystals that are trapped within the cell. Thus, a count of respiring cells can be made. The formazan can also be eluted with isopropyl alcohol and an indirect measure of cell concentration (number or mass) made from a spectrophotometric measurement of the intensity of the color released.

Filamentous bacteria (including actinomycetes, many cyanobacteria, and some gram negatives, such as *Sphaerotilus* and *Beggiatoa*), fungi, and algae can pose a problem in quantification in that a single "count" might be composed of only one cell a few micrometers long or might contain many cells extending to a length of 1 mm or more. In some cases the number of cells in the filament may be counted or the filament length measured, perhaps for conversion to a biomass estimate (similar to the calculations of Section 11.5.3). Similarly, counts of colonial protozoa and algae may require some enumeration of individuals or a measure of the size of the colony. For bacterial associations such as floc, individual counts may be impossible, but the size of the association may be measured.

Image analysis is a tool that can be combined with several of the techniques described above. A digital image of the microscope field is captured and can then be analyzed using computer software to pick out and quantify specific size (including filament length or floc size), shape, and/or color (including from stains and fluorescence) objects. This information can then be provided as total or differential counts, and can also be used to estimate cell volume (and hence mass; Section 11.5.3).

11.5.2 Indirect Methods

Several commonly used methods of estimating the number of microorganisms in a sample depend on **culturing** them in or on a medium that supports their growth. These methods require the use of **aseptic technique**, in which **sterile** materials (initially containing no organisms) are used along with a variety of measures to prevent contamination by other organisms (e.g., from the air, the analyst's hands, or the lab bench surface). They also involve a period of **incubation** (time for growth under designated conditions) of from one to several days for most common tests.

In one type of approach (used for most plate count and membrane filtration methods; see below), the organisms are grown on a solid surface until each initial cell produces a visible **colony** consisting of many millions of daughter cells. The number of colonies is then counted and gives an estimate of the number of colony-forming units (**CFUs**) originally present. It is usually assumed that each colony arose from a single cell (or spore), but in some cases it is likely that a small cluster of cells or a filament was present initially and gave rise to the colony. This is especially problematic for fungi and actinomycetes, whose counts can increase dramatically with sporulation even though little change in the actual biomass present has occurred.

Another approach (used in most probable number techniques and some plate count and membrane filtration methods) is to grow the cells in or on a medium until they produce some detectable change that indicates their presence. This might be a change in pH, increased turbidity, the disappearance of a substrate, or the appearance of a product. Often, color indicators are used to make the activity readily apparent.

One advantage of culturing techniques is that they often can be performed in a way that minimizes interferences from extraneous materials. They may also be used to count cells at lower densities than direct methods (without a concentration step). However, culturing requires a medium on which the organisms of interest can grow. Since no single medium is suitable for all prokaryotes, a true total count cannot be obtained. In fact, total

culturable bacterial counts in many samples are one to two orders of magnitude lower than total direct counts.

On the other hand, media can be made intentionally **selective**, so that only certain organisms will grow. For example, a selective substrate can be used, or a chemical that is inhibitory to nontarget organisms can be incorporated. Similarly, selective temperatures, salinities, or pHs can be used. Additionally, media can be **differential**, so that among the organisms that do grow, the ones of interest can be told apart from others. Often, this is done by a color reaction.

Some common approaches are described briefly below. By varying the medium used and the incubation conditions (e.g., temperature, aerobic vs. anaerobic), a wide variety of microorganisms can be enumerated.

Plate Counts Probably the most common methods of estimating bacterial numbers are plate counts. These use **agar**, or occasionally, some other solidifying agent, to convert liquid medium (**broth**) into a solid. Agar is particularly useful because it does not melt in water until heated to boiling, but it does not resolidify until it is cooled below 45 to 48°C. However, because some bacteria can digest agar (an organic polymer produced by certain red algae), it cannot be used in some special applications.

In the **spread plate** technique (Figure 11.11), an agar medium (often, 10 to 15 mL) is added to a small (commonly 100 mm in diameter by 15 mm high) sterile glass or plastic **petri dish** and allowed to solidify. A small (0.1 mL) amount of sample or dilution is then placed on top and spread over the surface of the medium using a bent glass rod called a **hockey stick**. Visible colonies that grow on the surface of the medium after incubation are counted.

Example 11.3 If an average of 38 colonies is counted for plates containing 15 mL of agar medium on which 0.1 mL of a 10^{-4} dilution of a sample was spread-plated, what was the concentration of CFUs in the original sample?

Answer

$$\text{CFUs/mL} = 38\,\text{colonies/plate} \div 0.1\,\text{mL} \div 10^{-4} = 3.8 \times 10^6\,\text{CFU/mL}$$

Note that the amount of agar is irrelevant.

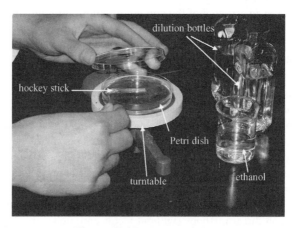

Figure 11.11 Spread plating.

For **pour plates**, a 1-mL sample or dilution is put on the bottom of a petri dish. Then 10 to 20 mL of the desired molten agar medium is added, and the dish is covered and swirled to mix the contents. The medium solidifies with the microorganisms distributed throughout it. After incubation, the visible colonies are counted.

Aside from the other limitations of methods involving culturing, incubation time is also an important factor affecting the accuracy of plate counts. If insufficient time is allowed, of course, some colonies will not yet have grown to visible size. On the other hand, if too much time is allowed, some colonies may grow so large that they cover up others, again leading to underestimation. Alternatively, secondary colonies may form (e.g., from motile bacteria swimming away to start new colonies), giving overestimations.

Plate counts are typically suitable for 20 to a few hundred CFUs per plate. Lower counts have poor precision, and at higher numbers they crowd each other and interfere with growth. If cell density is too high, the plate is recorded as **TNTC** (too numerous to count), and it is recognized that a greater dilution factor should have been used. Since it is often difficult to know ahead of time how many organisms are present, usually a range of dilutions are plated; some may end up with no colonies, others TNTC, but hopefully those in the middle will have countable numbers.

Membrane Filtration Methods   Microorganisms can be collected on the surface of a membrane by filtration (Figure 11.12). If the membrane filter is then incubated on the surface of a growth medium in a small (usually, ~50 mm in diameter) sterile petri dish, the colonies that form give an estimate of the original numbers present. The medium can either be a prepoured agar (similar to a spread plate) or a broth contained in an absorbent pad.

Figure 11.12   Membrane filtration.

Example 11.4 If an average of 41 colonies is counted per filter when 10 mL of a 10^{-3} dilution of the sample is filtered, what was the concentration of CFUs in the original sample?

Answer

$$\text{CFUs/mL} = 41 \text{ colonies/plate} \div 10\,\text{mL} \div 10^{-3} = 4.1 \times 10^3 \text{ CFU/mL}$$

Membrane filtration can be especially helpful in enumerating microorganisms present at low concentrations in water. However, particulate material present in the sample can be an interference.

Most Probable Number Techniques Suppose a sample that originally contained five bacteria per milliliter was diluted 10- and 100-fold (Figure 11.13). Then five tubes of media were each inoculated with 1 mL of the sample, another five tubes were each inoculated with 1 mL of the 10-fold dilution, and another five tubes were each inoculated with 1 mL of the 100-fold dilution (a total of 15 tubes inoculated, five at each of the three different concentrations). Further, assume that each cell could grow if it is placed in a tube. Which of the tubes are likely to show growth? Since on average the tubes receiving 1 mL of undiluted sample receive five bacteria each, it is likely they will all grow, although there is a statistical possibility that one or more tubes may not get any bacteria (and will not grow), whereas others will probably get more than five. However, for the 10-fold dilution, on average each tube gets 0.5 cell. Of course, a tube cannot receive 0.5 cell, so that some tubes will get one or more cells and be "positive" (show growth), whereas others will not get a cell and will thus be negative. Probably two or three of the

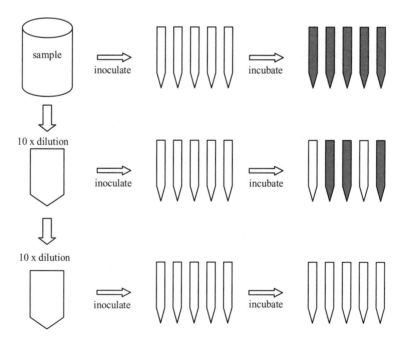

Figure 11.13 MPN method schematic. Positive tubes are shown as dark after incubation.

tubes will end up positive. At the 100-fold dilution, on average only 0.05 cell is added to each tube, and hence all may be expected to be negative.

The pattern of positive tubes that results (in this case, perhaps 5-3-0) is an indication of the cell numbers originally present. Through a mathematical procedure, in fact, the **most probable number** (MPN) of cells originally present resulting in that pattern can be estimated. These values have been tabulated for several different numbers of replicates (e.g., 3, 5, and 10 tubes per dilution), and also for different dilution factors (e.g., two fold as well as 10-fold serial dilutions). Note that a positive or negative result in each tube is used to arrive at a quantitative estimate of cell numbers. The estimate from a pattern of 5-3-0 is an MPN of 8.0 mL^{-1} of the original sample. (A response of 5-2-0 gives an MPN of 5.0, the starting value used in the example.)

An obvious disadvantage of this approach is that it produces a statistical estimate rather than an actual count. For example, with the 5-3-0 example, there is a 95% likelihood that the actual number is between 3.0 and 25, an undesirably broad range. On the other hand, the MPN is potentially useful for a wide variety of samples and can be adapted for many types of microorganisms, including some for which other methods have not been successful.

Turbidity and Absorbance   A method that can be used in laboratory cultures, especially pure cultures, grown on soluble (and preferably colorless) media is measurement of **turbidity** (cloudiness) or **absorbance**. Turbidity is the amount of light dispersed 90° from the path of incident light passing through a material. It is measured with a turbidimeter using a photocell at right angles to the light path. Absorbance is the reduction in the transmission of light along the light path and may occur due to both dispersion and absorption of light. Absorbance is also called **optical density** (OD) and is usually measured with a **spectrophotometer** (also called a **spectrometer**). Absorbance is used more often than turbidity because spectrometers are more commonly available than turbidimeters.

If a small amount of microbial suspension is placed in a sample tube, its turbidity, or the amount of light dispersed toward the photocell, is proportional, up to a point, to the number of particles in the suspension. Similarly, for absorbance, the amount of light that passes through the suspension will be inversely proportional to the concentration of organisms, provided that particle size does not change. This is a form of *Beer's law* (absorbance is proportional to concentration), which is the basis of most quantitative spectrophotometry (although in this case a substantial amount of the light is refracted rather than actually absorbed). Thus, increases in either turbidity or absorbance can be used as a surrogate measure of growth, or correlated through use of a calibration curve to microbial counts obtained by other methods.

One advantage of using turbidity or absorbance is that they are not destructive of the culture. In fact, special flasks with sidearms are available so that the turbidity of a pure culture can be determined over time without the need to open the flask and risk contamination. However, if particle size increases, through flocculent growth or filament elongation, turbidity and OD will underestimate the cell count or biomass. Also, this approach generally cannot be used with environmental samples.

Counting Viruses and Bacteriovores   Viruses are too small to see under any light microscope for direct counting. Also, since they grow only within cells of other organisms, they

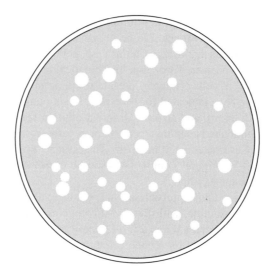

Figure 11.14   Sketch of petri dish with bacteriophage plaques. (Plate contains 43 PFUs.)

cannot be cultured directly like bacteria. Thus, alternative methods are needed to enumerate them.

If a suspension of the bacteria *Escherichia coli* is spread over the surface of a plate containing an appropriate medium, a uniform bacterial **lawn** will develop. If at the same time that the culture is added, a sample containing **coliphage** (viruses that attack *E. coli*) is included, "holes" in the lawn will develop as the viruses infect and kill cells and then spread to neighboring cells. These clear areas, where the bacteria have been lysed by the phage, are referred to as **plaques** (Figure 11.14). The number of such plaques is then an estimate of the number of plaque-forming units (**PFUs**) present in the volume of sample added. Similarly, other phage can be enumerated using appropriate host bacteria for the lawn. Analogous to CFUs, it is often expected that each PFU stems from a single initial phage.

Many animal and plant viruses can be counted in similar ways using appropriate host cell cultures. In these cases, however, individual infected cells may be counted rather than plaques. Alternatively, an MPN type of procedure might be used, or if necessary, the development of infection in exposed whole organisms.

Plaque formation can also be used to count some organisms that feed on bacteria, such as amoeba. Host cell lysis caused by *Bdellovibrio*, the bacteriovorous proteobacteria (Section 10.5.6), also forms plaques.

11.5.3   Relationship between Numbers and Mass

The size of a population of microorganisms can be expressed on the basis of their number or their mass. These are, of course, expected to have some relationship. In fact, if the size and shape of a particular microbe are known, its mass can usually be estimated easily.

Example 11.5   A rod-shaped bacterial cell is 0.8 μm in diameter and 2 μm in length. What is its mass?

Answer   The rod shape can be approximated by that of a cylinder, so that its volume, V, can be estimated as

$$V = \pi r^2 h = \pi (0.4\,\mu m)^2 (2\,\mu m) \approx 1.0\,\mu m^3 \left(\frac{1\,cm}{10^4\,\mu m} \right)^3 = 1.0 \times 10^{-12}\,cm^3$$

Most cells have a density only slightly greater than that of water, perhaps 1.05 to $1.10\,g/cm^3$, so that this cell would have a mass of $\sim 1 \times 10^{-12}\,g$. Since cells are typically about 85% water, this would be a dry weight of about $1.5 \times 10^{-13}\,g$ per cell.

If 10^9 of the rods in Example 11.5 were present in a sample, this could then be estimated to correspond to about 1 mg of the wet weight of cells (10^9 cells $\times 10^{-12}\,g/cell = 10^{-3}\,g = 1\,mg$). Alternatively, if the mass of these cells in a sample was known, the numbers present could be estimated.

Most bacteria probably have wet weights in the range 0.2 to $3 \times 10^{-12}\,g$ and dry weights of 0.3 to $4 \times 10^{-13}\,g$.

11.5.4   Surface Area/Volume Ratio

The surfaces of prokaryotic cells are not just protective wrappings. They often are highly active regions that may be required to transport incoming substrates, remove outgoing products and wastes, shed excess heat, regulate internal osmotic pressure, and support flagella. Thus, the surface area/volume (or mass) ratio (A_S/V) of prokaryotic cells plays an important part in their survival and metabolic success.

A_S/V is not commonly measured, but it is important to realize that it is very dependent on cell size. For cocci, for example, assuming that they are spherical with radius r,

$$\frac{A_S}{V} = \frac{4\pi r^2}{4/3\pi r^3} = \frac{3}{r}$$

Thus, this ratio is highest in small organisms and will decrease in an inversely proportional way with size. In fact, this is a factor that limits the size of single-celled organisms.

11.6   MEASURING MICROBIAL ACTIVITY

In some cases it may be desirable to quantify the general or specific activity of microorganisms in a system. This may then be used to estimate microbial mass or numbers, or may be of interest in its own right. What usually is meant by *microbial activity* is the rate at which microbially mediated reactions are occurring. Thus, the rate of activity might be determined by measuring the rate of disappearance of any of the reactants or the appearance of any of the products. In some cases, it may also be measured by looking at the rate of enzymatic activity in a sample.

11.6.1 Aerobic Respiration

A generalized overall reaction for aerobic respiration of organic material can be expressed as

$$\text{organic matter} + O_2 \rightarrow CO_2 + H_2O + \text{new biomass} + \text{energy}$$

Analogous reactions can be developed for inorganic substrates such as ammonia and hydrogen sulfide, which are used by lithotrophic organisms. The rate of change of each of the six quantities in this equation in theory could be used to measure the rate of aerobic microbial activity. This is typically done by making repetitive measurements of disappearance or accumulation over time. However, such repeated sampling may inadvertently affect the system itself. Alternatively, multiple replicate systems can be established initially and one or more sacrificed at each different sampling time. A disadvantage of this approach is the inherent variability between "replicate" systems.

Organic Matter Disappearance   In some situations the microbial activity on a particular compound is of interest. If this compound can be measured, its disappearance over time will be a direct indication of activity. However, proper controls are necessary to ensure that the activity is biological rather than the result of other mechanisms, such as volatilization, adsorption, or chemical transformation. **Killed controls**, in which any organisms present or added are destroyed by heat, ultraviolet radiation, or chemical means, are often best for such techniques. Also, disappearance of the parent compound does not necessarily mean that it is being completely biodegraded; some compounds are merely transformed to metabolic products that then persist and accumulate (see "Biodegradation" in Section 13.1.3).

Alternatively, some general measure of organic material might be used, and its disappearance monitored. **Biochemical oxygen demand** (BOD), **chemical oxygen demand** (COD), and **total organic carbon** (TOC) are three such measures commonly used for this purpose and are discussed further in the section "Quantification of Organic Carbon" in Section 13.1.3.

Measurements of organic material are often of great importance for measuring extent of disappearance but typically are not convenient for measuring rates. Good rate measurements usually require frequent sampling, which may be disruptive to the system or demand large numbers of replicate systems for sacrifice. In most cases the need for chemical analysis also places practical limits on the number of samples that can be processed.

Oxygen Utilization   In the headspace of a closed system of fixed volume and constant temperature, oxygen removed by respiration is replaced by an almost equal molar amount of carbon dioxide produced. However, carbon dioxide, a weak acid, is easily removed by absorption with an alkali, such as KOH solution. The pressure in the system will then drop as oxygen is consumed. Alternatively, if the pressure and temperature of the system are held constant, the volume will decrease. In both cases, frequent measurements can be taken without the need for expensive or disruptive analyses. Such methods are referred to as **respirometric** techniques.

Early respirometers typically required manual measurements of changes in pressure. Many modern systems can collect data automatically every few seconds, if desired, allowing careful monitoring of the pattern of oxygen utilization (Figure 11.15). In one device,

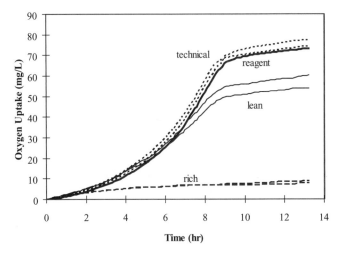

Figure 11.15 Example of respirometry data. Oxygen uptake with various grades (reagent, technical, lean, and rich) of monoethanol amine (MEA). (Data courtesy of Y. Lam, R. M. Cowan, and P. F. Strom.)

for example, oxygen is replaced automatically each time a slight pressure change occurs, and the amount of oxygen supplied is recorded.

With the development of dependable electrochemical **dissolved oxygen** (DO) probes, it has also become common to measure the **oxygen uptake rate** (OUR) directly in liquid culture. The liquid is first aerated, if needed, and then the drop in DO over time is measured and reported in milligrams of DO uptake per liter per minute (or hour). This may be further normalized to the biomass present, usually measured as SS or VSS, to give a specific OUR (**SOUR**), in units such as milligrams of DO uptake per minute per gram of biomass.

In systems in which the air flows through continuously, it may be possible to determine oxygen utilization by measuring the difference in the entrance and exit concentrations of oxygen in the airstream. Similarly, in continuous-flow liquid systems it may be possible to measure oxygen uptake by measuring changes in DO concentrations.

Production of Carbon Dioxide   The carbon dioxide produced by respiration might be measured directly (e.g., with an infrared spectrophotometer) or absorbed in an alkali and measured by titration with acid. For an aerated system, the exit gas might be bubbled through KOH solution in several flasks in series to be sure that it is all removed. Special **biometer flasks** (also known as *Bartha flasks* after their inventor, Richard Bartha) with sidearms containing KOH can also be used (Figure 11.16). Measurements of CO_2 can be particularly useful because they indicate mineralization of organic matter. Especially if appropriately ^{14}C-radiolabeled test compounds are available, [14]CO_2 production gives clear evidence of their complete biodegradation.

Production of Water   In most aqueous systems the amount of **metabolic water** produced is too small to measure compared to the amount of water already in the system.

New Biomass   Many of the methods described in Section 11.4 might be used to determine biomass over time. Alternatively, counts (Sections 11.5.1 and 11.5.2) might be used and converted (Section 11.5.3), if necessary, to biomass. However, the relationships

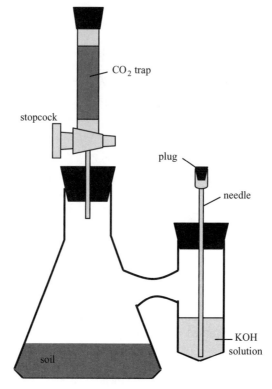

Figure 11.16   Biometer flasks.

between new biomass produced and other aspects of microbial activity may be complex. These are discussed more in Section 11.7.

Energy   Some of the energy released by metabolic activities is captured and used for growth and other cell functions; the rest is given off as heat. Special instruments known as **calorimeters** can sometimes be used to measure this release of heat experimentally. The amount of heat released per mass of oxygen utilized is ~14,000 J/g O_2 for oxidation of a wide variety of organic compounds. Also, in some systems of interest to environmental engineers and scientists, such as composting, sufficient heat is released that it may be possible to measure temperature changes or other signs of heat production as a means of determining the rate of activity [see the discussion of equation (16.5)].

11.6.2   Anaerobic Systems

In anaerobic systems it may also be possible to measure the disappearance of organic material or of electron acceptors such as nitrate or sulfate. Alternatively, monitoring the appearance of particular products such as ethanol or acetate may be feasible (if they accumulate). In methanogenic systems, the rate of formation of gaseous products can be measured volumetrically if they can be trapped in a gas collection column (Figure 11.17). This will typically be about 65% methane and 35% carbon dioxide (with traces of other gases). The total can be measured, or the carbon dioxide can be removed by scrubbing with KOH.

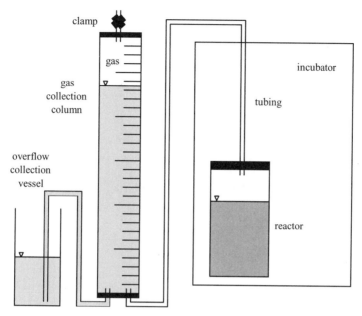

Figure 11.17   System for anaerobic gas production measurement.

11.6.3   Enzyme Activity

The amount of activity of a particular enzyme or group of enzymes in a sample can serve as an indication of the number and activity of microorganisms present. Probably the most widely used is the **dehydrogenase** assay, in which a tetrazolium salt is converted to a colored formazan product that is extracted and quantified. Other general enzyme assays include those for proteases, phosphatases, and esterases. Particular activities of interest, such as cellulose (cellulases), starch (amylases), or chitin (chitinases) degradation or nitrogen fixation (nitrogenases), also can be quantified.

11.7   GROWTH

Some microorganisms can reproduce asexually by budding, and many eukaryotes have a method of sexual reproduction. However, most microorganisms commonly reproduce asexually by **binary fission**, and that is the means of growth that we focus on here. In this process, a single "parent" cell physically splits into two genetically identical (in the absence of mutations) "daughter" cells. Since the original cell does not "die" in this process, terms such as *age* and *life span* have a different meaning than with plants and animals.

11.7.1   Exponential Growth

If they were able to grow freely without outside constraints (and no removal or death), cells growing by binary fission would follow the pattern depicted in Figure 11.18. Under these conditions the time between each successive split, producing a new

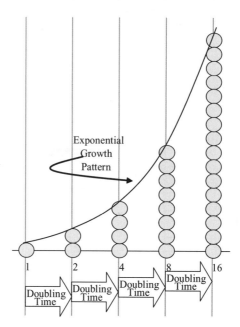

Figure 11.18 Theoretical microbial growth schematic.

"generation" of daughter cells, is constant and is called the **generation time** (t_g). Since the number of cells doubles each generation, the same period also is commonly referred to as the **doubling time** (t_d). This type of growth pattern is referred to as **exponential growth**. Note that whereas the time period is constant, the number of new cells added during that period doubles each time.

Most people do not have an intuitive feeling for exponential growth. As a simple mental exercise, picture folding a standard sheet of paper in half, then in half again, and again. Imagine that you could continue folding it in half 50 times. How tall would the folded stack of paper then be: a meter? 100 m ($\sim$ a football field)? New York to San Francisco ($\sim$5000 km)? After first trying to guess, calculate an approximate answer by assuming that the paper is 0.1 mm thick and doubles each time it is folded (or 0.1 mm $\times 2^{50}$). (*Note*: $2^{10} = \sim 10^3$.)

Minimum Doubling Times   Under ideal growth conditions, microorganisms will achieve their most rapid growth and hence their minimum t_d. This value is inherent to the particular strain of organism (for the system in which it is growing), and there is in fact a tremendous range of minimum doubling times in the microbial world. Some examples are shown in Figure 11.19.

Many common fast-growing bacteria (e.g., *Escherichia coli*) can have generation times of only 20 minutes. A few, such as the thermophile *Bacillus stearothermophilus* (found, for example, in composting piles), grow even more rapidly, with minimum generation times of around 10 minutes. This is about 1,000,000 times faster than a typical human generation of approximately two decades!

Prokaryotes in general have faster growth rates than eukaryotes, with their greater complexity. However, there are also slow-growing bacteria. Organisms tend to adapt to

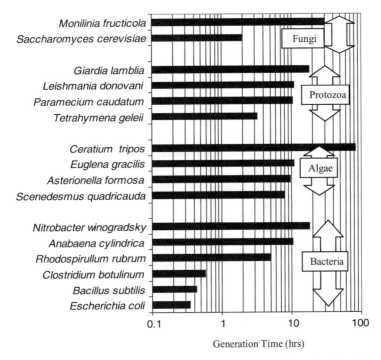

Figure 11.19   Representative microbial generation times under optimal conditions.

their environments, so that bacteria living in systems where substrates become available slowly will tend to be slow growing. Also, bacteria utilizing substrates that release little energy may have minimum doubling times of several hours or even days. Such substrates can include hard-to-degrade organic substances as well as some inorganic energy sources. For example, the chemolithotroph *Nitrobacter winogradskyi* derives only ~17 kcal of energy per mole of NO_2^--N that it oxidizes to nitrate, as compared to the ~600 kcal/mol potentially available to *E. coli* from the oxidation of glucose. This is reflected in the ~18-hour minimum generation time of *N. winogradskyi* (Figure 11.19).

Modeling Exponential Growth   If one cell begins growing exponentially, and cell number (N) is plotted vs. time, initially N increases in discrete steps: 1–2–4–8–16–32–64–··· (Figure 11.20). After a period of time, the cell divisions would no longer be synchronized precisely, so that the intermediate numbers (such as 50) would be present for brief periods. Still, only whole numbers of cells can occur (not 50.2), making the plot a series of small steps. However, once high enough numbers are reached, the rate of increase would appear to be smooth because the steps are so small.

On the other hand, biomass increases as a continuous function (except on the molecular level). Thus, biomass is theoretically more amenable than cell number to mathematical expression and is the focus of our discussion. In practice, however, it is also usually perfectly acceptable to use the same sorts of expressions with N, except when counts are very low.

Rather than biomass itself, more commonly it is biomass concentration (X) that is considered. This typically might be on a mass per volume basis for liquids (e.g., mg/L in water or media) and gases (mg/m^3 of air), or on a mass per mass basis for solids (mg/kg of soil or sludge).

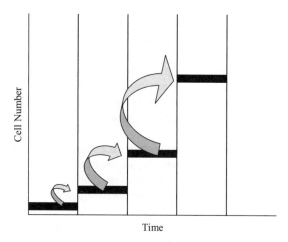

Figure 11.20 Theoretical stair-stepped batch bacterial growth curve.

If the logarithm of X is plotted vs. time (t) for an exponentially growing culture, a straight line results (Figure 11.21). Using natural logarithms ($\log_e$, or $\ln$), the equation for a straight line then gives

$$\ln X_t = \mu t + \ln X_0 \qquad (11.1)$$

where X_t is the value of X at time t, μ is the slope of the line, and X_0 is the starting value of X (at time 0). Differentiating equation (11.1) (assuming that μ is constant):

$$\frac{1}{X}\frac{dX}{dt} = \mu \qquad (11.2)$$

$$\frac{dX}{dt} = \mu X \qquad (11.3)$$

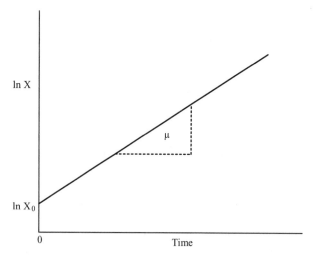

Figure 11.21 Similar plot of exponential growth.

From equation (11.2) it can be seen that μ is equal to the rate at which the biomass concentration is growing, divided by the biomass concentration at that time. Therefore, μ is referred to as the **specific growth rate** and has units of time^{-1} (inverse time, or "per time"). Growth rates are typically expressed per minute or per hour for bacteria in the laboratory, but more often per day in environmental systems and for other micro-organisms, since growth rates are lower.

Returning to equation (11.1) and rearranging yields

$$\ln X_t - \ln X_0 = \mu t$$

$$\ln \frac{X_t}{X_0} = \mu t$$

Exponentiating each side gives

$$e^{\ln(X_t/X_0)} = e^{\mu t}$$

$$\frac{X_t}{X_0} = e^{\mu t}$$

$$X_t = X_0 e^{\mu t} \tag{11.4}$$

There are thus three forms of the exponential growth equation: the differential form [equation (11.3)]; the logarithmic form [equation (11.1), which gives a straight line on a semilog plot]; and the exponential form [equation (11.4), useful for calculating the biomass concentration at a time in the near future].

Equation (11.3) is the most fundamental form of the three. This is because, unlike the other two, it is not dependent on an assumption that μ is constant. It simply states that the rate of biomass increase is proportional to the specific growth rate and to the current amount of biomass.

Assuming again that μ is constant, suppose that we look at X_t after one doubling time, when it would have doubled from its starting value to $2X_0$:

$$X_t = 2X_0 = X_0 e^{\mu t_d}$$

$$2 = e^{\mu t_d}$$

$$\ln 2 = \mu t_d$$

Thus, the relationship between specific growth rate and doubling time is

$$\mu = \frac{\ln 2}{t_d} \tag{11.5}$$

At the minimum doubling time, the **maximum specific growth rate** will be achieved. The symbol $\hat{\mu}$ (read "mu-hat") or μ_{max} is commonly used for this term. Using *E. coli*'s minimum t_d value of 20 min, for example:

$$\hat{\mu} = \frac{\ln 2}{20\,\text{min}} \times \frac{60\,\text{min}}{1\,\text{h}} = \frac{3 \ln 2}{1\,\text{h}} = 2.08\,\text{h}^{-1} = 50\,\text{day}^{-1}$$

Example 11.6   What is the meaning of the specific growth rate, μ?

Answer   It is the rate of change of biomass concentration in a batch culture (one with no inflow or outflow) at any instant, divided by the concentration at that point in time [equation (11.2)]. Suppose that $\mu = 0.05\,h^{-1}$ in a culture that at a certain time has a concentration of 100 mg/L. From equation (11.3):

$$\frac{dX}{dt} = \mu X = 0.05\,h^{-1} \times 100\,mg/L = 5\,mg/L \cdot h$$

In other words, at that instant, the concentration would be increasing at the rate of 5 mg/L per hour. At a later time, the concentration might reach 200 mg/L, and then

$$\frac{dX}{dt} = \mu X = \frac{0.05}{h} \times 200\,\frac{mg}{L} = 10\,\frac{mg}{L \cdot h}$$

So, as the concentration increased, the rate of increase also increased, even though the specific growth rate itself did not change.

The doubling time is easily computed from μ by rearranging equation (11.5):

$$t_d = \frac{\ln 2}{\mu} = \frac{0.693}{0.05\,h^{-1}} = 13.86\,h$$

Using equation (11.4), the biomass concentration at a future time for an exponentially growing culture can be calculated based on the present concentration and the specific growth rate. As an example, starting with a single cell with a mass of 10^{-12} g in 1 L of medium, and using the maximum specific growth rate of *E. coli*, after 10 hours the biomass concentration would be

$$X_t = X_0 e^{\mu t} = 10^{-12}\,g/L \times e^{(2.08/h)(10\,h)} \approx 1\,mg/L$$

However, this equation predicts that after 2 days, the biomass of the same culture would be $\sim 2 \times 10^{31}$ g, or 2×10^{25} metric tons. This is about 4000 times the mass of Earth; obviously, this exponential growth model has its limits! In the next section we show how an improved model can be developed.

11.7.2   Batch Growth Curve

A **batch** system is one in which all nutrients are present at the beginning and are not resupplied—there is no inflow or outflow, except perhaps for aeration. The flask of medium above in which *E. coli* was growing is a good example. In nature, a fresh deposit of cow manure on soil or the death of a fish in a pond would represent "batches" of nutrients made available at one time.

As the *E. coli* example demonstrated, exponential growth cannot continue for very long in a batch system. Depletion of substrates and/or buildup of inhibitory products will soon lead to decreasing specific growth rates. Thus, much of the time, microorganisms are likely to be growing at rates that are far below their $\hat{\mu}$ value. Equation (11.3) can still be used to describe this curve, but since $\mu \neq$ constant, it can no longer be integrated to give equations (11.1) and (11.4).

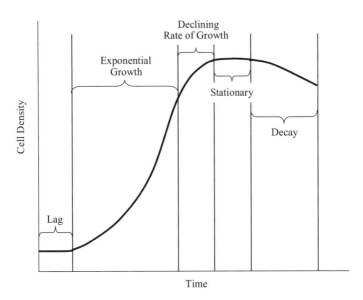

Figure 11.22 Typical batch bacterial growth curve.

Figure 11.22 shows a typical growth curve for a newly inoculated batch laboratory culture. For convenience, it can be broken up into the five phases shown and is discussed below. A similar response would be expected for other batch cultures.

Lag Phase   The transfer of a small inoculum of "seed" cells into a fresh batch of medium generally involves new conditions for the organisms, such as different organic substrates, form of nitrogen, pH, oxygen tension, and/or salt concentration. It will frequently take some time, referred to as a **lag**, for the cells to become adapted to this new environment before they start growing exponentially. This may involve production of enzymes to metabolize new organic substrates or utilize nutrients in another form, or simply replacement of cell constituents that had been depleted previously. Depending on how big a change is required and how rapidly the organism can respond, the lag period may vary from seconds to many hours.

Some scientists differentiate between a lag period, during which there is no growth ($\mu = 0$), and the lag phase, which lasts until exponential growth begins. For a short time between these two points, the culture may appear to be growing at a rate that is linear in a plot of X vs. t. This sometimes is referred to as an **arithmetic growth phase**. However, it should be noted that increases in biomass during exponential growth start off very slowly, and unless very precise and sensitive measurements are made, this may mimic a lag or arithmetic growth period.

Exponential Growth Phase   If an appropriate medium has been provided, once the organisms adapt to their new environment they will grow exponentially ($\mu = c$, where c is a constant greater than 0). Plenty of substrate and nutrients are available, and no harmful products of metabolism have yet accumulated. Thus, μ will approach the maximum specific growth rate $\hat{\mu}$ for the organism in that particular environment (in Section 11.7.6 we discuss some factors that affect $\hat{\mu}$). However, this phase can last for only a relatively short period of time before the ideal conditions deteriorate.

Declining Rate of Growth Phase   An old adage states that "all good things must come to an end," and this certainly applies to unrestricted growth within a batch reactor. Eventually, one or more factors (usually, substrate depletion, but perhaps product buildup, pH shift, nutrient deficiency, and/or something else) trigger an inevitable decline in cell growth rate. Biomass still increases, but at a continually decreasing rate ($\mu > 0$, but decreasing).

Most often this decrease in μ is the result of substrate depletion. This is commonly modeled by **Monod kinetics**, in which

$$\mu = \hat{\mu}\frac{S}{S + K_s} \tag{11.6}$$

where S is the substrate concentration (mg/L) and K_S is the half-saturation coefficient (mg/L). The **half-saturation coefficient** is the substrate concentration that allows for half of the maximum specific growth rate, as can be seen in Figure 11.23. At high substrate concentrations,

$$(S \gg K_s), \quad \frac{S}{S + K_s} \approx 1, \quad \text{and} \quad \mu \approx \hat{\mu} \tag{11.7}$$

At low substrate concentrations,

$$(S \ll K_s), \quad \frac{S}{S + K_s} \approx \frac{S}{K_s}, \quad \text{so} \quad \mu \approx \frac{\hat{\mu}}{K_s}S \tag{11.8}$$

When $S = K_s$,

$$\mu = \frac{1}{2}\hat{\mu} \tag{11.9}$$

At low S [equation (11.8)], the Monod expression approaches first order; that is, the rate is proportional to S to the first power ($S^1 = S$). Because S is often low in environmental

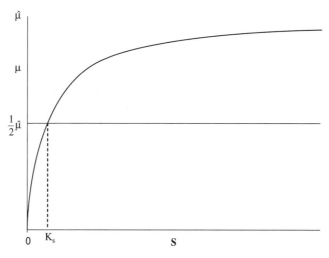

Figure 11.23   Monod kinetics.

systems, growth-related kinetic expressions are sometimes approximated by first-order equations of the form kS, where k is a constant for that system. However, at high substrate concentrations [equation (11.7)], Monod kinetics approach zero order (since $S^0 = 1$, the rate is unaffected by substrate concentration). Table 11.5 provides some half-saturation and other coefficients. (Although the discussion so far has focused on the carbon and energy source as the limiting nutrient, the same approach can be used for other requirements, such as dissolved oxygen, as indicated in the table; see "Multiple Substrates" in Section 11.7.4.)

Note that the Monod equation has the same form as the Michaelis–Menten equation used to describe enzyme kinetics (Section 5.3.1). However, unlike Michaelis–Menten kinetics for enzymes, which can be derived from fundamental chemical kinetic principles, the Monod expression for growth is empirical—an approximate fit to observed results.

Example 11.7 Assuming that an organism grows according to Monod kinetics with a maximum growth rate of 10 per day and a half-saturation coefficient of 5 mg/L, how fast would it be growing if the substrate concentration were 100 mg/L? What if it dropped to 20 mg/L?

Answer   At $S = 100$ mg/L:

$$\mu = \hat{\mu}\frac{S}{K_S + S} = 10\,\text{day}^{-1} \times \frac{100}{5 + 100} = 10\left(\frac{100}{105}\right) = 9.52\,\text{day}^{-1}$$

TABLE 11.5   Examples of Monod Kinetic Coefficients[a]

Organism	Substrate[b]	T (°C)	$\hat{\mu}$ (day^{-1})	K_s (mg/L)	K_{DO} (mg/L)	Y	b (day^{-1})
Mixed heterotrophic bacteria	Sewage COD	20	6	20	0.2	0.67	0.05
Mixed heterotrophic bacteria	Monoethanol amine	20	8.8	9	–	0.5	0.03
Sphaerotilus natans	Glucose	20	6.5	10	0.01	0.53	0.05
Citrobacter sp.	Glucose	20	9.2	5	0.15	0.55	0.15
Escherichia coli	Glucose	37	35	2	–	–	–
	Lactose	37	20	20			
Pseudomonas putida	Benzene	20	8.4	0.4	–	1.2[c]	0.5
	Toluene	20	8.9	0.2		1.3[c]	1.4
Ammonium-oxidizing bacteria	Ammonium-N	20	0.8	1.0	0.5	0.24[d]	n.a.[d]
Nitrite-oxidizing bacteria	Nitrite-N	20	0.8	1.3	0.7		

[a]$\hat{\mu}$, maximum growth rate at indicated temperature (T); K_s, half-saturation coefficient for the energy source; K_{DO}, half-saturation coefficient for dissolved oxygen; Y, yield (COD basis); b, decay coefficient.
[b]Substrate used as energy source. COD, chemical oxygen demand (see Section 13.1.3).
[c]Note that yields can be greater than 1.0, since during growth the organism incorporates oxygen and nutrients as well as the benzene or toluene carbon and hydrogen.
[d]Based on International Association on Water's Activated Sludge Model No. 1 (ASM1); ASM1 lumps ammonium oxidation with nitrite oxidation. A typical value for the decay coefficient for autotrophs was not available.

At $S = 20$ mg/L:

$$\mu = \hat{\mu}\frac{S}{K_S + S} = 10\,\mathrm{day}^{-1} \times \frac{20}{5 + 20} = 10\left(\frac{20}{25}\right) = 8.00\,\mathrm{day}^{-1}$$

Note that the growth rate is over 95% of the maximum with 100 mg/L of substrate, and is still 80% of the maximum when the substrate concentration is 20 mg/L.

Stationary Phase   Once the substrate is sufficiently depleted, growth nearly stops ($\mu \approx 0$). This plateau in biomass concentration is referred to as the **stationary phase**. Actually, the cells may still be growing slowly, but this is counterbalanced by the loss in mass through decay (the next phase).

Decay Phase   If a person stops eating, he or she gradually loses weight. This is also true of microorganisms. Their mass decreases, or **decay** occurs, if substrate is unavailable. The cell continues to carry on some metabolic processes, and if there are no external sources of energy, it must utilize internal ones. This is called **endogenous** metabolism. It will at first be based on storage materials but will then progress to include nonessential cell components, and eventually, essential ones. (The need to regenerate these components can be one cause of a lag phase when organisms from an old culture are transferred to fresh medium.) At some point, sufficient damage may be done so that the cell cannot recover and is thus no longer viable.

Energy is also used for cell **maintenance** during growth. Thus, the observed or **net specific growth rate** (μ_n) actually represents a higher true growth rate combined with some decay. Decay is often modeled as a first-order reaction with respect to biomass, with the **decay coefficient** constant for a given system. In the past, the symbol k_d was often used for this coefficient, but now b is more common, so that

$$\mu_n = \mu - b \tag{11.10}$$

Like growth rate, the decay coefficient has units of inverse time. Although it can vary considerably, in many wastewater treatment and other systems the decay coefficient may have a value of around 0.05 day^{-1}. This means that about 5% of the biomass will decay away each day. [Actually, the amount that decays in 1 day is $(1 - e^{(-1\,\mathrm{day})(0.05\,\mathrm{day}^{-1})}) \times 100\% = 4.88\%$. This is less than 5% because as the amount of biomass present decreases, so does the amount that decays.]

Specialized resting stages, such as spores or cysts, can be formed by some microorganisms. By greatly slowing down cell metabolism, they decrease decay rates, sometimes very dramatically.

Overall Equation   The growth equation (11.3) can now be rewritten incorporating Monod kinetics [equation (11.6)] and decay [equation (11.10)]:

$$\frac{dX}{dt} = \mu_n X$$
$$= (\mu - b)X$$
$$= \left(\hat{\mu}\frac{S}{S + K_s} - b\right)X \tag{11.11}$$

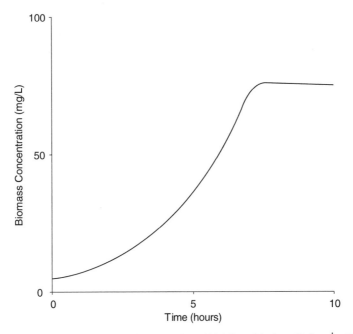

Figure 11.24 Batch growth curve using equation (11.11), with $\hat{\mu} = 10\,\text{day}^{-1}$, $K_s = 5$ mg/L, $Y = 0.6$ (explained below), $b = 0.1\,\text{day}^{-1}$, and starting values of S and $X = 120$ and 5 mg/L, respectively.

This expression will approximate the batch growth curve after the lag phase. The exponential phase occurs when $S \gg K_s$, so that $\mu \approx \hat{\mu}$. As substrate is depleted, μ decreases and the culture enters the declining growth rate phase. Once $\mu \approx b$, then $\mu_n \approx 0$ and the culture enters stationary phase. Finally, as S decreases further, $\mu < b$, so that $\mu_n < 0$ and the decay phase occurs. An example is shown in Figure 11.24.

Example 11.8   If $\hat{\mu} = 10\,\text{day}^{-1}$, $K_s = 5$ mg/L, and $b = 0.05\,\text{day}^{-1}$, at what substrate concentration will net growth equal zero?
   Answer

$$\mu_n = \hat{\mu}\,\frac{S}{S + K_s} - b = 0$$

$$S = K_s\,\frac{b}{\hat{\mu} - b} = 5\,\text{mg/L}\left(\frac{0.05\,\text{day}^{-1}}{10\,\text{day}^{-1} - 0.05\,\text{day}^{-1}}\right) = 0.025\,\text{mg/L}$$

If the substrate concentration falls below this value, the culture will be in the endogenous phase.

11.7.3   Death, Viability, and Cryptic Growth

In Section 11.7.2 we described the batch growth curve based on biomass (Figure 11.22). If cell numbers are used instead, the curve will have a similar shape. However, cell number does not decrease because of decay. It will, on the other hand, decrease as a result of cell death. Thus, this final phase is also sometimes referred to as the *death phase*.

Note that as decay does not necessarily lead to death, so death, or loss of viability, does not necessarily lead to decay. Also like decay, death does not necessarily occur only in the final phase; rather, it becomes more apparent there because it is not masked by high growth rates. Despite these similarities, the mechanisms of death (such as predation, lysis, acute toxicity, and lethal mutations) and decay are potentially very different.

When using biomass as a measure of microorganism concentration, viability may be of importance (although it is often ignored). In this sense, **viability** refers to the fraction or percent of the biomass that is still "alive," or capable of growth. Thus, if we consider X_L as the living biomass and X_D as the dead biomass, the total biomass (X_T) and viability (v) will be

$$X_T = X_L + X_D$$
$$v = \frac{X_L}{X_T} \tag{11.12}$$

The "growth" equation for each fraction can be considered using the same subscripts for the coefficients (although only μ_L is a real growth rate). First, the rate of change of the concentration of the living organisms is dependent on their rate of growth minus their rate of death:

$$\frac{dX_L}{dt} = \mu_L X_L - \mu_D X_L$$

The rate of change of the concentration of the dead biomass is dependent on the rate of formation from the death of living cells:

$$\frac{dX_D}{dt} = \mu_D X_L$$

What is typically observed is the rate of change of the total biomass:

$$\frac{dX_T}{dt} = \mu_T X_T \tag{11.13}$$

However, the rate of change of the total biomass can also be expressed as the sum of the rates for the two fractions:

$$\frac{dX_T}{dt} = \frac{dX_L}{dt} + \frac{dX_D}{dt} = \mu_L X_L \tag{11.14}$$

The right-hand side of equation (11.14) also makes sense, in that only the living organisms can grow to produce more total biomass. However, this means that equations (11.13) and (11.14) can be set equal:

$$\mu_T X_T = \mu_L X_L \quad \text{or} \quad \mu_T = \mu_L \frac{X_L}{X_T} \tag{11.15}$$

Substituting in equation (11.12) yields

$$\mu_T = \nu\mu_L$$

Thus, the growth rate observed for the total biomass of a culture with a 10% viability is only 10% of the actual growth rate of the living organisms. This means that the living organisms are in fact growing 10 times faster than the apparent overall growth rate, in order to produce all of both the living and the dead biomass.

If cells die and lyse (or lyse and die), their cell constituents become available as substrates for other microorganisms. This is sometimes referred to as **cryptic** growth.

11.7.4 Substrate Utilization

In many cases, to the environmental engineer and scientist it might actually be the substrate that is of greater direct concern than the microorganisms. For example, in biological wastewater treatment, the "substrates" are the constituents of the waste that are to be removed. Similarly, in polluted aquifers or streams, the contaminants may be the substrates. In such cases, the growth of the microorganisms themselves may be of only secondary interest.

Microorganisms may utilize substrates as a source of cellular constituents, to supply energy, and/or as an electron acceptor. Thus, the rate of substrate utilization often is considered to be proportional to the rate of growth:

$$\frac{dS}{dt} \propto -\mu X \tag{11.16}$$

where the negative sign indicates that substrate is decreasing and the Monod expression [equation (11.6)] is frequently used for μ.

Yield Coefficient   The proportionality coefficient for equation (11.16) is called the **yield**, Y. It is often reasonable to expect that for a given amount of substrate utilized, a given amount of biomass can be formed. Thus, Y has units of biomass produced per mass of substrate utilized, and the equation can be rewritten as

$$\frac{dS}{dt} = -\frac{\mu}{Y} X \tag{11.17}$$

The expression μ/Y, referred to as the **specific substrate utilization rate**, is given the symbol U, or sometimes k or q:

$$U = \frac{\mu}{Y} \tag{11.18}$$

Using Monod kinetics gives

$$\frac{dS}{dt} = -\frac{\hat{\mu}}{Y} \frac{S}{S + K_s} X$$

Since both $\hat{\mu}$ and Y are considered to be constants for a specific system, we also can define a new coefficient, the **maximum specific substrate utilization rate**:

$$U_{\max} = \frac{\hat{\mu}}{Y}$$

The value of Y can vary tremendously, depending on both the substrate and the microorganism. For many carbohydrates, common heterotrophic bacteria have yields of about 0.5 to 0.6 g of dry biomass produced per gram of carbohydrate utilized. A yield value can be greater than 1.0, and in fact typically is so for hydrocarbons as well as for the oxygen used as an electron acceptor. On the other hand, the yield for an autolithotrophic nitrifying bacteria growing on nitrite may be 0.05 or less. Various units may also be used; for example, the biomass and organic substrate could be expressed on a basis of dry weight, carbon, or chemical oxygen demand (see Section 13.1.3).

The yield described here is the *true yield*. The *observed yield* or *actual* amount of biomass production will be lower. This is because some of the biomass produced will be lost through decay. On the other hand, some of the substrate removed may simply be adsorbed, or even precipitated, rather than utilized. For this reason, some prefer the term **specific substrate removal rate**, or **specific substrate uptake rate**, rather than *utilization*.

Equation (11.18) can be rearranged as $\mu = Y_U$. Subtracting b from both sides and remembering that $\mu - b = \mu_n$ gives

$$\mu_n = YU - b \tag{11.19}$$

Multiple Substrates   It is common to focus on one substrate when examining microbial growth. In reality, of course, organisms need many substrates: an energy source; an electron acceptor; sources of carbon, nitrogen, phosphorus, and all the other essential elements; and perhaps organic growth factors. Most of these will normally be present in great excess of microbial needs, and hence can be ignored. Baron Justus von Liebig noted this for plants in 1840, leading to **Liebig's law of the minimum**: that the nutrient in shortest supply will limit growth. For microorganisms, Monod kinetics (discussed in Section 11.7.2) are generally used to describe this common situation, in which one substrate is or becomes **limiting**. However, what if the concentrations of two (or more) substrates are sufficiently low that they both will limit the amount and rate of growth?

One empirical approach used to describe this problem is an interactive, multiplicative Monod model. For the case of two potentially limiting substrates, A and B:

$$\mu = \hat{\mu} \frac{S_A}{S_A + K_A} \frac{S_B}{S_B + K_B} \tag{11.20}$$

Additional substrate terms can be added in the same way, as needed. Note that if $S_B \gg K_B$, the second term approaches 1, and the equation reduces to the basic Monod expression [equation (11.6)]. This can be used to give a quantitative definition to the term **limiting substrate**, as one that is not present in sufficient concentration to allow growth at more than some percentage (perhaps 90 or 95%) of the maximum rate.

Example 11.9   Suppose that a particular microorganism's growth can be described using the multiplicative Monod model. Given the substrate concentrations and kinetic coefficients below, what would its growth rate be? Would either of the substrates be considered limiting (using <90% of the maximum rate as the criterion)?

Answer

$$S_1 = 25 \, \text{mg/L glucose}$$
$$S_2 = 10 \, \text{mg/L ammonium-N}$$
$$K_1 = 5.0 \, \text{mg/L}$$
$$K_2 = 0.50 \, \text{mg/L}$$
$$\hat{\mu} = 6.0 \, \text{day}^{-1}$$
$$\mu = \hat{\mu} \frac{S_1}{K_1 + S_1} \frac{S_2}{K_2 + S_2}$$
$$= 6.0 \left(\frac{25}{5.0 + 25} \right) \left(\frac{10}{0.50 + 10} \right) = (6.0)(0.833)(0.952) = 4.8 \, \text{day}^{-1}$$

The glucose concentration allows growth at only 83% of the maximum (limiting by the stated criterion), whereas the ammonium-N concentration allows growth at 95% of the maximum (not limiting, even though it is at a lower concentration than the glucose).

Equation (11.20) for microbial growth has the same form as equation (5.38), which described enzyme kinetics for two substrates. However, whereas the enzymatic equation is mechanistic (derived from enzyme kinetics), the growth equation is again empirical.

A somewhat different problem is the presence of multiple substrates fulfilling the same nutritional need. In laboratory systems, microbiologists are able to study the uptake of a single substrate by a pure culture of bacteria. However, in most environmental systems, there are likely to be several utilizable organic substances present, for example, and perhaps several electron acceptors and nitrogen sources as well. Will an organism use only one at a time, or two or more simultaneously?

This turns out to be a very complex question, and the answer depends on the specific substrates and organism involved. On a practical level, this is often addressed by using some single general measure of organic matter, such as BOD or COD (see Section 13.1.3), rather than trying to account for the actual mechanisms involved. Electron acceptors are more likely to be used sequentially, with oxygen preferred when it is available. However, the ability to use a specific electron acceptor depends on the particular organism.

Of course, there are usually many different microorganisms present as well, not a pure culture of a single species. This, similarly, is commonly handled by using a general measure of biomass that lumps all or many different species together. Although these simplifications severely compromise any mechanistic basis for the models used, they have still been found in practice to provide useful information.

11.7.5   Continuous Culture and the Chemostat

Although some microbial systems can be represented as batches (Section 11.7.2), many others have flows of material entering and leaving, including streams, lakes, groundwater

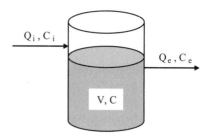

Figure 11.25   Schematic of a continuous culture system.

aquifers, and most biological wastewater treatment plants. Such **continuous culture** systems receive inputs of new substrates, but also lose substrates and even organisms in the outflow. Thus, to model a continuous culture, in addition to knowing the reactions that are occurring, it is necessary to keep track of the inputs to and outputs from the system. This is usually done by means of a **mass balance**.

A general mass balance equation for a particular component, A, in a continuous flow system (Figure 11.25) can be written as

rate of change of mass$_A$ in system = (rate of mass$_A$ input) − (rate of mass$_A$ output)

+ (rate of mass$_A$ production through reactions)

or if M_A is used for the mass of A and R_A for the reaction term, then

$$\frac{dM_A}{dt} = \text{(rate of } M_A \text{ input)} - \text{(rate of } M_A \text{ output)} + R_A$$

For a continuous-flow (Q) system of volume V, in which concentration $(C = M/V)$ of component A is measured in the influent and effluent (subscripts i and e, respectively) this can be expressed as

$$\frac{d(VC)}{dt} = (Q_iC_i) - (Q_eC_e) + (VR_C) \tag{11.21}$$

Note that in a batch system, $Q_i = Q_e = 0$ and $V = $ constant, so that

$$\frac{dC}{dt} = R_C$$

Thus, from equations (11.11) and (11.17), for biomass and substrate, respectively, we see that

$$R_X = \mu_n X \tag{11.22}$$

$$R_S = -\frac{\mu}{Y} X \tag{11.23}$$

Chemostat   One particular type of laboratory continuous culture system in which certain constraints are imposed is known as a **chemostat** (Figure 11.26). It is of great utility as a research tool but is also of interest because many observations about its functioning can be generalized to other continuous culture systems, such as activated sludge wastewater treatment.

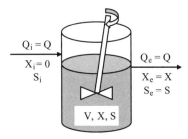

Figure 11.26 Schematic of a chemostat.

One requirement for a chemostat is that $Q_i = Q_e = Q$ = constant. If the influent and effluent flows are equal and constant, the chemostat volume will also be constant. It is also assumed that the substrate concentration in the influent (S_i) will be constant and that this flow will contain no biomass $(X_i = 0)$.

Another requirement for a chemostat is that it must be completely mixed (indicated by the "propeller" in Figure 11.26), so that the concentration of each constituent in every drop of water in the reactor is equivalent to its concentration in every other drop. A consequence of complete mixing is that the concentration in the effluent will be equal to the concentration in the reactor $(X_e = X, S_e = S)$, since a drop leaving the reactor is equivalent to a drop inside the reactor. This means that if the influent concentration is 1000 mg/L of substrate, for example, and the effluent concentration is 5 mg/L, the concentration everywhere within the reactor is 5 mg/L. Thus, a microorganism in this reactor "sees" only 5 mg/L of substrate, never the 1000 mg/L being fed to it. This concept may seem difficult to grasp at first, but keep in mind that every drop entering the reactor (at 1000 mg/L) is instantaneously mixed and diluted throughout the reactor volume, and that further, the substrate in it is being consumed by the microorganisms present.

The general mass balance equation [equation (11.21)] can now be rewritten for biomass in the chemostat as

$$\frac{d(VX)}{dt} = Q_i X_i - Q_e X_e + V R_X$$

$$V \frac{dX}{dt} = 0 - QX + V \mu_n X$$

$$\frac{dX}{dt} = -\frac{Q}{V} X + \mu_n X$$

$$= \left(\mu_n - \frac{Q}{V} \right) X$$

$$= (\mu_n - D) X \tag{11.24}$$

where $D = Q/V$. D has units of $(\text{time})^{-1}$ and is referred to as the **dilution rate**. It represents the number of times the contents of the reactor is replaced per unit of time. The inverse of D, referred to as the **hydraulic residence time** (HRT), indicates the amount of time the water (and hence any soluble materials) spends in the reactor. It is widely used in environmental engineering and science for a variety of reactor types and systems and is often represented by the symbol θ, so that $\theta = V/Q = 1/D$.

For substrate, the mass balance equation is

$$\frac{d(VS)}{dt} = Q_i S_i - Q_e S_e + V R_S$$

$$V\frac{dS}{dt} = QS_i - QS_e - \left(V\frac{\mu}{Y}X\right) \tag{11.25}$$

$$\frac{dS}{dt} = D(S_i - S_e) - \left(\frac{\mu}{Y}X\right)$$

One characteristic that makes the chemostat so interesting is that once it is inoculated with microorganisms that can grow in the system and utilize the substrate, it will naturally move toward a stable condition referred to as **steady state**. Steady state occurs when the concentrations in the reactor no longer change, that is, when $dX/dt = 0$ and $dS/dt = 0$. If the microorganisms grow faster $(dX/dt > 0)$, the substrate concentration drops $(dS/dt < 0)$, and this slows the growth back down. If the growth is too slow $(dX/dt < 0)$, organisms **wash out** of the system faster than they are growing; this drop in biomass allows the substrate to increase $(dS/dt > 0)$, and this leads to faster growth. Thus, the biomass concentration grows to the point at which it uses up substrate at the same rate that it is added.

At steady state,

$$\frac{dX}{dt} = (\mu_n - D)X = 0 \tag{11.26}$$

and X does not change. This can occur under two conditions: when $\mu_n - D = 0$ ($\mu_n = D$) or when $X = 0$ (no biomass in the system). If $X \neq 0$, substituting for μ_n and using $\tilde{S}$ for the steady state value of S, gives

$$\mu_n = \hat{\mu}\left(\frac{\tilde{S}}{\tilde{S} + K_s}\right) - b = D \tag{11.27}$$

Thus, for the steady-state chemostat reactor, the net growth rate is equal to the dilution rate. Since $D = QN$, the dilution rate can be controlled merely by controlling Q for a given reactor volume. Thus, the biological growth rate can be controlled directly by the chemostat operator. A similar relationship holds for some wastewater treatment processes, such as activated sludge (Section 16.1.3), in which the net growth rate can be controlled by the amount of sludge wasting. This gives the process operator a means of direct control over the biology of the process.

Equation (11.27) can be solved for the effluent substrate concentration:

$$\tilde{S} = \frac{K_S(D + b)}{\hat{\mu} - (D + b)} \tag{11.28}$$

Note that since $\tilde{S}$ is a function of only constants, it too must be a constant, confirming that $dS/dt = 0$:

$$\frac{dS}{dt} = D(S_i - \tilde{S}) - \left(\frac{\mu}{Y}\tilde{X}\right) = 0$$

Solving for $\tilde{X}$ gives

$$\tilde{X} = \frac{D(S_i - \tilde{S})Y}{\mu}$$

However, at steady state, $\mu_n = \mu - b = D$ (if $\tilde{X} \neq 0$), so $\mu = D + b$. Substitution then gives

$$\tilde{X} = \frac{D(S_i - \tilde{S})Y}{D + b} \tag{11.29}$$

Considering equation (11.26) and its meaning shows one value of the chemostat as a research tool. Once set up and inoculated, a chemostat will automatically run to a steady state in which X and S will be constant. If the setup conditions are reasonable, so that $\tilde{X} \neq 0$, the net growth rate will be equal to the dilution rate ($\mu_n = D$). Since $D = Q/V$, this means that any desired growth rate up to almost $\hat{\mu}$ can be chosen simply by adjusting the flow rate (Q) in a reactor of constant volume V.

Equation (11.28) shows further that the steady-state value of substrate will depend on the dilution rate but will be independent of the influent substrate concentration. A higher S_i leads to a higher steady-state biomass concentration $\tilde{X}$ [equation (11.29)], not a higher $\tilde{S}$. Thus, by selecting Q, chemostat operators can select the effluent substrate concentration; then choosing the influent substrate concentration allows them to pick the biomass concentration. This type of relationship holds true similarly for the activated sludge process.

It is also possible to calculate a minimum steady-state substrate concentration ($\tilde{S}_{min}$) that can be approached in a chemostat. This will occur when $\mu_n = D = 0$ (Figure 11.27). Substituting into equation (11.28) yields

$$\tilde{S}_{min} = \frac{K_S b}{\hat{\mu} - b}$$

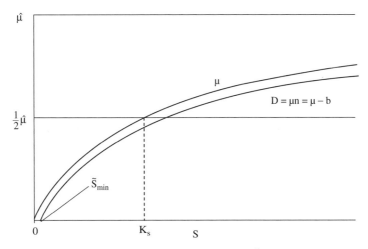

Figure 11.27   Steady-state substrate minimum, $\tilde{S}_{min}$, in a chemostat.

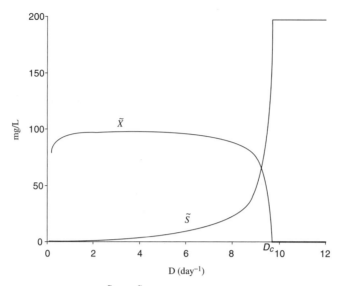

Figure 11.28   Theoretical plots of $\tilde{S}$ and $\tilde{X}$ vs. D in a chemostat. Values used in equations (11.28) and (11.29) were $\hat{\mu} = 10\,\mathrm{day}^{-1}$, $K_s = 5$ mg/L, $b = 0.05\,\mathrm{day}^{-1}$, $Y = 0.5$, and $S_i = 200$ mg/L.

This value can only be approached, since if $D = 0$, then $Q = 0$, and the system becomes a batch, not a chemostat.

On the other hand, if D is too high, the organisms cannot grow fast enough to maintain themselves in the reactor (Figure 11.28). At this critical dilution rate (D_c), or washout rate, $\tilde{X} = 0$ and thus $\tilde{S} = S_i$. In other words, as D (and hence, growth rate) increases, $\tilde{S}$ also increases until it reaches the influent concentration. Since it cannot go any higher, neither can growth rate, and the culture washes out. The highest net growth rate that can be approached (at steady state) is thus

$$D_c = \mu_n = \hat{\mu}\frac{S_i}{S_i + K_s} - b$$

Determining Microbial Kinetic Coefficients in Chemostats   One use of chemostats is for determining the microbial kinetic coefficients $\hat{\mu}$ and K_s. For this purpose, chemostats are run several times at different dilution rates and the steady-state substrate concentration is determined for each. A plot of $\tilde{S}$ vs. D is then made (Figure 11.28). Note in the figure that at values of $D > D_c$, $\tilde{X} = 0$ and $\tilde{S} = S_i$. Also, at low values of D, the importance of decay (b) leads to a decrease in $\tilde{X}$.

From the development of equation (11.28), we can see that

$$D + b = \hat{\mu}\frac{\tilde{S}}{\tilde{S} + K_s} \tag{11.30}$$

Such nonlinear equations are becoming readily usable on computers, but a traditional linearization method, the **Lineweaver–Burke plot**, may still be useful to know. (It can also be applied to data collected from batch cultures.) This involves inverting equation (11.30):

$$\frac{1}{D+b} = \frac{\tilde{S} + K_s}{\hat{\mu}\tilde{S}}$$

$$= \frac{1}{\hat{\mu}} + \frac{K_s}{\hat{\mu}}\frac{1}{\tilde{S}} \tag{11.31}$$

For values of $D \gg b$, a plot of $1/D$ vs. $1/\tilde{S}$ now gives a straight line with an intercept of $1/\hat{\mu}$ and a slope of $K_s/\hat{\mu}$ (Figure 11.29). The value of $\hat{\mu}$ is then obtained by taking the inverse of the y-intercept value. K_s can also be calculated from the slope once $\hat{\mu}$ is known; however, for imperfect data with their variability, it is usually better (particularly if b is not very small; see the "apparent slope" in Figure 11.29) to obtain it from the plot of $\tilde{S}$ vs. D (Figure 11.28). This is done by noting the D value that corresponds to $\frac{1}{2}\hat{\mu}$, then finding the corresponding S value that gives this growth rate (by definition, K_s). (To improve the quality of the estimates of the coefficients obtained, the slope $K_s/\hat{\mu}$ can then be calculated and used to better draw the Lineweaver–Burke plot, with the new intercept giving a better value of $\hat{\mu}$, which can then be used in turn to better estimate K_s from the plot of $\tilde{S}$ vs. D. After two or three iterations, the values will stabilize on the best estimate for the data.)

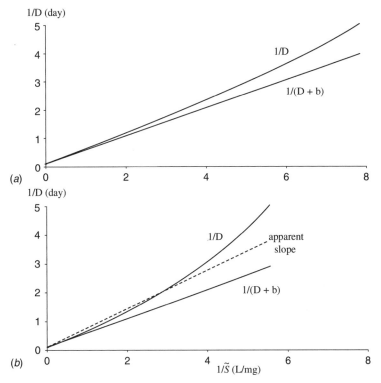

Figure 11.29 Lineweaver–Burke plot. In part (a) the same coefficients are used as in Figure 11.28 to generate the theoretical curves. In part (b) the value of $b = 0.15$, showing its effect on the nonlinearity of the $1/D$ plot.

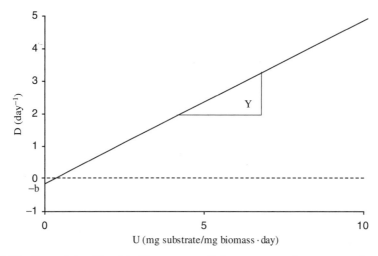

Figure 11.30 Determining Y and b. The same coefficients were used to generate the plot as in Figure 11.28, except that $b = 0.15$.

One criticism of the use of Lineweaver–Burke plots is that with some data they do not give very good estimates of the coefficients (particularly for K_s values obtained from the slope only rather than the $\tilde{S}$ vs. D plot). Nonlinear regression is now a preferred approach if sufficient data points have been collected.

The coefficients Y and b can also be found from chemostat data. Since $\mu_n = D$ in a chemostat, we can rewrite equation (11.19) as

$$D = YU - b \qquad (11.32)$$

Thus, by plotting D vs. U, Y (the slope) and b (the negative of the y-axis intercept) can be determined (Figure 11.30). For this purpose, U can be determined as

$$U = \frac{Q(S_i - \tilde{S})}{V\tilde{X}} = \frac{D(S_i - \tilde{S})}{\tilde{X}}$$

11.7.6 Environmental Factors

In addition to the concentrations of various substrates, other environmental factors, such as temperature, pH, pressure, moisture, and salinity, can clearly affect microbial growth. These can be viewed as influencing the kinetic coefficients, such as $\hat{\mu}$, K_s, Y, and b, that are used to describe growth and substrate utilization. (In fact, the observation that they vary is the reason that they have been referred to in this chapter as coefficients, rather than constants.) The effects of temperature and pH, which are usually the most important such factors in systems of interest to environmental engineers and scientists, are discussed briefly below.

Temperature   As discussed in Chapter 10, microorganisms are capable of growth at a wide variety of temperatures. However, each species has a particular **minimum** and **maximum** temperature at which it can grow, and an **optimum** temperature within that range.

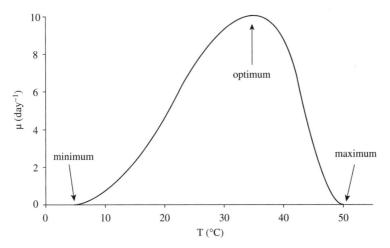

Figure 11.31   Dependence of microbial growth on temperature for a hypothetical mesophile.

These three values are referred to as the organism's **cardinal** temperatures (Figure 11.31). The minimum temperature for growth probably represents the point at which the cell's lipid-rich membranes become too rigid to be functional (somewhat like the increase in viscosity of an automobile's engine oil in winter), and/or reactions become too slow to maintain essential cell processes. The maximum temperature is reached when critical cell components are destroyed; in most cases this results from the **denaturation** (loss of three-dimensional structure) of proteins. The optimum usually is taken to be the temperature giving the highest $\hat{\mu}$ value, but it can also refer to a slightly different temperature (due to changes in yield or decay, for example) at which doubling time is shortest or biomass accumulation is most rapid.

The minimum and maximum temperatures for growth may form a narrow band or be widely separated. Similarly, growth rates may drop off rapidly as temperature departs from the optimum, or there may be a broad plateau of nearly optimal temperatures. However, the optimum temperature is always closer to the maximum than it is to the minimum. This is because increases in temperature will lead to increases in reaction rates, speeding growth, up to the point at which the enzymes or other cell constituents become unstable.

A common way to incorporate temperature effects into models is to use the following approximation, which is derived from work by von Arrhenius in the nineteenth century. Based on equation (5.48), the maximum specific growth rate $\hat{\mu}_T$ at the temperature of interest (T, in °C) is estimated from the rate $\hat{\mu}_{20}$ at 20°C as

$$\hat{\mu}_T = \hat{\mu}_{20}\Theta^{T-20} \tag{11.33}$$

where Θ is a **dimensionless** (no units) empirical coefficient, often taken to be 1.05 in the absence of an experimentally derived value. A related rule of thumb that is sometimes used is that reaction rates (and hence growth) will double for every 10°C increase in temperature. This is called the Q_{10} *rule* and corresponds to an Θ value in equation (11.33) of about 1.07. However, both of these approximations are suitable for mesophilic organisms only at temperatures between their minimum and optimum; outside this range (such as can occur in composting) they can give grossly misleading results.

The other kinetic coefficients (K_s, Y, b) also are affected by temperature. Decay rates (b) probably increase with temperature because of increased reaction rates, but the relationships with the other parameters are not well established. For the relatively few times in which it has been determined, K_s values have been observed to increase with temperature in some cases, but to decrease in others.

pH   Microbial growth occurs over a wide range of pH due to the existence of both acidic and alkaline extremophiles (Chapter 10). However, as with temperature, individual species have a much narrower range, with minimum, maximum, and optimum pH values. This stems from the effect that pH has on functional groups in essential microbial components, especially proteins. The pH may also affect the form (including shape and charge), and hence availability and toxicity, of organic and inorganic (e.g., NH_3/NH_4^+, H_2S/HS^-) substrates. In fact, pH is so important that even extremophiles maintain nearly neutral pH conditions inside the cell.

The overall pH range for a species may be narrow or broad. Unlike the case of temperature, the optimum can be near either end or, more likely, in the middle of this range; also, it may be narrow or broad (several pH units). Since there is no general pattern, the effect on kinetic coefficients (especially $\hat{\mu}$) must be determined individually for each situation of interest.

11.7.7   Inhibition

In addition to being limited by low concentrations of one or more substrates or unfavorable environmental factors, it is also possible for growth to be less than maximal due to the presence of toxic agents. The effect may be to slow down reactions or to cause damage that requires diversion of resources for repair. In general, this inhibitory agent could be (1) a substrate used for growth, if it is at high enough concentrations; (2) a product of cell metabolism; or (3) another substance or external factor.

Substrate Inhibition   Some growth substrates are toxic at higher concentrations. Potential mechanisms include reactions with critical cell constituents and dissolution of membranes. Several models have been developed to describe the effects observed. The most common is a modification of the Monod expression, based on the form of the Haldane model [equation (5.38)] for substrate inhibition of enzymes. In the context of microbial growth, it is usually referred to as the *Andrews model* (Figure 11.32):

$$\mu = \hat{\mu}\,\frac{S}{S + K_S + S/K_I}$$

If it is assumed that K_I is much greater than K_S, this equation can be approximately expressed as

$$\mu = \hat{\mu}\,\frac{S}{S + K_S}\frac{K_I}{S + K_I} \tag{11.34}$$

where K_I is the half-inhibitory coefficient (mg/L). Note that when $S \ll K_I$, the added term becomes 1, leaving the Monod expression. Also, when $S = K_I$, this term becomes $\frac{1}{2}$ in

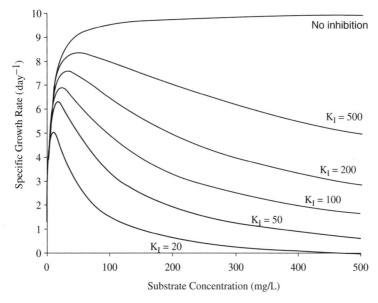

Figure 11.32   Andrews model of substrate-inhibited growth.

equation (11.34), and μ thus becomes half of the Monod value. In the Andrews model, growth rate approaches zero asymptotically as substrate increases. Some other models incorporate an inhibitory level above which there is no growth; this has been found to provide a better fit for growth on toluene, for example, which compromises the cell membrane at higher concentrations.

Product Inhibition   Products of cell metabolism can accumulate in the organism's environment and produce inhibitory effects. In some cases this may be a general effect, such as a decrease in pH. In other cases it may be a specific toxic product that is inhibitory, such as the release of ammonia and hydrogen sulfide from proteins, nitrite produced by ammonium oxidizers, or ethanol produced by fermentations. In the case of composting, accumulation of heat leading to excessively high temperatures is also, in a sense, an example of product inhibition.

Other Inhibitory Agents   Other toxic chemicals present naturally or as contaminants in the organism's environment may exert inhibitory effects but not serve as substrates. These might include heavy metals, toxic organics, or chlorine (perhaps added for disinfection). Ultraviolet light or other radiation can also be inhibitory. Depending on its mode of action, one way to model the effect of the concentration (I) of an inhibitory substance is to use another modification of the Monod expression:

$$\mu = \hat{\mu}\frac{S}{S+K_s}\frac{K_I}{I+K_I} \qquad (11.35)$$

However, other models might be more appropriate, if, for example, there is an inhibitory substance concentration above which the cell becomes inactive.

PROBLEMS

11.1. What would be the approximate dry weight of the cells in 1.0 L of medium containing 10^6 cells/mL if they are 1.0-μm-diameter cocci?

11.2. One rod-shaped bacterium has a diameter of 0.5 μm and a length of 1.5 μm, and another is 1.0×3.0 μm. **(a)** Assuming that they have equal densities, what is the ratio (smaller/larger) of their mass? **(b)** What is the ratio of their surface area/volume ratios?

11.3. If the minimum doubling time of *E. coli* in a particular system is 20 minutes, what is its maximum growth rate in units of **(a)** h^{-1}; **(b)** day^{-1}?

11.4. Given the Monod coefficients for heterotrophs and for nitrifiers below, find the substrate concentrations (mg/L of COD and NH_3-N, respectively) that result in a growth rate of 0.1 day^{-1}. Use the following maximum growth rate coefficients and the Monod half-rate coefficients (from the International Water Association's Activated Sludge Model No.1):

$$\text{For heterotrophs}: \quad \mu_{max,H} = 6.0\,day^{-1} \qquad K_S = 20.0\,mg\,COD/L$$
$$\text{For autotrophs}: \quad \mu_{max,A} = 0.80\,day^{-1} \qquad K_{NH} = 1.0\,mg\,NH_3\text{-}N/L$$

11.5. Given the two-substrate growth model and coefficients for the activated sludge process given below, compute the growth rate at a substrate concentration (S) of 5.0 mg/L and dissolved oxygen (DO) concentration (S_O) of 0.3 mg/L. What would the growth rate be if the DO term were excluded?

From Activated Sludge Model No.1: For aerobic heterotrophic growth,

$$\mu_g = \mu_{max,H}\frac{S_S}{K_S + S_S}\frac{S_O}{K_{O,H} + S_O}$$
$$S_S = 5.0\,mg\,COD/L$$
$$S_O = 0.3\,mg\,DO/L$$
$$K_S = 20.0\,mg/L \text{ (Monod coefficient for COD)}$$
$$K_{O,H} = 0.20\,mg/L \text{ (Monod coefficient for DO)}$$
$$\mu_{max,H} = 6.0\,day^{-1} \text{ (maximum growth rate)}$$

REFERENCES

Clesceri, L. S., A. E. Greenberg, and A. D. Eaton (Eds.), 1998. *Standard Methods for the Examination of Water and Wastewater*, 20th ed., American Public Health Association, Washington, DC. Updated regularly, and available at www.wef.org.

Grady, C. P. L., Jr., G. T. Daigger, and H. C. Lim, 1999. *Biological Wastewater Treatment*, 2nd ed., Marcel Dekker, New York.

International Association on Water, 2000. *Activated Sludge Models ASM1, ASM2, ASM2d and ASM3*, Scientific and Technical Report 9, IAW, London.

Metcalf and Eddy, 2003. *Wastewater Engineering: Treatment and Reuse*, 4th ed., revised by G. Tchobanoglous, F. L. Burton, and H. D. Stensel, McGraw-Hill, New York.

Murray, P. R. (Ed.-in-Chief), 1999. *Manual of Clinical Microbiology*, 7th ed., American Society for Microbiology, Washington, DC.

Young, J. C., and R. M. Cowan, 2004. *Respirometry for Environmental Science and Engineering*, SJ Enterprises, Springdale, AR.

12

EFFECT OF MICROBES ON
HUMAN HEALTH

Higher organisms evolved in a microbial world. From the perspective of microorganisms, plants and animals represent another environment to colonize. Thus, it is not surprising that humans, as well as other animals and plants, have a diverse community of microbes living on and in them. Some of these "passengers" are normal, beneficial, or even necessary (e.g., rhizobia in some plants; cellulose degraders in animals such as termites and ruminants), whereas others are abnormal, harmful, or even fatal.

12.1 MICROBIAL COLONIZATION OF HUMANS

The human body is a highly complex, multicellular life form whose structure and function depends on the coordinated interaction of roughly 10 trillion individual cells. On a microscopic level, the average human also carries a similar number of nonhuman visitors living opportunistically on and within the body. Freshly scrubbed and brushed, therefore, your body may outwardly appear clean and wholesome, but it is anything but sterile.

Your body is the mobile, warm-blooded equivalent of an ocean's coral reef, supporting a vast and highly divergent range of life. These microbes stretch from head to toe, spread across your skin, hide in the crevices of your mouth and nose, and follow your food from start to finish. Their presence is not just normal, but helpful or even necessary.

The life-style of these microorganisms is often described as a commensal relationship, in which they are tolerated by or even offer certain benefits to their host. For example, microbes are commonly found within the human intestinal tract, where they help to digest some foods and produce essential nutrients such as vitamins B_{12}, K, thiamin, riboflavin, and pyroxidine.

Environmental Biology for Engineers and Scientists, by David A. Vaccari, Peter F. Strom, and James E. Alleman
Copyright © 2006 John Wiley & Sons, Inc.

Perhaps even more important, the skin's normal biota actually offers a protective effect known as **colonization resistance**, which effectively safeguards the body against a hostile takeover by **pathogenic** (disease-causing) microbes. This effect can readily be appreciated during those periods when broad-spectrum antibiotics are used to combat a disease. The associated stress imposed on a body's normal biota can then lead to invasion by opportunistic, abnormal microbes (e.g., excessive growths of fungi such as *Candida*), which may lead to secondary health problems.

The composition of the human-associated microbial community will vary to some extent from one person to the next, and to some degree may also change with time. However, Table 12.1 provides a basic overview of the common locations and examples of the makeup of this normal microbial biota.

12.1.1 Abnormal Microbial Infection

Although the vast majority of microbes pose no threat whatsoever to human health, there are many forms that are outright hazards. The importance of the infectious diseases they cause is demonstrated by the rates of **mortality** (death) to which they can be linked. Even today, infectious disease is the world's leading cause of death, with fatalities exceeding 15 million per year (Figure 12.1). Rates of overall **morbidity** (illness, both fatal and nonfatal) are of course much higher and have a tremendous impact on the world's economy and each person's quality of life.

Respiratory and gastrointestinal diseases account for over half of these deaths. It is also sad to note that the majority of these deaths are of children below the age of 5 (the criterion used for **infant mortality**). Underdeveloped or developing countries are most heavily hit. Within industrialized countries such as the United States, for instance, the infant mortality rates fall below 1%, but these figures skyrocket to nearly 10% in developing countries and to more than 15% in the world's least developed countries. Furthermore, about half of the world's population is considered to be at risk to a wide range of infectious diseases.

Indeed, within the past millennia, microbial disease has proven to be a formidable adversary, one that has the potential to decimate the human population if left unchecked. During the Middle Ages and extending into the nineteenth century, diseases such as bubonic plague, cholera, and typhoid swept through Europe, causing massive mortality. The influenza pandemic at the end of World War I, for example, killed more people than the war itself.

Microorganisms are ubiquitous on and within the bodies of virtually all higher lifeforms (excluding those raised in laboratories under special "germ-free" conditions), and for the most part they are innocuous for their **host** (the organism that supports them). However, abnormal proliferation of indigenous microbes, or invasion from an external source, can lead to disease. Most of the important problems are infectious in nature, being spread by the dissemination of viable, pathogenic cells from one host to another by either direct or indirect means. In many instances, pathogenic microbial agents may also subsist within an environmental reservoir, lingering in wait for an opportunity to assert their influence.

A **parasite** is an organism that lives in a close relationship with another organism, benefiting at the expense of its host. **Pathogens** are thus parasites that do enough harm to their host to result in disease. However, it is also common to call disease-producing viruses, bacteria, and fungi *pathogens*, while referring to infective protozoans and worms as *parasites*.

TABLE 12.1   Examples of Normal Microbial Biota of Human Body Regions[a]

Region	Examples
Skin	*Acinetobacter*
	Corynebacterium
	Propionibacterium acnes
	Staphylococcus aureus
	Staphylococcus epidermidis
	Streptococcus
Nose, nasopharynx, and sinuses	*Haemophilus*
	Neisseria
	Staphylococcus aureus
	Staphylococcus epidermidis
	Streptococcus pneumoniae
	Streptococcus pyogenes
Mouth and throat	*Corynebacterium*
	Fusobacterium
	Neisseria
	Staphylococcus epidermidis
	Streptococcus mitis
	S. salivarius
	Treponema
Lower respiratory tract	Normally, few microorganisms
Stomach (pH $\sim$ 2)	*Helicobacter pylori*
	Lactobacillus
Small intestine (pH $\sim$ 4–5)	*Enterococcus*
	Lactobacillus
Large intestine (colon) (pH $\sim$ 7)	*Bacteroides*
	Clostridium
	Enterococcus faecalis
	Escherichia coli
	Eubacterium
	Lactobacillus
Urethra	*Escherichia coli*
	Proteus mirabilis
Vagina	*Candida* (yeast)
	Escherichia coli
	Lactobacillus acidophilus
	Streptococcus

[a]All are bacteria, except *Candida*, a fungus.

The Germ Theory of Disease and Koch's Postulates   We now take for granted that infectious diseases are caused by microbes, but this is actually a relatively new concept. Although Leeuwenhoek (Section 10.2.1) had first observed microbes in the seventeenth century, they were generally thought to be too small and unimportant to affect the health of higher organisms. In the early nineteenth century, fungal diseases of plants and later animals (silkworms) were first recognized. When Pasteur (Section 10.2.2) summarized his findings on the **germ theory of disease** in 1862, and referred to microbially caused spoilage as "diseases" of wine and beer, this influenced Englishman Joseph Lister to theorize that surgical wound infections might be the result of bacterial growth. His

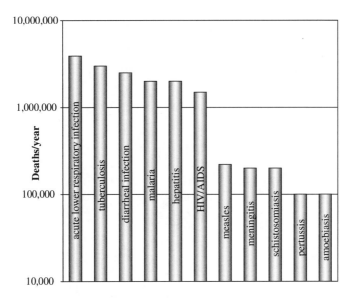

Figure 12.1   Infectious diseases with annual global mortalities of at least 100,000.

development of antiseptic surgical techniques in 1864 succeeded in greatly reducing the number of such often fatal outcomes. In 1873, Norwegian physician Gerhard Henrik Armauer Hansen, who oversaw a leper hospital, proposed that a specific bacterium (now called *Mycobacterium leprae*) was responsible for leprosy.

However, it was not until the definitive work of German Robert Koch (Figure 12.2) in 1876 that bacteria were proven to be agents of disease. About 10 years earlier, C. J. Davaine and others had shown that rod-shaped bacteria were present in the blood of animals suffering from anthrax (an important disease of animals, occasionally transmitted to humans), but not in that of healthy animals, and that an injection of infected blood would cause anthrax in previously uninfected animals. Koch provided the final proof of the bacterial **etiology** (causation) of anthrax by isolating the organism (now called *Bacillus anthracis*), growing it in pure culture in the laboratory, and injecting it into healthy mice, which then developed anthrax.

Koch was thus the first to meet the criteria for proving the causative relationship between a microorganism and a specific disease, as proposed in 1840 by German J. Henle (one of Koch's teachers). These are now referred to as **Koch's postulates**:

1. The microbial agent must be present in every diseased organism, but not in healthy organisms.
2. The agent must be isolated from the diseased host and grown in pure culture.
3. Inoculation of a healthy susceptible host with the pure culture must result in the same disease.
4. The same agent must be recoverable from the experimentally infected host.

These postulates, which Koch published in 1884 following his work on tuberculosis (which later won a Nobel prize), still represent an important benchmark used to judge whether there is a causal relationship between a particular microbe and a given disease.

Figure 12.2 Robert Koch. (Photo by F. H. Hancox, 1896.)

The methods that Koch developed also helped lead to rapid identification of numerous other causative agents of disease (Table 12.2).

Prevalence and Distribution of Diseases Diseases can vary widely in their **prevalence** (fraction of individuals infected) and geographic distribution. **Epidemiologists** (scientists who study diseases and their transmission) refer to an **endemic** disease as one that is constantly present in a specific area, usually at a low level (relatively few affected individuals). An **epidemic** refers to a disease with an unusually high prevalence in a specific geographical area. A **pandemic** is a widespread—nearly global—epidemic. The term **outbreak** is used to describe a sudden increase in the prevalence of a disease in a specific population; this may be associated with a single source, such as a contaminated water supply or food. Each victim may be called a **case**.

Disease Transmission A disease **reservoir** is a site in which an infectious agent remains viable so that it can serve as a source of infection for new hosts. Commonly, this is the pool of already infected hosts, so that humans are the primary reservoir for many human diseases. However, some disease organisms can also infect other species, and some go through a complex life cycle in which the host species alternate for the various life stages.

TABLE 12.2   Chronology of Some Major Disease Agent Discoveries

Disease	Microbial Agent[a]	Discoverers	Year
Anthrax	*Bacillus anthracis*	Koch	1876
Gonorrhea	*Neisseria gonorrhoeae*	Neisser	1879
Malaria	*Plasmodium* spp.	Laveran	1880
Tuberculosis	*Mycobacterium tuberculosis*	Koch	1882
Cholera	*Vibrio cholerae*	Koch	1883
Diphtheria	*Corynebacterium diphtheriae*	Klebs and Loeffler	1883–1884
Typhoid fever	*Salmonella typhi*	Gaffky	1884
Diarrhea	*Escherichia coli*	Escherich	1885
Tetanus	*Clostridium tetani*	Nicolaier and Kitasato	1885–1889
Pneumonia	*Streptococcus pneumoniae*	Fraenkel	1886
Meningitis	*Neisseria meningitidis*	Weichselbaum	1887
Gas gangrene	*Clostridium perfringens*	Welch and Nuttal	1892
Plague	*Yersinia pestis*	Kitasato and Yersin	1894
Botulism	*Clostridium botulinum*	Van Ermengem	1896
Dysentery	*Shigella dysenteriae*	Shiga	1898
Syphilis	*Treponema pallidum*	Schaudin and Hoffmann	1905
Whooping cough	*Bordella pertussis*	Bordet and Gengou	1906
Rocky Mountain spotted fever	*Rickettsia rickettsii*	Ricketts	1909
Tularemia	*Francisella tularensis*	McCoy and Chapin	1912

[a]All are bacteria, except *Plasmodium*, a protozoan.

A disease mainly of animals that also can be transmitted to humans (such as anthrax) is called a **zoonosis** (plural, *zoonoses*). Some infectious agents are able to survive in the environment, outside any host. This includes some species, referred to as **opportunistic pathogens**, which normally grow in soil or water or other environments but are capable of infecting a host that is **compromised** (weakened) by an injury, condition, or disease.

The human reservoir may include both people with the disease and those who are infected but show no symptoms. Individuals with such **subclinical** infections are referred to as asymptomatic **carriers**. The most notorious example of a carrier was Mary Mallon, known as "Typhoid Mary." Although she showed no symptoms herself, she was the source of a number of typhoid outbreaks in the New York City area in the early twentieth century. Because she refused to stop working as a food handler, eventually she was jailed.

One source of infection is the native microorganisms living on or in an individual. As a result of some change in it (increased **virulence**, the ability to cause disease) or (more commonly) the host, this usually innocuous **indigenous** microbe may become pathogenic. For example, the common intestinal *Escherichia coli* can acquire a virulence factor leading to severe diarrhea, the common skin inhabitant *Staphylococcus aureus* can cause serious illness after being transferred into formerly sterile tissue as the result of a subcutaneous wound penetration, or a fungal vaginal infection can result from suppression of the normally dominant bacterial populations because of antibiotic use.

However, in most cases an infectious agent must be **transmitted** from one host to another. This transmission may be relatively direct or may involve an intermediate. Direct transmission occurs through contact or exchange of bodily fluids between an infected host and the new, previously noninfected host. Rabies and sexually transmitted diseases such as syphilis are spread in this way. Many skin diseases, such as ringworm, are also transmitted via direct contact, but they may also be spread indirectly on objects (such as towels) because the causative agent can persist in the environment for a sufficient period of time. Most respiratory diseases are **droplet infections**, spread through **aerosols** (liquid or solid particles in air) resulting from exhaling, sneezing, or coughing. These remain suspended briefly in the air (during which time the agents remain viable), then are inhaled by the new host. Thus, the transmission is not truly direct, but is still generally considered to be so because the time in the air is so short (minutes or less). It is now believed that the common cold often is spread by direct contact through the following scenario: an infected person's hand becomes infectious when it touches the fluids of his or her mouth, nose, or eye; transmission occurs when he or she touches another person's hand; the new host becomes infected when touching his or her own mouth, nose, or eye.

Indirect transmission occurs through some other medium. A nonliving material that is capable of infecting a large number of individuals is referred to as a **vehicle**. Food and water are the two most common. Air can also be considered a vehicle, provided that the infectious agent is able to survive in this usually hostile (mainly because of drying and ultraviolet radiation) environment. Other nonliving means of transmission, such as clothing, furniture, toys, doorknobs, and bandages, are referred to as **fomites**.

A living intermediate for indirect transmission is called a **vector**. Many common vectors are biting insects (e.g., mosquitoes, malaria and some types of encephalitis; fleas, plague; lice, typhus; and flies, sleeping sickness) or ticks (Lyme disease and Rocky Mountain spotted fever). The vector picks up the infectious agent when it bites an infected host, then transmits it to a new host with another bite. In many cases the infectious agent reproduces within the vector (which is then considered an alternate host), thus increasing the likelihood of successful transmission to the next host. However, in some cases an organism such as a nonbiting fly is simply a mechanical vector, transporting the infectious agent from host to host on its mouth parts, feet, wings, or body hairs.

Not every transmission of a pathogen to a new potential host results in infection and disease. First, the host must be susceptible—a species that the pathogen can parasitize, and without previously developed **immunity** (resistance) from vaccination or prior exposure. Also, at least a minimum quantity of pathogens, called an **infective dose**, must be transmitted. Although for a few pathogens (perhaps the virus hepatitis A or the roundworm *Ascaris*) only a single viable particle (e.g., cell, spore, cyst, egg) may be sufficient, it more commonly takes at least tens (e.g., the bacterium *Shigella* and the protozoans *Entamoeba histolytica*, *Cryptosporidium parvum*, and *Giardia lamblia*), thousands (*Vibrio cholerae*), or even millions (e.g., *Salmonella*, *Clostridium perfringens*) of pathogens to overcome a healthy body's defense mechanisms and produce disease.

Example 12.1 While hiking in a national forest, a group of healthy young adults drinks from a clear mountain brook. Unknown to them, a short distance upstream there is a colony of beaver that has been infected with *Giardia lamblia*, and the water they consumed contains 2 viable cysts/mL. How much could a person probably drink without developing giardiasis, assuming that the infective dose is 20 cysts?

Solution If the number of cysts likely to result in disease is 20, a person could drink

$$20\,\text{cysts} \div 2\,\text{cysts/mL} = 10\,\text{mL}$$

or about 2 teaspoons. Thus the hikers are likely to develop giardiasis.

Not all people show equal resistance to disease, and individual resistance also varies over time. In general, the very young and very old have weaker immune systems, and some diseases may also severely compromise the ability to fight off other infections. This is of special concern in hospitals, since patients are typically in poorer health than is the general population, and thus more susceptible to contracting additional diseases. A **nosocomial** infection is one that is acquired in a hospital. In the United States alone, about 100,000 deaths per year occur from such infections, making them the fourth highest killer here (after heart disease, cancer, and strokes).

Parasites and their hosts evolve together over time. In general, it is not in the "interest" of a pathogen to kill its host. In fact, some fatal diseases, such as rabies and plague, are actually zoonoses, with humans an accidental (and therefore unadapted) host. Cholera, which is a human disease, appears to have grown somewhat milder over the last two centuries. Other human diseases remain fatal, however, perhaps because they have not sufficiently evolved yet. Also, previously unexposed populations may be particularly susceptible to a disease. This was the case, for example, among the indigenous peoples of North America when smallpox and measles were introduced from Europe.

Routes of Infection   There are several logical ways in which to classify diseases. One way is by the region of the body that is infected, such as respiratory disease, gastrointestinal illness, or urinary tract infection. Another is to group diseases by their causative agents, such as diseases caused by particular types of bacteria or viruses. However, for environmental engineers and scientists, it is usually most helpful to categorize diseases by their mode of transmission.

That is the approach that is taken here. In the following sections we look at diseases transmitted by water, food, air, vectors, sexual activity, and other direct contact. This is somewhat arbitrary, since a single disease may be spread by more than one route (e.g., water and food). The greatest emphasis is placed on waterborne disease, since this is where the role of the environmental engineer and scientist is greatest.

Table 12.3 provides an alphabetical listing of many diseases, along with their causative agents and mode of transmission. Although the list is certainly not all-inclusive, an effort was made to list all diseases mentioned in this and other chapters.

TABLE 12.3   Selected Diseases of Humans, Their Causative Agents, and Major Modes of Transmission

Disease	Microbial Agent	Taxa[a]	Transmission[b]	Comments
Abscesses, boils	*Staphylococcus aureus*	b-7	C-w; I	Most common agent
African sleeping sickness	*Trypanosoma gambiense, T. rhodesiense*	p-f	V-tsetse fly	Trypanosomiasis
Amoebic dysentery	*Entamoeba histolytica*	p-a	W-f/o	Amebiasis
Anthrax	*Bacillus anthracis*	b-7	C-d; A	Zoonosis—sheep, cattle; usually contracted through cuts in skin; airborne as biological weapon

(Continued)

TABLE 12.3 (*Continued*)

Disease	Microbial Agent	Taxa[a]	Transmission[b]	Comments
Aquired immune deficiency syndrome (AIDS)	Human immuno deficiency virus (Retroviridae)	v	S; C-b	
Ascariasis	*Ascaris lumbricoides*	w-n	F-f/o	Ingestion of contaminated soil or plants
Aspergillosis	*Aspergillus fumigatus, Aspergillus* spp.	f-d	A	Opportunistic pathogen
Athlete's foot, jock itch	*Epidermophyton* spp., *Trichophyton* spp.	f-d	C-d, f	
Bacterial dysentery	*Shigella dysenteriae, Shigella* spp.	b-6γ	C,W,F-f/o	Shigellosis
Botulism	*Clostridium botulinum*	b-7	F	Nearly 100% fatal food poisoning
Bovine spongiform encephalitis	BSE prion	pri	F	Mad cow disease
Chaga's disease	*Trypanosoma cruzi*	p-f	V-triatomid bug	South American sleeping sickness
Chickenpox	Varicella-zoster virus (Herpesviridae)	v	A; C-d	Varicella; same virus causes shingles (zoster)
Chlamydial urethritis and pelvic inflammatory disease	*Chlamydia trachomatis*	b-8	S	Most common sexually transmitted disease
Cholera	*Vibrio cholerae*	b-6γ	W,F -f/o	Gastrointestinal
Cold sores	Herpes simplex type 1 virus (Herpesviridae)	v	C-d	Fever blisters
Common cold	Rhinoviruses (Picornaviridae), coronaviruses (Coronaviridae), adenoviruses (Adenoviridae), others	v	A; C-d,f	Rhinoviruses account for ~75% of colds, coronaviruses ~15%; adenovirus colds may be more severe
Cow pox	Vaccinia virus (Poxviridae)	v	C-d	Spread between cows and humans during milking; source of smallpox vaccine
Creutzfeldt–Jakob disease	CJD prion	pri	F	Fatal brain disease
Cryptosporidiosis	*Cryptosporidium parvum*	p-s	W-ing	From animal feces; infectious oocyst
Dengue fever	Dengue virus (Flaviviridae)	v	V-mosquito	Usually not fatal; can be a hemorrhagic fever
Diphtheria	*Corynebacterium diphtheriae*	b-7	A	Respiratory disease— throat and tonsils

TABLE 12.3 (*Continued*)

Disease	Microbial Agent	Taxa[a]	Transmission[b]	Comments
Ear infections (otitis media)	*Streptococcus pyogenes*	b-7	C-d; I	Present in respiratory tract
	Pseudomonas aeruginosa	b-6γ	W-c	Swimmers' ear; opportunistic
Ebola hemorrhagic fever	Ebola virus (Filoviridae)	v	C-b,d,f	90% mortality
Ergotism	*Claviceps purpurea*	f-a	F	Infects cereals; severe central nervous system damage if ingested
Food poisoning	*Bacillus cereus*	b-7	F	Starchy foods
	Clostridium perfringens	b-7	F	Common; meats (7–15 h after eating)
	Staphylococcus aureus	b-7	F	Most common (1–6 h after eating)
Gas gangrene	*Clostridium perfringens*	b-7	C-w	Wound infection
Gastroenteritis	*Campylobacter jejuni*	b-6ε	F,W-f/o	Human and animal reservoirs
	Escherichia coli O157:H7	b-6γ	F	
	Norwalk virus (*Calvicindae*)	v	F,W,C-f/o	
	Rotaviruses (Reoviridae)	v	W	
	Vibrio parahemolyticus	b-6γ	W(F)-ing	Raw fish and shellfish
Infantile acute gastroenteritis	Rotaviruses (Reoviridae)	v	W	Major agent; major cause of mortality in children
Genital herpes	Herpes simplex type 2 virus (Herpesviridae)	v	S	Associated with cervical cancer
Genital warts	Papilloma virus (Papovaviridae)	v	S	Associated with cervical cancer
Giardiasis	*Giardia lamblia*	p-f	W-ing	From animal feces; infectious cyst
Gonorrhea	*Neisseria gonorrhoeae*	b-6β	S	
Hantavirus pulmonary syndrome	Hantavirus (Bunyaviridae)	v	A	Zoonosis—aerosolized mouse droppings
Hepatitis A	Hepatitis A virus (Picornaviridae)	v	W,F,C-f/o	Infectious hepatitis
Hepatitis B	Hepatitis B virus (Hepadnaviridae)	v	S; C-b	Serum hepatitis, liver cancer

(*Continued*)

TABLE 12.3   (*Continued*)

Disease	Microbial Agent	Taxa[a]	Transmission[b]	Comments
Histoplasmosis	*Histoplasma capsulatum*	f-d	A	Respiratory disease—inhaled spores germinate in lungs
Impetigo	*Staphylococcus aureus, Streptococcus pyogenes*	b-7	C; I	Skin sores
Influenza	Influenza virus (Orthomyxoviridae)	v	A	
Kuru	Kuru prion	pri	F	Ritualistic cannibalism
Legionellosis	*Legionella pneumophilia*	b-6γ	W-inh	Legionnaires' disease; milder infections called Pontiac fever
Leprosy	*Mycobacterium leprae*	b-7	C; A	Hansen's disease
Leptospirosis	*Leptospira interrogans*	b-9	W-c	Zoonosis—rodent, dog, or pig urine
Listeriosis	*Listeria monocytogenes*	b-7	F	Agent common in soil and water
Lyme disease	*Borrelia burgdorferi*	b-9	V-tick	Deer and mice are primary hosts
Malaria	*Plasmodium* spp.	p-s	V-mosquito	
Measles	Measles virus (Paramyxoviridae)	v	A; C	Rubeola
Meningococcal meningitis	*Neisseria meningitidis*	b-6β	A	50% mortality without treatment
Mononucleosis	Mononucleosis virus (Herpesviridae)	v	C-d,f	"Kissing disease"
Mumps	Mumps virus (Paramyxoviridae)	v	A	Epidemic parotitis
Peptic ulcers	*Helicobacter pylori*	b-6ε	W?,F?,C?	*H. pylori* found in drinking water
Pfisteria	*Pfisteria*	a-d	W-c	Skin disease in fishermen
Pinkeye	*Haemophilus aegyptius* (*H. influenzae*), *Moraxella lacunata*	b-6γ	C-d,f; A	Bacterial conjunctivitis
Plague, bubonic	*Yersinia pestis*	b-6γ	V-flea	Zoonosis—primary hosts are rats, other rodents
Plague, pneumonic	*Yersinia pestis*	b-6γ	A	Form of plague in which lungs infected; highly contagious
Pneumonia, pneumococcal	*Streptococcus pneumoniae*	b-7	A	
Pneumonia, viral	RS virus (Paramyxoviridae)	v	A	Respiratory syncytial disease

TABLE 12.3   (*Continued*)

Disease	Microbial Agent	Taxa[a]	Transmission[b]	Comments
Pneumonia, walking	*Mycoplasma pneumoniae*	b-7	A; C	Primary atypical pneumonia
Poison mushrooms	*Amanita* spp., others	f-b	F	Some fatal, others sickening or hallucinogenic
Poliomyelitis	Poliovirus (Picornaviridae)	v	W-f/o	An enterovirus; causes paralysis
Primary Amoebic meningoencephalitis	*Naegleria fowleri*	p-a	W-c	Swimming in warm ponds; enters through mucous membranes in mouth
Psittacosis	*Chlamydia psittaci*	b-8	A	Zoonosis—birds; 20% mortality without antibiotics
Rabies	Rabies virus (Rhabdoviridae)	v	C-animal bite	Zoonosis—mammals
Rheumatic fever	*Streptococcus pyogenes*	b-7	A	Autoimmune disease from untreated strep throat
Ringworm	*Epidermophyton* spp., *Trichophyton* spp.	f-d	C-d, f	
Rocky Mountain spotted fever	*Rickettsia rickettsii*	b-6α	V-tick	Tick-borne typhus
Rubella	Rubella virus (Togaviridae)	v	A; C	German measles
Salmonellosis	*Salmonella* spp.	b-6γ	F,W,C-f/o	Gastrointestinal disease
San Joaquin Valley fever	*Coccidioides immitis*	f-d	A	Coccidioidomycosis; respiratory; opportunistic
Scarlet fever	*Streptococcus pyogenes*	b-7	A	Systemic disease from untreated strep throat
Schistosomiasis	*Schistosoma* spp.	w-t	W-c	Bilharziasis; snails are intermediate hosts
Smallpox	Smallpox virus (Poxviridae)	v	C-d,f; A	Variola; eradicated
Staph infections	*Staphylococcus aureus*	b-7	A; C; I	
Strep throat	*Streptococcus pyogenes*	b-7	A; I	Respiratory disease
Swimmers' itch	*Schistosoma* spp.	w-t	W-c	Zoonosis—birds; snails are intermediate hosts
Syphilis	*Treponema pallidum*	b-9	S	
Tapeworm	*Diphyllobothrium*, *Taenia saginata*, *T. solium*	w-c	F	Fish, beef, pork

(*Continued*)

TABLE 12.3   (*Continued*)

Disease	Microbial Agent	Taxa[a]	Transmission[b]	Comments
Tetanus	*Clostridium tetani*	b-7	C-w	Lockjaw
Toxic shock syndrome	*Staphylococcus aureus*	b-7	C; I	Most common agent
Toxoplasmosis	*Toxoplasma gondii*	p-s	F; C	Undercooked meat, cat feces
Trachoma	*Chlamydia trachomatis*	b-8	C-d,f	Leading cause of blindness
Travelers' diarrhea (also see Gastroenteritis)	*Escherichia coli* (pathogenic strains)	b-6γ	W,F-f/o	
	Rotaviruses (Reoviridae)	v		
Trichinosis	*Trichinella spiralis*	w-n	F	Mainly undercooked pork, but now rare in United States
Trichomoniasis	*Trichomonas vaginalis*	p-f	S	
Tuberculosis	*Mycobacterium tuberculosis*	b-7	A	
Tularemia	*Francisella tularensis*	b-6γ	V-deer fly	Rabbit fever; zoonosis—rodents
Typhoid fever	*Salmonella typhi*	b-6γ	W,F-f/o	
Typhus fever	*Rickettsia prowazekii*	b-6α	V-louse	Epidemic typhus
Vaginal yeast infection	*Candida albicans (Monilia albicans)*	f-d	S; I	Vulvovaginitis, candidiasis, moniliasis; opportunistic
West Nile encephalitis	West Nile Virus (Flaviviridae)	v	V-mosquito	Zoonosis—birds
Whooping cough	*Bordetella pertussis*	b-6β	A	Pertussis; upper respiratory tract
Yaws	*Treponema pertenue*	b-9	C-d	Produces skin sores
Yellow fever	Yellow Fever Virus (Flaviviridae)	v	V-mosquito	"Yellow jack"; mortality up to 50%
Yersiniosis	*Yersinia enterocolitica*	b-6γ	F; W?	Acute gastroenteritis

[a]a, algae (-d, dinoflagellate); b, bacteria (- number, refers to bacterial kingdom, from Table 10.3; for Proteobacteria, Kingdom 6, class α, β, γ, δ, ε also indicated); f, fungus (-a, ascomycete; -b, basidiomycete; -d, deuteromycete); p, protozoans [-a, amoeba (Sarcodina); -f, flagellate; -s, sporozoan]; pri, prion; v, virus (family given in parentheses in preceding column after name of agent); w, worm [-c, cestode; -n, nematode (roundworm); -t, trematode (fluke)].

[b]Main routes of transmission: A, air; C, contact (-b, blood; -d, direct; -f, fomites; -w, wound); F, food; I, indigenous; S, sexual; V, vector; W, water (-c, contact; -f/o, fecal–oral route; -ing, other ingestion; -inh, inhalation).

12.2   WATERBORNE DISEASES

The major sanitary concern with waterborne disease classically has been contamination of drinking water with fecal material, commonly from wastewater. This important **fecal–oral route** is discussed in more detail below. However, it is now recognized that several other modes of disease transmission through water are also possible. We therefore look first more broadly at the types of water, sources of contamination, and routes of infection that can contribute to waterborne disease.

12.2.1 Types of Water

People may get their **potable** (drinking) water from a community source, such as a public or private water utility, or from an individual private source, such as a residential well. Water supplied to members of the public from private sources (usually, wells) by hotels, gasoline stations, camps, and similar small institutions are referred to as *noncommunity sources*. Potable water is also used for food preparation, cleaning dishes and clothes, and washing and bathing, as well as for direct ingestion. In most cases it is also used for flushing toilets and watering lawns and gardens, although in areas with limited supplies a separate nonpotable source may be used for these purposes. Industries commonly use potable water for their process water needs, sometimes following further purification. After use, much of the water from homes and industry becomes wastewater.

Other types of water that might be involved in disease transmission include recreational waters (such as rivers, lakes, and oceans) used for swimming and other water contact sports, swimming pool water, and irrigation water. Some natural waters are used for fish and shellfish harvesting. Industrial cooling waters may be from a potable source, or from ground or surface waters. Also, of course, there is precipitation (rain, snow, sleet, hail) and stormwater runoff.

12.2.2 Sources of Contamination

As indicated earlier, a major source of potential pathogens is human fecal material. People with gastrointestinal and some other types of infections may **shed** massive numbers of pathogens in their feces. Urine is also a potential source of contamination for some diseases. Thus, untreated or inadequately treated human wastes and sewage are a major sanitary concern.

However, there also are several other potential sources of water contamination. Water from activities such as bathing, showering, and toothbrushing, and from hand, dish, and clothes washing (often referred to as **graywater**, as opposed to the *blackwater* containing fecal material) may also contain pathogens, although typically in lower concentrations and of some different types. Urban and suburban stormwater contains fecal material and urine from pets and probably from rats, squirrels, and other wildlife. In some areas heavy concentrations of geese contribute large quantities of waste material. Agriculture may also be an important source of contamination directly from the animals being raised, or from the spreading of their manure; this may enter water through treated or untreated discharges, stormwater, or irrigation return flows (the portion of the irrigation water that flows off the field and back to a surface water, mainly to prevent salt buildup). Even in pristine areas water may not be safe to drink without treatment because of pathogens contributed by wildlife (e.g., beaver are a major reservoir of *Giardia*).

Human solid wastes may contain pathogens (e.g., from used facial tissues, diapers, and sanitary pads) that can enter water from litter, runoff, or landfill **leachate** (water that passes through the waste material in a landfill). Some industries (e.g., slaughterhouses, tanneries), institutions (e.g., hospitals), and other facilities (e.g., laboratories and doctor, dentist, and veterinary offices) are other potential sources. And especially in swimming pools, even the human body itself is a source, from skin sloughings and mouth, nose, and eye discharges.

12.2.3 Routes of Infection

Ingestion Ingestion is the major route of infection for waterborne diseases, primarily from drinking contaminated water. However, ingestion can also occur from eating foods containing or washed with the water, or from eating fish or shellfish harvested from contaminated water (which could also be considered transmission through food). Ingestion may also occur during swimming or other water-contact activities.

Most important waterborne diseases are transmitted through the fecal–oral route, discussed in Section 12.2.4. However, some pathogens occur naturally in water. *Vibrio parahemolyticus*, for example, is a marine bacterium that cause gastroenteritis, mainly in people who eat raw fish or shellfish that have accumulated it (or in some cases, been infected by it). Other diseases, such as giardiasis and cryptosporidiosis, often appear to result from contamination by the feces of wildlife or domestic animals, although the traditional (human) fecal–oral route can also play a role.

Contact It is not always necessary to ingest a waterborne pathogen for it to cause disease. For example, some bacterial eye and ear infections, such as those caused by the opportunistic pathogen *Pseudomonas aeruginosa*, can be transmitted through water in swimming pools that are overused or inadequately maintained. **Leptospirosis**, on the other hand, is caused by an obligate parasite, *Leptospira interrogans*, that is spread through contact with infected urine. The bacteria enters through mucous membranes or cuts in the skin, then attacks the kidneys and liver. It can be fatal in humans, but the main reservoirs are rodents, dogs, and pigs.

As another example, some free-living protozoans (e.g., *Naegleria*) in warm ponds can cause a fatal **primary amoebic meningoencephalitis** in swimmers. This organism also contaminated the hot springs of the famous Roman baths at Bath, England, and forced their closing for many years in the late twentieth century. The organism enters the body by penetrating the mucous membranes of the mouth and nose. (Thus, the route of exposure is not really ingestion, since the amoeba is not swallowed.)

Schistosomiasis is one of the most important debilitating parasitic diseases of the tropics, infecting over 200 million people. It is caused by several species of trematode flatworms of the genus *Schistosoma*. Eggs of these blood flukes are discharged in the urine and/or feces (depending on species) of the infected human. As part of their complex life cycle, the eggs hatch in water and the larva are ingested by certain aquatic snails, which then become infected and serve as the intermediate host. A distinct larval stage, the cercaria, develops, leaves the snail, and penetrates the skin of a new human host who walks through or bathes in the water. Building of dams has inadvertently led to increases in this disease by extending the range of the snails.

Fortunately, the appropriate snails do not live in the United States, so schistosomiasis does not occur here. However, related flukes do infect birds, again with snails as the intermediate host. Occasionally, one of these cercaria will inadvertently penetrate a human's skin, where it dies, unable to complete its life cycle. The resulting irritation is referred to as **swimmers' itch**.

Inhalation   Inhalation can be a route of exposure for a few waterborne diseases. **Legio-naires' disease**, or legionellosis, caused by *Legionella pneumophilia*, is perhaps the best known example. This bacterium grows in warm water, such as that in cooling towers, air-conditioning system evaporators, and hot-water tanks, as well as in natural aquatic habitats. If aerosols from such a system are inhaled, infection and potentially fatal pneumonia can result, especially in elderly or immunocompromised patients. (A milder infection of *L. pneumophilia* is referred to as **Pontiac fever**.) Unlike most other respiratory diseases, however, legionellosis is not spread from person to person, but rather, from aerosolized water. The first recognized outbreak resulted in 26 deaths at a 1976 convention in Philadelphia of the American Legion (hence its name). Later cases have included exposure while taking a shower.

12.2.4   Fecal–Oral Route

Fecal material is almost universally recognized as being unsanitary. How is it then that the fecal–oral route is such an important source of disease transmission (in food and from direct contact, as well as for water)? Part of the answer can be seen in Figure 12.3. Towns historically have been located along rivers or streams that served as their water source. Water typically is withdrawn upstream of the town and wastewater is discharged downstream. Beyond the obvious sanitary merits of this approach, it allowed both water delivery and wastewater collection to be by gravity: the water and wastewater flowing downhill. However, as areas became more densely populated, the wastewater discharge of one town soon became the water intake for the next community downstream.

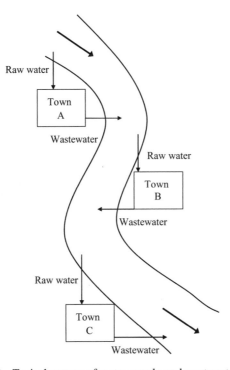

Figure 12.3   Typical pattern of water supply and wastewater disposal.

Most diseases transmitted by the fecal–oral route (whether through water, food, or direct contact) have their primary effect on the intestines and thus are referred to as **enteric**. To understand their importance, it is helpful to look at the historical impact of two such bacterial diseases, cholera and typhoid fever.

Cholera   Cholera is an acute intestinal disease marked by severe diarrhea and vomiting. The bacterium responsible, *Vibrio cholerae*, reproduces in the small intestine, releasing an **enterotoxin** (a toxin affecting the intestines) that triggers the resultant stress on the victim. Bodily fluids may be depleted so rapidly that the victim dies within hours unless preventive measures are taken, such as providing intravenous replacement of fluids and salts.

Cholera was originally endemic to South Asia, but there have been seven major pandemics (the first beginning in 1817), and it is now also endemic to South and Central America and perhaps to the Gulf coast of the United States. The early U.S. epidemics (first: 1832–1834; second: 1849–1854) produced widespread fear—not surprisingly, since there was no cure for this usually fatal disease. In 1832, 20% of the population of New Orleans died of cholera, while in 1849 another 5000 people died there, along with 8000 in New York City.

In 1854, **Dr. John Snow** (Figure 12.4) performed two studies of the incidence of cholera in London. These are now recognized as the first epidemiological studies ever

Figure 12.4   John Snow in 1857, one year before his death.

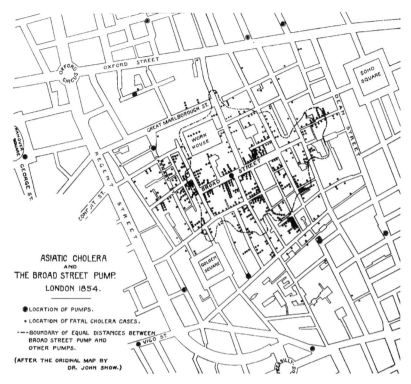

Figure 12.5 John Snow: cholera deaths and the Broad Street pump.

conducted, predating the formalization of the germ theory of disease (Section 12.1.1). In the better known of these studies, Snow plotted on a map the residence of each person who died of cholera in one area of town served by public wells with pumps (buildings there did not have indoor plumbing). He found 521 cholera deaths within 250 yards of the **Broad Street pump** (Figure 12.5). The incidence of cholera was much higher among people living close to, and therefore presumably using water from, this pump as opposed to others located in the area. He also found evidence that this well was contaminated with sewage, and in an example of an early public health measure, had the pump handle removed (rendering it unusable)! The adjacent pub is now named in his honor (Figure 12.6).

In the second study, Snow reviewed cholera deaths in another part of London served by two competing water companies. Customers of the Southwark and Vauxhall Company had 31.5 cholera deaths per 1000 houses served, whereas the rate for the Lambeth Company was 3.7 per 1000 houses served. Both companies took their water from the Thames River and delivered it through pipelines, without treatment. However, Southwark and Vauxhall withdrew water from the river near central London, where it was contaminated with untreated sewage, whereas Lambeth's source was upstream of the city, and relatively pure.

Attempts to control cholera also figured in some of the earliest efforts at water purification as a public health measure. During the German epidemic of 1892, the adjoining cities Hamburg (upstream) and Altona (downstream) were drawing their water from the Elbe. However, Altona practiced slow sand filtration, and despite the more contaminated

Figure 12.6   Snow's Pub in London's Soho district, adjacent to the infamous Broad Street pump.

water source, had a cholera death rate of only 2.3 per 1000 people (many traceable to drinking Hamburg water), compared to 13.4 for Hamburg (Table 12.4).

Medical treatment traditionally involved intravenous rehydration until the disease ran its course. However, this was not available to many in poorer countries. A simpler treatment called *oral rehydration therapy* (ORT) is saving many lives around the world (see Section 9.9).

Cholera was **eradicated** (eliminated) in the United States in 1911. However, it reappeared in 1973, and there have been small numbers of cases since, mostly associated with eating shellfish harvested along the coast of the Gulf of Mexico. Worldwide, cholera still accounts for more than 120,000 deaths per year.

Typhoid Fever   Although typhoid fever (caused by *Salmonella typhi*) was later transmitted mainly through food in the United States, originally it was also a major waterborne disease. Although it does not have as high a mortality rate, and thus did not provoke quite the same level of fear as cholera, it probably lead to more deaths overall: 500,000 cases with 40,000 deaths in 1909, for example. Water filtration also had a beneficial effect in reducing the incidence of typhoid fever, as can be seen in Table 12.4 for the cities of Pittsburgh, Cincinnati, and Louisville.

Another water treatment technology, disinfection with chlorine, was first utilized for an urban water supply in the United States in Jersey City, New Jersey, in 1908. Some of the early beneficial effects of chlorination in controlling disease are also shown in Table 12.4.

TABLE 12.4   Correlation between Sand Filtration (SF) or Chlorination (Chlor) of Water Supplies and Deaths from Cholera or Typhoid Fever

Disease	Date	City	Treatment	Deaths	Rate/100,000
Cholera	1892	Hamburg, Germany[a]	None	8606	1344
		Altona, Germany[a]	SF	328	230
Typhoid fever	1907	Pittsburgh, PA	None		125
	1908		SF[b]		49
	1902–1907	Cincinnati, OH	None		57
	1908–1913		SF		11
	1904–1909	Louisville, KT	None		58
	1910–1915		SF		24
	11–12/1914	Hull, Quebec[c]	None	200	1000
		Ottawa, Ontario[c]	Chlor	28	28
	1917	Wheeling, WV	None	200	
	1918		None	155	
	1–3/1919		Chlor[d]	7	

[a]Adjoining cities; Hamburg is immediately upstream of Altona on the Elbe River.
[b]Only part of the water supply was filtered.
[c]On opposite banks of the Ottawa River, Canada.
[d]Chlorination began late in 1918.

Figure 12.7 shows the drop in typhoid in New York attributable largely to improved water treatment.

In the United States, the annual incidence of typhoid fever has dropped to about 0.2 case per 100,000 population, and the majority of these are in people who acquired the disease while abroad. However, worldwide estimates for the incidence of typhoid fever are still on the order of 16 million cases and over 600,000 deaths a year. Although animals are a major source of other *Salmonella* species, the reservoir for *S. typhi* is usually man.

Other Fecal–Oral Route Diseases   As indicated in Table 12.3, a number of other waterborne diseases are spread through the fecal–oral route. This includes viral diseases such as poliomyelitis (now essentially eradicated in the United States through vaccination) and

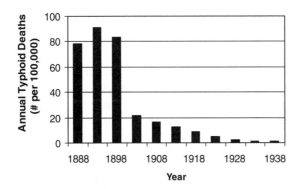

Figure 12.7   Annual typhoid cases (per 100,000 residents) in New York, 1888–1938.

hepatitis A. In one classic case (1955), 30,000 cases of infectious hepatitis were traced to drinking water in New Delhi, India, that was properly chlorinated but which had excess turbidity. It is believed that the viruses within the particulate material were protected from the chlorine. Waterborne viruses (especially rotaviruses) are also responsible for a portion of travelers' diarrhea, and contribute to the over 3 million yearly death toll worldwide from diarrheal diseases, mostly among young children (infantile acute gastroenteritis).

Other important bacterial enteric diseases include **salmonellosis** (caused by *Salmonella* species other than *S. typhi*) and bacterial dysentery, or **shigellosis** (caused by *Shigella* spp.). Salmonellosis is also easily spread through food, and *Shigella* by the fecal–oral route through direct contact, especially in settings such as child-care centers.

Amebiasis, or **amoebic dysentery**, is caused by the protozoan *Entamoeba histolytica*. This is the third most prevalent parasitic disease worldwide (after schistosomiasis and malaria) and has the eleventh highest annual mortality (100,000) among all infectious diseases. The infectious stage is a cyst.

Breaking the Chain   In most developed countries, waterborne disease from the fecal–oral route has been greatly reduced. Public health practices in such countries have been successful in breaking the chain of disease and its transmission. Through proper sewage treatment and waste disposal, fecal contamination of water has decreased dramatically. Water purification techniques, in turn, have been successful in virtually eliminating pathogens from drinking water. As a result, there are relatively few infected persons in the population who are shedding pathogens into the wastewater. This smaller reservoir further helps prevent disease transmission even when there are lapses in wastewater or potable water treatment. Quickly locating the source of an outbreak and treating infected persons further help minimize the spread of disease. Safe potable water also makes hand washing easier and more effective, reducing fecal–oral transmission through food and direct contact.

12.2.5 Modern and Recent Outbreaks

Table 12.5 shows the 108 known outbreaks of waterborne disease in the United States for 1993–2000. The six organisms responsible for more than one outbreak were the protozoans *Cryptosporidium parvum* and *Giardia* spp. and the bacteria *Salmonella* spp., *Shigella* spp., *Escherichia coli* type O157:H7 (Section 12.3.2), and *Campylobacter jejuni*. There were also two viruses. All produce gastrointestinal illness, and all but *Salmonella* and *Shigella* are relatively newly recognized as causes of waterborne disease. Many of the AGI (acute gastrointestinal illness from an unknown agent) cases are probably from viruses, although one was chemical.

Cryptosporidiosis and Giardiasis   The very large number of cases of cryptosporidiosis in the table results from a single major outbreak in Milwaukee, Wisconsin, in which 403,000 people were infected, with 4400 hospitalized and 100 fatalities. Worldwide it is estimated that there may be 500 million cases per year. A major outbreak of both cryptosporidiosis and giardiasis also occurred in Sydney, Australia, in 1999, threatening the Summer Olympics held there the following year. Both protozoans form resistant resting stages (oocysts for *Cryptosporidium*, cysts for *Giardia*) that are highly resistant to disinfection. This has led to a requirement in the United States for filtration of all community water systems that utilize surface water sources.

TABLE 12.5 Disease Outbreaks in the United States Associated with Drinking Water, 1993–2000[a]

Etiologic Agent	Type of Water System						Total		Percentages		
	Community		Noncommunity		Individual						
	Outbreaks	Cases	Outbreaks	Cases	Outbreaks	Cases	Outbreaks	Cases	Outbreaks	Cases[c]	Cases[d]
Giardia spp.	8	1,891	3	139	6	25	17	2,055	15.7	0.50	23.3
Cryptosporidium parvum	5	404,642	1	27	2	39	8	404,708	7.4	98.22	16.7
Shigella spp.	1	83	3	323	1	33	5	439	4.6	0.11	5.0
Escherichia coli O157:H7	3	208	2	37	3	12	8	257	7.4	0.06	2.9
Campylobacter jejuni	1	172	3	66	1	102	5	340	4.6	0.08	3.9
E. coli O157:H7/C. jejuni	0	0	1	781	0	0	1	781	0.9	0.19	8.9
Salmonella spp.	2	749	0	0	1	84	3	833	2.8	0.20	9.5
Non-O1 Vibrio cholerae	1	11	0	0	0	0	1	11	0.9	0.00	0.1
Plesiomonas shigelloides	0	0	1	60	0	0	1	60	0.9	0.01	0.7
Norwalk-like viruses	0	0	3	356	0	0	3	356	2.8	0.09	4.0
Small round-structured Virus	1	148	1	70	0	0	2	218	1.9	0.05	2.5
AGI[b]	8	85	17	1,465	10	208	35	1,758	32.4	0.43	20.0
Chemical	13	215	0	0	6	8	19	223	17.6	0.05	2.5
Total	43	408,204	35	3,324	30	511	108	412,039	100.0	100.0	100.0
Percentage[c]	39.8	99.1	32.4	0.8	27.8	0.12	100.0	100.0			
Percentage[d]		56.4		37.8		5.8		100.0			

[a]Compiled from data available on the Centers for Disease Control Web site; n = 108.
[b]AGI, acute gastrointestinal illness of unknown etiology.
[c]Percentage based on all 108 outbreaks or 412,039 cases.
[d]Percentage based on 8802 cases, excluding the Milwaukee outbreak of Cryptosporidium infecting 403,237.

TABLE 12.6   Drinking Water System Deficiencies Associated with Waterborne Disease Outbreaks in the United States, 1993–2000[a]

| | Type of Water System | | | | | | | |
| | Community | | Noncommunity | | Individual | | Total | |
Type of Deficiency	No.	%	No.	%	No.	%	No.	%
Untreated surface water	0	0.0	0	0.0	2	6.7	2	1.9
Untreated groundwater	5	11.6	17	48.6	14	46.7	36	33.3
Treatment	16	37.2	13	37.1	2	6.7	31	28.7
Distribution system	18	41.9	4	11.4	4	13.3	26	24.1
Unknown	4	9.3	1	2.9	8	26.7	13	12.0
Total	43	100.0	35	100.0	30	100.0	108	100.0

[a]Compiled from data available on the U.S. Centers for Disease Control Web site; $n = 108$.

Campylobacter   The first reported outbreak of waterborne gastroenteritis in the United States involving *C. jejuni* was in 1978, in Bennington, Vermont. Of a total population of 10,000 using the community water supply, 2000 became ill. *Campylobacter* (both *C. jejuni* and *C. fetus*) can also be spread by food, especially chicken and turkey (which are reservoirs).

System Deficiency   Table 12.6 indicates the breakdown of incidents by the type of deficiency in the potable water system. For community systems, the outbreaks are almost evenly divided between treatment and distribution problems, whereas for noncommunity and individual systems, consuming untreated water was the primary known cause.

12.3   FOODBORNE DISEASES

There are many diseases that can be caused by the consumption of contaminated foods. The responsible microbial agent may have been present in the original material (e.g., growth of *Salmonella* in chickens), or been introduced at some point during subsequent handling and preparation steps as a result of improper sanitation. In most instances, these problems can be linked to improper cooking or storage, particularly of foods containing meat, milk, eggs, cheese, poultry, fish, and shellfish. Table 12.7 provides an approximate percentile breakdown of various bacterial foodborne diseases in the United States by their causative agent and indicates the food products with which they are typically associated.

The sanitary aspects of commercial food distribution and handling within the United States have long been carefully regulated, but there are still serious lapses. Additionally, many foods are now being imported from countries around the world with far less stringent standards. However, most cases are associated with poor practices in homes, institutions (e.g., nursing homes), or small restaurants.

Foodborne diseases of microbial origin may be split into two categories. **Food poisoning** results from ingestion of preformed microbial toxins. In some cases, the microorganism itself may no longer be viable or may be incapable of infecting a human host, but the products of its previous activity result in disease. **Food-transmitted infection**, on the other hand (although sometimes also called food poisoning by the public), results when the causative organism is transmitted via food, then parasitizes the new host to produce disease. Some of the diseases of concern are discussed briefly below. Note that the specific symptoms, particularly the time to their onset, may help distinguish the particular agent.

TABLE 12.7   Bacterial Foodborne Diseases in the United States

Bacterial Agent	Type[a]	Commonly Contaminated Foods	(%)[b]
Bacillus cereus	P	Rice and starchy foods	1
Campylobacter spp.[c]	I	Chicken and milk	8
Clostridium botulinum	P	Home-canned foods; smoked fish	<1
Clostridium perfringens	P	Reheated meats/meat products	7
Escherichia coli[d]	I	Meats	12
Listeria monocytogenes	I	Dairy products, meat, fresh produce	1
Salmonella spp.	I	Chicken and other meats, egg products, fresh produce	55
Shigella spp.[e]	I	Meats, fresh produce	6
Staphylococcus aureus	P	Meats, desserts, salads with eggs and/or mayonnaise	8
Vibrio cholerae	I	Seafood	<1
Vibrio parahaemolyticus	I	Seafood	1
Yersinia enterocolitica	I	Pork and milk	<1

[a] I = infection; P = poisoning.

[b] Based on 196 outbreaks (8047 cases) of bacterial etiology reported in 2003 by the Foodborne Outbreak Response and Surveillance Unit, U.S. Centers for Disease Control and Prevention.

[c] Mostly *C. jejuni*.

[d] Mostly *E. coli* O157:H7.

[e] Mostly *S. sonnei*.

12.3.1   Bacterial Food Poisoning

Staphylococcus   *Staphylococcus aureus* is the most common cause of food poisoning. It is often present on the skin, so that it is easily introduced into food during preparation and handling. If dishes containing mayonnaise, meat, poultry, or creamy sauces or fillings are not properly refrigerated (e.g., at a picnic), the organism can grow and produce enterotoxins. These can provoke severe vomiting and diarrhea within 1 to 6 hours.

Clostridium perfringens   *Clostridium perfringens* is another major widespread cause of food poisoning throughout the world. It produces an enterotoxin in foods that 8 to 22 hours after ingestion creates severe diarrhea and intestinal cramps, but not vomiting. The resulting distress typically lasts for a period of about 1 day, beyond which point there is almost always a full recovery. *C. perfringens* is a normal resident in the intestinal tract but in numbers sufficiently low to avoid any negative impact. Thus, it is a common contaminant of meat products as well as soil and sewage. Because it is an endospore former, it most commonly causes problems in foods that are not heated sufficiently to inactivate the spores present, and then held unrefrigerated (at temperatures of 20 to 40°C), allowing for rapid growth and toxin production.

Botulism   *Clostridium botulinum* is also a sporeformer and is common in soil and sediment. Again, if contaminated food is not sufficiently cooked, growth can occur later. In this case, however, neurotoxins are produced, and the resulting illness, botulism, is often fatal. The toxins are also destroyed by adequate cooking, so that home-canned foods (which may be inadequately heated and then stored at room temperature) that are later eaten raw are a major source of this illness.

Unpasteurized honey may also contain spores of *C. botulinum*, although toxin is not produced there. However, germination and toxin production may occur in the intestines of very young (less than 2 months old) infants who are given raw honey, leading to sudden death from infant botulism. Babies should therefore not be given raw honey until their intestinal biota has developed sufficiently that it can resist such opportunistic infection.

Waterfowl are also sometimes killed by botulism. The organism can grow in the rich sediments of ponds, especially in parks where people throw bread in the water to feed ducks and geese. The birds may be poisoned when they later eat bread off the bottom.

12.3.2 Bacterial Infections

Salmonellosis   Based on their surface antigens, it is believed that there are actually 2200 different species of *Salmonella*, but in the United States, *Salmonella typhimurium* and *S. enteritidis* are generally considered to be the leading causes of salmonellosis (gastrointestinal illness other than typhoid fever caused by *Salmonella*). Many commercially raised chickens harbor *Salmonella*, leading to contamination of their meat and eggs. Proper cooking prevents infection, but consuming runny eggs, reusing surfaces or utensils that were improperly cleaned, and not refrigerating foods promptly all pose a risk. Other meats and milk can also be sources, and because *Salmonella* can grow in many such foods, contamination by a human carrier is also possible, typically from inadequate hand washing. Turtles and some other reptiles may also be carriers (which is a reason that their sale as pets is now banned in many areas).

Salmonellosis is characterized by the sudden onset of gastrointestinal distress, including fever, abdominal cramps, diarrhea (usually bloody), and vomiting, starting 12 to 48 hours after eating. Antibiotic treatment is usually highly effective, but massive infections may be fatal. It is estimated that there are 50,000 foodborne cases per year in the United States.

Toxigenic E. coli   There are many strains of *Escherichia coli* that live normally in the intestines of humans and animals without causing disease. However, in the late 1960s, medical researchers began finding a number of enterotoxin-producing strains. In 1973, an acute outbreak of gastroenteritis at Japan's Nagoya airport, which affected 956 people, was linked to the presence of several such strains in a restaurant's drinking water supply. Two years later, the presence of another hazardous strain (*E. coli* O6:H16) in a natural spring caused more than 1000 cases of acute diarrhea at Oregon's Crater Lake National Park.

As of the early 1980s, another new *E. coli* strain, type O157:H7, was linked to cases involving bloody and nonbloody diarrhea that were accompanied by abdominal cramps. The latter strain also has been found to foster debilitating blood and kidney illnesses and can affect the central nervous system. This virulent strain is now blamed for 250 deaths a year. Most cases are linked to eating undercooked ground beef or drinking unpasteurized milk. Dairy and beef cattle appear to be a major reservoir for this type of *E. coli*.

Yersiniosis   Yersiniosis is a severe gastroenteritis caused by infection with *Yersinia enterocolitica*, most commonly from contaminated food. These enterotoxin-producing bacteria are invasive pathogens that are able to penetrate the gut lining, entering the lymphatic system and blood and causing severe pain similar to that of appendicitis.

Listeriosis   Listeriosis is caused by *Listeria monocytogenes*. It is a widespread soil and water organism that is becoming an important pathogen in part because it is cold tolerant. Thus, its growth is not adequately controlled by refrigeration. It grows within and kills the white blood cells that try to control the infection, leading to a serious illness with a 25% mortality rate.

12.3.3   Other Agents

Viruses   Some viruses, such as hepatitis A, rotaviruses, and Norwalk virus, can be spread by food as well as by water. In part, this is expected, since water is used in the preparation of many foods. However, unlike bacteria, viruses cannot grow or replicate in food. (Animal viruses only replicate within living animal cells.) Thus, the initial contamination must be with a sufficient quantity of viruses to constitute an infective dose, or disease will not occur.

Fungi   Several types of fungi can cause food poisoning. Some wild mushrooms are themselves highly poisonous, but may be mistaken for edible varieties by inexperienced pickers. On the other hand, growth of fungi within a food material can produce toxins. Ergotism, as described in Section 10.7.4, for example, is caused by the neurotoxins produced by the ascomycete *Claviceps purpurea* in infected cereal grains (especially, rye). Similarly, the aflatoxins produced by some strains of *Aspergillus*, particularly during growth on stored grains and peanuts, are potential **carcinogens** (cancer-causing agents). However, infection by ingested fungi is rare.

Algae   Some of the marine dinoflagellates occasionally "bloom" in large numbers to produce "red tides" (Section 10.7.2). Certain strains, including many *Gymnodinium*, *Gonyaulax*, and *Pfisteria*, may release neurotoxins that can be taken up by filter-feeding shellfish, such as mussels, clams, and oysters. (Thus, this listing could also be under waterborne disease.) These concentrated toxins, which are not destroyed by cooking, can in turn produce disease in man when ingested, in some cases leading to respiratory failure and death. At least with *Pfisteria*, there are also reports of skin damage from contact with toxin-contaminated water or fish.

Parasites   **Toxoplasmosis** is caused by the protozoan *Toxoplasma gondii*, a sporozoan. It is ingested in inadequately cooked meat, using the same utensils for meat before and after cooking, or licking the fingers after handling raw meat. (It can also be spread through contact with cat feces.) In general, infection causes no symptoms, but it can cause birth defects if a pregnant woman becomes infected.

Tapeworms are members of the Cestoda class of flatworms (Platyhelminthes). The most important as human parasites are those acquired by eating raw or undercooked beef, pork, and fish. Encysted larvae in the muscle tissue of these animals are released from the cyst once they are ingested, and attach to the intestinal lining of the small intestine. There they quickly mature, reaching lengths of up to 15 m (50 ft). Tapeworms have no mouth or digestive tract, relying on the host to provide nutrients that it absorbs through its body wall. Over 1 million eggs per day may be produced in the worm's ripened body segments and shed with the human host's feces. If the eggs are ingested by the appropriate intermediate host (cattle, swine, or fish, respectively), the larvae develop and invade the muscle, where they encyst, completing the cycle.

The causative agent of **ascariasis** is the nematode (roundworm) *Ascaris lumbricoides*. The **ova** (eggs) are ingested on contaminated vegetables or from dirty hands, and hatch in the intestine. The larvae penetrate the intestinal wall, migrate to the lungs, climb up the respiratory tract to the throat, and are then swallowed again. The adults now attach to the intestinal wall and feed on the partially digested food, reaching a length of 25 cm (10 in.). After mating, a mature female may produce 200,000 eggs per day. These are shed in the feces and may remain infective in soil or sludge for several months.

Trichinosis is caused by another nematode, *Trichinella spiralis*. This roundworm is able to infect several species, but humans are exposed mainly by eating undercooked pork (or bear, in some places). The adult matures in the small intestine, with the female producing over 1000 larvae. These migrate throughout the body and then encyst in the muscle. Since human muscle tissue is not eaten, the larvae eventually die, and humans can be considered an accidental host. However, the disease may be very painful and is incurable. Fortunately, trichinosis has been virtually eradicated from the swine herds in the United States.

Prions   Prions (Section 10.8.2) are infective proteins that attack the brain of their host. The diseases they cause can be spread through ingestion of infected animal tissue, especially brain, containing the prions, and are always fatal. Cooking is generally considered inadequate to inactivate them. **Bovine spongiform encephalitis**, or **mad cow disease**, is probably the most widely known example, and apparently it can affect humans. **Scrapie** attacks sheep. Two human prion diseases are **Creutzfeldt–Jakob disease**, which some believe can be transmitted by ingesting beef products from "mad cows," and **kuru**, which is associated with ritualistic cannibalism among indigenous peoples of New Guinea. Perhaps as a result of the infected cattle herds in Great Britain, 81 people there and two in France have died.

12.4   AIR-TRANSMITTED DISEASES

Aerial disease transmission typically starts with an infected host's release of aerosol droplets (e.g., from a sneeze or cough) and subsequent reinhalation by a new host. Because of their mode of entry, it is usually the respiratory tract that becomes infected. Acute respiratory infections (e.g., pneumonia, whooping cough, meningococcal meningitis, and diphtheria) account for 4 million annual deaths worldwide just among children. Among adults, tuberculosis still is responsible for 3 million deaths and nearly 9 million infections a year, and influenza also takes a heavy toll.

12.4.1   Pneumonia

Streptococcus pneumoniae can cause infections deep within the lungs, attacking and inflaming the **alveoli** (the lung's minute, saclike surfaces at which oxygen transfer takes place). Without prompt treatment, this pneumococcal pneumonia may trigger a rapid accumulation of fluid in the lungs, severely impeding oxygen transfer and resulting in mortality rates of 30%. Even with treatment, up to 10% mortality is reported.

Pneumonia may also be caused by viruses, other bacteria (e.g., *Klebsiella pneumoniae*, an opportunistic pathogen), and some fungi (e.g., *Histoplasma*, *Aspergillus*).

12.4.2   Other Streptococcal Infections

Although *Streptococcus pyogenes* is found in the upper respiratory tract of many people, sometimes (virulent strains or weakened hosts) it causes the disease known as **strep throat**. In addition to a sore throat, this may lead to tonsillitis, and in some cases ear infections (otitis media). If not treated, some strains produce a toxin leading to damage of small blood vessels, a fever, and a rash, a disease known as **scarlet fever**. A few strains may produce **rheumatic fever**, which can lead to heart, kidney, and joint damage.

Figure 12.8   Bills of Mortality, published by the Company of Parish Clerks in London, England in 1665.

12.4.3   Tuberculosis

During the Middle Ages, tuberculosis was referred to as consumption, since it seemingly consumed its victims. England's routinely published Bills of Mortality (Figure 12.8) typically cited this disease as one of the leading causes of death. Even a century ago, tuberculosis was still one of the most important infectious diseases in the world, accounting for roughly 10 to 15% of all deaths.

The bacterial agent of this disease, *Mycobacterium tuberculosis*, was discovered in 1882 by Robert Koch, an achievement for which he was awarded a Nobel prize in 1905. Over the next several decades, this disease provided one of the principal motivations behind the drive to develop **antibiotic** drugs (a variety of substances, usually obtained from microorganisms, that inhibit the growth of or destroy certain other microorganisms).

Today, after massive efforts to eliminate this disease, the annual incidence in the United States remains in excess of 20,000 new cases, of which 10% may well be fatal. On a global basis, the toll is even more severe, with several million deaths per year. Furthermore, in recent years, there has been a dramatic reoccurrence of cases in the United States, which appears to be tied to the emergence of drug-resistant strains.

The incidence of tuberculosis is increased within areas in which numbers of people are living within tight quarters (e.g., military barracks, homeless shelters). An initial infection starts with the inhalation of airborne droplets or dust bearing viable cells, which then lodge deep within the lungs and begin their growth. Although acute **pulmonary** (lung) disease can develop, in the majority of cases these bacteria colonize the interior respiratory surface tissue in a vertical fashion, extending outward away from the lungs, building characteristic **tubercles** through which the victim's incoming oxygen must be channeled before reaching the lung's gas transfer surface. These tubercle formations ensure the availability of oxygen for these strict aerobes.

The World Health Organization has declared that tuberculosis presently qualifies as a global emergency. At its present rate of mortality, tuberculosis threatens to kill 30 million people in the next 10 years. Populations in developing countries are most at risk, but even in the United States some 15 million people are infected (although not all have the disease; ∼20,000 new cases are reported here each year). Aside from these frightful numbers, there is also evidence that drug-resistant strains of tuberculosis are multiplying in many areas, to the point where an incurable strain may eventually develop.

12.4.4 Influenza

Although generally perceived by the public to be a recurring, endemic, non-life-threatening problem, viral influenza ranks high on the list of diseases with epidemic potential. During years with high levels of influenza incidence, tens of thousands of fatalities may be experienced in the United States. During the World War I pandemic of 1918, it has been estimated that 40 to 50 million victims died worldwide, more than died directly from the war itself!

The traditional reservoirs for influenza include both humans and animals (especially chickens, ducks, and pigs). Close contact between these host species can provide an opportunity for mixing of the various strains in a single host. Unfortunately, the influenza virus readily undergoes genetic reassortment, leading to recombinant interactions that may develop as new strains of the influenza virus. The tendency in some parts of the Far East to maintain chickens, ducks, and pigs in or near residences is generally considered to be a reason why newer strains often develop in, and emanate from, this geographical area. Since the human immune system generally does not recognize the newer strains, they are able to cause a new epidemic of the disease.

However, virulent strains of influenza virus can also develop from genome mutation. One such event was documented in 1983 when a single mutation in a previously avirulent strain triggered a fatal chicken epidemic in Pennsylvania.

12.4.5 Diphtheria and Whooping Cough (Pertussis)

Diphtheria and pertussis, formerly important childhood bacterial respiratory diseases, have largely been eliminated in developed countries through widespread use of the DPT vaccine (effective against diphtheria, pertussis, and tetanus). Diphtheria is caused by *Corynebacterium diphtheriae*, infecting the nose, throat, and tonsils. In addition to producing an inflammatory response at the infected sites, surface lesions known as *pseudomembranes* characteristically form that can block the passage of air. Some strains also secrete toxins that kill the cells of the surrounding tissue.

Bordetella pertussis is the causative agent for whooping cough. This highly contagious disease, which attacks the upper respiratory tract, draws its name from the violent recurring cough that it triggers. Epidemics in developing countries have had fatality rates as high as 10 to 15%, and pertussis still has the tenth-highest rate of mortality in the world for infectious diseases, with roughly 100,000 deaths per year. This disease appears to be making a comeback in the U.S. due to dropping vaccination rates.

12.4.6   Meningococcal Meningitis

Neisseria meningitidis is part of the normal biota of the nasal cavity and throat in a fourth or more of the population, but these carriers show no symptoms of disease. For reasons not yet fully understood, these bacteria may invade the bloodstream and subsequently invade and colonize the **meninges** (the membranes that surround the spinal cord and brain). Early symptoms may include headache, fever, and vomiting, and death can quickly follow, due to the **endotoxin** produced (a toxin associated with the surface of the producing cells, usually gram-negative bacteria). Without proper care, the fatality rate can exceed 80%, but with treatment this can be reduced to below 10%.

12.4.7   Histoplasmosis, San Joaquin Valley Fever, and Aspergillosis (Respiratory Mycoses)

Mycoses (diseases caused by fungi) include several respiratory infections resulting from inhalation of viable spores of opportunistic pathogens that usually grow in soil. Usually, the host is immunocompromised, or in some cases is exposed to very high levels of airborne spores.

For example, *Histoplasma capsulatum*, the cause of histoplasmosis, is particularly common in areas contaminated with chicken or bat feces. The fungal growth develops inside the lung, in a fashion comparable to the tubercles formed during tuberculosis. Histoplasmosis is endemic in parts of the midwestern United States.

A similar lung disease known as San Joaquin Valley fever, caused by *Coccidioides immitis*, is seen more commonly in the southwestern regions of the United States. This fungus lives in desert soils, and in some localities the levels of asymptomatic infection may be as high as 80%.

Several species of *Aspergillus* can also cause a lung infection known as *aspergillosis*. *A. fumigatus* has been of particular concern around some sludge and yard waste composting sites, where it can apparently grow on woodchips, leaves, and other cellulosic materials. It can tolerate higher temperatures (50°C) than can most other fungi. However, the actual incidence of disease seems to be very low (few or no cases per year), although allergic reactions are more common.

12.4.8   Hantavirus Pulmonary Syndrome

Hantavirus pulmonary syndrome (HPS) was first noticed in the southwestern United States in 1993 by physicians treating victims with symptoms frightfully similar to those of Ebola (Section 12.7.9). Patients were suffering from an initial fever followed by the abrupt onset of acute pulmonary edema and shock. This outbreak involved 53 infections, with 32 fatalities. A rapid, systematic epidemiological study eventually determined that the responsible agent was a hantavirus, with the deer mouse as the principal reservoir. Infection usually stems from inhalation of aerosolized dried mouse feces or urine.

12.5   VECTOR-TRANSMITTED DISEASES

A number of diseases require transmission from host to host via another organism, often a blood-feeding insect or tick. This vector typically picks up the parasite when it bites an infected host, then injects it into the bloodstream of the new host with a subsequent bite. In many cases, the organism also replicates in the vector, increasing its chances of infecting the next human host successfully.

The environmental engineer or scientist may especially need to be aware of diseases transmitted by mosquitoes. Certain types of water projects (dams, canals) may increase or decrease the prevalence of the waterbodies in which various types of mosquito larvae develop. Also, the use of pesticides to control mosquitoes has both provided important public health benefits from reduced disease transmission and caused serious environmental impacts on nontarget species, including humans.

12.5.1 Malaria

Malaria is endemic to nearly 100 countries, and more than 40% of the world's population is at risk. This is the fourth most common infectious disease, with perhaps 300 million cases and 2 million annual fatalities. Many of the victims are children living in Africa south of the Sahara Desert who do not have access to prophylactic antimalarial drugs. Although malaria was once an endemic problem in the southeastern United States, the current U.S. incidence (∼1000 infections and one death per year) is largely associated with travel to infected areas.

Malaria has been recognized for centuries, being attributed originally to the dank, miasmic atmosphere associated with humid swampy areas. In fact, the name given to this disease, *mal aria* (literally, "bad air"), reflects this early hypothesis. The true culprit, however, is a sporozoan parasite, *Plasmodium*. The vectors are female mosquitoes of the genus *Anopheles*.

There are four species of *Plasmodium* linked with malaria; some lie dormant for several years, and others can be acutely lethal. Upon infection the protozoan proceeds to the liver, where it reproduces and infects red blood cells, eventually interfering with the body's ability to transfer oxygen. Symptoms, including fatigue, anemia, fevers or chills, and nausea, may be similar to those of flu or food poisoning, and in many instances they are mistakenly overlooked by travelers as jet lag. *Plasmodium vivax* can linger within a victim's liver for many years, causing recurring periods of incapacitating fatigue. Similar chronic problems can be caused by *P. ovale* or *P. malariae*. However, an infection by *P. falciparum* can be fatal within a single day following the initial onset of symptoms.

Drugs to counteract malaria were originally derived from a chemical known as quinine, extracted from the bark of trees found in the Amazon rain forest. For many years, the quinine derivative chloroquine provided a high degree of protection, but in recent decades its effectiveness has been compromised by the evolution of drug-resistant *Plasmodium* strains. A new generation of improved drugs has been devised, but resistance to these has also started to appear. The best control in many places, including the United States, has come from eradication of the insect vector, preventing transmission.

12.5.2 Trypanosomiasis (African Sleeping Sickness)

African sleeping sickness is a chronic and often fatal disease caused by the flagellated protozoans *Trypanosoma gambiense* (central Africa) and *T. rhodesiense* (eastern Africa). It is spread by blood-feeding tsetse flies and infects a variety of animals. The "sleep" is really a coma resulting from invasion of the brain. South American sleeping sickness, or Chaga's disease, is caused by *T. cruzi* and is spread by biting triatomid bugs.

12.5.3   Plague

There are three distinct clinical forms of plague caused by the bacterium *Yersinia pestis*: **bubonic** (affecting the lymph nodes), **pneumonic** (lungs), and **systemic** (entire body). In the bubonic form, which is the most common, the lymph nodes become painfully enlarged (particularly in the groin area), forming *buboes*. Symptoms include high fever, rapid and irregular pulse, hemorrhage, prostration, delirium, shock, coma, and death within 3 to 5 days. Another name for this disease, the Black Death, reflects the fact that plague hemorrhages often take on a blackish discoloration.

Although pneumonic plague is mainly spread through the air by droplets, bubonic plague transmission involves the rat flea as a vector. The disease is actually a zoonosis, with rats the primary host. Once enough rats die, however, the infected fleas look for other hosts, including humans.

Epidemics of this disease repeatedly killed more than a fourth of the population of Europe during the Middle Ages and thus is partly responsible for this period being referred to as the Dark Ages there. Only malaria and tuberculosis have been more deadly. The original onset of the Black Death midway through the fourteenth century has been linked to the Crusaders, whose return across the Mediterranean brought the black rat (*Rattus rattus*) as a hidden stowaway on their ships. The resulting first pandemic spread of plague throughout Europe and parts of Asia is believed to have killed as much as three-fourths of the population in a period of less than 20 years. The disease is still prevalent in some areas of the world (an outbreak occurred in India in 1994), but antibiotics have greatly reduced the mortality rate. The few cases that occur in the United States each year are mainly in the southwest, where the disease appears to be endemic (but usually not fatal) among ground squirrels and other wild rodents.

12.5.4   Typhus Fever

Typhus fever is caused by the intracellular bacterium *Rickettsia prowazekii*. It is spread from infected humans to new human hosts by lice. Before effective control of this biting insect vector, epidemics could be devastating. An epidemic in Europe during World War I, for example, killed 3 million people. Antibiotics are now another control method for the pathogen.

12.5.5   Rocky Mountain Spotted Fever

Rocky Mountain spotted fever (first reported in the Rocky Mountains, but actually more common in the east) is caused by *Rickettsia rickettsii*. This organism is related to the bacterium that causes typhus but is spread from human to human by ticks (hence its other name, tick-borne typhus), especially dog and wood ticks. Like typhus fever, it has a fairly high mortality rate (about 10%), but it does not produce epidemics.

12.5.6   Lyme Disease

Lyme disease (first reported in Lyme, Connecticut) also is spread through tick bites, but usually by the much smaller deer tick. It is caused by the spirochete *Borrelia burgdorferi*. It has a low mortality rate but can produce chronic disease if not treated properly.

12.5.7 Dengue

This is the world's most common mosquito-borne viral disease; almost half the world's population is at risk, with 20 million cases occurring a year in more than 100 countries. In 1995, the worst dengue epidemic in Latin America and the Caribbean for 15 years struck at least 14 countries, causing more than 200,000 cases of dengue fever and almost 6000 cases of the more serious dengue hemorrhagic fever.

Although this disease may occur in explosive epidemics, the rate of fatality is low. The disease is marked by acute onset, with subsequent fevers for 5 to 7 days, perhaps marked with headache, joint and muscle pain, and rash. Prolonged fatigue and depression are usually seen during the period of recovery.

12.5.8 Yellow Fever

Yellow fever is rare today in the United States and has been controlled in much of the world. However, this viral disease was once a major scourge, making many tropical and subtropical areas in the Americas virtually uninhabitable, with epidemics even reaching New York City. In 1793, Philadelphia (then the largest U.S. city, with 40,000 people) suffered nearly 5000 deaths, and 40% of the population evacuated the city. Victims suffer a high fever and jaundice (turn yellow) from infection of the liver. Most people considered yellow fever contagious, but finally in 1900 a team led by Walter Reed demonstrated that it was transmitted by mosquitoes. Disease incidence has been greatly reduced by control of this vector, and more recently by the development of a vaccine.

12.5.9 Rabies

Rabies is a fatal viral disease of animals that can also be transmitted to humans. Although it attacks primarily the central nervous system, infective particles are present in the saliva of rabid animals. Thus, it is most frequently transmitted by an animal bite, which injects the virus into the new host. About 35,000 cases a year are reported worldwide, all of them fatal, as there is no cure once symptoms develop. However, because of an effective vaccination program for dogs, cats, and other at-risk domestic animals and the availability of a vaccine for people bitten by wild animals or unvaccinated pets, there are typically fewer than five cases per year in the United States.

12.5.10 West Nile Encephalitis

A sometimes fatal encephalitis caused by the West Nile virus is endemic to areas such as Egypt and Israel, but in 1999 it also emerged in New York City. It is a zoonosis of birds, spread by mosquitoes that may also bite and infect humans. By 2004 it had spread to much of the United States.

12.6 SEXUALLY TRANSMITTED DISEASES

Most sexually transmitted pathogens are unable to survive in the environment for even relatively short periods (minutes). Thus, they can only be transmitted by close contact, such as occurs during sexual activity.

12.6.1   Syphilis

Syphilis is caused by the spirochete bacterium *Treponema palladium*. Historically, this has been one of the world's most frequent and widely spread communicable diseases, but its incidence in the United States and many other areas has been greatly reduced through a variety of public health measures (such as requirements for premarital blood tests) and the use of antibiotics. Still, there are an estimated 50 million new cases a year worldwide.

The usual mode of transmission for this disease is direct contact with an infective lesion during sexual activity (although a pregnant woman may also infect her fetus). The spirochete enters the new host through small breaks in the skin, multiplies at the site, and forms a sore known as a **chancre**. This primary stage is highly infective but heals on its own. However, by this time the bacteria have spread to other areas throughout the body, causing secondary lesions that are also infective. Eventually, sometimes after 20 or more years, the disease may enter a tertiary stage that can affect the heart or brain, and without treatment may be fatal almost 10% of the time.

12.6.2   Gonorrhea

Gonorrhea, which now is a much more common disease (250 million cases a year worldwide, 400,000 in the United States) than syphilis, is a bacterial disease caused by the gram-negative **diplococci** (spherical cells in pairs), *Neisseria gonorrhea*. It mainly affects the urethra, and in women also the reproductive organs [**pelvic inflammatory disease** (PID)], leading potentially to sterility. It is most often spread through contact with the mucous membranes of an infected person during sexual activity, although a newborn may also acquire it (usually, as an eye infection) from an infected mother during birth.

12.6.3   Chlamydial Infections

Chlamydia trachomatis is the most common cause of sexually transmitted disease (5 million cases per year in the United States). It can cause inflammation of the urethra and PID, with symptoms similar to a milder form of gonorrhea. It can also be transmitted to newborns during birth, leading to a potentially blinding eye infection and/or pneumonia.

12.6.4   AIDS

Acquired immune deficiency syndrome (AIDS) is a later stage of infection with the human immunodeficiency virus (HIV). The disease was first recognized in 1981, and the virus identified in 1984. A variety of cells can be infected, but helper T lymphocytes are a particular target. These cells play a critical role in the immune system, so that a major effect of the virus is a weakening of the body's immunity and hence greatly increased susceptibility to a variety of other diseases, including some cancers. It is transmitted through sexual activity or through contaminated blood (particularly from shared needles among intravenous drug users). It is believed that over 1 million people have been infected in the United States, with over 400,000 already dead. Worldwide HIV has reached more than 30 million people, and it is having a devastating effect in some areas, particularly in Africa. Although there is no cure or vaccination, a number of potent drug therapies have been developed that can control the disease, at least for now, in many persons. However, the treatment itself is debilitating, and it is too expensive for most people in developing countries.

12.6.5 Genital Herpes and Warts

Genital herpes is caused by the herpes simplex virus type 2, and genital warts are caused by papilloma virus. Both are incurable, although new drugs offer some control. Both also have been associated with cervical cancer.

12.6.6 Trichomoniasis

The flagellated protozoan *Trichomonas vaginalis* is the cause of trichomoniasis, which at 2.5 million cases per year is probably the second (after *Chlamydia*) most common sexually transmitted disease in the United States. It may also be spread by fomites, especially towels and clothing, but has no cyst stage and so cannot survive long outside a host. It mainly infects the vagina in women and the urethra in men.

12.6.7 Yeast Infections

A vaginal yeast infection, **candidiasis**, is caused by *Candida albicans*. Although it can be transmitted sexually, it also is often indigenous. Increased growth in the vagina, leading to symptoms, may occur during antibiotic treatment (of an unrelated bacterial infection) that inadvertently suppresses the normal vaginal bacteria (such as *Lactobacillus*). Birth control pills and immunosuppressant drugs are other possible factors.

12.7 OTHER DISEASES TRANSMITTED BY CONTACT

Injuries in which the skin has been cut, broken, or punctured can lead to direct infections by a variety of pathogens, as can contact with blood or other bodily fluids. Only some of those not listed above are included here.

12.7.1 Tetanus

The infectious agent for tetanus, *Clostridium tentani*, is an anaerobic sporeformer that is common in soil. It enters the body through a skin wound or puncture, then grows and releases a powerful **exotoxin** (a toxin released into the surrounding "medium" by the producing organisms, usually gram-positive bacteria) that specifically attacks the nervous system. If not properly treated, the damaged nerves can result in muscle spasms that are usually fatal. One symptom is clenched teeth, leading to the common name, **lockjaw**. The DTP (diphtheria, tetanus, and pertussis) inoculations given routinely in childhood, with 10-year boosters for tetanus, provide immunity to this disease.

12.7.2 Gangrene

Clostridium perfringens can also invade a wound or puncture site and grow in the surrounding dead tissue. Gas is released as a product of fermentation, and along with enzyme activity, further damages cells. The gas pressure may stop blood flow, leading to blackening of the tissue. The toxins released may prove fatal without treatment, and in severe cases, amputation of an affected limb may be necessary.

12.7.3 Trachoma

Chlamydia trachomatis, in addition to being the cause of the most widespread sexually transmitted disease (Section 12.6.3), also causes trachoma, an infection of the eye that is the world's leading cause of blindness. This form of the bacteria is spread by direct contact (often from touching infected liquid from the eye) or fomites, such as towels.

12.7.4 Bacterial Conjunctivitis

Bacterial conjunctivitis, commonly known as **pinkeye**, may be caused by several bacteria, especially *Haemophilus aegyptius* (also called *H. influenzae*) and *Moraxella lacunata*. The primary mechanism for its dissemination is finger-to-eye contact with discharges from an infected host's eye or fomites, although improperly disinfected swimming pools may also be implicated.

12.7.5 Anthrax

Anthrax is mainly a disease of animals. It is relatively uncommon in industrialized countries and is primarily an occupational concern for workers handling infected hides, bone, and wool of sheep, cattle, goats, horses, and pigs. It can be spread by direct contact through punctures or cuts in the skin and has a mortality rate of about 10%. As indicated in Section 12.1.1, this is the disease for which, in 1876, Koch's principles were first demonstrated.

However, because the causative agent, *Bacillus anthracis*, is a bacterial endospore-former and easily grown in culture in the laboratory, anthrax has been of increased concern as a biological warfare agent. If dispersed in the air and inhaled in high amounts, virulent strains cause a respiratory disease that is rapidly fatal in virtually 100% of unvaccinated victims.

12.7.6 Leprosy

Leprosy, or **Hansens's disease**, is caused by *Mycobacterium leprae*. This slow-growing bacteria may affect the body in several ways, especially causing disfigurement and nerve damage. Although it is not usually fatal, it can be severely disabling and has long led to discrimination against its sufferers. It is spread through prolonged close contact. Treatment is now possible with antibiotics.

12.7.7 Athlete's Foot and Ringworm

A **dermatomycosis** is a fungal infection of the skin, hair, or finger- or toenails. Athlete's foot and ringworm are two common examples. They may be caused by several species of deuteromycetes or ascomycetes, including *Epidermophyton* and *Trichophyton*. They can survive on moist surfaces (such as shower tiles) and in materials such as bath mats and towels for days or more. Ringworm also is spread by direct contact (e.g., among wrestlers) and by infected animals such as cats.

12.7.8   Hepatitis B

Hepatitis B, or **serum hepatitis**, is spread mainly through contaminated blood, often from unsterilized needles shared by drug users or used for tattoos or ear or body piercing. The virus can also be transmitted sexually. Over 100,000 people are infected yearly in the United States, but this number is decreasing due to the recent introduction of a vaccine. In addition to the initial disease, which is more severe than hepatitis A (more liver damage and fatality rate of 10%), those infected are at higher risk of liver cancer.

12.7.9   Ebola

The **index** (first noted) case for the Ebola virus took place in Zaire in 1976, with more than 300 victims of this severe **hemorrhagic** (bleeding) fever. A second outbreak in Sudan killed another 150 people, and other, usually smaller outbreaks have since been reported, mainly in east, central, and southern areas of Africa. Ebola virus spreads through the body, with rapid **necrosis** (death) of cells in the infected organs, particularly those of the liver, lymph system, kidneys, ovaries, and testes. Patients usually have sustained high fevers and may become delirious and difficult to control. After about a week, profuse bleeding occurs throughout the body, and 90% of patients die.

It is believed that animals, including monkeys and perhaps also bats, are the main reservoir for the virus. Ebola outbreaks result from person-to-person transmission involving close contact, probably with infected blood or other fluids, although the exact route of transmission is unknown. Unfortunately, nosocomial spread typically occurs before the disease is recognized, and hospital staff members are often among the early victims.

12.8   CONTROL OF INFECTION

A variety of methods have been used to help prevent the spread of disease. These involve a variety of physical steps taken outside the human body to prevent transmission, ways to make a person immune to a disease, and chemicals administered to control a microbial agent or its products once they are inside a host.

12.8.1   Physical Steps to Prevent Transmission

General sanitation is an important step in preventing transmission of many diseases. Proper disposal of sewage and adequate treatment of potable water, including filtration and disinfection, have gone a long way to limit waterborne disease in many areas. Adequate cooking, proper refrigeration, and thorough handwashing have been important steps in controlling foodborne outbreaks. **Pasteurization** (heat treatment to kill pathogens) or complete **sterilization** (killing all microorganisms) of milk and many other foods has also had a major role in reducing disease. Vector control has been effective for many diseases, and pesticides, despite their potential other risks, certainly have played an important role in this regard. Use of condoms reduces the risk of spreading sexually transmitted diseases. Sterile surgical equipment, bandages, and other hospital equipment and supplies and the use of surgical masks and gloves has greatly reduced infections from medical procedures. **Quarantine** (forced isolation of infectious persons) still plays an important role in the control of some diseases. These are only a few examples of physical changes in the environment that can help in controlling disease transmission. Maintaining a healthy, normal microbial biota in and on the body can also help prevent invasion by a pathogen.

12.8.2 Immunity and Vaccination

Immunity (the ability to resist infection based on mobilization of the immune system) to many diseases can result from a prior infection of the same agent. Getting the measles, for example, protects the host from being infected again later. Thus, a person can contract many diseases only once. Colds and influenza, on the other hand, stem from viruses that continue to produce new strains that avoid the body's predeveloped defenses, so that they may be contracted repeatedly. Prior exposure also does not protect against many microbial toxins, such as those involved in botulism food poisoning or against some parasitic infections, such as schistosomiasis, tapeworm, and athlete's foot.

Infection by a similar agent can in some cases provide protection from a disease. Contracting cowpox, for example, protected people from later getting the much more serious disease, smallpox. Similarly, it appears that having yaws (a disease caused by the spirochete bacterium *Treponema pertenue* and spread by direct contact) offers protection from syphilis (caused by *T. pallidum*).

Acquiring immunity through **vaccination** has played a major role in the control of a variety of diseases, particularly viral infections that are otherwise untreatable. This involves exposing a person to a material that will stimulate immunity to a particular agent. The material used may be a killed, inactivated, or weakened strain of the agent, a portion of its surface, one of its products, or a closely related agent.

One of the greatest successes of vaccination has been with **smallpox**. This once devastating disease killed millions in the Old World and scarred countless others. It was then brought to the New World, where it killed millions of Native Americans, in some cases wiping out entire cultures. In fact, the infection was spread intentionally to some tribes by European invaders who distributed infected blankets (an early example of biological warfare). Vaccination with material from the pustules (pox) of victims was practiced in Asia for at least several centuries, but sometimes lead to serious infection. In 1798, Edward Jenner reported on his experiments and observations in England involving cowpox, a related but mild disease of cows and milkmaids. He developed a vaccine based on this virus that provided immunity to smallpox. Vaccination was so successful that by 1966, the World Health Organization undertook a program to eradicate smallpox worldwide. In part because humans are the only known reservoir, the aggressive **surveillance** (tracking) and vaccination programs succeeded—the last "natural" case of smallpox occurred in 1977 (a small outbreak occurred among laboratory workers in England the following year). Since the disease no longer exists (the virus has been stored in two laboratories, in Atlanta and Moscow), vaccination has now been discontinued.

Another great success of vaccination has been with polio. This viral disease has now been eliminated in the Americas, the western Pacific, and Europe, and was scheduled for worldwide eradication by the end of 2002. (This target was missed, with 12 countries in Asia and Africa still not virus free as of early 2005; see www.who.org for updates.)

Table 12.8 shows the vaccinations approved for use in the United States, with those recommended for children and the general population indicated by footnote *a*. Some, such as DTP (diphtheria, tetanus, pertussis), are typically given as combined vaccinations, although individual inoculations or other combinations are possible (marked as footnote *b*). The others are designed for special populations that are at higher risk of a particular

TABLE 12.8   Licensed Vaccines and Toxoids Available in the United States

Vaccine	Type	Route of Administration
Adenovirus	Live virus	Oral
Anthrax	Inactivated bacteria	Subcutaneous
Bacillus of Calmette and Guerin (BCG)	Live bacteria	Intradermal/ percutaneous
Cholera	Inactivated bacteria	Subcutaneous or intradermal
Diphtheria–tetanus–acellular pertussis (DTaP)[b]	Toxoids and inactivated bacterial components	Intramuscular
Diphtheria–tetanus–pertussis (DTP)[a]	Toxoids and inactivated whole bacteria	Intramuscular
DTP–*Haemophilus influenzae* type b conjugate (DTP-Hib)[b]	Toxoids, inactivated whole bacteria, and bacterial polysaccharide conjugated to protein	Intramuscular
Haemophilus influenzae type b conjugate (Hib)[a]	Bacterial polysaccharide conjugated to protein	Intramuscular
Hepatitis B[a]	Inactive viral antigen	Intramuscular
Influenza	Inactivated virus or viral components	Intramuscular
Japanese encephalitis	Inactivated virus	Subcutaneous
Measles[b]	Live virus	Subcutaneous
Measles–mumps–rubella (MMR)[a]	Live virus	Subcutaneous
Meningococcal	Bacterial polysaccharides of serotypes A/C/Y/W-135	Subcutaneous
Mumps[b]	Live virus	Subcutaneous
Pertussis[b]	Inactivated whole bacteria	Intramuscular
Plague	Inactivated bacteria	Intramuscular
Pneumococcal	Bacterial polysaccharides of 23 pneumococcal types	Intramuscular or subcutaneous
Poliovirus vaccine, inactivated (IPV)	Inactivated viruses of all three serotypes	Subcutaneous
Poliovirus vaccine, oral (OPV)[a]	Live viruses of all three serotypes	Oral
Rabies	Inactivated virus	Intramuscular or intradermal
Rubella (German measles)[b]	Live virus	Subcutaneous
Tetanus[b]	Inactivated toxin (toxoid)	Intramuscular
Tetanus–diphtheria (Td or DT)[b]	Inactivated toxins (toxoids)	Intramuscular
Typhoid (parenteral)/ (Ty21a oral)	Inactivated bacteria/live bacteria	Subcutaneous/oral
Varicella (chickenpox)[a]	Live virus	Subcutaneous
Yellow fever	Live virus	Subcutaneous

[a]Part of standard recommended vaccination program for children.
[b]Alternative form of recommended vaccination for children.

disease based on occupation (including the military), travel to areas where the disease is endemic, or other factors. Most vaccinations are given as an injection, but a few are taken orally.

In some cases, the immune system can be "helped" by injection of an immune globulin, or antibody. Approved immune globulins are listed in Table 12.9.

TABLE 12.9   Immune Globulins and Antitoxins Available in the United States

Immunobiologic	Indication(s)
Botulinum antitoxin	Treatment of botulism
Cytomegalovirus immune globulin, intravenous (CMV-IGIV)	Prophylaxis for bone marrow and kidney transplant recipients
Diphtheria antitoxin	Treatment of respiratory diphtheria
Immune globulin (IG)	Hepatitis A pre- and postexposure prophylaxis; measles postexposure prophylaxis
Immune globulin, intravenous (IGIV)	Replacement therapy for antibody deficiency disorders; immune thrombocytopenic purpura (ITP); hypogammaglobulinemia in chronic lymphocytic leukemia; Kawasaki disease
Hepatitis B immune globulin (HBIG)	Hepatitis B postexposure prophylaxis
Rabies immune globulin (HRIG)	Rabies postexposure management of persons not previously immunized with rabies vaccine
Tetanus immune globulin (TIG)	Tetanus treatment; postexposure prophylaxis of persons not adequately immunized with tetanus toxoid
Vaccinia immune globulin (VIG)	Treatment of eczema vaccinatum, vaccinia necrosum, and ocular vaccinia
Varicella-zoster immune globulin (VZIG)	Postexposure prophylaxis of susceptible immunocompromised persons, certain susceptible pregnant women, and perinatally exposed newborn infants

12.8.3   Antibiotics and Antitoxins

Alexander Fleming in 1928 made a serendipitous discovery that *Staphylococcus* bacteria would not grow in the presence of the fungus *Penicillium notatum*. Eventually, the active ingredient, **penicillin**, was isolated and since 1941 it has been of great value in the control of certain bacterial diseases, especially syphilis, pneumococcal pneumonia, and some staphylococcal and streptococcal infections. Shortly thereafter, in 1943, Selman Waksman and Albert Shatz found streptomycin, produced by the actinomycete *Streptomyces griseus*, which was effective against tuberculosis and many gram-negative bacteria. In addition to many other antibacterials, compounds active against some fungi and protozoa have since been developed, and most recently there has been success with antiviral compounds. The term **antibiotic** dates from 1889, when Paul Vuillemin used it for a chemical compound produced by *Pseudomonas aeruginosa* that was inhibitory to other bacteria.

An antibiotic is a chemical that can be used to control the activity of selected microorganisms, especially within the body (as opposed to a **disinfectant**, which is used outside the body and which is not selective). Specific antibiotics are typically effective against a particular range of pathogens. A wide array of antibiotics now exist, with many produced by other organisms but some produced synthetically. Antibiotics have become a major tool in the control of disease. However, strains of pathogens that are resistant to our present antibiotics have emerged, and the search for new ones thus continues with a sense of urgency.

In some cases it is a toxin produced by the microbial agent that is a major cause of disease. Chemicals that are able to prevent or counteract these toxic effects are referred to as **antitoxins**. A few such compounds have been developed, including those effective against botulism and diphtheria toxin (Table 12.9).

12.9   INDICATOR ORGANISMS

How do we know that the treated potable water entering the distribution system or coming out of a residential tap is microbially safe to drink or bathe with? Has the disinfection of a wastewater been effective? Is the water at the beach "swimmable"? Can shellfish be harvested from this bed?

Ideally, we might like to answer these questions by determining the concentration of each specific pathogen and parasite in the water. However, this is clearly impractical for a number of reasons, including the very large number of possible disease agents and the resulting enormity of the job and tremendous expense. Further, it is likely that not all pathogens are known yet, and even for some of those that are recognized there is no method to quantify them. Perhaps in the future, molecular techniques or other advances may make this goal more nearly attainable, but for now it is necessary to rely on some other approach.

12.9.1   Physical–Chemical Indicators

As a practical matter, there is a need for some **real-time measures** of the microbial quality of drinking water and the effectiveness of disinfection. Tests that take several days, or in some cases even several hours, may be too slow to indicate that a problem has occurred during treatment. By that time a drinking water may already have been consumed, or a wastewater already discharged. The **turbidity** (cloudiness) of a water can serve as one useful measure, since it can **indicate** failure of previous treatment steps, and since the small particles imparting turbidity might harbor pathogens. The **chlorine residual** (concentration of active chlorine products remaining in the system) may also be of use where chlorination for disinfection is practiced. As another example, during autoclaving (moist heat sterilization) of some laboratory and hospital supplies, tape impregnated with a heat-sensitive chemical can be used to indicate that the appropriate temperature was maintained.

12.9.2   Microbiological Indicators

Where real-time measures are not essential, or for use in confirming their validity, it is logical to consider monitoring of **indicator organisms** as surrogates for pathogens themselves. For example, certain bacteria present in feces might be used to indicate the presence of fecal pollution and hence the likelihood that pathogens are present. Other organisms might be appropriate for other purposes, such as for evaluating swimming pools.

Because a major public health concern traditionally has been contamination of drinking, shellfish harvesting, and recreational waters with sewage, considerable effort has gone into developing methods for microbial indicators of fecal contamination in such waters. However, although several of these tests are used routinely and are generally reliable, it is important to remember the purpose for which they were developed. These standard indicators may not be appropriate for all applications, especially those for which they were not envisioned, such as microorganisms in aerosols.

Characteristics of an Ideal Indicator Organism   The ideal microbial indicator of fecal contamination would have a number of characteristics. (Note that similar characteristics could be enumerated for other types of indicators as well.) First, it should be present in human feces in high numbers, making the test sensitive and thereby avoiding **false negative** results (in which a test says that contamination is absent even though it is really present). In fact, the ideal would be that the indicator to pathogen ratio would be high and stable. However, it is recognized that a constant ratio, at least, is not feasible, since the concentration of pathogens will vary based on the rate of infection in the population.

On the other hand, the indicator organism should be absent from uncontaminated materials (avoiding **false positive** results). In many cases, though, its presence in the feces of other warm-blooded animals would be desirable, since some diseases can be spread from animal to humans, but this would depend on the particular application.

The indicator organism should also die off in the environment at a slightly slower rate than the pathogens. Dying too quickly could lead to false negatives, while dying too slowly would lead to false positive results.

In terms of the actual testing, it would be very nice if rapid, easy, and inexpensive methods were available. Also, it would be preferable that the indicator organism itself is not pathogenic. After all, we do not want a large number of labs growing up vast quantities of potentially disease-causing organisms.

As might be expected, no ideal indicators are known. However, a number of methods have been developed that meet the criteria above for indicators of fecal contamination or other purposes to a useful extent. Several of these are described below. An important source for the detailed procedures is *Standard Methods for the Examination of Water and Wastewater* (Clesceri et al., 1998).

Total Coliforms   **Total coliforms** are aerobic or facultatively anaerobic, gram-negative, nonsporeforming rods that ferment lactose (milk sugar) with acid and gas production (gas production not considered in an alternative procedure) within 48 hours (24 hours in the alternative test) at 35°C. This is an **operationally defined** group of microorganisms: that is, organisms that show up as positive in the test are defined as members of the group. The test is designed to include selected members of the bacterial family Enterobacteriaceae, especially many *E. coli*. This species is common (but not predominant) in the intestinal tract of humans and other warm-blooded animals. Also, it is closely related to *Salmonella* and *Shigella*, two of the bacteria historically of greatest concern for waterborne disease; thus its die-away in the environment is likely to be similar. However, not all *E. coli* are coliforms (e.g., some strains cannot ferment lactose), and not all coliforms are *E. coli*. Examples of other members of the family Enterobacteriaceae that may include coliforms are *Enterobacter*, *Klebsiella pneumoniae*, and *Citrobacter*. Additionally, some coliforms can grow and survive in soil or water. Also, while they serve as reasonably good indicators for *Salmonella* and *Shigella*, their survival of disinfection and various environmental factors may differ substantially from that of some other disease agents, including other bacteria, protozoa, and viruses.

Still, total coliforms are one of the most useful indicator groups for their intended purpose. About 10^6 mL^{-1} are commonly found in raw sewage. (In a practice dating from the early years of the field, microbial counts in water are often expressed "per 100 mL"; the same unit also is sometimes used for wastewater, so that this number of coliforms would be 10^8 per 100 mL.) They are used routinely to monitor the microbial safety of drinking water supplies. Total coliform counts also may be helpful in evaluating the fecal contamination of some foods and the disinfection efficiency for wastewaters. Both an MPN and an alternative membrane filtration procedure are available (Section 11.5.2).

Fecal Coliforms   Since not all coliforms are of fecal origin, it would be useful to have a test that was more selective and included only fecal bacteria. The **fecal coliform** test attempts to achieve this mainly by incubation at 44.5°C, hoping by the higher temperature to eliminate bacteria that are better adapted to the lower soil and water temperatures of the ambient environment. The test is partially successful in meeting its goals. It is estimated that about 90% of the coliforms that grow at 44.5°C are of fecal origin (10% false positives); however, only about 90% of the coliforms of fecal origin can grow at the elevated temperature (10% false negatives). Additionally, precise control of the temperature is critical, as even 0.5°C higher can prevent the growth of many fecal coliforms, while slightly lower temperatures allow the growth of more nonfecal organisms.

Fecal coliforms are considered good indicators for outdoor swimming pools and recreational waters. (See also the fecal coliform/fecal strep ratio, below.) Again, both MPN and membrane filtration methods can be used.

Fecal Streptococci/Enterococci   Although also operationally defined, the **fecal streptococci** consist mainly of *Streptococcus faecalis*, *S. faecium*, *S. avium*, *S. bovis*, *S. equinus*, and *S. gallinarum*; the **enterococci** include only *S. faecalis*, *S. faecium*, *S. avium*, and *S. gallinarum*. The first two species tend to be more common in humans and rats, whereas the others tend to be more common in chickens, cattle, horses, and domestic fowl (birds), respectively. However, they are not truly host-species specific.

The fecal streptococci are common inhabitants of the intestinal tracts of warm-blooded animals. In fact, humans are atypical in that they usually have a greater number of coliforms than fecal streptococci; in most other species, the fecal streptococci are more numerous. Thus, for fresh contamination (the ratio changes with age), a fecal coliform/fecal streptococci ratio of ≥ 4 may be considered indicative of human pollution, whereas a ratio ≤ 0.7 suggests an animal source.

The enterococci are now usually considered the best indicator for recreational surface waters, particularly ocean beaches. Once again, both MPN and membrane filtration methods are available. Incubation is at 35°C for 24 to 48 hours, depending on the test used.

Heterotrophic Plate Count   The **heterotrophic plate count** (HPC), sometimes called the *standard* or *total plate count*, uses a nonselective medium in an attempt to include as many of the bacteria present as possible. Incubation is at 35°C for 48 hours or at 20 to 28°C for 5 to 7 days. One use of the HPC is that increases can indicate problems in water treatment processes or potable water distribution systems. High counts in indoor pools indicate poor disinfection and/or overuse or inadequate cleaning. Available tests include pour plate, spread plate, and membrane filtration methods.

Specific Organisms   In some cases, tests for specific organisms may be desirable. For example, in indoor swimming pools and whirlpools or spas, tests for *Pseudomonas aeruginosa* and *Staphylococcus* are recommended. These organisms serve not only as indicators of microbial loadings from skin and mucous membranes, but are themselves pathogens of potential concern. (HPC serves as a recommended indicator of disinfection efficiency for these systems.) *Clostridium perfringens*, which as an endosporeformer can persist for long periods in the environment, has been used to monitor the dispersal of wastewater treatment sludges disposed of in the ocean (now banned in the United States). Some coliphage (viruses that infect *E. coli*) have been used to monitor disinfection. Coliphage F2 has been found to be very heat resistant and was thus used as a **conservative indicator** (one that provides an extra margin of safety) during the development of criteria for pathogen destruction during composting. In monitoring microbial aerosols generated during composting operations, *Aspergillus fumigatus* has sometimes been used, since its spores will survive longer than coliforms, and since it is itself of concern.

These examples are meant to be illustrative rather than exhaustive. Remember, the available indicators were developed for certain purposes and should not be adopted arbitrarily for other situations.

PROBLEMS

12.1. *Bulking* is a "disease" of activated sludge wastewater treatment plants (see Section 16.1.3) in which the excessive growth of filamentous organisms interferes with normal settling. *Escherichia coli* is commonly found at high concentrations (e.g., $10^6 \, \text{mL}^{-1}$) in activated sludge, and sometimes grows in short chains when grown in the laboratory on agar medium. This led some early investigators to suggest erroneously that *E. coli* was the cause of bulking. How might knowledge of Koch's principles have helped these researchers avoid this error?

12.2. The hot-water system at a motel contains $3.0 \times 10^5 \, \text{mL}^{-1}$ viable cells of *Legionella pneumophila*. A guest takes an 8-minute shower with the water running at 2.5 gal/min, 60% of it hot water. Assuming that 0.1% of the cells present are aerosolized and that 1% of those are inhaled, is this person likely to develop Legionnaires' disease if the infective dose is 100,000 cells? (*Note:* The actual infective dose for Legionnaires' disease is not known.)

12.3. With regard to the fecal–oral route of waterborne disease, what steps can be taken to break the chain of infection?

12.4. What factors make total coliforms a good indicator for monitoring of drinking water? What are some of the group's limitations as an indicator for this purpose?

12.5. What are some of the points that should be considered in selecting indicator organism groups to be used for monitoring a public swimming pool?

REFERENCES

Atlas, R. M., and L. C. Parks (Eds.), 1997. *Handbook of Microbiological Media*, 2nd ed., CRC Press, Boca Raton, FL.

Centers for Disease Control and Prevention. www.cdc.gov.

Clesceri, L. S., A. E. Greenberg, and A. D. Eaton (Eds.), 1998. *Standard Methods for the Examination of Water and Wastewater*, 20th ed., American Public Health Association, Washington, DC. Updated regularly and available at www.wef.org.

Murray, P. R . (Ed.-in-Chief), 1999. *Manual of Clinical Microbiology*, ASM Press, Washington, DC.

Scheld, W. M., W. A. Craig, and J. M. Hughes (Eds.), 2001. *Emerging Infections*, Vol. 5. ASM Press, Washington, DC. See also Vols. 1 to 4.

World Health Organization. www.who.org.

13

MICROBIAL TRANSFORMATIONS

In this chapter we focus on the role that microorganisms play in the cycling of elements important to life. Unlike the organisms themselves, these atoms never "die" or disappear. Instead, they continuously pass through a grand set of **biogeochemical cycles** whose mechanisms, although carried out at times on a minute scale, are vital to life. Quantitative aspects of these cycles, and their interactions with other parts of ecosystems, are described in Chapter 14.

These biogeochemical cycles include a variety of metabolic pathways as well as abiotic reactions that continuously replenish the chemical ingredients of life. Microbes are the agents of most of these reactions, often through adding electrons to and removing electrons from atoms within a sequence of redox reactions. Often, it is only microbes that can carry out a specific reaction. In some other cases, the reactions might still occur without microorganisms, but only at much lower rates.

Atoms found in living organisms, whether macro, micro, or trace ingredients, were at one time in inorganic, nonliving materials and eventually will be released from their assimilated organic forms and returned to an inorganic state. Thus, the microbial reactions not only create essential materials (e.g., reduced nitrogen compounds) but also eliminate unwanted, and potentially harmful, residuals (e.g., waste and decay products generated through the normal course of life and death).

In a few cases, essential elements occur almost entirely in one oxidation state. Phosphorus, which occurs almost solely (exceptions include some pesticides and nerve gases) in the form of phosphate, is a prime example. These elements are still potentially cycled between organic and inorganic forms, but through hydrolysis or other such reactions, not through oxidation or reduction.

Environmental Biology for Engineers and Scientists, by David A. Vaccari, Peter F. Strom, and James E. Alleman
Copyright © 2006 John Wiley & Sons, Inc.

Biogeochemical cycles are not maintained by any single species; rather, they represent the collective effects of diverse, yet inherently coordinated, microbial consortia as well as interactions of microbes with higher organisms. In the following sections we examine the specific elements carbon, nitrogen, sulfur, iron, and manganese and their cyclic metabolic transitions, while also discussing oxygen and hydrogen.

In each case, these cycles provide an endless exchange of oxidation states through which these atoms are transformed, both oxidatively and reductively (Section 5.1.2). The oxidation state assumed by any given atom reflects the level and mode of electrons that it shares with its neighbors. For example, carbon has a range of oxidation states from -4 to $+4$, as dictated by its **atomic number** (AN), 6. This means that carbon has six **protons** (positive charge) in the nucleus (equal to its atomic number) and six **electrons** (negative charge) outside the nucleus, balancing the atom's charge.

The nucleus of an atom (except most hydrogen nuclei) also contains neutral particles, called **neutrons**. The sum of the protons and neutrons is the **atomic weight** of the atom. Carbon usually has six neutrons, so that its usual atomic weight is 12 (six protons + six neutrons). However, some forms, or **isotopes**, of carbon have seven or eight neutrons, so that they have an atomic weight of 13 or 14. These forms are relatively rare, so that on average, carbon typically has an atomic weight of 12.011.

The electrons surround an atom's nucleus in **orbital shells**. The first shell for any element can only have a maximum of two electrons, while the next can have a maximum of eight. (The third shell can have up to 18, but is also stable with only eight.) Therefore, after subtracting the two electrons allocated to its first shell, carbon's second orbital contains the four remaining electrons. In this configuration (six electrons along with the six protons in the nucleus), elemental carbon's net electrical charge (oxidation state) would then be zero. In fact, in elemental form, the oxidation state of all elements is zero. This oxidation state will change, though, as the outer shell either gains or loses electrons while bonding with adjacent atoms. For example (Figure 13.1), the carbon atom found in a methane molecule, CH_4, effectively pulls four additional negatively charged electrons into its

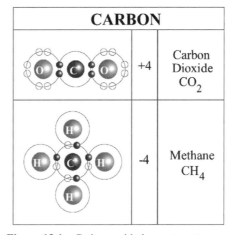

Figure 13.1   Carbon oxidation state extremes.

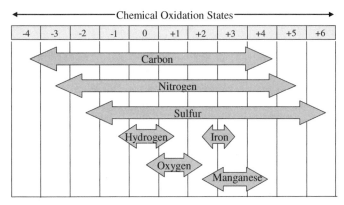

Figure 13.2 Redox ranges for some elements of major biochemical importance.

outer shell (one from each of the four hydrogens), thereby changing its oxidation state to −4 (10 negatively charged electrons versus six positively charged protons). Conversely, the carbon found in carbon dioxide (CO_2) essentially lends all four of its outer shell electrons to a pair of surrounding oxygen atoms, thereby shifting its oxidation state to +4.

Carbon, nitrogen, and sulfur are the three main elements of biological interest with a full eight-electron range in oxidation state. As shown in Figure 13.2, the ranges for these three elements (C, N, and S) are each, respectively, shifted one electron higher.

Table 13.1 shows the common oxidation states of some of the other elements of biological importance. Oxygen's (AN = 8) outer shell has six electrons, producing an aggressive tendency to add two additional electrons (thereby filling its outer shell) rather than releasing any to other adjacent atoms. This characteristic qualifies oxygen as having a high degree of electronegativity (Section 3.2). As a result, oxygen's metabolic range of oxidation states only includes −2 (e.g., H_2O), besides 0 (diatomic oxygen, O_2). Hydrogen (AN = 1) has possible states of −1, 0, and +1, since its single outer shell has only one electron of the two it would take to fill it. In living matter it typically donates its electron (thus, oxidation state = +1) to a more electronegative (e.g., oxygen, carbon, sulfur, nitrogen) receptor. Metabolic iron (AN = 26), occurs mainly in the +2 (ferrous) and +3 (ferric) oxidation states, while manganese (AN = 25) is found mainly as the +2 (manganous) and +4 (manganic) forms. Chlorine (AN = 17) as a disinfectant in water (HOCl, hypochlorous acid) has an oxidation state of +1, but as chloride or in organic compounds it is −1.

13.1 CARBON

Carbon can be converted from inorganic to organic form by a variety of microorganisms and by plants. It then is transformed, by anabolic and catabolic processes (Section 5.4), to

TABLE 13.1   Common Oxidation States of Some Biologically Important Elements

Element	Symbol	Atomic Number	Atomic Weight	Common Oxidation States
Aluminum	Al	13	26.98	0, +3
Arsenic	As	33	74.92	+3, +5
Bromine	Br	35	79.90	−1
Cadmium	Cd	48	112.40	+2
Calcium	Ca	20	40.08	+2
Carbon	C	6	12.01	−4, −3, −2, −1, 0, +1 +2, +3 +4
Chlorine	Cl	17	35.45	−1, 0, +1
Chromium	Cr	24	52.00	0, +3, +6
Cobalt	Co	27	58.93	+2, +3
Copper	Cu	29	63.54	0, +1, +2
Fluorine	F	9	19.00	−1
Hydrogen	H	1	1.01	−1, 0, +1
Iodine	I	53	126.90	0, −1
Iron	Fe	26	55.85	0, +2, +3
Lead	Pb	82	207.2	0, +2, +4
Magnesium	Mg	12	24.31	+2
Manganese	Mn	25	54.94	+2, +4, +7
Mercury	Hg	80	200.59	0, +1, +2
Molybdenum	Mo	42	95.94	+6
Nickel	Ni	28	58.69	+2, +3
Nitrogen	N	7	14.01	−3, −1, 0, +1, +2, +3, +5
Oxygen	O	8	16.00	−2, 0
Phosphorus	P	15	30.97	+5
Potassium	K	19	39.10	+1
Selenium	Se	34	78.96	−2, +4, +6
Silica	Si	14	28.09	+4
Sodium	Na	11	22.99	+1
Sulfur	S	16	32.06	−2, 0, +2, +4, +6
Tin	Sn	50	118.69	+2, +4
Vanadium	V	23	50.94	+5
Zinc	Zn	30	65.38	0, +2

a tremendous array of biologically important compounds in all living things. Some of these compounds are released as waste, or eventually the organism itself dies. They are then broken down, mainly by microorganisms, and eventually the carbon returned to inorganic form.

With carbon's central importance to all known forms of life, perhaps it is not surprising that the redox reactions through which these atoms are cycled (Figure 13.3) have such a high degree of breadth, complexity, and variability. In fact, its role as the dominant element in living organisms appears to stem from the great diversity of its bonding options. It typically forms four bonds with adjacent atoms. This is more than hydrogen (one bond) or oxygen (two), but not really different from nitrogen (three or four), and less than phosphorus (five) or sulfur (two to six). However, its great versatility also

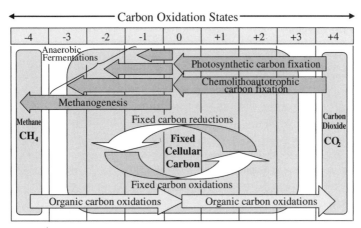

Figure 13.3 Biochemical carbon transformations.

stems from the fact that it can occur in all of the oxidation states from -4 to $+4$ (Table 13.2).

Organic Carbon   What exactly is meant by *organic carbon*? The term dates to the time that such compounds were considered to be produced only by organisms. German Freidrich Wöhler is generally credited with the first synthesis of an organic compound, urea, in 1928. Now, of course, there are many synthetic organic compounds. So again, what does *organic* mean?

Interestingly, most chemistry texts do not provide a real definition. A dictionary may state that organic chemistry is the study of carbon compounds, but carbon dioxide is not considered organic (although it is produced by organisms). Another definition that has been used is "compounds containing a C—C bond." This would exclude carbon dioxide, as desired, but would also exclude methanol, formic acid, and formaldehyde, which are generally considered organic. (Formic acid, for example, is produced by ants for their sting, as noted in the scientific name of the family, *Formicidae*.) Another definition is "compounds containing a C—H bond." This would exclude carbon dioxide and include methanol, formic acid, and formaldehyde. It would also include methane, which some people consider organic, but not carbon tetrachloride. Interestingly, neither definition would include urea!

Perhaps this explains why most texts avoid the question—there is not a simple answer. Once again, nature does not easily fit into human classification systems. For our purposes, we will consider carbon-containing compounds (not the elemental forms graphite and diamond) to be organics, with the exception of carbon dioxide, carbonic acid, bicarbonate, carbonate, and carbon monoxide. Methane we consider both organic, since it is the carbon and energy source for methanotrophs (Section 10.5.6), and inorganic, since it is a major end product of anaerobic mineralization and also the product of carbon dioxide respiration (Section 13.1.1).

13.1.1   Carbon Reduction

The reductive side of the carbon cycle is what makes continued life on Earth possible. Without this transformation, and specifically the biological **fixation** of inorganic carbon

TABLE 13.2   Simple Carbon Compounds and the Oxidation State of the Carbon[a]

Oxidation State of Carbon	Chemical Name	Formula	Structure
+4	Carbon dioxide	CO_2	$O=C=O$
	Carbonic acid	H_2CO_3	$HO-C{\overset{O}{\underset{OH}{}}}$
	Bicarbonate	HCO_3^-	$HO-C{\overset{O}{\underset{O^-}{}}}$
	Carbonate	CO_3^{2-}	$O^--C{\overset{O}{\underset{O^-}{}}}$
	Carbon tetrachloride	CCl_4	$Cl-\overset{\overset{Cl}{\vert}}{\underset{\underset{Cl}{\vert}}{C}}-Cl$
+3	Acetic acid[a]	CH_3C^*OOH	$H-\overset{\overset{H}{\vert}}{\underset{\underset{H}{\vert}}{C}}-C^*{\overset{O}{\underset{OH}{}}}$
+2	Formic acid	$HCOOH$	$H-C{\overset{O}{\underset{OH}{}}}$
+1	Acetaldehyde[a]	CH_3C^*HO	$H-\overset{\overset{H}{\vert}}{\underset{\underset{H}{\vert}}{C}}-C^*{\overset{O}{\underset{H}{}}}$
0	Formaldehyde	$HCHO$	$H-C{\overset{O}{\underset{H}{}}}$
−1	Ethanol[a]	$CH_3C^*H_2OH$	$H-\overset{\overset{H}{\vert}}{\underset{\underset{H}{\vert}}{C}}-\overset{\overset{H}{\vert}}{\underset{\underset{H}{\vert}}{C^*}}-OH$
−2	Methanol	CH_3OH	$H-\overset{\overset{H}{\vert}}{\underset{\underset{H}{\vert}}{C}}-OH$
−3	Ethanol[a]	$C^*H_3CH_2OH$	$H-\overset{\overset{H}{\vert}}{\underset{\underset{H}{\vert}}{\overset{*}{C}}}-\overset{\overset{H}{\vert}}{\underset{\underset{H}{\vert}}{C}}-OH$
	Acetic acid[a]	C^*H_3COOH	$H-\overset{\overset{H}{\vert}}{\underset{\underset{H}{\vert}}{\overset{*}{C}}}-C{\overset{O}{\underset{OH}{}}}$
−4	Methane	CH_4	$H-\overset{\overset{H}{\vert}}{\underset{\underset{H}{\vert}}{C}}-H$

[a]For the two-carbon compounds, the oxidation state given is for the C marked with an asterisk.

(carbon dioxide, carbonic acid, bicarbonate, and carbonate) to organic forms, life would be dependent on chemically fixed organic matter, and thus extremely limited.

Organisms that can grow on inorganic carbon, thereby converting it to fixed, organic form, are referred to as *autotrophs* (Section 10.3.2). Carbon reduction requires energy, a portion of which is stored as chemical energy in the fixed carbon that is then available to other life-forms. The primary source of energy utilized for this purpose, providing most of the new organic carbon that is metabolically synthesized on our planet, is light. Organisms able to utilize light as their energy source are phototrophs and are referred to as *photosynthetic*. This includes almost all plants and algae, and some bacteria. Some autotrophic bacteria and archaea, referred to as *chemolithotrophs* (Section 10.3.1), are able to use inorganic chemical energy to fix carbon.

Once carbon has been fixed, it can be further reduced by organisms to generate the cellular constituents that they need for growth. However, this reduction also requires energy, either directly or to regenerate intermediates.

Many organisms (including humans) also reduce carbon during a process known as **fermentation**. This is an anaerobic energy-yielding process during which organic carbon is both oxidized and reduced, either in different molecules or in different parts of the same molecule. Because it is the oxidation that provides energy, fermentation is discussed in Section 13.1.2.

Photosynthesis   As a new day starts and light intensity builds, photosynthetic reactions metabolically infuse new organic carbon back into the cycle of life. The net production of glucose and oxygen generated through this process by plants and algae can be summarized as

$$6CO_2 + 6H_2O + \text{light energy} \rightarrow C_6H_{12}O_6 + 6O_2 \qquad (13.1)$$

Cyanobacteria carry out the same reaction, but other photosynthetic bacteria utilize hydrogen, hydrogen sulfide, or elemental sulfur instead of water as their electron donor, and thus do not produce oxygen (see Table 10.4).

Almost all phototrophs use the same reaction sequence, referred to as the *Calvin cycle* after its discoverer, to invest energy drawn from sunlight into fixing inorganic carbon (Section 5.4.5). Exceptions are some green sulfur bacteria, which can run the Krebs cycle (Section 5.4.3) in reverse to fix carbon, and some green nonsulfur bacteria, which utilize a unique hydroxypropionate cycle to form the two carbon compound glyoxylate (CHO—COOH) from two carbon dioxide molecules.

The Calvin cycle itself was summarized in Figure 5.13 and is shown in more detail in Figure 13.4. It can be considered to consist of three parts. The first step combines the carbon dioxide with a five-carbon sugar containing a phosphate on each end [ribulose 1,5-diphosphate; also called ribulose 1,5-bisphosphate (RuBP)]. The unstable product breaks down to produce two molecules of a phosphorylated, three-carbon compound called 3-phosphoglycerate. This overall reaction is catalyzed by ribulose-1,5-bisphosphate carboxylase/oxygenase (**RubisCO**), which may be the most abundant enzyme on Earth.

During the second part of the cycle, the newly fixed carbon is reduced further, using additional energy, to glyceraldehyde 3-phosphate. In the final part of the cycle, a six-carbon sugar is produced, and the ribulose 1,5-diphosphate regenerated in preparation for subsequent cycles.

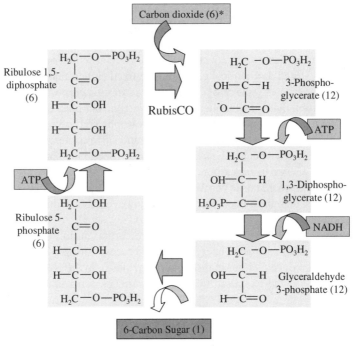

Figure 13.4   Calvin cycle carbon fixation process. *(No.) refers to number of molecules.

Photosynthetic organisms utilize part of the carbon they fix as an energy source, particularly in the dark (Section 13.1.2). However, the rest is used to produce the organic compounds they need for growth. It is this net fixation of carbon that sustains the ecosystems dependent on these **primary producers**, directly or indirectly, for their carbon and energy supply.

Of course, there are considerable variations in the rates of organic carbon assimilation around the world. Tropical rain forests, for example, have some of the highest rates of carbon uptake on Earth, with annual net fixation of up to $1.5 \, kg \, C/m^2$ of land surface area. Freshwater wetlands are almost as productive. Forests, grasslands, and cultivated plants in the middle latitudes typically assimilate 50% as much carbon, although intensive agriculture can more than double the production of the rain forest. The open ocean, deserts, and the Arctic tundra fix much less carbon (typically 0.05, 0.1, and $0.2 \, kg/m^2$, respectively, per year). Sunlight intensity, temperature, nutrient supply, and water availability are all factors (see Chapters 14 and 15).

The use of sunlight for photosynthesis is not very efficient. On average, only about 0.1% of the energy available is utilized. Even with intensive agriculture, this value is unlikely to exceed 1%. Each gram of carbon fixed represents about 10 kJ of stored energy.

It is estimated that photosynthesis provides our planet with a cumulative carbon assimilation rate of approximately 10^{11} metric tons per year. Based on the stoichiometry in equation (13.1), this means that at the same time, almost 3×10^{11} metric tons per year of oxygen is produced. The long-term maintenance of this essential activity, however, is sensitive to disruption. Whether measured in acreage of rain forest lost to deforestation, wetlands drainage, or accidental tanker spills and resultant oil slicks, humans certainly have the potential to cause severe environmental stress.

Chemolithoautotrophic Carbon Fixation   Although they utilize chemical energy stored in inorganic molecules rather than light, chemolithoautotrophs also utilize the Calvin cycle to fix inorganic carbon. Worldwide, a substantial amount of carbon is fixed through their action. In fact, in some environments in which sunlight is absent, they may be the major or only primary producers. Surrounding thermal vents in the deep ocean, for example, the release of hydrogen sulfide serves as an energy source for sulfide-oxidizing bacteria and archaea, which then form the base of an extensive ecosystem, including giant worms and clams. It also appears that the abiotic release of hydrogen gas in deep deposits of basalt (a volcanic rock) in the Pacific northwestern region of the United States may support specialized autotrophic bacteria that serve as the base of a microbial ecosystem.

Nitrifying (see more in Section 13.2.2) and sulfur-oxidizing (Section 13.3.2) bacteria may be important in waste treatment and aquatic and soil environments, and sulfur-and iron-oxidizing (Section 13.4.2) bacteria play a major role in acid mine drainage. However, in these cases it is not the carbon fixation that is of greatest environmental significance. It recently has been suggested, however, that in drinking water systems using chloramination (an ammonium/chlorine combination) for disinfection, the nitrification that sometimes occurs may actually contribute materially to the otherwise low organic carbon content of the water.

Methanogenesis   Methanogenesis, the production (genesis) of methane, is a form of carbon reduction carried out solely by a group of strictly anaerobic archaea referred to collectively as *methanogens* (Section 10.6.3). One reaction commonly utilized is actually an anaerobic respiration in which CO_2 [equation (13.2)], or occasionally, CO [equation (13.3)], is reduced to methane during oxidation of H_2. Many methanogens can ferment formic acid [equation (13.4)] or methanol [equation (13.5)], with some of the molecules being oxidized to CO_2 and others reduced to CH_4. A few methanogens can similarly ferment methyl amines or, less commonly, methyl sulfides. Many can also ferment acetic acid, oxidizing part of the molecule to CO_2 while reducing the other part of same molecule to CH_4 [equation (13.6)]:

$$4H_2 + CO_2 \rightarrow CH_4 + 2H_2O \qquad (13.2)$$

$$3H_2 + CO \rightarrow CH_4 + H_2O \qquad (13.3)$$

$$4CHOOH \rightarrow CH_4 + 3CO_2 + 2H_2O \qquad (13.4)$$

$$4CH_3OH \rightarrow 3CH_4 + CO_2 + 2H_2O \qquad (13.5)$$

$$CH_3COOH \rightarrow CH_4 + CO_2 \qquad (13.6)$$

Often under methanogenic conditions, such as in landfills and anaerobic digesters, the net gas production is almost two-thirds methane and one-third carbon dioxide, with small amounts of other gases. Additional details about methanogenesis are given in Sections 13.1.1 and 16.2.1.

13.1.2   Carbon Oxidation

The oxidation of organic carbon compounds provides energy to a cell. Even in photosynthetic organisms, it is the oxidation of the fixed carbon produced (with the energy from sunlight) that provides the cells with most of the energy they use.

The two overall types of energy-yielding metabolism are respiration (Section 5.4.3) and fermentation (Section 5.4.2). In respiration, an inorganic molecule acts as the **terminal electron acceptor**, whereas in fermentation, an organic molecule serves this function. Fermentation is an anaerobic process, whereas there are both aerobic and anaerobic forms of respiration.

Respiration   Respiration provides more energy to the cell then fermentation, so is favored whenever it is possible. **Aerobic respiration**, in which molecular oxygen (O_2) is the terminal electron acceptor, provides the most energy; thus, aerobic organisms are likely to dominate wherever oxygen is available.

Since our atmosphere is 21% oxygen, it might be expected that oxygen is readily available except in sealed systems. However, oxygen is poorly soluble in water, with only 9.1 mg/L of **dissolved oxygen** (**DO**) present at saturation (equilibrium with the atmosphere) at 20°C. Also, oxygen diffusion through pores in solid matrices (such as soil and composting material) occurs by a factor of 4 orders of magnitude more slowly in water than in air. Thus, in aqueous systems (such as wastewater, liquid sludges, surface waters, river and lake sediments, groundwater, and flooded soils) the oxidation of even small amounts of organic material can quickly deplete DO, leading to anaerobic conditions.

In the absence of oxygen, many bacteria can carry out respiration using nitrate as the terminal electron acceptor (Section 5.4.3). This **denitrification** (Section 13.2.1) provides about 85% of the energy available aerobically from oxidation of the same organic compound with oxygen. The nitrate is reduced first to nitrite, then eventually to nitrogen gas. Thus, nitrite can also be utilized, but yields still less energy. Denitrifying bacteria include a wide diversity of aerobic bacteria that can utilize nitrate and nitrite as alternative electron acceptors when oxygen is absent. The process itself is discussed in more detail as part of the nitrogen cycle (Section 13.2).

Ferric iron can also be utilized for respiration by some aerobic bacteria, providing a similar amount of energy to nitrite. Some other metals may also be utilized by facultatively or strictly anaerobic bacteria.

In the absence of oxygen, nitrate, nitrite, and ferric iron, some anaerobic bacteria and archaea can utilize sulfate as an electron acceptor, producing hydrogen sulfide. This is particularly important in salt marshes and other systems where sulfate is abundant (Section 13.3). Once the sulfate is also depleted, methanogens can utilize CO_2 as an electron acceptor, producing methane [equation (13.2)].

Under anaerobic conditions, particular bacteria, especially some sulfate reducers, may be able to utilize other "unusual" electron acceptors. Examples include the use of a number of chlorinated organic compounds, including important environmental contaminants such as polychlorinated biphenyls (PCBs), trichloroethylene (TCE), tetrachloroethylene [commonly called perchloroethylene (PCE)], and vinyl chloride, during which a hydrogen is substituted for a chlorine atom. This represents both a chemical reduction and a **dechlorination** of the organic molecule and results in the release of a chloride ion (Cl^-). A similar reaction may occur with brominated and perhaps other halogenated compounds. Thus, the more general term **reductive dehalogenation** is often used to describe this process.

Fermentation   Although it is usually bacteria, archaea, some fungi, and a few protozoa that are thought of as being anaerobic, higher organisms also carry out fermentations. During vigorous exercise, for example, not enough oxygen can be delivered to the muscles, leading to buildup of the fermentation product lactic acid. This is what makes your arms and legs feel heavy and tired. Gradually, as you recover, the lactic acid is transported to the liver and converted there for other uses.

Because an organic compound serves as the electron acceptor in fermentation, some of the organic material is reduced while oxidizing other organic molecules. In the examples above for methanogenesis, one reaction involves four molecules of formic acid [equation (13.4)], one of which is reduced to CH_4 while the other three are oxidized to CO_2. For the fermentation of the more reduced compound methanol [equation (13.5)], the ratio of these two products is reversed. With acetic acid [equation (13.6)], on the other hand, of the two carbons of a single molecule, one is oxidized while the other is reduced.

13.1.3   Carbon in Environmental Engineering and Science

It traditionally has been the organic carbon (along with pathogens) that was of the greatest concern in water pollution (Section 15.2.7), leading to the construction of wastewater treatment plants (Chapter 16) that focus on its removal. Management of wastewater treatment sludges often has **stabilization** of the organic material as a major objective. (Stabilization involves conversion of readily degradable materials to those that change only slowly; see later in this subsection). Municipal solid waste management also must stabilize the organic material (e.g., by incineration or composting), or else deal with the consequences (e.g., attraction of vermin, settling, and leachate and gas production during landfilling). Similarly, with soil and groundwater contamination, it is often organic carbon that is the target of remediation. Undesirable tastes and odors in drinking water, and the formation of cancer-causing compounds during disinfection, are traceable to organic compounds present in the water supply. Even air pollution control may involve organics, such as volatile organic compounds (VOCs) and soot (which includes organic particles), as important contaminants. Most individual toxic compounds of concern in water, soil, and air are also organics.

Thus, much of environmental engineering and science is directed at control of organic carbon or an understanding of its fate and effects in the environment. In particular cases the emphasis may be on a single compound, a particular class of compounds (such as petroleum hydrocarbons or chlorinated solvents), some broad fraction (such as oil and grease, or oxygen demanding biodegradable compounds), or the total organics. One concern might be rapid biodegradation, leading to depletion of oxygen, while another material might pose a hazard because it is very resistant to degradation. Slow-to-degrade compounds may be bioaccumulated (Section 18.7.2) in organisms, perhaps leading to toxic effects even if they are present only at low concentrations in the environment. Other slowly degrading compounds, such as many plastics, may pose aesthetic problems, or perhaps injure wildlife that eat (discarded plastic bags that are mistaken for jellyfish) or become entangled (abandoned fishing nets or six-pack rings) in them.

In some water bodies, contamination with excess levels of nitrogen and/or phosphorus may lead to excessive growth of aquatic plants, algae, or cyanobacteria (eutrophication; Section 15.2.6). In this case organic carbon is not added directly, but instead, becomes problematic after it is formed through photosynthesis.

On the other hand, for heterotrophic organisms, it is typically the availability of organic material that limits growth. This is particularly true for water and soil microorganisms, where survival is often dependent on an ability to subsist on very low levels of organic substrates, or to grow quickly when high concentrations of substrate suddenly become available (e.g., through death of a plant or animal, or depositing of animal waste products), followed by persistence until the next such event.

Thus, it is clear that the carbon cycle is of major importance in environmental engineering and science, and the next two subsections deal with measurement of organics and their biodegradation. However, it is also worth noting the potentially fragile nature of this essential cycle. This was demonstrated on a small scale with the Biosphere II facility in Oracle, Arizona (Section 14.2.2). That attempt to build a closed ecosystem including humans was unable to keep the carbon cycle in balance, producing too much CO_2 and too little O_2. On a broader scale, global warming, largely as a result of increased atmospheric CO_2 levels, represents another example of a carbon cycle that is not in balance, with potentially major effects. In considering such global change, it is perhaps instructive to keep in mind that Earth's atmosphere was not always as we know it today. In fact, the low CO_2 (380 ppm) and high O_2 (21%) that we consider "normal" is a result of a previous "imbalance" brought about by the advent of oxygen-producing photosynthesis (Section 10.1).

An important lesson to be learned from these observations is the crucial balance that comes into play with the life-sustaining biogeochemical cycling of an element such as carbon. Compared to our planet, the scale of Biosphere II was so small that it was very sensitive to an imbalance, and problems manifested themselves in a correspondingly short period of time. The biogeochemical cycles of our planet, on the other hand, were established over a far longer period (geologic time). Nonetheless, the success of the ecosystems that sustain us depends on a harmony in nature, one that has finite (if unknown) limits to the levels of human influence that it can tolerate.

Quantification of Organic Carbon   Because organic carbon is often the contaminant of greatest concern in water pollution and waste treatment, a number of approaches to measuring it have been developed. In some cases, the concentration of individual constituents must be known, especially for toxic substances. In these instances, sophisticated instrumental techniques such as gas chromatography (GC) or high-performance liquid chromatography (HPLC) may be used to separate and quantitate the compounds of concern, followed by mass spectroscopy to identify them with a high degree of certainty.

Another reason to know the amount of a compound present is to estimate the potential oxygen demand it might exert, for example, in a stream or during wastewater treatment. For a pure compound the **theoretical oxygen demand** (ThOD) can be calculated based on the stoichiometry of its complete oxidation.

Example 13.1   What is the ThOD of glucose, $C_6H_{12}O_6$?
  Answer

$$C_6H_{12}O_6 + 6O_2 \rightarrow 6CO_2 + 6H_2O$$

Thus, 6 mol of oxygen is needed to oxidize each mole of glucose completely. Since 1 mol of oxygen is 32 g, $6 \times 32 = 192$ g of oxygen is needed for 180 g (1 mol) of glucose, or 1.067 g of O_2 per gram of glucose.

More generally, for a hydrocarbon or carbohydrate, the equation can be written as

$$C_aH_bO_c + nO_2 \rightarrow aCO_2 + \frac{b}{2}H_2O \qquad (13.7)$$

where

$$n = a + \frac{1}{4}b - \frac{1}{2}c \qquad (13.8)$$

Hence, with MW representing the molecular weight of the compound,

$$\text{ThOD} = \frac{32n}{\text{MW}} = \frac{32a + 8b - 16c}{12a + b + 16c}$$

To calculate the ThOD of an organic compound containing other elements (besides C, H, and O), the final form of the elements after the reaction must be known. P can be considered to be in the form of phosphate and chlorine as chloride (Cl^-). However, nitrogen might end up as ammonia or nitrate, and sulfur as sulfide or sulfate. Often, it will be assumed that the products are in the reduced form (NH_3 and H_2S) so that the carbonaceous oxygen demand is calculated; any additional oxygen demand for oxidation of the inorganics can then be determined separately.

For mixed contaminants, such as in municipal wastewater or sludge, it is often necessary or sufficient to have an estimate of the total amount of organic matter present, or some fraction thereof, rather than trying to quantify each of the individual components separately. One of the older of such aggregate measures, still commonly used for sludges, is **volatile solids** (VS). In this test a sample is dried and its mass found to determine the **total solids** (TS). It is then ignited at around 500 °C, which destroys the organic material. (Note that this is also one weakness of the method, in that some inorganics also may volatilize, leading to overestimation of the organic content.) The VS is the difference in mass between the TS and the residual **ash**. Rather than use the total solids, the sample also can be filtered first and the test performed on the **suspended solids** (SS) trapped on the filter, yielding the **volatile suspended solids** (VSS). Organisms and their associated materials and waste products, including wastewater treatment plant sludges, often have a volatile fraction of around 85%.

This approach is also used to estimate the carbon content of soils. The volatile fraction is determined, and then multiplied by a factor, often around 0.55, since this is the typical fraction of the soil organic material that actually consists of carbon.

Total organic carbon (TOC) is an instrumental analysis method that gives a more selective measure of organic matter. Using combustion or chemical oxidation, the organic matter in a sample is converted to CO_2, which is then measured using infrared spectrometry. Correction must be made for the inorganic carbon present, either by acidification (to turn it all to CO_2) and purging (bubbling with nitrogen gas) to remove it before analysis (note that this also strips out volatile organic compounds, so that they are missed in the analysis), or by also measuring the CO_2 present before oxidation (inorganic forms) and subtracting it from the value obtained after oxidation (which includes both the organic and inorganic sources).

The TOC of pure compounds can also be calculated from the chemical formula.

Example 13.2 What is the TOC of glucose, $C_6H_{12}O_6$?

Answer There are 6 mol of C per mole of glucose.

$$\frac{6 \times 12}{6 \times 12 + 12 \times 1 + 6 \times 16} = \frac{72 \text{ g C/mol glucose}}{180 \text{ g glucose/mol glucose}} = 0.40 \text{ g TOC/g glucose}$$

For complex mixtures of unknown composition, a ThOD cannot be calculated. However, the **chemical oxygen demand** (**COD**) laboratory test empirically estimates this value. It utilizes heat and strong oxidizing agents to convert most organics to CO_2, then estimates the amount of oxygen that would have been necessary to carry out the same reaction. One limitation of the method is that despite the very vigorous conditions, a few organic materials are not oxidized completely.

However, the most common method for estimating the strength of the oxygen-demanding organic material in wastewaters—as required by many regulations—is the **biochemical oxygen demand** (BOD). This test determines empirically, under standard conditions, the amount of oxygen utilized during microbial oxidation of the organics present in the sample. Thus, it measures only the biodegradable portion of the total oxygen demand. However, this is often the fraction of interest in determining the impact of a discharge on a receiving water, or the amount of aeration capacity needed for an aerobic treatment process, since only the biodegradable organics are likely to be oxidized under such conditions.

Like many of the tests described above, the BOD is **operationally defined**; that is, the BOD is the number that results from conducting the test. Thus, it is important to use a standard protocol, including a specialized bottle (Figure 13.5), so that results will be reproducible. The sample is first oxygenated to bring the DO up to near saturation (9.1 mg/L), the DO is measured, and the bottle is then sealed. After incubation in the dark (to prevent photosynthetic production of oxygen) at 20°C, usually for 5 days, the DO is measured again. The difference represents the oxygen utilized biochemically.

The BOD of domestic sewage is typically around 200 mg/L, yet the initially aerated sample can hold no more than about 9 mg/L of DO. This means that substantial dilution is required (with a specified standard buffer), and the final oxygen depletion in the bottle is than multiplied by the dilution factor. To ensure that sufficient microorganisms are present initially, some samples are seeded (inoculated) with a small amount of wastewater treatment plant effluent or settled sewage, and the oxygen demand of this **seed** must be corrected for in the final BOD calculation.

Figure 13.6 shows a typical plot of the BOD exerted versus time for a sewage sample. The first surge of oxygen demand results from oxidation of organic materials, while the second, often starting at about day 7, is a result of the oxidation of ammonia (nitrification; see Section 13.3.2). The standard BOD test is run for 5 days (**BOD₅**), so that usually it includes much of the **carbonaceous** oxygen demand (**C-BOD**), but little of the **nitrogenous** demand (**N-BOD** or **NOD**). However, this makes the procedure difficult to use for short-term decision-making purposes, as the results will not be known for at least 5 days. (This choice is also unfortunate for the practical reason that tests started on a Tuesday must be finished on a Sunday!) In some cases the N-BOD begins to be exerted much earlier, giving a BOD₅ value that overestimates the C-BOD. If desired, the N-BOD can be suppressed by adding to the BOD bottle a compound that acts as a specific inhibitor of nitrification, preventing exertion of the N-BOD.

Figure 13.5   BOD bottle.

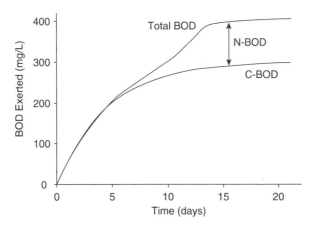

Figure 13.6   BOD exerted vs. time.

TABLE 13.3 Comparison of BOD, COD, and TOC Tests

	BOD$_5$	COD	TOC
Typical value for domestic sewage (mg/L)	200	400	150
Time for analysis	5 days	3 hours	10 minutes
Instrumentation cost	Very low/low[a]	Low	High
Chemical cost	Medium/low[a]	High	Low
Positive interferences[b]	Ammonia, sulfide, nitrite, ferrous iron	Chloride, sulfide, nitrite, ferrous iron	Carbonates, bicarbonates
Negative interferences[b]	Toxics, photosynthesis	Volatile organics, pyridines	[c]

[a]Depends on whether using chemical analysis or a probe for measuring dissolved oxygen.
[b]Corrective measures are possible for most interferences. Uncorrected positive interferences lead to overestimation; negative ones lead to underestimation.
[c]Using some TOC methods, not all organic carbon is reacted and measured.

The **ultimate BOD** (BOD$_u$) is the carbonaceous BOD that results from prolonged incubation, usually taken as 20 days. For domestic sewage, this value is often about 70% of the COD. This is because even for highly biodegradable compounds, some of the substrate is converted to cellular material, not respired for energy. Since the BOD$_5$ is often about 70% of the BOD$_u$, the standard BOD$_5$ test typically gives a value that is about 50% of the COD. However, for various compounds and mixtures, this ratio can vary from 0 (nonbiodegradable under the BOD test conditions) to more than 1.0 (biodegradable, but not chemically oxidized in the COD test). Table 13.3 compares BOD, COD, and TOC, showing their relative values for domestic sewage and some of the advantages and disadvantages of each test.

Because the BOD procedure relies on oxidation by microorganisms, it is inherently prone to variability despite efforts at standardization. Even slight variations in inoculation or other test conditions can lead to changes in oxidation rate and hence different oxygen depletions at the close of the test period. Despite its weaknesses, however, the BOD is still highly useful and widely used.

ThOD, COD, and BOD all give a good indication of the amount of energy available from oxidation of a compound or mixture. This is because the amount of oxygen consumed in an oxidation is linked very closely to the amount of energy released during that reaction [typically, 14,000 J/g O_2; see the discussion of equation (16.5) in Section 16.2.3].

Biodegradation   **Biodegradation** can be defined as the transformation of a compound through biological activity. It was first included in some dictionaries in the 1960s. **Degradation** is a broader term that would also include transformation as a result of other chemical or even physical changes. This might include **photolysis** (a chemical reaction in which light provides the activation energy), for example, or abiotic catalysis at the surface of a clay mineral.

It is interesting to note that some of the early tests of biodegradability were for plastics, and degradation was considered undesirable. Thus, even changes in color or opacity were considered degradation and might lead to rejection of a new polymer. On the other hand, some plastic bags now labeled "biodegradable" degrade only very slowly, and thus are

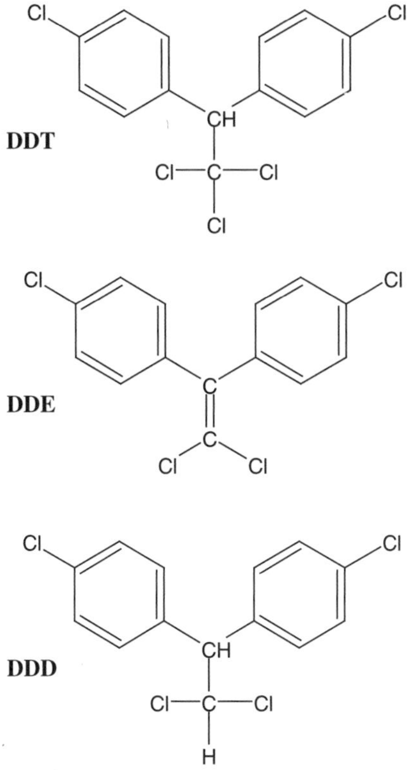

Figure 13.7 Structures of DDT (dichlorodiphenyltrichloroethane), DDE (dichlorodiphenyldichloroethene, and DDD (dichlorodiphenyldichloroethane).

inappropriate for use in composting operations. Similarly, the pesticide DDT was found to disappear fairly rapidly under some conditions; unfortunately, it was simply converted to DDD or DDE, which were at least as problematic as the parent compound (Figure 13.7). Thus, this biotransformation (Section 18.5) did not represent substantial biodegradation of the compound.

It is often desirable to be more specific when discussing biodegradation, and to consider both the rate and the extent of the transformation. **Mineralization**, or ultimate biodegradation, is the complete conversion of an organic molecule to inorganic products (mainly CO_2 and H_2O under aerobic conditions, CO_2 and CH_4 or H_2S under anaerobic ones). **Readily biodegradable** organics can be rapidly mineralized. **Recalcitrant** compounds (also called **refractory**), on the other hand, are difficult to degrade, and are mineralized slowly if at all. However, *biodegradability* refers to an inherent property of the material; whether it actually will be biodegraded depends on the environment in which it is present. For example, readable newspapers and intact pieces of fruit buried for decades have been recovered from landfills, particularly under dry conditions. Even more dramatically (Figure 13.8), the corpses of murder victims have been recovered from northern bogs (with cold, acid, and anaerobic conditions) after 2000 years with much of their clothing, skin, hair, and even stomach contents intact!

In general, conditions that promote growth are likely to speed biodegradation. Thus, warm, moist conditions often lead to higher rates than cold, dry ones. Also, most—but

Figure 13.8   Tollund man: photograph of a 2000-year-old body recovered from a Danish bog. (Photo by Lennart Larsen, National Museum of Denmark.)

not all—materials are more readily biodegradable under aerobic conditions than under anaerobic ones.

Most biodegradation occurs as a result of microbial activity in which the compound is utilized as a carbon and energy source. Typically, complex organics are broken down enzymatically into simpler molecules, which then enter the cell's basic biochemical pathways, such as glycolysis and the Krebs cycle (Section 5.4.3). Compounds containing nitrogen, sulfur, or other essential elements may also be utilized as sources of these nutrients.

All naturally occurring and most human-made organic compounds are biodegradable under appropriate conditions. However, some **xenobiotic** ("foreign" to biology; i.e., synthetic) compounds are extremely recalcitrant. In fact, many, such as polychlorinated biphenyls (PCBs) and most plastics, were designed to resist degradation. Long hydrocarbon chain molecules, such as polyethylene, are slow to degrade because microbial attack is mainly from the ends. Branching also makes biodegradation difficult, as it can hinder or prevent the molecule from fitting into the active site of appropriate enzymes. The early alkyl benzenesulfonate (ABS) detergents, used widely in the 1950s, for example, were highly branched and thus recalcitrant, leading to dramatic foaming in wastewater treatment plants and receiving waters (Figure 13.9). This problem was eliminated when the linear alkyl sulfonate (LAS) detergents were introduced as a replacement.

More recently, methyl tertiary butyl ether (**MTBE**) has become a problem in groundwaters. The history of the use of this branched compound represents an interesting lesson

Figure 13.9 Foaming from detergents in the 1950s at (*a*) a municipal sewage treatment plant and (*b*) the Passaic River, in New Jersey. (Photos by Joseph V. Hunter.)

in the interrelatedness of environmental problems. For many years, lead was added to gasoline as an antiknocking or octane-enhancing compound. However, the realization that this lead was contributing to permanent neurological problems (including decreased intelligence), especially in urban children, led to its reduction and eventual elimination from gasoline in the United States as an environmental health measure (although it is still used in many other countries, with the same unfortunate effects).

$$\begin{array}{c} CH_3 \\ | \\ H_3C-C-O-CH_3 \\ | \\ CH_3 \end{array}$$

MTBE

As a substitute for lead, MTBE was added to gasoline beginning in the mid-1970s at about a 5% level. Almost 20 years later, when a decision was made to increase the amount of oxygenate (oxygen-containing organic compounds) in gasoline as a means of reducing carbon monoxide emissions (an air pollution problem), MTBE was the compound of choice in much of the country. (MTBE was favored by the oil industry, which produces it, whereas ethanol, which can be produced from corn, was favored by agricultural interests.) Since it was already in wide use, little consideration was given to the environmental impacts of increasing its concentration to the range 10 to 15%. However, with the wide use of gasoline and the frequency of unintended environmental releases through leaks and spills, MTBE quickly became a major groundwater contaminant. Compounding the problem is MTBE's ready solubility in water and poor sorption to soil, leading to its rapid migration. A further complication is the relative ineffectiveness of carbon adsorption and air stripping (volatilization) as treatment technologies. Worst of all, both its branched structure and its ether bond make MTBE resistant (although not immune) to biological attack as well.

Molecules that are too large or otherwise not taken up by a cell are first attacked by **extracellular enzymes**, or **exoenzymes**, released by the organisms. (*Note*: The term *exoenzyme* is also used by biochemists to refer to enzymes that attack polymers from the end rather than in the middle.) Many of these are **hydrolases**, enzymes that react by adding or removing water from a molecule (without an oxidation or reduction). These often are relatively specific, as is the case with many cellulases (attack cellulose), amylases (starch), lipases (lipids), and proteases (proteins).

However, some organisms also produce nonspecific extracellular enzymes capable of degrading a wide variety of compounds. One such example is the lignin-degrading basidiomycete fungus *Phanerochaete chrysosporium*. Lignin is a complex, nitrogen-containing polymer. It is produced by woody plants for structural support and to protect cellulose from degradation. Although it is very common, it is relatively recalcitrant due to its heterogeneous, irregularly branched structure. However, under aerobic conditions it can be attacked by *P. chrysosporium* and other white rot fungi through an exocellular peroxidase enzyme system that produces free radicals. These highly reactive substances react to nonspecifically oxidize complex organic molecules (including lignin), releasing smaller subunits that are more readily biodegradable.

Extracellular enzymes represent a cost to a cell in terms of energy and materials released to the environment. The cell benefits if it is able to take up and assimilate (incorporate into biomass) or oxidize (for energy) some of the smaller molecules produced by the enzyme's activity. (Occasionally, another benefit of exocellular enzymatic activity may be to decrease the toxicity of a compound.) However, cells that do not release extracellular enzymes also benefit from the presence of these utilizable molecules (or decreased toxicity), but without paying the cost of the enzyme's production. Such "freeloading" organisms are thus in a sense parasitic on the extracellular enzyme-producing organisms.

Oxidation of many organics involves enzymes (mainly intracellular) known as *oxygenases*. Monooxygenases, which add a single oxygen atom to the molecule, are found in both eukaryotes and prokaryotes. Dioxygenases add both atoms of an elemental oxygen molecule to the organic and are most common in bacteria. Note that with oxygenases, oxygen is being used as a reactant, not as an electron acceptor.

Figure 13.10 shows a typical monooxygenase oxidation of a straight-chain hydrocarbon, leading in several steps to production of a fatty acid. The fatty acid then can be

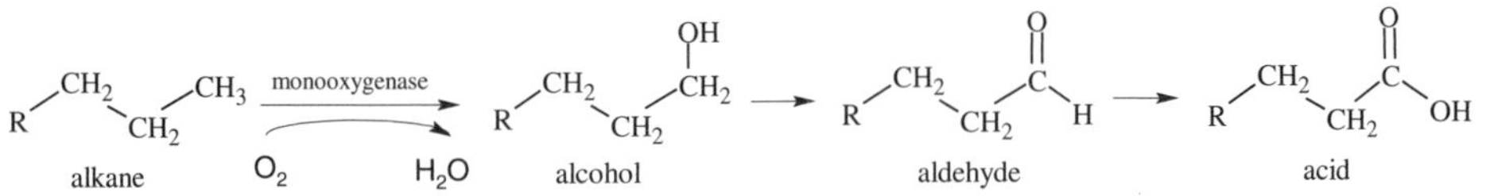

Figure 13.10 Monooxygenase activity on a straight-chain hydrocarbon.

mineralized via the standard β-oxidation pathway (Section 5.4.4). Note that the other atom from the O_2 molecule is reduced to H_2O.

Figure 13.11 shows five bacterial pathways for the initial steps in the biodegradation of toluene, an aromatic, naturally occurring hydrocarbon that is present in gasoline. The first four involve a monooxygenase, the fifth a dioxygenase. The first three of the

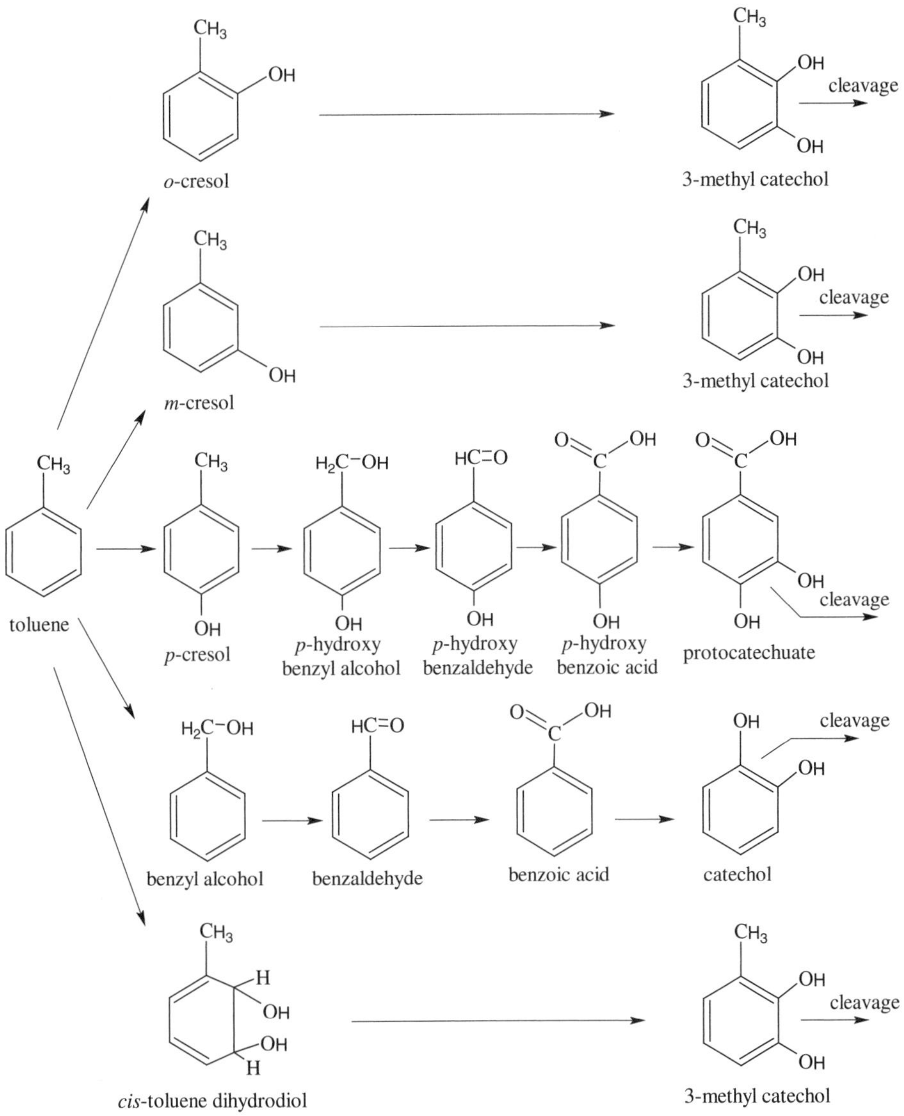

Figure 13.11 Five pathways for toluene oxidation.

monooxygenase pathways involve addition of a hydroxyl group to the ring in three different positions (ortho-, meta-, and para-), whereas the fourth involves oxidation of the methyl group. For all of the pathways, the next step will be **cleavage** (breaking) of the ring (usually between the two hydroxyl groups), leading to products (dicarboxylic acids) that after a few more transformations can enter the Krebs cycle. Detailed pathway information is now available for many degradation reactions (e.g., see the University of Minnesota Biocatalysis/Biodegradation Database Web site.)

Oxygenases, like most other enzymes, tend to be fairly specific. However, some degree of nonspecificity can be beneficial, in that it can allow an organism to utilize a newly available substrate, even if at a lower efficiency. Over an extended time, it can be expected that new forms of the enzyme, better adapted to the new substrate, will emerge (through random variation, genetic exchange, or rearrangement), giving the organism utilizing them a selective advantage when the substrate is available.

On the other hand, a nonspecific enzyme may transform a nontarget substrate, yielding a product that cannot be further metabolized by the cell. For example, a cell growing on toluene might accidentally cleave an aromatic ring containing a chlorine substitution. The resulting dicarboxylic acid may not be suitable for further metabolism because of the large chlorine atom still present. Then the cell has wasted the material and energy it invested in transforming a compound that provides it with no benefit. This type of activity, in which a cell growing on one compound "mistakenly" transforms another without any benefit, is referred to as **cometabolism** (Figure 13.12). In this case, over time it can be expected that the organism will develop more specific enzymes to prevent the drain of

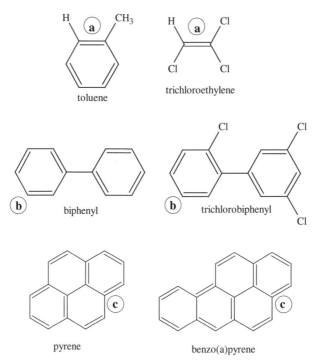

Figure 13.12   Possible cometabolism. The main (growth) substrate in each pair is located on the left, with areas of similarity indicated by a, b, or c.

cometabolism, or alternatively, it may develop new enzymes and pathways so that the **dead-end product** can be further utilized. Although it is not beneficial to the microorganism, cometabolism may be desirable from a human point of view because it may be possible to exploit it to degrade otherwise nondegradable compounds.

Even with energy-yielding metabolism, dead-end products may form. In other cases, intermediates may accumulate because their rate of production is greater than their rate of further transformation. Occasionally, this accumulating compound will be more toxic than the parent material, leading to at least a temporary increase in toxicity during biodegradation. The increased toxicity may have the effect of further slowing degradative activity, as well as having other undesirable effects on the ecosystem.

In aerobic systems it is common for a single species of bacteria to be able to utilize a single organic compound as its sole source of carbon and energy. However, occasionally, organisms of one species can degrade the compound only partially, and those of another species will further transform, and probably mineralize, the intermediate produced. Such a combination of organisms of different species "working together" to metabolize substrates is referred to as a **consortium** (plural, *consortia*).

The activities of consortia can make detection of cometabolism difficult, as a cometabolic product may be degraded by other organisms before it accumulates sufficiently to be noticed. Also, the fact that a compound can be completely degraded by a single organism does not necessarily mean that this is the way it will be degraded in a particular environment—the actual degradation still may be by a consortium.

Under anaerobic conditions, consortia are common. A wide variety of species may be involved in hydrolyzing, or solubilizing, complex organics, followed by fermentations of the subunits produced. Then sulfate-reducing or methanogenic organisms may utilize the fermentation products, leading to mineralization.

The organisms in a microbial consortium may be only loosely associated, or they may be so closely linked that they are difficult to separate. A classic example is the case of methane production from ethanol. For many years this was attributed to *Methanobacillus omelianskii*, an "organism" that could be grown in "pure" culture with ethanol as the sole carbon and energy source. However, it was later demonstrated that, in fact, the culture contained two species: a bacterium that converted ethanol to acetate, and a methanogen (Archaea) that utilized the acetate, producing methane and carbon dioxide.

Occasionally, the balance among organisms in an anaerobic consortium is disturbed and mineralization does not occur. In the absence of sulfate, anaerobic mineralization is dependent on methanogens. Compared to the wide variety of organisms that are able to hydrolyze and ferment organics anaerobically, methanogens are a relatively limited group of strictly anaerobic archaea. Although most methanogens utilize acetic acid as a substrate, as a group they are sensitive to low pH. If the acetic acid is produced too quickly, the pH of the system will drop, inhibiting methanogenic activity. In turn, this leads to a further buildup of acid, a greater drop in pH, and even lower rates of methanogenesis and acid destruction.

Some human foods, such as pickles and sauerkraut, are preserved in this way. **Ensilage**, the process of making **silage**, is a means of animal feed preservation utilizing acid anaerobic conditions that has been employed in agriculture for many years. Fresh corn, hay, or other feed crops are placed in a silo, where they quickly ferment. The acid anaerobic conditions then prevent substantial further degradation. However, if oxygen is introduced in large amounts, the organic acids are quickly mineralized, pH rises, and biodegradation resumes.

Acid anaerobic conditions are also responsible for the preservation of the large masses of spongy rich organic material in bogs. This waterlogged **peat** soil, composed mainly of dead sphagnum moss (an acid-loving plant), accumulates over hundreds or even thousands of years. If the water is drained from such systems, aerobic conditions develop and the organics are quickly oxidized.

During waste treatment, acid anaerobic conditions are undesirable because they slow degradation, leading to preservation of the waste. In poorly run leaf "composting" (Section 16.2.3) operations (really, leaf dumps), large piles of leaves may be formed and simply left unattended. Acid anaerobic conditions develop quickly, so that on opening such piles 10 years later, the tree species of individual leaves can still be determined. Similarly, in anaerobic digesters for sludge treatment (Section 16.2.1), overloading or toxic shocks may lead to acid anaerobic conditions that prevent methane formation. The digester under such conditions is referred to as having gone "sour."

Another problem with acid anaerobic conditions in waste treatment is the potential for odors. The organic acids themselves may be volatilized, leading to a sour smell. Often, some alcohols are also formed, and apparently react with the acids to give esters, which may have a sweet smell. Thus, such systems often have a sweet–sour smell. This is not necessarily unpleasant for silage in an agricultural setting, but it is often not appreciated near wastewater treatment, composting, and landfill sites.

If sufficient proteinaceous material is present in the waste, more severe odors may also develop. Under most conditions the amino acids released by protein hydrolysis are **deaminated**; that is, the amino group is removed, releasing ammonium (Figure 13.13). This can lead to an ammonia odor at high pHs where volatilization is favored. However, under acid anaerobic conditions, amino acids may instead be **decarboxylated**. This leaves an amine, some of which are highly odorous, with names such as *putrescene* and *cadavarene*.

Another odor concern with anaerobic conditions is hydrogen sulfide. This can be formed through the reduction of sulfate during anaerobic respiration (Section 13.3.1) or through the release of reduced sulfur from organic compounds such as the amino acids methionine and cysteine.

In some cases, under both aerobic and anaerobic conditions, rather than being mineralized, organics are **polymerized** to form products with long-term stability. One example is with nitroaromatic compounds such as trinitrotoluene (TNT). Because the aromatic ring

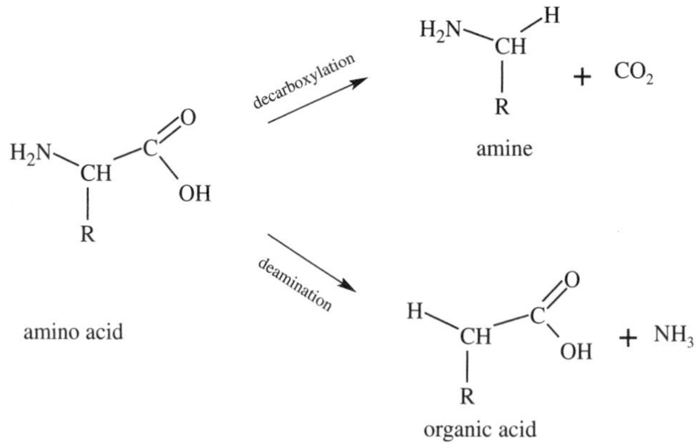

Figure 13.13 Deamination and decarboxylation of an amino acid. (R represents the specific amino acid side-chain.)

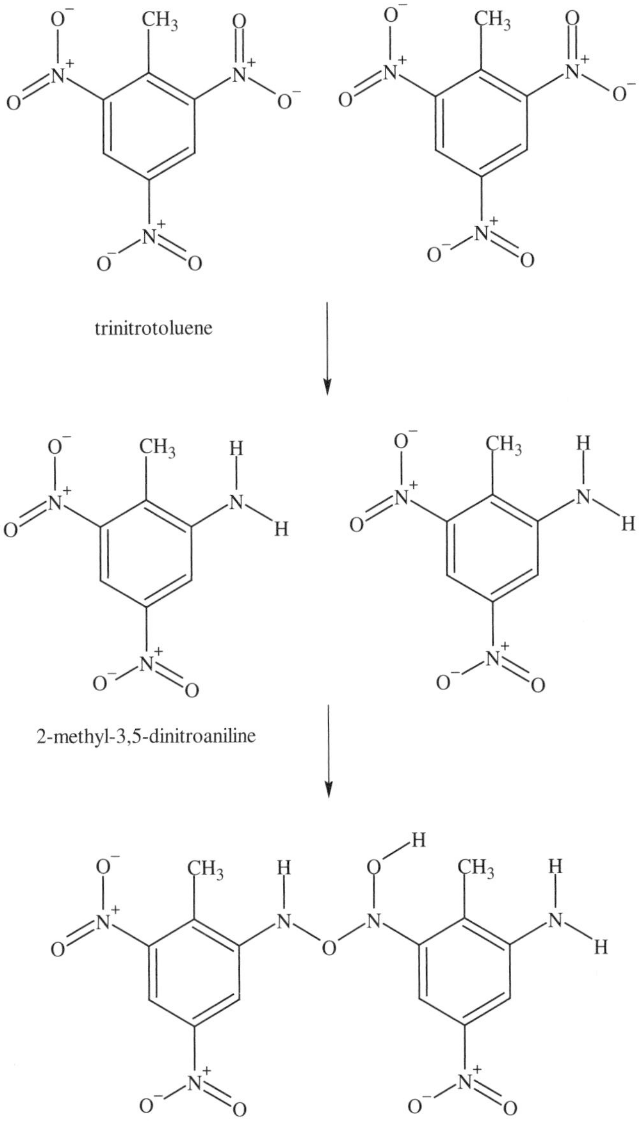

trinitrotoluene

2-methyl-3,5-dinitroaniline

Figure 13.14 Polymerization of trinitrotoluene.

is stabilized by the presence of the nitro groups, TNT is recalcitrant. Although it can be mineralized, biodegradation usually involves reduction of one or more of the nitro groups to an amine, which then reacts with a nitro group on an adjacent molecule (Figure 13.14). A series of such reactions leads to the disappearance of TNT through the formation of a polymer. The environmental significance of such polymers is still not thoroughly understood, as there is concern that the monomers may be released under some conditions.

Polynuclear aromatic hydrocarbons (PAHs or PNAs) are compounds containing two or more condensed (fused together) aromatic rings (Figure 13.15). They are common constituents of coal, oil, and creosote (a wood preservative) and are also products of incomplete combustion. They include the first known human carcinogen, benzo[*a*]pyrene

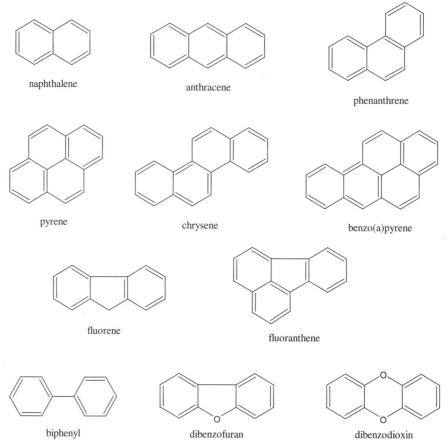

Figure 13.15 Polynuclear aromatic hydrocarbons and related compounds.

(BAP), which was found to be a cause (because of its presence in soot) of testicular cancer in English chimney sweeps.

Generally, the PAHs with two or three rings are readily biodegradable, at least under aerobic conditions. The four-ring compounds pyrene and chrysene are also usually mineralized, although chrysene more slowly. However, there usually seems to be little mineralization (with occasional exceptions) of the higher-molecular-weight PAHs. BAP does appear to be cometabolized to some extent. Probably as a result of initial cometabolic activity, it also appears to be incorporated into the soil humic materials. These complex, irregular soil organic polymers, such as humic acid (Figure 13.16), include phenolics, heterocyclics, sugars, and amino acid residues as building blocks, some probably derived from lignin degradation. Once a BAP or other PAH residue is incorporated into the polymer, it probably cannot be distinguished from other subunits, and thus is no longer expected to be of concern.

Over long periods of time, humic materials buried in anaerobic layers may be reduced further. Eventually, after millions of years, a combination of microbial degradation and abiotic processes at elevated temperatures and pressures has led to coal and oil formation. These organic materials have thus been held out of the active cycling of carbon for

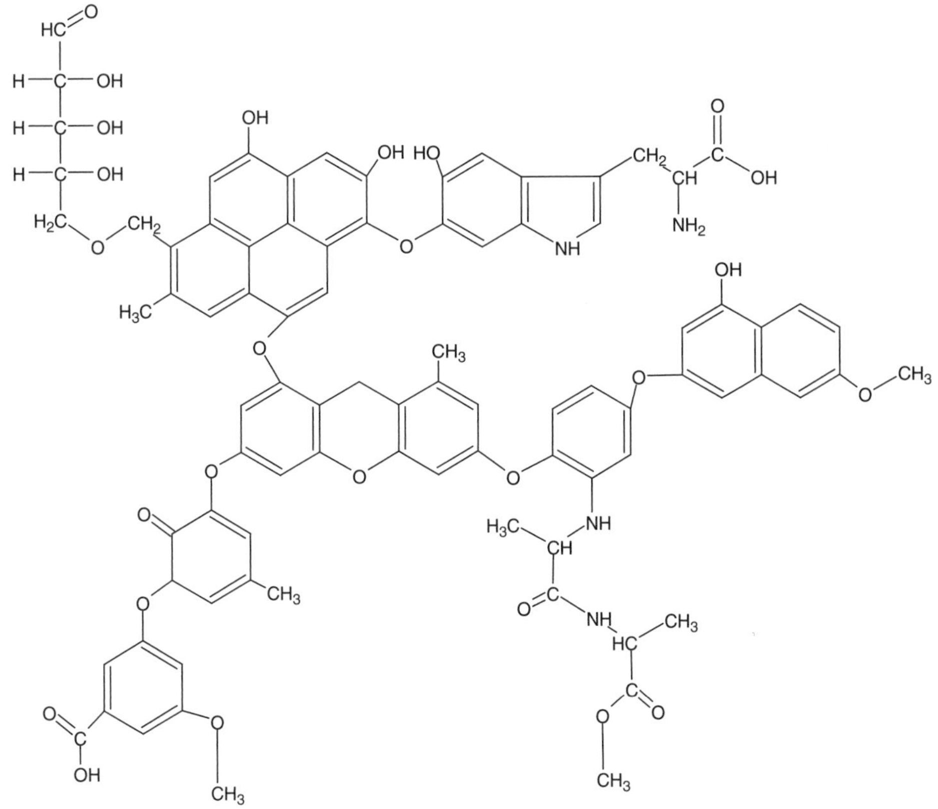

Figure 13.16   Example of a possible humic acid chemical structure.

100 million years or more. The rapidly increasing use of these fossil fuels over the last century is responsible for an increase in atmospheric CO_2 as this carbon reenters active cycling.

Note that although humic materials are not usually referred to as recalcitrant, they do not biodegrade rapidly. Rather, they are considered to be **stable** or to change only slowly. **Stabilization** of wastes thus refers to elimination of the readily degradable components through both mineralization and conversion to high molecular weight products such as humics.

Greenhouse Gases   In most cases, a desired major product of biodegradation is carbon dioxide. However, it is now recognized that atmospheric CO_2 concentrations have increased dramatically during the industrial age, from around 280 ppm prior to the late eighteenth century to 380 ppmv today. Most of this increase is from the combustion of fossil fuels, although some other human activities, such as deforestation, loss of soil organic matter, and draining of wetlands, may also have had a small role. The higher concentration of CO_2 acts to capture more of the infrared energy radiated from Earth's surface. This phenomenon, often referred to as the **greenhouse effect**, has contributed to an increase in average temperature worldwide, or **global warming**. Thus, some thought is now being given to slowing mineralization of certain organic materials and perhaps to rebuilding some stores of humic materials in soil, for example. This may be an area in which environmental engineers and scientists can make contributions in the future.

Carbon dioxide is the most abundant, but it is not the only greenhouse gas. In fact, methane has about 26 times the heat-trapping capacity of CO_2. Additionally, its concentration in the atmosphere has increased even more (proportionally) than that of carbon dioxide, from about 0.7 to almost 1.8 ppm. Much of this increase is from fossil-fuel extraction and use, but other major anthropogenic sources include cattle herds (digestive tracts), rice fields (flooded soils), and solid waste landfills. Even landfills with gas collection systems produce fugitive emissions. This is another area in which environmental professionals can expect to be called upon to help control greenhouse gas releases.

13.2 NITROGEN

Nitrogen is probably second only to carbon in the complexity of its cycling. Its roles in living things are likewise diverse, its organic compounds contributing to physical form (e.g., cells walls, exoskeletons, fingernails, hair), genetics (nucleic acids), and metabolism (enzymes).

With AN = 7, nitrogen has seven electrons, five of them in its outer shell. This results in a range of oxidation states from −3 to +5 (Table 13.1). These extremes are represented by **ammonia**- and **nitrate**-nitrogen, respectively (Figure 13.17). Many of the intermediate forms exist naturally, but −3 (ammonia and **organic nitrogen**), 0 (elemental N_2, also called **dinitrogen**), +3 (**nitrite**), and +5 (nitrate) are most common in soil, water, and biomass. Other fairly common states include −1 (**hydroxylamine**, NH_2OH, an intermediate in the oxidation of ammonia), +1 (**nitrous oxide**, N_2O, also known as *laughing gas*, which is an intermediate in nitrite reduction), +2 (**nitric oxide**, NO, another intermediate in nitrite reduction and also an important air contaminant), and +4 (**nitrogen dioxide**, NO_2, another air contaminant). In Figure 13.18 these forms are shown as part of the nitrogen cycle schematic.

The major accessible source of nitrogen in the biosphere is the elemental nitrogen gas (N_2) that makes up 78% (by volume) of the atmosphere. However, the ability to utilize elemental nitrogen is limited to a few groups of bacteria and archaea. Thus, nitrogen is

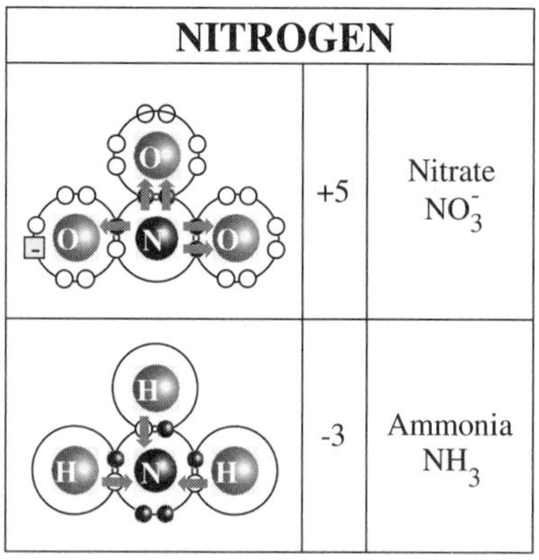

Figure 13.17 Nitrogen oxidation state extremes.

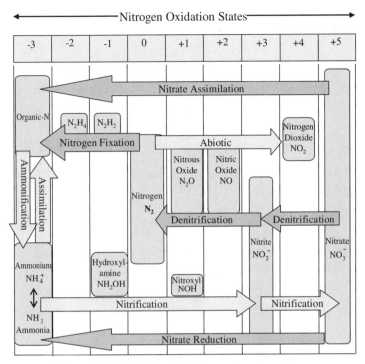

Figure 13.18   Biochemical nitrogen transformations.

a limiting nutrient in many environments, especially for photoautotrophs. Combined nitrogen, whether organic or inorganic, is referred to as **fixed-N**. In addition to microbial activity, nitrogen is naturally fixed in small amounts through photochemical reactions and electrical discharges (lightning) in the atmosphere. Beginning in the early twentieth century, large amounts of nitrogen have also been fixed through fertilizer production, and to a lesser extent, through high-temperature combustion (automobiles, electric power generation, and industrial processes) and explosives manufacturing. These human activities now account for almost 4×10^7 metric tons per year, more than 20% of all fixation. As of yet, this large increase in nitrogen fixation on a global scale does not seem to have "overloaded" other parts of the nitrogen cycle.

In addition to the essential role of microorganisms in nitrogen fixation, there are only a limited number of prokaryotes and a few fungi that can oxidize nitrogen. And although many prokaryotic and eukaryotic microorganisms and plants can utilize inorganic nitrogen in the form of ammonium and/or nitrate, many other microorganisms and all animals can obtain their nitrogen only from organic forms.

When referring to the concentrations of nitrogen-containing compounds, it is common to report them on the basis of the amount of N present rather than the amount of the compound. This makes it easier to keep track of the nitrogen as it is converted from one form to another. Thus, if 1.0 mg/L of NH_3-N is completely converted to nitrate, it yields 1.0 mg/L of $NO_3^- -N$. If this same conversion was reported as the concentration of the species themselves, it would instead be

$$1.21 \, \text{mg/L NH}_3 \rightarrow 4.43 \, \text{mg/L NO}_3^-$$

since for ammonia,

$$\frac{17}{14}(1.0\,\text{mg/L}) = 1.21\,\text{mg/L}$$

and for nitrate,

$$\frac{62}{14}(1.0\,\text{mg/L}) = 4.43\,\text{mg/L}$$

where the atomic weight of nitrogen $= 14$, the molecular weight of $NH_3 = 17$, and the formula weight of $NO_3^- = 62$.

Example 13.3   The drinking water standard for nitrate-N is 10 mg/L. How much is this when it is expressed as nitrate?

Answer   The formula weight for nitrate $= 1(14) + 3(16) = 62$ g/mol:

$$\frac{62\,\text{g/mol nitrate}}{14\,\text{g/mol N}}(10\,\text{mg/L N}) = 44.3\,\text{mg/L nitrate}$$

In addition to the oxidation and reduction reactions discussed in more detail below, several other reactions are important in the nitrogen cycle. One of these is the conversion of organic nitrogen to ammonia. As with any change from an organic form to an inorganic one, this process can be referred to as *mineralization*. However, a more specific name, **ammonification**, is often used in this case. The reverse reaction, **assimilation**, refers to the uptake of an element for incorporation in cell material. Ammonia assimilation does not represent a change in the oxidation state of the nitrogen. However, nitrate assimilation (as discussed below in Section 13.2.1) represents a reduction and requires energy.

Another important nitrogen transformation without oxidation or reduction included in Figure 13.18 is the acid–base reaction between ammonia (nonionized) and ammonium (a cation):

$$NH_4^+ \Leftrightarrow NH_3 + H^+ \tag{13.9}$$

When speaking of the ammonia-N concentration in water, it is usually the total ammonia-N, or NH_4^+-N $+ NH_3$-N that is meant, as this is what is measured by all of the commonly used analytical methods. The pK_a value ($-\log K_a$) at 25°C for this reaction, at which half of the total ammonia would be in each form, is about 9.3:

$$K_a = \frac{[NH_3][H^+]}{[NH_4^+]} = 10^{-9.3}$$

Since acid–base reactions are fairly rapid, at neutral pH values, only a small percentage of the total ammonia is expected to be present as NH_3 (0.50% at pH 7, 4.8% at pH 8).

Example 13.4   The pH of a water sample at 25°C is 7.5. If the measured total ammonia-N concentration is 3.0 mg/L, what is the concentration of nonionized NH_3-N?

Answer   Let $[N_T] = [NH_3] + [NH_4^+]$; then

$$K_a = \frac{[NH_3][H^+]}{[NH_4^+]} = \frac{[NH_3][H^+]}{[N_T] - [NH_3]}$$

$$[NH_3] = [N_T]\frac{K_a}{[H^+] + K_a}$$

$$NH_3\text{-}N = N_T\frac{K_a}{[H^+] + K_a} = N_T\left(\frac{10^{-pK_a}}{10^{-pH} + 10^{-pK_a}}\right) = N_T\left(\frac{1}{10^{pK_a - pH} + 1}\right)$$

$$= 3.0\left(\frac{1}{10^{9.3-7.5} + 1}\right) = \frac{3}{10^{1.8} + 1} = 0.047\,mg/L$$

13.2.1   Nitrogen Reduction

Important nitrogen reduction reactions include nitrogen fixation, dissimilatory nitrate reduction (including denitrification), and nitrate assimilation.

Nitrogen Fixation   Although a gaseous sea of N_2 blankets Earth, it offers nothing in the way of nutritional value except to a select few prokaryotes. The trivalent bond ($N\equiv N$) of this molecule is simply too strong for normal metabolic cleavage. Indeed, one of the names that Antonie Lavoisier originally considered in the late eighteenth century for this recently discovered gas was "*azote*," meaning lifeless or inert.

Since the vast majority of life-forms are unable to utilize nitrogen gas, a critical step in the global cycling of nitrogen is its fixation. **Nitrogen fixation** is the reduction of N_2 to organic or ammonia nitrogen. As indicated in Figure 13.18, some nitrogen is also fixed abiotically in natural systems through oxidation reactions involving combustion, photolysis, or electricity (lightning).

A variety of nitrogen fixers are known, scattered among a number of taxa, but all are prokaryotes. They are often separated into free-living vs. symbiotic forms, although this is not a phylogenetic approach. Although *Azotobacter* and *Beijerinckia* are probably the best known of the free-living, nitrogen-fixing aerobes among the proteobacteria, other examples are found among species of *Klebsiella* (but only when growing under anaerobic conditions) and *Citrobacter* (members of the Enterobacteriaceae family), *Methylomonas* (a methanotroph), and *Thiobacillus* (a sulfur-oxidizing autotroph). There are also aerobic gram positives (e.g., a few *Bacillus* and the actinomycete *Streptomyces)* and many cyanobacteria, such as *Anabaena* (see Figure 10.20) and *Nostoc*. Free-living anaerobic nitrogen fixers include members of the gram-positive spore-formers *Clostridium* and *Desulfotomaculum* (a sulfate-reducer); the sulfate-reducing proteobacteria *Desulfovibrio*; several phototrophs (Table 10.4), such as *Chlorobium* (green sulfur bacteria), *Chromatium* (purple sulfur), *Rhodospirillum* (purple nonsulfur), and *Heliobacterium* (another gram-positive spore-former); and a few methanogens, such as *Methanococcus*, which are archaea.

The best known symbiotic nitrogen fixers are **rhizobia**, proteobacteria such as *Rhizobium* associated with leguminous plants (e.g., beans, peas, soybeans, peanuts, alfalfa, and clover). Within the **rhizospere** (root zone in the soil), these bacteria infect the roots of the plant, forming **nodules** (knoblike growths). Root nodules help the host plant by providing fixed nitrogen (often, the limiting nutrient for plants), while providing organic substrates (often, the limiting nutrient for heterotrophs) produced by the plant for the bacteria. The actinomycete *Frankia* also produces root nodules, most commonly in nonleguminous woody plants such as the alder and bayberry.

Other nitrogen-fixing bacteria form associations with plants without nodule formation. *Clostridium* and *Desulfovibrio*, for example, grow in the root zone of eelgrass, a shallow saltwater plant, and *Azotobacter* and *Azospirillum* grow in the rhizosphere of some grasses, including corn. Nitrogen-fixing cyanobacteria may also form symbiotic relationships with fungi, as in some lichens, as well as with plants such as the water fern *Azolla* (used agriculturally to fix nitrogen in rice paddies). This great diversity of organisms and interactions further demonstrates the great adaptability of life!

Nitrogen fixation requires an uncommonly high metabolic investment of energy and reducing power to break molecular nitrogen's strong trivalent bond (almost twice the energy of oxygen's double bond). The conversion of a single molecule of N_2 into 2 mol of reduced nitrogen requires eight protons, eight electrons, and between 16 and 24 ATP (Figure 13.19).

The highly specialized **nitrogenase** enzyme complex, which includes dinitrogenase, employs cofactors containing iron and usually molybdenum or, occasionally, vanadium. Other iron-containing enzymes and ATP are needed to transfer the required electrons to the nitrogenase system. The reducing power (electrons) must come, in turn, from an electron-donating substrate (usually organic, although some nitrogen fixers can use sulfide, hydrogen, or carbon monoxide). Although the conversion of two nitrogen atoms from an oxidation state of 0 to -3 requires only six electrons, the process actually requires eight, producing a molecule of hydrogen gas (H_2).

There is some variation in the enzymes among different species, but even the nitrogenases of aerobic bacteria are typically highly sensitive to irreversible inactivation by oxygen. How, then, can these aerobes fix nitrogen? Obviously, they must have means of preventing the contact of oxygen with the enzymes. One approach, taken by *Klebsiella*, which is a facultative anaerobe, is to fix nitrogen only under anaerobic conditions. Other

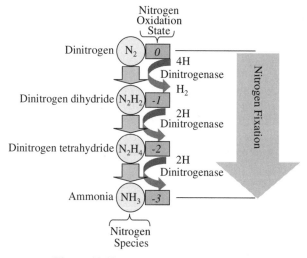

Figure 13.19 Nitrogen fixation pathway.

strategies include producing a physical barrier, forming a protective complex with another protein, and/or maintaining high rates of respiration to deplete oxygen near and within the cell. Thus, another advantage of being within a root nodule is the more restricted flux of oxygen. *Klebsiella* and some other free-living bacteria can also produce a thick slime layer. Being in the rhizosphere, where root exudates produce a higher concentration of respirable organic substrates, helps limit oxygen concentration.

Several of the filamentous cyanobacteria fix nitrogen only in a few differentiated cells (unusual in prokaryotes) with thick walls, called *heterocysts* (see Section 10.5.4 and Figure 10.20). The nitrogen is then distributed to the other cells in the filament, which in return provide organic substrates to the nonphotosynthetic heterocyst. Many cyanobacteria may also prefer reduced levels of oxygen (perhaps 10% of saturation).

Most nitrogenases will also cometabolize other triple bonds, such as those of acetylene (HC≡CH) and hydrogen cyanide (HC≡N). In fact, acetylene reduction to ethylene (H₂C=CH₂) is often used as an indirect means of measuring nitrogen-fixing activity.

Dissimilatory Nitrate Reduction, Including Denitrification   A wide variety of aerobic prokaryotes are also able to utilize nitrate (NO_3^-) as a terminal electron acceptor in the absence of oxygen. In many cases, including with strains of *Escherichia coli, Bacillus, Staphylococcus, Spirillum*, some actinomycetes, *Aquifex*, and some archaea, the product of this dissimilatory nitrate reduction is nitrite (NO_2^-). A few of these organisms are able to further reduce the nitrite to ammonia, but this type of ammonification (as opposed to the mineralization of organic nitrogen) appears to be of only minor importance in most environments.

However, under similar conditions, anaerobic respiration by a fairly broad array of otherwise aerobic prokaryotes, such as some *Pseudomonas, Thiobacillus denitrificans, Paracoccus*, and a few archaea, results in a sequential conversion (Figure 13.20) of nitrate

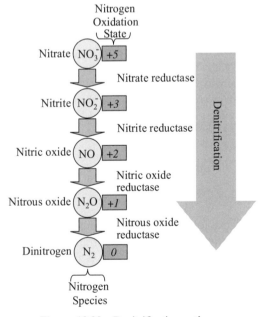

Figure 13.20   Denitrification pathway.

to nitrite to elemental nitrogen gas (N_2). The term **denitrification** is applied to this process because N_2 is unusable by most organisms, so that these reactions represent a loss of available (fixed) nitrogen to an aquatic or soil environment. (Similarly, the release of small amounts of the gaseous intermediates NO and N_2O represent losses of nitrogen from the system.)

A general reaction for oxidation of a carbohydrate through denitrification can be written as

$$\tfrac{5}{4}CH_2O + NO_3^- \rightarrow \tfrac{5}{4}CO_2 + \tfrac{1}{2}N_2 + \tfrac{3}{4}H_2O + OH^- \tag{13.10}$$

If the organic material oxidized is methanol, the reaction can be written instead as

$$\tfrac{5}{6}CH_3OH + NO_3^- \rightarrow \tfrac{5}{6}CO_2 + \tfrac{1}{2}N_2 + \tfrac{7}{6}H_2O + OH^- \tag{13.11}$$

Example 13.5   An anaerobic groundwater contains 100 mg/L of carbohydrate with an empirical formula of CH_2O. How much nitrate would have to be added to meet the stoichiometric requirement for complete oxidation of this contaminant?

Answer   From equation (13.10), 1.0 mol of nitrate would be needed per 1.25 mol of CH_2O, which has an apparent molecular weight of $1(12) + 2(1) + 1(16) = 30$:

$$\begin{aligned}
\text{required } NO_3^--N &= 100\,mg\ CH_2O/L\left(\frac{1\,mol\ CH_2O}{30\,g\ CH_2O}\right)\left(\frac{1\,mol\ NO_3^-}{1.25\,mol\ CH_2O}\right)\left(\frac{14\,g\ NO_3^-}{1\,mol\ NO_3^-}\right) \\
&= 37.3\,mg/L\ NO_3^--N
\end{aligned}$$

Under anaerobic conditions we often expect unpleasant odors to develop. When nitrate is present, however, this typically does not happen. Further, the active organisms are mainly aerobes that have the additional ability to respire using nitrate as an alternative terminal electron acceptor. Thus, sanitary engineers, and subsequently, environmental engineers and scientists, have traditionally referred to conditions in which oxygen is absent but nitrate is present as **anoxic** rather than anaerobic. However, please be aware that most biologists and others outside our field do not make this distinction, using anaerobic and anoxic as synonyms to refer to any system in which oxygen is absent.

Although most denitrifiers are organotrophs, some are lithotrophs utilizing hydrogen or sulfide. Recently, autotrophic nitrification (Section 13.2.2) also has been linked to denitrification, in a process dubbed **anammox** (*an*oxic *amm*onia *ox*idation). In oxidizing ammonia as an energy source, the organisms involved can reduce nitrate to nitrite and then reduce the nitrite to nitrogen gas:

$$NH_4^+ + NO_2^- \rightarrow N_2 + 2H_2O \tag{13.12}$$

This process does not seem to play a major environmental role in nitrogen cycling. However, it may be important to the organisms involved in some circumstances, particularly when aerobic conditions (under which the nitrate is produced) are followed by anoxic ones. This occurs in soils that are periodically flooded, as well as in some wastewater treatment plants.

In general, denitrification does not occur in the presence of oxygen. Since approximately 20% less energy is available to an organism when it is utilizing nitrate, there

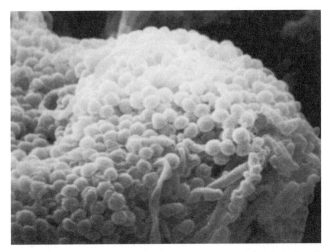

Figure 13.21   Clustered nitrifying cocci within activated sludge floc.

has been strong selective pressure for microbes to develop control mechanisms to suppress nitrate reduction in the presence of O_2. However, denitrification can occur in anoxic **microenvironments** (microscopically localized environments having different physical and chemical conditions) within heterogeneous aerobic systems, such as moist soils and microbial colonies.

Nitrate Assimilation   A wide variety of bacteria, archaea, fungi, algae, and most plants can utilize nitrate as their source of nitrogen. However, to do this, they must reduce it to an oxidation state of -3 (the level of ammonia and organic nitrogen). This is referred to as **assimilatory nitrate reduction**. While dissimilatory reduction can be linked to energy-yielding respiration, assimilatory reduction requires energy. As a result, the biomass yield of organisms growing on nitrate as their nitrogen source will be slightly lower than when ammonia is used. Hence, an organism that is able to use both ammonia and nitrate will utilize ammonia preferentially when it is available. On the other hand, the presence of oxygen does not inhibit assimilatory reduction.

13.2.2   Nitrogen Oxidation

The oxidation of ammonia to nitrite, and nitrite to nitrate, is known as **nitrification**. There is no known biochemical oxidation of nitrogen gas, although there are abiotic mechanisms (electrochemical, photochemical, thermal). Similarly, there is no known pathway for direct conversion of ammonia to nitrogen gas (the reverse of nitrogen fixation).

Autotrophic Nitrification   Most nitrification results from the activity of aerobic, autotrophic chemolithotrophs, referred to as **nitrifying bacteria** (Figure 13.21). They utilize reduced nitrogen (ammonia or nitrite) as their energy source, carbon dioxide as their carbon source, and oxygen as their electron acceptor.

The overall oxidation really consists of two separate steps, carried out by different bacteria (Figure 13.22). Ammonia oxidation involves the transformation of ammonia to nitrite, a six-electron transfer (-3 to $+3$), and is sometimes called nitrite formation or

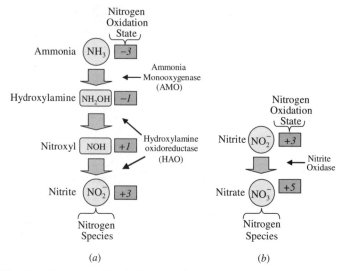

Figure 13.22 Steps in autotrophic nitrification: (a) ammonia oxidation; (b) nitrite oxidation.

nitritification. In the second step, nitrite oxidation or **nitratification**, nitrite is converted to nitrate, a two-electron change (+3 to +5). Similarly, the bacteria can be referred to as **ammonia oxidizers** (or **nitritifiers**) and **nitrite oxidizers** (**nitratifiers**). No known nitrifier can oxidize ammonia all the way to nitrate.

Balanced equations for the individual reactions and their sum can be written as

$$NH_4^+ + \tfrac{3}{2}O_2 \rightarrow NO_2^- + H_2O + 2H^+ \tag{13.13}$$

$$NO_2^- + \tfrac{1}{2}O_2 \rightarrow NO_3^- \tag{13.14}$$

$$NH_4^+ + 2O_2 \rightarrow NO_3^- + H_2O + 2H^+ \tag{13.15}$$

Note that the equations are written with ammonium and nitrite as the reactants, since it is these ionic forms that are expected to be predominant at neutral pH values. However, there is some evidence that the forms actually used by the bacteria are nonionized ammonia (NH_3) and nitrous acid (HNO_2). Also note that 2 mol of oxygen is used and that 2 mol of strong acid is produced from 1 mol of weak acid.

Known organisms carrying out the first (nitritification) step are the proteobacteria *Nitrosomonas* (β), *Nitrosospira* (β), and *Nitrosococcus* (γ). The second (nitratification) step is performed by the proteobacteria *Nitrobacter* (α), *Nitrococcus* (γ), and *Nitrospina* (δ), and by *Nitrospira*, a member of the Xenobacteria. Thus, although all of the nitrifiers were once included in the same family because of their activities, it is now recognized that they are phylogenetically diverse. Also, at one time most nitrifiers were assumed to be either *Nitrosomonas* or *Nitrobacter*. However, it is now recognized, based on genetic techniques, that although these are the most often cultured species, they are not necessarily the most common or most active in the environment. Hence, nitrifying activity should not be assigned to these genera unless they are actually identified.

As shown in Figure 13.22, ammonia monooxygenase (**AMO**) is the enzyme responsible for catalyzing the first reaction of nitrification, in which ammonia is oxidized to hydroxylamine. Hydroxylamine oxidoreductase then produces a transient intermediate (nitroxyl) while forming nitrite. Luckily, hydroxylamine itself rarely accumulates, as it is a potential mutagen. Nitrite oxidation is catalyzed by nitrite oxidase.

AMO shows some similarity to methane monooxygenase (**MMO**), the enzyme used by methanotrophs (Section 10.5.6) to oxidize methane. In fact, many ammonium oxidizers and methanotrophs can aerobically cometabolize each other's substrate as well as a number of other compounds, including trichloroethylene.

Compared to oxidation of organic compounds, relatively little energy is available to nitrifiers. It takes about 35 mol of NH_4^+ for ammonia oxidizers to fix 1 mol of CO_2. Nitrite oxidizers require even more substrate, about 100 mol of NO_2^- per mole of CO_2 fixed. Since the nitrite typically comes from the ammonia (100 mol of ammonia produces 100 mol of nitrite), this means that ammonia oxidizers are usually more abundant than nitrite oxidizers. Also, cell yields based on their energy source are much lower for nitrifiers than for most heterotrophs, often in the range 5 to 20% rather than 50 to 60%.

The fact that nitrifiers appear to belong to only a few genera suggests that there may be more limitations on their activity than would be true if they were more diverse. In fact, compared to many heterotrophs, nitrifiers are slow growing. Under ideal conditions, minimum doubling times are around 8 hours. Also, there are no known thermophilic nitrifiers, so that autotrophic nitrification does not occur in systems with temperatures above ~42°C. Furthermore, since there are no known sporeformers, elevated temperatures actually kill the nitrifiers; this means that activity is slow to return to a system (being dependent on reinvasion or reinoculation) even once elevated temperatures decrease. Optimum temperature is usually around 28 to 30°C, and activity is usually minimal at temperatures below 10°C.

Similarly, pH can be limiting. Optimum pH values are around 7.5 to 8, with almost no activity below pH 6. This may in part be because of the unavailability of nonionized ammonia at low pH values. Also, nitrite is more toxic at low pH, where it is present as nonionized nitrous acid. At high pH, toxicity from ammonia becomes a problem. On the other hand, although they are aerobic [with the exception of the anammox process, equation (13.12)], nitrifiers can survive for prolonged periods under anaerobic conditions and are effective at utilizing low concentrations of oxygen. In other words, they have a low K_s (half-saturation coefficient, Section 11.7.2) value for dissolved oxygen, typically below 0.5 mg/L. Similarly, they require only small amounts of their energy sources to approach maximum activity rates (K_s values for ammonia- or nitrite-N of 1 mg/L or less).

A close relationship between ammonia and nitrite oxidizers can be expected, since the product of the first group is the substrate for the second. Thus, the two groups are typically located in close physical association. Ammonia oxidizers are usually more abundant, since about three times as much energy is available from ammonia oxidation as nitrite oxidation. Typically, only traces of nitrite are seen in the environment. Thus, nitrification is often treated as though it was a single step, involving one group of bacteria. However, accumulations of nitrite can occur under transient conditions, particularly since the nitrite oxidizers appear to be a little more sensitive to low pH and high concentrations of ammonia and nitrite.

Heterotrophic Nitrification   Some heterotrophic bacteria and fungi are able to oxidize nitrite to nitrate, and/or occasionally, ammonia to nitrite. This does not appear to provide any benefit to the organism and hence is considered a type of cometabolism. Perhaps in some cases this represents assimilatory nitrate reduction enzymes working in reverse.

It is generally believed that heterotrophic nitrification plays only a small role in nitrogen cycling. However, in some environments, such as acid soils, in which autotrophic nitrification is severely inhibited, it may have a local effect.

13.2.3 Nitrogen in Environmental Engineering and Science

If nitrogen is second only to carbon in terms of the complexity of its cycling, it probably also is second only to carbon in terms of its importance in environmental engineering and science. The range of concerns includes wastewater and potable water treatment, surface and groundwater contamination, agriculture, sludge and solid waste management, human and environmental toxicity, and bioremediation. What is more, these problems may be interdependent. For example, in agriculture, nitrogen is often the limiting nutrient for many important crops (e.g., wheat, corn, cotton), so that loss of fixed nitrogen is a major concern. However, overapplication of chemical fertilizers, manures, or wastewater treatment sludges, which typically are high in nitrogen, can lead to groundwater (from leaching) and/or surface water (from runoff) contamination. In fact, nitrate is the leading groundwater contaminant in the United States, mostly as a result of agricultural practices.

Oxygen Demand   As noted in equation (13.15), complete stoichiometric oxidation of ammonium requires 2 mol of O_2 per mole of ammonium-N. This translates into $64/14 = 4.57$ mg O_2/mg NH_4^+-N oxidized to nitrate. Because the organisms also reduce CO_2 and assimilate some N for cell constituents, the actual ratio is a little lower, typically 4.33 mgO_2/mg NH_4^+-N (3.22 for ammonium oxidation to nitrite and 1.11 for converting nitrite to nitrate). This represents the N-BOD (Section 13.1.3).

Consider typical sewage (Table 13.4), with a BOD of 200 mg/L and a reduced nitrogen content of 30 mg/L (half organic-N and half ammonium-N). During secondary treatment (biological treatment required under federal law; see Chapter 16), the oxygen required to reduce the BOD to 20 mg/L (90% removal) is 180 mg/L. For the nitrogen, typically the organic-N is converted to ammonium, and about 5 mg/L is assimilated, leaving 25 mg/L. If this remaining nitrogen is oxidized, it would require $25 \times 4.33 = 108$ mg/L of oxygen, an increase of 60% of the original carbonaceous demand. Some industrial wastewaters may have much higher ammonium concentrations (even 1000s of mg/L) and thus represent even greater N-BODs.

TABLE 13.4 Typical Carbonaceous and Nitrogenous Biochemical Oxygen Demand of Treated[a] and Untreated Sewage

	C-BOD (mg/L)	TKN (mg/L)	N-BOD[b] (mg/L)	Total BOD (mg/L)	N-BOD (% of C-BOD)	N-BOD (% of total)
Sewage	200	30	130	330	65	39
Effluent nitrified	20	1	4.3	24	22	18
Effluent unnitrified	20	25	108	128	540	84

[a]Assuming secondary treatment with 90% C-BOD removal and assimilation of 5 mg/L NH_4^+-N. The remaining TKN is either oxidized to nitrate (nitrified) or remains unoxidized (unnitrified). Total Kjeldahl nitrogen (TKN) consists of organic-N (usually quickly ammonified) + ammonium-N.
[b]Based on 4.33 mg O_2 demand/mg NH_4^+-N oxidized.

Originally, biological sewage treatment plants produced nitrified effluents, since it was recognized that the presence of nitrate indicated thorough stabilization of the wastewater. However, once the BOD test became available, plants realized that they could reduce the energy costs associated with supplying oxygen (a major portion of the overall cost of treatment) by providing only carbonaceous removal. Thus, most U.S., plants for many years were designed and operated so as not to nitrify.

However, this meant that ammonium was being discharged to the receiving water, where it might also undergo nitrification and exert an N-BOD. Two classic papers in the 1960s, one examining the Thames estuary in London and the other the Grand River in Michigan, showed that this did in fact occur in some waters, seriously depleting DO. However, in some other receiving waters there was little evidence of nitrification. It now appears that nitrification can be expected in smaller rivers and streams, where the nitrifiers grow mainly attached to streambed surfaces (including vegetation), and in estuaries, where planktonic nitrifiers have time to grow. In larger rivers, where streambed surfaces are relatively small compared to the water volume, nitrifiers appear to settle out and to grow too slowly to oxidize most of the ammonia.

Ammonium may also enter streams from agricultural runoff, stormwater from suburban and urban areas (including erosion from fertilized lawns), and from natural organic inputs, such as falling leaves and bird droppings. Of course, this nitrogen also can represent an oxygen demand.

In denitrification, nitrate is used as an alternate electron acceptor to oxygen. Thus in systems in which carbon oxidation and nitrification take place, denitrification can result in "recovery" of some of the oxygen utilized for nitrification. For example, in equation (13.10), 1 mol of nitrate is "used" to oxidize 1.25 mol of carbohydrate. This nitrate required 2 mol of oxygen when it was produced from ammonia [equation (13.15)]. If the same amount of carbohydrate were oxidized by oxygen, 1.25 mol of O_2 would be required. Thus, using the nitrate saves using 1.25 mol of O_2, essentially "recovering" 62.5% ($100 \times 1.25/2$) of the oxygen that was used to form the nitrate. Some advanced wastewater treatment plants that nitrify take advantage of this to reduce aeration costs while removing nitrogen as N_2.

Toxicity of Nitrogen Compounds   If neither the treatment plant nor the receiving water nitrifies, the ammonia from discharges and runoff will be present in the stream. This can be a serious problem in that nonionized ammonia can be extremely toxic to aquatic organisms. It is recommended that nonionized ammonia concentrations in freshwater remain below 0.02 mg/L (20 ppb) to protect the most sensitive water uses (such as trout production), or below 0.04 mg/L in most other cases. (These values are an exception to the general reporting convention for nitrogen and are based on NH_3 rather than NH_3-N.) The amount of nonionized ammonia is strongly dependent on pH (Section 13.2), but at the slightly alkaline pH values of some streams, these levels can be exceeded. Thus, whether the stream nitrifies or not, there may be good arguments to require nitrification within the treatment plant, and it is becoming more common again.

Example 13.6   In Example 13.4, a water sample had 3.0 mg/L total ammonia- + ammonium-N. If that sample was taken from a stream below a wastewater treatment plant, does the stream meet an nonionized NH_3 standard of 0.04 mg/L?

Answer   The nonionized NH_3 concentration calculated for that example was 0.047 mg/L NH_3-N. Thus, the standard is not being met.

Ammonia is also toxic to microorganisms, including nitrifiers, and at high levels can interfere with treatment. As long as the nitrifiers are active, feed concentrations to a reactor can be high, but a sudden increase in concentration or pH can lead to failure. The increased concentration of nonionized ammonia slows down nitrifier activity, decreasing ammonia oxidation rates and leading to a rapid buildup in concentration and further increased toxicity. Typically, well-designed and well-operated reactors can handle up to 500 mg/L (or even higher) influent total ammonia concentrations, depending on pH, because the concentration within the reactor will remain well below 100 mg/L.

Nitrite is also highly toxic to aquatic organisms. U.S. recommendations at one time were to maintain concentrations of nitrite-N below 0.06 mg/L, and some European countries have suggested a 0.02 mg/L limit. Fortunately, nitrite rarely accumulates in the environment, since it is an intermediate for both nitrification and denitrification. Apparently for this reason, there is no longer a U.S. nitrite criterion.

Both nitrite and nitrate may be toxic to humans, and drinking water standards have been set at 1 mg/L for nitrite-N and 10 mg/L for nitrate-N. Upon ingestion, nitrate may be reduced to nitrite in the digestive tract. Once absorbed into the bloodstream, nitrite binds (preferentially over oxygen) with hemoglobin, leading to a mild to severe asphyxia known as **methemoglobinemia**. This is of particular concern for bottle-fed infants, in whom it is called "blue baby disease", as it can be fatal. Nitrite can also react with certain amines released during degradation of amino acids, forming **nitrosamines**, some of which are potent carcinogens.

Nutrient Enrichment   Usually, either nitrogen or phosphorus is the nutrient that limits the amount of photosynthetic growth that can occur in a water body. Typically, while phosphorus is limiting in lakes and reservoirs, nitrogen is the limiting nutrient in streams and coastal waters. Since a variety of organisms can utilize each form, addition of nitrogen as either ammonium or nitrate where N is limiting can lead to excessive growth of plants, algae, and/or cyanobacteria, speeding the process of eutrophication (Section 15.2.6). This is a major reason that some wastewater treatment plants are required to practice some form of advanced treatment to remove nitrogen, not merely to nitrify it.

Alkalinity Consumption   As can be seen from equation (13.13), each mole of ammonium oxidized to nitrite produces 2 mol of acidity. This uses up 2 mol of the **alkalinity** (acid-neutralizing, or pH-buffering, capacity) of the water in which it occurs. If the water does not initially have enough alkalinity, this will result in a drop in pH. This can, in turn, inhibit nitrification as the pH drops as low as 5.5, as well as resulting in other harmful effects in wastewater treatment or perhaps soil. (Ammonia concentrations are usually diluted enough that this level of acidification of a stream will not occur.) To prevent this, nitrifying treatment plants may have to add supplemental alkalinity to raise the pH to more desirable or permitted levels (usually, > 6.5). Acidity produced by nitrification can also be one of the reasons for adding lime (usually, it is really crushed limestone, $CaCO_3$) to agricultural fields or fertilized lawns and gardens.

Example 13.7   In a wastewater treatment plant, 20 mg/L of ammonium-nitrogen is nitrified to nitrate. How much alkalinity is consumed?

Answer   From equation (13.13), 2 mol of alkalinity is needed for every mole of nitrogen oxidized. By convention, alkalinity is expressed as $CaCO_3$ equivalents, with

2 mol of acid neutralized per mole (100 g). Thus 1 mol of $CaCO_3$ is needed for every mole of ammonium:

Alkalinity consumed

$$= 20\,\text{mg/L NH}_4{}^+\text{-N}\left(\frac{1\,\text{mol NH}_4{}^+\text{-N}}{14\,\text{g NH}_4{}^+\text{-N}}\right)\left(\frac{1\,\text{mol CaCO}_3}{1\,\text{mol NH}_4{}^+\text{-N}}\right)\left(\frac{100\,\text{g CaCO}_3}{1\,\text{mol CaCO}_3}\right)$$

$$= 20(7.14 \text{ mg CaCO}_3/\text{mg N}) = 143\,\text{mg/L CaCO}_3 \text{ alkalinity}$$

Denitrification produces 1 mol of alkalinity for each mole of nitrite or nitrate reduced to nitrogen gas [e.g., equations (13.10) and (13.11)]. Thus half of the alkalinity lost through nitrification can be recovered by denitrification while removing the nitrogen (as a gas) from the system. This will often eliminate the need for alkalinity addition in wastewater treatment, but would usually be undesirable in agriculture, where the nitrogen loss is typically a concern.

Disinfection   Nitrite and chlorine rapidly react to form nitrate and chloride, which has no disinfecting power. For example, in drinking water, where hypochlorous acid is the predominant active chlorine species;

$$\text{NO}_2^- + \text{HOCl} \rightarrow \text{NO}_3^- + \text{H}^+ + \text{Cl}^- \tag{13.16}$$

Thus, every milligram of NO_2^--N uses up $71/14 = 5.1$ mg of active chlorine (which is always expressed on the basis of Cl_2, with molecular weight of 71).

Ammonia also reacts with chlorine, but the products depend on the relative concentrations of the two. At low Cl/N molar ratios (≤ 1), mainly **monochloramine** (NH_2Cl), forms:

$$\text{NH}_4^+ + \text{HOCl} \rightarrow \text{NH}_2\text{Cl} + \text{H}_2\text{O} + \text{H}^+ \tag{13.17}$$

As the Cl/N ratio increases, **dichloramine** ($NHCl_2$) forms:

$$\text{NH}_2\text{Cl} + \text{HOCl} \rightarrow \text{NHCl}_2 + \text{H}_2\text{O} \tag{13.18}$$

Mono- and dichloramine are referred to as **combined chlorine**. Although they still represent active chlorine, they have much less disinfecting power than HOCl ($\sim$1 to 2% as effective). This is one reason that disinfection of wastewater (which usually contains ammonia) is less effective than disinfection of drinking water (which typically does not).

Finally, at still higher ratios, the chlorine oxidizes the ammonia to produce nitrogen gas or nitrate, or occasionally, trichloramine (NCl_3), depending on pH and the specific ratio.

$$\text{NHCl}_2 + \text{HOCl} \rightarrow \text{N}_2 \text{ and/or NO}_3^- \text{ and/or NCl}_3 \tag{13.19}$$

This is referred to as *breakpoint chlorination* (Figure 13.23), and these products have no disinfecting power. Thus, depending on the Cl/N ratio, adding more chlorine can actually result in a decrease in the remaining active chlorine, or **chlorine residual**. Once all of the ammonia is converted to nitrogen gas, nitrate, or trichloramine, any additional chlorine addition results in an increase in the chlorine residual.

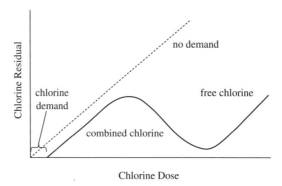

Figure 13.23   Breakpoint chlorination.

Other   In the soil, ammonium is often held at cation-exchange sites of clay minerals. Nitrification **mobilizes** this nitrogen, making it easier for plants to absorb if the ammonium in their rhizosphere has been depleted, but also easier for it to leach into the groundwater. On the other hand, assimilation of ammonium or nitrate by bacteria can make this nitrogen **immobilized** and unavailable to plants. Thus, if waste materials with a high carbon/nitrogen ratio, (such as some crop residues and some solid wastes) are applied to soil, the microbial activity on the carbonaceous materials can lead to a depletion of available soil N for plants. This is referred to as **nitrogen robbing**.

Some potable waters naturally contain ammonia, and others may add it intentionally. This results in the formation of chloramines, which although weaker disinfectants, are less likely to react with organic compounds in the water to produce trihalomethanes and other potential carcinogens. However, nitrifiers have been found growing as biofilms attached to the walls of the water distribution pipes in some of these systems, where their relative resistance to chlorination aids in their survival. Concerns arising from these observations include the production of nitrite (which inactivates chlorine), organic material (promoting growth of heterotrophs), and acidity (potentially contributing to corrosion, which might further increase concentrations of lead in the water).

During anaerobic digestion of sewage sludges, concentrations of ammonium (from ammonification), magnesium, and phosphate will sometimes become high enough to lead to the precipitation of magnesium ammonium phosphate, or **struvite** ($MgNH_4PO_4$). Heavy deposition of this mineral can clog pipes, coat heat exchangers, and otherwise interfere with digester operation.

Because of the limited metabolic diversity of nitrifiers, some compounds act as **selective inhibitors**, limiting nitrification while having little or no effect on other organisms. Allyl thiourea was one compound used early for this purpose in both agriculture (to prevent loss of nitrogen through nitrification followed by leaching or denitrification) and the BOD test (to prevent interference in the measurement of the C-BOD, Section 13.1.3), but it biodegrades fairly rapidly. Other compounds have also been used, most containing sulfur and/or nitrogen. The compound now recommended for the C-BOD test is 2-chloro-6-(trichloromethyl) pyridine, also called N-serve, nitrapyrin, and TCMP. For research purposes, nitrite oxidation has been inhibited using potassium perchlorate, $KClO_3$, but this compound may also inhibit nitrate reduction. It is also possible that a compound entering a wastewater treatment plant may interfere with nitrification without having a noticeable effect on other treatment processes.

13.3   SULFUR

The biogeochemical cycling of sulfur involves substantial abiotic chemical as well as the biochemical conversions. There are five sulfur oxidation states of appreciable biochemical importance in the environment. With an atomic number of 16, sulfur's outer orbital shell has six electrons. By either depleting or filling this shell, therefore, sulfur may reach extreme oxidation states of -2 (sulfide, S^{2-}) and $+6$ (sulfate, $SO_4{}^{2-}$) (Figure 13.24), as well as residing at 0 (elemental sulfur, S^0), $+2$ (thiosulfate, $S_2O_3{}^{2-}$), and $+4$ (sulfite, $SO_3{}^{2-}$). (*Note*: In pyrite, FeS_2, the sulfur has an average oxidation state of -1, although this may result from a combination of S^{2-} and S^0.)

As with reduced nitrogen in the form of ammonia, reduced sulfide can also be volatile (as hydrogen sulfide, H_2S). However, these two gases have opposite responses to pH. Whereas ammonia gas is a weak base and tends to form at higher pH levels ($pK_a = 9.3$), hydrogen sulfide is a weak acid ($pK_a = 7.2$ for first H^+, 11.9 for the second). Thus, it tends to be nonionized, and hence volatile, only at neutral or lower pH. Reduced sulfur in alkaline solutions tends to remain ionized, mostly as HS^- (or S^{2-} at very high pH). Sulfide also tends to react with metals to form insoluble precipitates. Ferrous sulfide (FeS), in particular, gives many anaerobic sediments and biofilms their black color.

The organic sulfur of the amino acids methionine and cysteine (Table 3.6) is also at an oxidation state of -2 and thus has two bonds with adjacent atoms (Section 3.2). One of these is to a carbon atom, but the other may be to a hydrogen (sulfhydryl bond) or to the sulfur (disulfide bond) of a second amino acid, an important factor in the three-dimensional configurations of proteins.

In comparison to some of the other elements, which are considerably more limiting, sulfur tends to be readily available in most environments. Fresh water usually contains at least 10 mg/L of sulfate, and sometimes much more (especially near the coast), and sewage usually has $\sim$30 mg/L more sulfate than the drinking water from which it is derived. The concentration in seawater is $\sim$2700 mg/L. The limit for sulfate in drinking water is 250 mg/L, but this is a secondary standard (protecting public welfare rather than health) because of taste and a laxative effect.

The reactions of the sulfur cycle are depicted schematically in Figure 13.25. As with the nitrogen cycle, some of the steps are carried out only by prokaryotes. Bacteria and

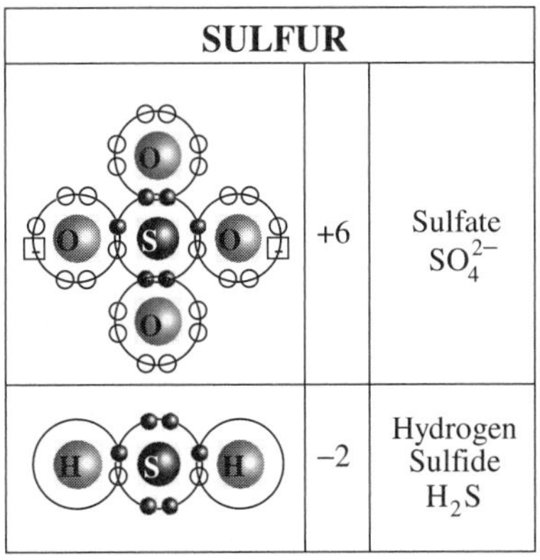

Figure 13.24   Sulfur oxidation state extremes.

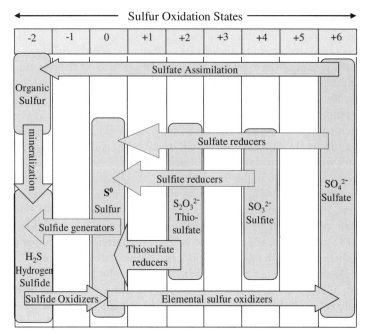

Figure 13.25 Biochemical sulfur transformations.

archaea associated with some of the nonassimilative sulfur reductions (use as an electron acceptor) and oxidations (use as an energy source) are listed in Table 13.5.

Also as with nitrogen, mineralization of organic sulfur does not involve oxidation or reduction of the sulfur, which is released as sulfide. However, assimilation of sulfide is uncommon, largely because of its toxicity. Instead, most organisms utilize sulfate (requiring a reduction) and/or organic sulfur.

13.3.1 Sulfur Reduction

Some bacteria and archaea, including chemoorganotrophs and chemolithotrophs, use sulfate or other oxidized forms of sulfur as an electron acceptor for anaerobic respiration, producing further reduced intermediates (sulfite, thiosulfate, elemental sulfur) or fully reduced sulfide. In addition, many organisms can assimilate sulfate through a reductive process.

Dissimilatory Sulfate and Sulfur Reduction The complete reduction of sulfate to hydrogen sulfide, during its use as an electron acceptor for anaerobic respiration, can be achieved by specialized but common and widespread groups of strictly anaerobic bacteria, referred to functionally as **sulfate-reducing bacteria** (SRBs). The best known of these are *Desulfovibrio*, a proteobacteria, and *Desulfotomaculum*, a gram-positive endosporeformer, although other genera are now recognized as well. Interestingly, one archaea, *Archaeoglobus*, is also known that can carry out this reaction. Some other anaerobic bacteria and archaea can reduce elemental sulfur to sulfide.

It was initially believed that SRBs could only utilize a few fermentation products, such as H_2, lactate, and pyruvate. However, it is now realized that some can utilize acetate and

TABLE 13.5 Examples of Prokaryotes Able to Oxidize or Reduce (Dissimilatory) Inorganic Sulfur

	Genus	Kingdom: Class	Forms of Sulfur Substrate
Reduce	*Desulfovibrio*	δ-Proteobacteria	SO_4^{2-}
	Desulfotomaculum	Firmicutes: Clostridia	SO_4^{2-}
	Thermodesulfobacterium	Aquaficae	SO_4^{2-}
	Pyrolobus	Crenarchaeota: Thermoprotei	$S_2O_3^{2-}$
	Pyrodictium	Crenarchaeota: Thermoprotei	S
	Thermoplasma	Euryarchaeota: Thermoplasma	S
	Archaeoglobus	Euryarchaeota: Thermococci	SO_4^{2-}
	Thermococcus, Pyrococcus	Euryarchaeota: Thermococci	S
Oxidize	*Chloroflexus*	Thermomicrobia: Chloroflexi (green nonsulfur)	H_2S
	Chlorobium	Chlorobia (green sulfur)	H_2S, S
	Rhodospirillum	α-Proteobacteria (purple nonsulfur)	H_2S
	Chromatium	γ-Proteobacteria (purple sulfur)	H_2S, S
	Thiobacillus	β-Proteobacteria	H_2S, S, $S_2O_3^{2-}$
	Beggiatoa, Thiothrix	γ-Proteobacteria	H_2S, S, $S_2O_3^{2-}$
	Aquafex	Aquaficae	S, $S_2O_3^{2-}$
	Sulfolobus	Crenarchaeota: Sulfolobi	H_2S, S
	Ferroglobus	Euryarchaeota: Thermococci	H_2S

other short- and long-chain fatty acids, ethanol and other alcohols, benzoate (which has an aromatic ring), and hexadecane (a hydrocarbon), among other compounds.

Sulfate reduction does not involve the electron transport chain utilized by aerobes and denitrifiers. In fact, unlike those reactions, it requires an initial activation of the sulfur with adenosine triphosphate (ATP) and thus involves an initial expenditure of energy. The sulfate within the adenosine phosphosulfate (APS) formed is then reduced (with the addition of two electrons) to sulfite. Sulfite reductase then catalyzes the sequential loss of three more electron pairs, yielding H_2S, which is released to the environment.

$$SO_4^{2-} + ATP \rightarrow APS + pyrophosphate \quad \text{(pyrophosphate is two phosphates joined)}$$

$$APS + 2e^- \rightarrow AMP + SO_3^{2-}$$

$$SO_3^{2-} + 6e^- \rightarrow H_2S$$

Assimilatory Sulfate Reduction The assimilatory reduction of sulfate can be carried out by most bacteria, archaea, fungi, algae, and plants, while many protozoa and animals require organic sulfur. This reduction also starts with the production of APS, but a second ATP is then used to attach another phosphate, forming phosphoadenosine 5′-phosphosulfate (PAPS). The sulfur is then reduced to sulfite, followed by further reduction to sulfide. To avoid toxicity, as the sulfide is formed it is quickly combined to form an organic sulfur compound such as cysteine.

Sulfur Disproportionation Some SRBs can simultaneously oxidize and reduce an intermediate sulfur compound such as sulfite or thiosulfate to provide energy. For example, three molecules of sulfite are oxidized to sulfate at the same time that one molecule is reduced to hydrogen sulfide.

$$4H_2SO_3 \rightarrow 3H_2SO_4 + H_2S$$

13.3.2   Sulfur Oxidation

A wide range of bacteria and archaea are able to oxidize reduced sulfur, including phototrophs and lithotrophs (Table 13.5). *Thiobacillus*, for example, is an aerobic or denitrifying (one species) lithotroph that can grow on sulfides. It is of particular interest because of the acid tolerance of several species; one has a pH optimum of 2! This is particularly useful to the organism, as the product of complete sulfide oxidation (sulfate) is actually sulfuric acid.

Beggiatoa, a filamentous proteobacteria, has a neutral pH range but is of interest because of its gliding motility. This allows it to move so that it can stay right at the boundary between an anaerobic environment (where the sulfide is formed) and the aerobic one from which it gets oxygen. When growing on sulfide, it deposits, internally, elemental sulfur that is readily visible as yellow granules under the microscope (see Figure 10.22). These may serve as a substrate reserve to be used if sulfide is depleted. However, they may also have the benefit of minimizing sulfuric acid production.

There are also a number of phototrophic bacteria that oxidize reduced sulfur (Table 10.4). The green and purple nonsulfur bacteria oxidize hydrogen sulfide but do not use or deposit elemental sulfur. The green and purple sulfur bacteria, on the other hand, utilize both H_2S and elemental sulfur, which they also deposit (externally for the green, internally for the purple).

13.3.3   Sulfur in Environmental Engineering and Science

Sulfur plays an important role in several areas of concern to environmental engineers and scientists, from corrosion to odors. Its important contribution to acid mine drainage is discussed below as part of the iron cycle (Section 13.4.3). Sulfur oxides are also a major component of acid rain, or acid deposition (see Section 15.7), but this stems mainly from combustion rather than microbial transformations.

Corrosion   Water reacts spontaneously with iron (Fe^0) to produce a thin layer of hydrogen, H_2, and ferrous hydroxide, $Fe(OH)_2$. Still, it was originally thought that iron and steel would not corrode under anaerobic conditions, since no oxygen would be present to oxidize the H_2 and thereby allow the reaction to continue. However, if sulfate is present, sulfate-reducing bacteria can utilize the H_2, perpetuating the reaction and leading to extensive deterioration:

$$4Fe^0 + 4H_2O + SO_4{}^{2-} \rightarrow FeS + 3Fe(OH)_2 + 2OH^- \tag{13.20}$$

The microbial reactions of the sulfur cycle are also responsible for the **crown corrosion** of sewers, particularly those made of concrete. Unlike water distribution pipes, which are pressurized, flow in sewers is usually by gravity, with the pipes often less than half full. During periods of low flow, wastewater velocities may drop to the point where organic matter deposits on the bottom. This is particularly common in flat areas, such as shore communities built on the coastal plain. After sufficient time, these bottom deposits and the wastewater itself may become **septic**, or go anaerobic, and if sufficient sulfate is present (as is also typical in shore communities), hydrogen sulfide will be formed by sulfate-reducing bacteria. A portion of this H_2S volatilizes from the liquid phase and is then absorbed in the moist biofilm at the crown (top) of the sewer. Here, under aerobic conditions, it can be reoxidized by bacteria such as *Thiobacillus* to sulfate, in the form of sulfuric acid (H_2SO_4), which then reacts with and dissolves the concrete. In fact, one of the sulfide-oxidizing bacteria involved is called *T. concretivorus* ("concrete eater"). Prior to an understanding of this process, it seemed mysterious that sewers corroded at the crown rather than at the bottom.

Odors   Decomposition of proteins can lead to the release of **mercaptans**, such as ethyl mercaptan (C_2H_5SH), from sulfur-containing amino acids. These volatile organic sulfur compounds are highly odorous. Both protein decomposition and sulfate reduction also lead to the production of H_2S, with its penetrating rotten egg odor. Thus, very strong and unpleasant odors can be released from a variety of "natural" anaerobic processes, including decomposition of vegetative debris in salt marshes and piles of manure. Odor problems from sulfides and mercaptans (as well as amines, organic acids, and other compounds) are also a major area of complaint for a variety of waste treatment facilities, including pumping stations on sewer lines, sludge tanks, composting facilities, land application sites, and landfills. In addition to aesthetic considerations, odors also appear to be associated with health effects in some people. Thus, control of odors is a major expense at some sites. Sulfides and mercaptans can be removed from gas streams by alkaline scrubbers (since at high pH the molecules ionize and are then nonvolatile), chemical oxidation (e.g., with chlorine), biological oxidation (e.g., in biofilters), or reaction with metals (e.g., iron). Alternatively, the formation or release of odorous compounds can be minimized by controlling oxygen levels, maintaining alkaline pH, and adding iron (to bind sulfides) or nitrate (leading to denitrification instead of sulfate reduction).

Toxicity   Hydrogen sulfide may be highly toxic, in both aquatic and soil systems, to both prokaryotes and eukaryotes. Once inside the cell, it binds rapidly to metals, inactivating cytochromes and other critical cell constituents. Low concentrations of H_2S in air are readily detectable by the odor. However, at higher concentrations it deadens the olfactory cells and can no longer be smelled. The buildup of hydrogen sulfide in confined spaces is one of the reasons (others include lack of oxygen and buildup of combustible gases) that entry into such areas without proper precautions is so dangerous, potentially resulting in death. Accidents involving H_2S frequently produce multiple fatalities, as would-be rescuers rush into contaminated spaces, only to become victims themselves. The gas is so toxic that even professional responders have been felled by gas leaking around their gas masks.

Wastewater Treatment   The sulfide present in wastewater can be oxidized to sulfate under aerobic conditions. Usually, this is a minor part of the secondary treatment process. However, some wastewaters may contain considerable amounts of sulfide, either from industrial inputs or from extensive sulfate reduction due to anaerobic conditions in the sewer lines or in earlier treatment steps. Under these conditions the sulfide may represent a considerable oxygen demand and in fact would be measured as C-BOD in standard tests. In activated sludge treatment (Section 16.1.3) under such conditions, filamentous sulfide-oxidizing bacteria such as *Thiothrix* may grow sufficiently to interfere with the settling that is an essential part of the process.

In another secondary treatment process involving rotating biological contactors (RBCs; Section 16.1.2), an anaerobic biofilm layer may develop that leads to H_2S production. The motile filamentous sulfide-oxidizing bacteria *Beggiatoa* may then grow at the interface between this region and the overlying aerobic zone. Growths of *Beggiatoa* under these conditions are sometimes so massive that they lead to the physical collapse of the system (e.g., shaft failure).

Methanogenic anaerobic treatment systems are commonly used for wastewater sludges (anaerobic digestion, Section 16.2.1) and occasionally, for high-strength wastewaters. However, if too much H_2S is produced (from amino acid mineralization or sulfate reduction), it can be toxic to the methanogens. One control measure is to add iron, which binds with the sulfide as FeS.

Sulfide can also interfere with disinfection. Active chlorine is rapidly consumed by reaction with sulfides and other reduced sulfur compounds.

Example 13.8   A wastewater contains 2 mg/L of hydrogen sulfide. How much chlorine will this consume?

Answer   Although the active form of chlorine in water is usually hypochlorous acid, HOCl, or chloramines, NH_2Cl (especially in wastewater), by convention (since 1 mol of either comes from 1 mol of molecular chlorine), chlorination calculations are based on the equivalent amount of Cl_2 (molecular weight $= 71$):

$$H_2S + 4Cl_2 + 4H_2O \rightarrow SO_4{}^{2-} + 8Cl^- + 10H^+$$

$$\text{chlorine consumed} = 2\,\text{mg/L } H_2S \left(\frac{1\,\text{mol } H_2S}{34\,\text{g } H_2S} \right) \left(\frac{4\,\text{mol } Cl_2}{1\,\text{mol } H_2S} \right) \left(\frac{71\,\text{g } Cl_2}{1\,\text{mol } Cl_2} \right)$$

$$= 2(8.35\,\text{mg } Cl_2/\text{mg } H_2S) = 17\,\text{mg/L } Cl_2$$

Dimethyl Sulfide   The biogenic production, fate, and impact of dimethyl sulfide (DMS), CH_3–S–CH_3, in remote oceans has only recently come under investigation. This reduced (oxidation state $= -2$) volatile compound is considered to be the most abundant form of organic sulfur in nature and the largest source of sulfur in the atmosphere. The environmental significance of DMS stems from its eventual contribution to the formation of atmospheric aerosols and cloud-forming particles that can impose a large-scale effect on the incidence of solar radiation over remote ocean areas.

Metabolic production of DMS has been estimated to amount to 4.5×10^7 metric tons per year. It results from the transformation of dimethylsulfonium propionate (DMSP), which is used by various marine algae, such as dinoflagellates, as a means of controlling their osmotic pressure. Bacteria metabolize released DMSP as a carbon and energy source, forming DMS as one of the end products. DMS then can be released into the atmosphere (Figure 13.26), where it undergoes photochemical oxidation to methane sulfonic acid (CH_3SO_3H), sulfur dioxide (SO_2), and non-sea-salt sulfate (as opposed to aerosolized sulfate from ocean spray).

$$\underset{H_3C}{\overset{H_3C}{>}}S^+ - CH_2 - CH_2 - C\underset{O^-}{\overset{O}{<}}$$

DMSP

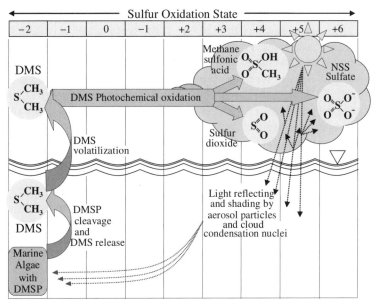

Figure 13.26   Dimethyl sulfide reactions over remote ocean waters. (DMSP, dimethylsulfonium propionate).

These aerosols themselves, along with the clouds that form when they serve as cloud condensation nuclei, affect the solar radiation reaching the ocean surface through scattering of incoming light and direct shading. In addition to the effects on the global radiation (and heat) balance, it may be that large-scale oceanic algal blooms promote atmospheric shading indirectly, thereby decreasing their own activity.

A variety of chemoorganotrophic and chemolithotrophic bacteria are able to oxidize DMS remaining in the water phase, producing dimethyl sulfoxide (DMSO), which is also frequently generated as a waste contaminant in paper manufacturing operations. Bacteria such as *Hyphomicrobium* are then capable of using DMSO aerobically as a sole source of carbon, energy, and sulfur. Alternatively, DMSO can be reduced under anaerobic conditions by organisms utilizing it as an electron acceptor, producing DMS again.

DMSO

13.4   IRON

In addition to its metallic elemental form (Fe^0), iron has only the $+2$ ferrous, Fe(II), and $+3$ ferric, Fe(III), oxidation states, separated by a one-electron difference. Most iron cycling involves oxidations and reductions between these two states, via both abiotic and biochemical mechanisms.

Iron is one of the most abundant elements on Earth, but it is confined primarily to the lithosphere (soil averages about 7% Fe). It is nonvolatile, and its concentration in air is virtually zero except for the small amounts present in airborne soil particles. In water, the solubility of ferric iron is low (<0.02 µg/L at pH 6), except under very acidic conditions or in some organic complexes. Ferrous iron is more soluble (about 20 mg/L at pH 6, but depending strongly on alkalinity). However, under the anaerobic conditions in which it is usually found, it is likely to form a precipitate with sulfide. Although iron is an essential element, it typically is present in organisms only at low concentrations (e.g., 0.2% of the dry weight of *E. coli*, Table 11.1). Thus, although iron is abundant and only small amounts are needed, it still may be available only at levels that limit growth in some environments, especially the open ocean.

13.4.1 Iron Reduction

In the absence of oxygen and nitrate, ferric iron can be used for respiration by a variety of aerobic bacteria. In the process, the Fe(III) is reduced to Fe(II). Both organotrophic and lithotrophic iron reducers are known. The process seems to be carried out by many of the same organisms that can reduce nitrate.

There are a number of natural oxygen-deficient conditions under which reduction might take place. These include anoxic biofilm layers, lake bottoms and sediments, water-saturated soils, groundwaters, and bogs. Reduction is a principal means of solubilizing iron-bearing solids. Iron reduction and resolubilization also occur in landfills and anaerobic digesters. Where sulfides are present, FeS will form, precipitating the iron while helping to limit odor and toxicity problems associated with H_2S.

13.4.2 Iron Oxidation

A diverse assortment of aerobic bacteria and archaea can oxidize ferrous iron, apparently using it as an energy source. Because only a small amount of energy is available from this transformation, large amounts of iron must be processed to support their growth.

At neutral pH values, ferrous iron is unstable under aerobic conditions and will be oxidized rapidly and spontaneously (abiotically) to the $+3$ form. Thus, microorganisms utilizing iron at such pH values must be located at the interface between aerobic and anaerobic zones, depending on this specialized niche so that they may capture this energy before it is lost. On the other hand, at low pH, ferrous iron is sufficiently stable that microorganisms can utilize it in aerobic systems.

Thus, iron oxidizers can be subdivided into two groups based on their pH preferences. The acidophiles include several *Thiobacillus*, *Leptospirillum* (a xenobacteria), and *Sulfolobus* (an archaea). Many of the same organisms can also oxidize reduced sulfur, an advantage since ferrous iron and sulfides often occur together. *T. ferrooxidans*, an autotroph, is the best known; it has a pH range of 0.5 to 6, with an optimum at 2! It utilizes a copper-containing protein, rusticyanin, for transferring the electron from iron during its oxidation (Figure 13.27). The amount of energy available from the oxidation is too small to use directly for ATP synthesis; instead, the proton gradient is utilized for this purpose.

Gallionella (Figure 13.28) grows at neutral pH, usually at a point where anaerobic groundwater comes in contact with oxygen (such as in wells). It is commonly heavily coated with ferric hydroxide, $Fe(OH)_3$. *Sphaerotilus* and *Leptothrix* also deposit oxidized iron on their sheath, but it is not certain that they are able to capture the energy from iron-oxidation. Other iron, oxidizing bacteria include *Planctomyces* and *Hyphomicrobium*;

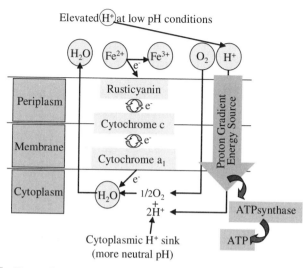

Figure 13.27   Energetic proton gradient for acidophilic iron-oxidizing chemolithotrophs.

interestingly, these bacteria both have stalks, although they are not related phylogeneti-cally. *Ferroglobus*, an archaea, is also of interest in that it oxidizes Fe(II) to Fe(III) under anoxic conditions, using nitrate as the electron acceptor.

13.4.3   Iron in Environmental Engineering and Science

Iron cycling plays a role in a number of issues of relevance to environmental engineers and scientists. These include problems of acidification, particularly acid mine drainage, and drinking water.

Figure 13.28   *Gallionella*-type bacteria causing iron precipitation in wells.

Acidification, Including Acid Mine Drainage   Oxidation of soluble ferrous iron appears to consume acidity:

$$Fe^{2+} + \tfrac{1}{4}O_2 + H^+ \rightarrow Fe^{3+} + \tfrac{1}{2}H_2O$$

However, the subsequent precipitation of ferric hydroxide has the effect of producing 3 mol of acid, so that the overall reaction can be written as

$$Fe^{2+} + \tfrac{1}{4}O_2 + 2.5H_2O \rightarrow Fe(OH)_3 + 2H^+ \tag{13.21}$$

During mining of coal and other minerals, anaerobic deposits containing pyrites (FeS_2) are commonly exposed to the air. This begins a complex process involving both abiotic and biochemical mechanisms that leads eventually to oxidation of both the iron and sulfur. Acidification initially results from the oxidation of reduced sulfur to sulfate (sulfuric acid), which then makes oxidation of the iron favorable. The first step, or **initiator reaction**, is typically abiotic at the starting neutral pH:

$$FeS_2 + 3.5O_2 + H_2O \rightarrow Fe^{2+} + 2SO_4{}^{2-} + 2H^+ \tag{13.22}$$

The production of acid leads to a drop in pH and the development of favorable conditions for bacteria such as *Thiobacillus ferrooxidans*. The overall reaction can be written as

$$FeS_2 + 3.75O_2 + 3.5H_2O \rightarrow Fe(OH)_3 + 2SO_4{}^{2-} + 4H^+ \tag{13.23}$$

Drainage from such areas can thus be highly acidic, with a pH of 2 not uncommon. This may have a drastic effect on the ecosystem of any nearby streams. An iron sulfate mineral known as **yellow boy** may also precipitate and color a stream bed dramatically.

Pyrites are also found in some clay minerals. This has been found to be a problem with the clay caps that are sometimes put over landfills—acid formation inhibits establishment of cover vegetation. Leachate from landfills containing reduced iron and/or sulfides may also become acidified through similar processes.

Drinking Water   With certain groundwater supplies, iron-oxidizing bacteria such as *Gallionella* in wells can cause clogging of the screens, resulting in a decrease in pumping capacity. Beyond the growth itself, the large amounts of precipitated iron they produce may be responsible for much of the problem. A similar effect can occur with the oxidation of reduced manganese (Section 13.5.3), but iron concentrations are typically much ($1000\times$) higher. Clogging of sand filters is also possible if the feed water contains sufficient reduced iron.

High iron concentrations can provoke consumer complaints about the aesthetic quality of drinking water. Dissolved iron can create taste problems, particularly with hot beverages such as coffee or tea. Also, the oxidation of reduced iron over time may produce a noticeable brownish discoloration (mainly from ferric hydroxide) on porcelain surfaces such as kitchen sinks, toilet bowls, and bathtubs, and even in clothes during washing. Additionally, there may be concern among customers over the safety of the "rusty" water itself. For these reasons, the secondary standard for iron in drinking water is 0.3 mg/L.

13.5   MANGANESE

Manganese can have an oxidation state as high as $+7$ in the form of permanganate ($KMnO_4$), which has some use in the environmental field as an oxidizing agent. However, within the biological realm, the natural states are $+2$ to $+4$, called, respectively, the "manganous" and "manganic" forms. Soluble Mn^{2+} and insoluble MnO_2 represent the principal forms that are found.

The dominant reservoir for manganese is the lithosphere; it forms $\sim 0.1\%$ of Earth's crust. Waters, both fresh and marine, rarely contain manganese concentrations above a few tenths of a part per million, and often far less. Like iron, manganese can also be chemically oxidized by atmospheric oxygen, but in this case pH levels of 8 or higher (as opposed to pH 4 to 5 for iron) are necessary to promote this spontaneous reaction. (This fact is used in the *Winkler test*, the standard chemical method of measuring dissolved oxygen; a solution of Mn^{2+} is added to the sample, followed by alkali, leading to formation of a stoichiometric amount of MnO_2 that is then analyzed iodometrically.)

13.5.1   Manganese Reduction

Manganese in the $+4$ state (i.e., MnO_2) can be reductively converted to Mn(II) by microbial activity under anoxic conditions. At least in some cases it appears that it is being used as an alternative electron acceptor for anaerobic respiration.

13.5.2   Manganese Oxidation

There is a relatively small group of bacteria capable of directly using manganese as an oxidative energy source, including some strains of *Arthrobacter*, *Bacillus manganicus*, *Corynebacterium*, *Flavobacterium*, *Gallionella*, *Hyphomicrobium*, *Pseudomonas*, and *Vibrio*. Several strains of iron-oxidizing bacteria, such as *Leptothrix* and *Sphaerotilus*, are also believed to be capable of cooxidizing manganese, precipitating it within their sheaths and metallic coatings.

13.5.3   Manganese in Environmental Engineering and Science

The cycling of manganese is not often regarded as a major environmental engineering and science concern, particularly since it is usually present only in the low parts per billion range. However, as with iron, higher levels of manganese may cause aesthetic problems in waters, or even difficulties with groundwater pumping. These issues tend to develop in relation to site-specific conditions.

Like iron, manganese can impose objectionable tastes in waters at quite low concentrations, particularly with hot coffee and tea. Discoloration with manganese generates a pinkish color. To address these concerns, the secondary standard for drinking water has been set at 50 parts per billion.

Typically, the problem is seen in communities using a lake or reservoir as their potable water source. Such water bodies tend to stratify (Section 15.2.1), with an aerobic zone (epilimnion) above an anaerobic one (hypolimnion). Seasonally, as the depths of these layers vary, the point of withdrawal for these waters (Figure 13.29) may extend

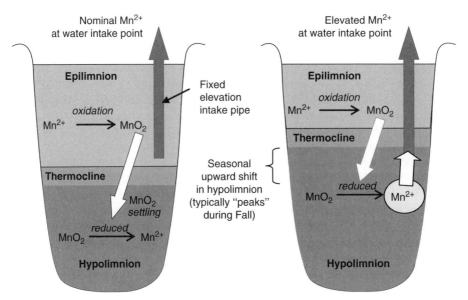

Figure 13.29 Seasonal shifts in lake stratification leading to increased presence of soluble reduced manganese (Mn^{2+}) in potable waters.

into the anaerobic zone. This can draw reduced manganese into the system, where it may later be oxidized.

Another potential problem occurs when groundwater is contaminated by landfill leachate or other degradable organic compounds, leading to anaerobic conditions. This can lead to mobilization of manganese and iron, potentially causing clogging of nearby wells (Figure 13.30).

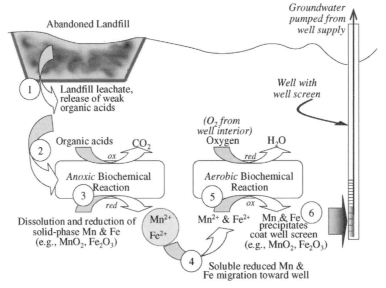

Figure 13.30 Mobilization of reduced manganese and iron by groundwater contamination.

PROBLEMS

13.1. What is the theoretical oxygen demand of a 20-mg/L solution of (**a**) pentane, C_5H_{12}; (b) ribose, $C_5H_{10}O_5$?

13.2. What are the total organic carbon concentrations of the solutions in Problem 13.1?

13.3. How much oxygen is utilized to nitrify 25 mg/L NH_4^+-N to nitrate? How much $CaCO_3$ alkalinity?

13.4. In a particular wastewater treatment plant, 200 mg/L of C-BOD and 20 mg/L of ammonium-N are oxidized using oxygen. Suppose that anoxic zones could be introduced so that all the nitrate produced was consumed by denitrification while oxidizing some of the wastewater C-BOD. By what percentage could the oxygen requirement potentially be reduced in this way?

13.5. A water quality limit of 0.04 mg/L NH_3 is placed on a stream receiving the effluent from a wastewater treatment plant. (**a**) How much total ammonia-N can be present in the water if the final pH of the stream is 7.3? (**b**) If the stream above the discharge point contains 0.2 mg/L total ammonia, and if the plant discharge is 20% of the total flow, how much total ammonia can the effluent contain without resulting in a violation of the limit?

13.6. In mg/L, how much methanol would be needed to denitrify 20 mg/L of nitrate-N stoichiometrically in an anoxic wastewater?

13.7. Suppose that a finished drinking water contains 0.5 mg/L ammonium nitrogen that becomes nitrified by biofilms in the water distribution system. How much chlorine residual will this nitrite consume?

REFERENCES

Atlas, R. M., 1997. *Principles of Microbiology*, 2nd ed., Wm. C. Brown, Dubuque, IA.

Atlas, R. M., and Richard Bartha, 1993. *Microbial Ecology: Fundamentals and Applications*, 3rd ed., Benjamin/Cummings, Redwood City, CA.

Madigan, M. T., J. M. Martinko, and J. Parker, 2000. *Brock Biology of Microorganisms*, 9th ed., Prentice Hall, Upper Saddle River, NJ.

Rittmann, B. E., and P. L. McCarty 2001. *Environmental Biotechnology: Principles and Applications*, McGraw-Hill, New York.

University of Minnesota Biocatalysis/Biodegradation Database, 2004. http://umbbd.ahc.umn.edu/. Updated regularly.

U.S. EPA, 1993. *Process Design Manual: Nitrogen Control*, EPA/625/R-93/010, U.S. Environmental Protection Agency, Office of Research and Development, Washington, DC.

Wackett, L. P., and C. D. Hershberger 2001, *Biocatalysis and Biodegradation: Microbial Transformation of Organic Compounds*, ASM Press, Washington, DC.

Young, L. Y., and C. Cerniglia (Eds.), 1995. *Microbial Transformation and Degradation of Toxic Organic Chemicals*, Wiley-Liss, New York.

14

ECOLOGY: THE GLOBAL VIEW OF LIFE

"**Ecology** is the scientific study of the interactions that determine the distribution and abundance of organisms" (Krebs, 1994). The effects of pollutants can be detected at all levels of the biological hierarchy. Our concern about these impacts falls mainly into two groups of organisms: humans and everything else. We are concerned directly with all types of impacts on humans. However, in nonhuman populations our primary concern is about loss of desirable species and the proliferation of undesirable ones resulting from our activities. Therefore, ecology provides an appropriate framework for examining our impact on the natural world.

The word *ecology* has as its root *oikos*, the Greek word for "house." The word *economy* is based on the same root. Thus, one can be thought of as the study of where we live; the other is its management. It is interesting that these two disciplines are commonly considered to conflict with each other. However, economics is now being brought to bear on environmental problems. Some think that if the true value of unspoiled ecosystems were taken into account, much destructive human activity would actually be seen not to be economically beneficial. In this view, the apparent economic advantages of polluting or destroying ecosystems comes from the costs being borne by others than those who receive the profits (the "tragedy of the commons").

Recall the following definitions from Section 2.2:

Population	A group of individuals of the same species living in the same environment and are actively interbreeding.
Community	A group of interacting populations occupying the same environment.
Environment	The physical and chemical surroundings in which individuals live.
Ecosystem	The combination of a community and its environment.

Environmental Biology for Engineers and Scientists, by David A. Vaccari, Peter F. Strom, and James E. Alleman
Copyright © 2006 John Wiley & Sons, Inc.

TABLE 14.1   Biomes of the Biosphere

Terrestrial biomes
    Tundra: arctic and alpine
    Boreal coniferous forests
    Temperate deciduous forests
    Temperate grassland
    Tropical grassland and savanna
    Chaparral: winter rain / summer drought
    Desert: herbaceous and shrub
    Semievergreen tropical forest: pronounced wet and dry seasons
    Evergreen tropical rain forest
Freshwater ecosystems
    Lentic (standing water): lakes, ponds, etc.
    Lotic (running water): rivers and streams
    Wetlands: marshes and swamp forests
Marine ecosystems
    Pelagic (open ocean)
    Continental shelf (inshore waters)
    Upwelling regions
    Estuaries (coastal bays, sounds, river mouths, salt marshes, etc.)

Source: Odum (1987).

These are the entities with which we are concerned directly in the study of ecology. A **biome** is a major type of ecosystem. Table 14.1 lists the most important biomes. The individual biomes are discussed in more detail in Chapter 15.

The range of environments in which a species lives is called its **habitat**. Each type of organism is thought to occupy a unique functional position, called its **niche**, in an ecosystem. The idea of a niche can be defined as a subset of the environmental and ecological variables that are favorable to a species. For example, two aquatic insects, the backswimmer (*Notonecta*) and the water-boatman (*Corixa*), occupy the same habitat among aquatic plants at the edge of a pond. But they have different niches because the former is a predator and the latter feeds on dead plant matter. Niche may be characterized by many variables, such as food source or tolerated physical conditions. The niches of two species may overlap but not coincide. It is said that two organisms in a particular ecosystem cannot occupy the same niche. However, they can coexist if there are even slight differences in their behavior.

Let us now systematize and categorize the ecological roles taken by organisms. Much of this is in terms that engineers will find familiar. For example, one important way to look at ecosystems is to examine their energy and material balances. Changes in sizes of populations are modeled in a way not very different from chemical or microbial kinetics. Most unique, perhaps, are the types of interactions between species such as competition or symbiosis. However, even these are studied in part using the mathematics of population dynamics. As you learn the ways in which we describe ecological roles and phenomena, think about how pollution could affect each.

14.1   FLOW OF ENERGY IN THE ECOSYSTEM

With minor exceptions, the biosphere is powered by the sun. The average amount of solar energy reaching Earth's surface (the **insolation**) in the United States ranges from 1250 to

2750 kcal/m$^2 \cdot$ day in January to 5250 to 7000 kcal/m$^2 \cdot$ day in July. For the entire year, averaged over all regions of the country, the input is about 3940 kcal/m$^2 \cdot$ day. Much smaller quantities originate from nonthermal sources. Geothermal energy derives from radioactive decay in the Earth and contributes about 0.5% of the solar input. Tidal friction extracts energy from the kinetic energy of the Earth–sun–moon system and is about 0.0017% of solar. A portion of the wind's energy comes from the kinetic energy of Earth's rotation. Oxidation of reduced inorganic minerals transported to the biosphere from deep in Earth's crust by hot springs and deep ocean vents provide energy for unique ecosystems adjacent to them. (However, this source is not completely independent of solar input, as it relies on oxygen produced mostly in the photosphere by solar energy.) These are the ultimate sources of energy for the biosphere. However, sensible heat and mechanical forms of energy cannot be used to form carbohydrates from CO_2. Organisms require high-quality energy in the form of photons or chemical bond energy in order to drive the reactions that produce biomass from inorganic precursors.

14.1.1 Primary Productivity

Most of the solar energy arriving at Earth's surface is converted to sensible heat or drives the processes of evaporation, wind, and waves. About 0.8%, however, is captured by photosynthetic conversion to organics. This autotrophic activity supplies practically all of the energy for the biosphere. The autotrophs responsible for the initial generation of fixed carbon from CO_2 are called **producers** in ecological terminology. The rate at which carbon is fixed in an environment is called its **primary productivity**. It can be measured either in energy terms (kcal/m$^2 \cdot$ day), or equivalently, in terms of biomass production rate (e.g., g/m$^2 \cdot$ yr). The basis area (m^2) refers to the area of the land. **Biomass** is usually measured as the total dry weight of the organisms under consideration. Carbohydrates have about 4.1 kcal/g (17.2 kJ/g).

Note that much of the activity of chemoautotrophs is not "primary" in the sense that the reduced minerals they are oxidizing were produced by other organisms. For example, hydrogen sulfide and ammonia are produced by anaerobic bacteria in marsh sediments. These can then be used by chemoautotrophic bacteria to fix CO_2. The energy for the anaerobes comes from organics that probably originated from plants. Other sources of these minerals, such as deep ocean vents or volcanoes, are abiotic in origin.

Primary productivity has two main components. **Gross primary productivity** (GPP) is the total amount of carbon fixed by the autotrophs. GPP may also be called the **rate of assimilation**. Only part of the gross productivity is available to other components of the ecosystem because the autotrophs themselves respire, consuming some of the fruits of their own labor. **Net primary productivity** (NPP) is the rate at which biomass or energy is accumulated by the autotrophs. It is equal to the difference between gross productivity and respiration. In practice, the net productivity is measured by harvesting and respiration measured by CO_2 production.

Gross primary productivity can be computed from measurements of net primary productivity and respiration. In some systems, such as the ocean, GPP is strongly correlated to the chlorophyll content. Chlorophyll can be measured by solvent extraction followed by spectrophotometry. Then one can apply a factor known as the **assimilation ratio**, which is the ratio of carbon fixation rate to chlorophyll concentration. For the ocean the assimilation ratio is fairly constant at 3.7 g of carbon fixed per hour per gram of chlorophyll. However, this ratio can vary widely for other ecosystems. Plants that are adapted

TABLE 14.2   Forms of Annual Primary Productivity for Several Ecosystems (kcal/m²)

	Alfalfa Field	Medium Age Oak/Pine Forest	Mature Rain Forest
Gross primary productivity	24,400	11,500	45,000
Net primary productivity	15,200	5,000	13,000
Net community productivity	14,400	2,000	Negligible
NPP/GPP	62%	43%	29%
NCP/GPP	59%	17%	0%

Source: Odum (1987).

to grow in shade usually have much higher chlorophyll concentrations in order to capture more effectively the energy available to them. This results in a lower assimilation ratio. Forests typically have about 0.4 to 3.0 g of chlorophyll per square meter, and an assimilation ratio that varies from about 0.4 to 4.0. Thinly vegetated areas such as new crops in the field, on the other hand, may have only 0.01 to 0.60 g of chlorophyll per square meter but fix from 8 to 40 g of carbon per gram of chlorophyll.

It is important to understand the relationship between the gross and net productivity, which are rates, and the biomass, or standing crop. Keep in mind that the amount of biomass present is not a measure of the gross or net productivity. As an analogy, consider a tank with water being pumped in, and with some flowing out through a hole in the bottom. The inflow is like the gross primary productivity, and the outflow is like respiration. If more is entering the tank than is draining out the hole, the difference accumulates. The rate of accumulation is like the net primary productivity. If the amount entering and leaving are equal, the amount in the tank stays the same and the system is at steady state. This would be equivalent to having zero net productivity and could occur with the tank full or almost empty. Consider a mature tropical rain forest (Table 14.2). The *gross* primary productivity is of such a forest is about 45,000 kcal/m²· yr. However, most of that is taken up by respiration to maintain the large biomass. Thus, the *net* primary productivity is essentially zero.

On the other hand, an alfalfa field is adding considerably to its biomass during the growing season. Its gross primary productivity is 24,400 kcal/m² · yr. Respiration accounts for 9200 kcal/m²· yr, leaving a net primary productivity of 15,200 kcal/m²· yr. However, the actual yield that farmers can obtain from the field will be lower because other heterotrophs (such as field mice) consume part of the crop. This leads to a third type of productivity, **net community productivity** (NCP), which is the gross primary productivity minus plant (autotrophic) respiration and minus the amount consumed by other organisms in the ecosystem (heterotrophic respiration). In the case of the alfalfa field, 800 kcal/m²· yr is consumed by heterotrophic respiration. This yields a net community productivity of 14,400 kcal/m²· yr, for an efficiency of 59%.

Primary productivity (GPP) can be limited by numerous factors, such as the amount of sunlight, availability of moisture and nutrients, soil properties, and so on. On land, moisture is the most serious limiting factor. In the open ocean, productivity is limited by nutrient availability and the ability of sunlight to penetrate the water. Ideal conditions can result in productivities that are as much as double the "typical" values (e.g., see Table 15.1); 50,000 kcal/m²·yr is thought to be the practical upper limit for any natural ecosystem. Changes in primary productivity, or comparisons to similar ecosystems, can be used as indicators of the health of an ecosystem. Worldwide, the total GPP is estimated to be about 10^{18} kcal/yr.

The primary productivity of the world's agriculture has only been keeping pace with increases in human population. Humans greatly increase the productivity of cultivated land by *energy subsidies*. This energy originates mostly from fossil fuels and takes the form of synthetic nitrogen fertilizer, pesticides, and mechanized agriculture (tilling, harvesting, etc.). Productivity of crops has also been increased by breeding new varieties. The greatly increased food production has resulted in an increase in the carrying capacity of the Earth for humans, and has been called the *green revolution*. However, it must be noted that the new crops are more dependent on energy subsidies than are traditional cultivars. Although the improved productivity has been important, underdeveloped countries have increased their food production as much by increasing the amount of land under cultivation. This results in destruction of forests and resulting loss in habitat for other species.

Humans have other uses for primary productivity besides as food. Cotton and flax fibers are used for clothing and wood for structures, paper, and fuel. Overharvesting of wood in poor regions of the world has led to deforestation and erosion.

14.1.2 Trophic Levels, and Food Chains and Webs

One of the most important relationships among organisms in an ecosystem is, of course, who eats whom. Ecologists discover these relationships by direct observation as well as by examining the stomach contents of animals captured from the wild. One species of bird may eat seeds, another might only eat certain types of insects, whereas a third may prey mostly on small mammals.

The energy and fixed carbon from the autotrophic producers are available to heterotrophs in the ecosystem, and these are called the **consumers**. Consumers that feed directly on the producers are called **primary consumers**, also called **herbivores** (plant eaters). Those that feed on herbivores are called **secondary consumers**, or **carnivores** (meat eaters). Next may come **tertiary consumers**, or **top carnivores**, which feed on carnivores.

This structure describes the **food chain**, the sequential feeding relationship from primary producers to higher consumers. The feeding levels in the food chain are collectively called **trophic levels**. The organisms in a particular trophic level have the same number of steps in their food chain between themselves and the primary producers.

The portion of each trophic level that is not used as food by a higher level will eventually die. A parallel food chain is based on this dead material. The organisms that rely on this are called **detritivores**, **decomposers**, **saprobes**, or **saprotrophs**, and a food chain based on them is called a **detritus food chain**. (In contrast, a food chain based on living biomass is called a **grazing food chain**.) The detritus food chain starts with organisms that feed directly on dead material, including bacteria, fungus, earthworms and other soil invertebrates, and some marine crustaceans, such as lobsters and crabs. These may, themselves, be fed upon. In fact, the grazing and detritus food chains are linked at all levels. For example, an owl may eat an insectivorous (insect-eating) shrew, which in turn has fed on saprobic soil invertebrates, and the same owl may also eat an herbivorous mouse (a grazer).

Ultimately, all the dead organic material is **mineralized**, that is, converted back to inorganic minerals such as CO_2, water, ammonia or nitrates, and salts. Along the way, the partially biodegraded biomass forms an important part of the soil and aquatic environments. The decomposers are the ultimate recyclers, making the material available for reuse by the primary producers. By releasing the minerals close to the roots of plants where they can be rapidly taken up again, they reduce the rate at which scarce minerals leach out of the ecosystem.

The energy for each trophic level is obtained from the level below. Some of the energy in a trophic level will be unavailable because it might be indigestible material, such as cellulose, lignin, chitin, or bone, or may be protected by defensive measures. Biomass may be stored, as in the case of peat formation in marshes, or it may be exported from the ecosystem, as in the case of migratory animals or tidal estuaries.

The energy input to a trophic level is the **gross energy intake**. Part of this is unused and is eliminated in the feces where it enters the detritus food chain. The remainder is the **assimilated energy**. Part of the assimilated energy is excreted with urine. A large part is used in respiration for maintenance and activity. The remainder is available for growth and reproduction, forming the **net productivity** of that trophic level. The **production efficiency** (also called **transfer efficiency**) of a species or of a trophic level as a whole can be defined as:

$$\text{Production efficiency} = \frac{\text{net productivity}}{\text{assimilation}} \tag{14.1}$$

Typical values for production efficiency are:

Birds	1.3%
Small mammals	1.5%
Other mammals	3.1%
Fish and social insects	9.8%
Other invertebrates	21–56%

Warm-blooded animals seem to pay a penalty in efficiency. The rate of respiration in animals can be related to body mass empirically by a log-log relationship. For example, for herbivorous mammals:

$$\log R = 0.774 + 0.727 \log M \tag{14.2}$$

where R is the respiration rate in kJ/day per individual and M is the body mass in grams (1 kJ = 0.239 kcal). Cold-blooded animals may use more than an order of magnitude less energy, but the value will be strongly dependent on ambient temperature.

Although they vary widely, a rule of thumb is that the production efficiency of trophic levels is typically about 10%. Thus, if the gross primary productivity is 60 kcal/m$^2\cdot$ day, the net primary productivity would be estimated at 6 kcal/m$^2\cdot$ day, and the herbivores would have a net productivity of about 0.6 kcal/m$^2\cdot$ day. Carnivores typically have a 20% efficiency, so net tertiary productivity would be estimated at 0.12 kcal/m$^2\cdot$ day.

As just described, the trophic structure of an ecosystem is commonly represented as an **energy pyramid**, which shows the relative proportions of energy or production in the various trophic levels. As Figure 14.1 shows, the pyramid can also be used to compare graphically quantities other than energy, such as population or standing crop. Note that the food chain is rarely longer than four trophic levels. Although this was originally thought to be due to low production efficiency, it now seems that the reason is due to the greater sensitivity of top carnivores to fluctuations in environmental conditions.

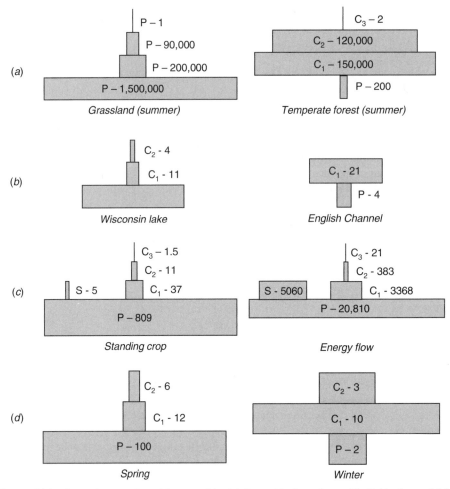

Figure 14.1   Several types of trophic pyramids. (*a*) Pyramid of numbers of individuals per 0.1 ha, not including microorganisms and soil animals. (*b*) Pyramid of biomass (grams dry weight per square meter). (*c*) Standing crop (kcal/m^2) vs. energy flow (kcal/m^2· yr) pyramids for Silver Springs, Florida. (*d*) Seasonal changes in the biomass pyramid in the water column of an Italian lake (planktonic organisms only) (mg/m^3). (Based on Odum, 1987.)

Of course, some organisms may feed at more than one trophic level. One example is the owl mentioned just above, which eats an herbivore and a carnivore. Humans are an example of an **omnivore**, which eats both plants and animals. However, this is unusual in natural systems. An exception is certain fishes that eat their way up the food chain as they grow. Other examples are the fox, skunk, and black bear.

The complexity of real systems can be described more precisely as a **food web**, which shows feeding relationships on a species-by-species basis. The food web shows the ecological dependencies of individual species more graphically than the energy pyramid. A web can be used to help predict the effect of a disturbance. For example, elimination of any one species from an ecosystem may increase the abundances of species that it feeds on and decrease the abundance of those that feed on it. Figure 14.2 shows two

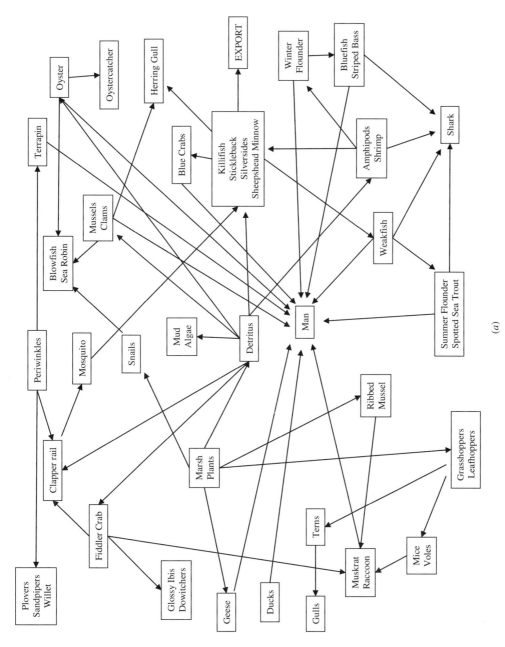

Figure 14.2 Food webs for (*a*) an unpolluted and (*b*) a polluted marsh/estuary. (Based on Mattson and Vallario, 1975.)

(*a*)

449

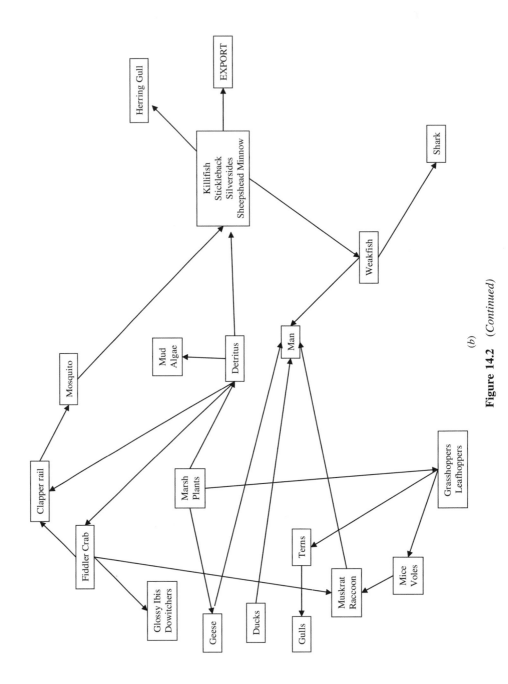

Figure 14.2 (*Continued*)

hypothetical food webs for a similar marsh, one that is unaffected by human activities, and a second that is affected heavily by pollution.

The more species there are in an ecosystem, the more possible interactions that could exist. Specifically, if there are s species, there could be up to $s(s-1)/2$ pairwise interactions. However, in reality, each species tends to have only about two interactions on average. Another generalization that can be made about food chains is that there tend to be about two to three prey species for each predator, no matter how many total species are present.

The flow of energy described in this section goes one way. It starts with the solar input, part of which is captured in gross productivity; what does not get used by the producers forms net productivity; successive trophic levels harvest part of this and produce their own biomass. At each step a majority of the energy is lost. Ultimately, all is returned to the environment as heat, after doing some useful work in producing biomass at each level. Although the ecosystem may try to optimize its use of energy, energy cannot be recycled. The second law of thermodynamics prevents the waste heat from being collected and reused for useful work.

14.2 FLOW OF MATTER IN ECOSYSTEMS

In contrast to the flow of energy in the ecosystem, materials can be and are recycled and reused. The patterns in which the elements are used, stored, made available again, and reused are called **biogeochemical cycles**. We have previously met biogeochemical cycles in Section 2.5 with reference to the effects of living things on the nonliving world, and in Chapter 13, since many of the important pathways in the cycles are catalyzed by microbes. Whereas Chapter 13 described the details of the microbial transformations involved, here we describe the overall cycles in quantitative terms as well as the abiotic parts of the cycles.

Each cycle consists of a network of **compartments** or **reservoirs**, which store forms of the element. The major compartments are the atmosphere, the biosphere, the hydrosphere, and the lithosphere. The **hydrosphere** is the liquid water on Earth, including rivers, lakes, groundwater, and the ocean. The **lithosphere** is the mineral part of the Earth, the crust with its soil, rocks and sediments, and mantle and core. The amount contained in each compartment is measured in units of mass or moles and is called the **standing stock**. The compartments are linked by flows or fluxes of the element, called **cycling rates**, which represent the rate at which the element moves from one compartment to the next. The units of the cycling rates are in mass per unit time or moles per unit time. The cycling rates may be the result of chemical or biochemical transformations, such as the fixation of carbon by photosynthesis, or simply the transport between phases such as the absorption of CO_2 by the oceans.

The elements constitute the basic nutrient requirements of the ecosystem. Of interest is the major storage compartment from which living things obtain each nutrient most directly. In the case of carbon and nitrogen, the inorganic source for living things is the atmosphere. Most of the other nutrients, such as phosphorus, sulfur, and potassium, originate in the lithosphere. However, weathering of rocks does not provide nutrients at a sufficient rate to replace losses of phosphorus and sulfur, and nitrification does not replace fixed nitrogen fast enough. Thus, the ecosystem must recycle its nutrients. This is one function of the detritus-based soil food chain. The decomposers release the nutrients in the dead organic matter. The partially decomposed organic matter increase the adsorptive capacity of the soil, preventing the nutrients from being leached out and lost to the ecosystem. Thus, a healthy soil environment is a key component of an ecosystem.

However, the soil is being compromised by human activities. Erosion causes a loss of organic-rich topsoil. Agricultural practices do more to replace lost nutrients than to replace soil organic matter, which contributes to favorable physical soil properties as well as chemical ones. Increased water runoff reduces infiltration, thereby reducing weathering of bedrock and the consequent liberation of minerals. As a result, agricultural fertilizers now need to include trace minerals in some areas.

When you examine biogeochemical cycles, an important consideration is whether or not the cycle is at steady state. **Steady state** is defined as the condition in which none of the variables are changing with time. A cycle is at steady state if all of its compartments are at steady state, which can be determined by a simple mass balance around each compartment (units of each term are mass or moles per unit time):

$$\text{accumulation} = \Sigma \text{ inputs} - \Sigma \text{ outputs} \pm \text{reactions} \qquad (14.3)$$

A compartment is at steady state if there is zero accumulation. The only reactions that create or destroy elements are nuclear ones, which we can ignore. (If we were doing mass balances on *compounds*, such as ammonia, we would have to consider reactions.) Thus, for a compartment to be at steady state, equation (14.3) reduces to

$$\Sigma \text{ inputs} = \Sigma \text{ outputs} \qquad (14.4)$$

Natural systems tend to be close to steady state, called the "balance of nature." An exception is when there is a net import or export from the ecosystem. Other exceptions operate at long time scales, such as when a long-term storage compartment is being formed. Examples of this are the deposition of carbonate sediments on the ocean floor, the formation of peat deposits in marshes, and burial of organic matter being converted over geologic time into coal or oil deposits. A system could also be disturbed from a steady-state condition by human activities.

Another important consideration is the size of various compartments relative to the total fluxes in or out of them. This leads to the concept of **turnover time**, θ, for a steady-state compartment without reaction:

$$\theta = \frac{\text{standing stock in a compartment}}{\Sigma \text{ input rates to that compartment}} \qquad (14.5)$$

For example, all the plants on Earth store about 600 Tg organic carbon ($1 \text{ Tg} = 10^9 \text{ kg}$). Photosynthesis forms about 120 Tg/yr, which is approximately balanced by respiration. Thus, the turnover time for organic carbon in the biosphere is five years. Compartments with short turnover times, on the order of days or weeks, respond more rapidly to disturbances and therefore may be more susceptible to pollution. Compartments with long turnover times change more slowly. However, it may take longer to recognize that a change is occurring, and it will also be slow to recover if a significant impact does occur.

Some cycles, such as phosphorus, tend to be local in scale. Others, such as carbon and nitrogen, include the atmosphere and the ocean as major compartments. These link all local cycles into a global cycle. Figure 14.3 shows some of the major compartments and fluxes associated with global cycles. In this chapter we discuss mainly global cycles. In the next chapter we look at local cycles for particular ecosystems.

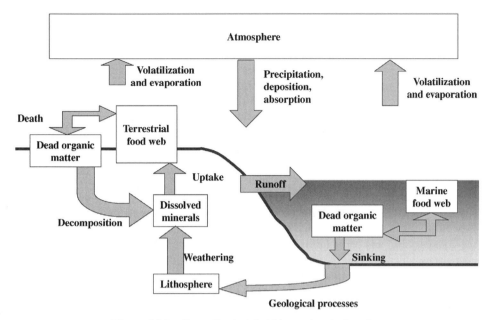

Figure 14.3 Generalized global biogeochemical cycle.

14.2.1 Sedimentary Cycles

All of the elemental cycles except for that of nitrogen are also linked to the **sedimentary cycle**, in which elements are cycled through Earth's crust (Figure 14.4). The cycling rates associated with this are small for most elements, but this only means that the turnover time is geologic in scale. Each of the cycles except that for nitrogen shows losses to

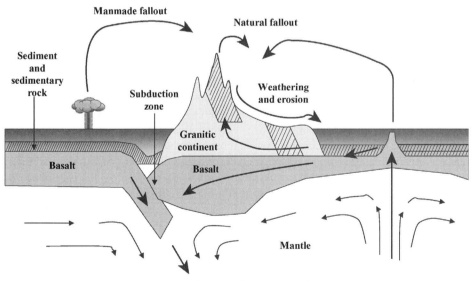

Figure 14.4 Sedimentary cycle.

the ecosystem as minerals are washed to sea and eventually buried in sediments. The losses are made up for by gains from volcanic activity or weathering of rocks. This part of the cycle is closed by processes driven by geologic rock-forming activity associated with continental drift. According to the theory of **continental drift**, Earth's crust is divided into sections called **tectonic plates**, which move around relative to one another on Earth's surface and upon which the continents ride. The motion is driven by thermal convection in Earth's molten **mantle**. The plates are formed continually at one edge by volcanic activity. The other edge, known as the **subduction zone**, is slowly pushed down into the mantle, carrying sediments with it. For example, the North American Plate is formed by volcanic activity at the mid-Atlantic ridge. Along the west coast of North America it is pushing over the Pacific Plate, forming a subduction zone there.

In the sedimentary cycle, buried minerals may follow several pathways to reenter the surface cycles. In the longest path they are carried into, and become part of, Earth's mantle. Eventually, they may resurface at a plate formation site. Other substances, especially the volatile sulfur and carbon dioxide, are expelled to the atmosphere by volcanic activity near the subduction zone. A third route is the crustal pathway. Buried sediments eventually consolidate to form **sedimentary rocks**, such as shale and limestone. Heat and pressure from deep burial may transform these into **metamorphic rocks**, such as slate and marble. Tectonic motion can cause parts of the continental masses to fold and crumple, forming mountains. This may raise their rocks above the surface, exposing them to erosion and making them available to the biosphere.

Living things act to reduce the rate at which minerals liberated by weathering return to the sedimentary cycle. Plants take up minerals faster than weathering makes them available. When plants and other living things die, the detritus food chain helps recycle them within the ecosystem. The saprotrophs release the minerals in close association with plants, so they can be reused immediately before they are lost. In this way most of the minerals in some ecosystems, such as the tropical rain forest, are stored in living things. Little is held in the soil.

14.2.2 Carbon Cycle

Carbon forms the backbone of biochemical compounds, and its fixation by primary producers coincides with the first biological step in the energy pyramid. Furthermore, the carbon cycle is at the center of one of the most important environmental impacts of human activities. Figure 14.5 illustrates some of the major parts of the global carbon cycle. The largest compartment shown is the ocean, which contains carbon mostly as dissolved carbonates. The ocean stores about 50 times as much CO_2 as the atmosphere does and has a very large turnover time of about 350 years. Terrestrial plants have a turnover time of less than five years. However, soil organic matter turns over about every 25 years. The biomass and soil organic matter form the major reservoir of reduced carbon, other than fossil fuels.

The importance of the carbon cycle reactions was dramatized by the experience of the Biosphere II project in Oracle, Arizona (the Earth is Biosphere I). Biosphere II is a 1.27-ha (3.15-acre) closed structure originally designed to support a crew of eight for several years without any food or other material supplies from the outside. It contained simulated ocean, desert, and forest ecosystems, as well as agricultural areas, plus an interconnected atmosphere. Thus, it was a highly visible, if only semiscientific, examination of a closed ecosystem. However, once the crew of eight was sealed inside in September 1991 for an initial two-year mission, it soon became apparent that oxygen was disappearing

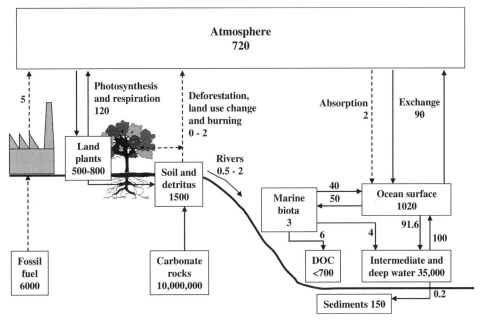

Figure 14.5 Global carbon cycle. Reservoirs units are 10^{15} g C, flux units are 10^{15} g C/yr. (From Odum, 1987.)

from the system. Eventually, the oxygen concentration dropped from the normal 21% to a dangerous 14%, and the total volume of the atmosphere decreased. (Huge, flexible "lungs" connected to the system gradually collapsed to maintain atmospheric pressure within.) It was suspected that the oxygen was being lost to respiration, but mysteriously, the increase in CO_2 from 330 ppmv to 4000 ppmv could not account for this. The high CO_2 levels lowered the pH of the crew's blood, which decreased the ability of their blood hemoglobin to carry oxygen. This, combined with the low oxygen level, led to complaints of fatigue among the crew. Ultimately, it was determined that the soil that Biosphere II was started with contained soil organic matter that was degrading, consuming oxygen and producing CO_2. The CO_2 then reacted with lime [$Ca(OH)_2$] in the concrete structure, producing calcium carbonate. Oxygen had to be brought into the facility, breaking the goal of achieving a closed ecological system. (*Note*: Biosphere II is now run as an educational and research facility, http://www.bio2.edu.)

Despite its low concentration, atmospheric CO_2 is a significant compartment. Its turnover time in Earth's atmosphere is about three years. Atmospheric CO_2 is an important "greenhouse gas" that affects the temperature of the Earth. In preindustrial times its concentration was about 280 ppmv (parts per million by volume). However, it is now about 350 ppmv and increasing by 1.5 ppmv per year. This level is already higher than levels going back 160,000 years, based on measurements in bubbles trapped in Antarctic ice core samples. Over that period, Earth's temperature correlated with atmospheric CO_2 concentration. At the current rate of increase, it will be double the preindustrial level in several decades.

It is well established that such an increase will trap an additional 4.35 watts (W) per square meter at Earth's surface, compared to the 235 W/m^2 normally absorbed by the surface. What is not so well settled is how much this extra heat energy will increase the temperature at the surface, called the **global warming** or **greenhouse effect**. A variety of mathematical climate models have been developed which predict an increase of from 1.5°C to 4.5°C, with a most likely value of 2.5°C. Besides potentially affecting temperature, climate, and sea level, the change in CO_2 could directly affect living things by fertilizing plant growth (Section 15.6). Potential ecological effects of global warming are described in Section 15.6.

Figure 14.5 shows primary productivity to be in equilibrium with plant and soil respiration. Fossil-fuel combustion contributes an extra 5.3×10^{15} g C/yr. In addition, about 1.0×10^{15} g C/yr comes from deforestation. However, only about 2.9×10^{15} g C/yr actually accumulates in the atmosphere. It is thought that the oceans absorb about 2.2×10^{15} g C/yr of the industrial contribution, about 42%. This leaves 1.2×10^{15} g C/yr unaccounted for, called "unknown sinks." It must be emphasized that global fluxes are difficult to measure. There is uncertainty whether terrestrial plants are increasing their removal of CO_2 from the atmosphere as its concentration increases, or whether such an effect is counterbalanced by human deforestation.

Atmospheric carbon is also present as carbon monoxide and methane. The turnover times for these are 0.1 and 3.6 years, respectively, compared to 4 years for CO_2. Methane is also a greenhouse gas. Although less important than CO_2, it is increasing faster, having already increased from about 700 ppbv (parts per billion by volume) in preindustrial levels to 1714 ppbv in 1992. Methane has many biogenic sources, including anaerobic degradation in wetland sediments and bacteria living symbiotically in the gut of termites. Anthropogenic sources contribute two-and-one-third times the natural sources. About half the anthropogenic sources are associated with food production, including rice paddies, biomass burning, and from livestock which belch gases formed by anaerobic bacteria in their gut. About one-fourth the anthropogenic source is from fossil-fuel use.

Two other greenhouse gases should be mentioned. Halocarbons are synthetic compounds used for refrigeration (e.g., Freon) among other things. They are highly stable, having half-lives in the atmosphere measured in decades. The most harmful forms have been banned because of their ability to destroy stratospheric ozone, and levels have stopped increasing. About two-thirds of the total nitrous oxide (N_2O) emissions are anthropogenic, and most of those are from tropical agriculture. N_2O has increased 13% over preindustrial levels.

At their current levels, these four gases trap an estimated 2.45 W/m^2 at Earth's surface. About 64% of this is due to CO_2, 19% to methane, 11% to halocarbons, and 6% to nitrous oxide. An increase in extreme weather events has been noted worldwide. Global temperatures have increased 0.3 to 0.6°C since measurements began in 1856. About half of that increase has come in the last 40 years, and the years 1990, 1991, 1994, 1995 were warmer than any of the years that preceded them. There are some who suggest that these changes may be natural, such as due to changes in solar output. Nevertheless, most scientists who have studied the data agree that these emissions have already produced detectable changes in Earth's climate and that their effects will have serious consequences for the Earth.

14.2.3 Hydrologic Cycle

The hydrologic or water cycle does not have any major flux components that are biological (Figure 14.6). Although water is consumed by photosynthesis and produced by

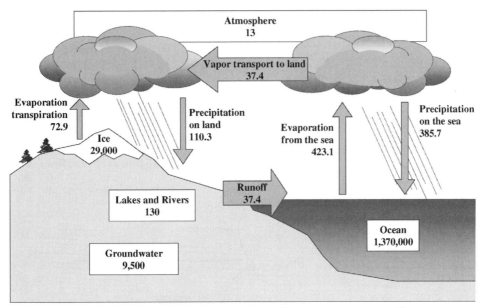

Figure 14.6 Hydrologic cycle. Reservoirs units are 10^{18} g C, flux units are 10^{18} g C/yr. (From Odum, 1987.)

respiration, these are very minor compared to other sources. The energy in rainfall does contribute an energy subsidy to ecosystems, as in rivers where it provides energy for transport of organisms and nutrients and for oxygenation of the water. And, of course, it provides water itself, the single compound that makes up the great majority of the mass of almost all living things. Water is the most common limiting factor in terrestrial productivity.

The biota affect the hydrologic cycle in several ways. In the process of **evapotranspiration**, plants withdraw moisture from the soil and evaporate it from their leaves. Plants also cool the soil by shading and evaporative cooling. Roots, detritus, and soil organic matter greatly increase the water-holding capacity of the soil.

14.2.4   Nitrogen Cycle

Nitrogen is the most common limiting factor in ecosystem productivity after water. Organisms require it to form all the amino acids and nucleic acids. Although N_2 is ubiquitous, it is not available to most organisms. The important role of microorganisms in catalyzing the fixation of N_2 to ammonia was discussed in Section 13.2.1. Nitrogen fixation also occurs abiotically in nature by lightning and fire, but this results in nitrogen oxides instead of ammonia. In fact, oxygen and nitrogen in the atmosphere are unstable thermodynamically. If they are heated together to a high enough temperature, they react to form nitrogen oxides. Thus, combustion processes from cooking fires to automobiles add to nitrogen fixation. Nitrogen oxides in the atmosphere can form nitrous and nitric acids, components of acid precipitation, described below. They also react with hydrocarbons to form photochemical smog, an irritating mixture containing ozone and nitrogenous organics. Another industrial source of fixed nitrogen is the Haber–Bosch

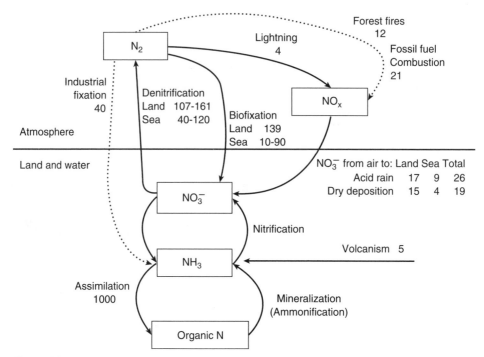

Figure 14.7   Fluxes in the global nitrogen cycle. Estimated fluxes in Tg/yr. Ammonia, organic nitrogen, and other forms also enter the atmosphere and oxidize or fall with rain. Dotted line arrows represent primarily anthropogenic fluxes. (From Odum, 1987.)

synthesis of ammonia. Much of this synthetic nitrogen is used in the manufacture of fertilizer, which goes directly into increasing the primary productivity of crops.

Many of the transformations that make up the nitrogen cycle are redox reactions. Nitrogen is present in its most reduced state as organic nitrogen or as ammonia. (The Kjeldahl nitrogen determination measures the sum of these two.) The most oxidized form is nitrate. Intermediate between these extremes are nitrite, NO, N_2O, N_2, N_2H_2, and N_2H_4, in order of increasing reduction (Figure 13.17). The primary biologically mediated steps in the nitrogen cycle are discussed next (Figures 14.7 and 14.8).

Fixation: the conversion of N_2 to ammonia. The prokaryotic aquatic blue-green bacteria (*cyanobacter*) are photosynthetic and get their energy directly from the sun. Bacteria such as *Rhizobium* and *Bradyrhizobium* beneficially infect the roots of leguminous plants such as alfalfa, clover, peas or beans, and some lichens and marine algae. Those plants sacrifice some of their growth to share the energy they captured from the sun with the nitrogen-fixing bacteria, in return for a rich source of available nitrogen. *Azotobacter* is a free-living bacterial nitrogen fixer. Fixation requires a considerable amount of energy. Legume bacteria use about 40 kcal (equivalent to about 10 g of glucose) to fix 1 g of nitrogen. Free-living nitrogen fixers require about 10 times as much.

Ammonification: the conversion of organic nitrogen to ammonia. This is performed by saprotrophic organisms which release nitrogen in excess of their own needs as ammonium.

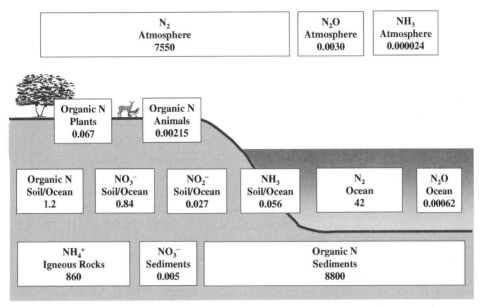

Figure 14.8 Another view of the global nitrogen cycle, showing storage reservoirs of nitrogen. Values are kg/m^2 of the Earth's surface. (Based on Whittaker, 1975.)

Nitrification: the chemoautotrophic conversion of ammonia first to nitrite, and subsequent conversion of nitrite to nitrate. The first step is accomplished by bacterial genuses such as *Nitrosospira* and *Nitrosomonas*. The second step is done by bacteria such as *Nitrobacter*, and in marine environments by *Nitrococcus*. Nitrification is also responsible for the production of nitric oxide (NO) and nitrous oxide (N$_2$O).

Denitrification: the conversion of nitrate and nitrite, through a series of steps, ultimately to nitrous oxide and N$_2$. This process is performed by numerous facultative heterotrophic bacteria when oxygen is limiting. Denitrification can occur even in nominally aerobic environments when anoxic microenvironments are present. This can occur, for example, in clumps of soil where organic matter is present to deplete the oxygen locally and provide energy for denitrification, or in the interior of flocs in the activated sludge wastewater treatment process. Nitrate serves as a substitute for oxygen as an electron acceptor for heterotrophic respiration.

Assimilation: the conversion of inorganic nitrogen to organic forms such as amino acids. Plants can absorb ammonia or nitrate, but mostly they use nitrate. Nitrite is toxic to plants. Plant fertilizers usually contain ammonia or urea, which breaks down to ammonia. The ammonia is then nitrified. After absorption, the plant converts the nitrate back into ammonia before converting it into amino acids or other organic nitrogen compounds.

The nitrous oxide produced by denitrification can be oxidized photochemically in the stratosphere to nitric oxide, which then reacts with ozone to form NO$_2$. Along with the NO$_x$ species from nitrification, these ultimately form nitrate again. This is deposited with precipitation and is part of the problem of acid deposition (Section 15.7). About 25% of the total NO$_x$-N production is thought to originate in lightning and in biological reactions in soil. The rest is anthropogenic, including fossil-fuel combustion and forest burning.

TABLE 14.3   Global Nitrogen Fluxes of Fixation and Denitrification (Tg N/yr)

Source	Fixation	Denitrification
Land (biological)	139	107–161
Ocean (biological)	10–90	40–120
Atmosphere (lightning)	4	
Forest fires	12	
Industrial, including fertilizers	40	
Fossil-fuel combustion	22	
Total	227–307	147–281

Source: Berner and Berner (1987).

Ammonia is also released by microbial activity in the soil. Much is reabsorbed by plants, which also absorb nitrate deposited on leaves. Ammonia in the air can neutralize some of the atmospheric sulfate, the other major acid rain component. Rainfall can also contain significant quantities of dissolved organic nitrogen (DON). The source of the DON may be from ocean spray tranported to the land.

Overall, rainfall contributes about 17 Tg N/yr in the form of nitrate to the land, plus 15 Tg N/yr as ammonia and 9 Tg N/yr as DON. Dry deposition contributes another 16 Tg/yr, mostly nitrate, for a total flux to land of 57 Tg/yr. The oceans receive a total of at least 20 Tg/yr.

The uncertainty in estimates for global nitrogen fixation and denitrification will need to be reduced before it can be determined if fixation and denitrification balance each other (Figure 14.7 and Table 14.3). The biological fixation on land includes an anthropogenic flux of 44 Tg/yr due to human cultivation of leguminous crops.

The nitrogen cycle is an interesting example of the Gaia hypothesis at work. Lovelock points out that one of the effects of life is the maintenance of chemical disequilibrium in the environment. If you turned off all biotic reactions, chemical equilibrium would eventually be achieved among the various species in the environment. In the case of nitrogen, the atmospheric O_2 would combine with its N_2 to form nitrate, which would be dissolved in the oceans. It is commonly understood that the oxygen in the atmosphere is maintained there by photoautotrophs. What is less well known is the fact that most of the nitrogen is in the atmosphere instead of the ocean, because of denitrifiers. Without life, Earth's atmosphere might resemble that of Mars, which is 95% CO_2.

Industrial fixation of nitrogen by the Haber–Bosch process has altered human ecology significantly. Each year, 175 million tons of nitrogen flow into the world's croplands, half of which is assimilated by cultivated plants. Synthetic fertilizers provide about 40% of all nitrogen taken up by these crops. The crops furnish about 75% of all nitrogen consumed by humans (the rest comes from fishing and from grazing cattle). Thus, about one-third of the protein in humanity's diet depends on synthetic nitrogen fertilizer. This has undoubtedly facilitated the tripling of the world's population that occurred in the twentieth century.

14.2.5   Sulfur Cycle

Roughly 90% of the world's supply of sulfur (about 1016 metric tons) is held in sediments and rocks and the remaining fraction is largely found dissolved within our oceans, which

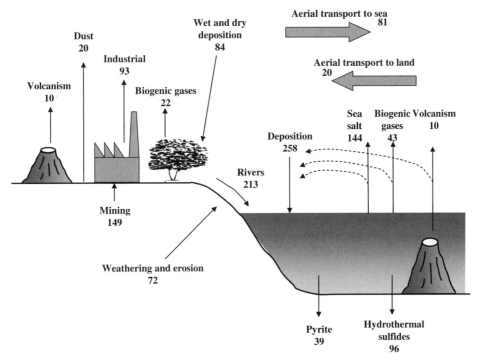

Figure 14.9 Global sulfur cycle. (Based on Krebs, 1994.)

in turn represents the dominant reservoir for the majority of sulfur being cycled within our biosphere. The cycle reactions for sulfur are depicted schematically in Figure 13.25, and the global cycle in Figure 14.9.

Sulfur is essential for life largely because it is a component of two of the amino acids, cysteine and methionine. The sulfur cycle may be the cycle that is most affected by human activities. Human emissions of CO_2 and nitrogen are about 5 to 10% of natural emissions. However, human activity releases about 160% as much sulfur as nature does.

Like nitrogen, sulfur cycles through a number of oxidation states. The most oxidized form, sulfate, serves as an alternative electron acceptor in respiration. It is generally used only if both oxygen and nitrate are absent, since it is less energetically favorable to organisms. Sulfate is the form of sulfur taken up by the primary producers, and thereby introduced into the food chain.

The most reduced inorganic form is sulfide. The toxic volatile compound hydrogen sulfide is commonly associated with anaerobic environments that have large amounts of organic matter, such as marsh sediments, anaerobic digesters in wastewater treatment plants, and petroleum deposits. In these the organic matter serves as a reducing agent, first reducing oxygen to CO_2, any nitrate is denitrified, and then all sulfate is reduced to H_2S. Sulfide is also found in deep groundwater aquifers, having been reduced by either soil organic matter or by ferrous iron. If ferrous iron is present, the sulfide will form the nonvolatile, low-solubility precipitate iron sulfide (FeS). Sulfide is also used by photoautotrophic bacteria as an electron donor in place of H_2O in a process similar to photosynthesis. Sulfur here plays the role of oxygen, and instead of O_2 being produced, elemental sulfur is the product. In aerobic environments, reduced forms of sulfur are rapidly oxidized to sulfate by chemoautotrophic bacteria. Thus, the sulfur cycle is closely linked to the carbon cycle.

The sulfur cycle is also linked to the phosphorus cycle in aquatic systems (Section 14.2.6). When iron sulfide is oxidized in aquatic sediments, sulfate is released and phosphorus precipitates with the iron, becoming unavailable to organisms. Under anoxic conditions, the iron is reduced, releasing the phosphorus to the water, and the iron precipitates again as a sulfide.

Because reduced forms of sulfur are often associated with deposits of fossil fuels and mineral ores, combustion and smelting of these resources often results in emissions of sulfur dioxide to the atmosphere. This is rapidly oxidized to sulfate. In the absence of alkaline species, this is present in the form of sulfuric acid aerosols. These are easily washed out of the atmosphere by precipitation, which as a result has greatly increased acidity and reduced pH. The result is called **acid rain** or **acid precipitation**. Acid precipitation has several ecological impacts on aquatic and terrestrial ecosystems where it falls. These are detailed in Section 15.7.

The ocean is also a significant source of atmospheric sulfur, with a biological origin. Marine phytoplankton produces volatile dimethyl sulfide [$(CH_3)_2S$]. In the air this oxidizes rapidly to sulfate, much of which is washed back into the sea.

14.2.6   Phosphorus Cycle

The phosphorus cycle differs from those of carbon, nitrogen, and sulfur in several ways: There are fewer steps, there is no change in oxidation state, there is no significant atmospheric component, and it tends to cycle locally. For these reasons, it is easier to study.

Phosphorus is present in three main forms: free or orthophosphate; polyphosphate, which is a polymer of orthophosphate; and organic phosphate. Although present in living things in much smaller quantities than carbon or nitrogen, its importance is clear from biochemistry, as phosphate forms the backbone of the DNA molecule and is central to energy metabolism in cells. Phosphate is the most common limiting nutrient in aquatic ecosystem. Humans disturb the phosphorus cycle in aquatic systems by discharging wastewater containing phosphate. This stimulates excessive cyanobacter, algae, and plant growth, which later die and deplete the water of oxygen. Aquatic systems with high nutrient loading are called **eutrophic** (Section 15.2.6).

Figure 14.10 shows one particular phosphorus cycle based on measurements in a salt marsh using radioactive tracers. Note that the various compartments do not have to be in steady state, as defined by equation (14.3). Each compartment may be importing or exporting phosphorus from the ecosystem. An energy flow diagram was also done for this ecosystem. An interesting conclusion stemmed from the observation that the filter-feeding mussels were more important for their role in nutrient cycling than for energy processing. In other words, they were more important for recycling phosphorus than as a food source for other organisms.

Sediments store and release phosphate to the water, depending on oxygen concentrations in aquatic systems. When oxygen is available, phosphate is absorbed by ferric hydroxide. When oxygen is limiting, the ferric iron is reduced to ferrous form, and the phosphate is released. This occurs seasonally in temperate-zone lakes.

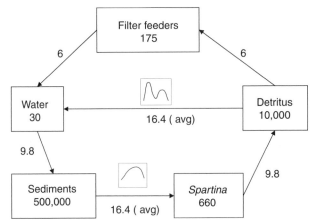

Figure 14.10   Phosphorus cycle from a Georgia salt marsh. Reservoirs are in mg P/m^2, fluxes are in mg P/m$^3 \cdot$day. Uptake by *Spartina* and release from detritus vary seasonally as shown. (Based on Odum, 1987.)

14.2.7   Cycles of Other Minerals

Numerous other minerals are required by living things, including potassium, calcium, and magnesium. Others are considered micronutrients, such as iron, chlorine, manganese, boron, zinc, copper, and molybdenum. Yet others, such as silicon, sodium, and cobalt, are required only by some organisms. Some are toxic, such as mercury and arsenic.

Iron is the sixth most common mineral element in the lithosphere (at approximately 5%), surpassed only by silica, calcium, magnesium, aluminum, and sodium. With an estimated total mass of 2×10^{17} metric tons of iron, there is roughly 10 times more iron than carbon in the lithosphere. However, the presence of iron in the biosphere is several thousand times lower than that of the macroelements, and its atmospheric fraction would be nil were it not for dust entrained by the wind. Its low solubility under aerobic conditions limits its availability. The two main forms are ferric (Fe^{3+}) and ferrous (Fe^{2+}). The solubility of ferric iron at pH 6.0 in waters of moderate alkalinity is about 0.13 µg/L. As a result, it is a limiting nutrient in some ecosystems, such as the open ocean. Ferrous iron is much more soluble, at about 20 to 30 mg/L, but this form is converted to ferric in the presence of oxygen. Ferrous iron may be found in deep groundwaters or wetland sediments where oxygen is depleted. When oxygen is present at pH levels above 6, ferrous iron is oxidized abiotically. At lower pH levels this occurs very slowly. But chemolithotrophic bacteria such as *Thiobacilus ferrooxidans* and *Gallionella ferruginea* can take over, harvesting energy in the process. Others, such as *Sphaerotilus* and *Leptothrix* perform the reaction but seem not to obtain energy from it.

One of the most thoroughly studied ecosystems in the world is the Hubbard Brook Experimental Forest in New Hampshire. A calcium budget developed for this forest found that it receives 3 kg/ha·yr from rainfall and 5 kg/ha·yr from weathering of bedrock. The output from the watershed via streamflow carries the sum of these, 8 kg/ha·yr. The biota and abiotic reservoirs exchange 50 kg/ha·yr between themselves. This shows how the biological community recycles the mineral many times faster than the overall throughput of the ecosystem.

Human activity has greatly affected the transport of many minerals. It has even created new ones. Radioactive strontium-90 is created by fallout from nuclear explosions. It behaves like calcium. Radioactive cesium-137 behaves like potassium and is rapidly recycled by organisms. Industrial activities have resulted in the release to the environment of otherwise scarce minerals, including mercury and chromium. Acid rain has increased leaching of aluminum into aquatic ecosystems, affecting fish life. Aluminum in soil water also decreases absorption of magnesium by plant roots. Recall that magnesium atoms are part of the chlorophyll molecule. Aluminum is added in the treatment of drinking water in the form of alum, although most is then removed.

14.2.8 System Models of Cycles

The cycling of nutrients among compartments can be modeled mathematically using simple rate expressions. Assuming steady-state nonreactive (conservative) systems, the expressions become even simpler. We illustrate with an example involving nitrogen balances, first as a linear steady-state system and then as a nonlinear system in both steady-state and dynamic conditions. The models can be used to explain and predict the distribution of nutrients in ecosystems.

Consider the system in Figure 14.11 which represents major nitrogen components in an aquatic system. The four compartments are: particulate nitrogen, which represents phytoplankton; dissolved organic nitrogen (DON), which is excreted by the phytoplankton; ammonia, produced by mineralization of the DON; and nitrate, produced by nitrification. Notice that the phytoplankton can utilize both nitrate and ammonia nitrogen.

The concentration of each species is represented by X and the flux between compartments by J. Each flux is assumed to be proportional to the concentration of the source species (first order kinetics). For example, the uptake of nitrate is assumed to be proportional to nitrate concentration: $J_1 = k_1 X_1$. This makes the system linear. The constant of proportionality is k. The system is set up by writing the mass balance equations for each

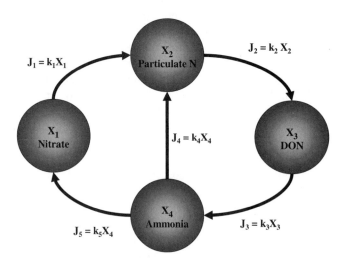

Figure 14.11 Simplified nitrogen cycle in the Bay of Quinte. (Based on Ricklefs, 1993.)

compartment based on equation (14.3):

Nitrate :	$-J_1 + J_5 = 0$	$-k_1X_1 + k_5X_5 = 0$
Particulate N :	$J_1 - J_2 + J_4 = 0$	$k_1X_1 - k_2X_2 - k_4X_4 = 0$
DON :	$J_2 - J_3 = 0$	$k_2X_2 - k_3X_3 = 0$
Ammonia :	$J_3 - J_4 - J_5 = 0$	$k_3X_3 - k_4X_4 - k_5X_5 = 0$

$$(14.6)$$

These equations form four equations with four unknowns. However, they are not *independent* equations. That is, one equation could be derived by combining the others. To see how this happens, write the equations for two compartments with two fluxes. You will see that the two mass balances are identical. Another way to see that the equations as given do not provide a useful solution is to recognize that one possible solution is to set all the concentrations to zero. In fact, an infinite number of solutions are possible, depending on the total concentration in the system. Thus, to find a unique solution we must specify the total concentration, M; then the individual X's will represent fractions of the total:

$$X_1 + X_2 + X_3 + X_4 = M \qquad (14.7)$$

If we replace one of the four mass balance expressions, say the nitrate equation, with this equation, we will have our four independent expressions. To solve we express them in matrix form:

$$\begin{bmatrix} 1 & 1 & 1 & 1 \\ k_1 & -k_2 & 0 & -k_4 \\ 0 & k_2 & -k_3 & 0 \\ 0 & 0 & k_3 & -k_4 - k_5 \end{bmatrix} \begin{bmatrix} X_1 \\ X_2 \\ X_3 \\ X_4 \end{bmatrix} = \begin{bmatrix} M \\ 0 \\ 0 \\ 0 \end{bmatrix} \qquad (14.8)$$

The coefficients must be obtained by experimental methods or by fitting the model to field measurements of the X's. The system could then be solved by standard matrix solution methods, such as Gaussian elimination. Matrix systems such as this can also be solved easily using spreadsheet software (see Table 14.4 and Problem 14.6).

The linear system described above is powerful but limited. Suppose that we tried to make it more realistic by improving the model for flux J_1, nitrate uptake by phytoplankton.

TABLE 14.4 Nitrogen Cycle Mass Balance Calculation

Process	Coefficient
Nitrate uptake (k_1)	16
Excretion (k_2)	4.1
Mineralization (k_3)	2
Ammonia uptake (k_4)	1.6
Nitrification (k_5)	3.1

Form of nitrogen	Fraction of total mass
Nitrate-N	0.081
Particulate-N	0.234
Dissolved organic N	0.481
Ammonia-N	0.204

In the linear model we assumed that this was proportional to nitrate concentration. However, it would be reasonable to assume that it would also be proportional to phytoplankton concentration. Thus, a better model would be $J_1 = k_1 X_1 X_2$. This is a second-order expression. When we insert it in our system, we can no longer create the matrix as we did above. We might make similar changes to the equations for other fluxes, and might even include Monod-type expressions.

These changes produce a set of nonlinear simultaneous algebraic equations. Such a system can be solved by Newton's method for systems of equations. An alternative and somewhat easier approach is to write the unsteady-state mass balances for each compartment, then simulate a sufficient period of time until steady-state is practically achieved. For example, in this case the equation for X_1 would be:

$$\frac{dX_1}{dt} = k_1 X_1 X_2 + k_4 X_4 - k_2 X_2 \tag{14.9}$$

The result is a system of nonlinear ordinary differential equations (ODEs), which must be solved numerically, such as by Euler's method or Runge–Kutta methods. Initial values must be specified for the X's. These can be estimates or solutions obtained from the linear case. Using the ODE model allows simulation of dynamic effects of disturbances. For example, the model could be used to determine the changes in phytoplankton populations that would result from a sudden addition of DON and ammonia, such as from a sewage spill.

Models of these kinds have been developed to simulate complete ecological systems. For example, models have been created for proposed closed ecological life support systems (CELSSs) to grow food and recycle wastes in support of human crews on long-term space missions. Global models have been used to make projections on economic resources such as food and minerals. Based on current scenarios for population growth and resource consumption, these models predict a catastrophic resource depletion some decades in the future. However, they do not take into account social responses to depletion, such as the development of new technologies, resource switching, or discovery of new resources. Nevertheless, the models make clear that our current world economy is not sustainable.

14.3   FACTORS THAT CONTROL POPULATIONS

We have just seen how energy and materials move through the ecosystem. Recall that our primary concern in ecology is to understand the distribution and abundance of organisms. Ultimately, this will relate to organisms' ability to gather and utilize resources. This, in turn, will depend on the availability of the resources in the environment, on the presence of other conditions favorable to organism growth and activity, and on their interactions with other species.

14.3.1   Limiting Factors and Interactions

Resources are factors that are consumed by an organism, resulting in their being removed from the environment. These include factors such as nutrients, but also less obvious factors, such as habitats. For example, barnacles require rock surfaces to grow on. As rock surfaces become colonized, growth and recruitment of new individuals is reduced by the crowding. Factors such as temperature are not resources because they are not depleted.

Historically, the effect of resources on growth has been described by **Liebig's law of the minimum**, which states that a population will increase until a single resource, called the **limiting factor**, becomes insufficient to support further growth. Thus, for every population in an ecosystem under steady-state conditions, Liebig's law predicts that each would be limited by a single factor. If that factor were known, the population could be increased by supplementing it.

Although it is still a useful concept and often applies to particular situations, Liebig's law is not generally true. For example, it may be possible to increase the growth of grasses in a meadow by supplementing either nitrogen, phosphorus, or water, and supplementing two or three of these together could increase growth even more. One of the factors could become limiting only if the others were present in excess.

Factors can also affect growth in a nonadditive way. For example, growth of *Impatiens parviflora* was increased 33% by adding nitrogen fertilizer and 19% by adding phosphorus. However, when both were added, the increase was 100%. If there were no interaction, adding both nitrogen and phosphorus would be expected to increase growth $33\% + 19\% = 52\%$. This is an example of interaction, which is not accounted for in Liebig's law. The concept of interaction is important: An *interaction* is when the sensitivity of a variable to one factor depends on the level of another factor. Here, **sensitivity** can be given a precise mathematical meaning. It is related to the rate of change of one variable with respect to another. One way to express the sensitivity of variable a to variable b (s_{ab}) is

$$s_{ab} = \frac{da}{db} \simeq \frac{\Delta a}{\Delta b} \qquad (14.10)$$

This expresses the sensitivity as the amount that a would change for a unit change in b.

Example 14.1 If increasing phosphorus concentration, P, from $10\,\mu g/L$ to $20\,\mu g/L$ results in an increase in chlorophyll concentration, c, in a water body from $3\,\mu g/L$ to $5\,\mu g/L$, the sensitivity of chlorophyll to phosphorus, s_{cP}, can be approximated as follows: By equation (14.10),

$$s_{ab} = \frac{5 - 3}{20 - 10} = 0.20\,\mu g\,\text{Chl-}a/\mu g P$$

That is, each 1-$\mu g/L$ increase in phosphorus produces a 0.20-$\mu g/L$ increase in chlorophyll a.

Example 14.2 Peak summer chlorophyll contents in lakes, C_a, have been empirically related to phosphorus concentration at spring turnover, P_w, as in equation (15.1): $C_a = 0.37\,P_w^{0.91}$ (both concentrations in $\mu g/L$). According to this relation, what is the sensitivity of chlorophyll to phosphorus at a concentration of $15\,\mu g\,P/L$?

Solution First, using equation (15.1), the expected concentration of chlorophyll a in the lake is

$$C_a = 0.37\,(15\,\mu g/L)^{0.91} = 4.35\,\mu g/L$$

Using equation (14.10), the sensitivity would be expressed as

$$s_{cp} = \frac{dC_a}{dP_w} = 0.3367P^{-0.09} = 0.264\,\mu g/\mu g$$

This indicates that a 1-$\mu g/L$ increase in phosphorus would be expected to produce a 0.264-$\mu g/L$ increase in chlorophyll a.

14.3.2   Resources and Environmental Conditions

The chemical resources include oxygen, CO_2 and water as well as other macronutrients and micronutrients. Other resources are light and habitat. The latter includes substrate, shade or sun, shelter, territory and space. Resources can also interact with environmental conditions. Some of the most important environmental conditions are temperature, humidity, pH, soil, salinity, fire, wind, and photoperiod.

The nutrients most often limiting are nitrogen and phosphorus. The ratio N/P in most biomass is about 16 : 1. In aquatic systems it is about 28 : 1. If phosphorus is supplemented, cyanobacter will fix nitrogen to bring the ratio back to nominal values. Therefore, phosphorus is the usual limiting nutrient. Cyanobacter are not abundant in marine environments. Thus, nitrogen may become limiting in saline systems. In the open ocean the micronutrient iron has been shown to be a limiting nutrient. Other micronutrient limitations may be significant in algal succession in lakes (see below).

The highest temperature that any organisms are known to tolerate is microorganisms at 88°C. Among animals, some fish and insects can go up to 50°C. Some organisms are known to perform better under temperature variations than at any particular constant temperature. Plants that use the C_4 photosynthesis outproduce the C_3 plants at high temperatures and light intensities. They also conserve water better. Conversely, the C_3 plants are more competitive under cooler, low-light conditions. The fraction of wild grasses that use the C_4 pathway varies along the Atlantic coast of North America, starting from 80% in Florida, 48% in the Carolinas, 34% in the mid-Atlantic region, 23% in New England, to 12% in the northern Maritime Provinces of Canada.

Water is, of course, one of the key determinants of the type of ecosystem that will exist in a terrestrial environment. A rough classification based on amount of precipitation can be made as follows:

0 to 25 cm/yr	Desert
25 to 75 cm/yr	Grassland or open woodland
75 to 125 cm/yr	Dry forest
125 cm/yr or more	Rain forest

Humidity affects water balance and cooling in plants and animals. As humidity falls below saturation, water is lost by evapotranspiration. Evapotranspiration is needed by plants to help them transport nutrients from the roots to the leaves. Too-low humidity can cause plant stomates to close to limit water loss. This prevents oxygen from diffusing out and CO_2 in, favoring photorespiration in C_3 plants. High humidity, on the other hand, both limits nutrient transport and encourages growth of plant diseases.

Many terrestrial ecosystems are adapted to periodic fires. Foresters have learned that too aggressive suppression of fires has resulted in shifts in population as less fire-resistant species outcompete the adapted natives. Wind can be a factor in dispersal of seeds, pollen, and insects. As a climatic factor it can affect survival of a species in harsh conditions. Pressure may be a factor only in the ocean, where it can reach 1000 atm. Pressure of that magnitude can affect the conformation and function of enzymes; organisms adapted to the deep ocean have evolved unique enzymes for that condition.

Photoperiod has a significant effect on the behavior of plants and animals. The best known example is that the flowering of many plants is controlled by the length of the day. Many physiological responses of animals are also tied to day length. Birds molt, migrate, and breed in response to day length. Insect eggs are stimulated to go into a resting stage by long days, so they will not hatch until the following spring, even if temperature and other factors are favorable.

14.3.3   Tolerated Range of Factors

Each species can tolerate a range within each factor. More precisely, their favorable range of factors can be defined as a region of the vector space defined by all the factors. Determining this space is very difficult. Laboratory tests can be misleading. The marine hydra *Cordylophora caspa* was found in the laboratory to have an optimal growth rate at 16 parts per thousand salinity. However, the species is never found at that salinity in nature, only at much lower levels. Evidently, some other factor in nature is limiting its distribution to only a part of the salinity range that it tolerates. Field and laboratory experiments are both needed to understand a species' range.

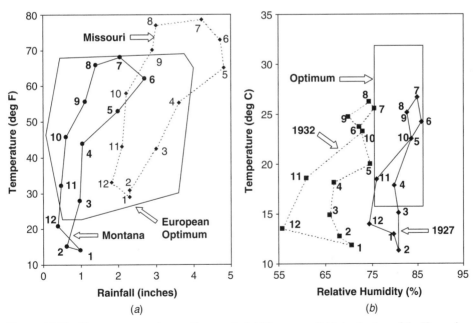

Figure 14.12   Temperature-moisture climograph. (*a*) The successful introduction of the Hungarian partridge to Montana, the unsuccessful introduction to Missouri, compared to average conditions in its native breeding range in Europe. (*b*) Conditions in Tel Aviv, Israel showing conditions favoring an outbreak of the Mediterranean fruit fly in 1927. (Redrawn from Odum, 1983; original from Twomey, 1936.)

An example of the combined effect of variables can be seen in Figure 14.12. The plot shows a **temperature–moisture climographs**, a phase-plane plot of temperature and rainfall. The numbers around each polygon indicate month of the year. The plot shows three climographs. One represents average conditions in the region of Europe where the Hungarian partridge naturally breeds. The other two are for the states of Montana and Missouri, where attempts were made to introduce the partridge. The introduction was successful in Montana but failed in Missouri.

The natural range of an organism can be examined to determine which levels of which factors are associated with the boundaries of the range. However, it is thought that different factors may operate within a range. In fact, it has been demonstrated that individuals near the edge of a geographic range may differ genetically from those in the center. The genetic variation may be due to natural selection, which in this context is called **factor compensation**. Thus, organisms living in the warmer region of a species range might not be able to tolerate the colder climate that other more adapted individuals of the same species can. The locally adapted populations are called **ecotypes**. For example, the jellyfish *Aurelia aurita* has a southern population that cannot swim outside the temperature range 11 to 36°C. A northern population is sharply limited to between 0 and 28°C. Although they are the same species, subpopulations have evolved to flourish over different temperature ranges.

A species is often limited by geographic factors such as mountain ranges or oceans, which in turn limit their dispersal. If transplanted outside their normal range, they reproduce and spread. This is seen most clearly in the many cases where humans introduced foreign species to an area, sometimes deliberately and sometimes inadvertently. Often, these species become nuisances and displace native species. Examples include the gypsy fly, the chestnut blight, and the starling in the United States, and the rabbit and the cane toad in Australia.

14.3.4   Species Interactions

Finally, a major factor that can limit the spread of a species is their interactions with other species. Within the ranges favorable to an organism, it is the interactions with other species that seem to dominate its distribution and abundance. Species interactions can be classified as beneficial, detrimental, or neutral in their effects on each of the interacting species. Note that a species that benefits from another may also become dependent on it. Thus, its range may be limited by the range of the species it depends on. A variety of major interaction types and subtypes have been identified (Table 14.5). We will give examples of several.

Competition arises when two species attempt to exploit the same resources. When the negative effect is due to the fact that one species' use of the resource reduces its availability to the other, it is called **resource competition**. In **interference competition**, one species has a more direct effect on the other's ability to compete for a resource. It may do so by chasing it away or by producing a chemical that inhibits competing species. Chemical inhibition of competitors is also called **allelopathy**. For example, sage plants (*Salvia leucophylla*) produce gaseous terpenes, which adsorb onto the soil near the plant and inhibit germination of seedlings. Some species of sponge produce a toxin that acts on other sponge species.

TABLE 14.5   Types of Two-Population Interactions[a]

Interaction Type	Species 1	Species 2	
Neutralism	0	0	No effect of either population on the other
Interference competition	−	−	Each species inhibits the other directly
Resource competition	−	−	Inhibition only when a resource is in short supply
Amensalism	−	0	One species is at disadvantage, other not affected
Commensalism	+	0	One species benefits, other not affected
Parasitism	+	−	A parasite is usually smaller than the host, and the host usually survives
Predation	+	−	A predator is usually larger than the prey, and the prey does not survive
Protocooperation	+	+	Favorable to both, but not obligatory
Mutualism	+	+	Favorable to both, and is obligatory

[a]0, no significant effect on that species; −, a detrimental effect; +, a beneficial effect.
Source: Odum (1987).

Predator–prey relationships result in the death of the prey. Most carnivorous organisms are predators (can you think of exceptions?). We usually think of this category in terms of animals that chase, catch, and eat other animals, but animals such as the blue whale simply seine the water for shrimp, small fish, and so on. In the microbial world, rotifers and stalked ciliates filter smaller organisms from the water. Populations of predator and prey often exhibit an out-of-phase oscillation. When the prey population increases, the predator population follows as its resource increases. This results in a drop in the prey population, followed a short time later by a drop in the predator population. This relationship is explored mathematically in Section 10.4.1.

Parasitism is distinguished from predation because the parasite is usually much smaller than its host, and the host will usually survive the interaction, although it may be harmed. Infectious disease falls into this category, although it is sometimes classified separately. Herbivores may are also be classified as separate +/− categories. Here we consider them to be predators if they kill the plant, or as parasites if they eat only part and the plant survives.

An example of predation involves the prickly pear cactus, which overran pastures and rangelands after it was introduced to Australia. The cactus moth was brought in from South America. Its caterpillar feeds on the cactus's new shoots, literally "nipping it in the bud". In a few years the cactus was no longer a problem. It wasn't eradicated, though, because the cactus could spread faster than the moth, although it would be followed soon after. It is interesting that it is difficult to find the moth in cactus stands either in Australia or its native South America. Its significant role might never have been discovered if not for the Australian experiment.

Predator–prey and parasitic relationships have resulted in what's been called an evolutionary "arms race" as hosts evolve ever better defenses and predators and parasites evolve to overcome them. Defensive tactics include escape, camouflage, body armor, and chemical defenses. Poisonous animals often have conspicuous coloration to warn potential predators away. Other species may mimic poisonous ones. Plants have been particularly adept at developing chemical defenses. They may be toxic or just noxious. Oaks and many other plants produce tannins, which complex with digestive enzymes of insects, rendering oak leaves nonnutritious. Nicotine from tobacco plants is a nerve toxin. Pyre-thrum is an insecticide that is extracted from the chrysanthemum plant.

Insects obtain food from flowers while pollinating them. Lichens are an association between a fungus and a flower; neither can live without the other. Bacteria in mammalian intestinal tracts are beneficial to the host. This is especially true for **ruminants**, animals such as cows who depend on bacteria to digest cellulose in their feed. Seed-eating birds help disperse seeds while obtaining food. Another term for mutualism is **symbiosis** or **synergy**. An example of **mutualism** that we have already met is the mycorrhyzal fungus (described in Section 10.7.4). This is a form of mutualism called **syntrophy**, in which two species provide some nutritional requirement for each other. Syntrophy is common among bacteria.

Pioneer communities such as are found in newly cleared areas tend to have more nega-tive interactions. Negative interactions tend to be less common in well-developed commu-nities. Even in laboratory predator–prey experiments, the population oscillations dampen as time goes on. Three-way interactions may also occur. An example is the scavenger–predator–prey relationship. The scavenger is a carnivore that eats remains of food killed by other animals.

14.4   POPULATIONS AND COMMUNITIES

Now we focus on the aggregate behavior of groups of individuals of one kind—species and groups of species—communities. The discussion here will emphasize animal popula-tions, although much of this applies to other kingdoms as well. The questions we address include: How will a given population change? Will it increase or stay the same? How high can it go? How can we compare the number of species in one ecosystem with those of similar ecosystems? How can we measure the effect of pollution on species distribution and abundance? How does the abundance of organisms of a species depend on such fac-tors as the geographical extent of its range, its trophic level, or its physical size? How does a community change in response to environmental influences such as seasonal changes or natural disasters?

14.4.1   Growth Models: Temporal Structure of Populations

Here we look with some depth at the methods that have been developed to predict changes in populations as they grow, use resources, compete, and feed on each other. This is one of the most active areas of ecology and is a prime example of the importance of mathematics to biology. For the most part, we deal with population as a continuous variable. The result-ing models are in the form of differential equations. Similar models can be developed for populations that change in discrete steps, such as annually. This would result in difference equations instead.

Vital Statistics and the Life Table   The size of a population will increase if the **natality** (rate of production of new organisms by birth, hatching, division, etc.) plus **immigration** (rate at which individuals join the population from outside) exceeds the sum of **mortality** (rate of death) and **emigration** (rate at which individuals leave the population). These four parameters determine the rate of change of a population. We focus on natality and mor-tality as the intrinsic determinants of growth.

For a given environmental situation, the average natality and mortality both depend on the age of the organism. Natality is usually zero for an initial period. Some organisms may then produce young at a fixed rate for their entire lives. Others, such as humans, are capable of bearing young only during a range of ages of the females. Some organisms, such as salmon, produce young only once in their lives, just before dying. The rate at which individuals in a given age group, x, produce young is called the **fecundity** or **fertility**, and is denoted b_x.

Mortality also depends on age. The age-specific rate is denoted q_x. Although each population is unique, three basic types of mortality vs. age relationships are noted:

Type I: fairly low and constant mortality over most of the life span, followed by increasing mortality with age. Humans follow this pattern.

Type II: constant mortality regardless of age. This is typical of birds.

Type III: high mortality among juveniles, followed by fairly low and constant mortality for the rest of the life span. This is typical of fishes and many lower animals such as zooplankton, which produce huge numbers of young, few of which survive to adulthood.

Species that go through several discrete stages, such as insects, may exhibit multiple plateaus of mortality in their lifetimes.

Mortality can be measured by following a group of individuals born in the same age group through their life span, until all have died. The life span is divided into age groups. The number that survive to enter each age group x, denoted n_x. A table of the values of n_x, called a **life table**, shows the age structure of the population. Table 14.6 is a life table for a population with a five-year life span. (Immigration and emigration are neglected.) The mortality is denoted q_x. Note that the midpoint of the age groups is used for calculations, so calculations for all organisms of age 0 to 1 are based on age 0.5.

This is called a **cohort** life table. It is based on a single group that goes through the age groups together. It is often difficult to follow a cohort for life. Instead, it is approximated by a synoptic, or **static**, life table. If the population is at steady state, the two will be identical. A steady-state condition means that natality equals total mortality and that none of the age groups is changing in size. A cohort life table can be used to compute the mortality as the fraction that die in each age group:

$$q_i = \frac{n_i - n_{i+1}}{n_i} \tag{14.11}$$

The information n_i, q_i, and b_i from the life table can be used to compute several things of interest, such as the steady-state age distribution changes in population and age

TABLE 14.6   Life Table for a Hypothetical Type I Population

i	Midpoint Age, x_i	Number, n_i	Natality, b_i	Mortality, q_i
0	0.5	1000	0.000	0.200
1	1.5	800	0.450	0.100
2	2.5	720	0.500	0.100
3	3.5	648	0.500	0.300
4	4.5	454	0.000	1.000

distribution when given a starting distribution and the intrinsic rate of increase of the population.

The steady-state age distribution is easily computed from mortality. A basis population must be assumed for the first age group, such as the number 1000 used for n_0 in Table 14.6. Then successive age group populations are computed by rearranging equation (14.11):

$$n_{i+1} = n_i(1 - q_i) \tag{14.12}$$

The age distribution is often presented as a histogram, as in Figure 14.13. This can be compared to a steady-state life table histogram. Populations with relatively large numbers of young and of individuals of offspring-producing age will tend to be increasing.

To compute unsteady-state changes, the mortality is simply applied to each age group at time step j to find the population entering the next age group:

$$n_{i+1,j+1} = n_{i,j}(1 - q_i) \tag{14.13}$$

The population of the first age group is obtained from the natality of each population, b_i:

$$n_{0,j+1} = \sum_{i=0}^{\infty} n_{i,j} b_i \tag{14.14}$$

Table 14.7 shows several generations that result based on the given starting population distribution and the life table in Table 14.6. Figure 14.14 is a plot of the total population over 45 generations of the organism.

Note the "baby boom" caused by the large number of young in the starting population. Eventually, the oscillations die out as the population assumes a stable age distribution. This approximates the steady state because natality exceeds mortality and the population is growing. Note that the population distribution at 45 days is still not at steady state. In fact, steady state is never reached. This population is growing exponentially, because birth exceeds mortality.

When the oscillations die out, the total population, N, increases in a constant proportion, λ, or geometrically, with each generation:

$$\lambda = \frac{N_{t+1}}{N_t} \tag{14.15}$$

$$N_t = N_0 \lambda^t \tag{14.16}$$

where N_t is the total population at time t and N_0 is the initial population. If population growth can be assumed to be continuous, growth can be described by the familiar first-order growth equation giving the rate of increase as being proportional to the current population:

$$\frac{dN}{dt} = rN \tag{14.17}$$

where r is called the **intrinsic rate of increase**. This is a the rate of increase normalized by the population $(dN/dt)/N$. The case of equation (14.17) in which r has a constant value is the **exponential growth rate** model.

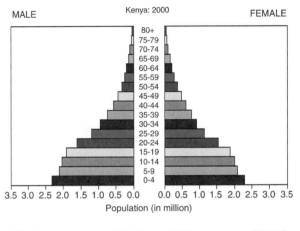

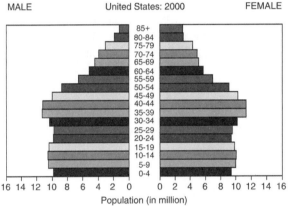

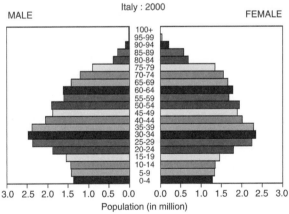

Source: U.S. Census Bureau, International Data Base.

Figure 14.13 Population histogram for three different growth scenarios. Kenya represents a country with a high growth rate, 2.5% per year. The U.S. growth rate is moderate, 1.0% per year. Italy is a country with a low growth rate, 0.3% per year, which is expected to begin to decline over the next several decades. Each bar represents a cohort of five years of age. (From U.S. Census Bureau, 2004.)

TABLE 14.7   Population Distribution Changes Based on Life Table from Table 14.6[a]

i	1	2	3	4	5	$\cdots$	44	45
0	1000	51	410	419	490		719	729
1	100	800	40	328	335		567	575
2	10	90	720	36	295		503	510
3	1	9	81	648	33		446	452
4	0	1	6	57	454		308	312
Total	1111	950	1257	1487	1606		2541	2578
λ	0.855	1.323	1.183	1.080	0.807		1.015	
$r(t)$	−0.156	0.280	0.168	0.077	−0.214		0.015	

[a]Each column represents one generation; each of rows 2 through 6 represents an age group.

Models based on differential equations such as this one are more appropriate to populations that breed continuously, with overlapping generations. On the other hand, equation (14.16) is better suited to species that breed at fixed intervals, once per season. The solution to equation (14.17), given an initial total population of N_0, is the exponential relationship

$$N_t = N_0\, e^{rt} \tag{14.18}$$

and by comparing (14.16) and (14.18), we obtain

$$r = \ln \lambda \tag{14.19}$$

The values of λ and r computed by equations (14.15) and (14.19) do not require age distribution information to predict growth—only data on total population and the assumption that the population distribution is stable. Examples of these calculations are given in Table 14.7. Notice that they vary widely until stability emerges.

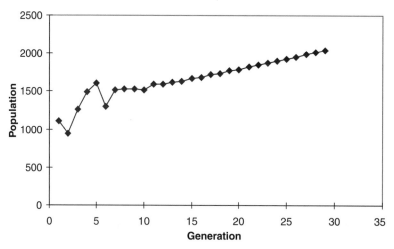

Figure 14.14   Predicted total population changes based on life table in Table 14.6.

An interesting parameter that can be computed from r is the doubling time, t_2:

$$t_2 = \frac{\ln 2}{r} = \frac{0.693}{r} \tag{14.20}$$

The values of r and λ could also be estimated directly from mortality and natality without having to simulate the population changes. Two intermediate values must first be computed: the **net reproductive rate**, R_0, which is the average number of offspring produced per individual over a generation; and the **mean length of a generation**, G, which is the average time between the birth of an offspring and that of its parents:

$$R_0 = \sum_{i=0}^{\infty} l_i \, b_i \tag{14.21}$$

where $l_i = n_i/n_0$ is the age distribution normalized to the first age group:

$$G = \frac{\displaystyle\sum_{i=0}^{\infty} l_i \, b_i \, x_i}{R_0} \tag{14.22}$$

$$r = \frac{\ln R_0}{G} \tag{14.23}$$

For the data from Table 14.6, we have $R_0 = 1.044$, $G = 2.47$, and $r = 0.0175$ per unit time, for a doubling time $t_2 = 39.6$ time units.

All populations are capable of exponential or geometric growth if favorable environmental conditions are maintained. However, this kind of growth is truly "explosive." Darwin gave an example of the elephant, which he predicted could grow from two individuals to 19 million over 750 years. This corresponds to a mere $r = 0.021$ per year, or a doubling time of 33 years. He also pointed out that although many parasites have a high intrinsic rate of increase, they are nevertheless rare. Clearly, the growth models described earlier need to take into account factors that limit growth, such as limiting factors, competition, and predation. These are the subjects of the next sections.

Single Species: The Logistic Equation   A simple strategy to place a limit on the exponential growth of equation (14.17) is to have r decrease as N increases. A simple way to do this is to have it decrease linearly from a **maximum specific growth rate**, r_0, when N is zero, down to zero when N increases to a particular level, K:

$$r = r_0 \frac{K - N}{K} \tag{14.24}$$

Since the rate goes to zero as N goes to K, the population cannot increase beyond K (except by immigration). Thus, K, called the **carrying capacity**, reflects the inherent ability of the ecosystem to support the population. Both K and r_0 can vary with environmental conditions; for example, they might decrease in time of drought. When the rate equation (14.24) is substituted into the growth equation (14.17), we get the important model of population growth in the presence of a limiting factor, called the **logistic equation**:

$$\frac{dN}{dt} = r_0 N \frac{K - N}{K} \tag{14.25}$$

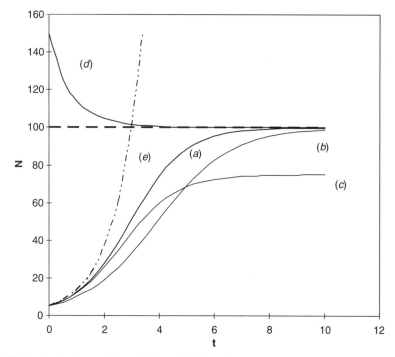

Figure 14.15   Logistic equation solution (14.26) with several parameter values, compared to exponential growth equation (14.18). The dashed line is $N = 100$. (*a*) logistic equation with $r_0 = 1.0$, $K = 100$, $N(0) = 5.0$; (*b*) logistic equation with $r_0 = 0.75$, $K = 100$, $N(0) = 5.0$; (*c*) logistic equation with $r_0 = 1.0$, $K = 70$, $N(0) = 5.0$; (*d*) logistic equation with $r_0 = 1.0$, $K = 100$, $N(0) = 150$; (*e*) exponential equation with $r_0 = 0.7$, $N(0) = 5.0$.

This equation has the following analytical solution for N at time t, $N(t)$, based on the initial population $N(0)$:

$$N(t) = \frac{K}{1 + be^{-r_0 t}} \qquad b = \frac{K - N(0)}{N(0)} \tag{14.26}$$

The solution has a sigmoid shape if $N(0) < K/2$ (Figure 14.15), sometimes referred to by ecologists as the **S-shaped** growth curve. This distinguishes the logistic growth curve from exponential growth, which is called the **J-shaped** growth curve (Figure 14.15*e*). Exponential growth is essentially like logistic growth but with an infinite carrying capacity. Curves (*b*) and (*c*) show the effect of varying the parameters, and curve (*d*) shows what happens if the initial population is greater than the carrying capacity.

Compare the logistic equation with the Monod model for microorganism growth [equation (11.6), Section 11.7.2]. The Monod model specifically includes the limiting resource needed for growth (substrate, S). In the Monod model, exponential growth is possible at any resource level as long as the resource concentration is held constant. Growth limitation will occur only if the disappearance of the resource is modeled separately. In the logistic equation, the organisms in a population are assumed to have to compete with each other for the resources, so as population increases, the growth rate decreases. The total amount of resource available to the population is assumed to be held constant.

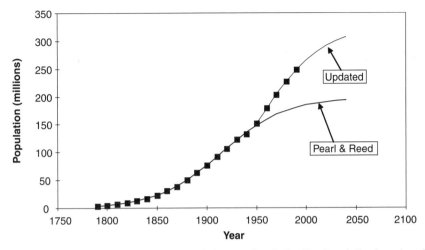

Figure 14.16   U.S. population data with logistic equation fit by Pearl and Reed, and updated logistic equation fitted to years 1950–1990.

The parameters of the logistic equation may be obtained by fitting the equation to laboratory or field data. For example, Pearl and Reed, who originally proposed the logistical equation in 1920, fitted it to U.S. population data from 1790 to 1910, when the population was about 92 million. They obtained $r_0 = 0.03134$ per year and $K = 197,273,000$. Thus, they projected a doubling of the U.S. population, followed by a leveling off. Notice that their prediction held for four more decades (Figure 14.16).

In 1950, however, the population started increasing much faster than predicted. The model fell victim to a basic problem with any model: The conditions on which it is based may change, resulting in either different parameter values or an altogether different model. In this case, the growth rate stopped decreasing as the logistic equation would have predicted. An updated model fitted to the data from 1950 to 1990 predicts almost the same $r_0 = 0.03281$ but a much higher $K = 324,796,000$. Thus, the carrying capacity of the United States seems to have increased, suggesting that the U.S. population will approach 325 million in the next century if things remain the same. (The population in 2005 is about 296 million.) The increased carrying capacity may have been due to the postwar prosperity and exploitation of new resources (raw materials, markets for our products). This result was a jump in the growth rate of the U.S. population called the "baby boom."

The logistic equation has led to a terminology to describe two basic reproductive strategies of organisms. An organism is called **K-selected** if it prospers by making efficient use of its resources, maximizing its carrying capacity. These organisms are more likely to be associated with crowded, stable ecosystems. They produce relatively small numbers of young but may nurture them to ensure their survival. Other organisms, called **r-selected**, adopt a strategy of rapid reproduction. These organisms are described as opportunistic. They can quickly take advantage of stressed conditions. They tend to produce large numbers of offspring but have much higher mortality. Weed species often fall into this category, as does that urban nuisance, the rat. The r-selected organisms will also have an advantage in temperate or arctic climates, where the short growing season favors rapidly growing species.

Laboratory experiments with single species populations have revealed behaviors that cannot be described by the models described above. Some species of *Daphnia* overshoot their carrying capacity and then oscillate around it. They are able to store food in the form of fat deposits, which delays the effects of resource limitations. This behavior can be described by using models with time-delay elements. For example, the logistic equation may be modified by using populations taken from \tau time units in the past, $N(t - \tau)$:

$$\frac{dN}{dt} = r_0 N(t) \frac{K - N(t - \tau)}{K} \tag{14.27}$$

Each of the foregoing models is **deterministic**, meaning that for each set of initial conditions, there is a unique prediction. However, the outcome in reality will be affected by unknown factors which produce random variation. A model that treats the variables as a random variable is a **stochastic** model. Each of the preceding models could be made stochastic by adding an independently and randomly distributed "error" at fixed intervals of time, before continuing the prediction to the next time step. The prediction of a stochastic model is not just a single value, but a value that has a probability distribution.

None of these models reflect interactions with other species. Whatever is the source of food for the organism will certainly be subject to its own population dynamics, in response to the predation. In the next several sections we address some of these issues.

Competition   The final term in equation (14.25) represents the effect of individuals from a single population competing among themselves. If another organism, species j, is using the same limiting resource as the population being modeled, species i, the logistic equation can be modified to take this into account:

$$\frac{dN_i}{dt} = r_i N_i \frac{K_i - N_i - b_{ij} N_j}{K_i} \tag{14.28}$$

where N_i is the population of species i, N_j the population of species j, and b_{ij} the positive number **coefficient of competition**. The coefficient of competition expresses how much the competing population effectively reduces availability of the resource to population i. A similar equation would be written for species j by switching the subscripts. If more than two species are interacting, additional terms of the form of $b_{ik} N_k$ would be included. A differential equation similar to (14.28) must be written for each interacting species, and the equations solved simultaneously.

A stability analysis of equation (14.28) reveals the conditions under which each species can survive. For two species we can form the dimensionless groups $a_{12} = b_{12}(K_2/K_1)$ and $a_{21} = b_{21}(K_1/K_2)$. If both $a_{12} < 1$ and $a_{21} < 1$, the species can coexist, although at a lower population than each by itself. If both $a_{12} \geq 1$ and $a_{21} \geq 1$, only one species can survive; which one depends on the exact dynamics and the initial conditions. If $a_{12} \geq 1$ and $a_{21} < 1$, species 1 will always win out. These dimensionless numbers can be interpreted as a measure of the competitiveness of each species. If both numbers are less than 1, competition is weak and the species can coexist. If both are greater than 1, competition is strong and only one species can survive.

Study of this model has led to the development of the **principle of competitive exclusion**, which states that species with identical resource requirements cannot coexist. Strong competition tends to result in the elimination of one species from the ecosystem. Thus, strong competition does not often occur in nature. In cases of apparent strong competition, it has often been found that the two species may have some significant difference. For example, five similar species of warblers in New England forests all eat insects. However, they feed at different parts of the forest canopy and in different ways, and they nest at different times. Strong predation by another species can keep several species sufficiently far below their carrying capacity that competition does not seriously limit the availability of the mutual resource. Darwin noticed this when he pointed out that grazing increases the diversity of plants in meadows. When the predation is limited, a single species tends to dominate. This is what has happened in many ecosystems affected by human activities. People have tended to remove top carnivores, such as grizzly bears and wolves in western North America. This actually had a negative impact on most of their prey, by increasing the role of competition among them. Thus, conservationists have the goal of reintroducing many of these predators to increase the diversity of the ecosystems.

The principle of competitive exclusion is linked to the concept of niche. The niche can be described as the resources and other factors favored by a species. Then the principle of competitive exclusion is equivalent to the statement made above that no two organisms can occupy the same niche.

Mutualism and Symbiosis   Mutualism and symbiosis are simply the opposite of competition. Instead of a species reducing the carrying capacity of another species, it increases it. This can be modeled simply by changing the sign of the interaction term in equation (14.28):

$$\frac{dN_i}{dt} = r_i N_i \frac{K_i - N_i + b_{ij}N_j}{K_i} \tag{14.29}$$

If $b_{12}b_{21} < 1$, there is a stable steady-state population for both species, which for each is greater than their respective carrying capacities. However, the model is not realistic for $b_{12}b_{21} > 1$, in which case the mutual benefit is so great that both populations grow without limit.

Predator–Prey: Lotka–Volterra Equations   One of the most interesting types of interactions are those between consumers (predators, parasites, and herbivores) and their hosts. Lotka and Volterra derived independently a model which although somewhat simplistic for modeling real populations, is nevertheless capable of describing some of the important behaviors found in predator–prey relationships. One important behavior is cycles in predator and prey populations (Figure 14.17). Intensive predation can cause a collapse in prey population. This is followed by a drop in predator population, due to the drop in food supply. As a result, the prey population can boom, followed by the predator to complete one cycle.

The Lotka–Volterra predator–prey model has the following form:

$$\begin{aligned}
\frac{dH}{dt} &= H(a - bP) \\
\frac{dP}{dt} &= P(cH - d)
\end{aligned} \tag{14.30}$$

where H is the host or prey population, P is the predator population, and a, b, c, and d are coefficients. Multiplying through, the host equation is seen to have two terms: The aH

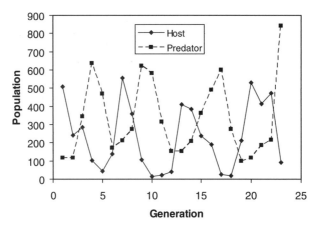

Figure 14.17 Oscillations in predator–prey populations: the predatory wasp *Heterospilus prosopidis* and its host the weevil *Callosobruchus chinensis*. (Data from Utida, 1957.)

term gives it exponential growth when the predator is absent. The $-bHP$ term is a negative interaction due to predation. The predator equation has a positive interaction corresponding to the negative one for the host. The $-dP$ term provides an exponential death rate without which the population could never decrease. There is no exponential growth term for the predator, and there is no carrying capacity for either equation.

The Lotka–Volterra model has a stable steady-state or equilibrium solution (other than $H = P = 0$) when $H = d/c$ and $P = a/b$. If values other than these are used as an initial condition, the solution will be a stable oscillation. The plot of host and predator populations vs. time (Figure 14.18*a*) shows that predator population peaks just after the host population. Examination of the phase-plane plot (Figure 14.18*b*) suggests that predator population peaks when the host population is at its greatest rate of decline.

More realistic models have been developed. For example, adding a term of the form $-eH^2$ to the host equation makes it equivalent to a logistic equation with a negative self-interaction. Other modifications place an upper limit on the amount of prey that each predator can consume. Other factors that affect predator–prey dynamics include the

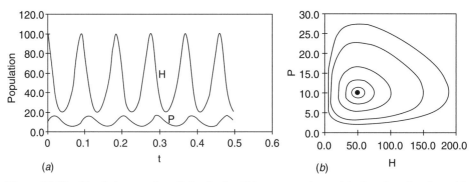

Figure 14.18 Simulation results of the Lotka–Volterra equations: (*a*) time-domain plot with $H(0) = 100$ and $P(0) = 10$; (*b*) phase-plane plot (P vs. H) for various initial conditions, and the equilibrium point.

availability of safe refuges, which limits predation when host population drops, and the presence of alternative food sources for the predator. Some of these factors make multiple steady-state values possible. This can lead to epidemics, when some environmental disturbance stimulates the system to jump to a different steady-state condition. Also, predator–prey populations can stabilize by some sort of accommodation to each other, reducing the amplitude of oscillations.

Empirical Models   Another way to model single populations is to use **autoregressive** models of the form

$$N(i) = \phi_1 N(i-1) + \phi_2 N(i-2) + \phi_3 N(i-3) + \cdots \qquad (14.31)$$

in which $N(i)$ represents the population at equispaced time intervals, and $N(i-k)$ is the population k time steps earlier (lagged by k). The coefficients can be obtained by multilinear regression techniques. Because this method is empirical, the coefficients cannot necessarily be identified with particular phenomena, such as growth, death, or predation rates. It is common to perform regressions of the form of equation (14.31) using logarithm-transformed variables instead of raw data. This can make the errors of the regression more normally distributed, and eliminates the possibility of negative predictions from the model. In addition, variables other than lagged population, such as the population of other species, can be used as dependent variables.

Multilinear regression software is capable of determining which coefficients, and therefore which independent variables, contribute significantly to the predictive ability of the model. Thus, it is possible to test which populations affect the dependent population and whether that effect is positive or negative.

Notice that the models discussed previously, including the exponential growth model (14.17), the logistic model (14.25), the competition model (14.28), the mutualism model (14.29), and the Lotka–Volterra predator prey model (14.30), can all be expressed as multivariable polynomials of the following form by combining terms and coefficients:

$$\frac{dN_1}{dt} = a_1 N_1 + a_2 N_1^2 + a_3 N_1 N_2 \qquad (14.32)$$

Effects of even more species can be modeled by adding appropriate terms to the model, such as $a_4 N_1 N_3$. By examining the signs of the coefficients of a pair of interacting species and comparing to the signs in Table 14.5, the type of interaction can be determined from population dynamics. This information can even help analyze the trophic structure of an ecosystem.

Terms such as $a_4 N_1 N_2 N_3$ test for three-way interactions. Statistical tests can show whether such effects are significant. Otherwise, they can be dropped from the model. As Krebs has stated: "Species interactions are rarely one-on-one in natural communities, and the untangling of complex sets of species interactions is an important focus in ecology today."

Stochastic Extinction   A population that is in equilibrium (birth rate = death rate = b) can nevertheless experience population fluctuations due to random variations. Thus, there is a finite probability that a population could go extinct just by chance. The

TABLE 14.8   Extinction Probability for Birth Rate = 0.5 per Year

Population	Time (yr)			
	1	10	100	1000
1	0.33	0.83	0.98	0.998
10	10^{-4}	0.16	0.82	0.980
100	10^{-48}	10^{-7}	0.14	0.819
1000	10^{-99}	10^{-79}	10^{-8}	0.135

Source: Ricklefs (1993).

probability, $p(t)$, of this occurring in a finite amount of time t can be estimated from b and from the average population:

$$p(t) = \left(\frac{bt}{1 + bt}\right)^N \tag{14.33}$$

For example, for a population with a birth rate of 0.5 per year, which is reasonable for terrestrial vertebrates, Table 14.8 shows the probabilities for various population sizes and periods of time. The table shows that the chance of extinction after a fixed period increases greatly as the size of the population decreases. For example, a population of 100 has a 14% chance of disappearing over the next century, whereas a population of 10 has an 82% chance.

This effect has been observed in nature in island ecosystems. On average the lost species are balanced by immigration or by evolution of new species. However, the role of humans is disturbing this balance. Urban and agricultural development may destroy only part of an ecosystem. Yet, even if a portion is preserved out of concern for the wildlife, the chances are increased that the species contained within will be lost. Simply putting a road through a wilderness divides it into two smaller parts. If movement of organisms of a particular species is inhibited from crossing, the overall chances of survival will have been decreased.

These concepts have led to the idea of **minimum viable population**, which is a level sufficient to protect a species not only from stochastic extinction but also from calamities such as fire or extreme weather. Furthermore, protection of a species requires that a species have multiple populations for two reasons: First, if one population is wiped out, it could be reintroduced from another. Second, having multiple centers of population maintains genetic diversity, which may be necessary to develop protection against evolving diseases.

14.4.2   Species Richness and Diversity: Synoptic Structure of Communities

Now we begin to look at communities instead of the individual populations that constitute them. Why do similar environmental conditions produce similar sets of species? The interactions between species are what stabilize a community. The more species that are present, the more interactions there can be. Removing any single organism eliminates its interactions, affecting those species with which it interacted. But the more overall species there are, the easier it will be for an affected organism to find a substitute. For example, suppose that mice are the primary food source for owls in a particular forest, and the mice are eliminated. If other small mammals, say rabbits, are available, the owls might be able to substitute them. Therefore, it is thought that species diversity is an indication of community stability and resistance to disturbance. The concept of species diversity bringing stability is used by conservationists to argue for the protection of ecosystems from human impacts.

There are a number of caveats to this idea. Disturbance, itself, can increase diversity. Well-developed ecosystems can settle into a lower-diversity state than they had during their stages of development. Also, it is observed that natural systems do not seem to have maximized their diversity as measured by the indices described below. This would require an even distribution of species. Instead, there are always some relatively rare species and others more dominant. Finally, it is possible that diversity is a consequence of ecosystem stability rather than a cause.

The Number of Species in an Ecosystem   The simplest measure of diversity is simply the number of species, S. It may be difficult to measure this in an ecosystem without impractically large samples. However, it can be estimated from a limited sample by assuming a lognormal probability distribution for species abundance. Generally, smaller geographic areas can be expected to have fewer species. An empirical rule relating S and area, A, is

$$S = cA^z \qquad (14.34)$$

where c is a parameter that varies with type of ecosystem and z is a parameter that has consistently been found to fall in the range 0.20 to 0.35 for islands. For large continental areas, the exponent ranges from 0.15 to 0.24, apparently due to the greater ability to import species. The number of species in ecosystems tends to increase toward the tropics, as does species diversity.

The **equilibrium theory of biogeography**, also called **island biogeography**, proposed by Robert MacArthur and E. O. Wilson, explains the relationship between area and number of species. They suggested that S is a result of a balance between immigration and extinction. Smaller areas would support smaller populations and thus would have higher rates of stochastic extinction [equation (14.33)]. Islands closer to the mainland have higher immigration rates and thus higher numbers of species. A similar balance occurs on the mainland itself, except that instead of immigration providing new species, it occurs by evolution. Of course, this acts more slowly, so the turnover of species is much slower.

Diversity Indices   A number of mathematical **diversity indices** have been developed to quantify species diversity better, giving more weight to more dominant species. Measurement of diversity may be subject to the sample-size limitation mentioned above. Also, measurements of diversity are usually limited to a subset of the community, such as a single taxa (birds, grasses, aquatic invertebrates, etc.) or a single trophic level. Despite these measurement problems and the caveats mentioned above with regard to their significance, diversity indices can be useful for making comparisons or for tracking trends.

Several of the most popular indexes can be computed from the proportions, p_i, of each of the S species. One is the **Simpson index**, D:

$$D = \frac{1}{\Sigma p_i^2} \qquad (14.35)$$

TABLE 14.9 Species Diversity Indexes for Hypothetical Communities Having Different Abundances

Proportion of Total Population					Simpson, D	Shannon–Weaver, H	exp H	Pielou's, Evenness, e
0.20	0.20	0.20	0.20	0.20	5.00	1.609	5.00	1.000
0.30	0.25	0.20	0.15	0.10	4.44	1.544	4.69	0.960
0.35	0.30	0.20	0.10	0.05	3.77	1.431	4.18	0.889
0.50	0.25	0.15	0.10	0.00	2.90	1.208	3.35	0.751
1.00	0.00	0.00	0.00	0.00	1.00	0.000	1.00	0.000

Source: Based on Ricklefs (1993).

The **Shannon–Weaver index**, H, is based on information theory. It gives more weight to rare species than does Simpson's index. It has the advantage of being normally distributed, making it possible to use statistical tests of significance to compare different values.

$$H = -\Sigma(p_i \ln p_i) \tag{14.36}$$

Table 14.9 shows examples for several different distributions of five species. Note that for H it is necessary to ignore zero values in the calculation (you can't take the log of zero).

The Shannon–Weaver index is often reported as its exponential form, since e^H is equal to S when the populations are evenly distributed. Both D and e^H vary from 1.0 when only one species is represented, to S if the species are evenly distributed. The Shannon–Weaver index can be used to compute an evenness parameter, **Pielou's index**, e. The evenness index varies from zero for maximally uneven, to 1.0 for perfectly uniform distribution:

$$e = \frac{H}{\ln S} \tag{14.37}$$

Because e depends explicitly on the number of species, its value is affected by including species with zero population in the count. Potentially, we could add any number of species which are absent from the ecosystem. Therefore, there would have to be some independent criteria for including such species, such as because they were formerly present in that ecosystem or because they are typically found in similar ecosystems.

A graphical approach that gives more information is to plot the probability distribution of the species. This plot is called the **dominance-diversity curve**. It is made by sorting the species by p_i, and plotting p_i vs. rank order. Evenness will be related to the slope of the curve.

Food web diversity has also been defined as a diversity index computed with abundances classified by trophic level instead of species.

14.4.3 Development and Succession: Temporal Structure of Communities

Previously, we looked at how to describe changes in individual populations. The basic population growth models often presuppose that environmental conditions, as reflected in carrying capacity, are constant. Community changes occur in nature as a result of disturbances that alter individual populations and environmental conditions. The most commonplace of these are the changes of seasons. More catastrophic changes include floods, fires, extreme weather, or volcanic eruptions. The time scales of human disturbances can range from practically instantaneous (nuclear explosions) to decades or more (global warming or soil depletion by agriculture). These disturbances affect the entire ecosystem, which responds as a unit. Although potentially, we could describe these ecosystem-wide changes using population models for all the individual species, we can note some holistic patterns.

A new habitat can be caused by a large event, such as a forest fire or clearcutting of a forest, or by a small local change such as the falling of a tree or a pile of dung left by a bear. In each case, the new habitat will be colonized by pioneers who are especially good at invading. These are often the *r*-selected organisms. Eventually, they will be replaced by more efficient *K*-selected species. In between, a variety of species may come and go in abundance. The process of sequential population changes initiated by a disturbance is called **succession**. The actual sequence of communities is called a **sere**. Barring further disturbances, succession ends when ecosystems develop a stable condition called a **climax**. The climax community is ultimately a community that can succeed itself. A climax community may form within a season or it make take decades, as in the case of forests.

As an example, an abandoned farm field may be colonized by grasses, which inhibit the germination of trees. The grasses attract herbivores, which create openings for shrubs by intense grazing. The shrubs provide shade, which enables pine to germinate and eventually to dominate. However, when the cover becomes too dense, the pine seedlings will not grow, and hardwood trees gain an advantage. Eventually, a climax community is formed as a hardwood forest. Populations of birds and other animals change as the food supply changes.

Another example is algal succession in temperate-zone lakes. As the water warms in the spring, the first species that dominate may be the cyanobacter. Some cyanobacter fix nitrogen and so are limited only by phosphorus. Eventually, they deplete the available phosphorus, removing it from the water column as they die and settle to the bottom. Other algae, such as diatoms, may flourish. But these require silicon for their shells, and when that is depleted, they may make way for green algae.

There are several forces driving succession. A community may change its environment, making it more suitable for others that succeed it as the shrubs made way for the pines. Succeeding communities are often those that tolerate a lower level of resources, as the diatoms that followed the cyanobacter. Sometimes the succeeding community must overcome inhibition by the previous residents, as the shrubs were inhibited by the grasses. Newly exposed soils may experience a gradual drop in pH from the accumulating products of plant decay.

Overall, succession seems to be related to a balance of colonizing ability of some species vs. the competitive ability of others. A particular sequence may not be uniquely determined by environmental conditions. One could wind up with one of several communities, depending on contingent factors during succession. However, chance factors seem to operate most strongly in earlier stages of succession.

Odum (1987) has listed a number of trends identified with succession in the absence of further disturbances (Table 14.10). Gross primary productivity tends to form a peak during succession and then declines as a community approaches the climax. Succession occurs in a biological wastewater treatment plant whenever conditions are changed. Plant startup is a prime example of this. Primary productivity is negligible in this system,

TABLE 14.10   Trends Expected During Undisturbed (Autogenic) Succession

Energetics
    Biomass (B) and organic detritus increase.
    Gross productivity (P) increases in primary succession, little change in secondary.
    Respiration (R) increases.
    P/R ratio moves toward unity (balance).
    B/P ratio increases.
Nutrient cycling
    Element cycles increasingly closed.
    Turnover time and storage of essential elements increase.
    Cycling ratio (recycle/throughput) increases.
    Nutrient retention and conservation increase.[a]
Population and community structure
    Species composition changes.
    Species diversity peaks in middle or end of sere.
    Species evenness peaks in middle or end of sere.
    r-Selected organisms replaced by *K*-selected.
    Life cycles increase in length and complexity.
    Size of organism and/or propagule (seed, offspring, etc.) increases.
    Mutualistic symbiosis increases.[a]
Stability
    Resistance increases.[a]
    Resilience decreases.[a]
Overall strategy
    Increasing efficiency of energy and nutrient utilization.[a]

[a]Trend based on theoretical considerations, not yet validated in the field.
Source: Based on Odum (1987).

but controlled energy input (organics in the wastewater) takes its place. This system is called *heterotrophic succession*. Respiration (per organism) peaks, then is reduced as substrate concentrations drop and respiration achieves balance with energy input.

As succession proceeds, nutrients are stored in biomass, and in soil in the case of primary succession. As these reservoirs build up, internal recycling increases, improving the efficiency of nutrient use. Species diversity may increase throughout the succession, although climax communities sometimes have less diversity than a peak. For example, relatively moist forests in the Great Lakes region showed a peak diversity in middle ages, whereas drier forests increase in diversity continuously.

The succession just described has a single direction of change. Some ecosystems exhibit cyclical changes even in the absence of external disturbance. The *Calluna* heath in Scotland has a cycle of 20 to 30 years. Heather establishes itself on bare ground and matures in about 7 to 15 years. After 14 to 25 years, this perennial loses its vigor, and other lichens and mosses invade. They eventually die out, leaving bare ground, and the cycle repeats.

14.4.4   Distribution vs. Abundance: Spatial Structure of Communities

Earlier sections treated the population of each species within an ecosystem as if it were a variable that depended only on time. However, their spatial distributions are interesting as well. Factors that control where an organism will be found include:

- Method of dispersal
- Physical factors
- Habitat selection
- Interactions with other species (predation, competition, disease)

Organisms disperse to colonize new areas in many ways. Plant seeds may be spread by wind, water, or animals, or they may be confined to the margins of the parent. Animals may disperse by their own motility or may be carried passively, like seeds, as in the case of insects. If the environment in the new area fits within their niche, they may thrive.

Some of the most general physical factors include climate, light availability, and the chemical environment. Geographically, these translate to such factors as latitude, altitude or depth (in soil or water), and position on a mountain slope. Soil or water chemistry may vary locally or regionally. Fire can alter the balance between grassland and forest, controlling the position of their border.

These factors vary in gradients throughout the world. For example, temperature, rainfall, depth of soil and light all vary with elevation along the slope of a mountain. Along the gradient various species rise and fall in abundance. In what are known as **open communities**, the population distributions overlap considerably, resulting in continuous variation in species along the gradient (Figure 14.19*a*). In the other extreme, known as **closed communities**, population abundances are strongly correlated (Figure 14.19*b*).

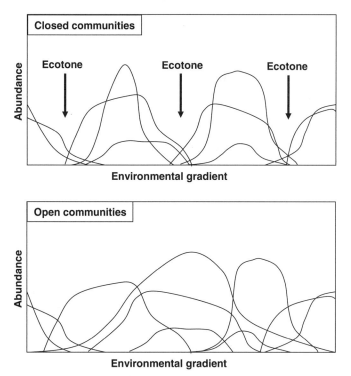

Figure 14.19 Population distributions along a hypothetical environmental gradient for both closed and open communities.

Closed communities are more easily defined as communities, since their boundaries are marked more clearly. Sharp boundaries between closed communities are called **ecotones**. They are especially prominent when the environmental gradient is fairly rapid, such as the transition between a wetland and adjacent woodlands. Nevertheless, they can also be present along gradual boundaries, such as the transition between grasslands and forest along long-range moisture gradients. In this case the boundaries are maintained by mutual inhibition, as when grasses prevent tree seeds from germinating and trees prevent grasses from growing by shading them.

Ecotones have greater species diversity not only because they may have representatives of two communities, but because some species benefit directly from both communities. This is called the **edge effect**, a concept that is attributed to Aldo Leopold. Many birds, such as the American robin, for example, nest in the forest but feed in fields. Human alterations to the landscape, such as by clearing fields or putting roads through wilderness areas, result in increasing the edges. This also has the effect of decreasing the area of each community, which can reduce the number of species each can support [equation (14.34)], and particular species with large area requirements may be eliminated.

We saw previously how the number of species tends to increase with the area of the ecosystem. However, each species may not necessarily range over the entire area, but rather, be limited by environmental conditions to a subset. It turns out, that most species have small ranges, and fewer have large ranges. Species with larger ranges also tend to have greater population density. These tendencies may be due to the differing dispersion capabilities of species or (as some think) to the artifacts of sampling, which are biased against rarer species.

Individuals within a population tend to distribute themselves more or less evenly due to intraspecies competition. Individuals or groups of vertebrates tend to restrict themselves to an area known as the **home range**. If the home range is defended aggressively and does not overlap with those of neighbors, it is called a **territory**. Species with complicated reproductive behavior, such as nest building and extended care of young, are more likely to be territorial. Birds form territories more often than other groups. Territoriality is a way to avoid the need for aggressiveness, since once a territory is formed and marked, others of the same species tend to respect it.

14.5   HUMANS IN THE BALANCE

As of the year 2005, the human population exceeds 6.4 billion people, versus 3.5 billion in 1950 and 1.6 billion at the start of the twentieth century. The growth rate in 2005 is 74 million per year (almost 1.2% per year). However, growth in food production is not keeping pace.

The amount of fish caught has increased 4.6 times from the 1950s to the late 1980s. Over the same period, world grain production has tripled. But since then, fishing has reached the sustainable-yield limit and has stopped increasing, and grain production has slowed its increase to 0.5% per year, far less than world population growth. Thus, the world's annual per capita grain production has dropped almost 10% from its highest value of 346 kg in 1984 to 313 kg in 1996.

Agricultural food production can be increased by either increasing the area under cultivation or by increasing the yield. The oldest means of increasing yield is to initiate irrigation. In the twentieth century, fertilizer use provided the most dramatic improvement. But fertilizer use is linked to water use. The more rain or irrigation that is applied, the more fertilizer the crops can utilize. World fertilizer use has actually dropped recently, as farmers have learned to apply it more precisely. A final way to improve yields was to develop better varieties of crops, but the high-yielding varieties require higher fertilizer doses.

Some of these practices cannot be extended or even be sustained. The per capita area under cultivation has been falling since the 1950s, and the total area has been dropping since 1981. Increasing the area now means bringing marginal lands under cultivation and/or destroying sensitive ecosystems such as tropical rainforests or wetlands. In the former Soviet Union the marginal land was subject to soil erosion. Over the 18 years leading to 1995, 26% of the land used for grain production had to be abandoned for this reason. Agricultural practices tend to increase erosion and do not contribute much organic matter by which natural systems rejuvenate the soil. Since 1950 some 20% of the world's topsoil has been lost. Industrial and urban development is also taking much prime farmland out of production, even in China.

The per capita amount of land under irrigation has been dropping since 1979. A principal reason is the depletion of underground aquifers. Twenty-one percent of U.S. cropland under irrigation gets its water from aquifers that are being drawn down. Many of the world's great rivers are pumped dry before they reach the sea. In the United States, these include the Colorado River and the Arkansas River. The river feeding the Aral Sea in Uzbekistan has run dry, and the sea itself is disappearing. China's Yellow River first ran dry in 1972, and does so now each year for longer and longer periods.

These dismal statistics call for an evaluation of the carrying capacity of Earth. The total world grain production is currently about 1.82 billion tons per year. This might be increased to 2 billion tons. Americans use about 800 kg per person per year, much of it indirectly in the form of meat and dairy products. Indians on average use only 200 kg per person per year, in a diet consisting mostly of grain only. Some think that the healthiest diet is what is known as the Mediterranean diet, which is lower in fat and protein than the American diet. It requires about 400 kg per person per year. Thus, the world's grain supply could support 2.5 billion people on the American diet, 5 billion on the Mediterranean diet, or 10 billion on the Indian diet.

The world population growth rate is 1.2% per year. The good news is that this is a slowing, down from a peak of 2.5% per year in the 1960s. Fertility has fallen significantly in most of the large developing countries. Modern contraceptives are becoming more accepted by these countries. For example, they are used by 66% of married women in Brazil. Indonesia is a model for population control in developing countries because it has linked it to health care and education for women and children, reducing infant mortality from 133 per 1000 births in the 1960s to 57 in the 1990s. Some think that feelings of economic security are an even more potent factor in controlling population than birth control education.

However, we are already beyond the number that could be supported on the Mediterranean diet. Even if population control is pursued aggressively by developing countries, the population will continue to grow because of the high proportion that is yet to reach childbearing age. Under this scenario the population is expected to level off at about 10 to 16 billion during the twenty-first century.

14.6   CONCLUSION

In the next chapter we study particular types of ecosystems. Based on what we have learned from this chapter, we can formulate a few questions to think about when examining any ecosystem:

- What is its primary productivity? Which organisms are responsible for most of its productivity?
- How much energy is stored and transferred at each trophic level? How much biomass?
- What do the biogeochemical cycles look like for this ecosystem? What are the amounts of nutrient storage and the flux rates between compartments?
- What are the important organisms and groups of organisms in this ecosystem? What adaptations make them abundant in this ecosystem?
- What is the niche of each of these organisms (habitat; food source; is it food for another organism; in what types of species interactions does it participate)?
- What limits the population of each species?
- What stage of development is the ecosystem in?
- How do human activities affect all of these factors?

In the next several chapters we travel from the land to the sea, examining several types of ecosystems with these questions in mind.

PROBLEMS

14.1. Decay of radioactive elements deep within the Earth produces a flux of heat energy toward the surface in the form of geothermal energy. Visible results include hot springs, geysers, and volcanoes. Humans can extract this energy to run machines. Could plants or microorganisms use this energy directly for primary productivity in nature? Explain.

14.2. The net primary productivity of a tall form of cord grass (*Spartina alterniflora*) has been measured to be 5800 $g/m^2 \cdot yr$ in a Georgia salt marsh. What is the efficiency of energy capture assuming an average solar energy input for the southeastern United States of 3888 $kcal/m^2 \cdot day$? Comment on the result.

14.3. The United States as a whole receives an energy subsidy in the form of fuels in the amount of about 1.8×10^3 $kcal/m^2 \cdot yr$. The world receives about 100 $kcal/m^2 \cdot yr$. Compare these figures to the gross and net primary productivity figures given in the text for the United States and for particular ecosystems.

14.4. Tommy's Pond averages 2.0 f deep and has an area of 0.40 acres (1 acre = 43,560 ft^2). The stormwater entering the pond was measured to have an average phosphorus concentration of 0.300 mg/L. The average flow rate of the stormwater may be assumed to be 5000 ft^3/day (1 ft^3 = 28.3 L). Thirty ducks live on the pond. Each duck is estimated to contribute phosphorus at the rate of 0.50 kg/y.

(a) What will be the steady state phosphorous concentration in the pond?

(b) What is the turnover rate for phosphorus in Tommy's Pond at steady state?

14.5. It can be difficult to visualize the huge quantities in global biogeochemical cycles. Earth has a land area of 148,429,000 km^2 and a water area 361,637,000 km^2. Normalize the fluxes and reservoirs in the carbon cycle as shown in Figure 14.5 on a per square meter basis. Separate the atmosphere into the portions over land (29.1%) and over water (70.9%). How does productivity on land compare to that on water?

14.6. Look at the magnitude of abiotic fixation of nitrogen. How long would it take to convert all the atmospheric nitrogen to nitrate if life suddenly disappeared from Earth? Use information from Figures 14.7 and 14.8, plus Earth's area from Problem 14.4.

14.7. Write a the balanced chemical equation for the reaction of water, oxygen, and nitrogen to nitric acid. How many moles of oxygen are needed for each mole of nitrogen? Would there be an excess left over of either, based on the current atmosphere of 79 mol% N_2 and 21 mol% O_2? Are there any other sources of oxygen?

14.8. Write a spreadsheet program to solve matrix (14.8). To do this in Microsoft Excel, first place the coefficients in a 4 × 4 array of cells, then the right-hand-side (RHS) vector in an adjacent column of cells. Say that the matrix is in cells A1 to D4 and the RHS vector is in cells E1 to E4. Use the cursor to highlight another empty 4 × 4 array: say, cells A6 to D9. Then type the following: "=minverse(A1:D4)." Now, hit "Cntrl-Alt-Enter." Cells E1 to E4 will display the inverse of the matrix. To get the solution, we must multiply the inverse by the RHS vector. To do this, highlight four empty cells in a vertical column. Then type "=mmult(A6:D9,E1:E4)" and hit "Cntrl-Alt-Enter." The cells will now hold the solution of the matrix system in order.

Notice that you can now change entries in the original matrix without reentering the functions. The solution changes automatically. The coefficients in Table 14.4 represent summer conditions. In winter the situation can be represented by the following values: $k_1 = 1.0; k_2 = 8.0; k_3 = 4.4; k_4 = 2.5; k_5 = 1.7$. Use your program to compute the resulting distribution of nitrogen forms in this ecosystem. Compare and explain the differences between summer and winter.

14.9. What type of species interaction describes the human–mosquito pair?

14.10. To see how doubling can rapidly cause populations to outstrip their resources, try the following math experiment. Suppose that you were to take a piece of paper 0.1 mm thick and fold it in half twice. This would quadruple the thickness to 0.4 mm. Suppose, instead, that you had folded it 50 times, if you could. Before computing the thickness, take a guess (a *big* guess). Then do the calculation. To what physical distance can you compare this?

14.11. A population of insects lives for two years. The first year they are a juvenile form that does not reproduce; the second year is an adult form that reproduces and then dies. Sampling finds 1000 juveniles and 10 adults per square meter. Create a life table for this insect. Compute the natality and mortality for each age group, and use the results to compute the net reproductive rate, the mean length of a generation, and the growth rate. Is the growth rate what you would expect?

14.12. For a carrying capacity K, what population, N, would result in a specific growth rate that is one-half of the maximum?

14.13. Both the logistic equation and the Monod model produce sigmoidal growth curves. (**a**) Why do we not use the logistic model for microorganism growth in wastewater treatment plants? (**b**) Conversely, why not use the Monod model for ecological population modeling?

14.14. (**a**) How many possible pairwise interactions are there among nine species? (**b**) If each species has, on average, only two interactions, how many interactions will there be in this case?

14.15. As of the year 2005, the human population exceeds 6 billion people vs. 3.5 billion in 1950 and 1.6 billion at the start of the twentieth century. Find the values of r and K that make the logistic equation fit these data. According to this, what is the carrying capacity of Earth? How does this compare with the estimates given in the text? What are possible explanations for the discrepancy?

14.16. In the Lotka–Volterra predator–prey model, the period of the oscillations depends on the initial conditions. But in the vicinity of the steady-state solution given in Section 14.4.1, it approaches $2\pi(ad)^{-1/2}$. Thus, a decrease in the host growth rate or the predator death rate would increase the period. Can you explain in words why this is so?

14.17. Suppose that someone were to enter a large temperate-zone hardwood forest and harvest all of the trees in a 100-ft circle. Describe the changes that might occur at the site if it were left undisturbed for the next 100 years.

14.18. Compute the Simpson and Shannon–Weaver diversity indices and Pielou's evenness index for a community consisting of four species that each comprises one-eighth of the organisms and a fifth species that comprises the other 50%. Which species distribution in Table 14.9 does this community most resemble?

REFERENCES

Berner, E. K., and R. A. Berner, 1987. *The Global Water Cycle: Geochemistry and Environment*, Prentice Hall, Englewood Cliffs, NJ.

Brown, L., C. Flavin, H. French, et al,, 1997. *State of the World 1997*, W.W. Norton, New York.

Krebs, C. J., 1994. *Ecology*, HarperCollins, New York.

Mattson, C. P., and N. C. Vallario, 1975. Hackensack Meadowlands Development Commission report.

Murray, J. D., 1993, *Mathematical Biology*, Springer-Verlag, New York.

Odum, E. P., 1987. *Basic Ecology*, Saunders College Publishing.

Odum, E. P., 1993. *Ecology and Our Endangered Life-Support Systems*, Sinauer Associates, Sunderland, MA.

Prigogine, Ilya, 1972. *Introduction to Nonequilibrium Thermodynamics* Wiley, New York.

Prigogine, Ilya, 1978. Time structure and fluctuations, *Science*, Vol. 201, pp. 777–785.

Prigogine, Ilya, G. Nicolis, and A. Babloyantz, 1972. Thermodynamics and evolution, *Physics Today*, Vol. 25, No. 11, pp. 23–38 and Vol. 12, pp. 138–144.

Raven, P. H., R. F. Evert, and S. E. Eichhorn, 1992. *Biology of Plants*, Worth Publishers, New York.

Ricklefs, R. E., 1993. *The Economy of Nature*: *A Textbook in Basic Ecology*, 3rd ed., W. H. Freeman, New York.

Smil, Vaclav, 1997. Global population and the nitrogen cycle, *Scientific American*, Vol. 277, No. 1, pp. 76–81.

U.S. Census Bureau, 2004. International data base, September, http://www.census.gov/ipc/www/idbsum.html.

Utida, Syunro, 1957. Population fluctuation, and experimental and theoretical approach, *Cold Spring Harbor Symposia on Quantitative Biology*, Vol. 22.

Whittaker, R. H., 1975. *Communities and Ecosystems*, Macmillan, New York.

15

ECOSYSTEMS AND APPLICATIONS

The major ecosystem types, or biomes, were listed in Table 14.1. Some of these are described in more detail here in a sequence from terrestrial to marine ecosystems, with stops along the way for aquatic and wetland systems. The gross primary productivity of the major ecosystems is listed in Table 15.1. The most productive ecosystems are estuaries, reefs, and tropical forests. Some of Earth's environments are practically sterile, such as the ice sheets of the Arctic and Antarctica and new land formed from volcanic ash or lava flows. The volcanic landscapes, will, however, soon be colonized if environmental conditions are favorable.

15.1 TERRESTRIAL ECOSYSTEMS

Terrestrial biomes can be distinguished on the basis of the dominant climax vegetation. The basic factors determining terrestrial ecosystems are temperature and moisture. Some of these determinations can be made based on annual temperature and precipitation (see Figure 15.1). Some biomes do not fit into this scheme, such as the chaparral, which depends on wet and dry seasons. Also note the overlap in regimes, since other factors may affect the ecosystem type that results. Fire affects whether a region will have forest or grassland, and the north slope of a mountain may have a different biome than the south-facing side.

The **tundra** is essentially a wet grassland that exists in cold climates at high latitudes (**arctic tundra**) or high altitudes (**alpine tundra**). The harsh climate excludes trees but supports shrubs and lichens. Although the growing season is short, arctic tundra has long photoperiods in the summer. This can result in primary productivity as high as 5 g of biomass per square meter per day. This can support a substantial animal population, the larger of which tend to be migratory. Only the upper few inches thaws during the

Environmental Biology for Engineers and Scientists,   by David A. Vaccari, Peter F. Strom, and James E. Alleman
Copyright © 2006 John Wiley & Sons, Inc.

TABLE 15.1   Gross Annual Primary Productivity of Some Major Ecosystems[a]

Ecosystem	GPP (kcal/m^2)
Marine	
Open ocean	1,000
Coastal ocean	2,000
Upwelling zones	6,000
Estuaries and reefs	20,000
Salt marsh	36,000
Terrestrial	
Deserts and tundras	200
Grasslands and pastures	2,500
Dry forests	2,500
Boreal coniferous forests	3,000
Cultivated lands without energy subsidy	3,000
Moist temperate forests	8,000
Cultivated lands with energy subsidy (mechanized agriculture)	12,000
Wet tropical and subtropical forests	20,000
Average for the biosphere	2,000

[a]Recall that 1000 g corresponds to about 4100 kcal.
Source: Odum (1987).

summer. Below this is the deep, permanently frozen soil called **permafrost**. Tundra is notoriously sensitive to disturbance and slow to recover. A footprint is said to remain visible for years.

The **taiga**, also called **boreal** (northern) **forest**, occupies vast areas of Canada and Russia and other areas of similar latitude. The dominant plants are conifers, including

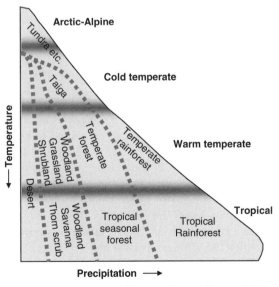

Figure 15.1   Temperature–moisture regimes of terrestrial biomes.

spruce, fir, and pine. Dense shade inhibits other plants, but the trees support small seed-eating animals. Larger animals are less common, as they require broadleafed plants for food. A form of coniferous forest called the **temperate rain forest** exists in humid coastal regions of North America from northern California to Alaska. The tree needles degrade slowly, building into a thick detritus mat on the forest floor.

The **temperate deciduous forest** occupies milder climate zones with ample rainfall. The canopy cover is less impenetrable, and smaller trees and shrubs (the **understory**) can grow. A large number of subdivisions exist, based on the type of dominant hardwoods that are present. Nuts, berries, and broadleaf plants support a well-developed food pyramid. The rate at which litter falls to the forest floor increases toward the equator, up to 1100 g/m^2 per year.

Where rainfall is insufficient to support the deciduous forest, a **grassland** will be found instead. A variety of types occur, depending on rainfall, as this type of ecosystem grades into desert. Moister regions support tall grasses and sod-forming grasses. More arid areas are characterized by short bunchgrasses. Grasses produce a large amount of litter and therefore organically rich soil. Most of the biomass is below ground. Large grazing animals are typically present, such as bison, gazelles, and kangaroos.

Grassland ecosystems are often maintained by fire, without which they would be succeeded by shrubland or forest. Forest can give way to grassland if fire follows an extended drought. Once the grasses are established in the resulting opening, fires become more frequent, due to the rapid production of leaf litter. This makes it difficult for trees to reestablish themselves.

Fire occurs naturally and can be a force in preserving an ecosystem. Even in forests, periodic fires confined to the understory (**surface fires**) eliminate deadwood, which otherwise could fuel a much more devastating **crown fire**. A crown fire destroys the mature trees as well as seeds and organic matter in the soil. A surface fire can release nutrients and stimulate germination of tree seeds. Pines tend to be more resistant to fire than hardwoods. Some mature coastal pine forests are called a **fire climax** ecosystem.

Woodlands tend to give way to shrublands under maritime influence. The **chaparral** is a shrubland that dominates in warm areas that have plenty of moisture in winter but are very dry in summer.

The **desert** includes regions with less than about 25 cm (10 in.) of rain per year, depending on the temperature. Primary productivity is proportional to rainfall, the limiting factor for this ecosystem. The dominant plants are either rapid-growing annuals, cacti, or specially adapted desert shrubs. The creosote bush dominates the hot deserts of southeastern North America. This plant exhibits **allelopathy**, in which it inhibits nearby competition chemically. In cooler northwestern deserts, sagebrush dominates. Desert animals are specially adapted to conserve water. Lizards are suited to the desert because they excrete dry uric acid instead of urea with water as mammals do. Nevertheless, some small rodents survive by excreting a very concentrated urine and by nocturnal life-styles. Irrigation of deserts can make them productive for agriculture. However, since the evaporation rate will be high, salts can slowly accumulate in the soil unless extra water is provided to remove the salts by runoff and infiltration. (In the Central Valley of California, selenium from irrigation and natural groundwater sources is accumulating in the soil. It is thought that the productive life of this important agricultural region may be limited to a matter of decades by this problem.)

Seasonal tropical forests occur in places such as south Asia, which has monsoon rains and a pronounced dry season, during which the trees lose their leaves. **Tropical rain**

forests occur where there is more than 200 cm of rain per year, distributed throughout the year. They are typically found at low elevations near the equator. They tend to be very rich in both plant and animal species. Some tracts of a few acres can have more species of trees than are found in all of Europe. A 6-square mile area of Panama was found to have at least 20,000 species of insects, compared to several hundred in all of France. Ants, termites, moths, and butterflies are important. A larger proportion of both plants and animals live in the upper layers of the forest, compared to temperate forests. In one forest, 50% of the mammals live in the trees. Plants that grow supported by other plants, with no roots reaching the ground, are called **epiphytes**. Many animals have formed symbiotic relationships with epiphytes. An important feature of tropical rain forests is the rapid cycling of nutrients within the ecosystem. Litter is decomposed rapidly, and the nutrients soon reabsorbed by the plants. Therefore, most of the nutrients are stored in the biomass, and the soil remains poor in nutrients. The rich growth of the rain forests made the early settlers think the soil must be rich. Many attempts at agriculture failed for a lack of this understanding.

15.1.1 Forest Nutrient Cycles

Besides participating in global biogeochemical cycles, ecosystems cycle nutrients internally. When the ecosystem is harvested, such as in logging, the nutrients must be replaced if the productivity of the ecosystem is to be sustained. For example, a forest ecosystem receives nutrient input from the atmosphere and from weathering of the bedrock. Atmospheric input includes nitrogen that is fixed by nitrification as well as nitrogen oxides from natural and human-made sources, and from dust and aerosols.

Forests export nutrients to the atmosphere and by streamflow. Streams leach mineral nutrients and carry away biomass that has fallen in. Microbes in anaerobic environments such as bog sediments eliminate nitrogen by denitrification and sulfur by reduction to H_2S. Trees also emit ammonia, H_2S, and organic compounds such as terpenes. The organic emission is famous for causing the haze in the Great Smokey Mountains of the southeastern United States and for giving President Ronald Reagan a reason to discount anthropogenic pollution, saying that "trees pollute." (Actually, it is true that plants may emit more hydrocarbons to the environment than comes from human activities. Aspen and oak leaves emit as much as 2% of the carbon produced by photosynthesis as isoprenes. The isoprenes may play a role in protecting photosynthetic membranes. However, photochemical smog found in urban air is caused not only by hydrocarbon emissions, but by nitrogen oxides produced together with them.)

The fluxes of nutrients within the forest are usually much greater than the external flows. These fluxes amount to an internal recycling. Nutrients taken up by trees are returned to the soil by leaf fall. Biodegradation by saprophytes releases the nutrients, whereupon they are again absorbed by the roots. The nutrient cycles operate faster in warmer climates, as biomass in the litter decays more rapidly. Plants develop thick root systems close to the surface to intercept nutrients before they leach away. As a result, more than half of the organic carbon in temperate forests is in the soil and litter; whereas in tropical rain forests the figure is 10 to 25%. Thus, tropical soils tend to be very poor in nutrients. As a result, harvesting the trees to convert a tropical forest to farmland often leaves the soil depleted of nutrients. Thus, ironically, these highly productive lands become unproductive in their new use. Temperate forest lands, on the other hand, make productive farmland.

TABLE 15.2   Annual Mineral Balances for Several Watersheds of the Hubbard Brook Experimental Forest, 1963–1969

Mineral	Input by Rain (kg/ha)	Output in Stream (kg/ha)	Net (kg/ha)	Stream Conc. (mg/L)
Calcium	2.6	11.8	−9.2	1.58
Magnesium	0.7	2.9	−2.2	0.39
Potassium	1.1	1.7	−0.6	0.23
Sodium	1.5	6.9	−5.4	0.92
Aluminum	Very small	1.8	−1.8	0.24
Ammonium	2.7	0.4	+2.3	0.05
Nitrate	16.3	8.7	+7.6	1.14
Sulfate	38.3	48.6	−10.3	6.40
Dissolved silica	Very small	35.1	−35.1	4.61
Bicarbonate	Very small	14.6	−14.6	1.90
Chloride	5.2	4.9	+0.3	0.64

Source: Krebs (1994).

When leaves and other detritus falls to the forest floor, nutrients are removed by leaching of soluble components, consumption by soil invertebrates, and mineralization by bacteria and fungi. Rate of degradation depends largely on the lignin content of the litter. For example, the highly digestible mulberry was found to lose 64% of its weight in a year on the ground in eastern Tennessee. Oak degraded by 39%, beech only 21%. Pines and conifers also degraded slowly. Lignins break down much more slowly than cellulose or hemicellulose, the other components of wood. Few bacteria can do the job. Most lignin decomposition is done by the **white rot fungi**, *Phanerochaete chrysosporium*.

Nutrient relationships have been studied using large-scale long-term experiments. One of the best known is the *Hubbard Brook Experimental Forest*. Stream and rainfall flow and composition were measured for a number of watersheds there (Table 15.2). A net loss of calcium, magnesium, potassium, sodium, aluminum, sulfate, silica, and bicarbonate are presumably balanced by inputs from weathering of bedrock. Ammonium, nitrate, and chloride showed greater inputs than outputs. The nitrogen species may also be lost by transformation.

About 60% of the rainfall that falls on undisturbed Hubbard Brook watersheds eventually leaves via the streams. The area is underlain by a tight bedrock formation, so losses to groundwater are negligible. Thus, most of the loss is by evaporation and transpiration. If a watershed is logged, streamflow increases drastically. Nitrate concentration in the streams also increases greatly, since nitrification proceeds normally in the soil, but uptake by trees is eliminated. As with anions in general, nitrate is weakly bound to soil and so is rapidly leached out in runoff.

15.1.2   Soil Ecology

Engineers are trained to think of soil just as a structural material. In fact, one of the first acts of developing a property is removal of the topsoil, the part most associated with the biosphere. The biomass in it makes it unstable and unpredictable from a structural point of view. To the biologist the soil is a living material, intimately connected with living things, and both created by and sustaining them.

TABLE 15.3   Average Turnover Time (years) for Organic Matter (OM) and Nutrients in Forest Soil

Region	OM	N	K	Ca	Mg	P
Boreal coniferous	353	230.0	94.0	149.0	455.0	324.0
Boreal deciduous	26	27.1	10.0	13.8	14.2	15.2
Temperate coniferous	17	17.9	2.2	5.9	12.9	15.3
Temperate deciduous	4	5.5	1.3	3.0	3.4	5.8
Mediterranean	3	3.6	0.2	3.8	2.2	0.9

Source: Krebs (1994); original source Cole and Rapp, 1981, in *Dynamic Properties of Forest Ecosystems*, D. E. Reichle (Ed). Cambridge University Press, Cambridge.

Soil is the part of terrestrial ecosystems with the slowest turnover time (Table 15.3). This makes it the part that regenerates slowest if disturbed. A forest fire can kill all aboveground life in an area, but if the soil is not deeply burned, the forest can soon grow back. But if erosion carries away the topsoil with its organic matter and nutrients, the area may remain desolate for a long time. Thus, one of the most important goals for conservationists should be preservation of an ecosystem's soil.

The mineral portion of soil is produced by weathering of bedrock, either locally or at a distant location followed by transport by wind or water. Weathering is caused by physical, chemical, and biological factors. **Physical weathering** is caused mainly by the expansion of water when it freezes in cracks. **Chemical weathering** is caused by water dissolving the more or less soluble components, and by acids from atmospheric carbon dioxide (which forms carbonic acid in water) or from acid rain. **Biological weathering** contributes to chemical weathering by adding more CO_2 plus organic acids to the water. Roots can help break up rocks by penetrating and forcing open cracks. Lichens produce acids that dissolve rocks.

A well-developed soil is divided into three main layers, or **horizons** (Figure 15.2). The **A horizon**, often called the **topsoil**, is the uppermost portion of the soil. The A horizon is rich in organic matter from decayed plants and animals. It is the site of most of the biological activity in the soil. Below this is the **B horizon**, which is lower in organic content and contains soil particles that are less weathered than the A horizon. The B horizon is a zone in which minerals, clay particles, and to a lesser extent organic matter are deposited after leaching from the A horizon. Next comes the **C horizon**, which is made up of partially broken-down and weathered rocks from which the soil of the upper layers is formed. Below this is the bedrock. Some classifications include the layer of decaying plant and animal matter above the A horizon, called the **O horizon**, which has little mineral matter. The portion of the soil that is subject to the influence of plant roots is called the **rhizosphere** or **root zone**.

When rocks weather, less soluble components such as silicates remain as solid particles. Particles larger than 20 μm are classified as sand; between 2 and 20 μm as silt, and particles smaller than 2 μm as clay. Soil with a preponderance of larger particles drain well, holding less moisture and nutrients. Clay particles tend to retain both. Mineral surfaces tend to have a net negative charge that enables them to adsorb cationic nutrient ions such as Ca^{2+}, Mg^{2+}, K^+, Na^+, NH_4^+, and H^+. Anionic nutrients such as NO_3^-, SO_4^{2-}, Cl^-, and HCO_3^- are not strongly adsorbed and leach more easily from soil. PO_4^{3-} is an exception since it forms very low solubility precipitates and adsorbs to mineral surfaces containing iron, aluminum, or calcium.

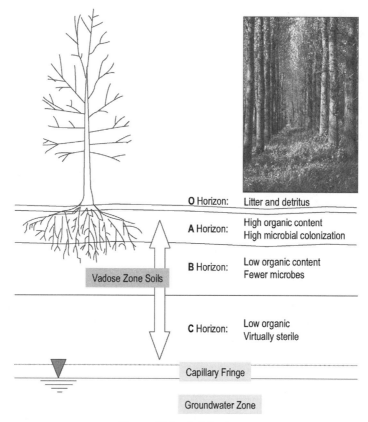

Figure 15.2   Soil horizons.

The capacity of soil to adsorb cations is called its **cation-exchange capacity** (**CEC**). The CEC describes the soil's capacity to serve as a reservoir for nutrients for plants. Because clay particles have a large surface area per unit mass, they have a relatively high CEC. For example, the CEC of kaolinite clay ranges from 3 to 15 milliequivalents (mEq) per 100 g. Montmorillonite clay, which has about an order-of-magnitude smaller particle size than kaolinite, has a CEC of 80 to 100 mEq per 100 g. The CEC of a whole soil will be strongly affected by its clay content as well as by organic matter (see below).

The relative strength of attraction of cations to clay particles is

$$H^+ > Ca^{2+} > Mg^{2+} > K^+ > Na^+$$

Thus, acidic conditions displace the other ions from the soil, allowing them to be leached. A low pH can also cause charge reversal on clay particles, enabling adsorption of anionic nutrients. In a process called **podsolization**, acidity breaks down the clay particles in the A horizon, precipitating their soluble salts in the B horizon. This reduces the CEC of the soil and therefore its capacity to hold nutrients. Podsolization commonly occurs in coniferous forests in cold regions, where the heavy fall of needles produce organic acids as they decay. This type of soil is called **spodisol**.

In low-lying tropical regions such as the Amazon basin, warm, moist conditions result in faster weathering than erosion. As a result, soil accumulates to great depths. Since the soil at the surface is so far from the parent mineral, the clay is leached out, leaving iron and aluminum oxides. These minerals do not retain nutrients as well as clay and can consolidate into a hard, concretelike material. This is called **laterite soil**, or **latisol**.

Soils with about 20% clay and 40% each of sand and silt are called **loam**. Increasing the fraction of one or the other component produces soils classified as "sandy loam," "clay loam," "silty clay loam," and so on. Soil with a clay content above 40 to 60% has a low permeability to water, which makes it a poor substrate for plants. A very sandy soil does not retain water well and has too low a CEC to store significant quantities of nutrients.

Thirty to fifty percent of soil volume is occupied by pore space. This fraction is called the **porosity**. If soil that is initially saturated with water is allowed to drain freely by gravity, a point is reached when the weight of the water is balanced by capillary forces holding the water in the pores. This point is called the **field capacity**. The moisture content can drop further by the action of air drying or by absorption by plant roots. The water content can be described in terms of the **soil water potential** or **suction pressure**, which expresses the pressure needed to extract the moisture from the soil. Free pure water has a soil water potential of zero.

The maximum pull most plants can exert on soil water is -1.5 MPa (about -15 atm). If the soil water potential falls below that level, it is considered to be below the **permanent wilting point** for that soil, and plants in that soil will be irreversibly wilted. A typical loam containing a fairly even distribution of sand, silt, and clay has a saturation of about 45 g of water per 100 g of dry soil. This corresponds to a porosity of 25%. Its field capacity is about 32 g per 100 g dry weight (d.w.; about 16% by volume), and the wilting point is about 7 g per 100 g d.w. (about 3% by volume). Thus, the amount of usable water in soil at the field capacity is about 25 g per 100 g d.w. (9% by volume). Soils with higher clay content have more of their water bound at the -1.5 MPa water potential, and therefore have a higher wilting point.

The bedrock is the ultimate source of most of the minerals for the ecosystem. Carbon and nitrogen, of course, come mostly from the atmosphere, and mineral input from the atmosphere, from dust, can also be significant. The bedrock provides the macronutrients phosphorus, potassium, calcium, magnesium, and sulfur, plus the micronutrients boron, copper, chlorine, iron, manganese, molybdenum, and zinc. The importance of bedrock is dramatized by the example of **serpentine soils**, which overlie serpentine rock, a magnesium iron silicate. These soils are deficient in calcium, nitrogen, phosphorus, and molybdenum, and high in magnesium, nickel, and chromium. Serpentine soils have poor plant coverage. What plants exist are characterized by specially adapted vegetation that tolerates these conditions. These plants cannot compete in normal soils, and normal plants cannot survive serpentine conditions.

Living things affect soil properties in other ways besides causing biological weathering. Living things in the soil mix and burrow, changing its structure, and living things contribute organic matter. **Soil organic matter** consists of plants and animals and the products of their degradation. Most of it is from plant material such as leaf litter in various states of decay. The conversion of dead biomass to soil organic matter is called **humification**. Biological materials that are degraded to the point of being unrecognizable are called **humic substances** (or **humus**). They represent about 5 to 10% of the dry weight of topsoil.

Humic substances are divided into three groups, based on whether they are soluble in acids, bases, or neither. The groups are called **fulvic acid**, **humic acid**, and **humins**.

Fulvic acid is the fraction that is soluble in strong acid and base solutions; humic acid is soluble in acids but insoluble in bases; humins are insoluble in both. None of these have a definite chemical structure. They are dominated by randomly polymerized phenolic rings with a variety of side groups, many of them acidic, and by numerous cross-links. Molar masses range from about 700 to 400,000, with fulvic acids tending to be the smallest molecules and humins the largest. Their structure makes them resistant to further biodegradation. The presence of large numbers of acidic groups, such as hydroxyl and carboxylate, makes humic substances a substantial component of the CEC of soil.

Humification occurs in several stages. First, organic polymers such as cellulose, hemicellulose, and lignin are broken down to monomers such as phenols, quinones, amino acids, and sugars. In the second stage of biodegradation, the monomers polymerize due to spontaneous reactions and due to enzyme catalysis, producing the humic substances. Following humification, the process of **mineralization** slowly converts humic substances to inorganic materials. In natural ecosystems the formation of humic substances tends to be in equilibrium with their mineralization.

Humification is caused mainly by saprotrophic bacteria and fungi. Sugars, fats, and proteins are decomposed fairly readily. Cellulose breaks down more slowly, and lignin even more slowly. Lignin may be the source of the phenolic rings that form the major building block of humic substances. As was mentioned previously, the fungus *P. chrysosporium* plays a major role in lignin degradation. The saprotrophic microbes are also important for the organic compounds they release to the soil environment. These include inhibitory compounds such as antibiotics, and stimulating compounds such as vitamins and essential amino acids.

In some ecosystems, organic matter accumulates to high levels. Grasslands produce large amounts of biomass. Humification proceeds rapidly, but mineralization is slow. Forests, on the other hand, recycle nutrients more efficiently. Thus, grasslands typically have five to 10 times as much soil organic matter as forests. This makes them particularly suitable for conversion to agriculture. Oxygen limitations in wetlands may slow both humification and mineralization. Biomass can accumulate to great depths in wetlands, forming **peat**. Coal is formed when peat is buried under heat and pressure over geologic time.

Experiments have shown that invertebrates play a major role in humification. Important groups include protozoans, nematodes, ostracods, snails and slugs, and earthworms. They act not primarily by digesting the biomass, but more by breaking up the biomass into smaller pieces, increasing its surface area, and by grazing the bacteria and fungi, which increases their activity. This is similar to the role of invertebrates found in biological wastewater treatment processes.

Humic substances are a major source of CEC in soil and increase its water-holding capacity as well as its nutrient capacity. Humic substances form chemical complexes, called **chelation**, with metal ions. This increases the bioavailability of the metals. Metals may be less toxic when chelated. For example, copper toxicity to phytoplankton depends on the free copper ion concentration. In marine environments the same copper concentration will be less toxic close to shore, where humic substances are more available, then in the open ocean. Soil organic matter also improves the soil structure, especially its friability. **Friability** is the ability of a soil to crumble. This property makes it easier for roots to penetrate. The increased CEC and friability are the primary reasons that we add composted manure to gardens and agricultural fields, not because of the nutrients they carry.

Agriculture removes organics by harvesting. This removes nutrients and reduces the generation of humic substances. Tilling increases erosion, removing existing topsoil. Adding fertilizers compensates for nutrient loss, but not for soil structure. A new development is **no-till farming**, in which weeds are controlled by the use of chemical herbicides instead of plowing. Crop residues and winter cover are left on the field as a mulch, which controls soil temperature and evaporation of moisture. The residues add to the production of humus and reduce erosion.

Agriculture, deforestation, and overgrazing all expose bare soil and increase the rate of erosion. Some estimate that as much as 1% of Earth's topsoil is lost each year. By one estimate, 25% of all farmland in the United States is losing topsoil at a rate exceeding a "tolerable" 1 in. every 34 years.

Humic and fulvic acids lower the pH of the soil. Another organic acid produced by degradation of biomass is oxalic acid, $(COOH)_2$. Oxalic acid can form complexes with otherwise insoluble iron and aluminum. The complex then leaches to lower soil horizons, where the metals are precipitated upon oxidation of the oxalic acid to CO_2. Oxalic acid also catalyzes the chemical weathering of rocks. Oxidation of sulfide minerals such as iron pyrite produces sulfuric acid. Soil pH was once thought to be a major factor controlling plant distribution. However, now it is known that although some plants have narrow pH tolerances, many are not strict. Even in some of those with strict limitations, it may be due to the effect of pH on nutrient availability rather than the effect of pH on the plant directly.

Animals are an important component of the soil community (Figure 15.3). Nematodes, annelids, and insects form the base of the detritus food web, along with fungi and bacteria. They aid in humification by digesting and reducing the size of biomass particles. Nematodes, or roundworms, are microscopic worms that are very abundant in soil. Up to 3 billion can be found 1 m^3 of soil. Ants and earthworms are important for their role in aeration of the soil, which they do by their burrowing activity. Mammals such as moles and shrews also form large burrows.

Earthworms are particularly important to the soil. Besides their function of aerating the soil by burrowing, they process the soil through their gut. Finally, they mix or till the soil by ingesting soil at depth, then discharging it at the surface when they come out at night. The discharged material can often be seen as twisted strands of soil, called **castings,** on the surface. Darwin spent a significant amount of time studying the role of earthworms in the years after he published *The Origin of Species*. He found that rocks in a field gradually became buried as soil was moved from underneath to atop.

Soil typically has 10^6 to 10^9 bacteria per gram. Most are associated with the decaying biomass in the O and A horizons. In the A horizon, more are associated with roots of plants than in free soil. Bacteria and fungi that perform the initial degradation of biomass include *Pseudomonas*, *Bacillus*, *Penicillium*, *Aspergillus*, and *Mucor*. About 10 to 33% of the bacteria are *Actinomycetes*, especially the genera *Streptomyces* and *Nocardia*. *Azotobacter* and some of the anaerobic *Clostridium* species can fix atmospheric nitrogen in free soil.

Chemolithotrophs include the nitrifiers *Nitrospira*, *Nitrosomonas*, *Nitrobacter*, and the *Thiobacillus*, the last of which oxidizes inorganic sulfur and ferrous iron. In agriculture, nitrification can increase leaching of nitrogen. Addition of nitrification inhibitors increases crop yield 10 to 15% and reduces groundwater pollution by nitrate.

Most kinds of fungi can be found in soil, usually in the top 10 cm, where they may constitute a major fraction of the soil biomass. Protozoans also occupy the upper layers;

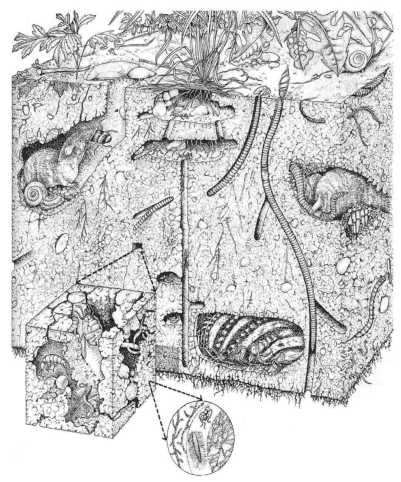

Figure 15.3   Soil community. (From Raven et al., 1992.)

populations are typically from 10^4 to 10^5 per gram of soil. They function as predators on bacteria and algae. Algae exist mostly on the surface, but they also have been found to exist heterotrophically at depths up to 1 m. Mycorrhyzal fungi are an important group, as has been discussed (see Section 10.7.4).

15.2   FRESHWATER ECOSYSTEMS

The three basic types of freshwater ecosystems are:

Lentic	Slow-moving water, including pools, ponds, and lakes
Lotic	Rapidly-moving water, such as rivers and streams
Wetlands	Areas the soil is saturated or inundated at least part of the time

Wetlands are considered separately in Section 15.3, where we also consider salt marshes. The science of inland waters is called **limnology**. The specialized science of rivers and streams is called **potomology** (which has the same root as *hippopotomus*, which in turn means "river horse.") Limnology and potomology include the physics and chemistry of the respective systems. Here, of course, we emphasize the biology.

15.2.1   Aquatic Environments

The **lacustrine** (lake) environment can be divided into categories by total depth and by the depth within the water column, by light availability, and by habitats. The shallow environment along the shore where rooted aquatic plants (including completely submerged plants) can grow is called the **littoral zone**. Deeper water is called the **limnetic** or **pelagic zone**. The upper layer reached by sunlight and therefore dominated by phytoplankton is called the **photic zone**. Below the photic zone, where light is attenuated to less than 1%, is the **profundal zone**. This is dominated by heterotrophic organisms subsisting on food that falls from above.

Deep waters may also be physically divided by thermal stratification in the winter and summer (Figure 15.4). Because water has a maximum density at 4°C, colder water in the winter and warmer water in the summer form a distinct layer that floats. The upper layer is called the **epilimnion**, the lower layer is the **hypolimnion**. Each of these layers is fairly uniform and, especially in the epilimnion, well mixed. These layers are separated by an area of rapid temperature and density change called the **thermocline**. The density gradient is very stable and prevents mixing between the two layers. The thermocline protects the profundal zone from the impacts of pollution. The littoral zone, on the other hand, is more susceptible because of its proximity to sources of pollution. The stratification shown in Figure 15.4 is typical of sufficiently deep lakes in summer in the temperate zone. Stratification varies significantly with season and local climate.

The basic characteristic of **riverine** (river or stream) environments is the velocity of flow. **Riffles** are relatively fast, shallow, and turbulent, whereas **pools** are laminar in flow, approaching lakes in their characteristics. Rivers generally lack the stratification found in lakes. Flow greatly affects dispersal of organisms and import and export of nutrients. Systems (whether rivers or lakes) that depend on their own productivity are called **autochthonous**. Those that depend on imported energy or nutrients are **allochthonous**. Lakes tend to be autochthonous. Rivers are more likely to be allochthonous, obtaining their organic matter from terrestrial sources such as leaf litter and other detritus in runoff.

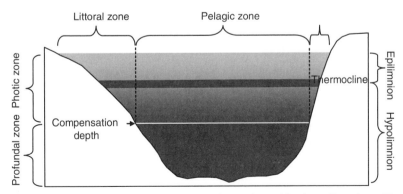

Figure 15.4 Lake zones classified by depth, light, and stratification. (Based on Horne and Goldman, 1994.)

15.2.2   Biota

Aquatic organisms are classified according to their habitats. The **plankton** are small free-floating organisms. Planktonic plants, called **phytoplankton**, include cyanobacter, green algae, and diatoms. Planktonic animals are **zooplankton** and include protozoans, small crustaceans such as *Daphnia* species and larval forms of other animals. Organisms such as duckweed that float on the surface are called **neuston**. Active swimmers that live in the pelagic zone, such as fish, are called **nekton**. The community of microscopic bacteria, fungi, and small plants and animals that live attached to submerged surfaces such as rocks are referred to as **aufwuchs**. A subset of these are the **periphyton**, which are algae that grow attached to surfaces. (Aufwuchs and periphyton are what makes submerged rocks so slippery to step on.) All organisms that live on the bottom or within the sediments, whether in the littoral or profundal zones, are called **benthic organisms**, or simply the **benthos**.

Lotic ecosystems have planktonic organisms only in the slowest of rivers. Periphyton dominate over phytoplankton, and benthos over zooplankton.

Prokaryotes   Three types of autotrophic bacteria are typically found in lakes: cyanobacter, chlorobacter, and chemoautotrophs. Cyanobacter (blue-green bacteria) are found in the epilimnion. Chlorobacteria peaks just below the thermocline, their numbers decreasing with depth. If the hypolimnion becomes anoxic, sulfate-reducing bacteria will appear to produce hydrogen sulfide. However, most of the bacteria are heterotrophs. Heterotrophic concentrations peak in the thermocline and near the sediment. In highly polluted streams the filamentous bacteria *Sphaerotilus* can form lush gray mats waving in the current.

One of the notable roles of cyanobacter, especially *Aphanizomenon*, is in the fixation of atmospheric nitrogen. This allows it to grow when nitrogen limitations apply to other algae. Some cyanobacter, such as *Oscillatoria* or *Microcystis*, cannot fix N_2. The latter depends on ammonia recycled by animal or bacterial excretion or by solubilization from the sediments. In rivers, *Nostoc* or *Rivularia* may fix considerable amounts of nitrogen.

Gas vacuoles enable cyanobacter to float to the surface during the calm night, where they can be found forming a scum at dawn. As the sun rises higher, the cyanobacter can be damaged by ultraviolet radiation, so they tend to sink by midmorning. *Anabaena* does this by collapsing gas vesicles by osmotic pressure which has increased as a result of glucose formed by photosynthesis. Colonies of *Microcystis* and *Aphanizomenon*, as well as filaments of *Anabaena*, also regulate their buoyancy by manufacturing glycogen, which is about 1.5 times as dense as normal cell material. At night they use up the glycogen and start to rise again. By moving up and down, the algae not only control their light exposure but take advantage of nutrients farther below the surface.

Cyanobacter (Figure 15.5) make chemicals that inhibit feeding by zooplankton. They make compounds that have been known to poison cows and pigs that drank from eutrophic sources of water. Cyanobacter and some actinomycetes also produce geosmin and methylisoborneol, the most common causes of taste and odor problems in drinking water. These compounds are contained in the cells, but released when the cells break down.

Green Plants   Algae are plantlike protists (Figure 15.6). They are distinguished from plants in that they do not have stems, roots, or leaves. They can be unicellular, colonial

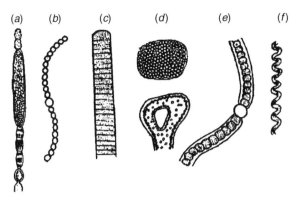

Figure 15.5 Cyanobacter: (*a*) *Aphanizomenon*; (*b*) *Anabaena*; (*c*) *Oscillatoria*; (*d*) *Polycystis*; (*e*) *Nostoc*; (*f*) *Spirulina*. (From *Standard Methods*, 12th ed. © 1965 American Public Health Association.)

(forming usually small groups of cells), or filamentous. Some, such as the dinoflagellates, are motile, enabling them to migrate in response to light. Periphyton are usually green algae, cyanobacter, or diatoms. The long, bright green filaments attached to rocks in streams are often cyanobacter. Other filamentous forms are green algae such as *Ulothrix* and *Cladophora*.

Diatoms (Figure 15.7) have an advantage over other plants in that their cell wall, being made of silica, requires about 12 times less energy to manufacture than cellulose. (Cyanobacter, which are prokaryotes, make cell walls of peptidoglycan, which requires more energy than cellulose. The ability of cyanobacter to control their light input by controlling buoyancy compensates them for this added cost.) However, diatoms are denser as a result, so under low-turbulence conditions they may settle out of the photic zone and be replaced by cyanobacter or dinoflagellates.

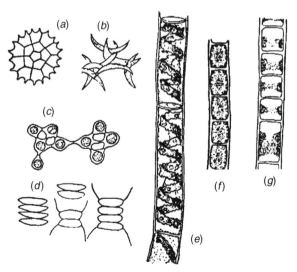

Figure 15.6 Green algae: (*a*) *Pediastrum*; (*b*) *Selenastrum*; (*c*) *Coelastrum*; (*d*) *Scenedesmus*; (*e*) *Spirogyra*; (*f*) *Microspora*; (*g*) *Ulothrix*. (From *Standard Methods*, 10th ed. © 1955 American Public Health Association.)

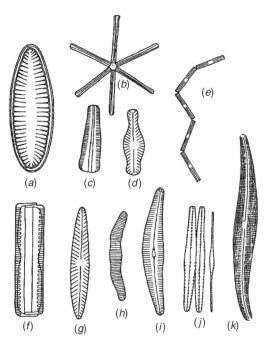

Figure 15.7   Diatoms: (*a*) *Surirella*; (*b*) *Asterionella*; (*c, d*) *Gomphonema*; (*e*) *Tabellaria*; (*f, g*) *Navicula*; (*h*) *Eunotia*; (*i*) *Cymbella*; (*j*) *Fragilaria*; (*k*) *Gyrosigma*. (From *Standard Methods*, 10th ed. © 1955 American Public Health Association.)

Dinoflagellates (Figure 15.8) can swim several meters per hour. They use their motility to swim upward in the morning to collect more light for photosynthesis, and then downward in the afternoon to avoid predation. In a condition called the **red tide**, dinoflagellates sometimes accumulate to nuisance levels in lakes, estuaries, and oceans, coloring the surface of the water blood-red. They can be toxic to fish and invertebrates and irritating to the skin of humans exposed by swimming. Examples of red tide genera include *Noctiluca* and *Gymnodinium* in marine systems and *Peridinium* and *Ceratium* in lakes.

In the open sea, the condition has been associated with salinity fronts where estuarine water converges with ocean currents. This leads to stratification between the estuarine water on top and the more saline water below. These waters may be complementary in the sense that each contains nutrients that may be limiting in the other. Algae grow preferentially at the boundary, where there will be some mixing. Dinoflagellates have the advantage of being able to move between the two layers, taking advantage of both.

Large plants are called **macrophytes**. Most are from the plant kingdom, although some are large algae (Table 15.4). Roots may be used for obtaining nutrition from the sediment or may be primarily for attachment. Nutrients can be obtained directly from the water. Besides their obvious role as primary producers, macrophytes are important as shelter and substrate for other organisms. Periphyton, which in this case are considered epiphytic, grow on the stems and leaves of the macrophytes. Protozoans roam the epiphytic cultures, and snails graze among them. Amphibians and small fish find shelter from predatory fish and birds. The macrophytes contribute greatly to the detritus food chain as dead plants settle to the bottom. Freshwater angiosperms will grow only at shallow depths. Deeper plants consist of ferns, mosses, and large algae. **Emergent** plants are those that grow

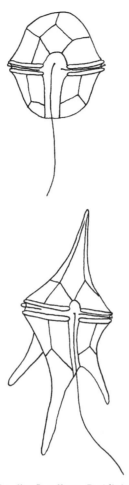

Figure 15.8   The dinoflagellates *Peridinium* and *Ceratium*.

out of the water. Some plants have only reproductive parts that are emergent. Many plants that are primarily emergent occur in marshes and are discussed in Section 15.3.

Zooplankton and Invertebrates   Aquatic ecosystems can have all the trophic levels at microscopic size, as we are accustomed to seeing on a large scale in terrestrial ecosystems. The zooplankton include protozoans, small crustaceans, rotifers, and other small invertebrates (Figures 15.9, 15.10, and 15.11). These may have roles of herbivore and primary and secondary carnivore. The nymph or larvae of many insects are aquatic and have similar roles.

Protozoans are widely dispersed, and tend to be most abundant in waters with large amounts of organic matter (Table 15.5). They feed on detritus and other single-celled organisms, whether bacteria, algae, or other protozoans. Some are obligate parasites, with hosts ranging from algae to fish. The presence of parasites can change the species composition of an aquatic ecosystem, such as by bringing an algal bloom to an end.

Rotifers are among the more interesting creatures to observe under a microscope as they move and feed. Most are **sessile** (nonmotile); many are motile but attach themselves

TABLE 15.4   Common Aquatic Macrophytes

Habitat	Taxonomic Group	Examples
Free-floating		
Subtropical and tropical lakes, slow streams	Monocot	*Pistia* (water lettuce)
	Monocot	*Eichhornia* (water hyacinth)
	Fern	*Salvinia* (water fern)
Temperate ponds	Fern	*Azolla* (water fern)
	Monocot	*Lemna* (duckweed)
Rooted		
Temperate marshes	Monocot	*Phragmites* (giant reed)
	Dicot	*Rorippa* (watercress)
Tropical marshes	Monocot	*Scirpus* (bulrush)
	Monocot	*Papyrus* (reed)
	Dicot	*Victoria* (water lily)
Ponds and slow streams	Dicot	*Nuphar* (water lily)
	Moncot	*Potamogeton* (pond weed)
	Dicot	*Ceratophyllum* (coontail)
Estuaries	Dicot	*Myriophyllum* (milfoil)
	Monocot	*Zostera* (eel grass)
	Monocot	*Ruppia* (widgeon grass)
Deep in oligotrophic lakes	Lycosida	*Isoetes* (quillwort)
	Large algae	*Chara* (stonewort)
	Large algae	*Nitella* (stonewort)
	Moss	*Fontinalis* (willow moss)

Source: Horne and Goldman (1994).

periodically to substrates by several "toes." A few have a ring of cilia that gives the impression of circular motion (accounting for the name) as it sweeps in suspended matter for food. Some are parasites or predators.

   Arthropods include two important aquatic groups: the crustaceans and the insects. Both have exoskeletons composed of chitin. Most of the larger zooplankton are crustaceans

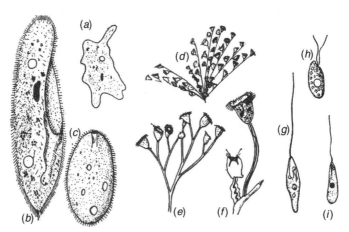

Figure 15.9   Protozoans. Amoebae: (*a*) *Amoeba*; ciliates: (*b*) *Paramecium*, (*c*) *Frontonia*; stalked ciliates: (*d*) *Carchesium*, (*e*) *Epistylis*, (*f*) *Vorticella*; flagellates: (*g*) *Peranema*, (*h*) *Chilomonas*, (*i*) *Astasia*. (From *Standard Methods*, 10th ed. © 1955 American Public Health Association.)

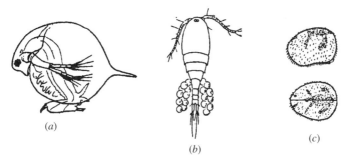

(a)

(b)

(c)

Figure 15.10 Planktonic crustaceans: (*a*) the cladoceran *Daphnia*; (*b*) the copepod *Cyclops*; (*c*) the ostracod *Cipridopsis*. (From *Standard Methods*, 10th ed. © 1955 American Public Health Association.)

(Table 15.6). Cladocerans, such as *Daphnia*, and copepods, such as *Cyclops*, combine with protozoans to form the majority of aquatic zooplankton.

The presence of some of the insect larvae or nymphs are indicators of a healthy aquatic environment (Table 15.7 and Figure 15.12). Examples include nymphs of stoneflies, mayflies, and caddisfly larvae. Recall that a **nymph** is an immature stage of that minority of insects that go through gradual metamorphosis. It is usually similar in form to the adult, except possibly lacking the adult's wings, coloration, and sexual maturity. The immature form of insects that go through complete metamorphosis, called a **larva**, may be very different in form (often, a worm or caterpillar) and habitat from the adult. The caddisfly larvae may be housed in a case made of bits of plant matter, as shown in Figure 15.12.

The insects occupy trophic levels between the primary producers and the fish. Hatching of adult insects may occur *en mass* and results in a feeding frenzy among the fish.

Benthos   The benthos contains organisms on and beneath the surface of the sediment. Their food web is based on detritivores: bacteria, fungi, protozoans, and invertebrate animals. The source of the energy is the primary productivity of the water column, for lakes, or the watershed, for rivers. Transport of pollutants through the pores of the sediments cannot be predicted using the diffusion methods from groundwater pollution hydrology. This is because burrowing organisms actively mix and transport chemicals in sediment (and the sediment itself), move water through the sediment, and provide channels for increased circulation and diffusion.

Two distinct benthic environments are the littoral and the profundal, according to whether they are above or below the thermocline, respectively. The littoral zone tends to have higher species diversity, although the profundal zone may have greater total numbers of organisms. The littoral benthos is subject to significant daily and seasonal changes. The profundal benthos may experience seasonal changes in oxygen in eutrophic lakes.

The littoral benthos consist of relatively more herbivores, grazers, and filterers than the profundal. Macrophytes can greatly increase the available habitat. Wave action and the proximity of the surface usually ensure an ample oxygen supply.

The profundal benthos consist mainly of worms and mollusks in four groups: oligochaete worms (only in cold waters), amphipods (crustaceans, including mysid shrimp), insect larvae, and clams (Figure 15.13). Most of these are detritus feeders, although there may be some predators, such as the midges. Most of the food input to the profundal benthos comes from the spring algae bloom, especially diatoms, which sink readily. The profundal benthos can store large amounts of lipids to get them through the rest of the year.

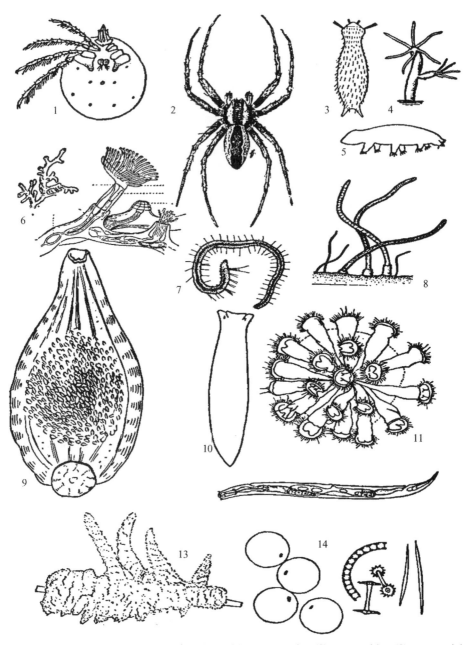

Figure 15.11 Other invertebrates: (1) hydrachnid, or water mite; (2) water spider; (3) gasterotrich, *Chaetonotus*; (4) coelenterate, *Hydra*; (5) tardigrade, *Macrobiotus*; (6) bryozoan, *Plumatella*; (7), bristle worm, *Nais*; (8) sewage worm, *Tubifex*; (9) leech, *Clepsine*; (10) flatworm, *Planaria*; (11) colonial rotifer, *Conochilus*; (12) nematode worm; (13) freshwater sponge; (14) gommules and spicules from the sponge. (From *Standard Methods*, 10th ed. © 1955 American Public Health Association.)

TABLE 15.5   Some Important Aquatic Protozoans

Phyla	Habitat	Examples
Flagellates (some are algae)	Lakes; oceans; some are parasites	*Synura* (Chrysomonad)* *Ceratium* (Dinoflagellate)* *Euglena* *Oikomonas*
Ciliates	Ponds, streams, detritus, some parasitic	*Paramecium* *Vorticella*
Amoeboid	Most waters	*Difflugia* *Globigerina* (Foraminifera) *Actinosphaerium* (Radiolaria) *Vampyrella* (algal parasite)
Sporozoans (all parasitic)	Aquatic organisms	*Henneguya* (on fish) *Plasmodium* (human malaria)

Source: Horne and Goldman (1994).

The differences in adaptation between littoral and benthic organisms can be represented by a comparison of the littoral amphipod *Gammarus pulex* and the benthic chironomid midge *Chironomus anthracinus*. Midges are larvae of predatory gnats. They feed on rotifers and burrow into the sediments at night. When the water is saturated, both species use oxygen at the rate of about 250 µg/g. But for *Gammarus* the rate decreases proportionally as oxygen saturation in its environment decreases. At 30% of saturation its respiration rate drops from to about 100 µg O_2 per gram of organism. The midge, on the other hand, maintains a fairly constant respiration rate down to 30% saturation; 100 µg/g

TABLE 15.6   Common Aquatic Crustaceans

Taxonomic Group	Most Common Habitat[a]	Feeding Method	Examples
Cladocera (water fleas)	FW, P	Predator	*Leptodora*
	FW, P	Filter	*Daphnia*
	FW, P	Filter	*Bosmina*
	FW, P, B	Filter	*Sida*
	FW, P, B	Filter	*Chydorus*
	FW, S, P	Predator	*Polyphemus*
Copepoda	FS, S, B	Filter and parasitic	*Nitocra*
	FS, S, B, P	Predator on animals	*Cyclops, Megacyclops*
		Predator on algae	*Thermocyclops, Eucyclops*
	FW, S, P	Filter	*Limnocalanus, Boeckella, Diaptomus*
Mysidacea (opossum shrimp)	FW, P	Predatory and detritovore	*Misis*
Amphipoda	S, FW, B	Predatory	*Gammarus* *Astacus*
Decapoda (freshwater crayfish)	S, FW, B	Predatory	*Pacifastacus* *Cancer*

[a]FW, freshwater; S, saline; B, benthic or attached; P, planktonic.
Source: Horne and Goldman (1994).

TABLE 15.7   Characteristics of Some Common Insect Larvae

Common Name and Order	Habitat	Food Habits	Examples
Nymphs			
Stoneflies (Plecoptera)	Rapids	Mainly carnivorous	*Nemoura*
Mayflies (Ephemeroptera)	All waters	Mainly herbivorous	*Baetis, Ephemerella*
Damselflies (Odonata)	Slow and stagnant	Carnivorous	
Dragonflies (Odonata)	Slow and stagnant	Carnivorous	*Libellula*
Water bugs (Hemiptera)	All waters	Carnivorous	*Notonecta* (backswimmer) *Gerris* (water strider)
Larvae			
Caddisfly (Trichoptera)	All waters	Carnivorous	
Water beetles (Coleoptera)	Slow or stagnant	Carnivorous	*Dytiscus* (diving beetle)
Order Diptera (and family)			
Mosquitoes	Pools at surface	Herbivorous	*Culex*
Blackflies	Rocks in rapids	Herbivorous	*Simulium*
True midges (chironomids)	All waters	Herbivorous	*Chironomus*
Horseflies	Beds in pools	Carnivorous	

Source: Needham and Needham (1962); Horne and Goldman (1994).

is not reached until about 5% of saturation. This is attributed to the presence of hemoglobin in the blood of the midge, which improves the efficiency of oxygen transfer.

Other invertebrates include various worms and mollusks (Table 15.8). Included are nematodes, oligochaetes, leeches, bivalves (clams), and snails. Most are detritivores. Leeches are predators that attach to fish. Some nematodes are internal fish parasites. Bivalves are becoming uncommon in American rivers. Snails feed on macrophytes and, along with insect larvae, can control their abundance. *Tubifex* can become so abundant in highly polluted streams that they form red mats on the bottom.

The zebra mussel, *Dreissena polymorpha*, is an exotic species imported to North America from Europe. It probably hitched a ride in the ballast water of a ship that released it near Detroit, Michigan, in 1985 or 1986. Since then it has been spreading rapidly, carried by currents and shipping. Zebra mussels cause great ecological and economic harm

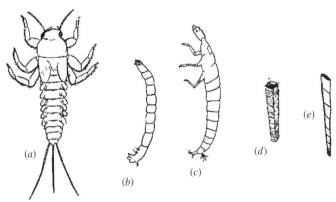

Figure 15.12   Insect larvae and nymphs: (*a*) Ephemeroptera (mayfly); (*b*) Diptera, *Chironomus*; (*c*) Trichoptera (caddisfly); (*d, e*) Trichoptera case. (From *Standard Methods*, 10th ed. © 1955 American Public Health Assoc., Inc.)

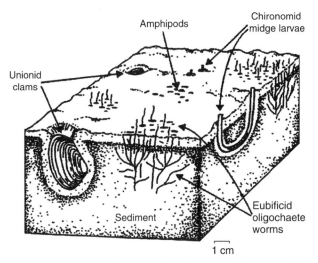

Figure 15.13   Zoobenthos of the profundal zone. (From Horne and Goldman, 1994. © McGraw-Hill, Inc. Used with permission.)

by occupying growing surfaces formerly used by other organisms and by growing on intake structures of power plants and drinking water plants, obstructing their flow.

Fish and Fisheries   Fish occupy and often dominate all the trophic levels from detritivore to herbivore to top carnivore (Table 15.9). They serve as prey to other fish and to aquatic birds or other animals. Since streams receive most of their nutrient input from terrestrial sources, harvesting of fish by birds, bears, or humans represents a return of some of those nutrients. Fish are the most motile of aquatic species; thus, they can live and breed in different places, and move in response to changes in food supply or other conditions.

Freshwater fish are hypertonic: Their bodily fluids are more saline than the water they swim in. Thus, they tend to absorb water by osmosis and lose salts by passive diffusion. To counter this, they discharge large volumes of dilute urine, and their cells absorb salt by active transport. They do not drink.

TABLE 15.8   Common Invertebrates Other Than Insects

Taxonomic Group	Feeding Method	Examples
Turbellaria (flatworms)	Carnivores	*Dugesia*
Nematodes (roundworms)	Carnivores, herbivores, parasites	*Dolichodorus*
Annelida		
Oligochaeta (segmented worms)	Sediment grazers	*Tubifex* (red worms)
Hirudinea (leeches)	Carnivores, detritivores	*Haemopis* (horse leech)
Mollusca		
Gastropoda (snails)	Grazer	*Limnaea* (pond snail)
		Planorbis (ramshorn snail)
Palacypoda (bivalves)	Filterer	*Anodonta* (swan mussel)
Crustacea		
Malacostraca	Detritivores	*Gammarus*
(crayfish, amphiphods)		*Astacus*

Source: Horne and Goldman (1994).

TABLE 15.9   Some Common Families of Fish

Family	Examples
Lampreys	Sea lamprey
Anguillids	Freshwater eel
Herrings	Alewife, herring
Salmonidae	Salmon, trout, whitefish
Cyprinids	Minnows, carp
Catostomids	Suckers
Ictalurids	Catfish
Cyprinodontids	Killifish
Gastersteids	Sticklebacks
Percichtyidae	Striped bass
Centrachids	Sunfish, largemouth black bass
Cichlids	Tilapia
Gadidae	Cod

Source: Horne and Goldman (1994).

Fish tend to feed selectively in the pelagic, littoral, or benthic zones. Food sources include phytoplankton, zooplankton, detritus, or other fish. Cannibalism is not uncommon, especially by adults on the young. Pelagic fish usually feed at the surface. Shad herring, whitefish, and minnows feed almost entirely on zooplankton. Some of the large predators will also depend on zooplankton in their early life stages. *Tilapia* feeds on the blue-green alga *Microsystis aeruginosa*. This makes it a candidate for producing meat on manned space flight, since the algae can also help close the nitrogen cycle on spacecraft.

Most fish live for many years, so the biomass is not subject to seasonal cycles. However, survival of the young produced in a single year, the **cohort**, depends on predation, the availability of food for the young (usually plankton), and environmental factors. In the presence of intense predation, juvenile fish may restrict their activity to macrophyte beds for protection. Others may venture into deep water to feed on zooplankton in the early hours of the day, retreating to the littoral zone when daylight makes them easy targets for predatory fish. Others move from the depths to the surface and back in a similar way.

Floods and droughts may affect the availability of nesting sites. In tropical rivers, which are much less affected by human activity than other rivers of the world, some fish depend on the floodplains to a great degree. They often breed there and feed heavily during floods.

When there are a large number of species with similar food requirements, species often evolve to avoid direct competition by specializing in a subset of the available food. This is called **resource partitioning**. In the case of fish, for example, several species that eat zooplankton may restrict their hunt to particular habitats. One species of fish may only eat organisms from the benthic zone, others only from the pelagic surface, and others only from the littoral zone.

Some fish are **anadromous**, meaning that they live most of their lives in salt water but spawn in fresh water. The best known example is the salmon, which remember the "smell" of the stream that they hatched in after several years in the ocean. They even remember the odor associated with each tributary they passed as a hatchling, and follow them back in reverse order. The six species of Pacific salmon die after spawning; Atlantic salmon and steelhead trout return year after year. The movement of grown fish can also be viewed as a transport of nutrients from the sea to the relatively nutrient-poor headwater streams.

Other fish are **catadromous**—they live in fresh water but return to the sea to spawn. The eels of the North Atlantic are a remarkable example of catadromous fish. For centuries it was observed that the adults all left their streams in the fall for the ocean, and in the spring a large upstream migration of matchstick-sized **elvers** appeared. Finally, in the early part of the twentieth century it was realized that all the eels from North America and Europe swam to the Sargasso Sea near Bermuda. There they spawned and the eels died. Over three years the larvae drift with the Gulf Stream and metamorphose into elvers and swim up the estuaries and streams of the North Atlantic. The males stay in the estuaries and the females continue upstream. They remain for 8 to 15 years before returning to their place of hatching. The North American and European species are distinct from each other, although they breed in overlapping areas of the Sargasso Sea and sort themselves out for the reverse migration.

Many human activities have disrupted fish populations around the world. The first apparent effect is usually on populations used for food supply. Direct chemical alteration by pollution can result in eutrophication or toxic exposures. Both can cause fish kills, the former by deoxygenation, the latter by direct toxic effect. Fish can tolerate dissolved oxygen levels down to about 4 to 5 mg/L. If not severe, eutrophication can increase fisheries. Exotic species disrupt the food chain. They may be introduced deliberately or inadvertently. The peacock bass, *Cichla ocellaris*, was introduced to a lake in Panama as a sport fish. Subsequently, 11 other species suffered serious declines, and gulls and herons lost their main food sources. The loss of several larvae feeders has resulted in an increase in mosquitoes in the area.

Inadvertent introductions may occur as a result of constructed waterways such as canals. The alewife and the sea lamprey were introduced to the Great Lakes in this way after having been limited by Niagara Falls. The sea lamprey, *Petromyzon marinus*, is a parasitic eel that attaches itself by its mouth to the side of a fish and feeds on the fish's body fluids. The alewife, *Alosa pseudoharengus*, competes with lake herring for zooplankton. The lampreys killed off the larger fish of the popular salmonid species, forcing fishermen to take smaller sizes. This gradually eliminated the breeding stock, until the fishery suddenly collapsed. Overfishing also caused depletion of some fisheries, such as sturgeon, which were not related to the introduction of exotic species.

On the other hand, construction of dams has created barriers to fish that need to migrate, such as the salmon. Dams may also flood breeding grounds. Flood control eliminates those species that depend on flooding to breed. Development of the river shore eliminates overhanging trees which shade and cool, as well as provide falling insects that some fish depend on, and leaf fall which supplies the detritivores.

Fisheries management is a set of strategies to ameliorate some of these problems. *Fish ladders* are artificial cascades that can be constructed alongside dams to provide routes for migrating fish. Reservoir releases are controlled with downstream fisheries in mind, controlling flow and temperature. Gravel beds are constructed in streams to furnish fish nesting sites. Competition by overabundant centrarchids in reservoirs can be controlled by lowering the reservoir level at the appropriate time to strand nesting areas. Log jams and fallen trees (snags) serve as sources of invertebrates for migrating fish, as well as resting places in migration. They used to be cleared routinely, but now that their role is understood, they are left undisturbed by fisheries managers. Finally, nutrients are sometimes introduced deliberately to increase primary productivity, ultimately increasing the productivity at higher levels of the food chain as well.

Another important tool of fisheries management is regulation of fishing. A rule of thumb for harvesting fish is that the maximum sustainable yield is obtained if about one-third of the population is taken in each reproductive period. This has been found in experimental systems to result in an equilibrium biomass that is slightly less than 50% of the unexploited fishery. The optimum was found to be independent of the carrying capacity. However, the situation is sometimes complicated by interactions with other species. Harvesting a single species can tip the balance toward a competitor. For example, heavy fishing of the Pacific sardine (*Sardinops caerulea*) resulted in an increase in the less valuable anchovy (*Engraulis mordox*).

Natural disturbances to ecosystems tend to show their strongest effects at the lowest trophic levels. The disturbance tends to dampen as it goes up the food chain, since each level often has alternative food supplies. On the other hand, removal of top carnivores tends to have impacts throughout the food chain, even disturbing phytoplankton populations. Fishing usually concentrates on carnivorous species and thus is an example of the latter type of disturbance.

15.2.3 Succession in Lakes

The epilimnion has plenty of oxygen since it is in contact with the atmosphere. It is also fairly productive since it has more sunlight available. However, biomass production gradually reduces nutrient availability during the growing season. Sedimentation of dead organisms transports the nutrients to the hypolimnion or to the sediments. Degradation reduces oxygen and releases nutrients in the hypolimnion. However, twice a year in the temperate zone, in spring and fall, these differences disappear as temperature changes destroy the stratification. The resulting mixing is called **turnover** (not to be confused with the idea of turnover time in nutrient cycling). The turnover returns nutrients to the surface and oxygen to the depths.

Spring turnover brings nutrients and light together with a warming environment. This initiates successional changes in populations. A sequence of species goes through boom-and-bust cycles. Each flourishes temporarily as it enjoys a competitive advantage over the others. Eventually, it depletes some necessary resource or comes under attack by zoo-plankton predators or fungal parasites, and the population crashes. This leaves the field for another opportunist, which can best compete in the resulting environment.

The largest total algal populations occur shortly after turnover. Large, sudden increases in phytoplankton populations are called **algal blooms**. During an algal bloom, cell counts on the order of millions per liter can color the lake green. The diatom *Asterionella* is often the first algae to bloom as longer days in spring increase the sunlight available. It is faster growing than its competitors *Fragilaria*, *Tabellaria*, and the cyanobacter, and it performs luxury uptake of phosphorus, up to 100 times its immediate requirement. However, it is susceptible to fungal attacks, which can give *Fragillaria* and *Tabellaria* a temporary advantage. If the filamentous cyanobacter *Oscillatoria* gains a foothold, it can shade *Asterionella*, outcompeting it for sunlight. Ultimately, silica limits *Asterionella* blooms (Figure 15.14). Unless some other nutrient becomes limiting, *Asterionella* stops growing when the silica concentration falls below 0.5 mg/L.

Because they can float on the surface, cyanobacter may gain an advantage when turbidity limits light to the green algae. A cyanobacter bloom may continue into the summer as nitrate and ammonia are depleted, since they can fix atmospheric nitrogen. Eventually, nutrients become depleted even for the cyanobacter. Iron is often one of the limiting

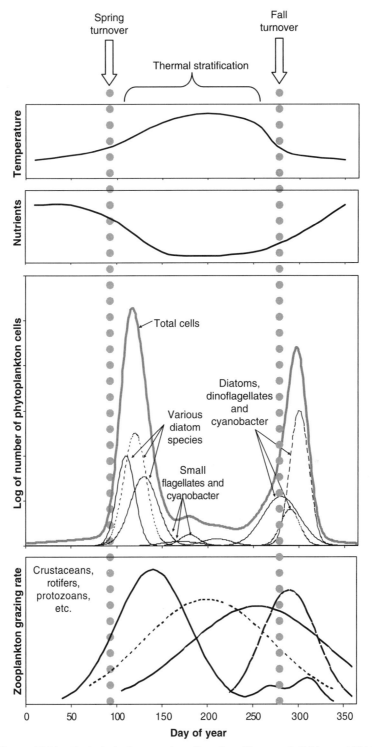

Figure 15.14   Typical algal succession. (Based on Horne and Goldman, 1994.)

nutrients. It is needed for the nitrogenase enzyme for nitrification. *Anabaena* and *Microcystis* often form late summer–autumn blooms as nutrients are recycled from sediments or the hypolimnion.

When the fall turnover comes, large numbers of the diatom *Melosira* can be resuspended from the sediments, becoming the dominant alga over the winter period. During the summer they settle back to the sediment before nutrient limitation begins to affect them. This, plus their tolerance of anoxic conditions, enables them to survive in the sediment.

Phytoplankton growth responds most strongly to light availability. But zooplankton feed at a rate strongly related to the temperature. Thus, they often peak later than the phytoplankton from the spring bloom, bringing it to an end. The elimination of phytoplankton brings about the **clearwater phase**, during which the lake is most transparent. This initial peak in zooplankton is usually dominated by cladocerans such as *Daphnia* or *Bosmina*. Other animals present in this peak include predatory rotifers, carnivorous and herbivorous copepods, and protozoans. Herbivorous rotifers follow the clearwater phase as they can eat larger and harder-to-digest algae, as well as toxic or repulsive cyanobacter rejected by the cladocerans.

15.2.4 Microbial Loop

Many of the aquatic organisms are omnivorous, complicating application of the concept of trophic levels. Figure 15.15 shows a food web for a small lake with organisms or groups arranged by trophic level. This lake is managed for fish production and thus receives an artificial energy input. Notice the large loss to the system from the emergence of chironomid insects.

As has been mentioned, most of the nutrients in the photic zone of lacustrine and marine ecosystems tend to be tied up in living things. This would place a severe limitation on further growth except for the existence of an efficient detrital recycling process. The

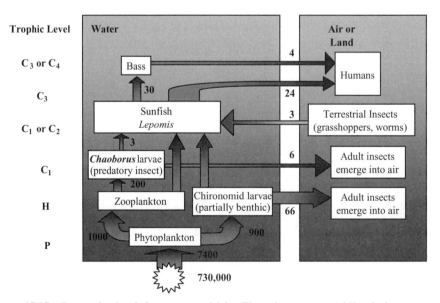

Figure 15.15 Energy food web for a managed lake. The values represent kilocalories per square meter per year. (From Horne and Goldman, 1994.)

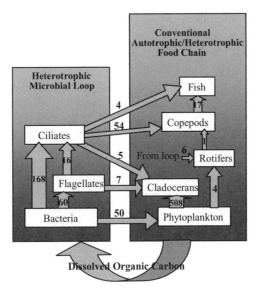

Figure 15.16 Carbon cycle in a lake during the summer. Values indicate grams carbon per square meter per day, ignoring respiration and cannibalism. (Based on Horne and Goldman, 1994.)

process known as the **microbial loop** involves the rapid utilization by bacteria and protozoans of organic matter liberated as waste products by all of the other aquatic organisms (Figure 15.16). This loop cycles nutrients largely independently of the phytoplankton, enabling growth to continue in the summer when the algal blooms have disappeared. The waste products are organics excreted by zooplankton and fish.

Another important source of soluble organic matter for the microbial loop are the **extracellular products of photosynthesis** (ECPP). These are organics such as glycolic acid that are produced by photosynthesis and leaked continuously by phytoplankton. The leakage has been measured to average from 7 to 41% of the carbon fixation rate. Nutrients are also recycled by bacterial and fungal degradation of dead organisms. These mechanisms are very important in the low-productivity areas of oligotrophic lakes and in the open ocean.

15.2.5 River Productivity

Riverine food webs are often dominated by allochthonous inputs from terrestrial sources. **Coarse particulate organic matter** (CPOM) from fallen leaves, dry grasses, and so on, are low in nitrogen. They require nutrients from the stream for their initial degradation. Otherwise, they are not readily available to the food chain. Nutrient availability, and therefore productivity, tends to be higher in hard-water areas (with high calcium and magnesium) than in soft-water regions.

Degradation of the CPOM is accomplished by a coating of fungi and bacteria that adheres to the surface. They break down the cellulose and lignin which cannot be digested by animals. Lignin is degraded mostly by fungi, including species adapted to function in the winter cold, to take advantage of the annual leaf fall. Invertebrates ingest the CPOM, but digest only the attached fungi and bacteria. The CPOM is excreted, and the coating regrows. In repeated cycles of degradation, ingestion, and excretion, the CPOM is gradually broken down into **fine particulate organic matter** (FPOM) and then to **dissolved organic matter** (DOM) until it has entered the food chain completely.

The invertebrates that feed on the allochthonous input fall into four groups. **Grazers** scrape bacteria, fungi, rotifers, and periphyton from surfaces. Examples of grazers are snails and caddis flies. **Shredders** cut and ingest pieces of leaves and stems which form the CPOM, reducing them in size. They include the stoneflies, amphipods, and crayfish. **Collectors** and **filterers** take FPOM from the water. Clams pump water through filtration organs, expending considerable energy. Midge larvae use fanlike collecting structures to ensnare particles. Caddisfly larvae spin long nets to capture FPOM from the current. Most benthic invertebrates simply use bristles on their appendages or mouthparts. Worms feed on FPOM that has settled to the bottom in more quiescent areas. **Predators** range from stonefly larvae and dragonfly nymphs to fish and birds of prey. Figure 15.17 shows a schematic food web for this ecosystem. This figure shows three kinds of inputs. Direct light energy fuels the autochthonous primary productivity. Dissolved organics may originate in decay products that run off from the land and provide an allochthonous input for microorganisms. Large organic matter feeds another allochthonous input to the cycle, involving the shredders that break the CPOM down to FPOM, ultimately providing food for the scrapers and the filter feeders. Finally, of course, each group of invertebrates is subject to predation by carnivores.

During the day, most invertebrates hide under or between rocks to avoid predation. However, at night many of these will join the current to migrate downstream. Together with FPOM they form **drift**, which is an important food source for larger collectors and fish. After a lifetime of drifting downstream, many insect larvae may emerge into the air, then fly upstream to lay their eggs, repeating the cycle.

According to the **river continuum concept**, the balance between allochthanous and autochthanous input changes gradually from small headwaters to large downriver condi-

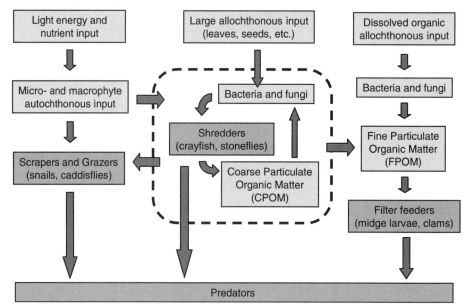

Figure 15.17 Allochthonous food web with functional groups. (Modified from Horne and Goldman, 1994.)

tions. As streams merge and move farther from their sources, FPOM increases relative to CPOM and the shredders give way to collectors and filterers. Farther downstream the river may acquire an autochthanous character based on attached algae and macrophytes, with grazers and collectors dominating. Finally, the river may become more like a lacustrine environment, dominated by collectors living off phytoplankton. The ratio of production to respiration exceeds 1, whereas respiration dominates at other parts of the river. Diversity is also greatest in this section. Figure 15.18 illustrates these changes with increase in **river**

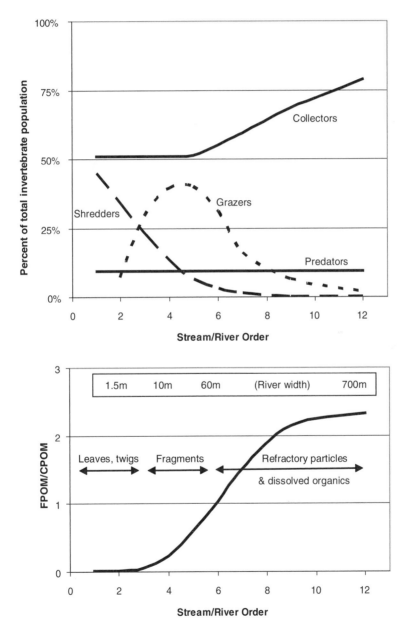

Figure 15.18   River continuum concept. (Based on Horne and Goldman, 1994.)

order, which can be roughly described as the number of branches in the river counting from the headwaters and incrementing the count whenever equal-order tributaries merge.

In large higher-order rivers, and especially in warmer climates, the low productivity in the main channel may be compensated for by a large input from the periodically inundated flood plain along the banks. Flooding results in a large input of nutrients and detritus. Many fish spawn in the floodplains, and rivers with large floodplains tend to have more diverse fish populations.

15.2.6   Nutrients and Eutrophication in Lakes

Aquatic systems are also classified according to their status with respect to nutrients or productivity (Table 15.10). Lakes that are relatively nutrient limited and low in productivity are called **oligotrophic** ("poorly fed"). Lakes with higher nutrient levels and higher productivity are called **eutrophic** ("well fed"). An intermediate designation is called **mesotrophic**.

Basically, an oligotrophic lake is one that would be perceived as "clean" vs. the relatively murky eutrophic lake. However, eutrophic does not necessarily mean polluted. Lakes proceed through a natural succession, starting as oligotrophic. Gradually, they accumulate sediments and nutrients. The shoreline grows more and more macrophytes as the lake becomes shallower. Eventually, they become marshes or wetlands, until they are finally transformed into terrestrial ecosystems.

Although it occurs naturally, addition of anthropogenic nutrients hastens this process. People contribute nutrients in the form of fertilizer runoff from lawns and agricultural lands, human waste from either treatment plant discharge or indirectly from septic tanks, and animal waste from pasture or feedlot runoff. In these cases the process is termed **cultural eutrophication**. Oligotrophic lakes are preferred for numerous human

TABLE 15.10   Comparison of Oligotrophic and Eutrophic Lakes

Criteria	Oligotrophic	Eutrophic
Phosphorus conc. (winter)	P < 10 µg/L	P > 20 µg/L
Organic matter	Low	High
Primary productivity	GPP < 150 g C/m^2·yr	GPP > 250 g C/m^2·yr
	Chlorophyll a < 3 µg/L	Chlorophyll a > 6 µg/L
Depth	<15–25 m	<10–15 m
Clarity	Clear, blue, Secchi depth > 5 m	Dark, turbid, Secchi depth < 3 m
Hypolimnion dissolved oxygen in summer	>50% saturation	<10% saturation (about 1 mg/L)
Bottom fauna	Diversified with deepwater fish	Low-DO-tolerant organisms; no deepwater fish; carp
Bottom sediment	Low organic matter content	Organic rich muck, high in N
Characteristic alga or genera	Green algae	Cyanobacter (*Anabaena,*
	Desmids (*Staurastrum*)	*Aphanizomenon, Microsystis*)
	Diatoms (*Tabellaria, Cyclotella*)	Diatoms (*Molosira,*
	Chrysophyceae (*Dinobryon*)	*Stephanodiscus, Asterionella*)
Examples	Lake Superior	Western Lake Erie
	Lake Geneva (Switzerland)	Lake Lugano (Switzerland)
	Lake Baikal (Russia)	

Source: Berner and Berner (1987); Conell and Miller (1984).

TABLE 15.11 Trophic Status of Several Lakes[a]

Lake	Status[a]	Avg Depth (m)	HRT (yr)	NO_3^{2-}-N	NH_4^+-N	P	SiO_2	Secchi (m)	Chloro-phyll a
Superior	O	149	184	220	<10	0.5	2000	11.3	0.9
Huron	O	59	21	180		0.5	800	5–9	1.8
Michigan	O–M	85	104	130		5	700	4.3	1.3
Erie	E	1.4	3	23		1	31	2–4.4	5.4
Ontario	M	86	8	40			100		
Tahoe	O	313	700	4	<2	~2			
Windermere (UK)		24	0.75	300–400	~10	5			
George (Uganda)		2.4	0.34	nd	<10	~2			
Baikal (Siberia)		730					3000		

[a]O, oligotrophic; M, mesotrophic; E, eutrophic; concentrations are summer measurements in μg/L; nd, nondetectable.
Source: Horne and Goldman (1994).

purposes, whether as a source of drinking water, food, or recreation. Among the problems caused by cultural eutrophication: Large plant biomass can cause oxygen depletion at night or upon death of the plants. High concentrations of algae cause taste and odor problems in water supplies. Some algal blooms can be toxic to animals that drink the water. Toxicity or deoxygenation can cause fish kills. Macrophytes can physically choke the lake, interfering with uses such as boating or swimming. Table 15.11 shows several examples of lakes in different categories of nutrient loading.

Primary productivity is limited by nutrient availability, especially phosphorus. Nitrogen is less often limiting because cyanobacter (blue-green bacteria) fix N_2. Nevertheless, dosages of nitrogen and phosphorus that individually stimulate a doubling in phytoplankton growth produce a fivefold increase when added simultaneously. Other potential limits to plant growth include silicon, CO_2, iron, and molybdenum.

Phosphorus concentration in a lake can be computed from material balance considerations. Loss of phosphorus from the water column by sedimentation is usually accounted for by a reaction term. The peak summer algae concentration, measured in terms of the average chlorophyll a concentration, shows a strong log-log correlation with the phosphorus concentration in the water during the spring turnover. A typical such relation is

$$C_a = 0.37 P_w^{0.91} \tag{15.1}$$

where the phosphorus concentration (P_w) and the chlorophyll a concentration (C_a) are both expressed in μg/L (see Example 14.2).

One source of phosphorus in the lake water column that is difficult to quantify originates from the sediment. Phosphorus that has been deposited in the past may be liberated again, producing what is known as **internal loading**. This phosphorus results from an interaction with sulfur and iron in the hypolimnion and the sediments of lakes (see Figure 15.19). When spring turnover brings oxygen to the depths, ferrous iron is oxidized to ferric, which precipitates with phosphorus and settles to the sediment. During summer stratification, oxygen is limiting and anoxic conditions may form in the sediment. The ferric iron and sulfate are reduced to ferrous and sulfide ions, respectively, liberating the phosphate. The sulfide precipitates the iron and phosphate is released to the hypolimnion.

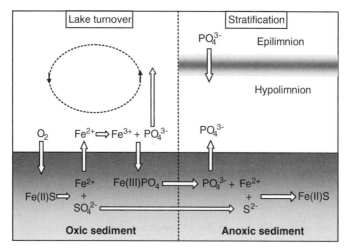

Figure 15.19   Phosphorus interaction with sulfur and iron in lakes. (Based on Horne and Goldman, 1994.)

In the fall turnover the phosphate is mixed back into surface layers, where it is available to plants and algae. With the transport of oxygen back to the depths, the iron and sulfur are oxidized again, and the cycle repeats. The process can be summarized by the following reaction, in which oxic conditions push the reaction to the right and anoxic conditions force it to the left.

$$4\,Fe(II)S + 4\,PO_4^{3-} + 4\,H^+ + 9\,O_2 \Leftrightarrow 4\,Fe(III)PO_4 + 4\,SO_4^{2-} + 2H_2O$$

The phosphate concentration falls in the epilimnion throughout the summer stratification as phytoplankton take it up, then die and settle toward the bottom. Phytoplankton are highly efficient at taking up phosphate, even at very low environmental concentrations. The Monod half-rate coefficient for phosphate uptake is about 1 to 3 µg/L PO_4-P. Thus, the uptake mechanism is rarely saturated unless pollution increases the concentration greatly. Differences in ability to compete for phosphorus may contribute to the succession of algal species during the summer stratification. Algae also exhibit **luxury consumption** of phosphorus, in which they store as much as 20 times the amount needed for cell division in cellular bodies called **polyphosphate granules**. Asterionella can store 100 times its immediate needs. By such hoarding they can continue their growth after the water column has been depleted of phosphorus.

Phosphorus is recycled rapidly in the water column. Zooplankton eat the phytoplankton, then excrete phosphorus at the rate of 10% of their body store daily. Half of this is phosphate and is readily reused. The other half is organic phosphate. Phytoplankton have the unusual capability to excrete the extracellular enzyme **alkaline phosphatase**. This acts in the water, outside the phytoplankton cell, to break the bond between the phosphate and the rest of the organic molecule. This represents a risky investment for the phytoplankton, since other species may benefit from the cost they bear for excreting the enzymes. When phosphorus is plentiful, the enzyme production slows considerably.

Diatoms need large amounts of silica (SiO_2) for their cell walls. About half of the dry weight of diatoms is silica. An absence of silica can result in replacement by cyanobacter.

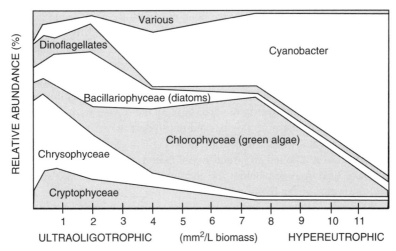

Figure 15.20   Changes in algal species composition with degrees of eutrophication. (From Connell and Miller, 1984.)

Rivers typically have about 13 mg SiO_2 per liter, derived from weathering of feldspars. Lakes range from 0.5 to 60 mg/L.

Iron, besides its role in the phosphorus cycle of lakes, is often a limiting nutrient. In the eplimnion much of the iron is chelated with organics. Some algae can only use inorganic iron, others only the chelated form, and others use both. Cyanobacter produce powerful iron-chelating organics, which limit their competitors' access to iron.

Eutrophication produces changes in the community species structure. As nutrient levels increase, the dominant algal species shift from diatoms to green algae to cyanobacter (Figure 15.20). In the more extreme forms of cultural eutrophication the phytoplankton growth exceeds the ability of zooplankton to control them. More bacteria and detritus feeders are found. Respiration by the plants at night or by bacteria degrading dead plants can deplete oxygen, affecting higher organisms including fish. Species diversity and richness may increase in moderate eutrophy, then decrease in the extreme form, although abundance continues to increase. At such high abundance these blooms may produce phytotoxins affecting higher forms of life, such as fish or cattle that drink the water.

Cultural eutrophication can be controlled by eliminating nutrient loading, reducing nutrients in the lake, or by addressing the symptoms, whether they be low dissolved oxygen or plant accumulation. Nutrients can be eliminated by using advanced wastewater treatment or by diverting the wastewater to another watershed. In the case of Lake Tahoe, a highly oligotrophic lake on the border between California and Nevada, the wastewater is both treated and pumped over the mountains surrounding the lake. This has eliminated enough phosphorus so that the lake has changed from being nitrogen limited to being phosphorus limited. However, significant nitrogen loading still results from atmospheric deposition of nitrogen oxides fixed by combustion in automobile engines and wood-burning stoves.

Dredging of lakes is used to remove phosphorus and nitrogen stored in sediments. Phosphorus can be precipitated from the water column by adding alum. This also removes suspended solids including bacteria, increasing the clarity of the water. Swimmers in small recreational lakes can tell if this treatment has recently been applied, because the alum flocs are easily resuspended by waving the hand over the bottom.

Algae can be killed or inhibited by application of 500 μg Cu/L using copper sulfate. Phytoplankton will be killed, although any taste and odor or toxic compounds may remain. Cyanobacter are especially sensitive to copper. Fish and zooplankton are relatively insensitive, but algae can develop resistance. In the case of macrophytes, physical removal both eliminates the nuisance and takes nutrients out of the system.

The problem may be low dissolved oxygen in the hypolimnion. In this case, mechanical mixing can be used to destratify the lake, mixing the oxygen-rich epilimnion and the deficient hypolimnion. Sometimes it is desired not to destroy the stratification: for example, to support cold-water fish such as trout. In this case a variety of **hypolimnetic aeration** techniques can be applied in which hypolimnion water usually is pumped to the surface, aerated, then returned below the thermocline. Oxygen can also be dissolved directly into the water of the hypolimnion.

15.2.7 Organic Pollution of Streams

The first successes of modern environmental engineers and scientists were in protecting human and environmental health from the effects of discharging human wastes (sewage) to streams. The concerns were transmission of waterborne disease and gross degradation of the receiving water body due to excessive organic loading. Sewage discharge affects streams by adding toxins, pathogens, suspended solids, nutrients, and readily biodegradable organic carbon. These produce a cascade of effects, mostly related to reduced oxygen concentration in the water column. The oxygen depletion, in turn, is due to augmented heterotrophic oxygen consumption in the water column and in the sediments and to chemoautotrophic oxidation of ammonia. Recall that the nitrogenous oxygen demand of typical domestic wastewaters can be very high, since nitrification of each milligram of NH_3-N to NO_3^--N consumes 4.57 mg of oxygen.

DO Sag Equation If the organic pollution comes from only a single point source, the stream gradually recovers with distance downstream in a fairly predictable way. The master variable in this situation is oxygen. The changes in oxygen can be described by a model involving the following variables with their typical units (note that g/m^3 is equivalent to mg/L):

C	**dissolved oxygen concentration** (also called **DO**; g/m^3).
C_s	**saturated DO**, or oxygen concentration the water would have if it were in equilibrium with the atmosphere. Values decrease with temperature, salinity, and altitude. Example values in pure water at sea level range from 14.6 g/m^3 at 0°C to 9.2 g/m^3 at 20°C.
L	**biochemical oxygen demand** (**BOD**; g/m^3).
D	**DO deficit**, the amount the DO is below saturation; equal to $C_s - C$.

The major processes affecting oxygen levels in streams are:

U	**oxygen uptake rate** (g/m^3 per day); a combination of heterotrophic (organic carbon) and autotrophic (ammonia) biochemical oxidation.
B	**benthic oxygen demand** ($g/m^2 \cdot day$), due to heterotrophic and autotrophic oxidation in the sediment. Also called **sediment oxygen demand** (**SOD**).

R **reaeration rate** (g/m^2 pre day); the rate at which oxygen enters the water from the atmosphere by physicochemical mass transfer.

P **specific photosynthetic oxygen production rate** ($g/m^3 \cdot day$), by green plants and algae.

At the point that a parcel of water mixes with the discharge, the mixture has a DO of C_0 and a BOD of L_0. The DO at this point corresponds to an initial deficit $D_0 = C_s - C_0$. As the parcel continues downstream, its DO changes in response to a balance between the sources of oxygen, R and P, and the depletion by U and B:

$$\frac{dC}{dt} = \frac{R}{h} + P - U - \frac{B}{h} \qquad (15.2)$$

where $h = V/A$ is the average depth of the water column, or the volume of a section of a stream, V, divided by its plan area, A. Note from the units that reaeration and benthic consumption are normalized to the area, since they represent transport of oxygen across the top and bottom surfaces of the water column, respectively.

At the same time that oxygen is disappearing according to equation (15.2), the BOD in the water column is being depleted as oxygen is consumed by the uptake process:

$$\frac{dL}{dt} = -U \qquad (15.3)$$

Much can be learned from a simplified model that neglects photosynthesis and benthic oxygen demand. Recall that rivers tend to be allochthonous to begin with, and this will be more so for polluted streams. Benthic demand may be quantitatively important, but it behaves similarly to the water column BOD and does not greatly change the qualitative behavior of the model. Thus, only U and R remain in the model. Oxygen uptake can be described as a first-order decay:

$$U = k_1 L \qquad (15.4)$$

The parameter k_1 is the rate coefficient for decay of BOD. This is sensitive to the temperature and biodegradability of the BOD. The k_1 values must be determined experimentally either in the laboratory or, preferably, in field experiments. To measure k_1 in the field, BOD bottles containing samples of river water are submerged directly in the stream being modeled for 2 to 6 hours. Initial and final DO measurements yield a rate of oxygen consumption, U. Dividing by the BOD yields a value for k_1. Biodegradation rate coefficients have been measured for numerous industrial chemicals in laboratory experiments (see Appendix B).

Usually, parallel experiments will be done with clear glass bottles and with black-painted ones. The difference between the light and dark bottles gives an estimate of phytoplankton photosynthesis, and the dark bottle measures respiration. This is done either to validate the assumption of neglecting photosynthesis or to quantify it for including in the model.

Combining equations (15.3) and (15.4) and solving the resulting differential equation produces the typical exponential or first-order decay of BOD in the stream (Figure 15.21a):

$$L = L_0 e^{(-k_1 \cdot t)} \qquad (15.5)$$

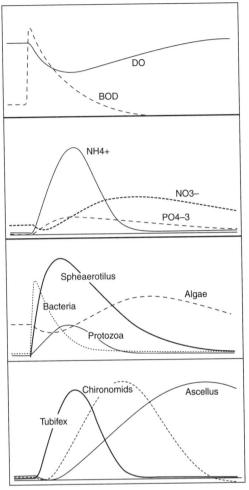

Figure 15.21   Changes downstream from a point of organic loading such as a sewage discharge. (Based on Connell and Miller, 1984.)

Note that the distance downstream, x, in Figure 15.21 corresponds to the distance traveled in time, t, at a stream velocity, v, by the transformation $x = vt$.

Reaeration is also first order, with respect to the DO deficit. That is, the rate at which oxygen enters the water is assumed to be proportional to how far the water is below saturation:

$$\frac{R}{h} = k_2 D = k_2 (C_s - C) \qquad (15.6)$$

The parameter k_2 is the coefficient of reaeration. It is controlled by turbulent mixing in the water. Correlations are available to estimate the value of k_2 from hydraulic parameters of the stream such as its depth, velocity, and hydraulic slope.

Combining (15.4) and (15.6) with (15.2) and (15.4) yields a differential equation which can be solved for initial conditions L_0 and D_0, resulting in the **Streeter–Phelps** or **DO sag equation** ($D = C_s - C$ is the DO deficit):

$$D = \frac{k_2 L_0}{k_2 - k_1}(e^{-k_1 t} - e^{-k_2 t}) + D_0 e^{-k_2 t} \qquad (15.7)$$

Figure 15.21a shows the shape of this curve. Qualitatively, the DO initially decreases rapidly when $U > R/h$. However, U is decreasing as L decreases (less food available for respiration), and R increases as C decreases farther below saturation. Ultimately, if the initial pollution is not too great, these two rates will approach each other, and the curve bottoms out. The DO deficit at this point is called the **critical deficit**, D_c, and occurs at a distance downstream called the **critical time of travel**, t_c:

$$t_c = \frac{1}{k_2 - k_1} \ln\left\{ \frac{k_2}{k_1} \left[1 - \frac{D_0(k_2 - k_1)}{k_2 L_0} \right] \right\} \qquad (15.8)$$

The critical deficit can then be found by substituting the result of equation (15.8) into (15.7).

As the BOD continues to decrease farther downstream, the stream starts to recover. If the critical DO concentration, $C_c = C_s - D_c$ is low enough, it can form a barrier to the movement of fish or other oxygen-sensitive organisms. The critical concentration is often used for regulatory purposes to indicate the seriousness of the pollution. In the regulatory process of **waste load allocation**, models of the river are manipulated to determine how much pollution must be reduced at the source to meet some minimum DO standard at the critical point.

Other Chemical Changes and Effects   Other changes occur as one travels downstream of the discharge. Biological effects are described below. Chemical changes include increases in nutrient concentrations. As shown in Figure 15.21b, ammonia increases due to mineralization of organic nitrogen and raw ammonia in the wastewater. The ammonia is gradually depleted as nitrification converts it ultimately into nitrate. The nitrate may be removed by phytoplankton uptake, and then either stays in the water column as organic nitrogen or is settled to the benthic zone. If the benthic zone is anoxic, nitrate may be removed from the water column to use to oxidize buried organics by denitrification.

Sedimentation of suspended solids near a discharge can smother benthic organisms. Sedimentation of suspended solids from the wastewater, plus biomass produced by growth on dissolved and colloidal material, can biodegrade in the sediment. The oxygen consumed in this way contributes to the benthic oxygen demand. Besides removing oxygen from the overlying water column, benthic oxygen demand reduces oxygen availability to the benthos. Benthic invertebrates are particularly sensitive.

Oxygen enters the sediment in a mass transport process similar to oxygen transport from the atmosphere to the water in reaeration. In this case the rate of transport is controlled by diffusion through the pores of the sediment and by the concentration in the water. If the oxygen demand is high enough or the rate of transport sufficiently low, the oxygen will only penetrate to a certain depth before it is consumed completely. Thus, there will be an oxic layer covering an anoxic layer, where denitrification replaces oxygen respiration. The depth of transition between the oxic and anoxic layers is controlled by a balance between the oxygen concentration in the water and the benthic oxygen demand.

At greater depths in the sediment, nitrate may be depleted and anaerobic conditions would prevail. Under these conditions respiration uses sulfate and carbonate as electron acceptors, producing H_2S and methane. The H_2S can diffuse back up through the sediment and enter the water column, where it can exert a toxic effect. The LC_{50} of H_2S (the concentration that kills half the organisms in a laboratory test) to several freshwater invertebrates has been measured to range from 11 to 1070 µg/L. Some zoobenthos can live in anoxic sediments because they circulate water containing oxygen mechanically from the surface of the sediment.

Ultimately, the organic matter becomes oxidized and no longer affects the river. However, many of the nutrients remain. In fact, degradation of organic-rich sediments may release previously stored nutrients. Thus, the high organic-loading condition of the stream can give way to a eutrophic situation, with photosynthesis exceeding the baseline level upstream of the discharge.

The Streeter–Phelps equation is a steady-state solution. If it is necessary to take photosynthesis into account, P should be modeled with a diurnal function. This would result in DO concentrations that oscillate, reaching a minimum in the morning. It is even possible to achieve supersaturation $(C > C_s)$ during sunny days if the amount of pollution is not great. The addition of organic pollution increases the ratio of heterotrophic to photosynthetic activity simply by providing food for the heterotrophs. However, photosynthesis may also be reduced directly by the effect of increased turbidity in the water, which can limit light availability.

Individual and Ecological Effects   Fish and larger invertebrates breathe through gills and have a circulatory system to bring oxygen-enriched blood to the tissues. When oxygen becomes depleted, these organisms can increase the pumping of water over the gills in order to extract more. Organisms such as fish that can readily move in response to oxygen have been found to avoid areas where concentrations have dropped below 3 to 5 mg/L.

Small invertebrates such as zooplankton obtain their oxygen by diffusion through the skin, without specialized organs; some do not even have circulatory systems. Some organisms are adapted to low-oxygen conditions. Annelid worms have special high-capacity hemoglobin. Chironomid larvae have a high concentration of hemoglobin, giving them a red color. Table 15.12 shows some of the oxygen limitations observed experimentally with aquatic animals.

Systems with large organic carbon input are called **saprobic**, as opposed to eutrophic, which refers primarily to excess nitrogen and phosphorus. In eutrophic waters primary production exceeds respiration. In saprobic waters the reverse is true. As organic pollution

TABLE 15.12   Dissolved Oxygen Limits for Several Organisms

Organism	Temperature (°C)	Type of Test	DO (mg/L)
Brown trout (*Salmo trutta*)	6.5–24	Limiting concentration	1.28–2.9
Coho salmon (*Oncorhynchus kisutch*)	16–24	Limiting concentration	1.3–2.0
Rainbow trout (*Salmo gairdnerii*)	11.1–20	Limiting concentration	1.05–3.7
Worm (*Nereis grubei*)	21.7–26.3	28-day LC_{50}	2.95
Worm (*Capitella capitata*)	21.7–26.3	28-day LC_{50}	1.50
Amphipod (*Hyalella azteca*)	—	Threshold	0.7
Amphipod (*Gammarus fasciatus*)	—	Threshold	4.3

Source: Connell and Miller (1984).

levels increase, population shifts occur. Clean waters are dominated by fish and green algae. The Shannon–Weaver diversity index for benthic invertebrates in clean water is greater than 3. At moderate saprobic levels, the diversity index drops to between 1 and 3 and the dominant population shifts to oligochaetes such as *tubifex*. Seriously polluted waters have these as well as increasing numbers of ciliates and flagellate protozoans, and a diversity index less than 1.

A succession occurs with time in saprobic lakes that is similar to what occurs downstream of a wastewater discharge. A pure lake begins in a **xenosaprobic** state, corresponding to oligotrophic. If a large pollution discharge occurs, it becomes **polysaprobic**. As the organic matter is degraded and the lake recovers, it passes through intermediate stages termed **mesosaprobic** and **oligosaprobic**. At this point the organic matter has been eliminated, but nutrients remain. Primary productivity may exceed respiration, and the lake passes into a eutrophic phase. The same terms are applied to rivers at various stages of recovery from pollution.

Besides tolerance of low oxygen levels, the animals that proliferate in saprobic conditions include detrital feeders that feed as filterers or collectors. Examples are certain caddis flies and mayflies, chrionomids, and clams. Grazers also are successful and include certain other caddis flies and mayflies, and snails. Predators such as leeches and midges respond to the increasing saprobic community. Table 15.13 shows some selected organisms that are found in various saprobic conditions. Although the term *mesosaprobic* is

TABLE 15.13   Tolerance of Organisms to Organic Pollution

Taxonomic Group	Oligosaprobic (Unpolluted)	Mesosaprobic (Moderately Polluted)	Polysaprobic (Highly Polluted)
Cyanobacter		*Oscillatoria*	*Chamasiphon*
Diatoms		*Nitzschia palea*	*Cocconeis placentula*
		Gomphonema parulum	
Green algae	*Chlamydomonas*	*Chlamydomonas*	*Chlamydomonas*
		Ulothrix	*Ulothrix*
		Scenedesmus	*Chaetophora*
Turbellaria		*Polycelis nigra*	*Polycelis cornuta*
Annelids	*Tubifex tubifex*	*Lubricus, Stylaria*	*Chaetogaster*
Hirudinea		*Erpobdella octoculata*	
(leeches)		*Helobdella stagnalis*	
		Glossiphonia spp.	
Crustaceans		*Daphnia*	*Daphnia*
		Cyclops	*Cyclops*
		Asellus aquaticus (water louse)	
		Sialis lutaria (alder flies)	
Insects	*Eristalis tenax* (rat-tailed maggot)	*Chironomous plumosus*	Stoneflies and mayflies (all species)
	Chironomous plumosus	*Caddis flies* *Hydropsyche* spp. *Anabolia, Molanna* spp.	*Chironomous* spp.
Fish		Eel, stickleback, carp, goldfish	Trout, pike, perch

Source: Connell and Miller, 1984.

associated with "moderately polluted," this condition can occur naturally if there are large allochthonous inputs.

15.3 WETLANDS

Wetlands have experienced a remarkable turnaround in the public mind. Not long ago they were derided as swamps and mires, sources of disease, impediments to travel, and a waste of real estate. If they could not be drained for agriculture or development, the only practical use for them was as a location for garbage dumps. If at all possible, they were eliminated by draining or filling. Between the late eighteenth century and the mid-1980s, it is estimated that over 50% of the wetlands in the lower 48 states of the United States have been lost.

Today they are regarded as natural treasures. They are protected by strict laws, and their importance is touted in newspaper articles and elementary school curricula. What has changed is public recognition of the ecological value of wetlands. Salt marshes are among the most productive ecosystems in the world (Table 15.1). They are the breeding ground for numerous animals that range far beyond the marsh, including shorebirds and pelagic ocean fish. Migratory birds use them as stopping-off points, where they feed and restore their energy reserves before continuing their journey.

Although wetlands occupy only about 2% of Earth's land area, they store 10 to 14% of its organic carbon in their soil. When wetlands are dewatered, such as by draining or by a decrease in precipitation, the organic matter is exposed to increased amounts of oxygen. As a result, the organic matter becomes oxidized to CO_2. This is one of the "positive feedback mechanisms" that may aggravate the effect of global warming.

Wetlands mitigate storm flows and flooding and serve as sources for groundwater recharge. Surface water passing through a marsh can be reduced in nutrient level and other contaminants. They have traditionally been, and continue to be, used as a source of natural resources, including lumber, moss, shellfish, waterfowl, and pelts. Finally, as can be attested to by the birders and hunters who venture into them, they have aesthetic value. It is fortunate that nature made its most productive ecosystem so inaccessible to humans, so that wetlands have survived even in the midst of the world's most densely populated cities.

Some consider wetlands to be merely an ecotone, an intermediate zone between terrestrial and deepwater ecosystems. They can also be viewed as a stage in the successional change from open water to terrestrial ecosystem (recall the discussion of eutrophication in Section 15.2.6). However, wetlands have a number of unique characteristics. Wetlands are defined by their water, soil, and plants. Specifically:

- Water is present at a depth either above the surface (but not so deep as to preclude the presence of rooted plants that emerge from the water) or below the soil surface but within the root zone.
- The soil is anaerobic within at least part of the root zone during a significant part of the growing season.
- The ecosystem is capable of supporting plants that are specially adapted for wet conditions and has no plants that are flood intolerant.

A more precise definition is necessary for legal purposes. This is discussed in Section 15.3.7. There are numerous terms for specific types of wetlands in both common use and scientific use. Sometimes, the terms *marsh* or *swamp* are used interchangeably with *wetland*. However, these terms both have more specific meanings. Some examples are:

Bog: a peat-accumulating wetland with little or no inflow or outflow. Most of its nutrients come from precipitation. Its waters tend to be acidic, and as a result it supports mosses primarily, such as sphagnum moss.

Fen: a peat-accumulating wetland that receives some flow and nutrients from surrounding uplands. It supports some vegetation similar to marshes.

Marsh: a wetland that is frequently or continuously inundated and has emergent plants that are specially adapted to saturated soil conditions. In Europe this term excludes wetlands that are peat accumulating.

Moor: in European use, any wetland that accumulates partially decayed organic matter.

Muskeg: large expanses of bogs, such as in northern Canada and Alaska.

Riparian wetlands (or floodplains): occasionally flooded lands along rivers that are dry during parts of the growing season.

Swamp: in the United States, a wetland dominated by shrubs or trees. In Europe, a forested fen or wetland dominated by reedgrass (*Phragmites*, or common reed).

Vernal pool: an intermittently inundated wet meadow that is usually dry during the summer and fall.

Table 15.14 shows how several of these wetland types fit along the continuums for plant type, water flow, soil organic matter content, pH, and nutrient condition. A **rheo-trophic** wetland is one that receives significant inflow vs. an **ombrotrophic** wetland that receives little flow. All of the characteristics in Table 15.14 relate to input of nutrients and whether the wetland accumulates peat. The oligotrophic wetlands become acidic because products of degradation are not washed away. Combined with anaerobic conditions in saturated soils, this limits biodegradation. Instead, the partially degraded biomass accumulates, forming peat.

Wetlands can also be classified as to whether they are coastal or inland, tidal or non-tidal, freshwater or saline, intermittent or permanent. **Tidal salt marshes** are found all along the U.S. Atlantic and Gulf coasts. They are dominated by the grass *Spartina* and the rush *Junca*. Moving inland the salt marsh gives way to tidal freshwater marshes. **Tidal freshwater marshes** are dominated by both grasses and broadleafed aquatic plants. In tropical and subtropical regions, such as southern Florida and parts of Louisiana and Texas, the tidal freshwater marsh, in turn, gives way to the mangrove swamp. The **mangrove swamp** is dominated by the salt-tolerant mangrove tree, such as the red mangrove *Rhizophora* or black mangrove *Avicennia*. All of these coastal marshes require protection from ocean storm action and are frequently found behind barrier islands.

There are four major types of inland wetlands in North America. **Freshwater marshs** occur throughout the continent and consist of shallow water, thin peat deposits, and diverse emergent plants, including cattails, arrowheads, pickerel-weed, reeds, and grasses and sedges. They include the prairie potholes of the Dakotas, the Great Lake marshes, and the Everglades of Florida. The **northern peatlands** include bogs and fens from the northern midwestern and northeastern United States and central and eastern Canada. Many are filled-in lake basins, in final stages of natural eutrophication (Section 15.2.6). The **southern deepwater swamps** have standing water through most of the growing season and are

TABLE 15.14   Comparison of Wetland Types and Their Characteristics

Bog	Peat-accumulating wetland; no significant inputs or outputs; supports acid-loving plants, especially sphagnum
Bottomland	Lowlands along streams and rivers; periodically flooded floodplains
Fen	Peat-accumulating wetland with inputs from surrounding uplands
Peatland	Any wetland with an accumulation of peat
Vernal pool	Shallow, wet meadow, intermittently flooded in winter and spring, but usually dry for most of the summer and fall
North American usage	
Marsh	Wetland that is frequently or continuously inundated and is dominated by emergent herbaceous vegetation
Muskeg	Extensive bogs or peatlands, mainly in Canada and Alaska
Playa	Marshlike ponds in the southwestern U.S.
Pothole	Marshlike ponds in the northern prairie, many formed by glaciation
Slough	Swamp or system of shallow lakes in north or midwestern U.S.; also a slowly flowing swamp or marsh in the southeastern U.S.
Swamp	Wetland dominated by trees or shrubs
Wet meadow	Grassland with water surface near but usually not above soil surface
Wet prairie	Intermediate between a *marsh* and a *wet meadow*
European usage	
Marsh	Non-peat-accumulating wetland with mineral soil
Mire	Any peat-accumulating wetland
Moor	Same as *peatland*
Reedswamp	Marsh dominated by *Phragmites* (common reed), especially in eastern Europe
Swamp	Forested fen or reedgrass-dominated wetland

Source: Mitsch and Gosselink (1993).

wooded with cypress (*Taxodium*) and gum/tupelo (*Nyssa*). The final freshwater type is the **riparian wetlands**. These are usually forested wetlands along the sides of rivers and streams. They are inundated on the average every one or two years in the United States, when the rivers overflow their banks.

Wetlands are not just passive products of the hydrological conditions in which they exist. In yet another example of the Gaia hypothesis at work, wetland plants exert control over the hydrology. By accumulating peat and trapping sediment, they create land with very little slope. The vegetation itself further obstructs the flow of water. These factors help dampen water-level variations. Trees reduce water loss by shading the water or land surface and by increasing storage. Animals alter the landscape as well. Beavers, notably, create new wetlands by damming streams. Alligators in the Everglades excavate "gator holes" that provide refuge for all kinds of aquatic organisms during the dry season.

15.3.1   Hydric Soils

The U.S. Department of Agriculture has defined the types of soil unique to wetlands as follows: "A **hydric soil** is a soil that is saturated, flooded, or ponded long enough during the growing season to develop anaerobic conditions that favor the growth and regeneration of hydrophytic vegetation." A hydric soil is considered a **mineral soil** if its organic matter content is less than 20 to 35% dry weight (12 to 20% organic carbon content). The exact percentage used depends on the frequency of saturation and the clay content. Otherwise, the soil is considered an **organic soil** (see Table 15.15).

TABLE 15.15   Properties of Mineral and Organic Soils in Wetlands

Property	Hydric Mineral Soil	Hydric Organic Soil
Organic matter, percent dry weight	Less than 20 to 35%	More than 20 to 35%
Organic carbon, percent dry weight	Less than 12 to 20%	More than 12 to 20%
pH	Approximately neutral	Acidic
Bulk density (g/cm^3)	1.0–2.0	Typically 0.2 to 0.3, although sphagnum moss peat can be as low as 0.04
Porosity	Low (45–55%)	High (about 80%)
Hydraulic conductivity	High	Low to high
Water-holding capacity	Low	High
Nutrient availability	High	Low
Cation-exchange capacity	Low, consisting of major cations	High, but consisting mostly of hydrogen ions
Example wetlands	Riparian forest, some marshes	Peat-accumulating wetlands

Source: Based on Mitsch and Gosselink (1993).

Organic matter gives soil lower density and higher cation-exchange capacity (CEC). The higher CEC does not translate into high nutrient content, due to the typically lower pH, which leaches mineral salts. Furthermore, organic complexation reduces the bioavailability of the nutrients that are present.

Hydric organic soil may be classified as a **muck**, in which two-thirds of the organic material is decomposed (or humified) and less than one-third of the plant fibers are identifiable as such. If less than one-third of the organic matter is humified and two-thirds of the plant fibers are still identifiable, the soil is considered a **peat**.

The molecular diffusivity of oxygen through water is about four orders of magnitude less than in air. Thus, inundation usually leads to anaerobic soil conditions, except possibly in a thin surface layer. Oxygen is usually depleted within hours or days of inundation. Further degradation of organic material can continue, although with reduced efficiency and velocity. Instead of oxygen, alternative electron acceptors are used, such as nitrate, manganic, ferric, sulfate, or carbonate ions, in order of decreasing metabolic efficiency.

Soil organic matter originates mostly from plant biomass. In most marshes, only a small portion (10 to 40%) of the plant matter is consumed by herbivores, entering the grazing food chain. The remainder fuels the detritus food chain. However, anaerobic conditions in the soil greatly reduces the rate of decomposition, even to the point where the rate of detritus mineralization is less than its production. Such wetlands are called **peat accumulating**. Peat can accumulate to great depths as the marsh builds up on previous layers of accumulation. Decomposition rates have been measured for *Typha* in freshwater marshes. Most show a half-life for degradation ranging from 67 to 365 days, with a median of 198 days.

The shift of equilibrium toward more reduced conditions can be measured by a parameter called the **redox potential**. It is analogous to the pH scale for acid–base equilibrium and is measured in a similar way using an electrode, usually connected to a standard pH meter. Redox potential is often measured in millivolts. When oxygen is present, the redox potential is around +400 to +700 mV. Under extreme reducing conditions it drops as low

TABLE 15.16   Redox Couples and Corresponding Redox Potentials

Element	Oxidized Form	Reduced Form	Approximate Redox Potential for the Transformation (mV)
Oxygen	O_2	H_2O	400 to 700
Nitrogen	NO_3^- (nitrate), NO_2^- (nitrite)	N_2O, N_2, NH_4^+	250
Manganese	Mn^{4+} (manganic)	Mn^{2+} (manganous)	225
Iron	Fe^{3+} (ferric)	Fe^{2+} (ferrous)	120
Sulfur	SO_4^{2-} (sulfate)	S^{2-} (sulfide)	-75 to -150
Carbon	CO_2	CH_4	-250 to -350

Source: Based on Mitsch and Gosselink (1993).

as -400 mV. Table 15.16 shows the redox potentials associated with redox transformations. For example, the table shows that iron will tend to be in the ferric form when the redox potential is above about 120, and in the ferrous form below that value. The exact redox potential for each transformation depends on the pH.

The sequence of redox reactions shows that the products of the most extreme reducing conditions are the gases hydrogen sulfide and methane. The latter forms **swamp gas** or **marsh gas**, which is formed only when the other electron acceptors are depleted. Methane production rates were measured as high as 440 mg $C/m^2 \cdot$ dayas an annual average in a tidal freshwater marsh in Lousisiana. A swamp in Michigan had an annual production of 100 mg $C/m^2 \cdot$ day. Rice paddies in various locations were measured at 36 to 385 mg $C/m^2 \cdot$ day $\cdot$ Salt marshes produce less methane, possibly because the seawater provides sulfate, which is a preferential electron acceptor over carbon dioxide.

Hydric mineral soils that are flooded all or most of the time form layers of bluish-gray reduced iron called **gleys**. If the flooding is intermittent, red or black spots called **mottles** form in the gley. Mottles are made of oxidized iron or manganese. Gleys and mottles are formed by microorganisms that require organic matter and anaerobic conditions.

Ferrous iron is very soluble and may diffuse to the roots of plants, where it is oxidized by oxygen supplied by the plant. This results in a coating of ferric hydroxide on the roots, which may limit nutrient uptake. If the dissolved ferrous iron seeps into an aerobic environment such as a pond, iron bacteria may oxidize it. It may then precipitate and form into concentrated deposits. This is the basis of the "bog-iron" ore deposits that are mined today.

15.3.2   Hydrophytic Vegetation

Living things can often be used as environmental indicators, due to their function as *integrators* of environmental conditions. That is, they respond to conditions over a period time rather than just reflecting current conditions. That is why wetland vegetation is an important clue to identifying a wetland. Wetlands are not always wet, and wetland soil is not always readily identifiable as a hydric soil.

Hydrophytic vegetation are macrophytic plants that grow in areas where soil saturation or inundation occurs with a frequency and duration sufficient to exert a controlling influence on the plant species present. They are distinguished from upland plants by having one or more adaptations to wetlands conditions. The adaptations have been classified as morphologic, physiologic, or reproductive.

Morphologic adaptations are changes in structure (see Table 15.17). Most are responses to anaerobic soil conditions. One of the most important of these is the formation

TABLE 15.17   Morphologic Adaptations to Wetlands Conditions

Buttressed tree trunks	Tree species (e.g., *Taxodium distichum*) may develop enlarged trunks in response to frequent inundation. This adaptation is a strong indicator of hydrophytic vegetation in nontropical forested areas.
Pneumatophores	These modified roots may serve as respiratory organs in species subjected to frequent inundation or soil saturation. Cypress knees are a classic example, but other species (e.g., *Nyssa aquatica*, *Rhizophora mangle*) may also develop pneumatophores.
Adventitious roots	Sometimes referred to as "water roots," adventitious roots occur on plant stems in positions where roots normally are not found. Small fibrous roots protruding from the base of trees (e.g., *Salix nigra*) or roots on stems of herbaceous plants and tree seedlings in positions immediately above the soil surface (e.g., *Ludwigia* spp.) occur in response to inundation or soil saturation. These usually develop during periods of sufficiently prolonged soil saturation to destroy most of the root system. (*Caution*: Not all adventitious roots develop as a result of inundation or soil saturation. For example, aerial roots on woody vines are not normally produced as a response to inundation or soil saturation.)
Shallow root systems	When soils are inundated or saturated for long periods during the growing season, anaerobic conditions develop in the zone of root growth. Most species with deep root systems cannot survive in such conditions. Most species capable of growth during periods when soils are oxygenated only near the surface have shallow root systems. In forested wetlands, wind-thrown trees are often indicative of shallow root systems.
Inflated leaves, stems, or roots	Many hydrophytic species, particularly herbs (e.g., *Limnobium spongia*, *Ludwigia* spp.) have or develop spongy (aerenchymous tissues in leaves, stems, and/or roots that provide buoyancy or support and serve as a reservoir or passageway for oxygen needed for metabolic processes.
Polymorphic leaves	Some herbaceous species produce different types of leaves, depending on the water level at the time of leaf formation: for example, *Alisma* spp. Produce strap-shaped leaves when totally submerged, but produce broader, floating leaves when plants are emergent. (*Caution*: Many upland species also produce polymorphic leaves.)
Floating leaves	Some species (e.g., *Nymphaea* spp.) produce leaves uniquely adapted for floating on a water surface. These leaves have stomata primarily on the upper surface and a thick waxy cuticle that restricts water penetration. The presence of species with floating leaves is strongly indicative of hydrophytic vegetation.
Floating stems	A number of species (e.g., *Alternathera philoxeroides*) produce matted stems that have large internal airspaces when occurring in inundated areas. Such species root in shallow water and grow across the water surface into deeper areas. Species with floating stems often produce adventitious roots at leaf nodes.
Hypertrophied lenticels	Some plant species (e.g., *Gleditsia aquatica*) produce enlarged lenticels on the stem in response to prolonged inundation or soil saturation. These are thought to increase oxygen uptake through the stem during such periods.
Multitrunks or stooling	Some woody hydrophytes characteristically produce several trunks of different ages or produce new stems arising from the base of a senescing individual (e.g., *Forestiera acuminata*, *Nyssa ogechee*) in response to inundation.
Oxygen pathway to roots	Some species (e.g., *Spartina alterniflora*) have a specialized cellular arrangement that facilitates diffusion of gaseous oxygen from leaves and stems to the root system.

Source: U.S. Army Corps of Engineers (1987).

of **aerenchyma**, which are airspaces within stem, leaf, and root tissues that provide a diffusional route of oxygen transfer from the surface to the roots. They are formed by cell separation during maturation or by breakdown of existing cells. The porosity of wetlands plants can be up to 60% vs. 2 to 7% in normal plants. A plant will increase its porosity in response to flooding.

Some trees produce **pneumatophores**, or **air roots**, which protrude above the ground some distance from the trunk of the tree and are thought to promote gas exchange with the atmosphere. An example are the "cypress knees" found in southern U.S. swamps. **Adventitious roots** are those that come from unusual locations, such as leaf nodes or in a circle around the base of the tree. The red mangrove (*Rhisophora* spp.) form arched adventitious **prop roots**. Adventitious roots and pneumatophores may have small pores called **lenticels** that provide a pathway for oxygen to reach the roots.

Some of the oxygen that plants transport to their roots affects the soil surrounding them. In a process called **rhizosphere oxygenation**, plants create a microaerobic environment around their roots in an otherwise anaerobic soil. *Spartina alterniflora* forms brown deposits around its roots because of precipitation of iron and manganese when they are oxidized. In the vicinity of mangrove prop roots or pneumatophores, the redox potential was higher and the sulfide concentration was three to five times lower than in mud deposits at a greater distance. Thus, the role of oxygen transport to the roots seems not only to provide metabolic oxygen but also to protect the roots from toxic minerals. Examples of plants with morphological adaptations are given in Table 15.18.

Physiological adaptations to anaerobic soil conditions are not readily identifiable in the field. However, they further illustrate the differences between wetlands and other plants (Table 15.19). Reproductive adaptations increase the chance of a plant becoming established in a wetlands area (see Table 15.20).

Salt marsh plants also need adaptations to survive in salt water. Exposure to salt water can be harmful both because of toxicity from salts such as sodium and because osmotic forces can cause lethal water loss. One protective mechanism is for their cell membranes to form a selectively permeable barrier, similar to an ultrafilter. The sap of some plants can have salt contents on the order of 3% of seawater salinity. However, no cell membrane is perfectly selective. Thus, the cells also selectively excrete harmful salts, especially sodium. Many salt marsh grasses, such as *Spartina*, actually form salt crystals on their leaf surfaces from these excretions. The cell prevents water loss by maintaining enough internal solutes in the form of potassium and organics. Although it has not been documented, it is thought that tidal freshwater marshes are even more productive than tidal salt marshes, because they receive the same nutrient and energy subsidies but do not have the salt stresses to deal with.

Salt marsh plants are more likely to use the C_4 photosynthesis pathway than upland plants. Recall that C_4 plants use CO_2 more efficiently than the more common C_3 plants. This allows them to keep their stomates closed more, with the result that they lose less water by evapotranspiration. The C_4 adaption helps plants survive arid conditions. The water potential of saline water can be as low as the soil in dry climates; despite the large amount of water present, it can be just as unavailable as in a desert. The C_4 mechanism uses malate to store CO_2. Thus, malate is involved in both respiration and photosynthesis in wetland plants, and its use is another indicator of hydrophytic vegetation. Examples of C_4 wetlands plants are *Spartina alterniflora, S. townsendii, S. foliosa, Cyperus rotundus, Echinochloa crusgalli, Panicum dichotomifloru, P. virgatum, Paspalum distichum, Phragmites communis*, and *Sporobolus cryptandrus*.

TABLE 15.18 Partial List of Species with Known Morphological Adaptations for Wetlands Conditions

Species	Common Name	Adaptation
Acer negundo	Box elder	Adventitious roots
Acer rubrum	Red maple	Hypertrophied lenticels
Acer saccharinum	Silver maple	Hypertrophied lenticels; adventitious roots (juvenile plants)
Alisma spp.	Water plantain	Polymorphic leaves
Alternanthera philoxeroides	Alligatorweed	Adventitious roots; inflated, floating stems
Avicennia nitida	Black mangrove	Pneumatophores; hypertrophied lenticels
Brasenia schreberi	Watershield	Inflated, floating leaves
Cladium mariscoides	Twig rush	Inflated stems
Cyperus spp. (most species)	Flat sedge	Inflated stems and leaves
Eleocharis spp. (most species)	Spikerush	Inflated stems and leaves
Forestiera acuminata	Swamp privet	Multitrunk, stooling
Fraxinus pennsylvanica	Green ash	Buttressed trunks; adventitious roots
Gleditsia aquatica	Water locust	Hypertrophied lenticels
Juncus spp.	Rush	Inflated stems and leaves
Limnobium spongia	Frogbit	Inflated, floating leaves
Ludwigia spp.	Waterprimrose	Adventitious roots; inflated floating stems
Menyanthes trifoliata	Buckbean	Inflated stems (rhizome)
Myrica gale	Sweetgale	Hypertrophied lenticels
Nelumbo spp.	Lotus	Floating leaves
Nuphar spp.	Cowlily	Floating leaves
Nymphaea spp.	Waterlily	Floating leaves
Nyssa aquatica	Water tupelo	Buttressed trunks; pneumatophores; adventitious roots
Nyssa ogechee	Ogechee tupelo	Buttressed trunks; multitrunk; stooling
Nyssa sylvatica var. *biflora*	Swamp blackgum	Buttressed trunks
Platanus occidentalis	Sycamore	Adventitious roots
Populus deltoides	Cottonwood	Adventitious roots
Quercus laurifolia	Laurel oak	Shallow root system
Quercus palustris	Pin oak	Adventitious roots
Rhizophora mangle	Red mangrove	Pneumatophores
Sagittaria spp.	Arrowhead	Polymorphic leaves
Salix spp.	Willow	Hypertrophied lenticels; adventitious roots; oxygen pathway to roots
Scirpus spp.	Bulrush	Inflated stems and leaves
Spartina alterniflora	Smooth cordgrass	Oxygen pathway to roots
Taxodium distichum	Bald cypress	Buttressed trunks; pneumatophores

Source: U.S. Army Corps of Engineers (1987).

15.3.3 Wetlands Animals

Animals exhibit specialized adaptations for wetlands, as well. Many benthic invertebrates, such as some nematodes, crabs, and clams, have either higher concentrations of oxygen-transporting pigments, or pigments with higher oxygen affinity than usual. Some organisms are effective at conserving oxygen. Fiddler crabs (*Uca* spp.) can maintain their resting respiration rates at oxygen levels down to 5 to 15% of saturation. At lower oxygen

TABLE 15.19   Physiological Adaptations to Wetlands Conditions

Accumulation of malate	Nonwetland species concentrate ethanol, a toxic by-product of anaerobic respiration, when growing in anaerobic soil conditions. Under such conditions, many hydrophytic species produce high concentrations of the nontoxic metabolite malate instead, and unchanged concentration of ethanol, thereby avoiding accumulation of toxic materials (e.g., *Glyceria maxima*, *Nyssa sylvatica* var. *Biflora*).
Increased levels of nitrate reductase	Nitrate reductase is an enzyme involved in conversion of nitrate nitrogen to nitrite nitrogen, an intermediate step in ammonium production. Nitrate ions can accept electrons as a replacement for gaseous oxygen in some species, thereby allowing continued functioning of metabolic processes under low soil oxygen conditions. Species that produce high levels of nitrate reductase include *Larix larcina*.
Slight increases in metabolic rates	Anaerobic soil conditions effect short-term increases in metabolic rates in most species. However, the rate of metabolism often increases only slightly in wetland species. Examples species include *Larix laricina* and *Sencio vulgaris*.
Rhizosphere oxidation	Some hydrophytic species (e.g., *Nyssa aquatica*, *Myrica gale*) are capable of transferring gaseous oxygen from the root system into soil pores immediately surrounding the roots. This prevents root deterioration and maintains the rates of water and nutrient absorption under anaerobic soil conditions.
Ability for root growth in low oxygen tensions	Some species (e.g., *Typha angustifolia*, *Juncus effusus*) have the ability to maintain root growth under soil oxygen concentrations as low as 0.5%. Although prolonged ($>$1 year) exposure to soil oxygen concentrations lower than 0.5% generally results in the death of most individuals, this adaptation enables some species to survive extended periods of anaerobic soil conditions.
Absence of alcohol dehydrogenase (ADH) activity	ADH is an enzyme associated with increased ethanol production. When the enzyme is not functioning, ethanol production does not increase significantly. Some hydrophytic species (e.g., *Potentilla anserina*, *Polygonum amphibium*) show only slight increases in ADH activity under anaerobic soil conditions.

Source: U.S. Army Corps of Engineers (1987).

levels (down to 2% of saturation) they can further reduce their respiration to survive. Clams can respire anaerobically for a time with their shells closed.

The simpler animals are **osmoconformers**; they adjust their cell osmolarity in response to the external salinity. More complex animals are **osmoregulators**, which maintain their internal environment in the face of external changes. The difference depends on the ability of the organism to excrete salt. In crabs it also depends on the permeability of the shell to salt. Crabs such as *Cancer* are always submerged; they have a more permeable shell and are an osmoconformer. Crabs that spend part of their time out of the water, such as *Uca* spp., are excellent osmoregulators. Their shells are relatively impermeable to salt. Other species fall in between.

The annelids are divided into two groups. The polychaetes are osmoconformers. Eggs and sperm are released into the water, where fertilization occurs. Oligochaetes and

TABLE 15.20   Reproductive Adaptations to Wetland Conditions

Prolonged seed viability	Some plant species produce seeds that may remain viable for 20 years or more. Exposure of these seeds to atmospheric oxygen usually triggers germination. Thus, species (e.g., *Taxodium distichum*) that grow in very wet areas may produce seeds that germinate only during infrequent periods when the soil is dewatered. (*Note*: Many upland species also have prolonged seed viability, but the trigger mechanism for germination is not exposure to atmospheric oxygen.)
Seed germination under low oxygen concentrations	Seeds of some hydrophytic species germinate when submerged. This enable germination during periods of early-spring inundation, which may provide resulting seedlings a competitive advantage over species whose seeds germinate only when exposed to atmospheric oxygen.
Flood-tolerant seedlings	Seedlings of some hydrophytic species (e.g., *Fraxinus pennsylvanica*) can survive moderate periods of total or partial inundation.

Source: U.S. Army Corps of Engineers (1987).

leeches, on the other hand, are osmoregulators. Fertilization is internal, and the resulting zygote is then enclosed in a cocoon.

15.3.4 Hydrology and Wetlands Ecology

The flow of water through a wetland provides many benefits. Foremost, it is often a source of nutrients from other ecosystems. It flushes toxins away. It facilitates dispersal of organisms. If the flow is tidal, it acts as a selection agent, inhibiting organisms that cannot tolerate the variation in water levels. In these functions, flow does work for the ecosystem and constitutes an energy subsidy to it. Wetlands with significant flow are termed **open**. Ecosystems without significant flow are called **closed**.

Flow can also have a negative effect as a source of stress and causing an export of biomass in the form of plant litter that is washed away by tides or floods. However, the evidence suggests that watersheds that include wetlands tend to export more biomass while retaining more nutrients. Salt marshes tend to export biomass in the form of plant litter at an order-of-magnitude higher rate than freshwater marshes, about 100 to 200 $g/m^2 \cdot yr$. This reduces the net primary productivity and provides an allochthanous input to downstream ecosystems.

Wetlands cycle nutrients similarly to terrestrial ecosystems. Two major differences are that the sediments tend to be anaerobic, and nutrients can be lost by export or burial in peat deposits. Anaerobic sediments cause a loss of nitrogen by denitrification, but phosphorus is made more soluble, and thus more available. However, if the wetland is open, it may not be as dependent on nutrient recycling as terrestrial ecosystems.

In some cases, primary productivity has been positively correlated to flow through a wetlands or to tidal range; in other cases, to phosphorus input. On the other hand, it has been found that freshwater marshes along Lake Erie exhibit an increase in productivity when they are isolated from the lake surface waters by dikes. It is probably best to think of wetland productivity as resulting from a combination of effects of nutrient input, biomass export and burial, degradation, and energy subsidies, and that the effect of flow is generally positive, although it might be an indirect effect.

Hydrology has a significant effect on species distribution. Since relatively few plant species are hydrophytic, the greater the duration or frequency of inundation, the lower the plant species richness and diversity. In fact, it is common to see a single species dominating a marsh, such as *Typha* (cattail) or *Phragmites* in freshwater marshes or *Spartina* in salt marshes. However, less intense flooding can create a higher species diversity than purely terrestrial ecosystems, since both hydrophytic and upland species may coexist in this situation.

15.3.5   Wetlands Nutrient Relationships

In closed wetlands, nutrient input and biomass output are reduced. Instead of exporting biomass, the wetland would tend to accumulate it in the form of peat. It is not true that all wetlands are highly productive. Peat-accumulating wetlands tend to be oligotrophic, with a consequently reduced gross primary productivity.

A major question about wetlands is their role in nutrient balances. A wetland is called a **source** if exports of an element exceed inputs. In particular, many wetlands are sources of organic carbon. A wetland is a **sink** if exports are less than inputs. Wetlands are considered **transformers** if they change the form of the element, such as by converting soluble phosphorus into particulate, or ammonia into nitrate. Few generalizations have been established to predict the nutrient balance of a wetland. Denitrification contributes to the role of wetlands as a nitrogen sink, and assimilation makes it either a sink or a transformer for phosphorus. However, many studies have also found wetlands to function as sources. Several multiyear studies found that wetlands used for wastewater treatment would act as sources in some years and sinks in others.

Because wetlands have aerobic and anaerobic environments in close proximity, they may be the major location for denitrification in the biosphere. The aerobic environment provides nitrate. This can leach into the anaerobic zone to be converted to N_2. Since such a large proportion of nitrogen fixation on Earth is anthropogenic, wetlands may play an important role in maintaining the balance.

As discussed elsewhere, wetlands also have an important role in global carbon cycling. Peat-accumulating wetlands are major sinks for carbon, providing negative feedback for global warming. On the other hand, they are a source of atmospheric methane, which is another greenhouse gas.

15.3.6   Major Wetland Types

Here we highlight several of the most extensive wetlands types, of the many that exist. Our interest is in those most likely to be affected by development activities.

Tidal Salt Marshes   Salt marshes tend to form in protected areas such as in the shelter of spits, bars, or islands, in bays, and in estuaries. Marshes may grow with sedimentation supplied by rivers or tidal action and influenced by sediment-trapping effects of the biota. In some places, sea-level rise results in an increase in the extent of salt marshes. For example, sea level along the Atlantic coast of North America has been increasing 1 to 3 mm per year for about 4000 years. Marshes kept pace by accumulating peat, in fact increasing their extent by growing both landward and seaward.

Salt marshes are primarily nitrogen limited. The main available form is ammonium. However, nonnutrient stresses may also be significant. These can include salt stress and lack of drainage. If evaporation exceeds rainfall, portions of the marsh away from the water may accumulate salt. Even salt-tolerant plants must expend energy to obtain water against the osmotic gradient.

Cordgrass (*S. alterniflora*) tends to dominate in the lower marsh, near the water. *S. alterniflora* has stiff leaves and occurs in two forms. The tall form is 1 to 3 m tall and dominates in the extreme lower portions of the marsh. The short form is 17 to 80 cm tall and is found in higher salinity conditions, farther from the water. Closer to land, *S. alterniflora* gives way to the salt-tolerant species *S. patens* and *Distichlis spicata* (spike grass). At the high-tide level will be found stands of *Juncus gerardi* or *J. Roemerianus*. Finally, at the part of the marsh reached only by the highest spring tides, if rainfall exceeds evaporation, salt-intolerant species such as *Panicum virgatum* and *Phragmites australis* are found.

Salt marsh fauna may be grouped into one of three habitats. The **aerial habitat** consists of the parts of the plants that are almost always above the water. It is similar to a terrestrial habitat and is dominated by grazing insects and spiders that live by chewing leaves or sucking the sap. Many birds feed off these insects as well as crabs, worms, and fish. Wading birds such as egrets and herons, and waterfowl such as the black duck, are residents of the salt marsh. The **benthic habitat** supports the detrital food chain and includes fungi, bacteria, and invertebrates.

The **aquatic habitat** includes some of the benthic organisms, but by convention may be defined to refer mostly to aquatic vertebrates, especially fish. Most salt marsh fish live at least part of their life cycles outside the marsh. Many fish and shellfish spawn upstream or offshore. The juveniles then find food and shelter in the salt marsh, eventually continuing their journey to their adult range. Over 90% of the commercially valuable fish and shellfish harvested off the Gulf and southeastern Atlantic coast of the United States spend part of their life-cycles in the salt marsh. This fact is one of the major utilitarian motives behind wetlands protection.

Freshwater Marshes   Freshwater marshes have a wide variety of types, so it is more difficult to make generalizations that apply to all. They range from tidal freshwater marshes to inland freshwater marshes. The latter include prairie potholes of the northern prairie of North America, the Everglades of Florida, the playas in arid areas of Texas and New Mexico, the Great Lakes marshes, and riverine marshes distributed in most regions. Hydrology is still a major controlling factor. The periodicity of inland marshes tends to be seasonal, in contrast to the tidal cycling of salt marshes.

The vegetation tends to grow in zones from upland to deep water (Figure 15.22). The edge that is periodically flooded is dominated by grasses. Where the water level is at or just above the soil, one can find sedges (*Carex* spp.), rushes (*Juncus* spp.), bulrush (*Juncus* spp.), and the broadleafed monocots, the arrowheads (*Sagittaria* spp.). In deeper water, cattails are common. The broadleafed cattail *Typha latifolia* is less floodresistant than the narrow-leafed cattail *Typha angustifolia*. The latter will grow in depths up to 1 m. Hardstem bulrushes (*Scirpus acutus*) occupies the deepest zone of emergent plants. At greater depth the floating plants are found. These include water lilies (*Nymphaea tuberosa* or *T. Odorata*) and water lotus (*Nelumbo lutea*). Also found at depths that preclude emergent vegetation are the submersed hydrophytes, including coontail (*Ceratophyllum demersum*), water millfoil (*Myriophyllum* spp.), pondweed (*Potamogeton* spp.) and waterweed (*Elodea canadensis*). Interestingly, many temperate aquatic ecosystems develop floating marshes, buoyed by trapped nitrogen and methane and plant material.

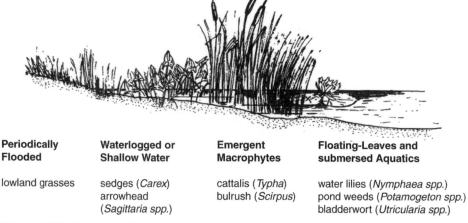

Periodically Flooded	Waterlogged or Shallow Water	Emergent Macrophytes	Floating-Leaves and submersed Aquatics
lowland grasses	sedges (*Carex*) arrowhead (*Sagittaria spp.*)	cattalis (*Typha*) bulrush (*Scirpus*)	water lilies (*Nymphaea spp.*) pond weeds (*Potamogeton spp.*) bladderwort (*Utricularia spp.*)

Figure 15.22   Zonation of vegetation in a freshwater marsh. (From Mitsch and Gosselink, 1993. Used with permission.)

Eichornia crassipes (water hyacinth) is an introduced species that grows very rapidly and has become a nuisance in the southeastern United States.

Phragmites and cattails are among the most productive wetlands plants, having photosynthetic efficiency from 4 to 7%. Measurements of primary production ranges from 1000 to 6000 g/m$^2 \cdot$yr for *Phragmites* and 2,040 to 3,450 g/m$^2 \cdot$yr for *Typha*. However, the pattern differs for these two plants. *Typha* grows fastest early in the growing season, then slows down. *Phragmites* grows at a fairly constant rate throughout the season.

The most obvious invertebrates in the freshwater marsh are members of the insect order Diptera, the true flies. These include the mosquitoes as well as midges and crane flies. Many are herbivores, although many others are benthic in their larval stages. Most of the mammals are also herbivorous. One of the most common mammals is the muskrat (*Ondatra zibethicus*).

Omnivorous and herbivorous waterfowl are plentiful. They nest in northern marshes and migrate between them and southern marshes in the winter. The loon frequents deeper ponds where there may be fish. Some ducks nest in upland locations; those that fish by diving nest over the water. Geese, swans, and canvasback ducks are the major herbivores in the marsh. Wading birds nest in colonies in marshes and fish in shallow waters. Songbirds and swallows usually nest and perch in uplands near the marsh, and fly into the marsh to feed.

Marshes may provide shelter for reproduction of fish from adjacent lakes. Common carp (*Cyprinus carpio*) are able to withstand great fluctuations in water temperature and dissolved oxygen that occurs in shallow marshes. They graze and uproot the vegetation, increasing turbidity in the water.

Freshwater marshes tend to be both nitrogen and phosphorus limited. Vegetation plays a role in nutrient cycling similar to that in forests. They remove nutrients from the soil, only to return them as dead plant material is degraded.

Riparian Wetlands   The riparian zone is the area adjacent to rivers that has a high water table and is regularly flooded by the river. Also called the **floodplain**, they are widely distributed, even in arid areas. They are open systems, with significant exchanges of energy and material with the adjacent upland and riverine ecosystems, as well as upstream and downstream areas. Because of their wide distribution and narrow geometry, they have been particularly vulnerable to draining and filling. Riparian wetlands provide water storage during flood stages of the river. The elimination of this "safety valve" has contributed to downstream damage due to floods.

Western riparian wetlands of the United States tend to be narrow and have steep geomorphology. The uplands tend to be clearly distinguished from the wetlands. Eastern riparian systems include those of the Mississippi valley, the southeastern and northeastern United States. They may have steep morphology along low-order streams near the high-elevation headwater. The higher-order streams at lower elevations tend to have gentler slopes, wider wetlands, and more gradual transition to uplands. A natural levee often forms by sediment deposition between the wetland and the river.

As flooding is intermittent, anoxic conditions in the soil occur only when inundated. The organic matter content is intermediate, about 2 to 5%. Clay deposits and import by flooding results in accumulation of a large quantity of nutrients.

Spatial and temporal scales are linked, as distance from the river (or actually, elevation) determines the frequency of flooding. Figure 15.23 shows an example of this in terms of a classification applied to southwestern U.S. bottomland forest wetlands. The type of vegetation correlates approximately with the zone. For example, among the oaks, overcup oak (*Quercus lyrata*) is typical of zone III, laurel oak (*Quercus laurifolia*) of zone IV, and water oak (*Quercus nigra*) in zone V.

In arid areas, riparian wetlands are easy to distinguish from upland ecosystems, since the latter are usually grasslands or deserts and the former are forested. They tend to form narrow strips along the river, snaking their way through the otherwise dry region. Obviously, they serve as a local oasis to animals and humans. These wetlands usually lack oak but include willow (*Salix* spp.) and cottonwood (*Populus fremontii*), which are also common throughout the United States. Western riparian wetlands also have sycamore (*Platanus wrightii*), ash (*Fraxinus pennsylvanica velutina*), and walnut (*Juglans major*), as well as alder (*Alnus tenuifolia*) at higher elevations.

Riparian wetlands often provide the only woodlands in an area, either because the uplands have been converted to agriculture or because it was that way in the first place, as in arid areas. This increases the importance of riparian wetlands for providing food and habitat, and as corridors for dispersal or migration.

Animal life in riparian wetlands is too diverse to describe here. Most are common to adjacent upland or aquatic systems as well. Some are strongly associated with wetlands, such as the beaver and the cottonmouth snake. In the western United States there are 88 species of birds that are strictly riparian. In southern California there are 45 riparian mammals. Some species of fish spawn in the floodplains during flood stage.

The intermittent nature of riparian flooding augments the ecosystem productivity by providing moisture without causing long periods of anoxic soil conditions, while providing nutrients and flushing wastes away. Typically, riparian ecosystems with regular wet and dry periods have an aboveground net biomass productivity exceeding 1000 $g/m^2 \cdot yr$.

Riparian ecosystems are open systems. Much of their productivity is exported to the adjacent waterway, and they can be sinks for nutrients from upland ecosystems.

15.3.7 Wetland Law and Management

In the United States, federal regulation of wetlands is not governed by any single piece of legislation, as is the case with air and water pollution. Instead, wetlands issues are covered

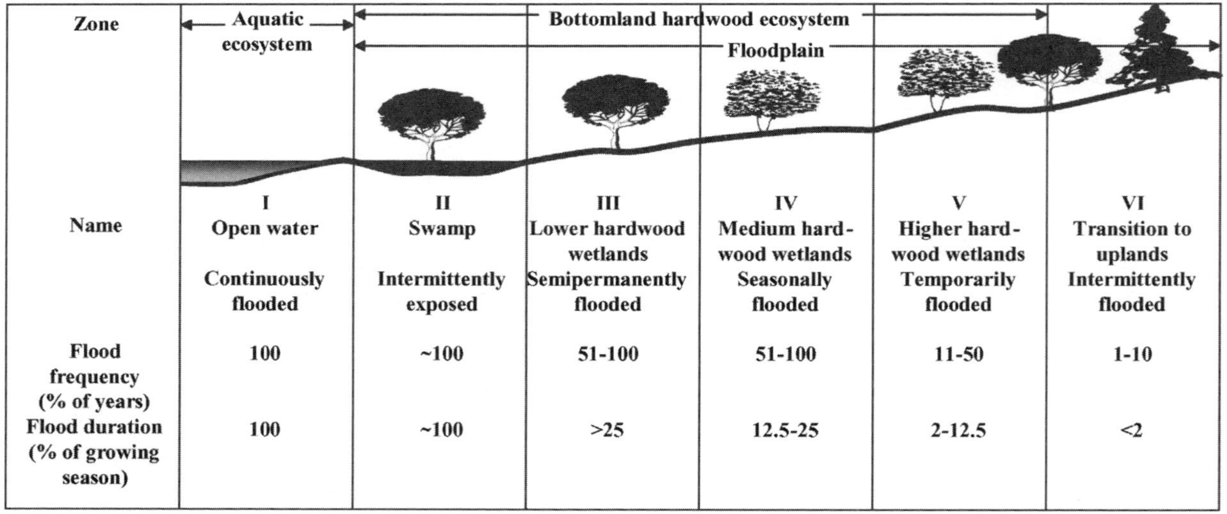

Zone	I	II	III	IV	V	VI
	Aquatic ecosystem	Bottomland hardwood ecosystem — Floodplain				
Name	Open water Continuously flooded	Swamp Intermittently exposed	Lower hardwood wetlands Semipermanently flooded	Medium hard-wood wetlands Seasonally flooded	Higher hard-wood wetlands Temporarily flooded	Transition to uplands Intermittently flooded
Flood frequency (% of years)	100	~100	51-100	51-100	11-50	1-10
Flood duration (% of growing season)	100	~100	>25	12.5-25	2-12.5	<2

Figure 15.23 Zones of a southeastern U.S. riparian wetland. (Based on Mitsch and Gosselink, 1993.)

by a variety of laws related to water pollution and flood control, plus Executive Orders. The major relevant law is Section 404 of the Federal Water Pollution Control Act of 1972 (the "Clean Water Act," PL 92-500). This section, without mentioning wetlands, gave the Army Corps of Engineers authority to regulate dredging and filling in US waters. Court decisions plus another Executive Order expanded this authority to wetlands.

This led to the problem of creating a legal definition of a wetland. **Wetlands identification and delineation** is the process of determining whether a property is a wetland and identifying the boundaries of the wetland. The primary guide for this purpose is the *Corps of Engineers Wetlands Delineation Manual* (U.S. Army Corp of Engineers, 1987). Identification and delineation are best done during the growing season. A site is classified as a wetland only if it possesses indicators of wetland hydrology, hydric soil, and hydrophytic vegetation. Hydrologic indicators include the actual presence of water at or near the soil surface, and stains from flooding on the trunks of trees. Hydric soil indicators include the presence of mottling below the surface.

For the purpose of wetlands identification and delineation, an area possesses hydrophytic vegetation only if the dominant vegetation have one or more of the special adaptations for survival in saturated soil conditions, such as were described in Section 15.3.2. The *Manual* places plants into several categories, based on the estimated probability that they occur in wetlands under natural conditions (Table 15.21). Species that have an indicator status of OBL, FACW, or FAC are considered to possess hydrophytic adaptations.

Exactly what is and isn't a wetland must be prescribed very precisely for the purpose of deciding whether legal restrictions apply. This is a matter of great importance to property owners, because if they cannot develop their property, the property loses much of its economic value. In 1992, the U.S. Supreme Court ruled that identifying a private parcel of land as a wetland could constitute a "taking" of that land for the public good, and that the landowner must therefore be compensated for the economic loss. Although this issue is not resolved entirely, it could have a major impact on the ability of the government to prevent further loss of wetlands.

The policy of "no net loss" has led to the strategy of creating wetlands to compensate for the loss of wetlands due to pressing economic needs, such as highway construction. Wetlands that are created where none existed before are called **constructed wetlands**. Besides their use to compensate for losses, wetlands may also be constructed to create

TABLE 15.21 Plant Indicator Status Categories

Indicator Category	Symbol	Wetlands Probability (%)	Nonwetlands Probability (%)	Examples
Obligate wetland plants	OBL	>99	<1	*Spartina alterniflora, Taxodium distichum*
Facultative wetland plants	FACW	67–99	1–33	*Fraxinus pennsylvanica, Cornus stolonifera*
Facultative plants	FAC	33–99	33–99	*Gleditsia triacanthos, Smilax rotundifolia*
Facultative upland plants	FACU	1–33	67–99	*Quercus rubra, Potentilla arguta*
Obligate upland plants	UPL	<1	>99	*Pinus echinata, Bromus mollis*

Source: U.S. Army Corps of Engineers (1987).

habitats fish and wildlife, for flood control, for nonpoint-source water pollution control, or for wastewater treatment.

Wetlands, whether constructed or natural, purify wastewater by physically entrapping particulate matter in the soil and organic litter, by the action of microorganisms on the plants and in the soil, and by uptake directly by macrophytes, especially of nutrients. Because the soil has both aerobic and anaerobic zones, wetlands can serve as a nitrogen sink by both nitrifying and denitrifying. Wastewater treatment systems tend to use either floating or emergent plants.

15.4   MARINE AND ESTUARINE ECOSYSTEMS

In contrast to the land, the oceans of the world are continuous with each other. Nevertheless, factors such as temperature, salinity, and lighting gradients isolate parts from each other. The **marine** environments are those with salt or brackish water. Included in these are the **estuarine** environments, which are bodies of water, partially enclosed by land, in which fresh and salt water mix. Estuaries communicate with rivers and the ocean. Saltwater marshes are often considered to be part of estuarine ecosystems.

A distinguishing feature of estuaries, besides having salinities intermediate between fresh and ocean waters, is the existence of a salinity gradient in one of several forms. In a **vertically mixed estuary** the salinity varies with distance from the mouth of the river, but not with depth. A **stratified estuary** has salinity that varies both along the axis of flow and with depth, to varying degrees. In the most extreme case, salt water flows upstream along the bottom while brackish water flows seaward along the surface. This affects the transport of nutrients and of planktonic organisms. The **salt wedge estuary** has salinity gradients with both depth and axially.

The oceans can be classified into two zones: the **pelagic** environment is occupied by free-floating organisms; the **benthic** environment is the bottom substrate. The benthic environment includes the **littoral** zone, which is the part between the high-tide and low-tide levels. Notice the difference from the use of this term for lakes, as defined in Section 15.2.1. Also, in oceans the term **abyssal**, which denotes the part of the ocean deeper than the continental shelves, takes the role of the term *profundal* in lakes.

The pelagic zones of the oceanic province can also be distinguished based on light penetration. Light penetration can be approximated as an exponential function with depth:

$$I = I_0 \exp(-\alpha_\lambda z) \tag{15.9}$$

where I is the light intensity at wavelength λ and at depth z, I_0 the intensity at the surface, and α_λ the empirical **extinction coefficient** corresponding to wavelength λ. For example, infrared wavelengths have $\alpha_\lambda = 20\,\mathrm{m^{-1}}$ in clear water. Using equation (15.9) to solve for the depth at which the intensity is 1% of the surface yields $z_{01} = 0.23$ m. Blue light is weakly absorbed, with $\alpha_\lambda = 0.004\,\mathrm{m^{-1}}$ and a resulting $z_{01} = 1150$ m. Red light is absorbed intermediately, with $\alpha_\lambda = 0.4\,\mathrm{m^{-1}}$ and a resulting $z_{01} = 11.5$ m. The total light energy versus depth will thus be a complex sum of exponentials for the various wavelengths (or, more accurately, an integral over wavelength). Overall, the extinction of total light energy reaches 1% in water as deep as 110 m in clear tropical seawater, or as shallow as only several meters in coastal waters.

Overall, photosynthesis occurs in the top 100 m of the ocean, a portion of the epipelagic zone called the **photic zone**. Below about 500 to 1000 m is the **aphotic zone**, in which essentially no light exists. In between is the "twilight" or **disphotic zone**.

15.4.1 Productivity and Nutrients

Primary production in the open ocean is thought to be due mostly to phytoplankton. Some 2 to 10% may be due to seaweeds, and less than 1% to chemoautotrophs. For all the oceans as a whole, the gross productivity is estimated to be 50 to 100 g $C/m^2 \cdot yr$. Due to surface scattering and light extinction, the efficiency of capturing incident sunlight is only about 0.1 to 0.2%.

The phytoplankton are dominated by diatoms, and secondarily by dinoflagellates. Although in the past most of the primary productivity of the oceans was attributed to the larger phytoplankton, evidence is growing that as much as 70% of all the photosynthetic activity in the oceans is due to smaller forms. These range from 2 to 20 μm in sizes and are termed **nanoplankton**. They include the small single-celled plants called *coccolithophores*. These are covered with plates of calcium carbonate. They can form deep beds of ooze on the ocean floor, which fossilize into chalk formations such as the White Cliffs of Dover in England. Another source of productivity is the phototrophic bacteria, such as cyanobacter (not necessarily nitrogen-fixing forms, however). Their abundance is of the order of 100,000 mL^{-1}. They are from 0.2 to 2 μm in size and are termed **picoplankton**.

Productivity has an interesting relationship with depth in the epipelagic zone. Typically, the photosynthesis rate, and the concentration of chlorophyll, peaks some tens of meters below the surface (Figure 15.24). Several possible reasons are given for the surprising observation that photosynthesis is not a maximum at the surface. One is that, ironically, the light intensity may be inhibitory, causing bleaching of phytoplankton. It may also be due to the effects of grazing by zooplankton. Mathematical modeling of phytoplankton growth dynamics suggests that nutrient limitations near the surface are responsible. Nutrients continue to increase with depth up to on the order of 1000 m, and then decrease modestly to the bottom. Oxygen reaches a minimum at similar depths, and then increases steadily toward the bottom.

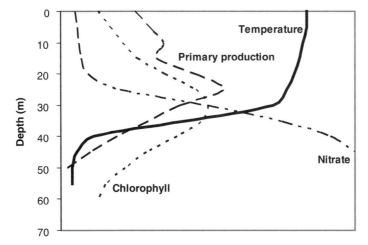

Figure 15.24 Typical tropical structure of the water column. (Based on Mann and Lazier, 1991.)

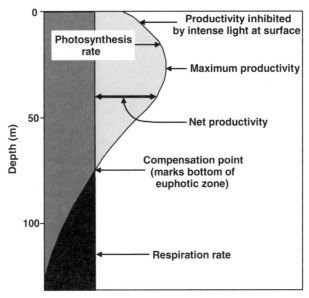

Figure 15.25 Productivity and respiration vs. depth, as would be measured using the light bottle/ dark bottle method. (Based on Garrison, 1993.)

If respiration is fairly uniform with depth, a point will occur where the rate of primary production becomes equal to the respiration rate, yielding a net primary production of zero. This depth is called the **compensation depth**. It tends to occur approximately at the depth where the light intensity has dropped to 1% of the surface intensity. This depth defines the bottom of the euphotic zone. Below the compensation depth respiration exceeds photosynthesis, and oxygen levels must be maintained by transport from surface waters by mixing and advection. Figure 15.25 shows these relationships, and Figure 15.26 shows how they can be measured.

There are some interesting seasonal and geographical patterns in productivity as well (Figure 15.27). One might expect that among open ocean waters, tropical oceans would have the highest productivity. However, nutrients are always being lost to the euphotic zone by the sedimentation of phytoplankton and zooplankton fecal pellets, and a permanent thermocline prevents the nutrients from being mixed back upward. At 20°N latitude the nitrate and phosphate concentrations are about 1% of the levels typical of temperate oceans during the winter. In the tropics the nutrients lie mostly below 150 m depth, and their concentrations peak at a depth of 500 to 1000 m. Productivity in most tropical waters is less than 30 g C/m^2 per year. Exceptions to this situation are upwelling zones and reefs, as noted below.

Polar zones, on the other hand, compensate for their low temperatures and low insolation by a lack of stratification. Thus, nutrients mix freely at the surface, except when snowmelt forms a shallow layer of relatively fresh water. The shallow stratified layer traps phytoplankton close to nutrients at the **pycnocline** (location of high vertical density gradient), producing a very intense, though short-lived algal bloom. In the Antarctic waters a unique circulation pattern produces upwelling of nutrients that stimulate productivity. However, even with these factors, the polar region productivity as a whole averages less than 25 g C/m^2 · yr.

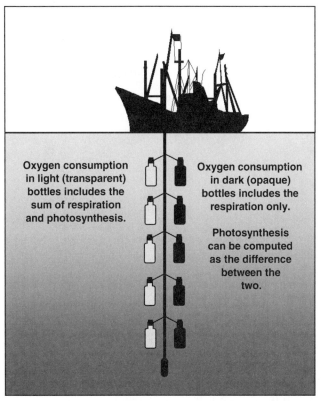

Figure 15.26 Light bottle/dark bottle method of measuring primary and net productivity vs. depth. At the compensation point the dissolved oxygen (DO) concentration in the light bottle will not change with time. Above that point, photosynthesis exceeds respiration and DO will increase. At greater depth the DO will decrease with time. (Based on Garrison, 1993.)

In temperate and subpolar oceans, on the other hand, productivity is at a higher average level, due to more dependable insolation and the nutrient replenishment that comes with seasonal stratification. Productivity peaks with a spring algal bloom, but it is moderately high yearround. Typical productivity is around 120 g $C/m^2 \cdot$ yr, but can be as high as 250 g $C/m^2 \cdot$ yr. Thus, these regions provide most of the productivity of the world's oceans.

In all climates, productivity as high as 1000 g $C/m^2 \cdot$ yr can be maintained where hydrodynamic factors bring nutrients to the surface. Places where this occurs are called **upwelling zones**. These are caused by wind-driven currents combined with the Coriolis effect. The **Coriolis effect** is the tendency caused by Earth's rotation for flows to be diverted to the right in the northern hemisphere and to the left in the southern. Thus, hemisphere, the trade winds around the equator cause a westerly current in the Pacific and Atlantic oceans. The Coriolis effect produces a divergence of flow, as the northern edge diverts toward the north and the southern toward the south. The net effect is that surface water flows away from the equator, and deeper waters are brought up to replace it. This phenomenon is called **equatorial upwelling**. The regions of high productivity that result from the upwelling nutrients can be seen in Figure 15.27. A similar divergence zone encircles the Antarctic, where ocean currents flow in opposite directions, due to

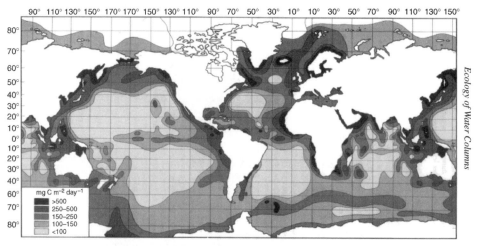

Figure 15.27   Global distribution of primary productivity in the world's oceans. (From Barnes and Mann, 1991. Used with permission.)

trade winds. The resulting enhancement to productivity partially compensates for the otherwise low productivity of the polar oceans.

Another important mechanism is **coastal upwelling**, in which winds parallel to the coast, in combination with the Coriolis effect, produce an offshore current. This brings the deeper waters to the surface with their nutrients. This occurs throughout the world but is particularly massive in four locations: the coasts of Peru, California and Oregon, northern Africa, and southwestern Africa. These upwelling zones form important food chains, with humans at the top. When climatic conditions interfere with their productivity, the human economy and food supply is severely affected. The most important example of this concerns the Peru upwelling, caused by a steady offshore wind at the equator, and is the basis of a huge commercial anchovy fishery.

Upwelling occurs on a smaller scale at the mouths of estuaries when relatively fresh water entrains deeper salt water. The mixture is still less dense than seawater and floats on the surface, carrying nutrient both from the river discharge and from deep entrained seawater.

Every two to 10 years, the winds responsible for the Peru upwelling reverse, in what is called the **southern oscillation**. Instead of the upwelling of cold, nutrient-rich water, warm, nutrient-depleted current appears, and the thermocline becomes deeper. The warm current is called the **El Niño**, and it creates havoc with the economy of Peru, as fishermen lose their anchovy harvest. (El Niño is also associated with climatic changes in regions far from Peru, such as in reducing the number of hurricanes in the Caribbean.) The phenomenon is referred to as ENSO, for El Niño–Southern Oscillation.

Phytoplankton extract carbon, nitrogen, and phosphorus from seawater in the ratio $116:16:1$, respectively. Carbon, of course, is readily available as carbonate. Nitrogen is more often limiting in marine environments than in freshwater environments, because nitrogen-fixing cyanobacter (blue-green bacteria) are less abundant in salt water. For this reason, and because phosphorus is recycled faster than nitrogen, phosphorus is less important than nitrogen as a limiting nutrient. Silicon can be a limiting nutrient for diatoms.

TABLE 15.22   Nutrient Enrichment Experiment Using Sargasso Sea Water

Nutrient Supplementation	Relative Uptake of ^{14}C Compared to Control (%)
Control (no supplementation)	100
N + P + metals, including iron	1290
N + P	110
N + P + metals except iron	108
N + P + iron	1200

Source: Krebs (1994); original source D. W. Menzel and J. H. Ryther, 1961, *Deep Sea Res.*, Vol. 7, pp. 275–281.

Since, as noted previously, silicon cell walls require less energy to form than those made of cellulose, diatoms are relatively more efficient, and therefore more productive, if silicon is sufficiently available.

Iron is often the limiting nutrient in the open ocean. As shown in laboratory results in Table 15.22, iron supplementation of water taken from the Sargasso Sea produced tenfold increases in productivity when combined with other nutrients. Iron alone produced a short-term increase in productivity, after which nitrogen and phosphorus limitations came into play.

A large-scale iron supplementation field experiment produced algal blooms in the Southern Ocean around Antarctica. The experiment was proposed to test a technology to increase the ocean's uptake of carbon dioxide, as a way to ameliorate anthropogenic CO_2 emissions that could lead to global warming. Since iron is a micronutrient, relatively small amounts would be necessary to eliminate it as a limiting factor. It could easily be distributed over a wide area. Although the field results support the technical feasibility of this idea, many other objections remain. These include the possibility of unintended ecological effects that are likely to result from such a large-scale manipulation of the environment.

15.4.2   Marine Adaptations

As with wetland animals (see Section 15.3.3), simpler animals tend to be osmoconformers. This is the case with most invertebrates, such as worms, mussels, and octopi. Their bodily fluids are isotonic with seawater. Therefore, they do not require special mechanisms to control their internal osmolarity.

Marine fish are hypotonic: Their body fluids are about one-third as saline as seawater. Without active mechanisms to counter it, their bodies would lose water by osmosis and absorb salt. The mechanism is the drinking of large volumes of water, coupled with the excretion of salt by special "chloride cells" in their gills and by the discharge of a highly concentrated urine. It is because humans do not have these adaptations that makes it dangerous for us to drink seawater.

Viscosity becomes an important factor for small organisms. **Viscosity** is a measure of the force required to create shear, or velocity gradient, in a fluid. Motion of differing objects can be compared in terms of the **Reynolds number**, which is the ratio of momentum to viscosity. If a human propels herself through the water with a few strokes of the arms, she can expect to coast a short distance because her momentum is great compared to the viscosity of the water. A microscopic zooplankter, on the other hand, stops within milliseconds if it stops propelling itself.

Salt water is more viscous than fresh water. Seawater is about 25% more viscous at 0°C than at 20°C. Figure 15.28 shows how two species of one genus react to this difference. The warm-water form has more elaborate appendages to slow its sedimentation rate in the less viscous water.

Many marine animals control their position not only by swimming but by buoyancy. Several mollusks have rigid air containers. These include the genus *Nautilus*, the cuttlefish *Sepia*, and the squid *Spirula*. Their air chambers just balance the buoyant weight of the rest of their bodies, making them neutrally buoyant. Some of the boney fish have a **swim bladder** to store air. The bladder is filled either by extracting gases from the blood or by direct ingestion through the esophagus. Active swimmers, such as the tuna, or fish that live on the bottom do not have swim bladders.

15.4.3 Marine Communities

Several marine communities are important or interesting enough to merit special attention.

Estuaries. Estuaries have high productivity because they receive a constant nutrient input from rivers, and tidal motions provide an energy subsidy that improves oxygen transfer

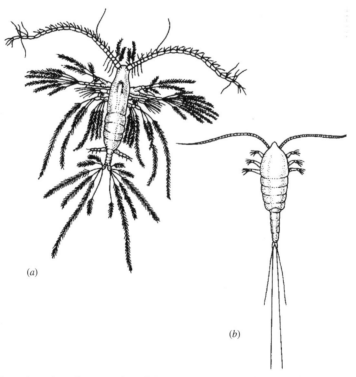

Figure 15.28 Adaptation of two species of the copepod genus *Oithona* to viscosity differences due to temperature: (*a*) warm-water species; (*b*) cold-water species. (From Garrison, 1993. © Wadsworth Publishing Co. Used with permission.)

and mass transfer of nutrients from the sediment. They receive detritus inputs from adjacent salt marshes. The organisms that populate estuaries are similar to those already described for salt marshes.

Rocky Intertidal Communities. This group constitute one type of littoral zone; others include the sand beach and the salt marsh. The wave energy that strikes these ecosystems would seem to make it an inhospitable place. Besides the crushing wave energy, they are subjected to sharp swings in moisture and temperature. In fact, they are highly productive ecosystems and have one of the highest species richness and diversities of any marine ecosystem. For this reason, and also because they are among the most accessible of the high-richness ecosystems, they are among the most interesting. There are few other environments where the casual observer could find so many different animals in one place.

Several reasons explain why. The wave energy provides mixing, gas transfer, and flushing of wastes. It also eliminates competition from species not specially adapted for this environment. The tides add to the energy subsidy. Detritus is brought in by the tides. Tides naturally create a range of habitats, classified by the fraction of the time that each is exposed. Areas near the low-tide mark are submerged almost all the time, high-tide areas only briefly. Since the intertidal zone spends time both above and below the water, organisms there are subjected to attack by predators from the land and the sea. Heavy predation is one of the factors that contributes directly to species diversity. Finally, the rocky intertidal environment itself provides many diverse habitats and niches: tide pools, crevices, exposed rock, and gravel beds.

Algae encrust rocks or grow in the form of large, tough, elastic, and slippery plants, such as kelp, to resist wave energy. Sessile filter-feeding animals such as barnacles and mussels attach themselves to rocks. Snails scrape algae from the rocks. Clams, octopi, starfish, crabs, sea urchins, anemones, sponges, shrimp, fish, and shorebirds are all present.

Coral Reefs. Coral reefs are unique and important communities formed from colonial types of coral. Corals are cnidarians with a polyp body plan that sits in a cuplike exoskeleton made of calcium carbonate. The opening at the top of the polyp has tentacles with stinging cells for capturing prey. The outer layer of the coral tissue has embedded within it thousands of dinoflagellates called **zooxanthellae**. There may be as many as 30,000 cells per milliliter of coral tissue, up to 75% by mass, and they provide the bright and varied colors for which coral reefs are famous. The coral and the zooxanthellae exist in a symbiotic relationship in which the coral provide shelter and nutrients and the algae provide food and removal of the coral's waste products (which are nutrients to them). The zooxanthellae also seem to give the coral its ability to secrete large amounts of calcium carbonate. The coral–algae association seems a most intimate form of nutrient recycling, which must be essential since coral tends to grow in nutrient-depleted waters.

Coral grows on a solid substrate. When the coral dies, the exoskeleton remains, and new organisms can grow on top of this. Over many hundreds and thousands of years the skeletons accumulate, forming reef structures. Coral can only grow in warm, shallow waters. If the land subsides or sea level increases (as has been happening over the last dozen millennia or so), the reef grows to maintain its position relative to the surface, forming offshore barrier reefs or, where a central island has become submerged, a ring-shaped atoll with a lagoon. Reef-forming coral grows almost exclusively in waters with a seasonal temperature minimum above 20°C.

The massive reef structure creates numerous ecological niches. The zooxanthellae, plus filamentous green algae and other algae, form the base of the coral reef food pyramid (Figure 15.29). The corals themselves are primary consumers, as well as fish, clams, sea urchins, crustaceans, and other invertebrates. Secondary consumers include sea urchins, sea anemones, sea stars, and fish. The top carnivores are eels, octopi, and barracuda. The animal diversity of coral reefs is the highest of all marine communities. Despite their high productivity, coral reefs do not support large fisheries. It is thought that high predation rates limit production at each level of the food chain. The overall effect is a low transfer efficiency.

Coral reefs will grow at depths no greater than 150 m, where light levels in the clear water they require falls below 4% of the surface levels. Too high a nutrient level hurts coral because other benthic plants and suspension feeders such as clams will outcompete them. Higher nutrient levels also brings increased plankton concentrations and resulting turbidity, which also inhibits coral growth.

Abyssalpelagic and Abyssal Benthos. With the exception of hot vents described below, the deep-sea bottom is sparsely populated. While the nearshore sediments may have 5000 g of biomass per square meter and the continental shelf could have 200, the abyssal

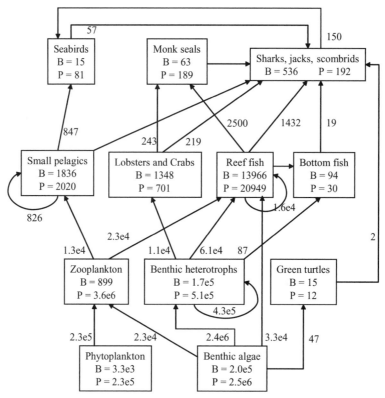

Figure 15.29   Food web of a coral reef with a biomass budget. *B* is average annual biomass in kg/ km^2; *P* is the production in kg/m$^2 \cdot$ yr. (Based on Barnes and Mann, 1991; original source is R. W., Grigg, J. J. Polovina, and M. J. Atkinson, 1984, *Coral Reefs*, Vol. 3, pp. 32–37).

benthos typically has less than 1 mg. Furthermore, the deeper it is, the less biomass there will be. This is because it is entirely a detritus-based food web, the energy input comes from sedimentation from the euphotic zone, and this input has increasing chances of being intercepted by pelagic organisms on its way down.

The environment is permanently dark, cold, and currents are weak. Adaptations include low rates of metabolism and growth, and enzymes that are specially optimized to function at the high pressures found on the ocean floor. Most organisms are blind, but some, such as the lanternfish, use **bioluminescence** (the biological production of light) to find and lure prey. All the animals are predators or scavengers. There are no plants, so there are no herbivores.

Mesopelagic fish have swim bladders, which makes them difficult to harvest without rupturing them and killing the fish. Fish from below 1000 m depth typically lack swim bladders. They and other organisms from this depth can survive in aquaria at the surface if the temperature and other physical factors are favorable. Some fish, such as the lanternfish, migrate diurnally to the euphotic zone at night to feed, then descend to between 700 and 900 m during the day. More than 2000 species of animals live in the aphotic zone, including copepods, ostracods, jellyfish, prawns, mysids, amphipods, swimming worms, and a group of strange-looking fish. The fish are typically small but have huge mouths. Since prey are scarce, they need not to have to reject larger victims.

Feeding strategies of deep-sea creatures are often bizarre. Some bury themselves in the sediment with their mouths open at the surface. Other creatures mistake them for caves and crawl in for shelter, only to be forced to crawl right into the stomach by downward-projecting spines. Others can smell dead organisms from kilometers away, then spend weeks crawling to the source of the scent. Some organisms feed less than once per year and live for hundreds of years.

Surprisingly, species diversity at the bottom of the sea is very high. This may be explained by the principle of competitive exclusion, which states that high predation pressure prevents any single species from dominating.

Hot Vent Communities. In 1977, scientists in the submersible *Alvin* discovered a new ecosystem northeast of the Galápagos Islands while searching for a source of heated water at a depth of 3000 m. There, at the rift where two of the tectonic plates that form Earth's crust are spreading apart, hydrothermal vents were spewing water at 350°C, laden with dissolved minerals, including H_2S.

But the astonishing thing about the vents were the abundance of benthic invertebrates surrounding them, many of them huge in size and previously unknown to science. Besides large crabs, clams, shrimp, and anemones, there were huge "tube worms," contained in parchmentlike tubes the diameter of a human arm and about 3 to 4 m. Three species have been found so far, forming a new genus called *Riftia*, in the phylum *Pogonophora*. Altogether, about 100 species live in vent communities.

The vent community had no plants and was too dense to subsist on detritus from above. What was the source of the energy to maintain it? Furthermore, the worms had no mouth, digestive tract, or anus. Instead, the worms and some clams contained "feeding bodies" packed with chemoautotrophic bacteria. The worms would absorb hydrogen sulfide from the water and transport it to the feeding bodies. These would then produce the carbohydrates to sustain the ecosystem. Thus, this unique ecosystem is based not on the energy from sunlight, as is every other ecosystem on Earth, but on geochemical energy from deep within the Earth itself.

Hot vent communities have been discovered in seas throughout the world, including sites on the mid-Atlantic ridge, the Gulf of California, offshore from the state of Washington, and near Okinawa. Similar communities have been found where hydrogen sulfide–rich water seeps from the base of the continental shelf off Florida, Oregon, and Japan. These communities also have tube worms. In a third type of deep ocean chemoautotrophic ecosystem, sites were found in the Gulf of Mexico and Baffin Bay, where ecosystems subsist on hydrogen sulfide and methane seeping from natural hydrocarbon deposits.

15.4.4   Adverse Impacts on Marine Ecosystems

Many years ago, the vastness of the oceans made it easy to assume that nothing that humans could do to it could have any long-term or widespread effect. Today, however, our species has become so successful that nothing in the biosphere can be held to be insensitive to our activities.

Overfishing. Improvement of fishing technology led to overfishing in many marine fisheries. There are many instances of this. Peru led the world in its catch with 4 million metric tons in 1960. In 10 years it increased the harvest to 13 million metric tons. But this high level, combined with a series of El Niño events, devastated the industry until it was reduced to 100,000 tons in 1985. Cod and herring are seriously depleted in the North Atlantic. The Pacific sardine has been reduced to **commercial extinction**, in which it is no longer profitable to harvest it. Of 11 species of large whales that have been hunted commercially, eight are now commercially extinct.

When a fishery becomes overfished, instead of prudently reducing its take, fishermen instead increase their efforts by adding more boats and more efficient methods. One of these methods is called **drift net fishing**, in which a 7-m-wide net up to 80 km long is deployed vertically overnight, then hauled in the next morning. This method ensnares turtles, birds, and marine mammals that are not intended for harvesting. Furthermore, about 1600 km of these nets are lost each season, becoming "ghost nets" that drift and kill for decades.

Oil Pollution. Oil pollution in the ocean is sometimes very dramatic. However, spills due to accidents are a small part of the total discharge: only 13% in 1985. A significant fraction, about 8%, is of natural origin. A major source (about one-third) is due to routine flushing of oil tankers when loading and unloading. Another third comes from use and disposal on land via rivers and streams. Once in the water, fractions of crude oil may disperse by evaporating, dissolving, forming emulsions, or settling to the bottom. Floating oil slicks may directly harm wildlife at the surface, especially birds, and can devastate benthic organisms of the littoral zone. These effects are most severe and may persist for many years. The dispersed fractions are biodegraded or assimilated by zooplankton and higher organisms, thus entering the food chain. Compounds absorbed by organisms have toxic effects at all trophic levels. Also, they may affect the commercial use of the resource by imparting an unpleasant taste to fish, shellfish, and so on. Spills of refined hydrocarbons can be more damaging than crude oil, since it contains lighter fractions and additives that are more easily taken up by organisms and are often more toxic.

Water Pollution. Coastal areas are affected by discharge of water pollutants from land sources via rivers, direct pipeline disposal, or dumping by barges. Barge dumping was practiced by communities in the New York Metropolitan Area up to the early 1990s. Originally, sewage sludge dumping was confined to a 10-km-square site some 20 km from the coast of New Jersey and Long Island. When it became known that sludge was accumulating at that site, an interim site was chosen in deeper waters at a distance of 196 km from New York Harbor. The new site provided enhanced dilution. Ultimately, ocean dumping was banned altogether in favor of land-based alternatives.

When these discharges are used as an alternative to treatment, they discharge toxic materials and nutrients to the marine environment. The nutrients can cause algal blooms in excess of what might occur naturally. Sometimes, the algae themselves are toxic to marine life. Other times, the algae have been responsible for fish or benthos kills due to deoxygenation, when the algae decompose. Discharge of sewage and sludge have also been blamed for diseases of finfish, including "black gill" and "fin rot."

Severe cases of **hypoxia** (low-dissolved-oxygen conditions) have been observed in the New York Bight and in the Gulf of Mexico. The New York Bight is the continental shelf area into which the Hudson River discharges. In one incident in the late 1980s, dissolved oxygen levels dropped to low levels over a 300 × 100 km area in the Bight, falling all the way to zero in a portion of the region off Atlantic City, New Jersey. A similar incident occurred in a 420-km band along Louisiana in the Gulf of Mexico. Even in the absence of such catastrophes, continuing sanitary pollution causes public health risks to swimmers in the ocean and eliminates many shellfish beds from use as a resource due to contamination by pathogens.

Coral Diseases. In 1983, scientists discovered a disturbing disease of coral called **coral bleaching**. For unknown reasons, the coral will expel its zooxanthellae, leaving the coral a pale color. Without their symbionts, coral cannot deposit calcium carbonate and are subject to erosion. If it does not regain its zooxanthellae, the coral dies.

This was first noticed in the Pacific but has since been found in Caribbean coral reefs as well. The cause of this disease is still unknown and is being sought with some urgency by marine biologists. Some blame a combination of pollution and high temperatures. The original observation occurred during a particularly severe El Niño event in the eastern Pacific, when sea surface temperatures averaged 30 to 31°C for 5 to 6 months.

Coral are vulnerable to other problems as well. The crown-of-thorns sea star *Acanthaster planci* is destroying coral reefs in the western Pacific. A number of other diseases are found on coral, including **white-band disease**. Anecdotal evidence points to many of these problems as having proliferated recently. Thus, human activities are being blamed, although there is no strong evidence for it.

Thermal Pollution. Industries, especially steam-powered electrical generating plants, often discharge large amounts of heated water to adjacent surface waters. At low levels these may provide an energy subsidy to the ecosystem, encouraging growth. The benefits may vary by organism, resulting in population shifts. Detrimental effects include stimulation of fungal organisms, which may cause disease for aquatic organisms. Cyanobacter have been found to dominate at temperatures above 35°C, green algae between 28 and 35°C, and diatoms at lower temperatures. Oxygen consumption rates increase with temperature, but oxygen solubility decreases, both effects increasing oxygen depletion. Stress may occur if the heated discharge is suddenly halted due to operating considerations of the source.

The sensitivity of animals to temperature may be measured by the **upper lethal temperature (ULT)**, the temperature at which 50% mortality occurs over long-term exposure. The ULT for the fishes *Salmo trutta* and *Salmo gairdneri* was measured at 25 and 24°C, respectively. Others, including the fish *Carassius auratus*, *Cyprinus carpio*, and *Lepomis macrochirus* and the invertebrates *Gammarus fasciatus*, *Procladius*, and *Cryptochironomus* had ULT between 30 and 35°C. At lower temperatures, growth and reproduction will be inhibited.

Use of large volumes of surface water for cooling can have a direct effect as the water is passed through pumps and heat exchangers. Shear and pressure forces can kill or damage organisms. Discharge of water from below the thermocline in reservoirs can produce a low-termperature anomaly. These usually have nonlethal effects, such as population shifts or inhibition of spawning of fish.

15.5   MICROBIAL ECOLOGY

The study of microorganisms too often treats each species as a pure strain. However, in the environment this is rarely the case. Many of the important behaviors depend on the interactions among microbial populations and between them and other organisms. Here we describe some of these interactions.

Cooperation is the positive interaction among organisms within a single population. Bacteria cooperate in numerous ways. Bacteria form flocs, biofilms, and other kinds of aggregates for the purpose of degrading solid substrates, whether lignin, cellulose, or rocks. The aggregates can benefit more efficiently from the investment of extracelllular products than a single individual can. In laboratory-batch microorganism growth experiments, a lag phase of growth is often missing when a large initial concentration is used. It may be that at a high initial density, the cells share growth factors with each other. A certain minimum number of microorganisms, called the **infective dose**, is required to transmit a disease to a host. The bacteria in this case cooperate to overwhelm the host's defenses.

Individuals also *compete* within a population. They certainly use the same resources. In the case of photoautotrophic microbes, there is competition for light.

Microorganisms exhibit most of the types of two-population interactions shown in Table 14.5. *Neutralism* (0/0) is uncommon among microorganisms if there is any overlap in their niches. *Commensalism* (+/0) is common. One microbial population may solubilize rock minerals, and other microbes benefit from the nutrients. One microbe may manufacture growth factors that others require. Biotin is a growth factor that is very important in marine habitats. Methanogenesis may result from commensal relationships. *Desulfovibrio* produces acetate and hydrogen by fermentation, which is then used by *Methanobacterium* to reduce CO_2 to methane. Several of the microbially mediated pathways in biogeochemical cycling involve one population using the products of another (e.g., *Nitrosomonas* and *Nitrobacter*).

Cometabolism is, by some definitions, commensal. *Mycobacterium vaccae* can cometabolize cyclohexane if it has propane available for its own benefit. The cyclohexane is converted to cyclohexanone, which can be used by other bacteria. The *Mycobacterium* does not assimilate the cyclohexanone itself.

One population may utilize a substance that is toxic to another population. An example is the oxidation of H_2S. Some marine bacteria produce organics that chelate heavy metals, reducing their bioavailability.

Synergism (+/+) refers to a loose two-way benefit, in which both populations can exist separately, but benefit when together. Sometimes two populations together can produce metabolic products that neither can produce by themselves. For example, neither *Sterptococcus faecalis* nor *Escherichia coli* can convert arginine to putrescine, but together, they can. Cyclohexane can be degraded by *Nocardia* and *Pseudomonas* together, but not separately. The reason is that *Nocardia* needs biotin and other growth factors produced by *Pseudomonas* to do the job. Together, *Arthrobacter* and *Streptomyces* can degrade and grow on the organophosphate pesticide diazinon, but not separately. *Pseudomonas stutzeri* combines with *Pseudomonas aeruginosa* to mineralize parathion. One form of synergism is called **syntrophy** or **cross-feeding**. In this situation one species provides a nutritional requirement for another, and vice-versa. For example, *Enterococcus faecalis* requires folic acid for growth, and *Lactobacillus arabinosus* requires phenylalanine. Together, they can grow in a medium that contains neither, because *E. faecalis* produces pheynylalanine and *L arabinosus* produces folic acid.

Mutualism, or *symbiosis* (+/+), is an obligatory relationship between two organisms that enables them to occupy a habitat that they otherwise could not. Lichens, an association between a fungus and a photoautotroph, are the best known example of symbiosis. Some protozoans form mutualistic associations with algae. For example, *Paramecium* can contain many cells of the alga *Chlorella* within its protoplasm. Under conditions of stress, the protozoan may digest its algae. *Paramecium aurelia* can also harbor a bacterium of the *Rickettsia* group. These paramecia have an advantage over a different strain from the same species that does not contain the bacteria. Apparently, the *Rickettsia* manufacture a toxin used to inhibit neighbors.

Some methanogenic cultures, once thought to be pure, have been found to consist of mutualistic associations. *Methanobacterium omelianskii* is associated with another strain, called "S." Methane formation involves an electron transfer reaction that actually starts in the S organism and ends in the *Methanobacterium*. It is thought that some other methanogenic mechanism may involve interactions among three organisms. Methanogens also participate in a mutualistic relationship with eubacteria. The methanogens require simple substrates such as acetate, CO_2, formate, and methanol. These are waste products for the eubacterial degradation of more complex organics. Thus eubacter benefits by having their wastes removed, and the archaean methanogens benefit by having substrates provided.

Bacteria can also enter into symbiotic relationships with viruses. *Corynebacterium diphtheriae* harbors a virus that enables it to produce a toxin and to infect host organisms, causing the disease diptheria. Without the virus, it cannot do either.

Competition (−/−) is also uncommon because of the principle of competitive exclusion. Unless competition is weak, two species cannot occupy the same niche. However, if each species has some distinguishing feature, they can compete in overlapping portions of their niches. The classic example of these behaviors is laboratory work by Gause (Figure 15.30). This shows two species of *Paramecium* that although they do not attack each other or secrete toxins, compete to the degree that one species is eliminated after 16 days. In a separate experiment, *P. caudatum* and *P. bursaria* were found to be able to coexist because they tended to occupy different areas of the laboratory flask.

Nutrient limitations can also produce coexistence. When the diatoms *Asterionella formosa* and *Cyclotella meneghiniana* were cultured together, the former would be limited by silica and the latter by phosphate.

Some organisms are relatively efficient at low substrate concentrations but are outcompeted at high nutrient levels. This causes the microbial populations in raw sewage to shift

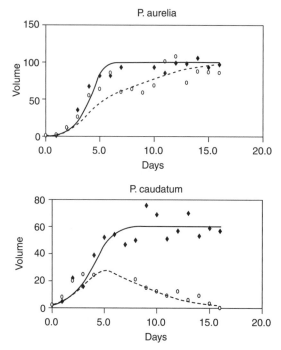

Figure 15.30   Competition between *P. aurelia* and *P. caudatum*. Solid diamonds show growth of each paramecium species in the absence of the other. Open circles show growth in mixed culture. The curves are for equation (14.28) fitted to the data by eye. For *P.aurelia*, $r = 1.5\,day^{-1}$, $K = 100$ volume units, and $b = 1.5$. For *P.caudatum*, $r = 1.0\,day^{-1}$, $K = 60$, and $b = 0.8$. (Original data from Gause, 1934.)

to the typical populations found in rivers and streams as substrate is consumed. This also explains why some organisms that cause filamentous bulking in the activated sludge process, such as *Sphaerotilus natans*, are favored by low-loading conditions. Figure 15.31 shows how this could occur by comparing the growth rate, r_g, as a function of substrate concentration, S, for two hypothetical microorganisms. Species A has a higher maximum growth rate, μ_m; species B has a lower Monod coefficient, K_S. As a result, species B has the higher growth rate at low substrate concentration (to the left of the dashed line).

Among bacteria, *amensalism* (*antagonism*) (0/−), usually comes in the form of allelopathy, the production of chemical inhibitors. In some cases it takes the form of acid production leading to pH changes, many produce toxic short-chain fatty acids, and some produce complex inhibitors that we exploit as antibiotics. Inhibitors allow the first population to gain a foothold in an environment to exclude newcomers.

Thiobacillus oxidans oxidizes sulfur, producing sulfuric acid, which lowers the pH of mine drainage to between 1 and 2, inhibiting most other microbes. Microorganisms on the skin produce fatty acids that are believed to prevent colonization by yeasts and other microbes. The fatty acids produced during methanogenesis are not just intermediates in the reaction but inhibit microbes that otherwise would disrupt the electron transfer between *Methanobacterium* and S.

Antibiotics are substances that kill or inhibit at very low concentration. Their role in natural habitats is unclear. They do not seem to accumulate to effective levels under natural conditions.

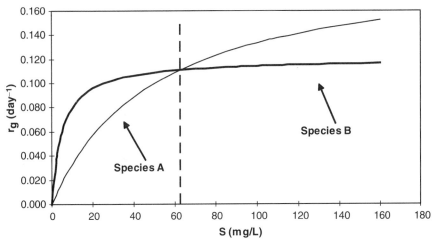

Figure 15.31   Effect of substrate concentration on competition. Maximum growth rate, μ_m, for species A and B are 0.2 and 0.12 day^{-1}, respectively, and K_S values are 50 and 5.0 mg/L, respectively.

Bacteria, fungi, algae, and protozoans all have their *parasites*. Many of the parasites are viruses, which are obligate parasites. *Vibrio cholerae* is thought to be eliminated from river water following fecal contamination by the action of viral parasites. *Bdellovibrio* is a bacterium that is parasitic on bacteria. It attaches to larger gram-negative bacteria, causing cell lysis. *E. coli* has been found to be partially protected from *Bdellovibrio* by the presence of clay particles. Other bacteria produce extracellular enzymes that can lyse bacteria, algae, or protozoans, making their cell contents available for uptake.

It can sometimes be difficult to distinguish *predation* from parasitism among microbes. The distinction can be made if predation is limited to cases where the prey is actually ingested by the predator. Some protozoans prey on other protozoans. *Didinium* preys on *Paramecium*. Many protozoans prey on bacteria. When bacteria are under strong predation pressure, they may change their growth habits: for example, growing in an attached form instead of dispersed. Again, clay particles have been observed to reduce the vulnerability of some bacteria.

Even algae and fungi can prey on other organisms. Some dinoflagellates consume bacteria, other algae, and other dinoflagellates. Slime molds such as *Dictyostelium* prey on bacteria. The **nematode-trapping fungi** capture nematodes by adhesives or by "lassoing" them with constricting rings. When it attempts to swim through a ring, the ring swells, trapping the worm. Despite thrashing motions, the nematode cannot free itself. Eventually, the fungus hyphae grow into the body of the prey. Interestingly, the fungi do not produce the trapping structures in laboratory culture unless nematodes are also present in the culture.

15.6   BIOLOGICAL EFFECTS OF GREENHOUSE GASES AND CLIMATE CHANGE

Our industrial economy is almost entirely heterotrophic, relying on energy stored in coal and petroleum deposits. These were formed from atmospheric CO_2 removed over the eons by primary production of ages past. The combustion of these resources is returning much of this to the atmosphere. The CO_2 and other anthropogenic gas emissions have already resulted in more than 1% increase in the amount of energy from the sun being trapped at Earth's surface (Section 14.2.2). Now the question is: What are the biological and ecological effects of the greenhouse gases and the global warming that they are expected to cause?

Because CO_2 is a reactant in photosynthesis, increasing its concentration would be expected to increase primary productivity. Assuming that moisture and other nutrients are readily available, a doubling of CO_2 increases aboveground biomass of plants by anywhere from zero to 50%. The C_3 plants will be more sensitive. Trees increase their productivity by an average of 40% with CO_2 doubling. The C_4 plants are less sensitive, since they have evolved mechanisms to concentrate CO_2 within the leaf. However, C_4 plants may benefit by being able to operate longer with their stomates closed, thus conserving more moisture.

Ecological effects have been more difficult to predict experimentally. One result tested the effect of decreased moisture in arctic tundra. An increase in global temperature is expected to have uneven climate effects at different locations around the world. Normally, biomass accumulates in these ecosystems because frozen permafrost is not available to decomposers. A drop in water table level would expose this biomass to degradation, converting the tundra from CO_2 sink to CO_2 source.

The leaves of plants grown at high CO_2 levels have a lower nitrogen content. Thus, the many insects that are nitrogen limited and feed on these leaves may need to eat more to satisfy their needs. As a result, a fall in insect numbers and an increase in plant damage could both result.

The effect of warming itself is usually to increase growth rates of organisms such as plants and insects. However, this is unlikely to occur evenly. Instead, competitive balances are likely to be altered, resulting in significant population shifts. Thus, the temperature does not have to move outside the normal range of a particular organism in order to harm it—it just has to favor a competitor or predator preferentially. Warm years in western Canada and in Alaska usually result in pest outbreaks.

In some crustaceans, fish, and reptiles, sex is affected by the ambient temperature at a critical time during gestation. For example, certain turtles, such as the loggerhead sea turtle, produce a higher proportion of females as the temperature increases. This could affect their population.

A small change in average temperature could produce a significant shift in the seasonal warming in the spring. This could disrupt timing of insect hatching, with consequences up the food chain. For example, the emergence of the larvae of winter moth in Scotland is more sensitive to spring temperatures than is the opening of the buds of the sitka spruce that they feed on. Thus, warming could reduce the overlap between egg hatching and bud opening, reducing the survival of the moth larvae. Birds or other animals that eat the moth would be affected. It is possible that the timing of the moth hatching could adjust itself by natural selection. Genetic variation within the moth may include individuals that hatch later. If so, these will be more likely to survive the changed climate conditions.

Migrations could also be affected by timing shifts. For example, many shorebirds migrate from the tropics to the arctic along the east coast "flyway" of North America. They arrive at the Delaware Bay just when spring warming triggers horseshoe crab egg laying. The egg feast provides the energy they need to complete their journey. When the birds arrive at the arctic, they lay their eggs. The eggs hatch just before the peak in certain insects that the birds feed on. Thus, these birds are locked into climate-driven cycles of other organisms. Ecologists fear that warming could disrupt one or both of these timings, with disastrous effects for the shorebirds.

A population may respond to changing climate by shifting its range. However, in some cases this may not work with changes caused by greenhouse gases. One reason is that the data and predictions indicate that the temperature change is coming faster than in natural climate changes of the past. Furthermore, human settlement has fragmented natural terrestrial ecosystems, isolating them into reserves or "ecological islands." Thus, spatial shifts may run up against barriers of inhospitable environments, lacking appropriate habitats or food sources.

It is thought that polar regions will experience greater warming than lower latitudes. Melting of the great ice sheets there could increase sea level several meters, flooding coastal regions around the world. Thermal expansion of the ocean due to warming can also produce significant changes. Already, the ocean level is increasing by about 2 mm per year, although the cause of this trend cannot yet be proved.

Although many areas may experience increased precipitation with warming, others may become dryer. Lake Manitoba, one of southern Canada's largest lakes, was completely dry between 4000 and 6000 years ago when temperatures were $1^\circ C$ warmer than present. Few wetlands existed below $53^\circ N$ latitude in this period.

In the Experimental Lakes Area of northwestern Ontario, the air and water temperature increased $2^\circ C$ from the late 1960s to the late 1980s. Although this increase cannot be blamed conclusively on greenhouse warming, it nevertheless serves as an example of what could happen to similar ecosystems if such a change occurred for any reason. In this case precipitation decreased and evaporation increased. Streamflow dropped, and forest fires increased, increasing runoff. These caused increases in dissolved solids and nutrients in streams and lakes. Increased wind stresses on lakes deepened the epiliminions, reducing the amount of cold hypolimnion on which many species, such as lake trout, depend. The combination of higher temperature, increased nutrient content, and reduced renewal rate resulted in increased algae blooms. Runoff from burned areas contains sulfuric acid, which produces a pulse loading of acidity to surface waters which may add to the background acid loading caused by acid precipitation.

15.7 ACID DEPOSITION

Acid deposition arises from nitrogen and sulfur oxides that are produced by combustion processes and fall on downwind ecosystems as nitric and sulfuric acids. The deposition can be in the form of precipitation, including **acid rain**, or as **dry deposition** carried by aerosols or dust particles. Even fog and dew may be acidified. Most of the nitrogen originates from the air used in combustion, although some may come from the fuel, such as from the burning of biomass. Sulfur oxides come exclusively from the fuel. Coal and some crude oils are high in sulfur. Refined fuels such as gasoline tend to have much less. Emitted as SO_2, at greater distances more and more of the sulfur is in the form of sulfuric acid. Sulfur is usually a more important cause of acid rain than nitrogen is. Unpolluted rainfall has a pH of about 5.6. The most seriously affected area of North America is in the northeast, where rainfall pH averages as low as 4.0 to 4.2. Individual storms can be as low as pH 2 to 3. Acid deposition is being blamed for damaging biological and chemical effects on forests, soils, and surface waters.

Some of the direct effects of acid deposition on plants include eroding the cuticular layer of leaves, interfering with the normal function of stomatal guard cells, causing necrosis of leaf tissue, disturbing photosynthesis, and producing premature senescence (aging). Indirect effects include increased leaching of minerals and organics from damaged surfaces, greater susceptibility to drought, parasites, disease and other stresses, and alteration of symbiotic associations such as with mycorrhyzae fungi, nitrogen fixers, or lichen fungi.

Large areas of central Europe are experiencing severe forest damage. Here there are thought to be interactions with other pollutants, such as ozone. The effect of acid rain on forests is largely indirect, caused by changes in soil chemistry. Acid deposition solubilizes aluminum, which in turn reduces absorption of calcium, magnesium, and potassium by roots. Nitrates absorbed from the air stimulate growth, but the lack of other minerals obtained from the soil results in this growth being weak and susceptible to disease.

In some cases, natural selection has resulted in an increase in tolerance of plants to pollution compared to similar plants in unpolluted areas. Other responses include shifts in populations. Several studies have shown correlation between the lichen species present and sulfur dioxide concentrations. Higher SO_2 levels also correlate with a reduced number of lichen species. However, many of these effects have been observed close to smelters or coal-burning power plants.

The ecosystems that are most sensitive to acid deposition are aquatic. Most of the water in aquatic systems first fell on the land. Soil contains minerals, especially calcium and magnesium, which neutralizes some of the acidity and provides buffering capacity. This is especially true in areas underlain by limestone. Lakes in these areas are unaffected even if rainfall is acidic. However, areas of granite bedrock have less soil buffering, and aquatic systems in these areas are affected more severely impacted. The Adirondack mountains of New York and the Scandinavian countries of Europe have a combination of this type of geology and acid precipitation. In the 1930s only 4% of Adirondack lakes had a pH less than 5. In 1975 the amount was 51%. Of the lakes in with pH less than 5 in 1975, 90% had no fish.

Lakes that have normal pH ranges most of the year may experience shock loading of acidity from snowmelt, when acidity accumulated throughout the winter is released at once. Some lakes are acid naturally, due to organic acids produced by decomposition of leaf litter. Lakes affected by acid deposition can be distinguished because they lack bicarbonate, and the major anion balancing the cations hydrogen, calcium, and magnesium tends to be sulfate. Eutrophic lakes may be protected from acidification because nitrates are destroyed and alkalinity is produced in the anoxic hypolimnion by denitrification and sulfate reduction. However, once pH drops below 5.4 to 5.7, nitrification is inhibited and ammonium accumulates.

Lowered pH causes aluminum to leach from soil into streams and rivers. In the Adirondack Mountains of New York State, 217 high-altitude lakes were found to have pH from 4.25 to 7.33 and aluminum concentration from 10 to over 2000 µg/L. However, almost all of the lakes with pH above 5.0 had less than 125 µg Al/L, whereas almost all with pH less than 5.0 were at or above 125 µg Al/L. Aluminum forms insoluble precipitates with phosphorus, removing this nutrient from the water column and reducing primary productivity. As a result, acid lakes are more oligotrophic than normal and may be exceedingly clear. Aluminum becomes toxic to fish below pH 6.2. Below pH 6 the aluminum forms a precipitate on the gills of fish, interfering with calcium and sodium

TABLE 15.23   Median Percent Reduction in Several Types of Organisms in Adirondack Lakes That Have Been Affected Severely by Acid Deposition, Compared to Unaffected Maine Lakes

Group of Organisms	Adirondack Lakes	Maine Lakes
Rotifers	−4	0
Crustaceans	−18	0
Mollusks	−51	0
Leeches	−60	0
Cyprinids	−17	0
Percids	0	0

Source: Barnes and Mann (1991).

transport. The fish secretes mucus on the gills in response, and this reduces oxygen transport.

Dissolved organic carbon (DOC) in water absorbs ultraviolet (UV) light, which is harmful to algae. In the Experimental Lakes Area of Ontario, UV-B light normally penetrates 20 to 30 cm from the surface. When the lakes were experimentally acidified, the DOC precipitated and settled. UV-B penetration increased to almost 1.5 m.

Fish are most susceptible to low pH just after hatching. Aquatic invertebrates which provide food for fish at all stages may be affected. Lakes in Ontario and the eastern United States lose their lake trout population when their pH falls below 5.6 to 5.2. The eggs and tadpoles of frogs and toads are sensitive, but other amphibians, such as newts, are less so and may increase in abundance due to reduced competition. However, if the pH remains below 5, all animals and many plants will be eliminated. Some attached green algae and the acid-loving *Sphagmum* moss may flourish, sometimes maintaining primary productivity. Table 15.23 shows the impact of acidification on a variety of animal groups. Many of the lakes experienced total elimination of mollusks and leeches.

15.8   ENDANGERED SPECIES PROTECTION

In 1966, the U.S. Congress passed the Endangered Species Protection Act. Initially, it listed only animal species in need of protection, but the 1973 Endangered Species Act (ESA) made plants and all classes of invertebrates eligible. It also implemented the Convention on International Trade in Endangered Species of Wild Fauna and Flora (CITES), which restricted international trade in endangered species.

The ESA defined a species as **endangered** if it is in danger of extinction within the foreseeable future throughout all or a significant portion of its range. In addition, a species is defined as **threatened** if a species is likely to become endangered within the foreseeable future throughout all or a significant portion of its range. The act prohibits federal agencies from any actions that would harm a listed species or its "critical habitat." Conservation plans and land acquisition were authorized. The ultimate goal of the act is recovery of the species and removal from the list.

Amendments in 1982 required that the determination to add species to the list must be made solely on the basis of biological and trade information. Economic or other effects are not to be considered. However, 1988 amendments required that all reasonably

TABLE 15.24 U.S. Fish and Wildlife Service, Division of Endangered Species, "Box Score" of Species Listings and Recovery Plans as of November 1997

Group	Endangered		Threatened		Total Species	Species w/Plans
	U.S.	Foreign	U.S.	Foreign		
Mammals	57	251	7	16	331	41
Birds	75	178	15	6	274	74
Reptiles	14	65	19	14	114	30
Amphibians	9	8	7	1	25	11
Fishes	67	11	41	0	119	78
Snails	15	1	7	0	23	19
Clams	56	2	6	0	64	45
Crustaceans	15	0	3	0	18	7
Insects	24	4	9	0	37	21
Arachnids	5	0	0	0	5	4
Total animals	337	520	114	37	1008	330
Flowering plants	525	1	113	0	639	390
Conifers	2	0	0	2	4	1
Ferns and others	26	0	2	0	28	22
Total plants	553	1	115	2	671	413
Grand total	890	521	229	39	1676[a]	743[b]

[a]Some species are listed more than once in different populations. Some entries represent entire genera or families (e.g., lemurs and gibbons).

[b]There are 477 approved recovery plans. Some cover more than one species, and a few species have different plans for different parts of their range.

identifiable expenditures be reported for each species undergoing recovery action by state or federal governments. Table 15.24 summarizes the endangered and threatened species list.

Tables 15.25 and 15.26 give examples of some of the plants and animals that are endangered or threatened over a range of several states or more. Note that plants tend to be endangered or threatened over narrower ranges than animals. The southern U.S. states of the tend to have the most listed species, in the range of 50 to 90 each. Except for Florida, most of them are animals. The large number may be due to the species richness there, resulting from the warm, moist climate. California has 166 species listed, possibly because of its size and diversity of ecosystems. Hawaii alone has 300 listed species, 263 of which are plants. One reason may be the threat of exotic species, with which native flora are not adapted to compete. Alaska, on the other hand, has only five listed species, and Maine has just eight.

A high proportion of the endangered and threatened animals depend on wetlands for some part of their life cycle. The whooping crane (*Grus americana*) nest in marshes in Canada's Northwest Territories in the spring and summer, in water between 0.3 and 0.6 ft deep. In the fall they migrate south, stopping at riverine wetlands along the way. Ultimately, they arrive at the Aransas National Wildlife Refuge in Texas for the winter. Hunting and habitat loss almost eliminated them. In 1941 there were only 15 individuals left. By 1993 they had recovered to a population of 75.

Two of the species in Table 15.25 are mentioned because of the controversial economic and political roles that they have played. These are the northern spotted owl (*Strix occi-*

TABLE 15.25   Examples of Endangered and Threatened Animal Species by Region in the United States

Species	No. of States	U.S. Region
Bat, gray (*Myotis grisescens*)	12	Southern and midwestern states
Bat, Indiana (*Myotis sodalis*)	21	East of the Rockies
Beetle, American burying (=giant carrion) (*Nicrophorus americanus*)	8	East of the Rockies
Beetle, northeastern beach tiger (*Cicindela dorsalis dorsalis*)	6	Mid-Atlantic states
Beetle, Puritan tiger (*Cicindela puritana*)	5	Mid-Atlantic and northeast
Butterfly, Karner blue (*Lycaeides melissa samuelis*)	8	Midwest to northeast
Crane, whooping (*Grus americana*)	12	South-central to north-central, including mountain states
Darter, snail (*Percina tanasi*)	3	Southern states
Eagle, bald (*Haliaeetus leucocephalus*)	48	Contiguous United States
Falcon, American peregrine (*Falco peregrinus anatum*)	48	Continental United States, including Alaska
Fanshell (*Cyprogenia stegaria*)	8	South-central to midwest
Frog, California red-legged (*Rana aurora draytonii*)		California
Mussel, dwarf wedge (*Alasmidonta heterodon*)	9	East coast states
Owl, northern spotted (*Strix occidentalis caurina*)	3	Northwest coastal states
Pearlymussel, all varieties	12	Midwest to southern states (Tennessee alone has 16 endangered or threatened species of pearlymussels)
Pelican, brown (*Pelecanus occidentalis*)	8	Gulf and west coast states
Plover, piping (*Charadrius melodus*)	34	East of the Rockies
Sturgeon, pallid (*Scaphirhynchus albus*)	13	South-central to northern tier states
Tern, least (*Sterna antillarum*)	18	South central to northern tier states
Tern, roseate (*Sterna dougallii dougallii*)	11	Atlantic coast states
Turtle, green sea (*Chelonia mydas*)	22	All coastal states except Alaska
Turtle, hawksbill sea (*Eretmochelys imbricata*)	21	
Turtle, Kemp's (=Atlantic) ridley sea (*Lepidochelys kempii*)	16	
Turtle, leatherback sea (*Dermochelys coriacea*)	27	
Turtle, loggerhead sea (*Caretta caretta*)	24	
Turtle, olive (=Pacific) ridley sea (*Lepidochelys olivacea*)	7	
Wolf, gray (*Canis lupus*)	10	Northern tier states
Woodpecker, red-cockaded (*Picoides borealis*)	13	Southern states

dentalis caurina) and the snail darter (*Percina tanasi*). Both of these have been used in efforts to stop or limit economic activity. The snail darter is a small fish that would have been threatened by the construction of a dam in Tennessee. Opponents of the dam used the ESA successfully to stop the project and protect the snail darter's habitat. However, many people did not feel that the survival of a tiny fish should have a higher priority than a project that would have huge economic advantages for the region.

TABLE 15.26 Examples of Endangered and Threatened Plant Species by Region in the United States

Species	No. of States	U.S. Region
American chaffseed (*Schwalbea americana*)	7	Southeastern states
Canby's dropwort (*Oxypolis canbyi*)	5	Southern Atlantic coastal states
Eastern prairie fringed orchid (*Platanthera leucophaea*)	8	Northern states east of the Rockies
Green pitcher-plant (*Sarracenia oreophila*)	4	Southeastern states
Harperella (*Ptilimnium nodosum (=fluviatile)*)	7	Southeastern states
Marsh sandwort (*Arenaria paludicola*)	3	West coast states
Mead's milkweed (*Asclepias meadii*)	5	Northern plains states
Northeastern (=Barbed bristle) bulrush (*Scirpus ancistrochaetus*)	8	Mid-Atlantic states
Pondberry (*Lindera melissifolia*)	9	Southeastern states
Running buffalo clover (*Trifolium stoloniferum*)	6	South-central to midwest
Sandplain gerardia (*Agalinis acuta*)	5	Mid-Atlantic states
Small whorled pogonia (*Isotria medeoloides*)	16	East of the Mississippi River
Swamp pink (*Helonias bullata*)	7	Southeast and Mid-Atlantic states
Ute ladies'-tresses (*Spiranthes diluvialis*)	5	Northern Rockies Mountain states
Virginia spiraea (*Spiraea virginiana*)	9	South-central states
Western prairie fringed orchid (*Platanthera praeclara*)	7	Plains states
Water howellia (*Howellia aquatilis*)	5	Northwestern states

A later controversy involved the northern spotted owl. This owl requires "old growth" forest for its habitat. Environmentalists tried to use the ESA to protect those forests. The result was an acrimonious battle. Loggers portrayed environmentalists as preferring owls to people. Environmentalists pointed out that without protection the old growth forest would soon disappear, eliminating the jobs that depend on that resource anyway. The issue was resolved with a compromise that was protective of the forests.

These controversies have resulted in attacks on the Endangered Species Act itself, which have the possibility of weakening the act in the future. The debate over old growth forests also brought the issue of sustainability into the public eye. **Sustainability** refers to a system in which resources are not used faster than they are produced. Nutrient use in steady-state ecosystems provides a good natural example of sustainability. Ecosystems achieve sustainability with respect to nutrients by limiting their consumption and by recycling.

PROBLEMS

15.1. (**a**) Suppose that you are given the task of determining the habitat of a benthic organism, say a worm or a clam. Make a list of questions you would need to know the answer to in order to specify the habitat fairly completely.

(**b**) Suppose, instead, that you were asked to specify the organism's niche. What else would you want to ask?

15.2. Adding composted manure improves the friability and nutrient-holding capacity of a soil. Why is it better for the manure to be composted instead of being added untreated?

15.3. A lake receiving conventionally treated wastewater is starting to show signs of eutrophication. You might be able to control the algal blooms by upgrading the treatment plant for nutrient removal, but you must determine whether to require nitrification/denitrification, phosphorus removal, or something else. How could you determine the limiting nutrient in a laboratory experiment? What limiting nutrients would you test for?

15.4. Name one or two advantages that each of the following types of phytoplankton has over the other for survival in lakes: cyanobacter, diatoms, dinoflagellates.

15.5. Draw an hypothetical energy-flow diagram for a river as it travels from its source to its mouth. Show the allochthanous inputs, the different kinds of dissolved and particulate organic matter, and exchanges with sediments, floodplains, salt marshes, and its ocean discharge. Perform a similar analysis for nitrogen.

15.6. What advantages and disadvantages do constructed wetlands for wastewater treatment have over conventional biological wastewater treatment?

REFERENCES

Atlas, R. M., and R. Bartha, 1981. *Microbial Ecology: Fundamentals and Applications*, Addison-Wesley, Reading, MA.

Barnes, R. S. K., and K. H. Mann (Eds.), 1991. *Fundamentals of Aquatic Ecology*, Blackwell Scientific, Boston.

Berner, E. K., and R. A. Berner, 1987. *The Global Water Cycle*, Prentice Hall, Englewood Cliffs, NJ.

Coler, R. A., and J. P. Rockwood, 1989. *Water Pollution Biology: A Laboratory/Field Handbook*, Technomic Publishing, Lancaster, PA.

Connell, D. W., and G. J. Miller, 1984. *Chemistry and Ecotoxicology of Pollution*, Wiley, New York.

Garrison, T., 1993. *Oceanography: An Invitation to Marine Science*, Wadsworth, Belmont, CA.

Gause, G. F., 1934. *The Struggle for Existence*, Williams & Wilkins, Baltimore.

Hickman, C. P., Jr., L. S. Roberts, and A. Larson, 1996. *Integrated Principles of Zoology*, Wm. C. Brown, Dubuque, IA.

Horne, A. J., and C. R. Goldman, 1994. *Limnology*, McGraw-Hill, New York.

Kent, D. M., 1994. *Applied Wetlands Science and Technology*, Lewis Publishers, Chelsea, MI.

Krebs, C. J., 1994. *Ecology*, Harper Collins, New York.

Mann, K. H., and J. R. N. Lazier, 1991. *Dynamics of Marine Ecosystems: Biological–Physical Interactions in the Oceans*, Blackwell Scientific, Boston.

Masters, G. 1998. *Introduction to Environmental Engineering and Science*, Prentice Hall, Upper Saddle River, NJ.

McKinney, M. L., and R. M. Schoch, 1998. *Environmental Science, Systems and Solutions*, Jones & Bartlett, Sudbury, MA.

Mitsch, W. J., and J. G. Gosselink, 1993. *Wetlands*, Van Nostrand Reinhold, New York.

Needham, J. G., and P. R. Needham, 1962. *A Guide to the Study of Fresh-Water Biology*, Holden-Day, San Francisco.

Odum, Eugene P., 1987. *Basic Ecology*, Saunders College Publishing, New York.

Raven, P. H., R. F. Evert, and S. E. Eichhorn, 1992. *Biology of Plants*, Worth Publishers, New York.

Sharkey, T., 1996. Isoprene synthesis by plants and animals. *Endeavour*, Vol. 20, pp. 74–78.

Standard Methods, 10th ed., 1955. American Public Health Association, Washington, DC.

Standard Methods, 12th ed., 1965. American Public Health Association, Washington, DC.

Taiz, L., and E. Zeiger, 1988. *Plant Physiology*, Sinauer Associates, Sunderland, MA.

U.S. Army Corps of Engineers, 1987. *Corps of Engineers Wetlands Delineation Manual*, Environmental Laboratory, Department of the Army, Washington, DC, January.

U.S. EPA, 1988. *Constructed Wetlands and Aquatic Plant Systems for Municipal Wastewater Treatment*, EPA/625/1-88/022, U.S. Environmental Protection Agency, Washington, DC.

16

BIOLOGICAL APPLICATIONS FOR ENVIRONMENTAL CONTROL

Applied biology plays an altogether vital role for all humans, in terms of sustaining both human life and environment quality. On the production side of these benefits, we are all highly dependent on the renewable genesis of consumable products, from farms around the world that generate cultivated crops and managed livestock, from managed forests that furnish new supplies of timber and pulp, and from extensive commercial and industrial operations that supply a wide range of biochemical products ranging from basic foods and beverages (e.g., bread, cheese, wine, beer) to bulk chemicals (e.g., acetic acid, ethanol) to advanced pharmaceuticals.

Given that waste generation inevitably follows our consumption of commodities, we also depend heavily on controlled biology for yet another layer of applied benefits, as the biochemical basis for sustained environmental management. Indeed, we routinely rely on applied biology to cover both source and sink roles in closing out the grand mass balance of waste residuals within all environmental realms, from wastewater treatment to biosolids processing to solid waste degradation to soil, groundwater, and air contaminant remediation.

These biochemical benefits, therefore, are global in coverage while being evolutionary in impact. Dating back roughly 10,000 years, applied biology had an absolutely seminal impact on the advent of civilized human life, enabling our nomadic hunter–gatherer predecessors to adopt an entirely new life-style tied to their newfound abilities with beneficially controlling and manipulating biology.

Rather surprisingly, though, the methods and motives of applied biology are not unique to humans. In fact, based on what we know about the approximate origin of ant life forms in the mid-Cretaceous Era 80+ million years ago and their sophisticated colonial lifestyles, it is quite likely that we trailed ants in this regard by a considerable margin. For example, there are ant colonies that "farm" fungal growths for food, and there are even

Environmental Biology for Engineers and Scientists, by David A. Vaccari, Peter F. Strom, and James E. Alleman
Copyright © 2006 John Wiley & Sons, Inc.

ants that "herd" so-called "ant-cattle" stocks of aphid and larvae life-forms. One such exemplary leaf-cutting ant group may be found in Central and South American regions (e.g., typically associated with the *Atta* genus), where individual chambers within their colonies are allocated to carefully managed farming operations. These particular ants nurture, and then feed exclusively from, a form of fungus found only within these unique subterranean systems. Leaf and grass cuttings gathered by ants on the surface are carried underground to nurture the fungus growths held in chambers inside the colony, even to the point of using protease-rich anal secretions to "fertilize" the cuttings. Multiple fungal growths are raised on staggered 3- to 4-week cycles, with the ants living off the cloned, reproductive fruiting buds that grow on the fungus surface.

Whether their mode of sustenance were that of farming, herding, or just grazing, however, ant cultures have also adopted remarkably advanced measures for applied waste management. Ants routinely remove residues generated by farming or herding, along with dead members of the colony and other debris, to remote dumping sites located either deep within the underground colony or on external surface spoil sites carefully chosen to provide a downhill, easy-discharge location. Furthermore, most ants exhibit a fastidious disdain for contact with their wastes and garbage, and in some colonies there are even special ant groupings solely assigned the task of policing waste.

At least when measured within a historical timeframe, ants appear to have far surpassed humans in terms of waste management concerns, extending back millions of years instead of a few centuries. In our own case, large-scale organized efforts to collect and remove wastes were not implemented until after the Industrial Revolution, roughly three centuries ago, and controlled use of biological waste treatment as an environmental management tool extends back barely half that time.

Of course, although humans may have been beaten by ants on a time scale, we have now advanced the sophistication of our applied biology efforts to a far higher level, and this book's environmental management theme certainly demonstrates this progress. Whereas ants have long been content with just collecting and discarding wastes, humankind's modern approach to environmental engineering and science has developed and adopted far more advanced measures.

Developing an appreciation for, and understanding of, the biological processes now used for treatment of contaminated air, water, soil or solid wastes involves the basic principles of biology that we discussed earlier. Basic biology, especially biochemistry, governs all the processes involved. Microbiology is critical, since microorganisms dominate these processes, but there are also processes that depend on higher plants and even trees to degrade various wastes and/or to remediate the land on which they live. Finally, an understanding of ecology is necessary, since all of these processes involve mixtures of numerous interacting populations of organisms, and in some cases the consortia of organisms are able to achieve treatment goals that could not have been achieved by individual populations (Atlas and Bartha, 1987).

In whatever fashion the various mechanisms of applied biology might be used, though, the resulting benefits of preserving and protecting our natural assets are readily obvious. Biological systems inherently qualify as environmentally friendly, and as a naturally renewable resource they tend to offer significant economic advantages. At the same time, backed by eons of evolutionary adaptation, biological mechanisms commonly provide a highly effective, and metabolically diverse, strategy for effectively transforming waste residuals into innocuous by-products. Should it be necessary, most of these systems can even self-adjust (i.e., **acclimate**) to their circumstances (environmental conditions,

TABLE 16.1 Biological Applications for Environmental Control

Applied Realm of Environmental Management	Specific Process Application	Biological Level	
		Micro-scale[a]	Macro-scale[b]
Wastewater	Attached growth	×	
	Suspended growth	×	
	Stabilization lagoons	×	
	Constructed wetlands	×	×
Sludge	Anaerobic digestion	×	
	Aerobic digestion	×	
	Composting	×	
Water	Potable water	×	
Disinfection	Water and wastewater	×	
Solid waste	Landfill degradation	×	
Air	Biofiltration	×	
Soil and groundwater	Phytoremediation	×	×
	Bioremediation	×	×

[a]Microscale applications involve various combinations of bacteria, protozoms, rotifers, fungi, algae, etc.
[b]Macroscale applications involve higher plants.

competitive pressures, etc.), in an attempt to maintain optimal performance while handling metabolically an increasingly complex array of industrial chemicals, which may be **anthropogenic** (human-made) or **xenobiotic** (foreign to living things).

The following seven sections (Table 16.1) will provide brief introductions to several of these engineered applications with biology and their associated technical features. In each case, these summaries explore the backgrounds of the applied technologies and present overlapping views of the biological and engineering principles involved. Upon reviewing these sections, and again in keeping with the critical role of microbiology, it will be readily apparent that the majority of these "biological" mechanisms operate largely at the "microbial" end of the spectrum of life. Indeed, bacterial biomass commonly represents the driving force behind the success of most engineered processes, both in terms of the shear numbers of cells involved and their metabolic contributions.

Granted, there are usually a number of complementary roles played by higher lifeforms in a number of these systems, but the true backbone of these operations typically hinges on the performance of the bacteria involved. Indeed, even in the case of wetland and phytoremediation systems, these plant-based processes depend heavily on microbial (and largely bacterial) mechanisms that operate on the surfaces or in the immediate vicinity of their root systems.

Although the following sections were written with heuristic confidence about the process of designing and operating these biological systems, a fair and honest acknowledgment of our inherent limits is warranted. Whether wastewater microbes or constructed wetland plants are to be used, it is rather rare that we can select and sustain the exact group of organisms that we might feel can best accomplish a desired treatment goal, particularly when more often than not, these systems are open to the influence of opportunistic biota living within the natural environment. Instead, we identify desired goals, establish the environmental conditions involved, and design an appropriately engineered system, at which point nature will play a significant, if not the dominant, role in selecting the resulting mix of operative species working within our processes. Even then, however, the most rewarding aspect of applied environmental biology is that by merging our scientific and engineering skills with the formative power of natural life forces, we can secure synergistic outcomes with truly global benefits.

16.1   WASTEWATER TREATMENT

Historically, wastewater processing qualifies as our first approach to using applied biology constructively on an environmental basis. Natural biological processes have, in fact, dealt with the world's various waste streams from the earliest days of microbial life eons ago, but our own efforts with respect to waste management were barely better than those of ants until well beyond the Renaissance, content as we were with primitive privies, latrines, and pits for mere containment rather than complete treatment (Asimov, 1989).

Remarkably, our initial motivation to use applied waste treatment roughly five centuries ago was neither that of resolving serious and escalating health concerns tied to disease transmission nor that of reducing widespread aesthetic degradation. Instead, our original goal was actually that of commercial manufacturing, whereby our residuals might be transformed from disagreeable wastes into truly useful products with monetary value.

The classic set of sixteenth-century lithograph prints shown in Figures 16.1 and 16.2 depict two such waste processing systems, located in England and Germany, respectively, both designed in a fashion that we might presently qualify as biological nitrification filters. Urine and other nitrogen-rich wastes (e.g., slaughterhouse effluents) were pumped or ladled onto the filters and allowed to trickle over and through the piles. In turn, nitrate-rich

Figure 16.1   Sixteenth-century English nitre farming.

Figure 16.2   Sixteenth-century German nitre farming.

salt crystals (i.e., long know to alchemists as *nitre*) would then be scraped from these filter surfaces and sold as a key ingredient for producing gunpowder.

These commercial operations predated the discovery of bacterial life by several centuries. It was not until the late nineteenth century that we finally understood the true lithotrophic basis for reliance on biochemical nitrification to convert reduced ammonia to oxidized nitrates. Despite this fundamental uncertainty, nitre pile waste processes continued to represent a significant chemical manufacturing industry for several centuries and was even used for critical nitrate production during the Civil War by Confederate states in the southern United States.

Midway through the nineteenth century, a number of larger English cities began to use a patented processing scheme to recover waste nitrogen, but in this case the desired product was a nutrient-rich fertilizer. These so-called "ABC" systems used an altogether unusual set of chemical additives, including alum, blood, and clay (i.e., hence the unique acronym) to clot and sorb a wastewater's unwholesome contaminants while generating a nitrogen-bearing agricultural fertilizer sold under the promotional trade name "native guano."

At about the same time, securing complementary environmental benefits with wastewater treatment began to reach beyond the product-oriented stage, catalyzed both by the latest discoveries in microbiology and medicine and a belated recognition of the inherent necessity for sanitary and environmental improvements. These early years in the field of wastewater treatment were exciting times for this fledgling niche of applied environmental biology.

Land application operations were the earliest large-scale systems used for wastewater treatment, with several such operations being in use in Europe. The first land-application "filters" were remote tracts of land onto which wastewater would be spread, being cleansed as it passed through the soil. As our understanding of, and appreciation for, the underlying microbial mechanisms evolved during the middle to late nineteenth century, simple in-ground filters were soon followed by a steady progression of ever more advanced, and more heavily loaded, intermittent soil filter, contact bed, and trickling filter wastewater systems. In each case, the soil, gravel, rock, slate, and slag media surfaces involved were found to be densely covered with an attached biofilm rich in waste-degrading biomass, following a wastewater processing mode now classified as *attached-growth* treatment.

Early in the twentieth century, another seminal discovery was made that coagulated and settled biomass suspensions such as those derived from either settled discharges taken from the early attached-growth biofilm reactors or from even newer *suspended-growth* (as opposed to attached-growth) biomass tanks, could actually be recycled back to an aeration chamber for subsequent reuse, thereby accelerating the overall metabolic efficacy of these biological wastewater processing systems. Largely developed at Manchester, England, by Sir Gilbert John Fowler in the period 1912–1915, with several large full-scale facilities being built around the world within a period of just a few years, this newly devised *activated sludge* concept rapidly evolved into yet another parallel mode of suspended-growth system, further escalating the sophistication and effectiveness of wastewater treatment (Horan, 1990).

In the twenty-first century, the applied science of biological wastewater treatment continues to evolve in terms of the level of technical sophistication employed. Attached- and suspended-growth options still comprise the two major options for biological treatment, but in either case these older technologies have been upgraded in technical complexity and hardware, including all manner of internal pumping schemes, advanced aerators, online instrumentation, and computer-based control and automation. Whatever the approach, these wastewater processing systems now arguably represent the largest controlled application of microbiology in the world, at a scale far outstripping that of commercial and industrial fermentation and pharmaceutical production.

Whether based on attached- or suspended-growth processing methods, the overall technology of wastewater treatment (i.e., as applied in large-scale fashion by cities and industries rather than that of small-scale septic tanks, etc.) includes an integrated collection of processing steps encompassing not only biological processes, but also a complementary set of physical and chemical mechanisms. Figure 16.3 depicts the processing steps used with these types of conventional four-step systems. The initial **preliminary treatment** step screens, settles, and separates coarse solids, and a **primary treatment** step provides further solids separation, either by floating or settling. Primary treatment may remove onehalf to two-thirds of the incoming suspended solids and perhaps half as much of the **biochemical oxygen demand** (**BOD**), but neither preliminary nor primary treatment is able to remove colloidal or dissolved contaminants.

A subsequent **secondary treatment** step is used to remove the remaining biodegradable contaminants and suspended solids. As reviewed in the following sections, secondary treatment is typically completed using one of four typical processing options, including attached-growth (e.g., trickling filters) and suspended-growth (e.g., activated sludge) systems, stabilization lagoons, and constructed wetlands.

16.1.1 Process Fundamentals

It is common to view biological treatments basically as biodegradation processes, but this is only partially true. To a great extent, they are contaminant-separation processes. In the case of domestic wastewater treatment, they convert colloidal and dissolved solids to a form that can be removed by gravity sedimentation into a concentrated by-product called

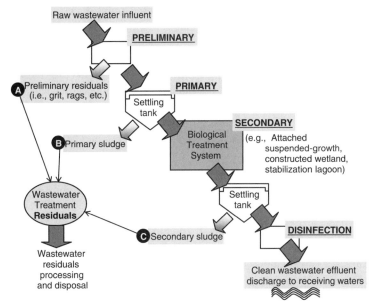

Figure 16.3 Conventional wastewater treatment processing scheme.

sludge or *biosolids*. They typically yield about 0.5 to 0.7 kg of biosolids (dry weight) per kilogram of BOD removed. Thus, they convert a large problem (the volume of wastewater to be treated) into a much smaller problem, but one that still needs an ultimate solution.

The dominant biochemical mechanism in the vast majority of secondary wastewater treatment processes is that of respiration, which is distinctly different from that of the fermentation reactions that come into play more often in the sludge processing systems (e.g., anaerobic digestion) used to process wastewater residuals. Both respiration and fermentation depend on a balanced set of redox reactions (i.e., coupling oxidation plus reduction steps), which are linked, respectively, to electron donor and acceptor compound conversions.

With respiration, both reduced biodegradable organics (i.e., carbonaceous BOD) and reduced, energy-rich inorganics (e.g., ammonia, nitrite, sulfide, ferrous iron) in the raw wastewater may be oxidized to secure energetic electrons, compounds referred to as **electron donors**. Conversely, the electrons will then be "respired" or consumed by an electron acceptor compound. In most conventional wastewater treatment, dissolved oxygen is preferred as the electron acceptor, so that the zero-valence gas-phase diatomic oxygen (O_2) species is reduced to water.

From a thermodynamic point of view, oxygen is the most favorable electron acceptor, and aerobic or facultative cells always opt for this aerobic respiration pathway if sufficient oxygen (i.e., greater than a few tenths of a milligram per liter) is present, largely ignoring other electron acceptors that might be available. Even in the absence of oxygen, respiration may be continued, following a pathway referred to as **anoxic respiration**, using a variety of alternative electron acceptors, such as nitrate, nitrite, sulfate, and oxidized iron and manganese. Whether aerobic or anoxic, the fact that respiration relies solely on inorganic electron acceptors is a notable feature for respiration and a key reason that fermentation is not used widely for wastewater treatment. Indeed, using anaerobic reactors rather than aerobic reactors would incur a far higher level of odor emission in these plants, brought about by volatile organic emissions generated reductively through fermentation.

Although characterized as biological processes, secondary treatment systems also involve a complex array of chemical transformations (e.g., acid–base, redox, chelation, sorption) and physical transfers (e.g., gas-liquid exchange, heat transfer). Furthermore, secondary biological operations involve biochemical mechanisms that couple removal and production pathways. Contaminants are removed via catabolic oxidization and anabolic assimilation while producing a concentrated waste product called **sludge** or **biosolids**, containing residual particulate solids at concentrations on the order of 1% or higher (i.e., $\geq 10,000$ ppm).

Notwithstanding their mutual reliance on aerobic respiration, there are considerable differences between the four main options for secondary wastewater processing in terms of their involved engineering and biological details, and there are also dramatic variations in their necessary footprint areas relative to high vs. low waste loading rates. Of course, the more heavily loaded, mechanically intensive attached- and suspended-growth systems would commensurately require the smallest land areas; the more natural slow-rate and lowly loaded design options (i.e., constructed wetlands and lagoons) would require far larger land areas. Figure 16.4 compares the typical organic loading rates (i.e., representing

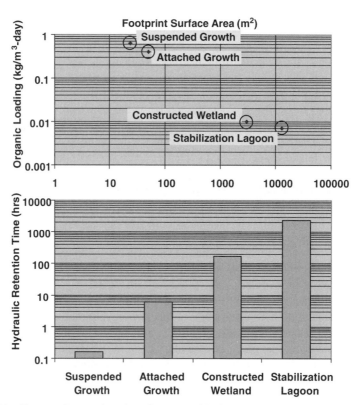

Figure 16.4 Comparative engineering features with four alternative 1000-person municipal wastewater design options.

the daily mass of applied organics per unit volume of tower, tank, wetland, or lagoon) that might be employed with each such design option, together with an estimate of the land areas (i.e., **footprint**) required to process the daily wastewater flow for an imaginary city of 1000 residents (assuming a daily per capita flow of 400 L and a BOD strength of 150 mg/L). As can be seen, there is a dramatic difference in the footprint size linked to these four alternatives, based on their varied loadings.

Two parameters can be used to compare wastewater treatment processes: hydraulic residence time and solids residence time. **Hydraulic residence time** (HRT) is the volume of the process divided by the flow rate through the process, and conceptually is equal to the average amount of time that a parcel of water spends in the process. This basic parameter relates the size of the process to the volume of material to be treated, and therefore is a key parameter describing the economics of the process. The smaller the HRT, the more economical the process is to construct.

Solids residence time (SRT), also known as the **mean cell residence time** (MCRT) or **sludge age**, is the average amount of time that the biomass resides in the process from when it is produced by microbial growth to when it is removed by deliberate wasting or by loss with the effluent. SRT is computed as the ratio of the mass of biomass within the process divided by the mass wasted or lost from the process per unit time. The SRT is basically the replacement growth rate of the biomass. For example, an SRT of 10 days means that one-tenth of the biomass is replaced by growth each day. Therefore, a low SRT implies a high growth rate, and vice versa. In terms of microbial kinetics, the SRT for a steady-state process is the same as the net growth rate described in Chapter 11. Because setting the SRT controls the net growth rate, it is the key parameter that controls the biology of the process.

Under optimum growth conditions (in particular, high substrate concentrations) microorganisms can reproduce as fast as once every 20 minutes. A common misconception among wastewater treatment plant operators is that high growth rate (being in the log growth phase in terms of the batch microbial growth curve) is a desirable operating goal. However, this is false for several reasons. One is that high growth rate implies high substrate concentration, and therefore high effluent BOD. On the contrary, it is essential to operate in the declining growth phase, which implies low substrate concentration. Under this condition the microbes are "starved" for substrate. This causes low growth and high SRT. The second reason for low growth rates is that the suspended or attached growth biological wastewater treatment processes depend on the ability of the microbes to produce exocellular polysaccharides which enable them to stick together (to form "flocs") or to solid surfaces (to form "biofilms"), making it possible to separate the solids from the wastewater and thus to decouple HRT from SRT.

The floc- or biofilm-forming behavior is important because it makes it easy to retain the biomass in the wastewater treatment process much longer than the wastewater itself. For example, consider a process operated with an SRT of about 10 days, to achieve a low enough effluent BOD. If this process were operated in a completely mixed reactor, it would have to also have an HRT of 10 days. That is, to treat 1 million gallons of wastewater per day would require a tank with a volume of 10 million gallons. However, if it were possible to keep the biomass in the reactor longer than the wastewater, it would be possible to build a much smaller tank. The attached-growth processes accomplish this by growing the biomass attached to a solid material as a biofilm. Suspended growth processes retain the biomass by using a sedimentation tank to trap the solids and then pump them back to the reactor. The biomass in the reactor requires only a matter of hours to absorb or otherwise remove the substrate from the wastewater. The purified wastewater can then be discharged as effluent. By separating the biomass it is possible to have an HRT on the order of several hours and an SRT of days. As a result, the treatment process can be made to be a small fraction of the size it would otherwise have to be. In the example just given, if the HRT were about 6 hours, the volume needed to treat 1 million gallons per day would be just 250,000 gallons instead of 10 million gallons.

The microorganisms that make up the floc or biofilm structure in biological wastewater treatment produce the exocellular polysaccharide only under conditions of "starvation" and low net growth rate, corresponding to the low substrate concentrations that go with high SRT. If the process is operated close to the minimum SRT, the substrate concentration in the process will increase toward the influent concentration. At high substrate concentrations, the microbes will show less tendency to form flocs or biofilm. Without the polysaccharide to help the bacteria to agglomerate into flocs, they will grow dispersed in the wastewater, increasing effluent turbidity, suspended solids, and BOD.

Flocculation behavior makes sense for bacterial survival. When an organism emits some substance outside its boundaries, there is a cost that must be accompanied by a return. Only under conditions of scarcity is this investment worthwhile for the microbe. The benefits are: (1) Entrapment of particulate substrate is enhanced; and (2) secretion of exocellular digestive enzymes by a single cell could result in most of that enzyme diffusing uselessly away from the cell, and a large fraction of any dissolved substrate formed by digestion of the particle would also diffuse away. Within a floc or biofilm, the bacteria are cooperating by releasing enzymes that benefit them collectively.

16.1.2 Attached-Growth Systems

The first large-scale success with biological wastewater treatment was achieved midway through the nineteenth century using an attached-growth process now referred to as **trickling filter** technology (Figure 16.5). The operative biochemical agent within these

Figure 16.5 Attached-growth wastewater treatment system.

Figure 16.6   Attached growth biofilm media options: (*a*) rock, (*b*) random packed plastic, and (*c*) nested modular plastic bundle.

systems develops as a thin microbial layer, commonly known as a **biofilm**, which grows in an attached fashion on the surface of stationary support media (see the media shown in Figure 16.6) and over which the incoming wastewater is then continuously distributed and streamed (see Figure 16.7). In turn, these attached-growth **biofilters** will then sorb, filter, and degrade waste contaminants as they trickle across their extensive attached-growth surface.

Compared with most other waste-processing strategies, trickling filters are fairly simple in design and operation, and as a result they still find considerable use (Water Environment Federation, 1991). While their overall effectiveness may be somewhat lower than that of other processing strategies (e.g., suspended-growth systems), the attached nature of these biofilms does provide several complementary benefits. Since the attached biofilm stays inside the reactor, there is less need to provide any sort of recycling process to return solids from the clarifier (as must be done in suspended-growth systems). Trickling filter systems also encounter fewer problems retaining their attached biomass during high-flow periods, whereas a suspended-growth reactor might overload the settler hydraulically, resulting in solids loss in the effluent. In many instances, and particularly in nitrifying

Figure 16.7   Attached growth wastewater treatment distributor.

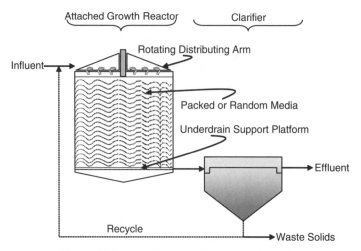

Figure 16.8   Attached-growth process schematic.

towers, where the attached biofilm is extremely thin and tightly bound, these reactors may not even require a follow-up clarifier.

The typical hardware components used with a conventional system (Figure 16.8) include a reactor filled with media, an underdrain platform at the base of the reactor on which the medium rests, a rotating influent distribution arm (or, less commonly, a collection of fixed spray distributors), and the necessary valves, pipes, and pumps required to route, and possibly recirculate, this flow through the reactor. In addition, there may be a downstream settling tank to clarify the outgoing waste stream prior to discharge, in which solids pulled or released (i.e., **sloughed**) from the biofilm surface can be settled.

As for the design strategies used by environmental engineers to configure and size these processes, hydraulic and substrate loadings typically represent the key parameters, as depicted in Figure 16.9. Table 16.2 provides an overview of the associated design

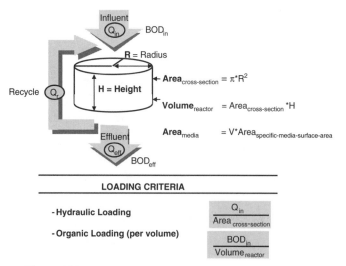

Figure 16.9   Attached-growth process design considerations.

TABLE 16.2   Typical Attached-Growth System Design Criteria

Fixed-Film Design Factor	Hydraulic Loading Range $(m^3/day \cdot m^2)$	Substrate Loading Range $(kg/m^3 \cdot day)$	Media Depth (m)	Recirculation Ratio Q_r/Q, Relative to Influent Flow
Low-rate biofilm system	1.2–3.6	0.08–0.4 (as CBOD)	2–2.5	0
High-rate ("roughing") biofilm system	48–150	1.6–8 (as CBOD)	5–12	0–1
Nitrification biofilm system	1.2–4.8	0.0015–0.002 (as N)	3–12	0–1

Source: Adapted from Metcalf and Eddy (2003).

values typically used for attached-growth systems, as well as information regarding expected reactor depths and recirculation ratios. Hydraulic loading is a primary design factor based on the flow of influent wastewater (which may or may not include recycle flow) applied per cross-sectional surface area of the medium (with typical units of, $m^3/day \cdot m^2$, or when simplified, m/day). One of the more subtle aspects of these systems is that biofilm surfaces will naturally retard the movement of liquid across a media's surface in proportion to the amount of dynamic drag imparted by the physical roughness of its liquid–biofilm interface. As a result, for any given system, an applied **total hydraulic loading** (THL), inclusive of influent plus recycle flow, could well result in considerably different operating values for the amount of time the flow actually spends in transit through the reactor, as would be quantified using a **hydraulic residence time** (HRT).

Without an attached biofilm, wastewater flow over smooth, clean media surfaces would experience little hydraulic resistance, and in turn would have a low hydraulic residence time. Conversely, mature biofilms have an inherent degree of surface roughness associated with microbes extending physically into the bulk fluid layer, which would then slow down the flow and increase the HRT. By analogy, a pail of water splashed across a floor will travel much faster across smooth and recently waxed tile (i.e., a clean medium) rather than a fibrous carpet (i.e., a highly nonhomogeneous microbial surface). As a result, real-world observations of attached-growth systems have shown a considerable difference in hydraulic residence times at equivalent THL loadings, by roughly a factor of 10 (e.g., ~1 minute for 3 m of clean media vs. ~10 minutes when passed over mature biofilm) (CH2M-Hill, 1984).

Attached-growth trickling filter systems used with wastewater treatment implicitly benefit from this phenomena, since it gives the biofilm more time to effectively treat an overflowing waste stream. Granted, compared to the sort of hydraulic residence times used with suspended-growth wastewater treatment (e.g., measured in hours) or lagoons and wetlands (e.g., measured in days, weeks, and possibly even months), biofilm systems have far lower contact times (e.g., ~10 to 20 minutes) in which the desired treatment can be completed. On the other hand, this sort of biofilm growth and surface roughness would be considered problematic in many other instances of attached-growth formation, including that of cooling towers, ship hulls, or your own teeth. In the particular case of cooling tower operations, for instance, biofilm growth on heat transfer surfaces would reduce the overflow velocity while increasing the depth and temperature of the overlying water stream, such that the tower would then have a lower thermal gradient and slower rate of heat release.

As for the second, substrate loading, design parameter, this variable considers the mass of substrate applied per unit time and per unit volume of the media (typically, kg substrate/day·m^3). The magnitude of these terms, of course, varies in relation to the type of substrate and the degradation rates expected. For example, even the low end of the organic loading rates typically used (e.g., ~0.08 kg CBOD$_5$/day·m^3) is far higher than that of nitrification rates (e.g., usually less than 0.002 kg ammonia-nitrogen/day·m^3) (Characklis and Marshall, 1990).

Example 16.1: Preliminary Trickling Filter Design   A newly proposed suburban community near Chicago, Illinois (i.e., Deer Creek, with 400 homes in a standard three-bedroom configuration) will require a dedicated wastewater treatment facility given the remote, unsewered location. During a preliminary public utilities hearing, the community's consulting engineer is asked to present a preliminary design estimate for sizing a standard attached-growth trickling filter, recommended as one possible option for the biological secondary stage in this proposed system. With an expected per-home family size of four and a projected average per-person wastewater discharge rate of 378.5 L/day (i.e., 100 gal/day-person), the daily wastewater flow for the entire community would be 605,700 L/day [(150 homes) (4 people/home) (378.5 L/person·day)] or 605.7 m^3/day. In turn, and given that the carbonaceous organic waste [i.e., typically quantified as a carbonaceous biochemical oxygen demand (CBOD)] in this sort of municipal wastewater is typically around 200 mg CBOD/L, the daily organic wastewater loading generated by this community would be 121.2 kg CBOD/day [i.e., (605.7 m^3/day) (0.2 g CBOD/L)], of which approximately 40% (or 48.5 kg BOD/day) would be removed by primary treatment, leaving a final CBOD load on this trickling filter system of 72.7 kg CBOD/day.

Preliminary Design Details

Hydraulic loading, system sizing, and retention times

Design hydraulic loading rate: 8 m^3/m^2·day (see Table 16.2)

Design media specific surface area: 98.4 m^2/m^3·day (see Table 16.2)

Design media depth: 3 m (see Table 16.2)

Cross-sectional tower area: (605.7 m^3/day)/(8 m^3/m^2·day) = 75.7 m^2

Circular tower diameter: [(4) (75.7 m^2)/π]$^{0.5}$ = 9.82 m diameter

Projected tower volume: (3 m media depth) (75.7 m^2) = 227.13 m^3

Projected total media surface area: (98.4 m^2/m^3) (227.13 m^3) = 22,350 m^2

Projected average liquid trickling film depth: 0.2 mm

Projected total tower liquid volume: (0.2 mm) (22,350 m^2) = 4.47 m^3

Projected hydraulic retention time: (4.47 m^3)/(605.7 m^3/day) = 10.6 min

Biomass values
Estimated active biomass depth: 0.15 mm (or 150 μm)
Estimated total biomass depth: 1 mm
Estimated biomass density at 5% solids $= 50$ kg/m^3
Estimated active biomass mass:

$$(0.15\,\text{mm})(22{,}350\,\text{m}^2)(50\,\text{kg/m}^3) = 167.7\,\text{kg}$$

Estimated total biomass mass:

$$(1\,\text{mm})(22{,}350\,\text{m}^2)(50/\text{kg/m}^3) = 1118\,\text{kg}$$

Organic loading
Projected volumetric organic loading:

$$(72.7\,\text{kg CBOD/day})/(227.13\,\text{m}^3) = 0.32\,\text{kg/m}^3 \cdot \text{day}$$

Related note
Although this determination was simplistically developed for a single tower, multiple towers are typically designed by engineers to provide beneficial system redundancy. In similar fashion, while this design estimate was based on an average daily flow, engineering designs are typically based on higher likely flows, such as the facility's daily peak flow, in order to physically accommodate these types of hydraulic transient events.

The third and fourth design factors listed in Table 16.2, which are somewhat secodary in importance relative to the preceding loading terms, refer to the typical depths and levels of recirculation provided with these fixed-film systems. Older, rock-media designs were seldom built with media depths beyond 2 to 3 m, but many newer, plastic-media units are considerably taller (with depths up to 10+ m) and are designed to handle much higher loads. As for the notion of incorporating a recycle stream, this capability is considered necessary with influent flows that tend to have extreme lows (e.g., smaller towns, day-time-only industrial operations), such that the biofilm can be kept wetted during periods with influent lulls rather than having it dry out.

Extending beyond these heuristic design criteria, there are also a number of empirical models used to qualify the reaction-rate kinetics and expected reactor performance levels with attached-growth reactors. These models, typically named in honor of the responsible author or group, include the following variations: Eckenfelder, Rose, NRC, Howland, Schulze, Germain, Velz, and modified-Velz. Each of these models is considered to have its own merits, but the modified-Velz model, given in equation (16.1), is one of the more widely used strategies. As is the case with many of these competing models, the modified-Velz model follows a first-order format:

$$C_{\text{in}} = \frac{C_{\text{out}}\exp(kD\Theta^{T-20})}{\text{THL}^n} \tag{16.1}$$

where

$$C_{\text{out}} = \text{outlet substrate concentration (mg/L)}$$
$$C_{\text{in}} = \text{inlet substrate concentration (mg/L)}$$
$$k = \text{empirical reaction-rate constant}$$

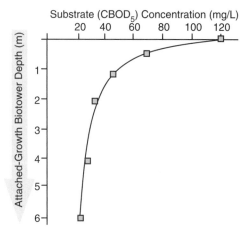

Figure 16.10   Representative pattern of exponential substrate removal relative to attached-growth biotower depth.

$$D = \text{media depth (m)}$$
$$\Theta^{T-20} = \text{temperature correction factor (relative to } 20°C)$$
$$\text{THL} = \text{hydraulic loading (based on influent plus recycle flow)}$$
$$n = \text{empirical loading constant}$$

Figure 16.10 accordingly presents a representative substrate removal profile derived with this model, as might be derived for a particular range of media heights relative to a given set of constant values for k, n, and Θ, In turn, this type of model does provide an informative characterization of biofilm performance, at least on a qualitative basis. However, on closer review, it would also appear that these types of models bear an inherent degree of fuzziness on several important issues.

The parameters of the modified Velz equation are sometimes reported with high precision (e.g., $n = 0.4332$; CH2M-Hill, 2000). However, it should be recognized that these types of documented constants reflect cumulative, averaged values derived from real-world biofilms whose actual site- and point-specific metabolic activities undoubtedly vary to a significant degree, depending on time, location, season, and so on. Indeed, these constants could well vary from reactor top vs. reactor bottom locations, and even from one substrata depth within a biofilm to another.

Yet another level of "fuzziness" exists with the total hydraulic loading rate (THL). The THL depends on influent flow and cross-sectional media area, but it does not actually account for the manner in which the influent is applied. For example, two parallel systems of equal design, construction, and hydraulic loading could have identical THLs, yet have sizably different levels of performance. In this case, the issue of "how" the influent is applied represents an important factor, as opposed to simply "how much," relative to both the rate of movement of the rotating influent distributor and the corresponding pulsing of flow across the media. With each successive radial pass, this rotating arm distributes a liquid stream whose downward passage across the media appears as a moving wave whose cyclic repetition follows that of the arm's travel speed. This pulsing and wavelike movement of liquid into and across the depth of the media would have an amplitude (i.e., height of the wave) that increased as the rotating arm was slowed. In turn, higher-amplitude waves would impose higher levels of physical shear and drag against the face of the biofilm, to an extent that would probably yield thinner attached-growth depths.

Undoubtedly, though, one of our largest levels of uncertainty with understanding attached-growth systems is our appreciation of the biofilm complexity in terms of biological makeup, physical conformation, and related metabolic behavior. Attached-growth system models akin to the modified-Velz equation implicitly suggest that biofilms are built and behave as homogeneous surfaces, with uniform levels of depth, kinetic activity, and structural composition, but this is hardly the case. Instead, biofilms comprise a highly nonhomogeneous, three-dimensional matrix with a decidedly complex physical structure. In short, attached-growth biofilms are configured as a densely arrayed microcosm of life, composed largely of bacteria but with several higher life-forms, including protozoans, rotifers, and nematodes, and possibly even fungus and insect forms as well.

This complexity within biofilm structures starts at an early stage during an initial period of startup and adhesion, where a preliminary layering of bacterial cells (i.e., a monolayer) binds to a clean media surface within a relatively short period of time (i.e., minutes to hours). This adhesive phenomenon stems from the presence of a sticky coating of exocellular polymeric materials secreted by bacterial cells growing under conditions of limited substrate availability, thereby allowing them to bind not only to media surfaces but also to other cells. In time, this monolayer coverage subsequently extends in thickness, progressively melding new and old cells as well as various enmeshed inorganic precipitates and particulate solids, to a point where the monolayer eventually reaches depths ranging from a few hundred to several thousand micrometers.

An apt analogy for mature biofilm growth is that of a forest canopy, with overarching branched limbs (newly grown cells) stretched across open glades (internal cavities gouged out by ongoing microbial sloughing and fluid shear) and underlying brush (deeper deposits of dead or inert cells). However, even then, this canopy will be apt to collect an added assortment of intertwined solids and cells, including a variety of higher life-forms living within the biofilm as well as entrapped wastewater solids. Given the chemical complexity of this film, yet another group of enmeshed, precipitated solids could well be wrapped into this matrix, including such materials as sulfide-, hydroxide-, phosphate-, or carbonate-based precipitates of iron, manganese, calcium, magnesium, and aluminum.

Of course, commensurate with this increased microbial depth, substrate and product transport through the film also becomes considerably more constrained, to a point where most biofilms eventually take on a vertically stratified layering. For those microorganisms living on the outer, aerobic edge of this biofilm (depicted schematically in Figure 16.11), their proximity to energy-rich substrates, nutrients, and oxygen received from the overlying bulk solution will allow them to reach the highest rates of aerobic metabolic activity. Figures 16.6 and 16.7 depict the type of bacterial growth found at this topmost, aerobic layer, including the considerable presence of filamentous forms, which give the biofilm its inherent outer roughness.

In particular, Figures 16.12 and 16.13 depict progressive enlargements of an extensive overlying growth of a filamentous *Beggiatoa*-type growth. Organisms such as this are often found living at this top-level biofilm region, catabolically using sulfide diffusing to the surface from the underlying anoxic–anaerobic (i.e., sulfate-reducing) zone. Strictly aerobic higher life-forms (e.g., most protozoans, rotifers, worms) are also commonly seen in this outer region, grazing steadily through this relatively active region.

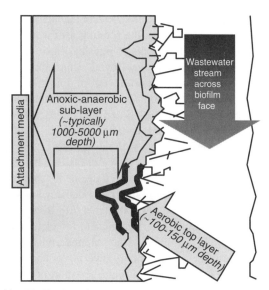

Figure 16.11   Biofilm layering (aerobic top vs. anaerobic bottom) schematic.

Moving downward into the biofilm, however, metabolic uptake will progressively deplete the available substrates and nutrients, imposing metabolic limitations that progressively retard the activity of these lower cells. Although no exact threshold can be predicted for the point at which the oxygen will be exhausted, interstitial measurements taken with microscale probes suggest that this point will occur roughly in the neighborhood of 150 μm.

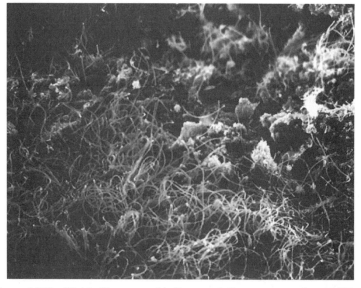

Figure 16.12   Highly filamentous biofilm morphology at top surface (∼800×).

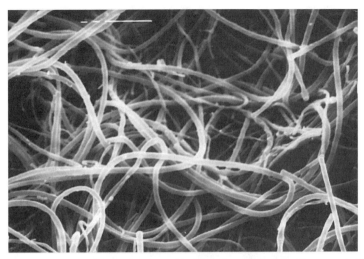

Figure 16.13   Biofilm surface close-up with dense Beggiatoa-type filaments (~2000×).

Moving deeper into the biofilm, therefore, the aerobic metabolic viability of the cells will decline as a direct consequence of diminished substrate and oxygen availability. Oxygen, in particular, declines steadily in availability as depth increases, to the point where it is depleted and anaerobic conditions begin. One such zone of transition from an overlying aerobic top layer to an underlying anoxic and/or anaerobic zone is depicted in Figures 16.14 and 16.15.

With oxygen reaching limiting conditions within the subsurface biofilm strata, anoxic respiration will accordingly deplete whatever alternative electron acceptors (i.e., nitrates, nitrites, etc.) might be present, eventually leading to sulfate reduction and a metabolic release of sulfides. Tucked just below the overlying *Beggiatoa* filaments shown in Figures 16.14 and 16.15 is a distinctly different type of bacteria, *Desulfovibrio*, which

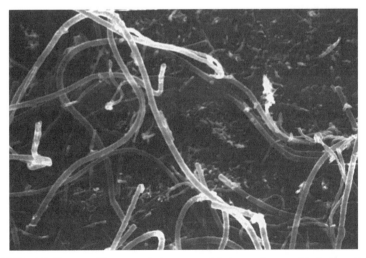

Figure 16.14   Biofilm transition zone between aerobic top and anaerobic bottom (~2000~).

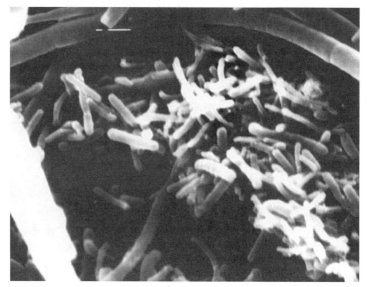

Figure 16.15   Biofilm transition zone (aerobic to anaerobic) close-up (~4000×).

is responsible for this sulfate reduction. These lower cells are more tightly packed and have a characteristic comma-shaped curvature. The last photograph in this series, shown in Figure 16.16, provides a close-up look at this group of cells.

The sulfides produced at this lower level will often scavenge and precipitate a variety of metals (probably dominated by iron sulfide, FeS, but undoubtedly including many other metal species), leading to a sort of cementation behavior that further inhibits chemical transport at these lower depths. Whereas optimal activity would be achieved with limited biofilm depths, therefore, the aging process experienced with biofilms leads to progressive metabolic constraints at the lower depths. Buried at a depth that limits their access to key

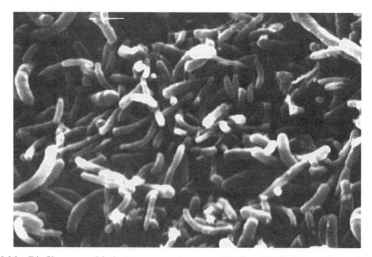

Figure 16.16   Biofilm anaerobic bottom zone close-up with Desulfovibrio-type bacteria (~4000×).

substrates and nutrients, as well as hampered by transport problems brought about by entrapped and precipitated solids, the underlying cells must inevitably shift to anaerobic (or at least anoxic) life-styles.

However, when viewed in terms of microbial diversity and potential treatment efficacy, the circumstance of having adjacently layered aerobic and anaerobic strata within a biofilm does appear to provide a potentially beneficial arrangement. Indeed, these strata could well complement one another, such as a reductive transformation of what might otherwise comprise recalcitrant organics (e.g., chlorophenolics, nitrophenolics, aniline dyes) within the anaerobic sublayer, followed by a concluding oxidation within the overlying aerobic layer.

At least in theory, it would also appear that it might hypothetically be possible to develop a beneficially layered sequence of nitrifying and denitrifying biofilms, such that nitrates produced in the oxidative overlying strata would then be directly inside reduced the lower anoxic–anaerobic depth. However, this conceptual scheme would be difficult to attain since the slow-growing nitrifiers are prone to overgrowth and submergence by faster multiplying heterotrophs, at which point competition for oxygen and other essential nutrients severely constrains their viability. As a result, nitrifying biofilm systems designed for oxidation of wastewater ammonia-nitrogen are usually configured as stand-alone processes, supplied with a pretreated wastewater whose biodegradable organic levels available for heterotrophic growth have been largely removed, and provided with specially designed media offering a far higher specific surface area more conducive to these slow-growing lithotrophs.

Once these attached-growth systems have been placed into use, though, and their multilayered biofilm configurations have been established, operating personnel have surprisingly few monitoring and control options with which they might try to improve or optimize performance. Merely measuring the attached mass or depth of biofilm within a reactor is quite difficult, although in rather rare instances, provisions have been made to track biofilm mass using removable sample coupons removed from inside the reactor.

Other than shifting reactor loading patterns by taking biofilm towers off- or online, the most significant control factor is probably that of altering the rotation speed for the distributors. In recent years, there has been a distinct shift to slow down the rotational speed of these distributing arms, dropping the number of rotations per minute (rpm) from values of 1 to 2 by a factor of 10 or more (i.e., to 0.1 or even lower rpm rates). Slowing down these distributing arms while maintaining the same influent waste flow means that the hydraulic loadings applied at the point of release are considerably higher. This increase subsequently raises the level of hydraulic shear as it moves downward across the face of the biofilm, and the added shear helps to keep the biofilm thinner and more active than would be the case with a much slower-moving water film.

Notwithstanding these operational constraints with control and monitoring, though, several significant developments have occurred with attached-growth media materials and configurations over the past several decades. In the realm of media options, rock media largely gave way to lighter plastic media with far higher specific surface areas midway through the twentieth century, and today's vendor options include a wide range of modular and random-packed versions. Table 16.3 provides specific technical details regarding several of these newer media forms compared against the older rock option. These newer plastic media provide several times more surface area per unit volume, which in turn yields considerable improvements in biofilm growth and loading capacities. In addition, plastic media are distinctly lighter and have far higher void fractions (i.e., the

TABLE 16.3 Attached-Growth Media Options

Biofilm Media Option	Nominal Size (cm)	Specific Media Surface Areas (m^3/m^2)	Media Void Fraction (%)	Media Density (kg/m^3)
Old rock media	10–12	30–60	40–60	800–1400
Standard plastic media	[a]	60–100	95–97	32–96
Standard random packed media	3–10	60–120	92–95	48–96
High-high plastic media	[a]	100–200	95–97	32–48
Ultrahigh random packed media	1–3	200–600	95–97	40–55

[a] Plastic media sheets are typically configured in ~60 × ~60 × ~120 cm bundles.
Source: Adapted from Metcalf and Eddy (2003).

portion of bed volume open to liquid and air flow rather than being occupied by the media itself), which subsequently helps to avoid problems with solids clogging that historically tended to affect the older rock media units.

As compared to the original stationary media mode, there have also been a number of recent developments based on rotational or fluidized media movement options. One such version shown in Figure 16.17 is called the **rotating biological contactor** (RBC). RBC

(a)   (b)

Figure 16.17 Attached-growth rotating biological contactor system: (*a*) covered tanks and gantry; (*b*) interior media view.

reactors are built as cylindrical media bundles with densely packed, rotating plastic sheets which may either be partially or fully submerged into the waste stream. During rotation of the partially submerged units, the attached biofilm is exposed sequentially to the wastewater and to the air, while submerged units may be operated in an aerated or even an anoxic or anaerobic mode.

Over the past two decades, several new strategies for attached-growth processing have been developed, including two options where the involved biofilms are maintained as suspended, millimeter-scale microbial clusters rather than using a traditional sheetlike configuration. The first such approach is typically aerobic, with small-diameter, inert media particles (typically, plastic or low-density fired-clay beads, sand, or activated carbon granules, etc., in the size range 1 to 4 mm) used as a biofilm carrier, with varied media densities used alternatively to secure either settleable or buoyant behavior. Because of the elevated biomass densities found in the latter options, their elevated oxygen uptake rates will require high-level aeration, such that these operations are commonly referred to as **biological aerated filter** (BAF) systems. These types of small media biofilm systems are typically operated in an up-flow operating regime, possibly with partial or full fluidization of the media during its operation on either a continuous or an intermittent basis and with intermittent backwashing being used (possibly with coaeration) to flush and extract entrapped solids. There are several advantages with the latter type of BAF reactor. First, the media usually has a cumulative surface area that is far higher than would be attainable had the same bed been packed with standard plastic media, and this increased surface area raises the mass of biofilm able to grow inside the reactor. To some extent the magnitude of this increase is tempered by the fact that physical rubbing of granules against one another keeps the biofilm thinner than what is usually seen in conventional trickling filter systems. However, the cumulative effect of having a denser and more active biomass is that the reactor loadings can be increased dramatically. In fact, volumetric organic loadings several times higher than that of standard fixed-film systems (Rittman and McCarty, 2001) have been reported with many of these newer BAF biofilm systems (e.g., at levels measured in the range 5 to 10 kg $CBOD_5/day \cdot m^3$).

Yet another innovative quasi-attached-growth concept uses suspended granular biofilm clusters whose dense, multimillimeter microbial matrix develops in a self-aggregated bead configuration rather than relying on the preceding inert support materials. These granulated biofilm clusters are then suspended in an *upflow anaerobic sludge blanket* (UASB) process, where incoming soluble COD is rapidly fermented to gaseous methane and carbon dioxide, which then rises and is released from the reactor through an overlying inverted-cone degasification cover. Internal recirculation of the granular biofilm cluster blanket is then achieved by upwelling lift provided via wastewater flow and initial gas buoyancy followed by subsequent downward settling of the denser microbial beads after gas detachment. Here again, these UASB systems carry extremely high granular biofilm densities, to the point where their permissible loadings with industrial-type soluble organics are among the highest possible with wastewater operations, ranging from 12 to 20 kg $CBOD_5/day \cdot m^3$ (Metcalf and Eddy, 2003).

Yet another hybrid approach for these biofilm systems couples attached-growth with suspended-growth processes using a **trickling filter/solids contactor** (TF/SC) scheme. These designs include a solids aeration tank placed between the biofilm tower and clarifier, with settled solids drawn from the clarifier underflow being recycled back to the intermediate solids contactor unit to enhance the uptake and separation of fine particulates. These intermediate tanks are typically not all that large, with HRTs usually below

1 hour, and their suspended solids levels are usually not all that high compared to conventional suspended-growth reactors at 700 to 1500 mg/L, but their ability to facilitate improve bioflocculation elevates the overall efficiency of the TF/SC process beyond that of standard biofilm processes, with expected removal efficiencies for BOD and suspended solids 10 to 15% higher than those of standard biofilm processes.

16.1.3 Suspended-Growth Systems

Suspended-growth processes used for wastewater treatment, such as that shown in Figure 16.18, have largely surpassed the attached-growth options over the past half-century in terms of general application, to the point where they have become the accepted state-of-the-art, best available practice for wastewater processing. There are several reasons for this preference, covering both performance and economic issues. These systems do have a higher energy demand to aerate and mix their suspended biomass, but they have earned a higher degree of confidence in terms of expected performance levels. These suspended-growth processes, commonly referred to as *activated sludge systems*, also tend to have better economy of scale in construction costs, whereas attached-growth is often used in smaller installations, due to its ease of operation and simpler mechanical design.

Figure 16.18   Suspended-growth (i.e., activated sludge) wastewater treatment facility.

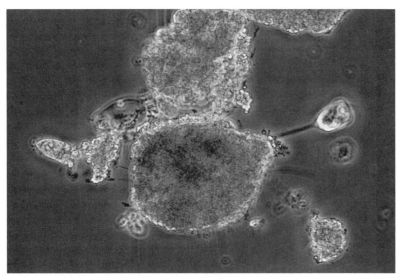

Figure 16.19 Typical light microscope image of suspended-growth (i.e., activated sludge) floc conformation.

The key biological structure within suspended-growth processes is that of an amorphous, clustered structure (see Figure 16.19) known as **floc**, whose constituent microbial population sustains a desired biochemical oxidation of wastewater contaminants. Like its biofilm counterpart, floc is composed of a complex aggregate of various bacterial and higher life biota, plus an added quantity of nonliving organic and inorganics material. Suspended-growth floc particles exist as free-floating structures whose physical, chemical, and biological exchange with the surrounding bulk solution functions in three dimensions, as compared to a biofilm sheet which is confined against the inert surface to which it is attached. Floc suspensions are referred to as **mixed liquor**, and the suspended solids concentration of this is denoted mixed liquor suspended solids (**MLSS**). These floc concentrations are usually maintained at total volatile suspended solids concentrations ranging from 1 to several grams per liter.

As was the case with biofilm development, the circumstance of having bacterial cells stick together in floc particles depends on starvation conditions which trigger an excellular release of adhesive exocellular polysaccharides. In turn, individual cells are able to self-adhere or **flocculate** into larger, enmeshed floc clusters. Floc dimensions are considerably smaller than multimillimeter biofilms, though, with typical diameters of 250 to 500 μm, and this limit is probably controlled by shear forces encountered during routine mixing, aeration, and pumping. Given that floc biomass is composed largely of bacterial cells, the specific gravity of these particles also tends to be only slightly higher than water (about 1.03 to 1.05), such that these relatively light floc particles can be kept in suspension with only moderate agitation and aeration.

Flocculation behavior also plays a role in terms of bacterial survival as a whole. When an organism metabolically releases any substance outside its boundaries, there is a cost that must be accompanied by a return. Only under conditions of scarcity is this investment worthwhile for the microbe, and in the case of floc physiology the benefit is that of enhancing the entrapment of particulate substrates. The secretion of exocellular digestive

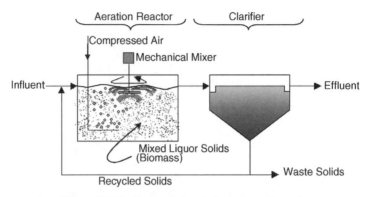

Figure 16.20 Suspended-growth design schematic.

enzymes by a single cell could result in most of that enzyme uselessly diffusing away from the cell, such that a large fraction of any dissolved substrate formed by digestion of a particle would also diffuse away. While clustered within an aggregated floc community, though, bacteria are cooperating by releasing enzymes that benefit them collectively.

The physical structures used in suspended-growth or activated sludge systems include both a reactor basin and a settling tank (Figure 16.20). Within the reactor basin, biodegradable substrates are then absorbed by the floc particles, along with oxygen and other key metabolic elements (e.g., nutrients, trace metals), and used to sustain a complex set of catabolic and anabolic reactions by which the wastewater is then cleansed biochemically. In turn, a subsequent settling tank is used to settle suspended floc biomass in preparation for recycle or wastage.

To maintain desired aerobic conditions inside the reactor, oxygen must be added routinely either by introducing compressed air or oxygen via **dispersed air** bubbles or through various **mechanical aeration** mechanisms. Extending beyond these reactor and clarifier components and their affiliated aeration equipment, suspended-growth systems will also require a complementary set of valves, pipes, and pumps used to control the reactor feed, as well as additional hardware for clarifier underflow recycling and wastage plus effluent discharge.

The settling tank, also called the **secondary clarifier**, has one basic function, to separate the floc biomass from the effluent. This operation can be divided into two complementary functions: clarification and thickening. **Clarification** is intended to produce a clear effluent that is nearly free of suspended solids. **Thickening** is the production of a concentrated underflow sludge which is then pumped from the bottom of the clarifier. Most of the thickened sludge is returned to the aeration tank [as a **recycle** or **return activated sludge** (RAS) stream], but a portion of the sludge will also be **wasted** and removed from the process for further sludge processing. In turn, this wastage turns out to be the key control parameter for a suspended-growth process, which by varying the age of this biomass (its SRT) controls net growth rate, will then regulate overall operational efficiency.

In this context of qualifying discrete reactor and settling operations, though, it should be noted that some suspended-growth systems do not follow this convention. For example, **sequencing batch reactor** (SBR) systems use individual tanks as both reactors and clarifiers, where aeration and settling phases are then sequenced during batchwise operating cycles. Another suspended-growth system, which is just beginning to see wider utilization in full-scale systems, is that of the **membrane bioreactor** (MBR), where semipermeable synthetic membranes are used for biomass separation in lieu of a settling tank. In these units, mixed liquor is passed across and through the membranes to obtain an effluent that is essentially free of all suspended solids, thereby directly retaining floc biomass within the aeration tank.

Numerous design options exist with respect to the manner in which the waste is introduced to these suspended-growth reactors, how the reactor system might be arranged in physically or temporally separated stages, the condition (i.e., aerobic, anoxic, anaerobic) of these stages, and the rates employed for hydraulic throughput and solids discharge. The vast majority of systems are maintained in a fully aerobic state for oxidative biodegradation of incoming organics, and in recent years an escalating number of these aerobic systems have also been used to maintain an additional oxidation of reduced nitrogen (ammonia and organic nitrogen) to nitrate by way of nitrification. In addition, there are also a variety of suspended-growth design options by which full nitrogen removal, including denitrification, as well as biological phosphorus removal can be achieved by, respectively, using an additional complement of suitably designed and controlled anoxic and anaerobic phases.

Six design variables are considered in the development of suspended-growth systems, including hydraulic residence time, solids residence time, specific substrate loading (mass substrate applied per mass cells per time), volumetric loading, reactor configuration, and recirculation rate (recycle flow relative to raw influent flow) ratio. Table 16.4 provides a general synopsis of the typical design criteria relative to four of the basic options for configuring these reactors, including conventional, high-rate, extended aeration, and nitrification systems. In addition, further engineering consideration must be given to the anticipated reactor solids levels, excess sludge production rates, solids–liquid separation methodologies, and overall environmental conditions (e.g., operating temperature).

The HRT and SRT were described in Section 16.1.1. The third variable, specific substrate loading, is often referred to in terms of a **food-to-microorganism** (F/M) **ratio**. Of course, the form of the food term will vary according to the purpose of the reactor and whether its intended goal was that of carbonaceous BOD, COD, ammonia-nitrogen, and so on, removal. Similarly, there are several techniques for quantifying "mass," but the most common procedure is to use **mixed liquor volatile suspended solids** (MLVSS). At steady state, the F/M ratio is inversely related to SRT. Specifying one is equivalent to specifying the other, and either one can be used for design purposes.

TABLE 16.4 Typical Suspended-Growth System Design Criteria

Suspended-Growth Design Factors	Hydraulic Residence Time (h)	Solids Residence Time (days)	Specific Substrate Loading (F/M) Rate (kg/kg VSS·day)	Volumetric Substrate Loading Rate (kg CBOD/m^3·day)	Recirculation Ratio $Q_r Q$, Relative to Influent Flow
Conventional	4–8	5–15	0.2–0.4 (as CBOD)	0.32–0.64	0.25–0.75
High-rate	2–4	3–5	0.4–1.5 (as CBOD)	1.6–16	1–5
Extended aeration	18–36	20–30	0.05–0.15 (as CBOD)	0.16–0.4	0.5–1.5
Nitrification	6–12	>8*	0.05–0.15 (as N)	0.2–0.4 (as N)	0.5–1.5

[a]Operating SRTs necessary for nitrification vary with temperature: >4 days in summer, >8 days in winter.
Source: Adapted from Metcalf and Eddy (2003).

The control strategies used with suspended-growth systems are more sophisticated than those used in attached-growth wastewater processes (Bailey and Ollis, 1977), including that of controlling the aeration plus clarifier recycle and wastage flow rates. The major operational objectives for suspended-growth systems, though, are to maximize effluent quality while minimizing costs for disposal of **waste biomass** or **sludge** (the production of which decrease with SRT) and energy used with mixing and aeration (which increase with SRT), and in this context, controlled biomass wastage represents the key control factor. In turn, there is a delicate balance with controlling SRT to maximize effluent quality while minimizing operating costs, to an extent that considerable experience is necessary for optimal process control given the dynamic impacts of seasonal temperatures, changes in wastewater character, and so on.

In general, control regimes designed to sustain higher rather than lower SRT conditions are often preferable, in that their correspondingly reduced (i.e., starvation) growth rates would correspond with low-level effluent substrate concentrations. As was the case with similarly starved biomass growth within adhesively bound biofilms, suspended-growth biomass maintained at these elevated SRT levels will also tend to produce beneficial, floc-forming exocellular polysaccharides.

In turn, the development of improved floc-forming biomass would be similarly beneficial in terms of desired settling performance. It is imperative that the suspended biomass be able to settle and compact effectively, forming a **sludge blanket**, so that these solids can be recycled back to the aeration tank rather than being lost in the settling tank's effluent. Settling behavior is, consequently, tracked in most systems. The oldest and still likely to be the most common method of monitoring settling is that of the **sludge volume index** (SVI). The SVI is defined as the volume in milliliters occupied by 1 g of sludge after 30 minutes of settling. A settling test is completed by adding mixed liquor to a **settleometer** (a vertical glass or plastic cylinder) or standard laboratory graduate cylinder (Figure 16.21). Although there are a number of variations, the basic test consists of allowing the sludge to settle for 30 minutes, then noting the volume of settled sludge [settled volume (SV)]. This value is then divided by the MLSS concentration:

$$SVI\,(mL/g) = \frac{SV(mL/L)}{MLSS(g/L)} \tag{16.2}$$

An acceptable SVI value (i.e., producing good settling) is considered to be in the neighborhood of 100 mL/g. An SVI greater than about 200 indicates bulking conditions, where settling performance is less than satisfactory. Values in between must be judged depending on the design and operating conditions of a particular plant. The settled 1-L sample shown on the right-hand side of Figure 16.21, for instance, has a 30-minute settled volume of 250 mL, and with a solids concentration of 2.5 g/L, the corresponding SVI would be 100 mL/g. Although rather crude, this simple test provides an approximate feel for the settleability of these solids, and when tracked on a recurring basis, the operating staff can predictably qualify "good" vs. "bad" behavior.

Beyond the SVI test, though, suspended-growth reactors are monitored in a fashion that pays a higher level of attention to the biological aspects of the involved biomass than is the case with the various attached-growth processes, taking into account such issues as expected rates of cell growth, decay, and respiration, as well as the overall morphology and consortia maintained within the involved biomass. For example, dissolved oxygen levels in the aeration reactors are monitored routinely, both to ensure appropriate

Figure 16.21   Sludge volume index testing for suspended-growth biomass.

oxygen transfer and to track biomass activity qualitatively. More precise measurements of mixed-liquor **oxygen utilization rate** (OUR) levels are also completed at many facilities on a frequent basis, using **respirometers** to evaluate biomass oxygen uptake such that they might quickly pinpoint potential problems with incoming wastewater toxins or other problematic metabolic conditions.

In yet another attempt to upgrade their monitoring efforts, many suspended-growth operations have also adopted routine measures to examine their floc with a light micro-scope on a recurring basis. This effort can be used to track chronological changes in its overall appearance, conformation, and approximate average diameter of a system's floc particles. At the same time, density counts taken with other higher life-forms, including those for protozoans and rotifers, may be recorded to evaluate their relative presence, or absence, as this factor may also signal symptomatic changes.

Floc particles may, in fact, have less microbial and metabolic diversity than biofilms (which often contain both aerobic and anaerobic zones), but they do enjoy several appar-ent advantages in terms of potentially securing maximal degradation rates. First and fore-most, the combined effects of submillimeter floc diameters and their inherently open structure as a whole means that their cells will have far closer proximity, and easier access, to substrates and oxygen coming from the surrounding bulk solution. In turn, it

Figure 16.22 Light microscope image of suspended-growth floc structure with Vorticella protozoans.

is highly likely that a far larger fraction of the biomass will be able to sustain oxidative, aerobic activity compared to that found in biofilms. Furthermore, this enhanced transport of substrates and nutrients into the floc would be mirrored by a comparable improvement in the release of unwanted metabolic products. Finally, the overall surface area of these floc particles, on a cumulative basis that factors in both outer surface and internal subsurface exposure, will be considerably higher than that of a comparably sized (similarly loaded) biofilm system. Here again, the higher surface area not only provides for better kinetics but also contributes to improvements in a range of supplemental impacts, including sorption, filtration, and chelation.

Most floc structures will also include any number of higher life-forms, including attached and free-swimming protozoans (Figures 16.22 and 16.23) and rotifers (Figure 16.24). These organisms are highly beneficial and symptomatic of desirable floc conditions in that their presence sets up a tiered food chain which at the bottom end helps to maintain a clean effluent by scavenging finer, unattached solids and free-swimming bacteria that might otherwise linger during clarification as a fine, turbid haze. The stalked protozoan species, *Vorticella*, seen in Figures 16.22 and 16.23 represents one such desirable scavenger, whose life-style and mode of operation is similar to that of a free-wheeling vacuum cleaner. This organism's contracted cilia, seen in top-down view in Figure 16.23, will open and extend outward into the bulk solution and then whirl rhythmically to create rotational fluid currents that pull unattached particulates in the bulk solution toward, and then into, this organism's mouthlike opening, where they are then ingested.

Problems with floc conformation and behavior may, however, be encountered, including various difficulties with settling or clarification. Table 16.5 provides a synopsis of these conditions. In fact, a considerable number of suspended-growth process problems are related to poor settling and clarification. Although settling is a purely physical step, it is largely floc microbiology and conformation that determines the settling characteristics of the sludge.

Figure 16.23   SEM image of suspended-growth floc with Vorticella protozoans.

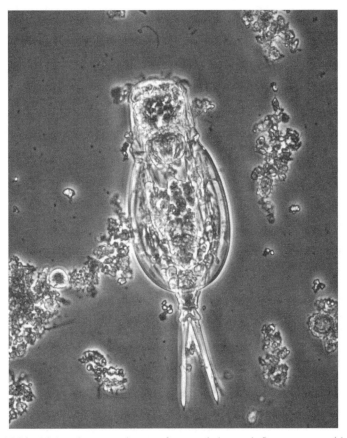

Figure 16.24   Light microscope image of suspended-growth floc structure with rotifer.

TABLE 16.5   Suspended-Growth Floc Problems and Possible Contributing Factors

Type of Floc Problem	Operational Difficulties	Possible Contributing Factors
Bulking floc	Poor thickening, resulting in secondary settler overloading	Low dissolved oxygen concentration Low F/M loading Insufficient soluble BOD Nutrient deficiency Low pH Completely mix reactor mode Septic wastewater High wastewater sulfide concentration
Zoogleal floc	Branched, fingerlike floc filaments	High F/M loading Low SRT
Jelly floc	Excessive floc exocellular slime	Nutrient deficiency
Dispersed floc	Floc destabilization and breakdown Hazy, cloudy effluent	High influent salt concentration High influent surfactant concentration Extreme high or low SRT
Pin floc	Elevated effluent suspended solids	High SRT
Rising sludge	Floc flotation vs. settling	Denitrifying sludge blanket and gas release Anaerobic sludge blanket and gas release
Foaming	Stabilization of bubbles Floc solids bubble entrapment	High influent surfactant concentration Hydrophobic cell surfaces Possible oil and grease recycle in skimmings

Certainly, one of the most frequent problems with suspended-growth systems is excessive filament presence, which leads to a condition referred to as **bulking**. Healthy, well-structured floc usually contains a nominal-to-moderate level of filament presence (Figures 16.25 and 16.26). In fact, it has been proposed that filaments may serve as a "backbone," giving strength to floc and allowing them to grow larger. Recent advancements with confocal microscopy have helped to confirm this backbone role with the filaments found in floc particles. The image shown in Figure 16.27 was clipped from an animated video sequence whose sequential frames depict the progressive rotation of a floc particle with a loosely arrayed, and highly nonhomogeneous, structure bearing several backbone-type filament threads. However, these filamentous forms are sometimes present in excessive amounts, extending out from the floc into the bulk solution, thereby leading to bulking behavior. Under these conditions the filaments can impede sedimentation in the secondary settling tank by two mechanisms. First, they act as a sort of parachute, increasing viscous drag and reducing the settling velocity of the flocs. Second, the filaments act to fend the flocs off from each other, preventing them from compacting

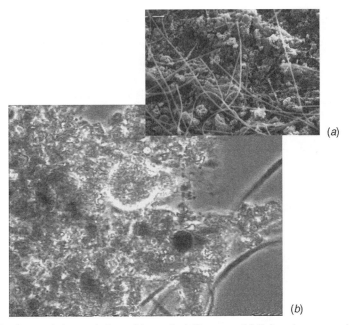

(a)

(b)

Figure 16.25   Suspended-growth floc with nominal filaments: (*a*) light microscope view at 400×; (*b*) SEM view at ∼600×.

toward the bottom of the tank, such that the sludge then takes on a "bulky" condition, with SVI values in excess of 150 mL/g.

Taken to an extreme, filaments may replace almost completely any semblance of a normal floclike structure (Figures 16.28 to 16.30). For many years, one particular sheathed bacteria, *Sphaerotilus natans*, was blamed routinely as the responsible agent for filamentous bulking problems, but further study demonstrated that several other bacterial forms, and occasionally fungi, could be involved. During the early 1970s the work of D. H. Eikelboom in The Netherlands developed a systematic microscopic method which

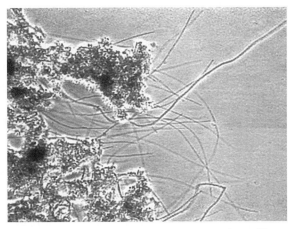

Figure 16.26   Suspended-growth floc (400×) with moderate filament growth.

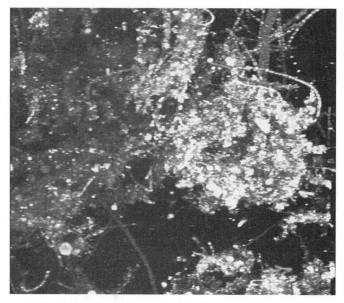

Figure 16.27 Confocal microscopy image of suspended growth floc structure.

recognized more than two dozen different filament types that could contribute to bulking events (Eikelboom and van Buijsen, 1981). Eikelboom's classic work, which was then continued and expanded by a group working with David Jenkins in the United States, strongly suggested that several causes were possible, each as a result of the excessive growth of a different type of filamentous organism. Table 16.6 provides an overview of

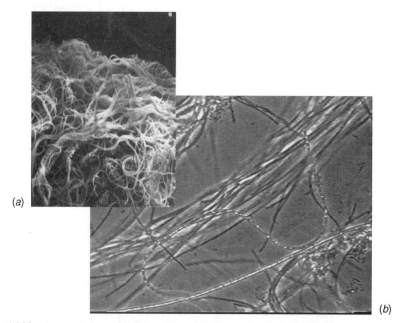

Figure 16.28 Suspended-growth floc with heavy filament growth: (*a*) light microscope view at 400×, (*b*) SEM view at ~800×.

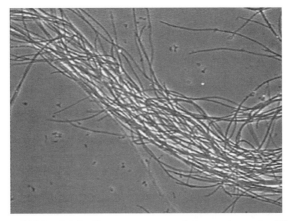

Figure 16.29 Solely filamentous suspended-growth matrix (400×).

the various filamentous forms they typically found, including some forms that have only been distinguished with numbers (as originally assigned by Eikelboom) since they have not been formally identified and given official genus and species names (Jenkins et al., 1993). Table 16.7 provides a synopsis of the correlation observed between the various microbial filaments and their contribution to bulking problems.

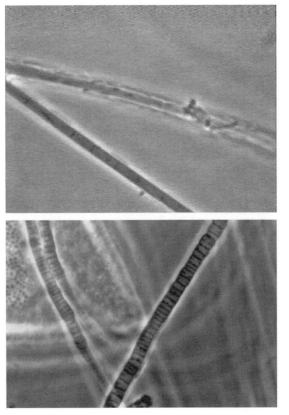

Figure 16.30 Magnified images of floc filaments.

TABLE 16.6 Microbial Factors with Filamentous Bulking and Foaming in Activated Sludges

| Rank | Bacterial Agent | Percentage of Facilities Experiencing Designated Problematic Bacterial Growth | |
		Primary	Secondary
1	*Nocardia* sp.	31	17
2	Type 1701 bacteria	29	24
3	Type 021N bacteria	19	15
4	Type 0041 bacteria	16	47
5	*Thiothrix* sp. bacteria	12	20
6	*Sphaerotilus natans*	12	19
7	*Microthrix parvicella*	10	3
8	Type 0092 Bacteria	9	4
9	*Haliscomenobacter hydrossis*	9	45
10	Type 0675 bacteria	7	16
Other forms	Bacterial types 0803, 1851, 0961, 0581, and 0914, plus *Nostoicoida* sp., *Beggiatoa* sp.; undefined fungi		

Source: Adapted from Grady et al. (1999).

Knowing which condition is the cause of a particular bulking episode—by determining which filamentous types are present—is of course very helpful in developing control strategies. Bulking often appears in relation to extremely low organic loadings (extremely low F/M or high SRT) or low dissolved oxygen concentrations in the aeration tank. Occasionally, high sulfide concentrations, either from the influent wastewater or produced in anaerobic zones within the treatment plant, can promote growth of filamentous sulfur-oxidizing bacteria. Low pH can also lead to fungal bulking, but this is a rather rare occurrence when pH control is available. Inadequate nutrients, especially nitrogen and probably phosphorus, can also lead to a form of bulking or to **jelly sludge**, in which excessive levels of exocellular slime materials are produced. However, this behavior is more apt to occur with unique industrial wastewaters (e.g.,

TABLE 16.7 Correlations Between Filamentous Organism Types and Causes of Bulking

Cause of Bulking	Indicative Filament Types
Major (occurring commonly)	
Low organic load	0041, 0092, *Microthrix parvicella*, *Nocardia*-like spp. 0581, 0803
Low dissolved oxygen	1701, 021N, *Sphaerotilus natans*, 1863
Minor (occurring infrequently)	
Low pH	Fungi
Sulfides	*Thiothrix*-like, *Beggiatoa*, 021N
Low N (or P)	*S. natans*

with canning operations), since domestic sewage usually has a considerable excess of nitrogen and phosphorus.

Yet another type of bulking, which is rather less common than the filamentous version, is caused by a particular type of floc structure largely composed of the bacterium, *Zoogloea ramigera*, and for this reason this problem is referred to as *zoogloeal bulking*. These flocs are characterized by branched, fingerlike projections that can have an effect on settling similar to that of other filaments. Whereas zoogloeal flocs are seen in small numbers in perhaps 10% of suspended-growth systems, they are present occasionally in excessive amounts and cause bulking. They are favored by high organic loading rates (high F/M, low SRT). At one time, early investigators thought that these zoogleal species were the dominant, and perhaps only, floc-forming microorganisms in suspended-growth systems. This assumption is now known to be incorrect, but the (improper) use of the term *zoogloea* in reference to all floc forms (rather than the specific unusual type of floc) persists among some in the field.

Note that not every filamentous type has been linked directly with a specific cause of bulking. Also, there is still some debate about some of the relationships. For example, one filament, generically referred to as type 021N, in addition to being a sulfide oxidizer, has been associated by some with low dissolved oxygen (DO) and by others with low F/M. Part of this disagreement may stem from the likelihood that some of the filamentous types established using only the microscope may actually be composed of more than one species. Future efforts to apply molecular biology tools will undoubtedly help to clear up some of this confusion.

Reactor design may also influence bulking. For example, with completely mixed reactor designs, wastewater and recycled sludge are vigorously mixed and rapidly distributed throughout the tank, such that the incoming wastewater is rapidly diluted as it enters the reactor. This circumstance then leads to uniformly low substrate concentrations within the aeration tank, a condition that favors the low-F/M filamentous bacteria. Bacteria living within these flocs would be at a disadvantage to the extended filaments, which reach out into the bulk solution to an extent where they have far higher access to substrates. By comparison, an alternative plug-flow reactor configuration has no such initial mixing, with wastewater and return sludge being introduced at the head end of the linear reactor and then flowing together down the length of the tank, or through several tanks in series. In turn, there is far less initial dilution of the incoming substrates, and the elevated initial substrate concentration facilitates a higher rate of diffusion into the floc to an extent which then negates the physical advantage of extended filament growth. However, plug-flow systems are more apt to experience higher initial oxygen uptake rates at the head end of their operations, and unless care is taken to meet this higher demand, there may still be a contrary opportunity for filament growth due to low-DO bulking. Yet another innovative approach to the control of filament bulking is to use a special reactor configuration designed specifically to optimize and select for the growth of floc-forming bacteria at the expense of filaments. These **selector** systems with suspended-growth processes are typically quite small, usually less than 1% of the total aeration tank volume and placed just ahead of the main aeration tank and then mixed and/or aerated to promote aerobic or anoxic conditions favorable to floc formers.

Another approach for control of nuisance filaments would be to apply chlorine or other strong oxidant (e.g., hydrogen peroxide) to the return sludge on an intermittent basis. Because the filaments are more exposed to the solution than are the floc formers, being extended outward from the floc into the surrounding bulk solution, they are also more susceptible to toxins such as these oxidants. However, the recurring necessity for using this method can be costly, and in the case of high-level chlorination there are also second-ary concerns stemming from the potential formation of toxic trihalomethane compounds within the mixed liquor.

Contrasted with the circumstance of excess filaments and bulking, though, there are also problems with settling stemming from other difficulties tied to unusual floc confor-mations or problems. One such condition, referred to as **pin floc** or **pin-point floc**, stems from the absence of any filaments whatsoever. These small-diameter floc forms are affected particularly by high-level aeration, which imposes excessive shear forces leading to floc disaggregation. Since there is no "net" of filaments that would otherwise strain out these particles physically, pin floc end up getting left behind during settling and thus con-tribute to a higher-than-desired effluent suspended solids concentrations.

Undesired floc breakdown and **dispersed growth** can also be encountered in a variety of instances. High-level influent concentrations of salt or surfactants have been found to destabilize floc structures, and short SRT systems with high-rate bacterial growth rates have shown a tendency toward dispersed rather than flocculated growth, due to a lack of exocellular polymer release. **Toxic shock** impacts have also been proposed as a factor behind dispersed growth, where a toxic shock initially kills off a portion of an existing biomass, and then the remaining organisms grow extremely fast, in an unflocculated fash-ion, because of the reduced competition for substrates.

Foaming is yet another problem that can occur in mixed liquors. Occasionally, this may be the result of a nonbiodegradable **surfactant** (a surface-active agent) that enters the system. Other times, degradable detergents and other foaming materials (including proteins and DNA) that are present in the wastewater may not be degraded quickly enough and may lead to surface tension reductions that trigger the release of foam. This typically occurs either during startup of a plant and with very short SRTs, or after washout of sludge during a storm. Both of these types of foam are usually white and can be controlled with defoaming compounds or water sprays. The most common and serious foaming problems, however, are the result of excessive growths of certain gram-positive filamen-tous bacteria. These seem all to be actinomycetes, and most are highly branched. They were originally called *Nocardia*, but it is now recognized that they also include several related genera, including *Gordonia, Skermania, Rhodococcus*, and *Tsukamurella*. Occasionally, the bulking organism *Microthrix parvicella* (which is not branched) also causes this type of foaming. These bacteria seem to have a waxy coating, which makes it hard to rewet them once they get lifted into the foam layer. They are slow-growing and therefore are favored by longer SRT conditions. However, once they form a foam, they may accumulate on the surface of the aeration and/or secondary settling tanks to an extent that control of waste to achieve a desired SRT is extremely difficult. In severe cases, half of the system's biomass may actually be in the foam, since it gets so thick. The foam may even overflow the tank, or get pushed out by the wind, creating a slippery safety hazard on walkways as well as an aesthetic problem. The foaming bacteria appear to be able to utilize hydrocarbons and other fats and oils that may not readily be available to many of the other organisms present. Defoaming compounds are not effective in their control, and in some cases may even be utilized by the organisms as a carbon and energy source. Instead, a combination of decreased sludge age and physical removal of the foam (by skimming or vacuum) is generally used. Improved oil and grease removal in primary treatment, and spray chlorination of the foam, may also be beneficial.

Occasionally, sludge will settle well in the clarifier and then suddenly bob back up to the top. This results from gas bubble formation in the sludge blanket while it is sitting at the bottom of the settling tank. This is called **rising sludge** if the bubbles are composed of nitrogen gas (N_2), formed as the result of denitrification. Since denitrification can take place only if nitrate and/or nitrite are present while oxygen is absent, this means that sludge rising can occur only if nitrification has first occurred in the aeration tank and then the sludge has gone anoxic in the settling tank. In cases where no nitrate or nitrite are present, **anaerobic sludge** can float to the top as a result of the production of other gases, especially hydrogen (H_2). These problems can be controlled by increasing the rate at which sludge is returned to the aeration tank, so that settled biomass within the clarifier's sludge blanket does not have a chance to go anoxic or anaerobic.

Another important aspect in the microbial makeup of a suspended-growth reactor's biomass involves the absence, or presence, of specialized bacteria uniquely acclimated for the metabolic uptake or conversion of specific contaminant species, including not only nutrients such as nitrogen and phosphorus but also a wide range of potentially inhibitory, and yet still biodegradable, compounds (e.g., phenolics, thiocyanates) (Bitton, 1999). The desired acclimation and retention of these specialized bacteria would be analogous to that of their development within fixed-film systems, and there is limited evidence to suggest that floc particles may also provide, at least partially, a comparable sequencing of cometabolic aerobes and anaerobes located within the exterior and interior regions respectively.

In the particular case of pursuing complementary oxidative and reductive nitrogen removal steps, design and control measures to secure the presence and vitality of nitrogen-oxidizing and nitrogen-reducing bacteria within biomass will determine whether nitrification and denitrification were able to occur. Given the low growth rates commonly maintained by nitrifying bacteria, the key to achieving nitrification within an activated sludge is that of retaining reactor biomass for periods (maintaining an SRT) in excess of 4 days (and perhaps as much as 8 to 10 days during cold-weather periods). At this point, a sufficient (albeit, even then rather small—at levels of a few percent or lower of the total viable cells) inventory of these nitrifying lithotrophs can be kept to effect the desired oxidation of ammonia to nitrate.

This strategy for securing nitrifiers has subtle aspects that are inherently similar to that associated with fixed-film systems. A competitive advantage for nitrifiers in fixed-film systems is maintained by limiting the amount of incoming organics available to competing heterotrophs; on the other hand, suspended-growth nitrifying reactors maintained at high solids retention times are able to produce similarly low bulk solution levels of available organic carbon, such that the faster-growing heterotrophs are energetically constrained and unable to impose competitive stress. In fact, the process of nitrification has now become a rather commonplace requirement for a considerable number of wastewater treatment facilities given the inhibitory impact caused by low-level (sub-ppm) free ammonia concentrations on fish living within downstream receiving waters. Taken one step further, many newer activated sludge systems are also being designed to encompass full nitrogen removal, thereby matching ammonia oxidation (via aerobic nitrification) with nitrate reduction (via anoxic denitrification).

This combined nitrification–denitrification practice provides a number of potential benefits, including (1) the ability to achieve enhanced levels of overall nitrogen removal, thereby reducing the downstream impact of this prospective nutrient; (2) the ability to secure a recovery of the oxidizing "power" provided by nitrates, thereby conserving

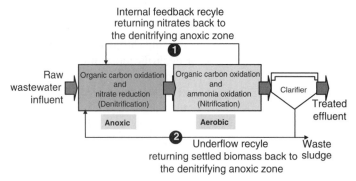

Figure 16.31   Representative two-tank plus two-recycle design scheme for biological nitrogen removal. (*Note*: This particular arrangement is referred to as the *modified Ludzack–Ettinger system*, based on its modified addition of an internal feedback recycle stream to the original Ludzack–Ettinger design scheme.)

the energetic investment previously made to create these oxidized products during aerobic nitrification; and (3) the ability to achieve a similar recovery of alkalinity during denitrification, thereby partially offsetting the alkalinity loss realized during prior nitrification.

Suspended-growth systems designed for combined nitrification and denitrification systems tend to fall into one of two alternative design categories, including zoned and temporally sequenced (i.e., flip-flopped aeration and mixing) arrangements to secure the necessary aerobic and anoxic phases. One such popular, zoned design strategy, known as the *modified Ludzack–Ettinger* (MLE) *scheme* is depicted in Figure 16.31. The first, anoxic stage in this MLE system mixes incoming raw wastewater with a nitrate-bearing flow brought back from the trailing aerobic nitrification reactor by way of an internal feedback recycle stream, thereby facilitating the conditions necessary for efficient reduction and removal of this recycled nitrogen. Within this initial anoxic zone, therefore, denitrifiers employ incoming organics as their electron donor while reductively using nitrates as an electron acceptor, thereby producing nitrogen gas.

Example 16.2: Preliminary Activated Sludge Design   After presenting their preliminary design estimate for a standard trickling filter with the Deer Creek, Illinois, community (see Example 16.1), the community's consulting engineer was asked to develop yet another preliminary design estimate for a standard secondary activated sludge system which might be used alternatively for this facility's secondary treatment process. As explained in Example 16.1, the expected average daily wastewater flow for this community will be 605.7 m³/day, and the carbonaceous organic loading (CBOD) entering this secondary activated sludge reactor system will be 72.7 kg BOD/day.

Preliminary Design Details

Hydraulic	Design hydraulic retention time (HRT): 4 h (see Table 16.4)
loading,	Design solids retention time (SRT): 8 days (see Table 16.4)
system	Design reactor depth: 5 m
sizing,	Projected reactor volume: $(605.7 \text{ m}^3/\text{day}) (4 \text{ h}) = 100.9 \text{ m}^3$
and retention	Projected reactor surface area: $100.9 \text{ m}^3/5 \text{ m} = 20.2 \text{ m}^2$
times	

Biomass content	Estimated theoretical biomass yield (Y_T): 0.5

Estimated biomass endogenous decay and death rate: 0.05 day^{-1}

Estimated active reactor biomass concentration:

$$(SRT/HRT)(Y_T/(1 + k_d \cdot SRT)(CBOD_{influent} - CBOD_{effluent}) =$$
$$1920\, g/m^3$$

Estimated active biomass mass: (1920 kg/m^3) (100.9 m^3) = 193.8 kg

Estimated total suspended solids mass: (active biomass/0.75) = 258.4 kg

Organic loading

Projected organic loading:

$$(72.7\, kg\ CBOD/day)/(100.9\, m^3) = 0.73\, kg/m^3 \cdot day$$

Projected F/M (food/microorganism) ratio:

$$(72.7\, kg\ BOD/day)/(258.4\, kg) = 0.38\, day^{-1}$$

Related notes

1. As with Example 16.1, this preliminary design was based on a single reactor receiving an average daily flow, as compared to multiple reactors and peak etc. design flow conditions.
2. The total suspended solids vs. active biomass differences was based on a typically observed biomass fraction of ~75% vs. an additional inert solids content of ~25%.
3. The biomass concentration derivation given above takes into account composite biomass growth for CBOD depletion, plus decay and death, with an assumed effluent BOD of 10 mg/L; further details regarding this analysis can be found in *Wastewater Engineering* (Metcalf and Eddy, 2003).
4. In comparison to the prior trickling filter design estimate, there are several important design differences, including:
   a. *Considerably different contact times* (~hours at activated sludge vs. ~seconds or minutes for the trickling filter).
   b. *Considerably different total reactor volumes* (the activated sludge system is ~2+ times smaller).
   c. *Roughly comparable active biomass levels but sizably different total biomass levels* (with a total of ~4+ times higher in the trickling filter, due to attached underlying anaerobic and inert biofilm).

Going beyond nitrogen removal, biological phosphorus removal can also be achieved in full **biological nutrient removal** (BNR) systems designed specifically to provide sequenced anaerobic and aerobic zones which would encourage the growth of a highly unique bacterial group largely believed to fall within the *Acinetobacter* genus. These unusual cells consume low-molecular-weight volatile fatty acids (VFAs) during an initial anaerobic zone while releasing internal reserves of polymerized phosphate (i.e., a polyphosphate known as **volutin**). Moving from the anaerobic to aerobic zones, though, these same cells reverse the process, balancing their accelerated uptake of soluble phosphorus against the oxidation of previously stored VFAs. The end result, therefore, is that of a biochemically mediated phase transfer of phosphate, shifting incoming soluble

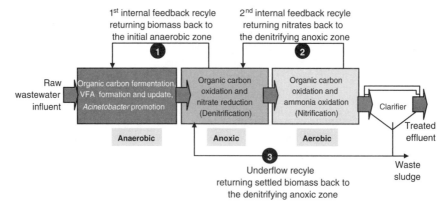

Figure 16.32   Representative three-tank plus three-recycle design scheme for combined biological nutrient removalm including nitrogen and phosphorus removal. (*Note*: This particular arrangement is often referred to as *UCT* or *VIP systems*, based on their respective geographic development at the University of Cape Town and Virginia; there are numerous alternatives using similar sets of anaerobic, anoxic, and aerobic reactors.)

phosphorus contained within the raw wastewater stream into phosphate-rich solid-phase intracellular storage granules contained within cells eventually wasted from the system.

Further design refinements to secure an overall goal of complete biological nutrient removal, encompassing both nitrogen and phosphorus, typically involve the addition of yet another separate, anaerobic zone or phase at the head end of the reactor train, as shown schematically in Figure 16.32. The second and third reactors in this sequence, offering anoxic and aerobic environments, respectively, will again play much the same roles, including oxidation of organic carbon in both stages as well as nitrification in the aerobic stage and denitrification in the anoxic stage. By adding an additional internal feedback recycle loop, though, biomass moved from the anoxic stage to the new head-end anaerobic stage is able to shift rapidly into an anaerobic condition (since it carries little, if any, nitrate or dissolved oxygen). In turn, fermentative transformation of incoming organics into volatile fatty acids sets up the initial conditions necessary for *Acinetobacter*'s phosphorus-related metabolism.

16.1.4 Stabilization Lagoon Systems

Stabilization lagoons represent one of the simplest possible options for wastewater treatment, with a level of engineering sophistication and cost below that of the preceding attached- and suspended-growth processes. As explained in an oft-quoted anonymous description—that "the design engineer drives a bulldozer and the rest is left to Mother Nature"—most lagoons are built as simple earthen basins, albeit with liners, and only limited operational oversight is provided. Given their simplicity, though, and ease of operation, lagoon systems are widely used in small rural communities and industries throughout the world.

There are three basic lagoon design options, including two that are solely aerobic or anaerobic, and a third, facultative version that has both overlying aerobic and underlying anaerobic layers. Whereas the aerobic and facultative lagoon options can be used for complete treatment of most municipal waste streams, anaerobic lagoon systems are more commonly used with high-strength industrial wastes (e.g., rendering plants, slaughter-houses, food processing residuals), and even then as a means of pretreatment prior to a following polishing step (e.g., followed by an aerobic lagoon, spray irrigation over land, or some other means of treatment).

All three of these basic lagoon design options have yet another subgroup set of design alternatives, depending on their projected organic loading rates and the means by which they will secure their necessary oxygenation and/or mixing.

Aerobic lagoons of the sort shown in Figure 16.33 rely on natural wind-induced mixing and/or surface aeration, plus algal photosynthesis, for oxygen resupply, and tend to be fairly shallow (with typical depths of about 2 m) and less heavily loaded. On the other hand, forced mixing and aeration using either mechanical agitation or subsurface compressed-air distribution (Figure 16.33) is also practiced with many aerobic lagoons, such that their design depths can be increased (to the range 3 to 4 m), along with receiving higher organic loadings. Anaerobic systems will be even deeper, often reaching depths of 4 to 5 m, to provide sufficient depths for the lower strata of the lagoon to fully reach a

(a)

(b)

Figure 16.33 Stabilization lagoons: (*a*) unaerated; (*b*) aerated.

point of oxygen depletion, and in many instances an overlying floating scum blanket will be allowed to collect on the lagoon surface to further constrain surface reaeration.

There is also a considerable range in the hydraulic detention times used with these various systems. Design HRTs below 1 week in length, at 3 to 6 days, may be used with high-rate operations, or for anaerobic or facultative pretreatment units ahead of a second, downstream aerobic polishing lagoon. However, many stand-alone lagoons, particularly those not provided with supplemental aeration hardware, are sized to provide detention times in excess of a week and possibly extending to month-plus periods. Pretreatment focused largely on screening solids from the influent waste prior to entering the lagoon may also be provided, particularly in the case of high-strength industrial or commercial operations for which prescreening might effectively pull out a substantial level of incoming contaminant solids loading.

Example 16.3: Preliminary Stabilization Lagoon Design   After completing the trickling filter and activated sludge preliminary designs, suppose that a third design overview is requested for the same Deer Creek, Illinois, community, with its 400 homes and 1600 residents. Given that their expected discharge point into a local creek has seasonal low-flow conditions that warrant high-level ammonia removal levels (i.e., to avoid summertime ammonia inhibition of sport fish in this Midwest city's regional area), it is assumed that a controlled-release, unmixed aerobic oxidation pond with a 6-month retention time will be required to handle the daily average wastewater flow of 605.7 m³/day and its commensurate total BOD loading of 121.2 kg BOD/day (*Note*: This design will be based on the community's full BOD load, since primary settling units are not typically included with these types of stabilization pond designs.)

Preliminary Design Details

Hydraulic loading, system sizing, and retention times	Design hydraulic retention time (HRT): 6 months

Hydraulic loading, system sizing, and retention times

Design hydraulic retention time (HRT): 6 months

Note: HRTs for these systems typically range from 1 to 6 months, depending on discharge restrictions into the receiving water body (e.g., a stream).

Design reactor depth: 2 m

(*Note*: This depth is appropriate for unmixed lagoons; facultative and mechanically mixed aerobic lagoons have depths of 3 to 4 m, while anaerobic lagoons handling higher strength wastes are considerably deeper.)

Projected reactor volume: (605.7 m³/day)(6 months) = 110,600 m³

Projected reactor surface area: (110,600 m³)/(1.5 m) = 55,300 m²

(*Note*: This area is equivalent to 13.7 acres, which would correspond to a per-area loading of approximately 116 people/acre.)

Organic loading

Projected organic loading:

$$(121.2\,\text{kg CBOD/day})/(110,600\text{m}^3) = 0.0011\,\text{kg/m}^3 \cdot \text{day}$$

Related notes

1. This lagoon is rather large in surface area given its long HRT and shallow, unmixed depth; deeper, mixed ponds with lower HRTs often have per-capita loadings of 1 acre per 300 to 600 people.

2. As compared to either the activated sludge or trickling filter design options, this controlled (i.e., low-rate seasonal) discharge oxidation pond design would have a physical footprint roughly 1000 times larger.

3. Although it does require significantly more space, this type of facility should be less costly to build and apt to require less operational attention, although the stability of this system's effluent quality will probably be much lower.

To some extent, it is reasonable to classify these lagoons as quasi-suspended-growth systems given their content of suspended bacterial and algal biomass, although in this case the level of entrained biomass concentration tends to be several fold lower (i.e., generally measured in 100s of mg/L as opposed to 1000s). At the same time, most aerobic and facultative lagoon volumetric organic loading rates (ranging from 0.001 to 0.01 kg/ $m^3 \cdot$ day) are also much lower than the more advanced attached- and suspended-growth processes, such that the biomass found in these lagoons tends to have a greater proportion of higher life-forms (e.g., scavenging protozoans, rotifers) and a lower net mass fraction of lower bacteria than what would be seen in a more highly loaded attached- or suspended-growth process.

Yet another unique aspect with open, aerobic lagoons (and to a lesser degree in mixed lagoons, due to the higher entrained solids levels and lower opacities) is that they have considerably more involvement with algal growth, due to the reduced density of their suspended solids and the fact that light can penetrate much deeper than is the case with most other waste treatment systems. With the incoming waste stream continuously introducing a fresh supply of readily available nutrients, algal and other small-scale plant growth can undergo large seasonal swings and will have a correspondingly significant impact on suspended solids levels in the effluent as well as the soluble oxygen and carbon dioxide levels.

The preferred circumstance for lagoons in terms of their algal composition it that of having suspended, submerged algal cells, where their metabolism often has a distinct, diurnal impact, resulting in sizable swings in dissolved oxygen and pH [i.e., higher DO and higher pH during the day, when algal photosynthesis and oxygen release offset bacterial respiration (at the same time, algae also removes CO_2); lower DO and lower pH at night, while both bacteria and algae are respiring and jointly releasing CO_2]. These algae are also effective scavengers of soluble nutrients, including both nitrogen and phosphorus, assimilating them routinely into solid-phase algal cell mass and then removing them once these cells die and settle onto the lagoon bottom.

However, there is also a smaller number of lagoons whose algal or plant content is established in a free-floating fashion, such as that of the classic duckweed plant form shown in Figure 16.34, where their presence tends to result in a more extensive mat stretched across the lagoon surface. These mats may appear quite green and healthy, but their activity and presence are somewhat contrary to conditions normally considered optimal for these lagoons. Free-hanging duckweed roots, called *fronds*, extend only a few centimeters in depth, and they contribute little, if any, dissolved oxygen via photosynthesis. Even worse, they physically block wind-induced surface reaeration, to a degree that their subsurface conditions are more likely anoxic to quasi-anaerobic. BOD and suspended solids removal can still be secured in these matted lagoons, albeit with

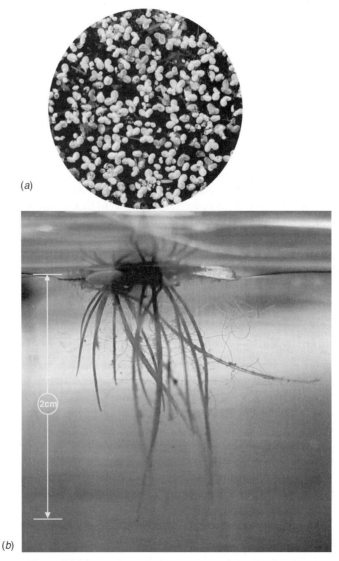

Figure 16.34 Duckweed plant: (*a*) overview; (*b*) closeup.

lower loading rates, and nutrient uptake will still occur, but effective long-term operation may require surface harvesting measures.

As for the issue of exactly what happens to the accumulation of dead algae, bacteria, and other waste solids once they settle and accumulate on the bottoms of these lagoons, this topic still represents a highly variable circumstance for most lagoons. Ideally, the input of biodegradable solids settling into a lagoon's bottom sludge blanket would then undergo decay at a rate not much different from that of their addition. If, indeed, this rate of accumulation were to be low, the sludge blanket depth would not increase, there would be no need to clean (i.e., remove sludge from) the lagoon periodically, and there would be no reduction over time in the lagoon's active (as opposed to settled sludge) volume. Unfortunately, the history with most lagoons is that they tend to accumulate settled solids at a rate that warrants their cleaning at multiyear to decade-long intervals, without which there would be a debilitating, progressive drop in effective lagoon volume.

Of course, there is also a negative side to large, exposed lagoon surfaces, particularly in regions faced with colder seasons, in that they will experience sizably higher levels of evaporative cooling that could lead to significant drops in temperatures during cold-weather periods. In turn, these reduced temperatures will then affect bacterial efficacy in the lagoon, particularly that of ammonia oxidation by highly sensitive nitrifying bacteria. From a regulatory perspective, this circumstance of decreased nitrification during cold-weather periods is most important for lagoons that discharge into creeks and streams with a high degree of variation in their seasonal flow, where low-level dilution of a high-level effluent ammonia discharge might then cause downstream ammonia toxicity problems for fish. In fact, many sport fish, such as trout, which might be found in streams, rivers, and lakes downstream of these lagoons, are highly sensitive to quite low ammonia levels, to an extent where 0.1 mg NH_3-N/L concentrations might incur serious, possibly fatal, stress. To resolve this problem, therefore, there is yet another lagoon option which incorporates a controlled-discharge release, with far longer HRTs extending up to, and even beyond, 6 months in time, to ensure that the effluent release can be delayed until the receiving water body provides an acceptable level of dilution that would negate this type of ammonia toxicity concern.

Lagoon effluent quality as a whole is not usually comparable to that of the more advanced waste treatment processes (e.g., activated sludge), with effluent BOD levels typically in the range 40 to 60 mg/L and even higher solids levels, typically 60+ mg/L. Algae represent a large fraction of these solids, however, and as a result, regulatory criteria levels with lagoon suspended solids concentrations are either set higher to accommodate this circumstance or filtration hardware (i.e., mechanical screens or sand, etc. filters) will need to be installed if deemed necessary to remove the latter solids prior to discharge. Therefore, disregarding discharge sites whose low-level dilution might incur nitrification concerns, lagoon effluents are commonly considered acceptable for general discharge.

16.1.5 Constructed Wetland Systems

Wetland systems have played an important role in the process of cleansing surface waters since the original evolution of Earth's hydrological and biological ecosystems eons ago. However, our own efforts with extrapolating this natural concept into controlled treatment systems for remediation of wastewater and stormwater runoff have been far shorter, roughly covering only the past half-century.

The involved processing strategy is that of densely clustering water-loving plants within shallow (~60 to 75 cm deep) basins and then allowing the incoming waste or runoff flow to experience a prolonged period of contact with these plants and their extensive root matrixes. Rather naturally, the public's qualitative perception of these types of environmentally "green" plant-based systems inherently tends to be more positive than is the case with traditional, concrete-intensive wastewater treatment options, and as such, these systems have recently drawn considerable appeal (Cole, 1998).

Indeed, the purported benefits of green wetland systems, compared to the conventional approach of hardware-intensive wastewater treatment, are as follows (U.S. EPA, 2000):

Figure 16.35   Constructed wetland wastewater treatment systems.

- They can be constructed in highly attractive, environmentally friendly configurations (see, e.g., Figure 16.35).
- They may offer lower capital costs (at least for smaller applications).
- They will almost certainly have lower operating costs.
- They should require relatively less operating attention.

Clearly, a primary benefit of constructed wetlands, in terms of their engineered design, construction, and operation, is that of their simplicity. Wetland cells are typically constructed to use a gravity-based flow scheme that has few, if any, moving parts; most systems use only solar energy; and their demands on operator attention and care are quite low (being limited to occasional flushing of the influent and effluent lines to avoid plugging, and seasonal removal of dead, senescent plant debris).

However, these systems can also have a number of related disadvantages that must be recognized, particularly in terms of performance. One key issue is that their best levels of effluent quality (e.g., BOD, TSS, ammonia) are seldom realized in a matter of weeks or months after a new facility is installed, but instead, will probably require a year-plus period for system maturation. In addition, these systems are also apt to experience a higher degree of seasonal variability than what is typically the case with conventional wastewater systems, and these seasonal variations will be difficult to control. At the same time, these natural systems also tend to be susceptible to problems of a sort that would otherwise not be an issue with standard wastewater treatment operations. For example, the performance of the plants could be partially disrupted or completely curtailed by plant infestation and disease. Given the inherent plug-flow mode of wastewater passage through these units, an incoming slug of potentially harmful household chemicals (e.g., paint thinner, lawn care chemicals, bleach) could also kill a substantial front-end area of this biological operation for an extended period. In the particular case of wetlands with which their waste stream is open to the atmosphere, they have been known to encounter troublesome problems with mosquito growth as well as occasional odor generation.

These constructed wetland designs can be roughly subdivided into two major options, including those that rely on emergent plants and those that use submerged plants either alone or in combination with emergent plants. The names given to these two options are, respectively, that of a **subsurface flow** (SSF) and a **free water surface** (FWS) constructed wetland. In either case, it should be noted that some form of pretreatment (e.g., using a septic tank) is routinely provided ahead of wetlands receiving wastewater flows, to remove setttleable solids and floating oils and grease that might otherwise pass into the wetlands, where they might degrade system performance. In general, submerged-flow wetland systems tend to require somewhat less operational attention than free-water systems, and they may also have a slight advantage in terms of required land areas for comparable flows.

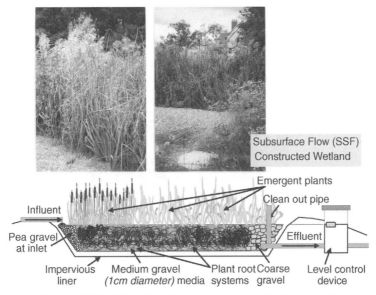

Figure 16.36   Subsurface flow constructed wetland schematic.

The SSF options (see Figure 16.36) are commonly constructed with one or more cells filled with various forms of a porous media, such as coarse and fine gravel washed previously to remove fine solids that would otherwise reduce necessary void volume, and planted with various forms of emergent wetland plants. Larger-diameter media (i.e., 4- to 5-cm coarse gravel) is normally placed across the head-end face of the cell to spread laterally and distribute the incoming waste uniformly, and then again spread across the back of the cell to recollect the effluent and direct it smoothly to the system's outlet piping. Within the wetland itself, a small-diameter medium (e.g., typically a washed, 0.75- to 1.5-cm-diameter river gravel) is used to sustain plant rooting. Wastewater introduced into these cells subsequently flows horizontally through the open, void space of the porous, subsurface medium at a relatively low velocity, during which it comes into contact with both the plant's extensive root system and its affiliated rhizosphere consortia of microorganisms.

The FWS option (depicted schematically in Figure 16.37) for constructed wetlands is built much the same as those in the SSF mode, with single or serial open cells that are similarly lined with much the same barrier material. These free-water cells, however, are only partially filled (10 → 15 cm deep) with soil or a small-diameter (0.5- to 1-cm) gravel medium to support the plant root systems, such that the remaining depth is then filled with water and with floating plus rooted plant mass. Both emergent and submergent plant varieties can be grown in these cells; emergent plants will root themselves in the bottom medium, while the submerged plants can either exist in a free-floating mode whose roots dangle downward or are again embedded into the bottom medium. When given a means of supplemental aeration (e.g., using compressed air diffusers), the depths used with free-water systems are sometimes extended somewhat beyond 1-m depths.

Hydraulic residence time (HRT) is a primary design factor for all of these constructed wetlands, with minimal recommended values of 5 to 7 days (Hammer, 1989). Conservative designers may opt for considerably longer times, and many systems are built with HRTs extending into multiweek periods. However, when provided with supplemental aeration, or in the case of most subsurface flow designs, these residence times are typically reduced downward to values around 1 week.

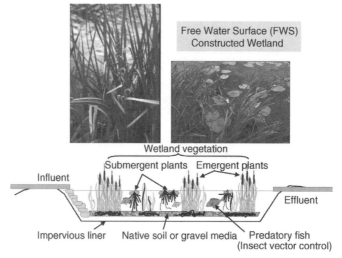

Figure 16.37   Free-water surface constructed wetland schematic.

Example 16.4: Preliminary Constructed Wetland Design Having completed their three prior preliminary designs, one final process design overview is requested for the Deer Creek, Illinois, community, in terms of a constructed wetland system. Given the prolonged subfreezing winter temperatures typically encountered at this northern midwest U.S. community, a subsurface-flow (SSF) constructed wetland will be sized preliminarily to handle a daily average wastewater flow of 605.7 m^3/day and its commensurate total BOD loading of 121.2 BOD/day (*Note*: Although these constructed wetlands are commonly built with preliminary septic tank units, this example design will be based on the community's full BOD load as a conservative measure in light of potential septic tank upset conditions).

Preliminary Design Details

Hydraulic loading, system sizing, and residence times	Design hydraulic residence time (HRT): 7 days Design wetland media depth: 0.61 m (i.e., 24 in.) Projected wetland liquid depth: 0.46 m (i.e., 18 in.) (*Note*: Typically, only the lower two-thirds of the media depth is saturated.) Expected media porosity $= 0.4$ Projected wetland volume:

$$(605.7 \, \text{m}^3/\text{day})(7 \, \text{days})(1/0.4) = 10,600 \, \text{m}^3$$

Projected wetland surface area: $(10,600 \, \text{m}^3)/(0.46 \, \text{m}) = 23,043 \, \text{m}^2$ ($\sim$1.9 ha)

Projected wetland total media volume: $(23,043 \, \text{m}^3)/(0.61 \, \text{m}) = 14,046 \, \text{m}^3$

(*Note*: The latter volume would include both coarse inlet and outlet media plus smaller 0.75- to 1.5-cm river gravel.)

Organic loading   Projected organic loading:

$$(121.2 \, \text{kg CBOD}/\text{day})/(10,600 \, \text{m}^3) = 0.0069 \, \text{kg/m}^3 \cdot \text{day}$$

Hydraulic loading   Projected hydraulic loading:

$$(605.7 \, \text{m}^3/\text{day})/(1.9 \, \text{ha}) = 318 \, \text{m}^3/\text{ha} \cdot \text{day}$$

Related notes

1. As with the stabilization lagoon option, a constructed wetland system will need to be considerably larger than the activated sludge or trickling filter designs for this community, in both surface area and volume.
2. Construction costs for constructed wetlands can escalate rapidly with larger facilities (e.g., beyond a daily flow of $\sim$500 m^3/day) due to the capital cost of the media (e.g., washed coarse and fine gravel) and its on-site delivery.
3. Constructed wetlands are also apt to require less operational attention, but their cold-weather ability to secure nitrification may require additional treatment steps depending on regulatory requirements for effluent ammonia.

In the case of a free-water-surface operation, the determination of minimal hydraulic retention time is based simply on the ratio of the system's open water volume to the expected peak incoming wastewater flow. For subsurface-flow designs, however, this calculation must take into account the available pore volume of the medium through which the wastewater will be flowing, as follows:

$$\mathrm{HRT} = \frac{ADP}{Q} \tag{16.3}$$

where A is the wetland surface area (m^2), D is the wetland depth (m), P is the porosity (dimensionless), and Q is the incoming peak wastewater flow (m^3/day).

Despite widespread use of the latter hydraulic retention-time parameter, there is considerable uncertainty as to its validity and usefulness. Much of this uncertainty hinges on the fact that the pore space volume of newly installed gravel will change once it begins to fill with root material, interstitial microbial growth, and entrapped suspended. As seen in Figure 16.35, the density of plants placed within a typical wetland tends to be quite high, to the point where their root systems undoubtedly fill a sizable fraction of the soil's available pore space.

Once the anticipated waste stream extends much above a few thousand liters per day, constructed wetlands are usually designed with multiple cells arranged to receive the incoming flow in serial-flow fashion. Within each cell the ratio of length to width used for the actual layout of constructed wetlands is known as the **aspect ratio**. Low aspect values, particularly those below 1 : 1 (*L/W*, length to width), are prone to short-circuiting hydraulic problems, where the passing flow is not distributed equitably across the entire cross section of the cell. This problem would then be reflected in poor and uneven levels of treatment (i.e., erratic levels of BOD, TSS, and ammonia removal). Performance will typically improve with larger aspect ratios, but these benefits level off around a figure of 3 : 1. Even at this ratio, attention must still be given to ensuring that the incoming waste stream is distributed uniformly across the face of the wetland.

Various technical references recommend that these organic loadings be kept below 80 to 110 kg BOD/ha·day [approximately 8 to 11 g (0.008 to 0.011 kg) of carbonaceous BOD (CBOD) applied per square meter per day] in order to achieve acceptable levels of treatment. Hydraulic loading rates should also considered in the design of most constructed wetlands. These values are inversely related to HRT, whereby longer retention-time systems will be maintained with a loading range of 150 to 300 m^3/ha·day, whereas the highly loaded, shorter HRT designs will operate at hydraulic loadings from 300 to 600 m^3/ha·day.

Most wetlands will experience a significant loss of water on a seasonal basis due to evapotranspiration, especially during a warm late spring through early fall time frame. This behavior is beneficial in the sense that it reduces the necessary volume of discharge, but evapotranspiration may also complicate the process of discharge permitting. Indeed, the current practice of establishing effluent discharge limits on BOD, suspended solids, ammonia, nitrate, and so on, for any wastewater treatment operation is almost always based on contaminant concentrations rather than total mass. If, for instance, a warm-weather SSF wetland operation were to release one-half of its incoming wastewater flow via evapotranspiration, the remaining 50% flow rate subject to actual discharge might well experience seemingly unacceptable effluent concentrations despite the reality that the total mass of contaminants had actually been reduced. Consideration is therefore being given to an arrangement whereby these permits might be written to secure a necessary reduction in influent mass of BOD, and so on, rather than concentration.

The next set of design issues covers installation of a subsurface liner as well as cell depth and media placement. SSF and FWS cell bottoms will need to covered with an impervious, low-permeability liner, typically constructed with overlapped, seamed sections of high-density polyethylene (HDPE) sheets rolled across the cell floors and sidewalls. The peripheral edges of these sites are typically edged with an elevated soil berm, with the wetland's impervious underlying liner rolled up and over the top of the berm, where it is then tucked into a stone-filled trench to secure it in place. This exposed section of the liner is then covered with a layer of gravel that protects the liner against direct physical wear as well as blocking sunlight irradiation that could degrade the liner. The key benefit of installing these berms is that they restrict stormwater runoff from directly entering the wetland, where it would impose an unnecessary additional hydraulic loading.

Balanced against the aforementioned engineering details and recommendations applied to the design and construction of these facilities, though, our technical understanding of wetlands is frankly still evolving, given our relatively brief experience with their use. In fact, the current state of the art for constructed wetlands is still far less established than that of conventional wastewater treatment options (e.g., attached- and suspended-growth), to the point where the engineered practice of designing and operating constructed wetlands remains a rather inexact science.

Of course, a key factor with wetlands is that of maintaining a healthy population of true "wetland" plants and ensuring their ability to grow under truly wet conditions. Plantings introduced to these wetland systems during construction will typically encompass a variety of plants as opposed to just one type, as a means of promoting a more ecologically stable population (i.e., a diverse population whose varying levels of water uptake, root depths, root densities, and so on, inherently provided a higher degree of environmental tolerance, disease resistance, contaminant sensitivities, etc.).

Free-water-surface systems will usually include a combination of free-floating and rooted plants and they will naturally distribute themselves across the surface of the wetland. However, in the case of subsurface-flow units, a preplanned layout is typically used for the installed plants, where they are initially placed in successive layers or bands stretched perpendicularly across the face of the direction of flow. Second, the plants selected for each such band are often staggered to take advantage of their various root forms. Shallower rooted plants tend to be placed earlier in the bed to give the incoming wastewater flow more of a chance to pass downward into the underlying medium, which helps to negate short-circuiting across the top edge of the bed. Subsequently, deeper-rooted plants may be placed farther back in the bed as the flow reaches a point where it has been fully distributed across the vertical profile of the bed.

As for the types of water-tolerant, if not water-loving plants actually planted within these facilities during initial construction, a number of options are commonly associated with each of these subsurface-flow and free-water-surface systems Table 16.8. Interestingly, fragmites plants are banned in some regions because of their aggressive nature, because their seeds might travel away from the site and subsequently colonize areas outside the original site. For that matter, each of the various plant types tends to have good and bad features.

Figure 16.38 depicts schematically the aboveground and root-zone conformations observed with a few of the more widely used plants associated with subsurface-flow design. Bulrush plants (such as river bulrush, *Scirpus fluviatilis*) and sedges (such as

TABLE 16.8   Typical Constructed Wetland System Plant Types

Recommended Plant Type	Subsurface Flow	Free-Water-Surface
Arrowhead (*Sagittaria* spp.)	×	×
Bulrush (*Scipus* spp.)	×	×
Coontail (*Ceratophyllum* spp.)	×	
Pondweed (*Potamogeton* spp.)	×	
Rush (*Juncus* spp.)	×	×
Sedge (*Carex* spp.)	×	×
Water plantain (*Alisma* spp.)	×	×
Water hyacinth		×
Duckweed		×

fox sedge, *Carex vulpinoidea*) are generally considered to offer superior levels of nutrient removal, with ideal levels of root-zone penetration (typically 0.6 m) and high levels of root mass and surface area. Conversely, although cattails (*Typha latifolia*) are often observed in natural wetland settings, they are generally regarded as a nuisance plant in constructed wetlands since their shallow-rooting tendency leads to relatively low treatment properties (i.e., much of the wastewater passing through an SSF bed would pass beneath the cattail root zone).

Duckweed and water hyacinths (see the insert image on the top of Figure 16.34) are commonly used in free-water systems. Duckweed has a far smaller root mass than that of hyacinth, and it extends far less deeply into the FWS system water depth, but even then it is often considered to be an important contributor to free-water systems, either by itself or as a secondary plant form. Water hyacinths are even more favored in free-water

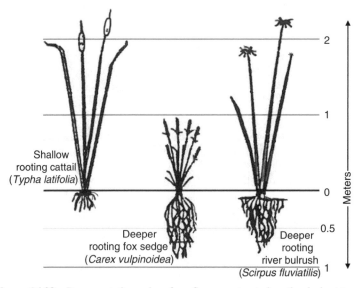

Figure 16.38   Representative subsurface-flow constructed wetland plant types.

systems given the sizable extent of their root systems, which extend deeply into the depth of these ponds. Hyacinths can also sustain high nutrient uptake rates, and as such they are also used in applications for tertiary-level wastewater polishing wetlands. Large, rapidly growing plants such as hyacinths, though, can clog a free-water system unless they are harvested on a regular basis, and in turn this residual plant mass could constitute a sizable waste material. Even then, however, these waste plants may still be put to beneficial use; for example, Disney World in Florida mulches their waste hyacinth plants and digests the residual organics to produce methane gas for use as an energy-rich fuel.

One of the key uncertainties with wetland systems, though, is that of the true mechanism(s) behind which contaminants are removed within wetlands, and whether the responsible agent be that of plant- or microbial-based pathways. Much the same argument could also be raised for phytoremediation systems (see Section 16.2.4), in that our understanding of these processes bears a similar degree of lingering fuzziness.

Undoubtedly, plants can contribute directly to the observed uptake and degradation of contaminants, using a complementary range of plant-based sorption, uptake, accumulation, volatilization, and degradation mechanisms. Furthermore, the medium itself will provide some measure of complementary treatment, through either simple filtration or more complex physical–chemical mechanisms (e.g., sorption, ion exchange). Plants will also play a significant role in the metabolic passage of water through a wetland, and in some instances their rates of water loss through evapotranspiration will prove to be the dominant path for water movement, even to the point where the wetland may have little, if any (e.g., 5% or less), actual free-flowing discharge.

However, microbes extensively colonized onto and around plant root surfaces will clearly play an important, perhaps even dominant role with contaminant removal in plant-based wetlands. Figure 16.39 depicts this sort of extensive bacterial attachment, using an acridine orange cell stain to highlight living cells against the darker, unstained root surface.

Wetland plants nurture this sort of microbial colonization actively on several accounts. First, plant root and root tip surfaces inherently offer an extensive surface area for this sort of microbe growth. Second, plants will pump photosynthetically generated oxygen

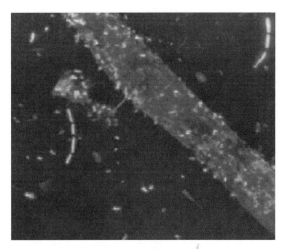

Figure 16.39 Representative image of bacterial colonization on plant root surface. (Courtesy of Jay Garland.)

through these same roots, such that this release will nurture the growth and colonization of aerobic microbes (i.e., largely bacteria plus fungi) within the root's immediate vicinity. Third, plants actively release an organic-rich mix of simple sugars from their root, known as root **exudates**, which will again sustain and promote active microbial growth.

On a collective basis, therefore, these plants will promote the extensive growth of microbial biomass within the immediate vicinity of their root structures and, in turn, these microbes will contribute to the wetland system's overall treatment efficiency. Cyclic oxygen release and uptake by wetland plants will actually occur on a diurnal cycle commensurate with the metabolic swing between daytime photosynthesis and nighttime respiration. FWS systems may consequently experience daily shifts in oxidizing and reducing conditions within the bulk wetland fluid, which intermittently promotes aerobic, anoxic, and anaerobic bacterial metabolism. For SSF systems, though, these same shifts in day–night plant behavior will similarly affect microbial growth in the root rhizosphere zone. Microbes immediately adjacent to root surfaces will probably be capable of sustained aerobic activity during daytime periods, but farther back from this oxygenated zone, the soil and its microbes will routinely follow a more anaerobic life-style. As such, there both aerobic and anaerobic microbial mechanisms will be involved in the degradation process, as well as a variety of interrelated precipitation, sorption, ion-exchange, filtration, and even volatilization reactions.

Regional climactic conditions will represent another significant design factor, both in terms of selecting an SSF vs. FWS design strategy and in selecting the types of plants to be used. Whereas free-water wetlands are not suited to geographical areas in which prolonged winter weather would freeze much of the system, SSF operations have proven to be useful in cold-weather regions extending even into the Arctic Circle. During the winter, an ice layer will develop across the top of SSF wetlands, extending perhaps as much as 20 cm or more in depth. However, overlying ice and snow layers act as an insulating cover so that the underlying medium and roots remain at a temperature above freezing. Granted, at these ambient air temperatures the plants themselves will have shifted from an active to a dormant state, but their remaining vascular stalks are believed to act as physical channels or conduits that carry oxygen through the ice and into the medium where degradative metabolism is still maintained by an active microbial consortia. The reduced temperature, however, results in slower metabolic rates that must be taken into account by providing longer hydraulic retention times. Conversely, summertime wetland operations not only enjoy considerably faster rates of microbial degradation, but in many instance the plants involved will maintain far higher rates of evapotranspiration. In fact, depending on which plants are used, there may even be periods during summer season operation at which a wetland's incoming flow is largely released into the atmosphere rather than being released as a conventional discharge.

Many of the emergent plant species used in SSF systems tend to be fairly tolerant of cold-weather exposure. Conversely, hyacinths used in FWS applications are not hardy (able to tolerate freezing weather), and therefore their use is limited to warm climates. Duckweed, on the other hand, is able to withstand colder climates, but even then this plant form is still not widely used in these areas, given the general reluctance to use free-water systems in cold-weather localities.

Notwithstanding the apparent biological benefits of constructed wetlands, negative factors can be associated with these operations that must be addressed. For example, free-water systems, particularly those without supplemental aeration, have been known to encounter troublesome problems with mosquito growth as well as occasional odor generation. In both instances, though, these types of problems can be alleviated or avoided with either aerated free-water systems or subsurface-flow operations.

Annual maintenance for wetlands is, again, a rather inexact science. Harvesting is practiced infrequently with plants grown within subsurface systems. Instead, carefully controlled burning is used more commonly to remove the burden of dead, senescent plants and to help with eradicating invasive plants that have moved into the wetland during the past growth season. Dandelions and goldenrod are two of the more common invasive species.

With free-water wetlands, the plants maintained within lowly loaded operations will probably have to be thinned out with harvesting on an annual basis. Commensurate with increased loading levels, these harvesting rates will then have to be increased, to the point where monthly or biweekly plant thinning may be necessary with the more highly loaded (e.g., aerated) systems.

As for the regulatory requirements and monitoring policies currently stipulated by various states for constructed wetlands, there is considerable variation. For example, different stipulations may be placed on the location of flow monitoring equipment, but in most instances this equipment is situated at the effluent end of the last wetland cell. In some instances, a second flow monitoring device (i.e., typically a flow totalizer) is installed at the head end of the system, by which the operations personnel can then quantify the level of water loss maintained within the overall operation due to direct evaporation plus plant-based evapotranspiration.

FWS and SSF systems both typically exhibit fairly high level BOD and SS removal efficiencies, ranging from 750 to 90%, albeit somewhat below the performance levels that might be achieved with either fixed-film or activated sludge processes. For many wetland applications, particularly those faced with tight surface or subsurface discharge limits, nitrification will probably represent one of the most critical issues, and the relative rates, and efficiencies, of removing nitrogenous contaminants in these wetlands can vary immensely with time and from one site to the next in relation to plant type, organic loading rate, hydraulic loading, and climate. It is certainly possible for nitrification to occur in these wetlands, but the results quantified to date regarding wetland nitrification have been rather erratic, probably due to the fact that the temperatures and dissolved oxygen levels in these wetlands can vary quickly. One recent upgrade on this account has been that of using parallel SSF cells maintained in a flip-flop sequence of cyclic fill-and-draw stages, where these periodic, diurnal-type drain steps would hopefully provide increased oxygenation of nitrifiers working within the rhizosphere region.

16.2 SLUDGE TREATMENT

Wastewater treatment systems are designed to remove contaminants from the influent waste stream, thereby producing a clean liquid effluent as well as a concentrated, semi-solid residual that contains the extracted contaminants plus newly generated biological cells. Rather understandably, the latter concentrated residuals, particularly those in raw form, can have a rather unpleasant, if not altogether foul nature if not handled and managed properly. In fact, one of the most worrisome aspects of these sludge residuals, in terms of human health, is that of their potential contamination with a wide range of pathogenic, disease-causing organisms, ranging from viruses and bacteria and extending upward in scale to protozoans, as shown in Table 16.9.

TABLE 16.9 Principal Pathogens of Concern Associated with Municipal Wastewater Biosolids

Organism Type	Disease/Symptoms
Bacteria	
Salmonella	Salmonellosis and typhoid
Shigella	Bacillary dysentery
Yersinia sp.	Acute gastroenteritis
Vibrio cholerae	Cholera
Campylobacter jejuni	Gastroenteritis
Escherichia coli	Gastroenteritis
Enteric viruses	
Hepatitus A virus	Infectious hepatitis
Norwalk and	Epidemic gastroenteritis
Norwalk-like viruses	
Rotaviruses	
Acute gastroenteritis	Enteroviruses
Polioviruses	Poliomyelitis
Coxsackieviruses	Meningitis, encephalitis
Echoviruses	Gastroenteritis and respiratory infections
Reoviruses	Gastroenteritis and respiratory infections
Astroviruses	Epidemic gastroenteritis
Caliciviruses	Epidemic gastroenteritis
Protozoans	
Cryptosporidium	Gastroenteritis
Entamoeba histolytica	Acute enteritis
Giardia lamblia	Giardiasis
Balantidium coli	Diarrhea and dysentery
Toxoplasma gondii	Toxoplasmosis
Helminth worms	
Ascaris lumbricoides	Ascarids
Toxocara sp.	Ascarids
Trichuris sp.	Whipworms
Necatur americanus	Hookworm
Hymenolepis sp.	Taeniasis (tapeworms)
Taenia sp.	Taeniasis (tapeworms)

Up until the past decade, all forms of these concentrated residuals, whether raw or treated, were commonly described with the same "sludge" label in deference to the typically negative context associated with words whose spelling begins with "sl..." (e.g., slap, sleet, slum, slam, etc.). However, despite the unwholesome connotations implied by *sludge*, the nature and makeup of many properly treated residuals (i.e., suitably degraded, stabilized, disinfected, and dewatered) can actually be rather positive, if not altogether even valuable: harkening back, as it were, to the earliest days of wastewater processing as a means of generating nutrient-rich fertilizers.

In fact, many properly treated sludges can be used as beneficial soil amendments on agricultural lands once their negative attributes (i.e., pathogen content, high-level degradability, etc.) have been effectively resolved. A more positive label for these properly treated sludges, **biosolids**, was consequently chosen in conjunction with a national campaign by a U.S. association representing wastewater operations (i.e., the Water Environment Federation) during the 1990s to promote a more proactive image for a finished product whose prior standing as raw sludge carried a poor reputation among the general public.

The efficient and economical management of these wastewater residuals correspondingly represents an important engineering task, one that usually relies on the alternative use of several different biosolids treatment schemes. Granted, physical (e.g., wet air oxidation, pasteurization, gamma irradiation) and chemical (e.g., lime stabilization, silicate-based solidification) methods are used to stabilize and/or disinfect these residuals, but the majority of sludge processing systems employ biological mechanisms whose goal is similarly focused on degrading, stabilizing, and disinfecting the **putrescent** (easily degradable) and pathogenic (i.e., bacteria, viruses, parasites, etc.) problems associated with raw sludges.

The putrescible fraction of wastes can produce both nuisance (e.g., odor generation) and hazardous (e.g., hazardous gas generation, such as hydrogen sulfide) conditions, and a major goal of sludge treatment is **stabilization** with these putrescible materials. **Digestion** is a biochemical process by which stabilization can be achieved, such that physical reductions will then be realized in both the original solids mass and volume. Both anaerobic and aerobic digestion options are used with liquid-phase sludge processing systems, and aerobic digestion can also be completed in a more concentrated slurry- or solids-type mode using **composting** procedures.

One additional goal with any of these processes would be that of **disinfection**, by which another reduction will be secured in the original pathogen content. Of course, extending beyond any of these digestion schemes there is usually a final processing step, called **dewatering**, which is intended to reduce the water content of the biosolids and produce a semisolid material for subsequent disposal with a total suspended solids concentration above about 25%. The latter dewatering step is usually completed by physical processes such as filtration or centrifugation.

Before addressing specific biological options for sludge processing, further clarification is warranted as to the fact that two versions of raw residuals are produced at most primary and secondary wastewater plants. In either case, these raw residuals are then referred to as sludge, while their biochemical transformation and treatment subsequently generates a finished biosolids product. The primary sludge fraction includes heavy, particulate organic solids as well as various combinations of grit and inorganic fines extracted from the raw wastewater during its primary sedimentation. On the other hand, a large fraction of secondary sludge residuals is simply that of bacterial cells that grew within, and were then wasted from, the secondary biological treatment operations. Table 16.10

TABLE 16.10   Raw Primary and Secondary Sludge Characteristics

Organic Grouping	Primary %	Secondary %
Protein content	20–30[a]	32–41[a]
Grease and fat content	6–35[a]	Low
Cellulose content	8–15[a]	Low
Volatile solids	60–80[a]	59–88[a]

[a]By weight of dry total sludge solids.
Source: Adapted from Metcalf and Eddy (2003); Crites and Tchobanoglous (1998).

provides a general characterization of primary and secondary sludge forms. Before using any one of the various biological digestion and stabilization schemes, these primary and secondary streams will typically be blended and concentrated to achieve a desired starting point with a total solids content of 1.5 to 6% solids (by weight). In the following sections we provide additional details regarding strategies for converting sludge into biosolids.

16.2.1 Anaerobic Digestion

Although each of the alternative sludge digestion and composting processes have its own degree of metabolic complexity and operational peculiarities, anaerobic systems are widely regarded as the most difficult to maintain given the interlinked sensitivity of their biochemical pathways. Furthermore, where the aerobic options (i.e., aerobic digestion and composting) involve only an oxidative breakdown of the sludge solids and a single separation (liquid–solid) step, the anaerobic digestion process entails a sequential series of complex metabolic pathways and concluding separations of gas, liquid, and solid products. As a result, this digestion strategy has historically tended to require more attention and care while being more prone to upset. Conversely, though, there are a number of potential benefits to be gained with this technology. Most notably, the anaerobic digestion process tends to have a far lower energy demand (since it does not require aeration); at the same time, it can be used to generate a useful, energy-rich by-product in the form of methane gas that is created reductively by these metabolic reactions.

The underlying biochemical mechanism with anaerobic digestion is that of fermentation rather than the respiration reactions maintained in most aerobic wastewater treatment systems, as discussed previously. In either case, these options again involve a balanced set of oxidative and reductive (i.e., redox) reactions that are linked to electron donor and acceptor compound conversions, respectively. With fermentation reactions, the electron donor and acceptor compounds are typically both organic in form, although there are also critical pathways in which inorganic hydrogen and carbon dioxide gas species can fill the associated roles (Zehnder, 1988). Yet another unique feature of fermentation is that of a metabolic flexibility which allows a single organic compound to fill both roles, not only accepting electrons but also serving as the electron donor. For example, during the fermentative anaerobic transformation of acetate, this single substrate can simultaneously be oxidized to CO_2 ($CH_3COO^- + 2H_2O \rightarrow 2CO_2 + 4e^+ + 7H^+$) and reduced to methane, CH_4 ($CH_3COO^- + 4e^+ + 9H^+ \rightarrow 2CH_4 + 2H_2O$).

Figure 16.40 provides an overview of the three progressive metabolic steps that take place during the complete anaerobic digestion process: hydrolysis, acidogenesis, and methanogenesis. During the first, **hydrolysis** step, macromolecular organic compounds entering the system are transformed hydrolytically from large, and in many instances solid-phase macromolecular materials (e.g., cellulose, grease, protein, microbial cells) into their smaller, soluble building blocks (e.g., amino acids released from protein, carbohydrates from polysaccharides, fatty acids from lipids and fats). During subsequent fermentation maintained by the anaerobic digestion process, one fraction of these hydrolyzed organics will eventually be oxidized to carbon dioxide, another fraction will be reductively converted to methane, and a third, comparatively smaller fraction will be assimilated anabolically into a new anaerobic cell mass.

The two fermentation reactions following hydrolysis sequentially are acidogenesis and methanogenosis. **Acidogenesis** involves the formation of acid products by which

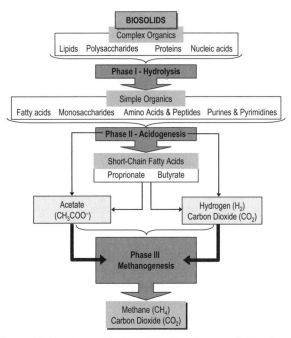

Figure 16.40   Anaerobic biosolids digestion metabolic phases.

the previously hydrolyzed organics are transformed into three substrate precursors (i.e., CO_2, H_2, and acetate) in preparation for the concluding methanogenesis transformation. Two different types of acidogenesis reactions are involved: one- and two-step conversions. The first such mechanism produces acetate primarily (i.e., by means of an **acetogenic** conversion, encompassing reactions that produce organic acids directly), whereas the second fermentative conversion involves intermediate production of volatile fatty acids (e.g., butyric and proprionic acid) and alcohols, which are converted subsequently to acetate, hydrogen, and carbon dioxide. The latter conversions are mediated by a wide range of gram-positive and gram-negative bacteria, whose classifications depend on the primary form of their fermentative products (Table 16.11).

TABLE 16.11   Fermentative Bacterial Types and Products

Fermentative Bacterial Groups	Representative Bacterial Forms	Proprionate	Butyrate	Acetate	Formate	Alchohols	H_2	CO_2
				Anaerobic Reaction Products				
Proprionic acid bacteria	*Proprionibacterium* *Chlostridium propionicum*	×		×				×
Butyric acid bacteria	*Chlostridium butyricum*		×	×			×	×
Mixed acid bacteria	Enteric gram-negative bacteria			×	×	×	×	×

The concluding step in this three-step sequence is **methanogenesis**, by which methane is produced. Methanogenesis converts low-molecular-weight organic precursor species into gaseous end products at both ends of the carbon oxidation state range, including fully reduced methane and fully oxidized carbon dioxide. At this point, therefore, the overall process of anaerobic digestion will have biochemically converted a sizable fraction of an initial sludge residual into a far more benign, possibly even energetically useful gaseous product.

Whereas the hydrolytic and acidogenic participants in this process are all members of the eubacteria domain, methanogens are classified as archaea. Methanogens include 17 genus members, typically subdivided into seven different groups based on their gramstain reaction status, morphology, DNA makeup, and methanogenic substrate form. On the one hand, these methanogens exhibit considerable variety in terms of their genetic structure (with $C + G\%$ values ranging from 26 to 62%), shape (short and long rods, plate-shaped cells, various cocci arrays, and filamentous forms), and temperature tolerances (including both mesophilic and thermophilic preferences).

Contrasted with their apparent physiological diversity, though, methanogens rely on three fairly narrow metabolic options for usable substrate forms, as depicted in Figure 16.41. The dominant substrate option, which includes a large number of methanogenic genus members, involves CO_2-type substrates where carbon dioxide serves as an acceptor for electrons donated by H_2, carbon monoxide, or formate. The second option, which includes only two genuses (*Methanosarcina* and *Methanothrix*) uses acetate in a metabolic process known as **acetoclastis**, a form of fermentation whereby this singular substrate (i.e., acetate) serves as both an electron acceptor and an electron donor. The third, **methylotrophic**, substrate option generally follows an electron flow scheme similar to that of the acetoclastic group, with substrates such as methanol or methylamine both

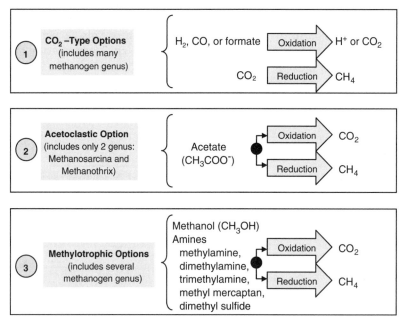

Figure 16.41 Methanogenic substrate options.

supplying and scavenging electrons. However, there are several more methanogenic genuses capable of pursuing this third "methyl" substrate option than are involved with acetoclastis, and in some cases they may also use hydrogen gas as a source of reducing power. Spread among these three catabolic options, most methanogens are only able to use one or two substrate forms, although *Methanosarcina* is far more facile given its ability to metabolize seven different substrates. In addition, there are methanogens whose usable substrate options lie outside the standard norm, including those able to subsist with various alcohol forms, including ethanol, 1- and 2-proponal, and 1-butanol.

Methanogenic metabolism naturally requires a highly reduced environment and is sustained only by microbes whose life-style is strictly anaerobic. Although the reductive metabolism of carbon dioxide practiced by most methanogens can be viewed as an anaerobic respiration pathway (using CO_2 as an electron acceptor in lieu of O_2), methanogens do not employ the same sort of cytochromes, quinones, and flavoproteins as are involved in a normal respiratory electron transport chain. Instead, methanogens use a number of highly specialized reducing enzymes and coenzymes for their reductive metabolism, sequentially including coenzyme factor F_{420}, methanopterin, methanofuran, coenzyme M (2-mercaptoethanesulfonic acid), coenzyme factor F_{430}, and a terminal coenzyme HS-HTP (7-mercaptoheptanoyl threonine phosphate).

Aside from the important biochemical roles played by these metabolic compounds, two of these enzymatic forms also provide a unique means of selectively identifying the presence of methanogen cells. Methanopterin, which resembles folic acid structurally, exhibits a bright blue fluorescence when illuminated with light at 342 nm, and coenzyme F_{420} projects a similarly distinct fluorescent blue-green hue when exposed to 420-nm epifluorescent illumination. Coenzyme factor F_{430} also absorbs light when irradiated at 430 nm, but unlike F_{420}, this coenzyme does not fluoresce. An environmentally important aspect of F_{430} activity, though, is that this coenzyme requires nickel at a level that gives methanogens a distinct anabolic requirement for this trace element.

The multistep and metabolically complex nature of these anaerobic mechanisms, therefore, introduces a set of concerns when applied to the pragmatic business of sludge digestion, particularly those systems designed for single-stage processing. Indeed, there is an extremely delicate balance that must be achieved in these single-stage anaerobic reactors between the involved acidogenic and methanogenic sequences in terms of their respective production and use of the acidogenic fermentation intermediates. Specifically, two different forms of key intermediates are involved, including both the low-molecular-weight volatile fatty acids (e.g., acetic, butyric, proprionic, etc.) and hydrogen gas.

The first issue with fatty acid intermediates revolves around system pH and the negative impact that acidic pH levels have on these anaerobic reactions. Methanogens are sensitive to pH outside the range 6 to 8 (Figure 16.42). A decrease in pH below 6 reduces the activity of the methanogens more than that of the acidogens. This causes a buildup of organic acids, further reducing pH. Thus, anaerobic digestion is unstable when confronted with pH disturbances. An important part of the inherent instability of the acidogenic-to-methanogenic linkage, therefore, stems from the fact that short-term lags in methanogenic activity (e.g., perhaps triggered by other environmental stress factors) can lead to upset pH transients from which the methanogens cannot readily recover.

The second crucial aspect with the metabolism of an anaerobic digester is that the reductive intermediate production of reduced hydrogen gas via the fermentative acidogens must be similarly balanced by its consumption during methanogenic metabolism. Should hydrogen gas levels rise within the digester much above trace values (i.e., above

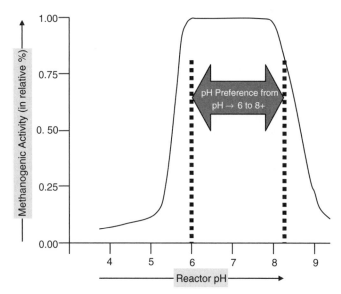

Figure 16.42 Approximate pH response by anaerobic methanogens. (Adapted from Speece, 1996.)

$\sim 10^{-3}$ atm), the metabolic harmony between acidogenic and methanogenic metabolism will again be disrupted, although in this case the negative impact is on the acidogenic cells.

This phenomenon reflects yet another delicate thermodynamic circumstance with the second set of fatty acid fermentation reactions, where the ambient hydrogen gas level must be kept sufficiently low to effect a net release of free energy. The free-energy values given in le 16.12 aptly qualify this dependence. In the absence of a sink (i.e., methanogenic consumption) for hydrogen gas during anaerobic digestion, the free-energy release for the fermentation of either proprionate or butyrate would be positive in value (with standard values of $+76.2$ and $+48.2$ kJ for these respective conversions), at which point these reactions would simply not occur.

Here again, a successfully maintained anaerobic digester system must achieve a balance between fermentative and methanogenic reactions. By holding the intermediate hydrogen gas levels at an acceptably low level (just as the fatty acid buildup had to be

TABLE 16.12 Fermentation Free-Energy Changes for Standard vs. Typical Reactor Values (kJ/Reaction)

Fermentation Type	Reaction	ΔG^a	ΔG^b
Proprionate fermentation to acetate, CO_2, and H_2	Proprionate $+ 3H_2O \rightarrow$ acetate $+ H^+ + H_2 + HCO_3^-$	$+76.2$	-5.5
Butyrate fermentation to acetate, CO_2, and H_2	Butyrate $+ 2H_2O \rightarrow 2\,\text{acetate} + H^+ + H_2$	$+48.2$	-17.6

[a] ΔG^0, Standard free-energy release with 1 M fatty acid and H_2 at 1 atm.
[b] ΔG, free-energy release with typical reactor conditions (1 mM fatty acid and H_2 at 10^{-3} atm).

kept in check) the acidogenic conversions will realize energetically favorable (i.e., negative free-energy release) conditions during fatty acid fermentation.

This interplay between H_2-producing and H_2-consuming anaerobes, known as *interspecies hydrogen transfer*, represents an exceedingly vital biochemical alliance, without which the overall process of anaerobic digestion would not be feasible. In this context, the bacteria responsible for fatty acid fermentations and preparative H_2 production are also referred to as *syntrophs* (literally translated as "eating together"), whereby their cooperative life-style synergistically supports the necessary metabolic exchange of this key electron donor.

Sulfur metabolism, particularly that of sulfide release and transformation, also represents an important biochemical aspect with the fermentation reactions found in anaerobic digesters. These systems have three possible sulfur sources, including that of incoming soluble sulfates being converted directly to sulfides, due to the highly reducing conditions within the digester. Sulfides may also be introduced to these digesters via the sludge streams, either as constitutive cellular elements (e.g., present within sulfur-bearing amino acids, etc.) released directly from lysed microbial cells found in the secondary sludge or as reduced metallic-sulfide (e.g., FeS) precipitates found within reduced primary sludges. Whatever their source, sulfides within anaerobic digestors may then undergo a variety of important transformations. Sulfide-based precipitation may occur to an extent that could reduce the net solubility of many metal species (e.g., iron, copper, nickel, zinc), possibly even to a degree that would reduce their availability as metabolically necessary trace elements. Hydrolysis of the sulfide may also take place given the slightly acidic conditions found in most anaerobic digestors (i.e., due to the low, $10^{-7.25}$, acid ionization constant for sulfide hydrolysis, $H_2S \rightarrow HS^- + H^+$), thereby leading to gas exchange (i.e., via stripping) into the headspace and contamination of the predominant methane product to an extent that could necessitate subsequent removal prior to using the methane as an energy-rich feed gas. The highly reduced environment of an anaerobic digester may also transform inorganic sulfides into various malodorous organic sulfides collectively known as **mercaptans**. le 16.13 provides a general synopsis of these various biochemical products and their associated chemical structures, ranging from simply methyl mercaptan to dimethyldisulfide. These chemical composites of reductively generated organic methyl groups and inorganic sulfide collectively bear an intensely malodorous nature whose repugnant smell, especially when blended with that of classic "rotten egg" smell of hydrogen sulfide, is well known to anyone who has ever dealt with any of these anaerobic materials.

Perhaps in part due to the latter complexity of these metabolic pathways and transformations, anaerobic digestion systems are designed and operated in a wide range of physical configurations. These variations can be subdivided into one of three generic options: conventional high-rate, phase-separated, and anaerobic contact reactors. Even then, there may be further refinements with regard to mesophilic ($\sim$20 to 35°C) and thermophilic ($\sim$45°C and higher) operating temperatures.

Conventional high-rate anaerobic digesters are typically built in either single- or dual-tank (i.e., a mixed first tank and stratified second tank) forms, which provide mixing, and possibly heating, to improve their overall effectiveness (Figures 16.43 and 16.44). These operations receive smaller, semicontinuous loads of incoming sludge rather than larger batch-type loadings spaced farther over a longer period, with the overall intent of fostering a more stable balance with the acidogenic and methanogenic reactions. The latter

TABLE 16.13   Mercaptan Species Generated by Anaerobic Sludge Metabolism

Chemical Name	Chemical Structure
Methyl mercaptan	H_3C-SH
Isopropyl mercaptan	H_3C \ $CH-SH$ / H_3C
Isobutyl mercaptan	H_3C \ $CH-CH_2-SH$ / H_3C
N-Butyl mercaptan	$H_3C-CH_2-CH_2-CH_2-SH$
Dimethyl sulfide	$H_3C-S-CH_3$
Dimethyl disulfide	$H_3C-S-S-CH_3$

approach toward infrequent, intermittent loadings might impose a "feast-to-famine" regime that would then tend to disturb the desired metabolic harmony considered optimal for anaerobic digestion.

Taking into account the intermittent nature of these incremental inputs, the recommended loading rates [in terms of the daily applied volatile suspended solids (VSS) load per unit tank volume] range from 1.6 to 4.8 kg $VSS/m^3 \cdot$ day. Based on the typical

Figure 16.43   Conventional anaerobic sludge digestor: (*a*) with internal mixing; (*b*) without mixing.

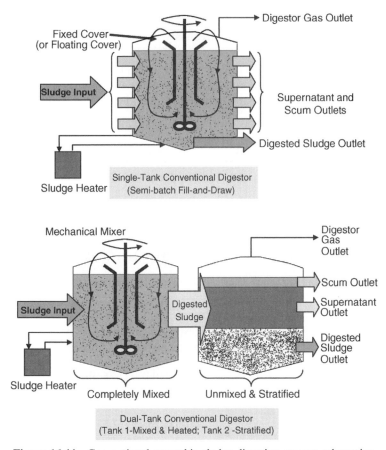

Figure 16.44 Conventional anaerobic sludge digestion process schematics.

characteristics of municipal wastewater sludge, these loading rates correspond directly to hydraulic retention times of 15 to 45 days, at which point these systems should be able to maintain volatile solids reductions on the order of 45 to 50% while producing 0.75 to 1.2 m^3 of methane-rich gas per kilogram of volatile solids removed. Federal stipulations for the operation of these systems in a fashion to reduce their pathogen content significantly mandates minimal solids residence times of 15 days at a temperature of 35 to 55°C, and 60 days for sludge digested at 20°C.

Example 16.5: Preliminary Anaerobic Digester Design   After completing the preceding series of four optional wastewater treatment designs, the Deer Creek, Illinois, consulting engineer is then asked for another set of preliminary design estimates relative to sludge processing. In this example, a preliminary design is developed for an anaerobic digester as might be used in conjunction with a conventional activated sludge system. Completion of this preliminary design involves four basic assumptions regarding sludge production and character. First, conventional wastewater treatment facilities typically generate about 0.25 kg total dry suspended solids per cubic meter of processed wastewater. Second, in its original wet state, raw primary plus secondary sludge generally contains a rather small solids fraction, approximately 1.2% by weight, or 12 kg/m^3. Third, raw sludge is commonly prethickened prior to digestion, and for this example a reasonable value of ~6% (60 kg/m^3) will be assumed. Fourth, blended primary plus secondary wastewater sludge typically has a volatile solids fraction of ~70%, the remaining ~30% being inert solids.

Preliminary Design Details

Sludge solids mass and volume, reactor sizing, and retention time

Design average wastewater flow: 605.7 m^3/day

Estimated daily total sludge solids mass: 151.43 kg TSS/day

$$(605.7\,\text{m}^3/\text{day})(0.25\,\text{kg/m}^3)$$

Estimated raw wet sludge volume: 2.5 m^3/day

$$(151.43\,\text{kg/day})/(60\,\text{kg/m}^3)$$

Estimated daily volatile sludge solids mass: 106 kg VSS/day

$$(151.43\,\text{kg/day})(0.7)$$

Design anaerobic digester solids loading rate: 1 kg VSS /m^3 · day

Design anaerobic digester volume: 106 m^3

$$(106\,\text{kg/day})/(1\,\text{kg/m}^3 \cdot \text{day})$$

Design anaerobic digester HRT: 42 days

$$(106\,\text{kg/day})/(2.5\,\text{m}^3 \cdot \text{day})$$

Related notes

1. The estimated raw wet sludge volume represents more than a 200-fold concentration of contaminants from the original wastewater flow into this residual sludge stream (i.e., 2.5 vs. 605.7 m^3/day, or a ratio of 1:~242).
2. The design solids loading rate represents a typical norm for conventional low-rate anaerobic digesters.
3. Here again, this single reactor design does not afford the desired system redundancy associated with multiple units.

Digester off-gas generated by these conventional high-rate operations typically contains about two-thirds methane and one-third CO_2, with a net heating value of approximately 22,400 kJ/m^3. By comparison, natural gas (i.e., a mixture of butane, methane, and propane) has an energetic value roughly 50% higher. Moderate-to-large wastewater treatment operations will, therefore, typically collect and use this gas to run boilers and internal combustion engines that then supply heat and power, although many smaller plants just burn their gas in a flare as a waste product. Prior to using this gas within a boiler or diesel engine, though, low-level gas contaminants such as water vapor and H_2S will routinely be removed to minimize their corrosive impact.

Although the vast majority of anaerobic systems currently in use for sludge digestion qualify as high-rate operations, experimental study and limited full-scale evaluation have demonstrated that further improvements in process efficiency and reliability can be obtained with appropriate revisions in the reactor staging, temperature and/or solids contact scheme. For example, even as early as the 1930s, it was noticed that thermophilic reactor temperatures from 50 to 55°C might provide distinctly higher metabolic rates for anaerobic fermentation, thereby resulting in significant reductions with necessary detention times. Unfortunately, though, these early thermophilic studies also revealed corresponding problems with process instabilities (e.g., higher levels of off-gas odor, as well as elevated foaming tendencies) that would largely negate any metabolic gains.

Recognizing these relative trade-offs (i.e., benefits vs. shortcomings) with mesophilic and thermophilic systems, several new schemes have been developed in recent years to combine the advantages of both. Two such **temperature-phased** processing options are shown in Figures 16.45 and 16.46, with one using a thermophilic-to-mesophilic sequence and the other using a mesophilic-to-thermophilic arrangement. In either case, the operational premise is that the first phase serves as a solids preprocessor, which then expedites the efficiency of the trailing, second phase. The resulting improvement in processing efficiency achieved with these temperature-phased designs is reflected in their reduced HRT requirements, which on average (at approximately 15 days) are both comparable to the lowest permissible values with standard high-rate designs.

Yet another version of these phase-separated design schemes was developed not only to provide different sequential reactor temperatures but also to secure a beneficial separation of the acidogenic and methanogenic reactions. These types of **acid-to-gas phased** processing options (Figure 16.45), are designed with even smaller hydraulic retention times (e.g., ~1 to 3 days) for the initial, "acid" phase, such that the slower-growing methanogens simply cannot be retained within the first reactor. At that point, therefore, the first reactor's hydrolytic and acidogenic roles are uncoupled from that of the concluding, second-phase methanogenic transformation, thereby promoting optimal conditions for each of these subdivided reactions.

This type of mesophilic-to-thermophilic acid-gas scheme has been used with considerable success on a full-scale level (i.e., at the Woodridge facility in DuPage County, Illinois) with less than a two-week total retention time. Such acid-gas-phased anaerobic digestion processes typically carry a far higher level of volatile acids (i.e., commonly ranging from 6000 to 8000 mg/L, and sometimes as high as 18,000 mg/L) within the first tank, and as a result the average pH in this reactor (~5.6) is considerably below the desired, let alone tolerable level for the methanogenic conversion desired. Another telltale indicator to the success of this two-phase digestion scheme is that the vast majority (typically, ~95%) of the system's overall gas production takes place within the concluding thermophilic reactor. Conversely, gas production, particularly that of hydrogen, takes place at only a nominal level within the first reactor, such that the acidogens are not constrained by the thermodynamic difficulties (i.e., at higher hydrogen levels) that might otherwise be imposed.

Further efforts to secure better, or more stable, performance with anaerobic digestion systems have also been made using changes in the mode of initial contact (e.g., with upflow passage through a fluidized bed of anaerobic granules) and/or final separation of the solid–liquid matrix (e.g., using mechanically induced separation as opposed to natural settling and flotation). In whatever fashion these systems might be constructed, energy savings represent a significant benefit with anaerobic digestion, including the

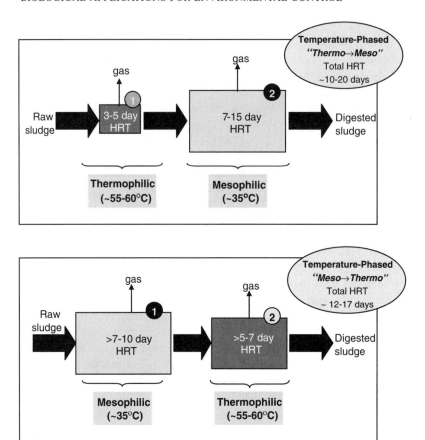

Figure 16.45 Temperature-phased anaerobic sludge digestion process schematics.

twin economic advantages of eliminating the aeration energy cost of aerobic processes and of recovering a good fraction of the energy content of the organic waste as gas-phase methane fuel.

16.2.2 Aerobic Digestion

Whether collected within wastewater treatment plants, solid waste landfills, or simply on a forest floor, organic solids will naturally degrade and decay under aerobic conditions through the natural process of oxidative breakdown. The microbial cells found in raw sludge will therefore undergo much the same progressive sequence of cell lysis, release of cytoplasmic materials, and final oxidative conversion within an aerobic digester, whereby the organics released from lysed cells serve as a source of substrates for other living aerobic cells. The metabolic conversion of wastewater sludge through aerobic digestion, such as the unit depicted in Figure 16.47, subsequently converts an initially organic-rich sludge matrix into carbon dioxide and residual solids that include not only both inorganic and recalcitrant organic forms but also the remaining complement of meta-bolically active aerobic oxidizers. The resulting product has the advantage not only of

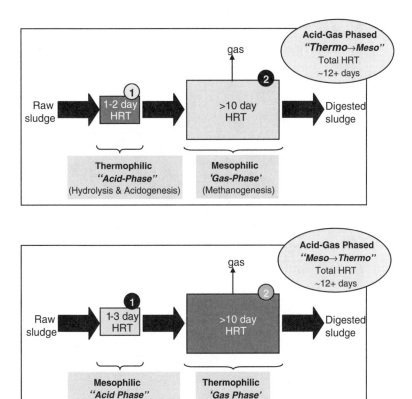

Figure 16.46 Acid-gas phased anaerobic sludge digestion process schematics.

being well stabilized, but produces much less of an odor problem than that of anaerobically digested sludge. The main disadvantages of aerobic digestion is the significant
energy cost of operation; the loss, compared to anaearobic digestion, of the energy
value of the waste; and the production of high levels of oxidized nitrate–nitrogen,
which may complicate subsequent land application options.

Compared to the complicated process of anaerobic digestion, aerobic sludge digestion
represents a far more simplistic and in large measure a single-step process. However, there
are still several important biological issues with aerobic digester operations that must be
considered and accommodated. The level of oxygen uptake maintained within these reactors provides a useful indication of the level of organic solids stability and digestion efficacy. This parameter is quantified as a **specific oxygen utilization rate** (SOUR) by which
the measured oxygen uptake rate is divided by the existing volatile suspended solids concentration to derive the *specific* (i.e., per mass of solids) value. The generally accepted
benchmark for suitably stabilized biosolids is a SOUR value of no more than 1.5 mg
of oxygen consumed per gram of total solids per hour.

The pH levels with standard, mesophilic aerobic sludge digestion operations can typically be expected to drop with time given the fact that the organic nitrogen released from

Figure 16.47   Aerobic sludge digestion system.

the degrading cells will hydrolyze to ammonia and be nitrified. This oxidative transformation releases 2 mol of hydrogen ions for every mole of oxidized nitrogen, which corresponds to an alkalinity consumption rate of 7.14 mg alkalinity (as $CaCO_3$) per milligram of fully oxidized ammonia-nitrogen. With incoming solids levels measured in percentile figures, and with commensurate ammonia-nitrogen releases of 1000+ mg N/L, it conceivable that this sort of pH drop could well shift the reactor to a sufficiently low level (i.e., <pH 5.5) that this nitrification process would actually be discontinued.

Without this sort of pH disruption, though, the effluent nitrate-nitrogen levels commonly observed in aerobically digested biosolids residuals would be quite high (measured in 100s if not even the 1000+ mg NO_3^--N/L range). The problem posed by these high residual nitrate levels is that unlike ammonium-nitrogen, NO_3^--N has no cationic affinity for soils. As a result, subsequent land application of these aerobically digested solids will require close attention to the potential migration of these nitrates to the groundwater (for which the U.S. EPA's Safe Drinking Water Standard is 10 mg NO_3^--N/L).

The regulatory stipulations on aerobic digester holding times (i.e., hydraulic retention times) relative to temperature stem from the necessity to maintain suitable reductions in pathogens. However, at the mesophilic conditions under which most aerobic digesters are maintained, retention times on the order of multiweek to multimonth periods are not uncommonly necessary to achieve volatile solids destruction levels up to or beyond the 38% range [i.e., the benchmark national standard for **vector attraction reduction** (VAR) with stable biosolids established by 503 Rule federal regulations (i.e., Code of Federal Regulations, 1993)]. Furthermore, during cold-weather periods, and with a corresponding drop in aerobic digestion efficacy, even longer periods will often be necessary.

As a proactive approach to escalating the standard aerobic digestion process, therefore, this technology has followed a trend analogous to that of the newer anaerobic schemes whereby reactor temperatures are increased to secure higher metabolic rates. This new design strategy, commonly known as **autothermal thermophilic aerobic digestion** (ATAD) (Figure 16.48) involves reactor temperatures starting in the mid-50°C range and in most cases reaching the 60+ °C range, at which point these processes experience far faster rates of lysis and oxidation. In addition, these ATAD operations also realize a sizable acceleration in their rates of overall disinfection.

Relatively little is known as yet about the microbial character and makeup of these types of systems, but it does appear that they may offer several new avenues for degrading a number of organic, possibly even hazardous compounds in addition to that of conventional wastewater sludge. In both cases, the levels of oxygen tension at which these high-temperature systems appear to operate tends to be much lower than what is usually seen with standard aerobic treatment reactors (e.g., ~0.5 mg of dissolved oxygen per liter with ATAD vs. 2+ mg/L in standard mesophilic aerobic digesters). In fact, at these DO levels, it is rather likely that the microbial consortia involved includes both aerobic and quasi-anaerobic microbes which live and work in metabolic harmony. Rather interestingly, the issue of nitrate buildup and release is also a moot concern at these thermophilic levels. Not only do the autotrophic bacteria responsible for nitrification effectively stop working at temperatures of about 40°C, but there is also a pronounced tendency to volatilize free

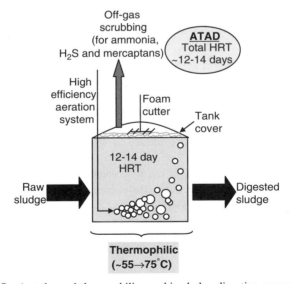

Figure 16.48 Autothermal thermophilic aerobic sludge digestion process schematic.

ammonia (NH_3) at this thermophilic level. Off-gas treatment (e.g., biofiltration) may, however, be required to deal with this released ammonia, let alone the release of other odorous compounds (e.g., reduced sulfur gases such as mercaptans and hydrogen sulfide).

Aerobic digestion is widely used as a sludge stabilization process at smaller wastewater plants (i.e., at flow rates below about 20,000 m^3/day). Batchwise operating regimes are common, but it is also possible to use intermittent batch (with cyclic settle, decant, and refill steps) and continuous-flow formats as well. Compared to anaerobic digestion, this aerobic option tends to be easier to maintain, and in the case of mesophilic systems has considerably less potential to release troublesome odors. These aerobically digested solids are also biochemically stable (i.e., resistant to further decay) and low in residual ammonia-nitrogen.

The key metabolic factor with aerobic digestion is that of the endogenous decay rate of the solids involved, which for most cells is in the neighborhood of 0.05 day^{-1} (i.e., 5% solids decay/day) at a temperature of 20°C. Of course, this rate of decay varies according to temperature and the degradable characteristics of the processed solids.

The design criteria typically used for sizing these units is therefore that of the retention time for these solids, for which values of 20+ days would accordingly provide volatile solids reductions of approximately 63% (37% remaining) when completed under batch conditions:

$$X_{20} = X_0 e^{-(0.05 \ \text{day}^{-1})(20 \ \text{days})}$$
$$\frac{X_{20}}{X_0} = 0.37 \tag{16.4}$$

where X_0 and X_{20} are, respectively, the initial and final (after 20 days) volatile solids concentrations found within the reactor. However, despite this theoretical level of performance, 503 Rule regulations require solids retention times (SRTs) of 40 days at 20°C and 60 days at 15°C before the product of this mesophilic aerobic digestion process is considered suitable for widespread use as a soil amendment based on desired VSS destruction. Here again, the latter increase in SRT reflects the fact that metabolic solids degradation slows down at colder temperatures.

Yet another important factor with aerobic digestion is that of its higher energy requirement, for both mixing and aeration, vs. that of anaerobic digesters, which not only require far less energy for mixing but also generate an energy-rich methane gas product. Aeration and mixing within an aerobic digester must therefore be provided using either mixers that entrain oxygen mechanically or compressed air blowers that diffuse oxygen into these tanks through bubble transfer. Mixing intensities with mechanical mixers are usually designed to provide 10 to 100 W/m^3 (0.4 to 0.5 hp per 1000 gal), while the diffused aeration systems are typically sized to provide 20 to 40 m^3/min of air per 1000-m^3 tank volume.

As mentioned previously, one variation to conventional aerobic digestion that has recently drawn considerable attention due to its inherent ability to pasteurize sludge effectively is that of the autothermal thermophilic aerobic digestion (ATAD) process, by which the heat released exothermically from the digesting sludge naturally causes the reactor temperature to rise into the thermophilic range (i.e., to values typically between 55 and 65°C, and sometimes well in the 70s). Federal regulations specifically stipulate a holding time of only 24 hours for those systems able consistently to maintain a temperature at or above 55°C.

In comparison to standard mesophilic operations, these ATAD systems are able to take advantage of a considerable increase in the available rate of endogenous decay, with values believed to be many times greater than those experienced at mesophilic temperatures. A properly designed and operated ATAD system does not need external heating. Heat is provided by the exothermic biodegradation reaction. A clear advantage of these ATAD systems is that their solids loading rates are considerably higher than either the mesophilic aerobic or anaerobic options, with values typically in the range 4.5 to 5 kg TSS/m^3·day. In turn, the design and operating SRT values used with ATAD systems can be decreased sizably, to values of a few weeks and possibly even less.

Example 16.6: Preliminary Autothermal Thermophilic Aerobic Digester Design Due to the volume estimated for the standard anaerobic digester system in Example 16.5, yet another request is then presented to the Deer Creek, Illinois, consulting engineer to prepare a second preliminary sludge processing design for a more advanced, and assumedly smaller, autothermal thermophilic aerobic digester. This ATAD design makes use of all four assumptions regarding sludge production and character (that were presented in Example 16.5).

Preliminary Design Details

Sludge solids mass and volume, reactor sizing, and retention time

Estimated daily total sludge solids mass: 151.43 kg TSS/day
Estimated raw wet sludge volume: 2.5 m^3/day
Design anaerobic digester solids loading rate: 4.8 kg TSS/m^3· day
Design ATAD digester volume:

$$31.5\,\text{m}^3 = (151.4\,\text{kg/day})/(4.8\,\text{kg/m}^3\cdot\text{day})$$

Design anaerobic digester HRT:

$$12.6\,\text{days} = (31.5\,\text{m}^3)/(2.5\,\text{m}^3\cdot\text{day})$$

Related notes

1. ATAD loading rates are sizably higher than those of even the high-rate anaerobic digester, and as a result these systems will require a distinctly smaller volume (i.e., in this case, the projected ATAD sizing is only one-third that of the anaerobic digester).
2. ATAD solids loading rates are generally based on the total vs. the volatile solids content of the sludge.
3. Prethickening of the incoming to sludge to values at or above 6% total suspended solids levels is very important with ATAD systems in order to secure autothermal (i.e., self-heating) operations.
4. Here again, this single-reactor design does not afford the desired system redundancy associated with multiple units.

To generate the level of heat output required to incur this temperature increase (i.e., raising the reactor temperature autothermally), the incoming total solids content of the raw sludge supply must be routinely prethickened (using thickeners, gravity belt thickeners, etc.) to values of 6% or higher. The corresponding density of these solids, and the fact that oxygen solubility drops considerably at higher temperatures, presents a distinct challenge in terms of providing the oxygen necessary to maintain aerobic conditions without unacceptably stripping heat away from the reactor at a rate that would negate the thermophilic condition desired. However, high-efficiency aerators (e.g., jet-type mixing units) have proven to be a suitable aeration technology for this type of application.

Despite the apparent technical benefits afforded by ATAD processing, a number of important operating details have yet to be fully resolved. First, instrumentation for measuring dissolved oxygen levels at these thermophilic operating temperatures has only recently been developed and has limited field experience. Preliminary testing has therefore been conducted using measurement of **oxidation–reduction potential** (ORP) as an alternative indication of the apparent aerobic nature of these reactors, and it does appear that holding this parameter within an approximate range of about −50 to −350 mV can subsequently be used to regulate aeration rates in a fashion that will obviate, or at least minimize, undesired shifts toward fully anaerobic conditions (with ORP dropping much below −400 mV), especially following intermittent loading events. Second, excessive foaming events have been observed at a number of early full-scale plants. Here again, this phenomenon is not fully understood, but it does appear that low-SRT operations are particularly prone to this problem, perhaps due to load-related swings in the cyclic level of soluble proteins. On the one hand, a limited amount of foam (i.e., ∼10 to 30 cm) is actually beneficial, in that it helps to provide an insulating blanket across the top of the tank. However, excessive foam production can lead to undesirable reactor overfoaming conditions, to the point where foam-cutter (essentially a coarsely toothed disk rotating just above the desired foam height) or foam-aspirator mechanisms are used to minimize this condition. Finally, it is important to note that the microbial behavior, temperature range, and environmental circumstance of ATAD processes is equivalent to that of biosolids composting operations, or at least the interior zones of the actively composting piles, as described in the following section.

16.2.3   Composting

Sludge residuals may also be biochemically digested, stabilized, and perhaps even partially or fully pasteurized using composting operations of the sort shown in Figure 16.49. When exposed to moisture and appropriate environmental conditions, organic, nutrient-rich surfaces will quickly become colonized by bacteria and other microorganisms. Since no organism can be 100% efficient in its metabolism, during the ensuing degradation of the organics, some chemical energy is wasted and given off as an **exothermic** heat release. Ordinarily, this release of heat would not be noticed, since it quickly dissipates into the environment. When solid-phase organic materials are held in a large pile, however, the pile itself acts as insulation and traps some of the heat. This effect can then lead to a noticeable increase in temperature of the material and is thus referred to as biological **self-heating** or **autothermal** metabolism. The intensity of self-heating can be surprising, particularly if the pile is sufficiently porous to allow oxygen penetration and if the available moisture content remains sufficiently high to sustain continued biodegradation. Temperatures of up to 80°C (176°F) can be reached with many materials, at which point subsequent chemical self-heating and eventual combustion may occur if moisture is still present. In fact, important early research on this process was completed in New Zealand during the 1960s, leading to documented reports of spontaneous ignition in piles of wool! In much the same fashion, spontaneous ignition is a familiar phenomenon

Figure 16.49 Windrow biosolids composting piles.

to farmers, who must, for example, carefully aerate and cool hay stored in barns to prevent disastrous fires.

This sort of heat production is, in fact, a natural occurrence with any biochemical system, and in our own case provides the thermochemical basis by which we intrinsically maintain an optimal body temperature. Exothermic biochemical energy is also released within aqueous-phase wastewater treatment processes, whether they be suspended-growth activated sludge or attached-growth biofilm processes, but the high thermal mass of the water and the cooling effect of aeration (by evaporation and conduction) typically offset the metabolic heat release to such a degree that it does not affect system temperature. In rather rare instances, high-strength industrial wastewater operations have experienced elevated temperatures due to their elevated levels of heat release. Similarly, the ATAD systems described previously are also designed to benefit specifically from this mechanism, in which aeration is maintained inside insulated reactors at rates close to stoichiometric levels such that their exothermic heat release is not excessively offset by off-gas heat loss.

As applied to the high-temperature degradation of wastes bearing high-level biodegradable solids, **composting** is an engineered process that utilizes self-heating for waste treatment purposes. By definition, it would be described as a mainly aerobic, self-heating, solid-phase biological treatment process. The goals of composting include stabilization, reduction, drying, and pathogen destruction. Traditionally, composting has been used to treat agricultural wastes such as crop residues and animal manures, and large-scale composting of separately collected yard wastes is a common practice in the United States, particularly in the northeast and midwest, where the fall leaf collection can be sizable (often 10 to 20% of the total MSW for the year!). In the broader context of environmental waste management, composting also has wide applicability for wastewater treatment sludges, **municipal solid waste** (MSW) fractions, and some industrial (including hazardous) wastes.

The primary objective of composting is to stabilize the waste material being treated. This results in a reduction of mass and volume as well as stabilization, destruction of **putrescibles** (rapidly degrading, odor-producing compounds). Another potential benefit of composting is that in many cases the final residue is a loamy, soil-like material with an earthy smell, called **compost**, which may be used beneficially as a soil amendment or surface mulch. The complementary fact that composting is capable of substantial pathogen and weed seed destruction is also quite beneficial. As a soil amendment, compost adds to the organic content of a soil, increasing its friability and its ability to adsorb water and nutrients. Uncomposted wastes are not as suitable for this purpose because they will degrade in the soil, depleting soil oxygen, which can harm plant roots.

Waste materials subjected to composting will typically undergo a succession of microorganisms in relation to progressive changes in waste character and environmental conditions. At first, mesophilic organisms originally present in the material, along with early invaders, will be dominant. This community may be highly diverse, including fungi, protozoans, and even invertebrate animals, such as earthworms, insects, and sow bugs, in addition to bacteria. However, as the ongoing heat release moves temperatures above 40°C, many of these original inhabitants are inhibited, and eventually, most are killed by the heat. At this point, having shifted into a thermophilic realm, small numbers of thermophiles that were present find suitable conditions and grow rapidly. This is a more select group, as few eukaryotes can survive at temperatures above 50°C. Above 62°C, the last fungi are unable to grow, and only bacteria (and perhaps archaea) are left. The known organisms in thermophilic composting include mainly *Bacillus* species, such as *B. stearothermophilus* and *B. coagulans*, and some actinomycetes, although there is evidence from molecular techniques that other groups may also be present. Eventually, as substrate is used up, the material will cool again, be recolonized by the germination of spores that survived the high temperatures, and be reinvaded by mesophiles.

Earlier, some composting enthusiasts seemed to believe "the hotter, the better." However, it is now well established that if not controlled properly, most composting materials will overheat, killing or severely inhibiting even the thermophilic microorganisms. This has led to the failure of many composting facilities, since the subsequent rate of degradation slows dramatically, and the remaining putrescibles lead to odor problems. Thus, an important goal of modern composting technologies is to maintain temperatures at a desirable level of ~60°C (140°F) to maintain the desired high rates of microbial activity. As a point of comparison, most home hot-water heaters are set at ≤130°F; reaching into the interior of an actively compost pile would lead to a serious burn!

It is also important that the composting material be kept mainly aerobic. Only aerobic metabolism releases energy rapidly enough to sustain this degree of self-heating. Also, avoiding extensive anaerobic conditions helps to minimize odor production. Usually, maintenance of at least a 10% oxygen partial pressure in the pore spaces within a pile (compared to the 21% oxygen in ambient air; Figure 16.50) will be adequate, depending

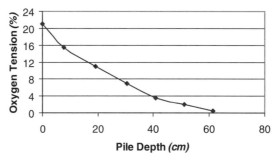

Figure 16.50   Biosolids composting pile oxygen tension relative to pile depth.

on the material and method used. To facilitate the desired transport of oxygen into these actively composting systems, therefore, the composting solids must be maintained in a suitably porous form. If the starting waste materials are too wet, as is the case with most raw sludges, they may need to be partially dewatered and may also need to be mixed with other materials to improve porosity (**bulking agents**, such as finished compost, MSW, or wood chips). If the pile is too wet, the water filling the pore spaces severely limits oxygenation, and anaerobic conditions develop. On the other hand, materials such as municipally collected leaves commonly start out too dry for rapid microbial growth and thus benefit from water addition. Generally, moisture contents in the range 50 to 70% will give the best results, but this depends on both the material and the type of composting system used. Many of the bulking agents used in composting (e.g., wood chips, shredded tire chips, bark) have an original bulk cost that warrants an attempt to secure their recovery and reuse. As shown in Figure 16.51, it may be possible to screen and recover these

(a)

(b)

Figure 16.51   Sludge composting: (*a*) aerated piles; (*b*) bulking agent screening and recovery.

supplemental bulking agents from the finished, composted product so that they might then be reused yet again with fresh sludge. This effort not only saves money by reducing the necessary volume of bulking material but also provides an initial seeding of the coblended material with thermophilic microorganisms, which then promotes faster startup times.

The relative masses of carbon and nitrogen (the C/N ratio) can be important in composting, with a ratio of about 30 : 1 often considered desirable. Materials such as dry leaves may have C/N ratios of 80 : 1; under such conditions, nitrogen becomes limiting and composting rates are slowed. This is normally acceptable for leaf composting, and N addition is not recommended, but this might be an issue for some industrial wastes. Municipal sewage sludge, though, tends to have a C/N ratio of $\leq 8 : 1$, indicating excess nitrogen. This will not decrease composting rates, but can lead to the release of nitrogen as either ammonia gas (making odors more of a problem) or as a potential water pollutant. Addition of a carbonaceous material may therefore be desirable. However, although a bulking agent such as wood chips will increase the calculated C/N ratio, most of the C is unavailable to microorganisms and hence may not produce the full beneficial effect expected.

As with other microbially based treatment systems, the presence of a large number and variety of microorganisms is desirable for rapid composting. However, waste materials typically already contain a high concentration and diversity of appropriate organisms, and under proper conditions their growth will be very rapid. In the few cases where the deliberate addition of microorganisms may be warranted, such as for some pasteurized food-processing residues, this is usually best accomplished by adding small amounts of finished compost or soil to the initial mixture. There is no scientific evidence that the addition of commercially available inocula or "compost starters" is beneficial.

On the other hand, waste materials may start out with numerous undesirable biological agents present, such as pathogens, parasites, and weed seeds. Composting can be extremely effective (better even than chemical disinfection, and probably second only to incineration among treatment processes) at inactivation of these undesirables. This is a result mainly of the high temperatures achieved (e.g., *Salmonella* will be reduced significantly within an hour or so at temperatures much above 60°C), but is also aided by the vigorous microbial activity that occurs. Thus, decreases of well above 99.99% are expected in properly run systems. In fact, a common criterion, maintenance of 55°C for 3 days, is predicted to give a minimum of 15 "9's" (i.e., 99.9999999999999%) reduction of even the most resistant pathogens. Federal (i.e., 503 Rule) regulations in the United States for pathogen control with composting systems stipulate specifically that the temperature of these piles must be held above 40°C or higher for a period of 5 days. This standard also requires that a temperature of 55°C or higher must be reached for a period of at least 4 hours in such piles to maintain the necessary reduction in pathogens.

One special concern with pathogens is that a few may actually grow during some composting processes. The best known example is *Aspergillus fumigatus*. This thermotolerant fungus is cellulolytic (degrades cellulose) and thus very common in nature and agriculture in soil and decaying vegetative material. However, it produces large numbers of spores that can cause a mild to severe allergic reaction in susceptible people. In a few cases, it is also able to opportunistically invade people with severely weakened immune systems, leading to potentially lethal infections. Sludge composting operations using wood chips as a bulking agent may release very high levels of *A. fumigatus* spores during the final screening step to remove the wood chips from the compost. Reuse of the wood chips then serves to heavily reinoculate the new pile with *A. fumigatus*. Elevated levels have also been observed at some leaf composting sites during the turning of windrows.

The overall composting reaction can be described using a modified form of the basic equation for aerobic respiration:

$$\text{organic matter} + O_2 \rightarrow CO_2 + H_2O + \text{compost} + \text{heat} \tag{16.5}$$

From this expression it can be seen that the rate of organic matter stabilization is proportional to the rate of heat production. Thus, maximizing the rate of heat production will maximize the rate of stabilization. However, if the material becomes too hot, rates slow dramatically. Once active self-heating occurs, therefore, it is necessary to remove heat at approximately the same rate it is released, so as to avoid exceeding 60°C in the material. This should be a major concern in system design, as the amount of heat requiring removal can be substantial. The amount of heat released per mass of oxygen consumed is approximately 14,000 J/g and is very nearly constant for a wide variety of different organic materials. At the same time, it would be necessary to provide adequate oxygen to reach this oxidative heat release, and to keep moisture and other parameters within a desirable range.

The goal of a composting system is to stabilize the particular material being treated in an efficient, economical, and environmentally sound manner. For some materials, such as leaves, a low-cost system can be used, even though the conditions it provides do not come close to maximizing composting rates. This is because the facility, if it is large and isolated enough, can simply allow extra time for completion (e.g., 6 to 18 months), and the materials can be managed so as not to cause problems during this time. Other materials, such as sludges, however, usually demand closer control and require composting systems that much more nearly achieve maximum rates. Otherwise, problems are likely to occur, and costs may soar.

Although there are many variations, there are really three approaches to large-scale composting: mixing; forced aeration; and both. Any of the three potentially can be done out in the open, under a roof, or in an enclosed reactor, although some combinations are more common and/or logical. Interestingly, almost all systems are operated as batch processes rather than continuous feed, as is the case with most other waste treatment systems.

Probably the most common type of composting, such as that shown in Figure 16.49, is that of *windrowing*. The sludge-plus-bulking agent windrow piles are constructed in an elongated, haystack shape (in cross section) up to perhaps ~1.3 to 2 m (4 to 6 ft) high and ~4 to 5 m (12 to 15 ft) wide, and lengths reaching up to and beyond 100 m (~300 ft). Periodically (e.g., twice a week initially, monthly later) it is mixed, or turned, using a front-end loader or specialized turning machine (as can be seen in the background of Figure 16.49). Windrowing is virtually the only method used for yard waste composting and is also used occasionally for sludges and solid wastes in areas that are sufficiently isolated, handling small volumes, or with other special circumstances. Also, it is commonly used in a curing stage, a low-rate finishing step after a more active composting phase. In some cases, the material is enclosed in bins, and mixed there: essentially, "windrowing" within an isolated reactor.

One disadvantage of static piles and other unmixed systems is that stratification occurs with gradients of temperature and moisture within the pile. Thus, some portions of the material may not heat sufficiently for pathogen kill, whereas others may dry or overheat and become inactive. Periodic windrow mixing or turning, therefore, helps to ensure uniform solids breakdown. Although some heat is lost during turning, and some oxygen is incorporated in the pile, the increase in microbial activity spurred by the mixing quickly (within hours) reheats the material and depletes this added oxygen. The height and width of these windrows must therefore be held to values of less than a few meters, such that oxygen may adequately diffuse into the pile interior from the surface. In fact, the ongoing release of heat from windrow piles helps to facilitate this aeration process, whereby the physical air current induced by heat rising upward and out of these hot piles essentially drafts or pulls in cooler air, and oxygen, at the pile base.

In lieu of intermittently mixed windrow systems, there are also units designed to provide nearly continuous mixing and/or aeration for composting wastes. Forced static pile aeration, as shown in Figure 16.51, represents yet another method developed by the Agricultural Research Service (U.S. Department of Agriculture) at Beltsville, Maryland. An aeration system (ducts, or a perforated false floor) in the lower portion of, or under, the pile is used, and control systems can be provided to either increase aeration (if the pile starts to get too hot) or decrease aeration (if the pile starts to get too cool), as needed. In some cases (including that of the piles seen in Figure 16.51), this supplemental airflow is drawn into and through the pile by applying a vacuum to the ductwork, with the exhaust air then being routed through a secondary smaller pile of aged compost material in order to screen out potentially problematic gas-phase odors, fungal spores, and so on. However, forced blower aeration directly into the pile interior has also proven to be effective, particularly in terms of securing direct oxygen entry to the pile's hottest and most active zone. These types of simple, feedback aeration schemes, based on a thermostatically regulated vacuum or blower, have a number of interesting features. First, the pile itself demands the amount of aeration it needs to remove enough heat to keep temperatures in the desirable range. This is important because the rate of activity (heat generation) changes with time as the readily degradable organics are depleted. Second, most of the cooling results from evaporation of water within the pile into the supplied air, at which point the pile starts to dry out. With some materials pile drying may be so extensive that composting rates decrease and supplemental water will need to be added. Often, however, the drying is beneficial, as it substantially reduces the remaining mass and volume and leads to a more stable final product that is more easily stored, transported, and ultimately used with gardening, land application, and so on, measures.

A third feature stems from the fact that the amount of oxygen required for biodegradation of the organic material is closely related to the amount of heat released [as can be seen from equation (16.5)]. Thus, an **air function ratio** can be defined for forced aeration systems as the amount of air required to remove the heat produced compared to the stoichiometric amount required to provide the oxygen necessary for the oxidation reaction that releases it. This ratio will vary slightly based on materials, ambient conditions, and pile temperature, but is typically about 8.5 to 9.0. This ensures that sufficient—in fact, considerably excess—oxygen will automatically be provided by the aeration required for cooling. It is interesting to note that the air function ratio must in fact be greater than 1.0 for self-heating to occur. As a practical matter, some aeration, typically provided by a timer, also is needed to supply oxygen before and after the phase of the process during which aeration for cooling is required: the come-up and cool-down stages.

The most advanced composting systems combine forced aeration for temperature control and oxygenation with mixing to increase rates and uniformity. Such systems include agitated beds and "mushroom tunnels" (so-called because they were developed in the mushroom industry to provide the high-quality compost needed for the commercial growing of mushrooms). Tunnels are enclosed, with a small headspace above the material, and a portion of the used air can be reused for aeration. This requires some cooling of the air and condensation of the water vapor present but can greatly reduce the amount of fresh air needed (since based on the air function ratio, little of the oxygen is utilized). This in turn reduces the amount of spent gas that must be vented, making control of odors and other volatile compounds simpler.

Example 16.7: Composting Air Function Ratio Sample Calculation   Based on equation (16.5) and the typical energy release of 14,000 J/g O_2 consumed, an air function ratio can be calculated for specific composting conditions. Assume, for example, that the ambient air is at 20°C and 50% relative humidity (RH), and the air exiting the composting pile is at 60°C and 100% RH. From thermodynamic data (found in psychrometric charts) it can be determined that this represents a change in enthalpy (heat energy) of 362 J/g dry air (part from the increase in temperature, and part from the increase in the amount of water vapor). Thus, the amount of dry air required to remove the heat released from 1 g of O_2 consumption is

$$\frac{14,000\,\text{J/g}\,O_2}{362\,\text{J/g dry air}} = 38.7\,\text{g dry air/g}\,O_2 \text{ consumed}$$

The amount of dry air required to provide 1 g of O_2 is only 4.31 g. Thus, the ratio is

$$\text{air function ratio} = \frac{\text{air required for cooling}}{\text{air required for providing oxygen}} = \frac{38.7\,\text{g/g}}{4.31\,\text{g/g}} = 8.98$$

The air function ratio will vary slightly depending mainly on the temperature and RH of the inlet and outlet air. However, it is usually between 8.5 and 9.0 for conditions that are likely to be encountered during thermophilic composting.

A number of other composting system designs have also been of initial interest because of their materials handling approaches, including semicontinuous feed systems such as silos (in which materials were added to the top and finished material was removed from the bottom) and rams (in which material was pushed along in a "tunnel"). However, these particular approaches greatly compacted the material, destroying the porosity that was essential to movement of air through it. It is important to keep in mind that although efficient materials handling approaches are important for cost-effectiveness, the system will fail if the basic biological requirements of the composting process are not sufficiently met.

16.3   POTABLE WATER TREATMENT

There are relatively few engineered applications for biological treatment within potable water systems. The rare exception to this is that of denitrifying processes intended to reduce excessive nitrate levels (i.e., above the regulated level of 10 mg/L), as might be used in association with shallow groundwaters affected by regional farming and fertilization activities. In contrast to beneficial instances of biological treatment with potable waters, though, biology can, and in many instances often does, become a troublesome

TABLE 16.14   Taste- and Odor-Producing Chemicals Generated by Bacteria and Algae

Chemical Name	Chemical Structure	Generating Microorganisms
Methylisoborneol (MIB)	OH	Actinomycetes *Oscillatoria*
Geosmin	OH	Actinomycetes *Anabaena* *Oscillatoria*

issue either in terms of affecting raw water quality or during subsequent dissemination of these waters via distribution piping.

Excessive biological growth of any sort, whether algal, bacterial, or other, can lead to elevated levels of turbidity (i.e., cloudiness) that would complicate subsequent treatment efforts. However, one of the largest biological problems in potable processing of surface waters is that of the formation of various taste- and odor-inducing chemicals. Two such compounds, geosmin and methylisoborneol (often referred to as MIB), are depicted in le 16.14, and it is these particular biochemical products that are most frequently cited as the chemical culprits behind serious aesthetic complaints. These compounds are produced biochemically by a variety of cyanobacteria (including *Oscillatoria* and *Anabaena*), actinomycetes (including *Streptomyces*, *Micromonospora*, and *Nocardia*), and algae (e.g., *Asterionella*), leading to taste and odor complaints that have been likened to "musty" or "fishy" (Darleym, 1982).

The ability of either compound to create problems is readily demonstrated by their respective aqueous-phase "threshold odor" values, which quantifies the concentration value at which humans may perceive their presence. As shown in Figure 16.52, geosmin and methylisoborneol have remarkably low "threshold" values, in the single-digit parts per trillion range. This level is nearly 1000 times lower than that of the next-highest compound (i.e., methyl sulfide, otherwise known as methyl mercaptan, which as described earlier, has a considerable odor of its own!).

The typical circumstance of these two products creating a problem is that of seasonal blooms of cyanobacteria and algae in reservoirs, lakes, and rivers, tied to changes in water temperature and nutrient availability. One such commonplace trigger is that of diurnal fall and spring overturns in deeper reservoirs and lakes, at which point the underlying phosphate-rich hypolimnion waters are returned to the surface to stimulate phototrophic microbial growth. Even without the bacterial release of geosmin or MIB, algal activity stimulated in much the same fashion may also result in the release of various phenolic-type organics, which can then react with chlorine during the treatment process to produce the strong taste- and odor-causing chlorinated phenols.

A second biological problem in potable water systems that may occur is that of bacterial colonization along the interior surfaces of distribution piping (Bitton and Gerba,

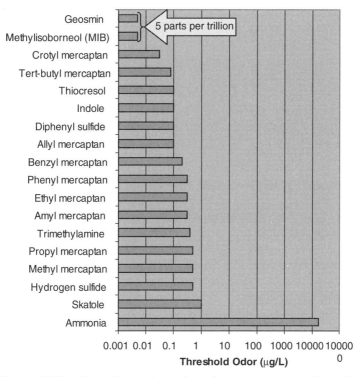

Figure 16.52 Threshold odor levels for various aqueous-phase chemicals.

1984). These microbes, such as *Aeromonas*, use low-level substrates in the form of biodegradable and assimilable organic carbon that in many instances are produced by the oxidative conversion of otherwise recalcitrant materials (e.g., naturally occurring humic acids) by chlorine. More often than not, these growths do not reach population densities that would unduly affect the healthful quality of the water. Their presence can nonetheless lead to potential corrosion problems, particularly with cast iron piping. Isolated patches of bacterial biofilm can accordingly create microsite niches whose isolated underlying strata may effect a reducing environment that, in time, corrosively degrades the pipe wall.

Neither geosmin nor MIB can be removed easily during the treatment of surface waters, at least using the conventional technology (e.g., coagulation, filtration, disinfection) employed at most municipal operations. As a result, the only approach to dealing with this problem is that of attempting to eliminate the microbial source of the problem, either by switching water sources or by working to eliminate the biological agent responsible. Reservoirs are often treated with copper salts, which inhibit algae growth at levels of 0.5 to 1.0 mg Cu/L.

The key to preventing growth of the organisms in the first place is restricting or eliminating the presence of nutrients, such as phosphate, which contribute to the offending phototrophic bloom. Within many reservoir and lake settings, these nutrients are introduced through a variety of nonpoint sources that are not particularly easy to control or eliminate, including fertilizer use on lawns bordering homes and cottages, septic tank leachates, or agricultural inputs received via upstream drainage. However, should the dominant phosphorus source be that of diurnal overturn and exchange, this mechanism may be controlled using low-level aeration of the hypolimnion region in a fashion that effectively ties up the available phosphate ions through precipitation with iron or calcium. This strategy can be relatively simple to employ, using flexible plastic piping, laid across the deepest reach of a lake, that is drilled with holes and supplied with compressed air such that it releases a steady current of bubbles. Another approach is somewhat more complicated, using skirted mechanical mixers or turbines that create hydraulic drafts to routinely mix the upper and lower reaches of the lake, but in either case the end result is that of inducing a degree of supplemental aeration that reduces or eliminates the presence of soluble phosphates.

16.4   WATER AND WASTEWATER DISINFECTION TREATMENT

One of the most significant public health advances over the past century was that of developing, and then routinely applying, suitable engineering methods for disinfecting potable waters that could retard, and ideally obviate, the transmission of waterborne disease. Rudimental disinfection measures based on water filtration (used by the ancient Egyptians) and heat treatment have long been practiced, but the advent of commercially available chlorine during the late nineteenth and early twentieth centuries effectively revolutionized the wide-scale utility and efficacy of this practice (Baker, 1948).

Indeed, chlorination has subsequently served as the dominant disinfectant "tool" for potable waters since the early twentieth century, but there are now at least six strategies by which waters, wastewaters, or sludges might be duly processed to achieve a desired reduction in their microbial content (Bryant et al., 1992; U.S. EPA, 1999):

1. Physical filtration (e.g., using ultrafiltration or reverse osmosis)
2. Heat treatment (e.g., boiling and pasteurization)
3. Physical sonication
4. Strong oxidant chemical treatment (e.g., using halogens such as chlorine, bromine or iodine, or ozone)
5. Nonoxidizing chemical treatment
6. Radiation treatment

These disinfection strategies can largely be subdivided into the following four options, although in many instances there are likely to be significant overlaps with the causative impacts imposed by any one disinfection procedure:

1. Remove cells physically
2. Alter and disrupt cell membrane permeability
3. Alter and disrupt metabolically essential proteins and enzymes
4. Alter and disrupt genetically essential nucleic material

The first such approach to using physical filtration to remove microbes depends on cell size, which is rather fortunate since chemical-based disinfection tends to become problematic with large cells or cysts, such as might be experienced with protozoan forms of *Giardia lamblia* and *Cryptosporidium*. Conventional slow- and rapid-sand filters have been used for more than a century to effectively reduce, if not completely eliminates, these larger microbes, and as a result these types of filters have been widely used for processing surface water sources susceptible to microbial contamination. However, the success of these operations can be compromised by human and technical shortcomings (e.g., inadequate filter backwashing regimes), as has been demonstrated in several U.S. cities over the past few decades. One such well-publicized incident with mismanaged water filter operations in Milwaukee, Wisconsin, during 1994 affected 400,000+ residents. The effectiveness of these operations has been improved with tightened stipulations (i.e., the U.S. EPA Interim Enhanced Surface Water Treatment Rule) on routine filter monitoring (U.S. EPA, 1998). In recent years, the disinfection effectiveness of media-based filtration has been eclipsed by the use of micro- and ultrafiltration systems with pore sizes in the double-digit nanometer range (e.g., typically 30 to ~100 nm), which are small enough to prevent the passage of any pathogens.

As for those disinfection strategies intended to negatively change the permeability or perhaps water content of cells, numerous examples can be seen with foods prepared in percentile-level salt, sugar, or organic-acid-rich conditions (e.g., pickles, candied fruits, cheeses, vinegar, tomato catsup). These preservation environments, many of which yield osmotic pressures intolerable to active cell growth, facilitate a bacteriostatic condition in which microbial growth has been effectively stopped without specifically killing the original cells. Large-scale adjustments to the osmotic pressure within water, wastewater, or sludge treatment processes would, of course, be infeasible given the necessary chemical dosage requirement (i.e., at expensive, high percentage levels).

Chemical disinfection agents that alter the form and function of membrane-bound transport enzymes could disrupt the transmembrane passage of essential substrates or nutrients. Whether or not the latter membrane-specific impact is realized in conjunction with the use of the more widely used antimicrobial chemicals (e.g., chlorine, ozone), enzyme disruption and denaturation are widely considered to be their dominant disinfection mechanism. These agents, and particularly those involving either strong-oxidant (e.g., chlorine) or superoxide (generated by means of irradiation) chemicals, readily disrupt a cell's hydrogen- and covalently bonded three-dimensional enzyme conformation. Having lost the enzyme's catalytic contribution to energy-yielding catabolism, these disinfected cells subsequently lack sufficient energetic resources to reproduce effectively.

Four different strong oxidants are widely used for disinfection: (1) halogens (chlorine, bromine, and iodine), (2) halogen-containing compounds (e.g., chlorine dioxide, chloramines, bromochlorodimethylhydantoin), (3) ozone, and (4) hydrogen peroxide (H_2O_2). In each case, the standard engineering application is that of dosing the applied chemical into a short-term contact chamber (i.e., typically designed for 15 minutes' retention) using a chemical delivery system preset to achieve a desired disinfectant concentration relative to measured flow. As shown in Figure 16.53, these disinfection contact chambers are often designed with a serpentine configuration in an attempt to secure a quasi-plug-flow regime.

Chlorine has been, and remains, the dominant disinfectant chemical with waters and wastewaters in the United States, applied either in gas (Cl_2), liquid (NaOCl), or solid [$Ca(OCl)_2$] form at what is likely to be the least possible cost (i.e., in the range of pennies per pound) for any disinfection option (White, 1992). Aside from cost, chlorine's advantages include its range of delivery options and expected efficiency. However, there are also shortcomings with its use, including the fact that there are significant safety issues to be addressed when storing and metering chlorine gas.

One key aspect of chlorine use is that of its sensitivity to pH. Above pH 7.5 the desired hypochlorous acid (HOCl) species found in aqueous environments (e.g., produced by the

Figure 16.53   Serpentine chlorination reactor for wastewater effluent disinfection.

hydration of chlorine: $Cl_2 + H_2O \rightarrow HOCl + H^+$) will disassociate into a hypochlorite (OCl^-) anion ($HOCl \rightarrow OCl^- + H^+$) whose antimicrobial efficiency is far lower than that of the hypochlorous acid (i.e., HOCl) form.

Another important fact is that hypochlorous acid reacts readily with reduced ammonia (NH_3), leading to a series of amination reactions and products [i.e., monochloramine, NH_2OCl; dichloramine, $NH(OCl)_2$; and nitrogen trichloride, NCl_3] whose bacteriocidal efficacy is again less that of the original HOCl.

Bromine use as a disinfectant also has its own set of unit features in terms of chemistry, benefits, and shortcomings (Water Environment Federation, 1996). First, the operative hypobromous acid (HOBr) species does not disassociate until it reaches a pH of about 8.5, so offering a wider range of serviceability them chlorine. Second, bromine tends to have a higher level of efficacy at equivalent concentrations, so lower dosage levels might be used. Third, the bromamine forms are all far better disinfectants than are the chloramines. Indeed, a solid-phase, bromine-bearing compound called bromochlorodimethylhydantoin (BCDMH) is widely marketed for hot tub and spa applications, given the particular prevalence of urinary ammonia release. One clear disadvantage, though, is that of cost, which tends to be several times higher than that of chlorine if the latter's efficacy attributes are ignored.

Hydrogen peroxide has little, if any, credible role as a disinfectant in environmental engineering systems, but ozone has found widespread acceptance, particularly in Europe, as a potable water disinfectant. As compared to any of the halogen-based options, ozone has the unique ability to dissipate shortly after its addition, without any semblance of a lingering residual. This lack of a residual is considered a disadvantage in the United States, where residual chlorine levels are maintained routinely in potable water delivery systems as a safeguard against subsequent contamination that can occur within the distribution system. However, the conventional wisdom in Europe is that disinfecting chemical residuals are both unwarranted and undesired. Yet another important aspect of ozone use is that it must be produced on-site, using electrical hardware that is not all that simple and at a cost that is considerably higher than that of chlorine.

A wide range of nonoxidizing organic and inorganic chemicals are used for, or are able to provide, disinfecting effects, including aldehydes (formaldehyde and glutaraldehyde), phenolics, alchohols (ethanol and isoproponal), cationic detergents, nitrites, and heavy metals (e.g., mercury, silver nitrate, tin, arsenic, copper). Although most of these chemicals have little relevance for the disinfection of waters, wastewaters, or sludges, there are two noteworthy exceptions. Specifically, silver-impregnated filters are sometimes marketed for point-of-use water conditioning devices, such as those that are sometimes screwed onto the outlets of sink faucets. In this instance, the silver is intended to be slowly leached from the filter medium (typically, activated carbon) at a rate that, hopefully, will retard the opportunistic formation of microbial biofilms intent on using sorbed organics as their energy source. A second nonoxidizing chemical disinfectant option is that of using cationic detergents in the form of quaternary ammonium compounds (formed as organic salts of ammonium chloride and commonly referred to as *quats*). One such common application is that of controlling biofilm growth on cooling tower surfaces, in which aggressive (i.e., oxidative) disinfection agents such as chlorine, bromine, and ozone would unacceptably attack exposed wood or metal heat-transfer surfaces.

One of the unique disinfection features of the bactericidal quaternary compounds is that they have a distinctly higher level of effectiveness with many gram-positive bacteria, probably due to the added depth and complexity of their membrane structure. Conversely, gram-negative cells as a whole are often similarly considered to be somewhat more resistant, perhaps due to the added depth and complexity of their membrane structure. Indeed, *Pseudomonas* probably tops the list in terms of durability, under conditions that would foil the vast majority of other cells (e.g., growth in distilled water). Similarly, gram-positive *Mycobacteria* species, as well as spore formers, also tend to exhibit this resistant nature when challenged with quaternary disinfectants, apparently based on the protective capacity of their respective outer cell coatings. Finally, given their chemical nature, these quat compounds also bear a unique sensitivity to inactivation when exposed to complexing soaps, detergents, and organic materials.

The fourth and final disinfection mechanism is that of altering and disrupting a cell's genetic makeup so that the cell is prevented from reproducing even though it may still have the energy to do so. This disinfection effect is largely associated with the use of ultraviolet irradiation, and its consequent high-energy cross-linking of adjacent nitrogen base groups poised side by side at various points within stranded DNA (see the thymine dimer reaction in Figure 16.54). The resulting impact of polymerizing DNA and formation of thymine dimers follow much the same path, such that cells are effectively sterilized by this UV exposure.

The range of wavelengths associated with ultraviolet irradiation actually has considerable breadth, from 4 to 300 nm, but the highest level of absorbance by DNA appears to fall nearly coincident with the maximal output value (i.e., at approximately 254 nm) for the emission spectrum of light emitted by mercury-vapor light bulbs. The standard engineering application of UV irradiation involves an array of mercury bulbs placed inside an irradiation chamber, through which the water or wastewater is passed, with each such tube being jacketed inside a UV-transparent quartz jacket. These tubes are aligned vertically or horizontally in a fashion where the hydromechanics of the operation provides maximal opportunity for exposure of cells flowing through the chamber while obviating

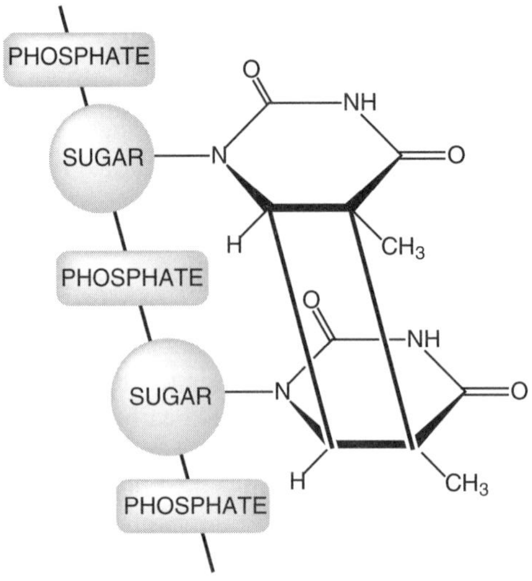

Thymine dimer

Figure 16.54   Thymine dimer formation along DNA strand produced by polymerizing ionizing or UV irradiation.

short-circuiting pathways that would degrade process efficiency. Figure 16.55 depicts one such array used for disinfection of a wastewater effluent immediately prior to discharge. UV irradiation with mercury bulbs offers an effective means of sterilization for those engineering applications involving fairly clear or optically transparent waters and wastewaters. Conversely, sludge is not amenable to UV disinfection given the unacceptably shallow degree of light penetration into this material.

Figure 16.55   Ultraviolet irradiation reactor for wastewater effluent disinfection.

Ionizing radiation using x-ray and gamma-ray beams with even higher energy levels than that of UV irradiation, can also be used in water, wastewater, and sludge disinfection. These ionizing mechanisms displace electrons during beam bombardment (i.e., at which point they are said to *ionize*), and in the presence of oxygen these displaced electrons electrochemically form a type of free radical (called *hydroxyl radicals*) that is highly toxic to microbial cells. Free radicals are highly reactive, having essentially no activation energy for their reaction. Given the acutely reactive nature of these radicals, they readily attack and destroy hydrogen bonds, double bonds, and ring structures essential to the metabolic utility of various cellular molecules. Yet another operative mechanism, and perhaps the key factor behind disinfection with ionizing irradiation, is that of a polymerizing impact (e.g., DNA thymine dimerization) whereby the biochemical effectiveness of complex molecules is degraded or terminated. This technology bears a degree of complexity and technical hazard that is not typically appropriate for most municipal applications, however, so that only a limited number of sites, the majority of which involve sludge disinfection presently rely on its use. On the other hand, ionizing radiation is used widely for the disinfection of pharmaceuticals and of disposable dental and medical supplies (e.g., syringes, gloves).

In reviewing the various options for disinfection, the mere fact that there is such a diverse range of disinfecting mechanisms and associated effects demonstrates that there is no perfect solution for all engineering applications. Excluding heat treatment and osmotic pressure, which have little if any pragmatic utility for large-scale engineered disinfection, the remaining options exhibit unique differences in their associated costs, technical complexity, residual impacts, environmental sensitivity, and relative performance, such that an appropriate assessment of many such site-specific issues have to be completed prior to making a final decision.

As for defining the precise goal of disinfection, the normal benchmark for potable water is that of zero residual fecal coliform presence. In the case of wastewater disinfection, though, effluent standards tend to be targeted for a somewhat higher level of residual bacterial presence in a fashion that implicitly reflects the latter phenomenon of microbial resistance. Compared against the typical levels of bacterial presence within raw, undisinfected wastewaters (such as those shown in le 16.15), therefore, effluent criteria for fecal coliform usually tend to be somewhat more tolerant (i.e., typically at 200 fecal coliform colony forming units per 100 mL). However, compliance with this sort of effluent fecal coliform standard would still require a sizable four- to five-log reduction (i.e., based on influent fecal coliform densities in the range 10^6 to 10^7 cells per 100 mL).

Extending beyond the theoretical aspects of these disinfection mechanisms there are pragmatic uncertainties and inherent regulatory concerns tied to the nature and degree

TABLE 16.15 Bacterial Levels Present in Representative Wastewaters

Wastewater	Observed Bacterial Densities (10^6 viable cells/liter)			
	Total Coliform	Fecal Coliform	Fecal Streptococci	FC/FS Ratio
A	172	172	40	4.3
B	330	109	24.7	4.4
C	19.4	3.4	0.64	5.3
D	63	17.2	2	8.6

Source: Adapted from Water Environment Federation (1996); Crites and Tchobanoglous (1998).

of the consequent biological response. For example, it is quite rare for 100% of the exposed cells to actually be killed following high-level UV exposure, and we now know that certain microbes have an inherent degree of disinfection insensitivity or resistance. One such microorganism that exhibits a definite tolerance for this ionizing impact, *Micrococcus radiodurans*, was named to reflect this remarkable resistance to UV irradiation, thereby offering yet another demonstration of the facile nature of bacterial cells. Furthermore, and although presently not well understand, there is evidence that some microbes are able to rehabilitate themselves following disinfection events. One such occurrence involves a mechanism known as dark-field repair, through which cells previously exposed to UV light are able to autorepair DNA sites affected by thymine dimer damage.

Balanced against the health benefits derived by the routine use of disinfectant chemicals and measures, significant biological concerns have also been raised regarding the formation and consequent human health-related impacts of potentially carcinogenic disinfection by-products (DBPs) during disinfection of water and wastewater. DBP compounds can be formed from halogen-based disinfection activities. For example, chlorination in the presence of natural organic compounds can lead to the formation of *trihalomethane* (THM) compounds, including chloroform and bromoform. In such cases, the desired disinfection benefits of using halogens are offset by problems among long-term consumers tied to cancer and reproductive effects. After recognizing this risk, the U.S. EPA identified a maximum contaminant level (MCL) for total THMs within community water systems serving at least 10,000 people that added a disinfectant to the drinking water during any part of the treatment process (whether it be for disinfection, filter cleaning, prechlorination, etc.). Specifically, the U.S. EPA has stipulated the Stage 1 Disinfectants and Disinfection Byproducts Rule (in compliance with the Safe Drinking Water Act Amendments) (U.S. EPA, 1998), which limits total THMs to an annual average value below 0.08 mg/L and total haloacetic acids to values below 0.06 mg/L, as well as setting additional limits on chlorite (1 mg/L) and bromite (0.01 mg/L) designed to further reduce carcinogenic risks from halogen-disinfected waters. The inherent irony is that ozone has been promoted partly in the hope of avoiding the DBPs from chlorination. Another irony is that some systems have switched to chloramination, which has a lower tendency to form DBPs. However, this has led to another health effect by increasing corrosion in plumbing systems, resulting in elevated lead concentrations in tap water.

16.5 SOLID WASTE TREATMENT

Biodegradation of dead plant, animal, and other residues has played an important role on our planet's surface since the inception of life eons ago, recycling valuable nutrients back into our biosphere while negating or at least reducing the shear physical burden of this never-ending natural debris stream. At least in theory, therefore, nature provides a highly useful example of a means by which we might manage our own solid waste residuals, taking full advantage of much the same aerobic and anaerobic degradative mechanisms.

The level of success presently realized in using biology effectively as a management process for human solid waste residuals has, however, considerable room for improvement. At the low end of this spectrum of biochemically engineered solid waste management strategies, the majority of municipal detritus generated by the world's current leader in per-capita solid waste production (i.e., the United States) follows a least-cost disposal

Figure 16.56   Solid waste landfill operation.

pathway with little, if any, thought given to promoting, let alone maximizing, its biological degradation, resulting in the consequent loss of potential resources, including nutrients, organic material, and energy. Indeed, expedient burial in subsurface landfills (Figure 16.56) has in this case become the U.S. norm, with emphasis given to prior source reduction and recycling of various fractions of the original waste as a means of reducing the shear magnitude of the problem.

However, there are areas in the world that exhibit clear and escalating evidence of a national shift toward biochemically engineered solid waste processing systems involving controlled biological strategies. Europe and Asia represent two regions where escalating land constraints that limit landfilling options are matched both by a national motivation to pursue environmentally friendly disposal strategies and the financial wherewithal and willingness to accept technical solutions considerably more expensive than mere burial.

Commensurate with the continuing escalation of our world's population and its attendant increase in human solid waste production, future generations will no doubt be aggressively challenged to pursue a better universal means of managing their solid waste problems. Although biology may, at best, be presently considered a minor, hidden aspect of solid waste processing, at least for much of the world, the opportunity at hand to employ renewable metabolic strategies beneficially will probably continue to attract increased attention (Palmisano and Barlaz, 1996).

One of the fundamental issues with solid waste residues is that of characterizing its nature and source, principally in terms of how degradable or recalcitrant it might be. The term *solid waste* truly covers a wide range of high-volume residues, including not only municipal wastes of the sort that might be generated in your own home to industrial (e.g., food-processing residues, spent foundry sands, slag), agricultural (e.g., manure, bedding, plant residues), power (e.g., fly ash, bottom ash), and even mining (e.g., overburden) wastes. Only two of these fractions, the municipal and agricultural groups, include putrescible materials, and even in the latter case, a large segment of these wastes (e.g., manures) are already being biochemically recycled for the beneficial purpose of rejuvenating farmland productivity.

The municipal solid waste (MSW) segment itself is composed of many different fractions whose putrescibility varies widely. At one end of this spectrum, food scraps have the highest level of potential degradability, due to both their organic makeup and their high water content. Cellulose-rich residues, whether discarded as lawn clippings or paper products, would also be amenable to biochemical degradation, although perhaps at a somewhat slower rate. Extending beyond these two segments, though, municipal solid waste includes many other materials whose composition will not be amenable, and possibly even antagonistic or inhibitory, to biochemical degradation. For example, MSW generated within affluent countries includes a sizable proportion of plastic, for which the vast majority will have no susceptibility to biochemical breakdown. Similarly, affluent countries generate municipal solid wastes with a proportionately higher percentage of metals (e.g., cans, batteries, used appliances) whose presence may actually lead to the release via leaching of soluble heavy (e.g., cadmium, chromium, nickel, lead) and transition (e.g., arsenic, selenium) metal ions that are detrimental to desired biochemical activity.

Yet another key issue is that of available moisture, since there must be adequate water present to facilitate and sustain the growth of biological populations. Contrasted against this necessity, though, modern landfills are provided with overlying physical caps (i.e., built with impervious clay and/or plastic liners) that intentionally limit the influx of water into their buried wastes, such that limited moisture presence within the subsurface waste matrix could well become a constraining factor. Indeed, solid waste materials excavated from decades-old landfills have in many instances proven to contain a remarkable amount of undegraded putrescible residuals (e.g., corncobs, newspapers) whose lingering structural integrity reflects what appears to be a desiccating condition within the waste vs. a moisture-bearing environment conducive to microbial degradation.

Extending beyond water presence, a number of additional ambient environmental conditions come into play. Nutrient availability will certainly be an issue, not only in regard to the macroscale distribution of carbon and nitrogen (i.e., for which a C/N ratio of $\sim 20 : 1$ is typically considered optimal), but also in terms of the available presence of lower-level essential elements (e.g., phosphorus, sulfur, potassium, iron, calcium). While the pH of the constitutive moisture must also be suitably conducive to microbial activity, fermentative metabolism will progressively release weak organic acids that could well shift the pH to a more acidic, and less optimal, state. In fact, a downward shift in pH of this sort might accordingly escalate the undesired rate of trace metals leaching from the waste matrix, thereby leading to inhibitory metal toxicity.

Finally, the relative combination of high solids and low moisture found in solid waste streams effectively yields a high-level specific energy content (i.e., cal/kg) that is higher than that of most other biodegraded waste (e.g., greater than wastewater and sludge), with the sole exception of agricultural manures. Commensurate with effective biodegradation of these wastes, therefore, the heat release per mass of degraded solids would not only be considerably higher, but also apt to be trapped inside the high-density waste, due to its insulating nature. On the one hand, this heat release and temperature increase could help to accelerate the ongoing biochemical process. However, as is the case with composting, this thermal buildup could accelerate evaporative water loss to a degree that eventually retards the desired activity.

There are several engineering and operational aspects of the current design and management of MSW landfills that are frankly contrary to a goal of optimizing their biological degradation. For example, restricting water egress into and out of this buried waste is routinely considered a beneficial environmental goal for landfill operations, even though this effort could well end up slowing down biochemical degradation. By reducing the hydraulic pressure gradient that might accrue with an interior water buildup, the transport of waterborne contaminants into the adjacent groundwater table can subsequently be reduced or obviated. At no point during this process, either during active filling of the landfill or after final closure, will the moisture content of the material ever come anywhere close to an optimal value relative to maximal microbial activity. Unlike composting, therefore, where moisture control is addressed routinely, landfills have no management plans that involve steps to proactively increase moisture content. Furthermore, efforts taken to heavily compress the waste to minimize its bulk volume, or of using a daily cover of $\sim$15 cm soil to restrict rodent and other vector access, work against desired microbial degradation, as these measures impose physical constraints that degrade process homogeneity. In fact, the mere circumstance of allowing homeowners to isolate much of their daily MSW within plastic bags contributes yet another complication, further exacerbating the waste's isolation into discrete, semi-isolated packets.

Given these inherent shortcomings with landfills in the context of biological MSW degradation, several alternative engineering strategies have been developed to secure higher levels of success. These systems are typically charged with a presorted waste stream whose degradable content has been enhanced by way of separating out a large fraction of the relatively nondegradable content (e.g., plastics, glass, metal). The engineering features of these systems subsequently vary in relation to operational solids content. In-vessel digester reactors are maintained with the lowest solids content and the highest water contents, including both low-solids units at 4 to 8% solids and high-solids units with solids values in the low- to mid-20% range. MSW is shredded and slurried before introduction to these reactors. These digesters include both aerobic and anaerobic options, much like sludge digestion vessels.

A second MSW processing option, composting systems, is again quite similar to the equivalent strategies used for sludge treatment, with waste solids levels approximately double that of high-solids MSW digesters (i.e., 40 to 50%). Both aerobic and anaerobic composting strategies are available, maintained in static and windrowed piles held either in open or in-vessel configurations. Yet another option is that of codisposal systems, treating a blended mixture of MSW and sludge. As compared to MSW digestion options, though, and their relative focus on volatile solids destruction, MSW composting reactors provide an overall volume reduction tied both to volatiles destruction and moisture loss allowed to accelerate at the end of the processing cycle during a final curing phase. Volume reductions on the order of 50% have been observed with both aerobic and anaerobic composting systems at processing times on the order to 3 to 4 weeks.

Finally, it is worth mentioning a process in which animals are the basic organism. Vermiculture (raising of earthworms) or vermicomposting uses earthworms to help decompose organic material, especially food-processing wastes (Datar et al., 1997). Vermicomposting produces stabilized waste similar to ordinary composting, although the process must be maintained at lower temperatures. The process has been applied mostly at small scale, including household use, although large-scale facilities exist that process up to 20,000 metric tons of organic waste per year.

16.6   AIR TREATMENT

Several abiotic engineering methods are commonly used to treat contaminated airstreams, including those of incineration, carbon absorption, condensation, and scrubbing (i.e., either with water, chemicals, or combinations thereof). However, within the past few decades the idea of using biofiltration columns has proven to be a highly attractive,

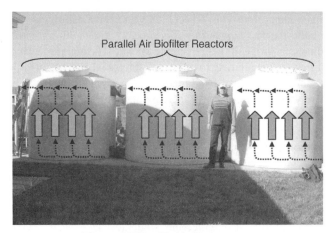

Figure 16.57   Air biofilter system.

complementary treatment strategy for air quality remediation. **Biofiltration** consists of a closed vessel containing a porous support medium with an attached biofilm through which a contaminated gas is passed. The attached biofilm will first sorb, and then degrade, the incoming contaminants. This technology first gained significant attention in Europe during the late 1970s, and there are now a number of relatively recent applications in the United States. However, as with constructed wetlands and phytoremediation systems, the involved technology is still in the process of being clarified and optimized.

Biofiltration technology is, in fact, rather similar to that of the attached-growth media-filled biofilters (e.g., trickling filters) described earlier, although in this case the stream to be treated is that of contaminated air rather than wastewater. Similarly, the mode of construction and operation used for biofiltration columns is also comparable to that of biosolids composting using aerated piles, although with a goal of treating the incoming gas rather than a solid-phase (sludge) material blended into the pile.

In some cases, the main objective of gas-phase biofiltration processing may simply be that of remediating a malodorous gas stream by removing the chemicals responsible (e.g., hydrogen sulfide, mercaptans, free ammonia). Figure 16.57 depicts one such full-scale application used to effectively cleanse these types of contaminants from the off-gas discharge generated by an aerobic thermophilic biosolids digester. Several comparable wastewater-related applications exist for handling contaminated, odor-bearing off-gas streams, including those released by attached-growth wastewater treatment towers, activated-sludge reactors, biosolids digesters, and even sewer force mains.

However, biofiltration may also be used to biochemically degrade one or more regulated volatile organic compound air contaminants such as those generated within a specific industrial airstream. High-level removals have, in fact, been observed with a number of industrial organic contaminants, including phenol, styrene, formaldehyde, BTEX, and methanol as well as various aldehydes and ketones. Many of these biofiltration applications have been tested within industrial applications in which the current abiotic off-gas treatment procedures are too mechanically complex, too expensive, or simply too inefficient.

Figure 16.58 depicts a typical biofiltration system. The various design strategies for biofiltration systems cover a wide range of options, some of which depend solely on

Figure 16.58   Air biofilter system.

the biofilter for contaminant removal (e.g., typically involving low-level contaminant concentrations), while in other instances the process train will include a pretreatment abiotic scrubber. Abiotic scrubbers are, in fact, often included as an integral part of many systems, helping to moderate influent gas temperatures and ensuring full humidification as well as securing preliminary sorption and partial removal of highly soluble contaminants such as ammonia. Four options are typically seen with these scrubbers: (1) water-only scrubbing, to humidify and cool an incoming gas stream as well as to initiate desired gas-to-liquid sorption; (2) acid spray scrubbing, to adjust pH in a downward fashion that shifts gaseous ammonia to ionic, liquid-phase NH_4^+; (3) caustic spray scrubbing, to adjust pH upward both to shift gaseous hydrogen sulfide to ionic, liquid-phase HS^- and to buffer both the scrubber water and gas-phase moisture against the acidity of incoming gaseous CO_2; and (4) oxidant (i.e., typically bleach) spray scrubbing, to oxidatively attack any chemically amenable contaminants.

Once this prescrubbed gas stream reaches the biofiltration zone, the targeted contaminants must be readily transferable from the gas phase to the water phase within which the biomass is growing. The success of this sorptive phenomenon depends on several factors, including the Henry's constant (K_H) of the involved contaminant species, ambient temperature, media-film pH, and the media-biofilm surface area available for interfacial gas–liquid exchange. Biofiltration-based degradation of contaminants with high K_H values may therefore be difficult, since it is likely that these species will remain in the gas phase rather than being sorbed and degraded.

Media temperatures between ~20 and $40°C$ appear to work best, avoiding low or high extremes that would retard the desired metabolic activity of the biofilm or escalate water evaporation to an unacceptably high level. For cold incoming gas streams, steam might be used to raise a reactor's operating temperature, and hot incoming gases can be cooled by means of the initial water scrubbing.

In terms of flow schemes and design retention times, both up- and down-flow regimes have been used, although the up-flow option is probably the most common. The contact time for gas throughput also represents another key factor, with most units sized to provide ~45 seconds. Pile moisture levels in the neighborhood of ~50% are considered optimal, and at the same time, the relative humidity in the gas stream will need to be above ~95+% to avoid excessive levels of water evaporation from the pile. Many biofiltration vessels are fitted with internal humidifying spray misters as a means of ensuring desired wetting and moisturization of the biofilm, and if necessary, supplemental nutrient or buffering chemicals may be blended into these wetting streams to enhance biological activity.

Extending beyond these heuristic details, though, there are at present relatively few specific criteria for designing or operating biofilter units, such that their design and construction represents, more a heuristic art than a codified engineering science. Indeed, the state of the art for this technology is still being actively refined, covering a wide range of design and operational factors (e.g., optimal bed depths and configurations relative to minimizing undesired short circuiting, airflow per unit volume of media, throughput velocities, necessary rates of water addition and scrubbing, nutrient and buffer requirements, bed media longevity and replacement practices, tolerance levels with loading variations, microbial seeding and startup requirements).

Undoubtedly, though, one of the most important variables with any biofiltration system is that of its media form and composition. In general, three different types of media are used in biofiltration reactors: (1) synthetic random-packed plastic media, (2) inorganic rock-type media (e.g., lava rock, limestone, clay, shale), and (3) organic plant- or tree-derived media (e.g., shredded or chipped roots and stems, as well as bark). The issue of media packing itself can impose a considerable resistance to the passage of the gas stream, such that it will have to be effectively pushed through the bed using a mechanical air pump or rotary blower. In many instances, though, coblended columns (see Figure 16.59) have been constructed successfully with various blends of organic and inorganic materials (e.g., shredded roots, stems, and bark plus limestone and clay admixtures), with the overall goal of providing an enhanced surface area conducive to biofilm growth, as well as a higher moisture retention capability and enhanced buffering capacity. In addition, these blended natural ingredients may contribute secondary benefits that further enhance metabolic efficiency. For example, the concomitant breakdown of plant- and tree-based support materials may beneficially serve as secondary organic substrates that promote complementary microbial (e.g., fungal) growth as well as potentially supplying a fraction of the metabolic nutrients required. Hardwood materials passed through shredders and grinders probably appear to represent one of the more desirable natural options, given their perceived durability and resistance to rotting associated with prolonged wetting while favoring biofilm attachment and retention. In addition, inorganic media additives, such as limestone and clay, may also provide a beneficial buffering capacity, which in the case of ammonia or sulfide hydrolysis may comprise a particularly important benefit.

With time, though, the historical pattern observed with all of these natural media is that they tend to degrade in terms of media depth and overall performance, at which point they would have to be supplemented with fresh media to rejuvenate the bed. During large-scale bed replacement with new media, partial blending with some of the older media would help to secure beneficial recycling of microbial seed. Here again, the science behind media replacement is still in a research stage, but 5 years is generally believed to be a reasonable time frame.

Yet another aspect that has not yet been fully explored is that of the presence and role of microscale metabolic niches that develop inside the biomass contained within biofilters. It

Figure 16.59   Natural shredded root and stem mass used for biofilter support media.

is highly likely that metabolically complementary anaerobic–aerobic mechanisms will develop in these sites, whereby special biodegradation pathways could be developed to attack normally recalcitrant chemicals (e.g., reductive dehalogenation of TCE), followed by aerobic degradation of the partially dehalogenated products).

16.7   SOIL AND GROUNDWATER TREATMENT

16.7.1   Phytoremediation

Phytoremediation is the use of plants to remove and/or biotransform contaminants. The process of phytoremediation is comparable to that of constructed wetlands. Both applications make use of macro- and microscale biology, and both concepts have captured considerable attention, even though they each still qualify as evolving technologies given their relatively short histories. However, phytoremediation also has several important distinctions (Schnoor et al., 1995). First, the focus of phytoremediation is that of biologically remediating contaminated soils, sediments, and waters (both surface and ground water) as opposed to wetland applications focused solely on wastewater treatment. Second, phytoremediation systems may employ trees as well as smaller plants as the primary biological agent. Indeed, for those applications in which they are suitable, phytoremediation systems may offer a highly attractive "green" means of decontaminating lands and waters.

Successful applications have been achieved experimentally with a wide range of inorganic (e.g., metals, ammonia) and organic [e.g., petroleum hydrocarbon, BTEX, creosote wood preservative, polycyclic aromatic hydrocarbon (PAH), refinery waste, organophosphate insecticide, chlorinated pesticide, chlorinated solvent, explosive, cyanate] contaminants (Jackson, 1997).

The potential benefits are again much the same as those of constructed wetlands, including that of an apparently simple, solar-driven, aesthetically pleasing, in situ "green" technology with few, if any, complex or energy-intensive hardware or operational requirements (i.e., as compared to conventional treatment operations that employ pumps, mixers, aerators, etc. that routinely use energy and require careful operator attention). These systems can also be self-sustaining in terms of procuring nutrients, they can make a beneficial contribution to the balance of water in their soils, they can establish a highly evolved complement of degradative enzymes, and they tend to be inexpensive both in their initial startup and in subsequent maintenance. Granted, this remediation approach will not work in all situations, and in even when it is successful, the remediation process will operate on a time-scale measured in years rather than in hours or days. The public perception of phytoremediation is extremely high, though, as a natural means of promoting the restoration of chemically contaminated sites.

However, the seemingly simplistic notion of using plants and trees to clean up these contaminated sites actually involves a far more sophisticated process than what is apparent to the eye. As was the case with constructed wetlands, the visibly "green" above ground portions of these systems are but a part, and in some cases perhaps even a lesser part, of an integrated remediation scheme that encompasses a complex array of physical, chemical, and certainly biological treatment factors. The type, density, and nurturing of the plants and trees is important as well as the nature of the soils (e.g., soil type, conductivity, depth to groundwater, nutrient availability) and climate (e.g., rainfall frequency and duration, radiation, seasonal climate, windspeed, humidity) in which they are grown. Finally, the character, concentration, location, and form (e.g., whether it is sorbed, soluble, solid) of the contaminating materials are also important factors.

Before delving into the underlying sophistication of these phytoremediation systems, though, one must develop a background understanding and appreciation of the vertical layering of the soils in which these bioremediating plants and trees grow, and their corresponding physical and chemical characteristics. Figure 15.2 provides a schematic overview of a representative soil–plant system and its associated horizontal layers. There are two major regions shown in this schematic, situated above and below the groundwater table, respectively, and known as the unsaturated and saturated zones. Within the **unsaturated zone** the soil water volume does not entirely fill the pore space in the soil. The unsaturated zone is also called the **vadose zone**. The remaining (partially or fully) open void space, however, will also facilitate better levels of aeration and oxygen transfer, which will then promote more aerobic microbial activity. The **saturated zone** is where the pores are completely filled with water. The **groundwater table** is the point in the saturated zone where the hydraulic head is equal to zero. Water is drawn by capillary action into the **capillary fringe** slightly above the groundwater table. The top of the capillary fringe marks the division between the saturated and unsaturated zones.

The level of the groundwater table may vary considerably from one location to another, and also according to temporal changes in precipitation and climate, but in most instances it lies many meters below the surface and at a level not usually reached by plant root systems. As a result, phytoremediation was developed with systems whose remediating activity took place almost totally within the uppermost layer of the unsaturated zone (i.e., at shallow depths). However, subsequent developments with the nurturing and use of deep-rooting plants and trees have now expanded this technology down to the saturated region, at which point phytoremediation could then deal with contaminants extending fully down to the groundwater table.

With most soil columns there will also be considerable variation in the horizontal and vertical homogeneity of the top and bottom layers. The top, unsaturated zone may be further divided into another series of layers vertically from top to bottom:

- *O horizon*: consisting mostly of leaf litter and other dead organic matter
- *A horizon*: the topsoil, with a high content of mostly degraded organic matter, well populated by microorganisms and invertebrates, often coincident with the **root zone**
- *B horizon*: less weathered mineral matter plus organic and inorganic matter leached from the A horizon
- *C horizon*: partially weathered material from the bedrock below, very low in organic matter and living organisms

Once the remediating plants and/or trees have been introduced successfully into a site, the means by which they can attempt to degrade or remove a group of involved contaminants can be quite diverse. An important, yet all too easily overlooked aspect of phytoremediation is that the plants are generally not the sole means of contaminant treatment. Granted, the plants themselves may contribute many different and important remediating effects (e.g., phytovolatilization), as discussed below, but in most instances their remediation role is metabolically complemented, perhaps even dominated by that of the microbes that are motivated correspondingly to live within the same soils. Figure 16.60 provides an overall synopsis of these prospective plant and microbial mechanisms, and in the following synopsis we examine each of these metabolic contributions relative to the overall process of phytoremediation.

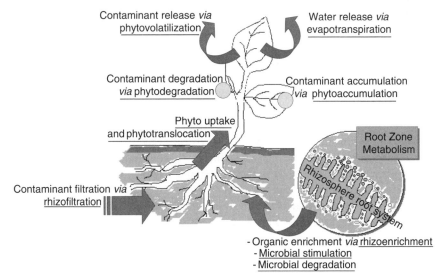

Figure 16.60 Phytoremediation mechanisms.

The first such mechanism, **rhizoenrichment**, stems from the fact that plants release into soils a number of exudates that are rich in organic carbon and that, in turn, effectively nurture the growth of many soil microorganisms. A sizable fraction of the carbon fixed through photosynthesis is released into soils, with estimates ranging from 10 to 30%. This material includes a range of readily biodegradable materials with small to moderately sized molecular weights, including sugar, protein, alcohol, and acids. Yet another group of organic carbon residuals are also released into soils by the senescence (aging) and decay of plant tissue, particularly that of fine-root biomass. There is also a beneficial physical impact with the growth and aging of plant roots, in that they tend to loosen the soil during both their growth and death, forming new paths for transporting water and aeration. This process subsequently tends to pull water to the topsoil surface while drying the lower saturated zones.

The latter enrichment of soils with organic carbon compounds exuded by plants subsequently promotes and maintains a significant enhancement in the growth of microbes within the immediate vicinity of the roots (i.e., **microbial stimulation**). There are actually two mechanisms by which plants provide this stimulation: by feeding the microorganisms with their exudates and by promoting the availability of oxygen. Here again, the photosynthetic activity of the plant is important, with at least some of its newly created oxygen being effectively pumped through the roots into the soil. Channeling created by roots, both alive and dead, also provides a means of physically opening the soil matrix and improving its porosity. The net effect of the added substrates and improved oxygen availability leads to levels of microbial activity and density that are considerably higher than those of barren, unvegetated soils, by several orders of magnitude. In a fashion analogous to that shown in Figure 16.39, between 5 and 10% of root surfaces will tend to the colonized by various forms of bacteria and fungi, and the adjacent rhizospheric soils will commonly experience considerable increases in their microbial density. Population densities of 5×10^6 total bacteria, 9×10^5 actinomycetes bacteria, and 2×10^3 fungi per gram of air-dried soil have been observed.

This symbiotic relationship between plants and their adjacent microbial consortia stimulates microbes, which in return assist the plants in securing nutrients and essential vitamins. Given the diversity of the substrate forms available to these microbes and the variable nature of the rhizospheric environment (i.e., with dynamic changes in oxygen content, soil water presence, pH, etc.), a wide range of microbial types is found in these soils. In turn, the metabolic breadth of these bacterial and fungal forms encompasses a considerable range of enzymatic mechanisms and pathways. The resulting, collective effect of these microbes is that they can be expected, either directly or indirectly, to play a significant role in degrading organic contaminants present in the soils (i.e., **microbial degradation**). Compared to readily biodegradable compounds, recalcitrant organic contaminants found in soils may not be directly oxidized by these root-zone microorganisms as an energy source, but they may nonetheless be converted. Indeed, in the presence of other biodegradable root exudates, the catabolic enzymes generated to catalyze these reactions sometimes cooxidize the recalcitrant materials through a cometabolic conversion. The relative contribution of plants vs. microbes to degradation no doubt varies from one situation to the other, and there are those who would argue that one or the other typically plays a more dominant role. However, irrespective of which contribution might be dominant, the fact remains that the efficacy of phytoremediation commonly involves a coordinated and harmonious set of biological mechanisms that span the micro- to macroscale of life.

Contaminants not degraded by rhizospheric microbes are available for plant uptake by roots, where they may then either be retained or translocated farther upward into a plant's shoots and leaves (i.e., **phyto uptake**). Some plants simply uptake contaminants and store them in their roots, whereas others both uptake and translocate contaminants. There is, admittedly, a degree of uncertainty about the nature of these combined processes and the conditions under which they may each take place, but it appears that the polarity of the contaminants is an important factor. One of the most frequently mentioned criteria in this regard is that of the *octanol-to-water partition coefficient* (K_{OW}). K_{OW} is used as an indicator to represent the relative ability of a compound to concentrate in lipids vs. water and is therefore more likely to concentrate in biological materials. K_{OW} tends to be inversely related to the polarity and therefore the aqueous solubility of a compound. If a compound has a K_{OW} value of 1, it has the same affinity for water as it has for octanol (and by implication, for lipids). If the K_{OW} value were 100, it has 100 times greater affinity for octanol (and lipids) than for water. Because of the wide range of K_{OW} values, it is often reported as log K_{OW}, so the range of K_{OW} from 1 to 100 corresponds to log K_{OW} from 0 to 2. Compounds amenable to plant uptake reportedly have logK_{OW} values in the range 0.5 to 3. Those materials having higher values, extending beyond the range of moderately nonpolar would simply be too tightly sorbed onto root surfaces to be translocated any farther within a plant. On the other hand, lower values (under 0.5) have such a high preference for water (i.e., being quite polar) that they would tend to remain within soil water rather than be taken in by roots in the first place.

Some plants not only transport contaminants across their cell membranes (i.e., through phyto uptake) but also move these materials internally beyond their roots (i.e., **phytotranslocation**). Here again, the polar vs. nonpolar nature of the contaminant is an issue, as well as the rate of transpiration being maintained by the plant. This rate of transpiration is, in fact, a key variable for translocation, with an apparent direct correlation between these two factors. Once a contaminant has been taken into a plant through these sequential processes of phyto-uptake and phytotranslocation, this contaminant could theoretically then be removed from the site by harvesting and subsequent disposal of the plant's aboveground biomass.

Following uptake and translocation, many plants have evolved compound-specific detoxification pathways that involve subsequent conjugation and compartmentation reactions that effectively bind contaminants into their structural makeup (i.e., **phytoaccumulation**). These biotransformed and phytoaccumulated compounds can either be deposited into vacuoles or converted into insoluble (and frequently covalent) complexes within cell wall components through a process known as *lignification*. In some cases, the accumulated compounds are passed unchanged into these deposits; in other instances the material being accumulated is that of degradation fragments produced through preceding biochemical conversions that transformed the contaminants into nonphytotoxic metabolites. However, it is also possible that some plants may accumulate contaminants internally to a level where an ecotoxicological hazard develops that would severely restrict subsequent consumption or disposal of the plants.

Phytoremediation plants can also produce a number of enzymes that may promote the internal metabolism and degradation of contaminants (i.e., **phytodegradation**). For example, nitroreductase enzymes can initiate the breakdown of nitroaromatic munitions; dehalogenase enzymes will promote the degradation of chlorinated compounds; nitrilase will contribute to the degradation of herbicides; phosphatases will facilitate the catalysis of organophosphates; and peroxidases will promote the destruction of phenols.

The next process, **phytovolatilization**, theoretically involves the uptake and translocation of contaminants into leaves; plants may then release these compounds into the atmosphere through a volatilization mechanism. One particular plant, *arabidopsis* (in the mustard family) has been found to produce a specific enzyme, mercury reductase, which reduces mercury to elemental mercury, which is then amenable to volatilization and release. Yet another known volatilization sequence involves the treatment of selenium-contaminated soils by rice, broccoli, and cabbage through the production of volatile *dimethylselenide* and *dimethyldiselenide*. In addition, there are a number of low-molecular-weight VOC-type organic molecules that appear to be easily translocated and volatilized by various plants. The extent to which the latter reactions actually take place under real-world conditions, however, is not well established.

Although roots generally cannot be harvested in a natural environment, another phytoremediation process, **rhizofiltration**, can be used where plants are raised in greenhouses and transplanted to sites to filter metals from wastewaters biochemically. As the roots become saturated with metal contaminants, they can be harvested and disposed of. Phytoremediation plants have also been used in this fashion to concentrate radionuclides via rhizofiltration in the Ukraine and Ashtabula, Ohio.

Extensive water uptake and release rates can also be maintained by a number of phytoremediation plants, including poplars, cottonwoods, and willows, in a fashion that will effectively pull contaminated groundwater plumes toward and through these phytoremediating tree roots (i.e., **evapotranspiration**). A single, mature willow tree, for example, can transpire more than 19 m^3 of water each day ($\sim$5000 gallons, or about 3.5 gal/min), and 1 ha (10,000 m^2) of a herbaceous plant such as saltwater cord grass has been found to evapotranspire even four times as much. There are several interrelated issues, including plant type, leaf area, nutrient availability, soil moisture, wind conditions, and relative humidity.

le 16.16 provides a general correlation of the plant types that have been tested for the various contaminant forms believed to be amenable to phytoremediation treatment. Extending beyond the matter of a plant's potential suitability for any given contaminant, there are also a wide range of characteristics for plants in terms of their relative environmental preferences. Some plants tend to have shallow roots (i.e., cottonwoods and willows roots), whereas **phreatophyte** plants have produced far deeper roots (i.e., aspens and alders). The family of *salicaceae* trees (including poplars) tends to have very high water uptake rates and is usually able to tolerate high organics. Some plants are rather salt-intolerant (e.g., hybrid poplars), whereas others have a high tolerance for salts (e.g., mesquite, salt cedar). Some plants prefer hot humid climates (e.g., bald cypress), whereas others prefer cold and dry climates (e.g., greasewood). Alfalfa plants, are often used due to their high nitrogen uptake rates and ability to maintain nitrogen fixation in the absence of available nitrogen.

The principles and practice of phytoremediation systems involve several important engineering aspects, but in reality the procedures still qualify as an emerging technology. The issues that must be considered include those of the involved soil characteristics, the targeted contaminants and current concentrations, and the relative depth of the existing residuals.

Concerns regarding soil type stem from the fact that various plants have different preferences for either fine- or coarse-grained soils, which probably reflects the ability of the soils to hold and transfer varying amounts of moisture, air, and nutrients. The site-specific and perhaps seasonally fluctuating depth to the groundwater table is also important, as it affects the means by which a plant can draw water.

TABLE 16.16 Phytoremediation Agents Relative to Contaminant Species

Organic Contaminants	Phytoremediation Agent	Inorganic Contaminants	Phytoremediation Agent
Atrazine	Alfalfa	Arsenic	Bluebells
			Maple tree
			Cattail
			Water lily
DDT	Pine tree	Lead	Alpine pennycress
	Mangrove		Indian mustard
			Alyssum
			Duckweed
			Brass buttons
Total petroleum hydrocarbons	Alfalfa	Cadmium	Duckweed
	Sorghum		Brass buttons
	Rye grass		Alyssum
	St. Augustine grass		Cabbage
	Hybrid poplars		Sunflower
			Turnips
BTEX	Alfalfa	Cesium	Sunflower
	Sorghum		
	Rye grass		
	St. Augustine grass		
	Hybrid poplars		
Gasoline	Daisy	Chromium	Alyssum
	Alfalfa		Cabbage
	Rye grass		Turnips
	St. Augustine grass		
	Hybrid poplars		
Chlorinated solvents (TCE, PCA, etc.)	Hybrid poplars	Cobalt	Sunflower
	Horsetail		
Polychlorinated biphenols	Crabapple	Copper	Duckweed
	Osage orange		Brass buttons
	Mulberry		Alyssum
			Cabbage
			Sunflower
			Turnips
Polycycloaromatic hydrocarbons	Mulberry	Lead	Sunflower
	Hackberry		
	Rye grass		
	Mulberry		
Pentachlorophenol	Rye grass	Mercury	Arabidobsis
	Mulberry		Duckweed
			Brass buttons
Nitroaromatics (TNT, etc.) (via nitroreductase)	Hornwart	Selenium	Duckweed
	Parrot feather		Brass buttons
	Eurasian water milfoil		Brassica
			Rice
			Broccoli
			Cabbage
Ammonia	Alfalfa	Strontium	Sunflower
Nitrate	Alfalfa	Uranium	Sunflower
		Zinc	Sunflower

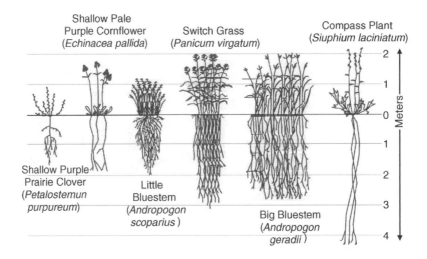

Figure 16.61   Representative phytoremediation plant variations.

The majority of phytoremediation plant systems, such as those depicted schematically in Figure 16.61, apply to soil depths extending to the first 2 to 3 m. In turn, most of these plants probably draw their water either from roots closely aligned to the surface or water drawn from vadose-zone pore moisture, as depicted in Figure 16.62. Most poplars, for instance, tend to have shallower root systems, and as a result, these types of plants have an inherent level of reliance on water being precipitated into, and then passed through, the surface soils.

However, there are also deep-rooting plants that maintain a more water-loving life-style, in which their root systems extend into the capillary soil region or underlying saturated groundwater zone (see Figure 16.63). These phreatophytic plants tend to have remarkable high summertime water uptake rates, possibly as a competitive means of trying to restrict the growth of their fellow plants. These deep-rooting plants generally have higher levels of plant biomass as well as higher overall growth rates.

With plants such as poplars, although they may not normally pursue this sort of phreatophytic growth mode, it is possible to induce these plants to form deep root systems using special drip-irrigation practices that effectively train them to pursue progressively deeper levels of water uptake. Aside from extending the potential reach of these phytoremediating plants to lower levels, this induction of a deep-rooting preference helps the plant with its future water demand. However, at these deeper strata the matter of oxygen availability subsequently becomes a concern, as does the presence of nutrients. Furthermore, deeper soils usually tend to be more tightly packed, such that root impedance may also be an important factor.

When motivated to adopt this phreatophytic mode, though, alders, ash, aspen, river birch, and poplar have proven to be fast growers and rapid water users, with daily uptake rates during peak summertime periods ranging from 100 to as much as 1000 L/day per tree. The resulting uptake of water from the deep, saturated soil zone may actually produce a sizable depression of 5 to 10 cm in the water table within the capture zone where water is being used by the trees. Field studies of this sort have been able experimentally to

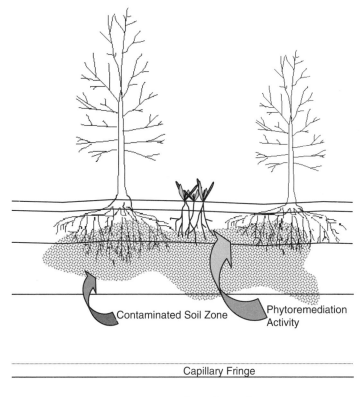

Figure 16.62   Surface and vadose-zone contaminant remediation via phytoremediation.

develop hydraulic barrier strips using deeply rooted trees planted in rows aligned perpen-
dicular to the direction of travel for the contaminated plume (as in Figure 16.64).

Of course, the actual process of evapotranspiration depends not only on the location
and depth of the roots, but also on the number of plant and atmospheric parameters.
The rate of water use by plants depends on the conductance of water through the plant
stoma as well as the cumulative surface area of the leaves through which the water
will finally be released into the atmosphere. The air temperature, wind speed, humidity,
and radiation intensity will also play a part in the final rate of this release.

For those phytoremediation systems that employ trees, the spacing provided with the
original plants can also prove to be an important factor. Of course, trees spaced too far
apart will not be able to provide nearly as much of a remediating impact, but plants spaced
too tightly together can also experience problems, particularly as they reach maturity. In
particular, tightly bunched trees block each other's light, and they will probably end up
contending for limiting nutrients. Recommended estimates for seedling plantings are in
the neighborhood of several thousand seedlings per hectare, which will drop to a level
of a few thousand after natural thinning takes place during the first few years of growth.
Clearly, there are instances in which phytoremediation systems can play a significant role
in site restoration, but it is not a universal solution for every contaminated soil. In some

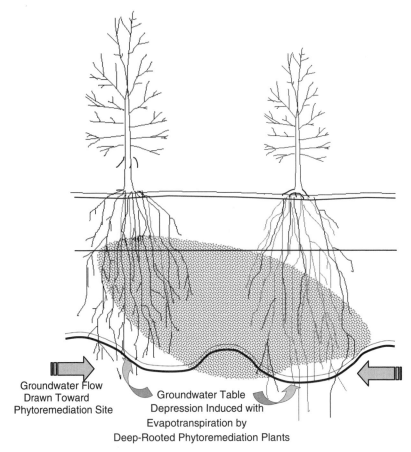

Groundwater Flow
Drawn Toward
Phytoremediation Site

Groundwater Table
Depression Induced with
Evapotranspiration by
Deep-Rooted Phytoremediation Plants

Figure 16.63   Groundwater contaminant remediation via phytoremediation.

cases, the contaminants may simply be toxic to the plants, and in others the soils will not be amenable to plant growth. Many strongly sorbed contaminants, such as PCBs or PAHs, may be too tightly adhered onto soils to be amenable to rhizospheric degradation or uptake. There will also be instances involving contaminants that would be bioaccumulated into the plants (either directly or as metabolic intermediates) in forms and at concentrations that are unsafe and unacceptable.

16.7.2   Bioremediation

The process of removing, or at least significantly degrading, the presence of contaminants in soils and groundwater using biology at the microbial level alone is known as *bioremediation* (Anderson et al., 1993). Bacteria and fungi play the dominant role in these systems. They metabolically degrade a variety of industrial, and in some instances even hazardous, wastes into less toxic, or perhaps nontoxic, products. There are two basic options to the circumstances under which a bioremediating activity might develop: the intrinsic and the engineered options. The underlying differences between them are related to the source of the responsible microbial agents and whether their presence stems from

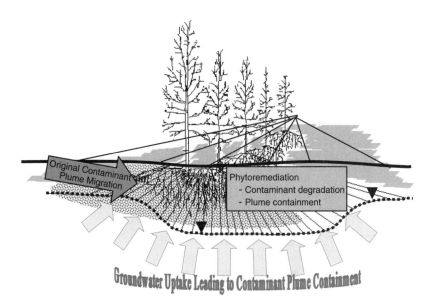

Figure 16.64 Groundwater uptake leading to contaminant plume containment.

naturally occurring organisms or a manipulated, and possibly even externally seeded, colonization.

At those sites where intrinsic bioremediation has been found to occur, the environmental conditions and contaminant characteristics of the site (i.e., types of contaminants, relative concentrations, etc.) are such that microbes already present in the soil and/or groundwater will naturally begin the remediating process of chemical degradation. In fact, it is likely that many contaminated sites could experience some degree of intrinsic bioremediation, given the biochemical diversity and metabolic flexibility of the microbial world. However, the unfortunate reality of intrinsic bioremediation is that it often fails to develop at a rate or degree that is sufficiently robust to fully remediate a given site within a time frame that is acceptably short. There are many reasons for these shortcomings. The chemistry of the soils might not have been conducive to microbial growth, as would be the case with pH extremes or nutrient limitations. There may also be physical limitations or constraints with these soils, created perhaps by extreme (whether high or low) moisture contents or unduly low soil permeabilities that hinder the movement of the bacteria or the necessary substrates and nutrients. Yet another set of problems could develop due to the contaminants themselves, which might either be recalcitrant in form or present at levels sufficiently high to inhibit, or possibly even kill, the existing microbes.

The strategy of invoking engineering methods was developed to enhance the attainable rate and efficacy of bioremediation. A wide variety of engineered bioremediation procedures have now been developed, including plans that attempt to optimize the environmental conditions under which these microbial reactions might occur and also to instigate and promote the growth of suitable microorganisms. These engineered bioremediation systems can largely be subdivided into two primary options (see Figure 16.65) In situ *treatment* means, literally, "in place," as opposed to ex situ *treatment*, in which soil is excavated for treatment aboveground.

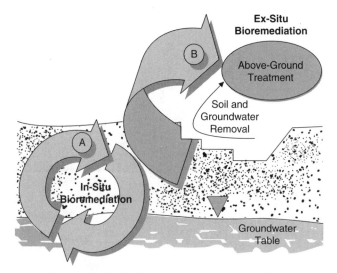

Figure 16.65 Engineered bioremediation options.

The synopsis given in le 16.17 provides an overview of the basic contaminant categories with which bioremediation has already proven to be a potentially suitable procedure, as well as information regarding the relative characteristics of the various chemical forms and the appropriate mode (aerobic or anaerobic) of their biochemical transformation. Several factors play an important role in this regard, including that of the

TABLE 16.17 Contaminant Forms Relative to Potential Bioremediation Treatment

Contaminant Compounds	Volatility[a] ($\propto K_H$)	Solubility[b] ($\propto K_{OW}$)	Biodegradability Aerobic	Biodegradability Anaerobic	In Situ Treatment Potential
Hydrocarbons					
Gasoline (C_4–C_{12})	+	+	+	−	Yes
Kerosene (C_6–C_{15})	+/−	+	+	−	Yes
Domestic fuel (C_9–C_{24})	−	−[c]	+/−	−	Yes
Lubricants (C_{15}–C_{40})	−	−	−	−	No
Aromatic hydrocarbons (BTEX)	+	+	+	+/−	Yes
Polycyclic aromati chydrocarbons					
Light (2-3 rings)	+/−	+/−[c]	+	−	Yes
Heavy (4-5 rings)	−	−[c]	−	−	No
Chlorinated hydrocarbons					
Aliphatics	+	+	−	+/−	Yes
Chlorobenzene	+	+	+	−	Yes
PCB	−	−	−	−	No
Pesticides	−	−	+/−	−	No
Heavy metals	−	+/−[c]	−	−	Yes

[a]Contaminants that tend to be volatile (with high K_H values) are listed as +.
[b]Contaminants that tend to be soluble in water (with low to moderate K_{OW} values) are listed as +.
[c]Contaminant solubility can be increased by detergents (for hydrocarbons) or acidification (for heavy metals).
Source: Adapted from Staps (1990).

compound's molecular weight and configuration, solubility, volatility, and degree of halogenation. For example, the list of suitable hydrocarbon contaminants extends from lighter fuels such as gasoline well into the heavier fuel range of 10 to 20 carbon atoms (Kostecki and Calabrese, 1991). However, beyond this level, the heavy carbon compounds (e.g., lubricants) are considerably less amenable to successful biodegradation. This pattern extends to the family of aromatic hydrocarbons, where smaller two- to three-ring forms may be biodegradable, whereas larger, four- and five-ring compounds are likely to be highly recalcitrant.

The biochemically resistant nature of the larger aliphatic and aromatic hydrocarbons stems not just from their size but also from their corresponding shift toward lower levels of water solubilities, as quantified by a determination of their octanol-to-water partitioning coefficient (K_{OW}). This variable qualifies the extent to which any given compound is attracted to (and miscible within) water as compared to its corresponding solubility in a nonpolar solvent, octanol. Contaminants regarded as water-loving or hydrophilic will have negative log K_{OW} values, by which their solubility naturally ensures a maximal opportunity for metabolic uptake and degradation. However, as the K_{OW} increases, the prospects for successful bioremediation begin to drop.

Moderately hydrophobic compounds with log K_{OW} values extending up to 1 to 2 will still be suitably available. The data given in le 16.18 for the **BTEX** (benzene, toluene, ethylbenzene, and xylene) **compounds** exemplify the upper end of these permissible values. Even though their log K_{OW} figures range from 2.12 (benzene) to 3.26 (xylene), these contaminants are all amenable to full aerobic catabolism.

Nonpolar organic liquids that have low aqueous solubility, and thus form immiscible phases, are commonly referred to as **nonaqueous phase liquids** (NAPLs), of which there are two versions, **dense NAPL** (DNAPL) and **light NAPL** (LNAPL), depending on whether they are more or less dense than water, respectively. Most hydrocarbon solvents are LNAPLs (e.g., benzene or mixtures such as gasoline), whereas most chlorinated hydrocarbon solvents (e.g., trichloroethylene, perchloroethylene) are DNAPLs. The sorptive preference of these NAPL forms to coalesce on subsurface soils substantially limits their available surface area and correspondingly constrains their effective bioavailability. There are engineering strategies to counter this behavior, however, by which surfactant (e.g., soaplike agents) chemicals can be introduced to increase the relative solubility of low-solubility contaminants (e.g., NAPLs) in a fashion which then enhances their prospective availability for metabolic degradation.

The volatility of a contaminant may also affect its rate of biodegradation, particularly given the fact that many bioremediation systems require aeration to enhance the

TABLE 16.18   Representative Aromatic Hydrocarbon (BTEX) Properties

Property	Units	Benzene	Ethylbenzene	Toluene	Xylene
Empirical formula		C_6H_6	C_8H_{10}	C_7H_8	C_8H_{10}
Molecular weight		78.12	106.18	92.15	106.18
Density at 20°C	g/cm^3	0.88	0.87	0.87	0.87
Water solubility	mg/L	1750	152	535	198
Henry's constant (K_H)	atm·m^3/mol	5.59×10^{-3}	6.43×10^{-3}	6.37×10^{-3}	7.04×10^{-3}
Octanol–water (log K_{OW}) Partitioning coefficient	Log K_{OW}	2.12	3.14	2.73	3.26

Source: Adapted from Verschueren (1983).

subsurface availability of oxygen. The key variable in this case is Henry's constant, K_H. Low-molecular-weight hydrocarbons, such as the BTEX compounds (see le 16.18), tend to have high K_H values, and as a result these types of contaminants can be removed from contaminated soils merely be providing thorough aeration or sparging [i.e., using an approach known as *soil vapor extraction* (SVE)]. The difficulty encountered with bioremediating high K_H compounds stems from their motivation to leave the soil by means of physical volatilization at a rate beyond that of its biochemical degradation. Although the soil is remediated, the contaminant is not degraded or concentrated, and may be emitted to the atmosphere.

As for the level of halogenation, the degree of substitution also tends to progressively reduce the biodegradability of a compound. Carbon tetrachloride offers a good example of this phenomenon Figure 16.66. Starting with methane as the carbon precursor, the successive halogen-based oxidations of this parent compound with chlorine will eventually shift the carbon oxidation state from a fully reduced (-4) to a fully oxidized ($+4$) level, at which point this CCl_4 compound no longer offers any energy whatsoever as an electron donor during aerobic degradation. Carbon tetrachloride consequently represents a highly recalcitrant compound in terms of its resistance to aerobic degradation. Halogenated compounds with lower levels of substitution may, however, be amenable to aerobic metabolism. At the same time, many of these compounds can also be degraded through reductive anaerobic pathways in lieu of aerobic oxidations.

Theoretically, soil and groundwater bioremediation systems can be engineered and maintained to provide a considerable variety of different biochemical mechanisms, the specific mode and implementation of which may then be tailored on a site-specific basis according to the target chemical(s) and the soil properties involved (Rittman and McCarty, 2001; National Research Council, 1993). Although the engineering requirements of these systems are rapidly becoming better established, the fact remains that this technology remains a rather inexact science. Simply put, our efforts to manipulate microbial reactions remotely within soils whose characteristics usually have a high degree of spatial inconsistency are seldom easy and are often quite difficult to complete at a desired rate or degree.

To date, the majority of bioremediation operations have been developed with the intent of using aerobic degradation mechanisms, but here again the reality of soil and pore space

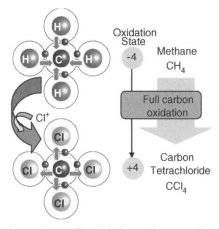

Figure 16.66   Oxidative methane (C − 4) halogenation to carbonte trachloride (C + 4).

ecosystems is that they probably experience a considerable range of aerobic, anoxic, and anaerobic reactions. At least for readily oxidizable contaminants such as the lighter hydrocarbons, this interplay is probably a moot issue given the fact that the end result usually involves a successfully remediated soil. These operations are engineered to provide aeration in the hopes of promoting aerobic remediation, and any commensurate subsurface reactions involving anoxic or anaerobic metabolism are essentially masked by the dominant effectiveness of the aerobic reactions.

In this case, therefore, it is aerobic bacteria and fungi that are relied upon to catalyze the degradation of the contaminants successfully using an array of hydroxylation, dealkylation, decarboxylation, and epoxidation reactions. The underlying thrust with these reactions is that of using the contaminants as a primary energy source, drawing electrons from the targeted contaminant and then releasing them back to oxygen as the final electron acceptor.

However, in certain instances there may well be nonaerobic reactions that offer degradative enzyme pathways that are better suited to a variety of specialized contaminants. For example, the nitroreductase enzyme typically associated with denitrification can reductively attack the nitro groups found on nitroaromatic compounds. In this case, therefore, an in situ remediation process designed to deal with soils contaminated with these types of otherwise recalcitrant energetic munitions residuals would benefit from the introduction of nitrates rather than oxygen, such that subsurface denitrification could be promoted. Extending beyond this particular application with nitroaromatics, denitrifying *Pseudomonas*, *Thiobacillus*, and *Bacillus* bacteria have also been shown to be effective metabolic oxidizers of xylene isomers.

The bioremediation of halogenated compounds involves yet another set of potential remediation pathways covering both aerobic and anaerobic conversions. One such aerobic option for dehalogenating these types of contaminants can be maintained by a cometabolism sequence facilitated by a special group of aerobes using monooxygenase enzymes. This generic group of bacteria includes **methanotrophic** (methane-oxidizing) and ammonia-oxidizing forms that employ **methane-monooxygenase** (MMO) and **ammonia-monooxygenase** (AMO) enzymes, respectively. For those sites at which this cometabolic scheme was selected, the soils would have to be supplied with the suitable substrates (e.g., dissolved methane or ammonia, and oxygen) to induce and promote the growth of the involved bacteria at levels sufficiently high to enrich the targeted soils with their monooxygenase enzymes.

Figure 16.67 schematically depicts one such cometabolic conversion of TCE, a widely observed soil and groundwater contaminant used historically in many industries as a solvent and degreasing agent. This particular reaction is known to be mediated by a number of bacteria, including *Alcaligenes* and *Pseudomonas* (Harker and Kim, 1990). These types of monooxygenase reactions convert the ethylene backbone ($-C=C-$) of TCE into an epoxide structure (i.e., TCE-epoxide) by oxidatively introducing a single mole of oxygen. This product is chemically unstable and subsequently breaks down into smaller fragments, which are then amenable to direct microbial breakdown.

The actual circumstance of these cometabolic reactions is that they occur due to inherent inefficiencies of the involved enzymes. In this case, these catalysts inadvertently attack and oxidize secondary compounds (e.g., TCE) other than the energy-releasing primary substrate (methane or ammonia) for which the enzyme had originally been produced. The difficulty with this process is that the level of inherent inefficiency for these cometabolic reactions is so low that far more primary substrate has to be introduced

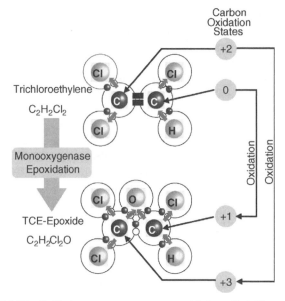

Figure 16.67   Oxidative monooxygenase epoxidation of trichloroethylene.

to the soil than the mass of secondary contaminant. In turn, the level of growth by the induced bacteria can be so high that it can clog the soil's pore space and restrict the migration of the added substrates away from the point of injection.

Dehalogenating pathways are also used with bioremediation systems that involve anaerobic reactions, including those of hydrolytic dehalogenation and reductive dehalogenation. The latter process of reductive dehalogenation has drawn the most attention, as it has proven to be widely applicable to the uncoupling of attached halogen species. One such process is depicted schematically in Figure 16.68, where the backbone carbon unit is

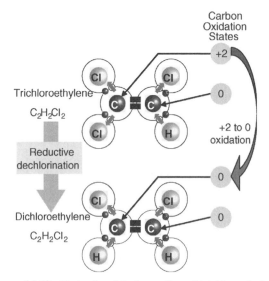

Figure 16.68   Reductive dehalogenation of trichloroethylene.

reductively converted during the course of its dehalogenation. The end products created by this latter sort of anaerobic transformation may not have been suitably transformed to qualify the bioremediation sequence to this point as an outright success. With a highly halogenated compound such as pentachlorophenol (PCP), several such reactions may be necessary. In the particular case of TCE, a second anaerobic repetition of the reductive dehalogenation reaction can lead to the formation of vinyl chloride ($H_2C{=}CHCl$). Given the fact that the latter product has been linked to carcinogenic impacts, therefore, this reactive sequence is not considered to be a suitable bioremediation mechanism. However, in many instances, a coordinated linkage of cyclic, sequential anaerobic–aerobic conversions could well achieve overall conversions of the targeted contaminants to a degree of transformation that is considerably higher than either option used independently.

The structural character and environmental conditions of a soil can play a sizable role by promoting or restricting the success of bioremediation. Certainly, one of the most important factors is that of a soil's **hydraulic conductivity** (K) (le 16.19), measured in velocity units (e.g., cm/s). This property ranges from highly permeable gravel and sand whose open void space is readily conducive to water transfer (i.e., with a K value of 10^2 cm/s or higher) to extremely tight, heavy clay soils with permeabilities (at or below 10^{-5} cm/s) which are so low that they restrict the necessary hydraulic throughput of moisture, nutrients, and substrates required for metabolic activity. In fact, the majority of successful in situ bioremediation systems have tended to fall with a hydraulic conductivity range of 10^{-3} to 10^{-1} cm/s (Staps, 1990).

Soil bioremediation depends on providing the necessary factors for microbial activity: moisture, nutrients including substrate and electron acceptors, and the presence of species of microbes capable of conducting the desired reaction. Moisture should be at a level of at least 40% of saturation (i.e., 40% of the soil pore space is filled with water and the remainder by air). Higher moisture levels would be acceptable for anaerobic systems, and may even be helpful, but in the case of aerobic bioremediation, a blockage of oxygen movement in soils with high pore water percentages would be decidedly harmful to the process. The oxidation of a readily biodegradable aliphatic hydrocarbon would, for example, require a supply of oxygen many times larger (in mass) than that of the oxidized contaminant, to the point where the soil would have had to be readily open to oxygenating air movement. The presence, or absence, of this soil moisture in remediation systems may also be affected in those aerobic systems that employ an aeration header or well, where the continuous addition of atmospheric air could induce an evaporative water loss that

TABLE 16.19 Soil Hydraulic Conductivity Relative to Bioremediation Potential

Soil Category	Hydraulic Conductivity (cm/s)	In Situ Bioremediation Potential
Very coarse sand and gravels	$\sim10^1$–10^2	Good
Coarse sand	$\sim10^0$–10^1	Good
Fine sand	$\sim10^{-1}$–10^0	Good
Loam	$\sim10^{-2}$–10^{-4}	Possibly poor
Peat	$< 10^{-5}$	Usually poor
Clay	$< 10^{-6}$	Usually poor

would have to be countered with a makeup water addition. Oxidation need not be provided only by air. Fully saturated systems can be provided with electron acceptors by way of several strategies. Oxygen can be dissolved in groundwater through sparging. It can also be introduced indirectly using hydrogen peroxide dissolved in water, which decomposes to produce oxygen. Alternative electron acceptors such as nitrate may be used in specialized circumstances.

As for the chemical nature of a soil subjected to bioremediation, the pH should be neither excessively high or low. Values between 6 and 8 are generally considered as a desirable operating range. The availability and supply of nutrients also represents an important contributing factor that will affect microbial activity, relative to the availability of key nutrients, including both organic and inorganic (e.g., primarily nitrogen and phosphorus). Ideally, the ratio of available carbon to nitrogen to phosphorus (C/N/P) should fall within the range $100 : 10 : 1$ to $100 : 1 : 0.5$. As with water availability, these pH and nutrient parameters may also be subject to change during the course of a remediation activity, such that it will need to be checked on a routine basis and corrected if necessary.

Finally, the temperature at which a remediating microbial reaction is maintained will probably have a corresponding impact on the metabolic rate. At depths much below about 1 m the temperature of soil quickly moderates to a fairly stable level of about 12 to 13°C, at which point most bioremediation microbes will experience little negative impact. However, it is possible for temperature to become an important factor, with both ex situ systems and near-surface soils, which are considerably more vulnerable to sizable temperature changes. In both instances, there may be seasonal extremes during which time the rates of microbial activity will be reduced substantially. Frost penetration during winter periods in cold-climate regions could, for example, extend for depths even beyond a meter, such that microbial activity could be commensurately depressed during these periods.

Extending beyond the characteristics and properties of the soils involved, many engineered bioremediation systems also rely on wells to deliver key materials, or air, to the active subsurface site. There are three different types of wells commonly used with bioremediation operations: those used for ventilating soils (i.e., **aerating wells**); those used to introduce water into the ground (i.e., **dosing wells**), possibly mixed with nutrients, and those used for accessing groundwater (i.e., **monitoring wells**) for sampling or monitoring purposes. The majority of these wells are installed as rigid pipe bodies constructed of inert plastic or metal, the lower reach of which has a series of slits cut into the face of the pipe to act as a screen. The lower screen face of any well serves as the interface between the well and the soil into which it was placed, and the narrowness of the screen openings is sufficiently small to prevent the unwanted migration of all but the smallest soil grains.

Well pipes used for groundwater sampling are commonly installed into a vertically aligned bore hole drilled into the soil to depths of a few meters to 10+ m (sufficiently deep to reach the groundwater table), using fairly narrow diameter (e.g., 2.5 cm) stainless steel or PVC casing. However, in the case of ventilating or dosing wells, there is considerably more variation in the types of wells and their installation, including changes in the alignment of the well, the materials employed, and the means by which they are put into place. The depth of the aerating and dosing wells can vary considerably, from shallow depths extending only a few meters to deep wells that reach 100+ m. To facilitate the desired flow of air through the aerating wells, they are much larger in diameter than dosing or monitoring well lines, with diameters of 7.5 to 10+ cm. Pipes installed into vertically aligned wells are typically installed into drilled wells, although many dosing wells are placed in a horizontal alignment using a predug ditch rather than being placed with drilling. In some cases, particularly with loosely packed loamy or sandy-loam soils, it may also be possible to install shallow-depth vertical wells with a cone penetrometer. However, in this case stainless steel well lines must be used to carry the bearing force during the installation and also to avoid subsequent problems with oxidization of the metal and consequent clogging of the screen.

For aerating wells, the depth of installation is important since air may escape from a shallow screen when dealing with highly permeable soils (i.e., that would offer a path of least resistance back to the soil surface rather than having the incoming air pass through the soil). Diffusers have been suggested, with the idea that they might improve aeration efficiency by creating smaller air bubbles with a higher surface area, but these diffuser bubbles quickly coalesce as the air enters the soil, to the point where there is no overall gain in transfer efficiency.

The type of aeration equipment to be selected depends on the necessary pressure required to push or pull the air through the wells. For remediation systems that require lower pressure [e.g., 10 to 50 kPa (1.5 to 7.5 psi)] they will commonly use rotary air pumps or blowers called *positive-displacement* (PD) *blowers*. These types of blowers effectively "push" a relatively constant volume of air drawn into the inlet through to the discharge. The air released is usually oil-free and can then be introduced directly into the soil. Experimental installations have also been constructed to employ passive aeration systems in which the motive power is supplied not by mechanical pumps or blowers but by wind or natural air exchange. The latter strategy depends on barometric-driven changes in airflow in a fashion analogous to that of the air exchange found inside caves.

In the case of more highly compacted and denser soils, high-pressure air compressor systems may be necessary to inject air at higher pressure levels of 50 to 200 kPa ($\sim$7.5 to 30 psi), and this airflow often contains a level of oil that needs to be filtered and removed prior to its introduction into soils. Outlet gas temperatures with the latter high-pressure systems may also be considerably higher than would be the case with a lower pressure operation. To some extent, this heat may be helpful in raising the aerated soil temperature and escalating the biokinetic reaction rates, but these higher temperatures may also require a shift in piping materials to alternatives (e.g., to non-HDPE forms such as stainless steel) able to handle these higher temperatures. Aerating wells maintained at high airflow rates and pressures also run the risk of fracturing tightly packed, less permeable soils, at which time problems with short-circuiting might subsequently arise. However, air extraction may also pull water and small soil fines into their well screens, which might then clog their openings and restrict subsequent fluid and air movement.

In general, there are three different configurations for aerating wells on an in situ basis with bioremediation sites, including those that inject air into a contaminated soil zone using pressure (i.e., **bioventing**), those that use combined air injection and extraction [i.e., **soil vapor extraction** (SVE)] to mobilize and extra contaminants from soils, and those that inject air into a contaminated groundwater zone using pressure (**air sparging**).

With bioventing, the incorporated low-level airflow not only oxygenates vadose soil with the goal of promoting enhanced aerobic metabolism of biodegradable contaminants within these soils but may also result in some measure of gas-phase release of the soil's more volatile contaminants. However, the primary goal of bioventing (Figure 16.69) is that of securing metabolic breakdown as opposed to volatilization; on the other hand, these goals are essentially reversed with SVE systems.

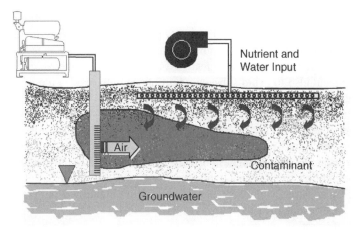

Figure 16.69   In situ engineered bioremediation using bioventing.

Bioventing has been used successfully to treat a variety of soil contaminants, including petroleum hydrocarbons and nonchlorinated solvents as well as a number of pesticides, wood preservatives, and other organic chemicals. Bioventing may also be effective at removing heavier, semivolatile petroleum hydrocarbons such as diesel, fuel oil, kerosene, and jet fuel. The key technical issues with bioventing all focus on the means of promoting optimal microbial activity. The desired ventilating airflow rate has to be kept sufficiently low to avoid significant losses of the more volatile contaminants, with typical application pressures in the range 10 to 50 kPa.

Moisture and nutrient additions are often provided using a companion dosing well system as replacements for their losses incurred due either to aeration or metabolic uptake. Intermittent pulsed additions of nutrients (NH_4Cl, KNO_3, NaH_2PO_4, etc. solutions) and water are often provided, with the intent of spreading these additions into the soil in a fashion that will distribute the zone of microbial activity away from the immediate reach of the well face. Cyclic aeration cycles have also been studied in terms of developing sequential aerobic–anaerobic degradation schemes that couple a far wider, and potentially synergistic, range of metabolic pathways. For example, the anaerobic sequence might enable the reductive dehalogenation of halogenated solvents otherwise unavailable to aerobes, at which point the dehalogenated products could be reoxidized and mineralized during the subsequent aerobic period.

Monitoring provided during the course of a bioventing operation includes measurements of soil moisture content, nutrient presence, and contaminant concentrations. The in situ soil vapor oxygen content can also be used an indicator of metabolic activity, whereby the oxygen uptake rate can be tracked following intermittent shutdowns of the aeration system. Higher respiration rates will, of course, tend to correspond to those areas where the contaminants are still present at the highest concentrations, and with time these rates will drop in parallel with contaminant depletion. Yet another option for tracking the existing metabolic activity is that of the in situ, vapor-phase carbon dioxide content.

The rate and degree to which soil contaminants can be remediated using bioventing depends on a number of factors, including the permeability of the soil and the form and concentration of the contaminants. In some cases, sizable levels of cleanup with easily degradable materials can be experienced in a matter of months, but successful

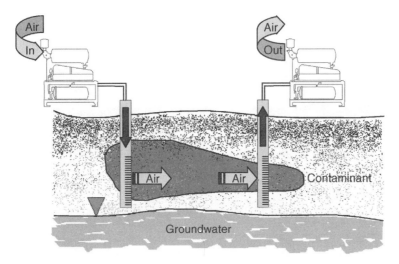

Figure 16.70   In situ engineered bioremediation using soil vapor extraction.

bioventing operations usually entail operational periods measured in years rather than months. Even then, the long-term prognosis for reaching desired cleanup endpoints will typically extend into a time frame measured in decades. These actual cleanup levels are usually derived on a site-specific basis and developed to achieve desired levels of risk reduction for each of the existing contaminant species.

There is also considerable variation in the size of these operations, ranging from applications with a single aerating well to fields involving dozens of wells. Yet another option is that some bioventing locations opt to add hydrogen peroxide (H_2O_2) in lieu of delivering oxygen, but this approach may face certain problems. It is possible for this H_2O_2 chemical to react with the soil and subsequently decompose, and it may also be that catalase-secreting bacteria could evolve at the vicinity of its release that would then lead to rapid decomposition and loss of the desired oxygen.

As the name implies, and as shown in Figure 16.70, **soil vapor extraction** (SVE) uses both positive and negative pressure to promote in situ bioremediation, essentially melding vacuum extraction to a bioventing system. Here again, both such aeration and vacuum operations are completed within a contaminated soil zone.

The third method of in situ groundwater treatment is that of **air sparging** (Figure 16.71), by which air is injected into the groundwater to promote the release and removal of contaminants. The key difference in this regard is that of aerating the groundwater zone rather than the overlying soils (as practiced with either bioventing or SVE). Two removal mechanisms will be involved, the first of which involves outright aerobic biodegradation within a reach of the oxygenated saturated soil called the **biologically active zone** (BAZ). The second mechanism involves volatilization and stripping of contaminants from the saturated to the unsaturated zone (i.e., from the groundwater into the vadose zone). As such, soil vapor extraction is typically used in conjunction with air sparging to ensure that contaminants released from the groundwater will be duly pulled from the overlying soils. Overall, though, sites amenable to air sparging must have soil hydraulic conductivity that is neither so low that it traps the incoming gas, nor too high that it allows short-circuiting of the incoming gas away from (as opposed to being diffusely spread across) a biologically active zone.

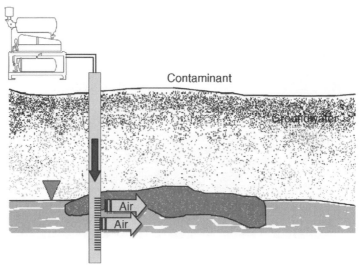

Figure 16.71   In situ engineered bioremediation using groundwater air sparging.

In lieu of directly aerating the groundwater, there is yet another engineering approach to in situ groundwater bioremediation, called **pump and recycle**, which involves the pumping of groundwater back to the soil surface, at which point these waters are then injected and infiltrated back into the overlying soils (see Figure 16.72). This single technology can provide an effective means of remediating contamination found not only in the groundwater but also in the overlying soils. To enhance and promote the metabolic efficiency of the process, an aboveground water-conditioning step may also be incorporated

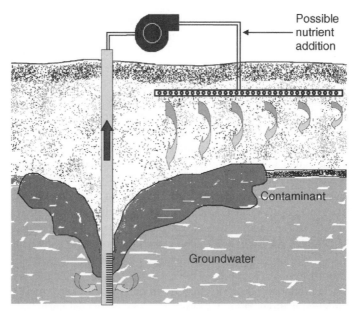

Figure 16.72   In situ engineered bioremediation using groundwater pump-and-recycle process.

into this overall process, whereby nutrients and oxygen (using either air or hydrogen peroxide) may be added to the contaminated water prior to its release back to the soil. One limitation with this technology, however, is that subsurface changes in soil layering and density may have a distinct impact on the physical course that this reinjected groundwater prefers to follow, such that these conditioned waters could well fail to reach the targeted areas of contamination.

Rather than attempting to complete a bioremediation effort with groundwaters on an in situ basis, there is also an ex situ bioremediation option in which the applied treatment steps are completed above the soil surface. This concept of **pump and treat** basically follows the general plan as that of the in situ groundwater bioremediation strategy mentioned previously, but in this case an engineered treatment process (i.e., activated sludge, trickling filter, etc.) is added to the aboveground flow scheme. The treated water is then recycled back to the ground, essentially flushing contaminants from the soil and moving them into the groundwater, where they are then extracted and pumped to the aboveground treatment unit (Figure 16.73). Yet another variation with this approach is that of using soil extraction agents to enhance the effectiveness of this soil washing effort. This process, known as **solvent extraction**, differs from conventional soil washing in that it uses surfactant-type organic chemicals as the solvent instead of water. The principal application of this process has been with the removal of polychlorinated biphenyls (PCBs), but the technology can be effective for petroleum removal as well. The disadvantages with this process are that these detergents may, at least in some instances, interfere with the desired separation and biotreatment processes.

There are also three other ex situ techniques that can be used to remediate contaminated soils, and to some extent these schemes can be faster, easier to control, and more amenable to a wider range contaminants and soil types than is the case with in

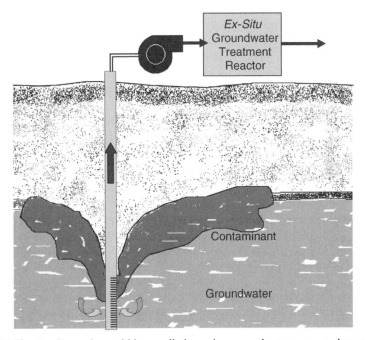

Figure 16.73 Ex situ engineered bioremediation using groundwater pump-and-treat process.

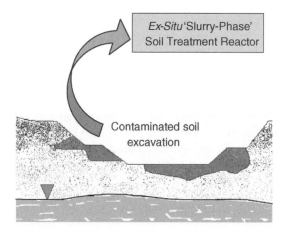

Figure 16.74 Ex situ engineered bioremediation using soil excavation and slurry-phase soil treatment.

situ techniques. However, the key problem with these ex situ procedures is that they require the initial excavation and relocation of the contaminated soils, and this effort can rapidly become prohibitively expensive. The first such option (Figure 16.74) is that of **slurry-phase soil bioremediation**, where the extracted contaminated soil is combined with water in an aboveground reactor and then mixed in batch fashion as a suspended slurry to facilitate subsequent microbial degradation.

The microbes involved may not only include a variety of indigenous forms but also specially acclimated organisms seeded into the reactor or recycled from prior treatment batches. Nutrients and oxygen can, of course, be added to the reactor to optimize the environmental conditions for the microorganisms that are degrading the contaminants. After the desired level of contaminant removal is completed, the water is extracted from the solids, and these water and solid residuals are then placed back into the ground. This slurry-phase biological treatment scheme can provide a fairly rapid means of remediating soil contaminants given its ability to maintain a homogeneous soil matrix and optimal microbial contact, particularly when compared to the sort of slow-rate bioremediation processes that might be achieved with in situ treatment of tight clay soils. However, there is clearly a trade-off with this technology in terms of the benefits of accelerated treatment vs. the costs associated with soil extraction.

The second ex situ soil bioremediation scheme (Figure 16.75) is completed as a solid-phase process in which the extracted, contaminated soils are placed in carefully engineered, typically enclosed, aboveground mounds known as **soil biopiles**. These soil piles are built with several meters of soil stacked vertically over an impermeable base constructed with either well-packed clay, plastic liners, or geomembranes, as well as an associated leachate collection system that ensures that the contaminants are not able to leave the site. Air is then drawn through the pile using a vacuum pump tied into an underlying air extraction piping network. Moisture, temperature, nutrients, and oxygen levels can then be controlled within these mounds to enhance their available rates of biodegradation, and in some cases the mounds may also be mixed or turned on an occasional basis to promote homogeneous remediation.

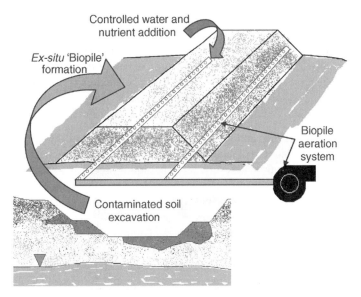

Controlled water and
nutrient addition

Ex-situ 'Biopile'
formation

Biopile
aeration
system

Contaminated soil
excavation

Figure 16.75   Ex situ engineered bioremediation using soil excavation and biopile process.

These solid-phase systems are somewhat easier to maintain and operate than is the case with slurry-phase units, but they do require more space, and their metabolic rates are typically lower, due to their reduced effective microbial intensity. There is also a variation in the original makeup of these piles in which the contaminated soils are mixed with various forms of bulking material (i.e., straw, hay, corncobs, wood chips, etc.) to open the soil and to make it more permeable to the passage of water, air, and nutrients. Here again, these piles will often be developed with some form of internal aeration header, or alternatively, mixed on a routine basis with mechanical turning to ensure that the pile contents are kept aerated.

The third option for ex situ soil remediation is that of **land farming**. This relatively simple process again involves the excavation of the contaminated soils, which are then spread across land areas which are then tilled periodically much like farmlands to entrain air and remix the actively remediating soil matrix. Moisture and nutrients are also monitored and added as needed to promote the desired bioremediation. Given the fact that most of these land-farming operations are maintained on unenclosed lands, there may be instances in which the contaminant removal process could actually be attributable more to volatilization than to biodegradation. If this is expected to be the case, where the rate of contaminants would probably be unacceptably high, this process may have to be shifted to an enclosed operation.

PROBLEMS

16.1. Call your state's regulatory department for environmental management, protection, or health and identify a local municipal or industrial wastewater treatment facility for which you could then obtain a 12-month span of compliance reports covering each of the following process streams: effluent (liquid) discharge and biosolids (residuals). Identify the stipulated parameters and values in each such case (e.g., effluent BOD, suspended solids, ammonium-nitrogen, fecal coliforms), and then qualify the relative compliance or noncompliance maintained with each such stream during the period of record. If, indeed, the facility has had any noncompliance events, characterize the nature and extent of these periods. Finally, obtain the necessary contact information (name and phone numbers) for this facility by which you can then arrange a subsequent on-site guided tour.

16.2. Obtain permission for, and then complete, a guided tour of the previously facility identified in Problem 16.1 and then complete the following steps: (**a**) prepare a schematic and technical description of the liquid processing stream involved; (**b**) identify the sources, characteristics (e.g., influent BOD, suspended solids), and quantities (i.e., percent municipal versus commercial versus industrial) of this facility's various wastewater sources; (**c**) identify this facility's typical removal efficiencies for BOD, suspended solids, and ammonium-nitrogen; (**d**) quantify the hydraulic retention times provided by each successive liquid processing step; and (**e**) qualitatively describe the apparent environmental condition of this facility's receiving water body and characterize the probable impact of this incoming discharge stream.

16.3. In conjunction with the site visit and tour completed for Problem 16.2, do the following: (**a**) prepare a schematic and technical description of the sludge processing streams involved; (**b**) qualify and quantify the daily wet sludge volume, moisture content, and volatile solids fraction, as well as the daily dry solids mass; (**c**) describe the involved process of sludge processing (e.g., stabilization, digestion, dewatering); and (**d**) describe the procedure(s) used for final disposal of this facility's biosolids materials.

16.4. In conjunction with the site visit and tour completed for Problem 16.2, do the following: (**a**) prepare a schematic and technical description of the disinfection process involved; (**b**) characterize the involved disinfection strategy and desired efficacy; (**c**) quantify the hydraulic retention time provided by this disinfection step; and (**d**) quantify the requisite levels of energy and chemical consumption.

16.5. Explain how the various processes used for water and wastewater disinfection (e.g., chlorination, bromination, ultraviolet irradiation) produce a desired pathogen inhibition or mortality, and then explain how these "disinfected" organisms might effectively be considered "dead" (i.e., unable to reproduce viably), although they exhibit catabolic activity (i.e., such that they still appear to be "alive").

16.6. Call your state's regulatory department for environmental management, protection, or health. Identify the location of a nearby constructed wetland used for either wastewater treatment or stormwater management, and obtain the necessary contact information (name and phone numbers) for this facility. Arrange and complete an on-site guided tour at this facility, and then do the following: (**a**) prepare a schematic and technical description of the flow scheme involved; (**b**) characterize the distribution and forms of plants found within this system; and (**c**) qualitatively characterize the apparent quality of the final effluent.

16.7. Prepare written answers to the following questions regarding subsurface-flow constructed wetlands: (**a**) Why are these systems commonly built with washed "river" gravel as the internal mediaum rather than unwashed sand, soil, or gravel; (**b**) What is the porosity of a washed gravel (in percent)? [*Note*: you might consider running a simple test using a small bucket filled with washed gravel set on a bathroom scale and then weighed with and without water filling the voids). (**c**) What happens to the latter void space within this gravel as the constructed wetland "ages"? (**d**) Is the latter change in porosity associated with aging of the wetland and plant root biomass growth considered during initial design and retention-time analysis?

16.8. Complete a literature review (e.g., using a Web of Science search) regarding the options for, and uses of, molecular biology procedures and tools (e.g., DGGE, PLFA, FISH) in terms of environmental biology research and assessment (i.e., covering both fundamental and applied applications), and then prepare a written synopsis of this information.

16.9. Complete a Web-based search for an available online case study report covering a successful soil bioremediation operation, and then prepare a written synopsis of this information, covering the following details: a) original site background, conditions, and scale; (**b**) site ownership and regulatory factors; (**c**) involved technical procedures; (**d**) stipulated remediation efficacy criteria (if any); (**e**) project time frames (start date and expected closure date); (**f**) postremediation characteristics and conditions; and (**g**) involved costs and funding sources.

16.10. Explain why it is often necessary to supply nutrients, including nitrogen and phosphorus, during in situ soil bioremediation operations.

16.11. Complete a Web-based search for an available online case study report covering a successful biofilter operation for air-phase contaminant treatment, and then prepare a written synopsis of this information, covering the following details: (**a**) facility background and conditions [e.g., airflow rates, contaminant(s) and concentration information]; (**b**) involved technical procedures (i.e., process scheme); (**c**) biofilter media form(s) and retention times; (**d**) stipulated efficacy levels; and (**e**) performance details observed (e.g., effluent contaminant levels).

16.12. Describe the benefits that might be achieved with prehumidifying an airflow stream passed through a gas-phase biofilter, using a water spray scrubber, in terms of physical, chemical, and biochemical attribute factors.

16.13. Complete a Web-based search for an available online case study report covering a successful phytoremediation operation, and then prepare a written synopsis of this information, covering the following details: (**a**) original site background, conditions, and scale; (**b**) site ownership and regulatory factors; (**c**) applied plant forms and growth patterns; and (**d**) chronological remediation efficacy.

16.14. Prepare a table qualitatively covering the various mechanisms by which biological factors and metabolism can negatively affect the aesthetic (e.g., taste, odor, clarity) quality of our environment, and in each such case identify possible remedial or corrective strategies potentially able to negate or obviate these problems.

16.15. Review the global biochemical nitrogen cycle and then describe the corresponding impacts and/or benefits of the following environmental situations: (**a**) oxidative nitrification completed within wastewater treatment facilities; (**b**) reductive denitrification completed within wastewater treatment facilities; (**c**) reductive nitrogen fixation maintained by algal growth within a polluted lake; (**d**) concentrated urine release, storage, and hydrolysis relative to confined animal feedlot operations; and (**e**) anhydrous ammonia application associated with farm fertilizer operations.

REFERENCES

Anderson, T. A., E. A. Guthrie, and B. T. Walton, 1993. Bioremediation, in In the rhizosphere, *Environmental Science and Technology*, Vol. 27, No. 13, pp. 2630–2636.

Asimov, I., 1989. *Asimov's Chronology of Science and Discovery*, Harper & Row, New York.

Atlas, R. M., and R. Bartha, 1987. *Microbial Ecology*, 2nd ed., Benjamin/Cummings, Menlo Park, CA.

Bailey, J. E., and D. F. Ollis, 1977. *Biochemical Engineering Fundamentals*, McGraw-Hill, New York.

Baker, M. N., 1948. *The Quest for Pure Water*, American Water Works Association, New York.

Bitton, G., 1999. *Wastewater Microbiology*, Wiley-Liss, New York.

Bitton, G., and C. P. Gerba, 1984. *Groundwater Pollution Microbiology*, Wiley, New York.

Bryant, E. A., G. P. Fulton, and G. C. Budd, 1992. *Disinfection Alternatives for Safe Drinking Water*, Van Nostrand Reinhold, New York.

CH2M-HILL, 1984. *A Comparison of Trickling Filter Media*, CH2M-Hill, Inc., Denver, CO.

Characklis, W. G., and K. C. Marshall, 1990. *Biofilms*, Wiley-Interscience, New York.

Code of Federal Regulations, 1993. *Standards for the Disposal of Sewage Sludge*, 40CFR, Title 40, Parts 257 and 503, U.S. government printing office, Washington, DC.

Cole, S., 1998. The emergence of treatment wetlands, *Environmental Science and Technology/News*, May 1, pp. 218A–223A.

Crites, R., and G. Tchobanoglous, 1998. *Small and Decentralized Wastewater Management Systems*, WCB McGraw-Hill, New York.

Darleym, W. M., 1982. *Algal Biology: A Physiological Approach*, Blackwell Scientific, Oxford.

Datar, M. T., M. N. Rao, and S. Reddy, Vermicomposting: a technological option for solid waste, 1997. *Management Journal, of Solid Waste Technology and Management*, Vol. 24, No. 2.

Eikelboom, D. H., and H. J. J. van Buijsen, 1981. *Microscopic Sludge Investigation Manual*, TNO Research Institute for Environmental Hygiene, Delft, The Netherlands.

Fox, J. C., P. R. Fitzgerald, and C. Lue-Hing, 1981. *Sewage Organisms: A Color Atlas*, Metropolitan Sanitary District of Greater Chicago, Chicago.

Gaudy, A. F. Jr., and E. T. Gaudy, 1988. *Elements of Bioenvironmental Engineering*, Engineering Press, San Jose, CA.

Grady, C. P. L., G. T. Daigger, and H. C. Lim, 1999. *Biological Wastewater Treament*, 2nd ed. Marcel Dekker, New York.

Hammer, D. A., 1989. *Constructed Wetlands for Wastewater Treatment: Municipal, Industrial and Agricultural*, Lewis Publishers, Chelsea, MI.

Harker, A. R., and Y. Kim, 1990. Trichloroethylene degradation by two independent aromatic-degrading pathways in *Alcaligenes eutrophus JMP134, Applied and Environmental Microbiology*, Vol. 56, No. 4, pp. 1179–1181.

Horan, N., 1990. *Biological Wastewater Treatment Systems: Theory and Operation*, Wiley, New York.

Jackson, L., 1997. Why choose phytoremediation: looking through the eyes of the customer, *IBC's 2nd Annual Conference on Phytoremediation*, Seattle, WA, June 18–19.

Jenkins, D., M. G., Richard, and G. T. Daigger, 1993. *Manual on the Causes and Control of Activated Sludge Bulking and Foaming*, Lewis Publishers, Boca Raton, FL.

Kostecki, P., and E. Calabrese, 1991. *Hydrocarbon Contaminated Soils and Groundwater*, Vol. 1, Lewis Publishers, Chelsea, MI.

Metcalf and Eddy, 2003. *Wastewater Engineering: Treatment and Reuse*, 4th ed., revised by G. Tchobanoglous, F. L. Burton, and H. D. Stensel, McGraw-Hill, New York.

National Research Council, 1993. *In-Situ Bioremediation: When Does It Work?* National Academy Press, Washington, DC.

Palmisano, A. C. and M. A. Barlaz, 1996. *Microbiology of Solid Waste*, CRC Press, Chelsea, MI.

Rittman, B. E., and P. L. McCarty, 2001. *Environmental Biotechnology: Principles and Applications*, McGraw-Hill, New York.

Schnoor, J. L., L. A. Light, S. C. McCutcheon, N. L. Wolfe, and L. H. Carreira, 1995. Phytoremediation of organic and nutrient contaminants, *Environmental Science and Technology*, Vol. 29, No. 7, pp. 318A–323A.

Speece, R. E., 1996. *Anaerobic Biotechnology*, Archae Press, Nashville, TN.

Staps, J. J. M., 1990. *International Evaluation of In-Situ Biorestoration of Contaminated Soil and Groundwater*, Report 738708006, National Institute of Public Health and Environmental Protection, Bilthoven, The Netherlands.

Strom, P. F., and M. S. Finstein, 1995. *New Jersey's Manual on Composting Leaves and Management of other Yard Trimmings*, New Jersey Department of Environmental Protection, Trenton, NJ, 81pp. [Available free online at http://www.state.nj.us/dep/dshw/rrtp/compost/front.htm]

U.S. EPA, 1998. *Interim Enhanced Surface Water Treatment Rule*, Office of Water, U.S. Environmental Protection Agency, Washington, DC.

U.S. EPA, 1999. *Alternative Disinfectants and Oxidants Guidance Manual*, Office of Water, U.S. Environmental Protection Agency, Washington, DC.

U.S. EPA, 2000. *Constructed Wetlands Treatment of Municipal Wastewaters*, National Risk Management Research Laboratory, Office of Research and Development, U.S. Environmental Protection Agency, Cincinnati, OH.

Verschueren, K., 1983. *Handbook of Environmental Data on Organic Chemicals*, Van Nostrand Reinhold, New York.

Water Environment Federation, 1991. *Design of Municipal Wastewater Treatment Plants*, Manual of Practice 8, WEF, Alexandria, VA.

Water Environment Federation, 1996. *Wastewater Disinfection*, prepared by the Task Force on Wastewater Disinfection, Technical Practice Committee, Municipal Subcommittee, WEF, Alexandria, VA.

White, G. C., 1992. *The Handbook of Chlorination and Alternative Disinfectants*, Van Nostrand Reinhold, New York.

Zehnder, A. J., 1988. *Biology of Anaerobic Microorganisms*, Wiley-Interscience, New York.

17

THE SCIENCE OF POISONS

Toxicology is the study of physical or chemical agents that produce adverse effects on biological systems. This definition is a bit overly broad for our purpose, since, for example, it could include nontoxic oxygen-demanding substances that can kill fish by depleting oxygen in a stream. In this chapter and the several that follow we only examine agents that affect individual organisms directly. Indirect effects, such as eutrophication or deoxygenation, are described in Chapter 15.

Furthermore, we are mainly concerned here with environmental pollutants. The field of toxicology also deals with other toxins, such as pharmaceuticals, food additives, and those that occur naturally. Of particular interest are xenobiotics. Various forms of radiation, if capable of depositing enough energy to break chemical bonds, can also produce toxic effects. Radiation with sufficient energy is called **ionizing radiation** or just **high-energy radiation**. Examples include ultraviolet, x-ray, and gamma radiation from the electromagnetic spectrum, and high-energy particles such as alpha or beta radiation (helium nuclei and electrons, respectively) from radioactive decay.

This chapter focuses on the general principles of toxicological effects at and below the organism level (e.g., biochemical, cellular, organ systems). In subsequent chapters we detail higher-level effects such as ecosystem-wide changes, or organism effects that are specific to particular groups of organisms, such as aquatic or mammalian.

Toxicology is an interesting combination of the qualitative and the quantitative. A major activity in toxicology is examining exposed organisms to determine the "how" and "what" of a toxin's effect: ultimately, it is hoped, to the molecular level of understanding. Another large area of activity is the measurement of toxic responses, in either laboratory experiments or in field measurements. These responses are usually studied in probabilistic terms using the tools of statistics. Toxicity tests involve exposing organisms

Environmental Biology for Engineers and Scientists, by David A. Vaccari, Peter F. Strom, and James E. Alleman
Copyright © 2006 John Wiley & Sons, Inc.

to varying amounts of toxicants and measuring the probability of a particular response at each dosage.

Detection of cause-and-effect relationships between toxicants and possible effects in human populations is the job of the intersecting field of epidemiology. **Epidemiology** is the study of disease occurrence. (Epidemiology is concerned not only with chemical or physical agents of disease, but also other causes, such as infectious agents.) Epidemiology uses statistical tools such as analysis of variance (ANOVA) and regression to detect such relationships empirically. Empirical evidence, however, is not proof of cause and effect. Proof of a direct effect requires confirmation by independent evidence, such as is provided by the other two toxicological activities. If all three point in the same direction, this makes a strong case. Quite often they are performed in the following order:

Epidemiology	Detection of a correlation between exposure to a toxin and some adverse effect, from field observations
Bioassay	Laboratory verification of the cause-and-effect relationship found by epidemiology
Biochemical/physiological	Determination of the mechanism of toxic effect

The sequence above is, however, a reactive situation. We are increasingly using a proactive stance, in which the bioassay comes first. In this way we will know if a chemical is toxic, and how, and prevent exposure before it can occur in the field.

17.1   MECHANISMS OF TOXICITY

We can distinguish between the *mechanism* of toxicity and the toxic *effect*. The toxic **effect** is the adverse reaction that is observed, whereas the **mechanism** is the underlying process that leads to the effect. Put another way, the mechanism is the cause, usually at the molecular level, and the effect is the result. In some cases, as in biochemical effects, the distinction can be arbitrary.

Ultimately, toxins act by reacting with specific compounds in the organism. The result may be an alteration to a metabolite, enzyme, or cell structure. The site where the toxin acts is called the **target tissue** or **target organ**. The specific compounds in the target tissue that the toxins react with are called **receptors**. Organs vary in their sensitivity because they may differ in numbers or types of receptors, toxin transport, or organ function. For example, neurons depend on mitochondrial activity for their ATP and are therefore more sensitive to substances, such as carbon monoxide, that affect oxygen transport. The liver and kidney receive more blood than do other organs, making them a frequent target. Rapidly dividing cells such as bone marrow and intestinal mucosa are more sensitive to genotoxins.

Some toxins are relatively nonselective; they exert their effect on any tissue they contact. This tends to be the case for more highly reactive toxins, such as strong oxidizers, irritants, or free-radical producers. Even nonselective toxins, however, can exert their damage selectively either at the tissue first contacted or at other susceptible tissues. This may be the mode of action for chemical disinfectants such as chlorine.

Biochemical changes may be **reversible** or **irreversible**. Physicochemical interactions tend to be reversible. Examples include carbon monoxide poisoning of hemoglobin and narcosis induced by hydrophobic solvents. Covalent bonding and degradation reactions between toxins and cellular substances are more likely to be irreversible. Examples are the bonding of mercury, lead, and cadmium to sulfhydryl (−SH) groups in proteins, and the damage to lung cell components by ozone.

Covalent Bonding   Examples of toxicity due to covalent bonding include protein denaturing by heavy metals and the interference of cyanide and sulfide with the cytochrome system. Mercury, lead, and cadmium bond to sulfhydryl (−SH) groups in proteins. Since these groups are important in protein folding, this causes denaturing of proteins, leading to profound effects. HCN and H_2S bond to iron in cytochromes. This halts the electron transport system, which is responsible for most of the energy yield of respiration. Others examples, which are described further below, are the reaction of pesticides with the enzyme acetylcholinesterase and reactions involving cellular DNA.

Physicochemical Bonding   Bonds between toxins and cellular substances involving van der Waals, hydrogen, and polar bonds also create toxic effects by modifying normal biochemical function. Examples are the complexation of metals and the effect of solvents on plasma membranes.

Multiple physicochemical bonds with a single molecule or complex is called **chelation**. Chelation can alter the distribution of substances, especially heavy metals, by effectively changing their physicochemical behavior (solubility, lipophilicity, etc.). This can affect excretion or can sequester substances within the organism. Enzyme activity can be susceptible to this chelation since many enzyme cofactors are inorganic metals.

For example, the solvent carbon disulfide, CS_2, reacts with amino acids to form thiazolidone and thiocarbamate compounds, which chelate zinc. Rats exposed to CS_2 rapidly lost Zn in the urine due to the increased solubility of the complex. Thiozolidone is also thought to make copper less available as an enzyme cofactor. The complexing agent ethylenediaminetetraacetic acid (EDTA) is used in cases of lead poisoning to hasten excretion of that toxic metal.

A phenomenon called **metal shift** is the effect in which a toxicant causes metals to decrease in one set of organs and increase in another. For example, feeding vanadium to rats at levels below 150 ppm resulted in iron moving into the liver and spleen, whereas above 250 ppm it caused iron concentrations in those organs to fall as low as one-third of the normal content. Metal shift might be caused by chelation or competitive absorption.

Enzyme Disruption   The normal action of enzymes can be disrupted in a number of ways. A toxin can bind, reversibly or irreversibly, to the active site or the binding site of a cofactor, or it may bind elsewhere, causing conformational changes affecting the active site. The toxin can also have its effect by reacting directly with a cofactor, changing its reactivity or availability. For example, fluoride inhibits enolase, one of the enzymes of glycolysis, presumably by complexing with its Mg^{2+} cofactor. Finally, a toxin may mimic a metabolite or cofactor, competing with the normal metabolite for the active site. This is the case with fluorocitrate, as described in Section 18.5.

An interesting comparison of effects occurs with two classes of insecticides, both of which affect the same enzyme, one irreversibly and the other reversibly. Both of these classes form covalent bonds with the neurotransmitter-degrading enzyme acetylcholinesterase, inhibiting it (Figure 17.1). Symptoms produced are muscular twitching, weakness, and ultimately paralysis. One class, the organophosphates, forms bonds that are practically irreversible. Carbamate insecticides, on the other hand, bond to the same enzyme

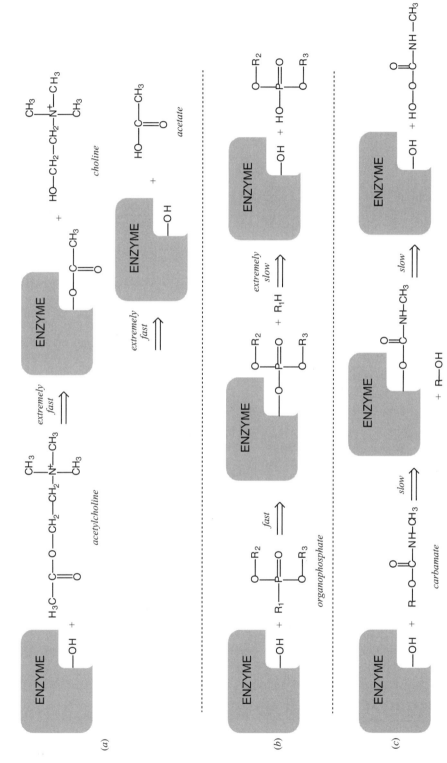

Figure 17.1 (*a*) Normal hydrolysis of acetocholine by the enzyme acetocholinesterase; (*b*) reaction of organophosphate pesticides with enzyme; (*c*) reaction of carbamate pesticides with enzyme.

707

reversibly. Both have similar effects, but recovery from carbamate poisoning is faster, making this a somewhat safer material to use. Another critical difference is that organophosphate exposure, being irreversible, is cumulative. A series of small exposures successively depletes the useful supply of acetylcholinesterase until a level is reached that causes detectable neurological effects. In fact, experiments with chickens have demonstrated that the cumulative dose of an organophosphate pesticide that produces an effect can be less than the single dose that would produce the same effect.

Membrane Changes   A common target of industrial pollutants is the plasma membrane. This is because many industrial chemicals, especially the solvents, are lipophilic. Once absorbed into the organism, they preferentially partition into the lipid bilayer, changing its physicochemical properties. These properties include its fluidity, permeability, and physicochemical interactions with embedded proteins. The effect on proteins affects their function, which includes acting as receptors for normal biochemical functions or as agents in active transport of ions and other substances across the plasma membrane. These actions are the presumed cause of narcosis, one of the major symptoms of solvent toxicity. In extreme cases, exposure to solvents can leach lipids out of the membrane, or cause the membrane to rupture, destroying the cell or organelle.

Irritants   Some of the more severe modes of action lead to **irritation** (damage to cells of a tissue). These modes include oxidation, severe pH changes, dehydration, or precipitation or denaturing of proteins. Dehydration can be caused by concentrated salt solutions. Proteins can be precipitated or denatured not only by changes in pH or ionic strength, but also by organics such as aldehydes or ketones. Examples of strong oxidizers are ozone and chlorine.

Free Radicals   Oxidizers, ionizing radiation, and metabolic transformations of some organic pollutants, can result in the formation of free radicals. **Free radicals** are compounds with unpaired electrons. They are very reactive and electrophilic and therefore tend to oxidize compounds. However, the reactions are nonspecific; that is, the radicals tend to react with almost any compound with which they come into contact. Some of the simpler radicals are the superoxide, peroxide, or hydroxyl free radicals:

$$\cdot HO_2 \qquad \cdot H_2O_2 \qquad \cdot OH$$

The dot symbolizes the unpaired electron. Some toxins themselves react to become free radicals, such as in the biotransformation of carbon tetrachloride:

$$CCl_4 \Rightarrow \cdot CCl_3$$

When radicals react with cellular compounds, they may either oxidize them, or they may transform them into radicals, such as lipid or DNA free radicals, which undergo further reactions. Several enzymes and the dietary antioxidants vitamin E, vitamin C, and glutathione act to protect the cell by destroying radicals.

17.2   ABIOTIC FACTORS THAT AFFECT TOXICITY

A number of abiotic factors affect the toxicity of substances. Notable examples in aquatic tests include the pH, temperature, dissolved oxygen content, salinity, alkalinity, and water hardness. The fact that all of these can have a simultaneous effect greatly complicates the application of toxicity data to the field. In most of the following cases, the research has been done on acute toxicity (i.e., those with rapid onset). Much less is known about the effect of these factors on chronic toxicity (i.e., those with delayed onset).

Temperature affects organisms directly as well as by changing the response to toxins. Thermal pollution of surface waters, often associated with cooling-water discharge from electric power plants, can affect fish behavior. By reducing the solubility of oxygen and increasing microbial oxygen uptake rate, it can apply a stress to many aquatic organisms. However, one cannot make generalizations about the effect of temperature on toxicity. The threshold toxicity of zinc to Atlantic salmon was higher at 19°C than at 3°C, but rainbow trout, which are in the same genus, were more tolerant to zinc at 3°C than at 20°C. The undissociated fraction of hydrogen cyanide shows one of the strongest temperature effects. Its LC_{50} (the concentration that is lethal to 50% of test organisms) increases with temperature in a log-linear relationship. The LC_{50} increases (i.e., it becomes less toxic) 12-fold, from 44 μg/L at 6.5°C to 530 μg/L at 25°C.

The effect of temperature can depend on other abiotic factors, thus forming a three-way interaction between the toxin dosage, temperature, and the other factor. In soft water, napthalenic acids are twice as toxic to snails at 20°C than at 30°C. However, no differences have been reported in hard water. One rule of thumb that can often be used is that the time it takes for an organism exposed to a lethal concentration to succumb follows a van't Hoff relationship with temperature: survival time decreases by a factor of 2 or 3 with each 10°C increase. For this reason, shorter-term acute toxicity tests could show an increasing toxicity with temperature, while the same test conducted for a longer time could show decreasing toxicity.

Dissolved oxygen (DO) in water is saturated with respect to the atmosphere at concentrations ranging from 7.5 to 14.6 mg/L, depending on temperature, salinity, and total pressure (and therefore, altitude). Most fish will tolerate DO levels as low as 2.0 mg/L, although their activities may be reduced. DO can affect aquatic toxicity in several ways: (1) by increasing exposure at a given activity level, since at low DO levels fish must circulate more water through their gills; (2) indirectly by increasing organism activity; and (3) by causing stress to organisms at low DO. Very commonly, there is no discernible effect of DO, or the effects observed are due to the oxygen stress and not the toxin. One case of strong interaction has been found in the effect of pulp mill effluent on young salmon. Only 2% of effluent mixed with half-strength seawater caused the fish to gasp for oxygen at the surface when the DO was reduced to 36% of saturation.

The *pH range* 6.5 to 9.0 is considered harmless to fish. The pH interacts strongly with the toxicity of weak acids and bases, such as cyanide, hydrogen sulfide, or ammonia, by shifting the balance between ionized and nonionized forms (Sections 3.3 and 18.1). The LC_{50} of total sulfide is 64 μg/L at pH 6.5 and 800 μg/L at pH 8.7. Thus, the undissociated form seems to be about 15 times as toxic as the ions. Metals are usually more toxic in the ionic form. Although phenol is an acid, its toxicity does not change much with pH. Alkalinity may have some effect. High-alkalinity waters are more buffered against changes in pH. The CO_2 released by gills can cause a local decrease in pH, affecting the toxicity of weak acids and bases. High-alkalinity waters are more buffered against changes in pH, and therefore have less of an effect from CO_2. Thus, at the same pH, the toxicity of ammonia to fish would be expected to be less at lower alkalinities.

Water hardness is due primarily to dissolved calcium and magnesium. A soft water is one with less than about 75 mg/L expressed in terms of $CaCO_3$, and above 300 mg/L as

CaCO$_3$, it is considered very hard. Surface waters tend to be soft; the world's rivers have a median hardness of 50 mg CaCO$_3$/L. Groundwater from limestone regions range from 350 to 380 mg CaCO$_3$/L.

Metal salts are less toxic in hard waters than in soft. For example, a very good correlation was found between the LC$_{50}$ of copper in mg/L to salmonid fish vs. hardness (H) in mg CaCO$_3$/L:

$$LC_{50} = 0.0034\,H^{0.91} \tag{17.1}$$

Based on such simple correlations, it was thought that membrane permeability was being affected by calcium. However, if the tests were done in water with combinations of pH and alkalinity different from those typically found in nature, the simple relationship disappears. It seems that the relationship was observed because in natural waters the hardness and alkalinity vary together. Now it is thought that it is the alkalinity that interacts with pH, and that this is due to the various complex ions formed by the equilibrium among the metal ion, hydroxide ions, and carbonate ions. For example, copper forms at least seven species in solution, each of which may have its own toxicity. Figure 17.2 shows the effect of hardness and pH on the toxicity of copper to rainbow trout. The curve represents a mathematically smoothed response surface. The low toxicity at pH 8 is thought to be due to most of the copper being in the form of carbonates and nonionized hydroxides, which may be less toxic than the other forms.

Salinity is not an important variable in toxicity.

Temperature and humidity strongly affects the sensitivity of plants to toxins in the air. An increase in either stimulates plants to open the leaf stomata, giving gaseous toxins admittance to the interior of the leaves. Light intensity also affects plant response, but the effect may be positive in some situations and negative in others.

Even *time of day* has been shown to have an effect on toxicity to animals. In rats and mice the cytochrome P450 enzyme system, which metabolizes many toxins, is most active just after dark. Factors relating to the care and handling of test animals, such as whether they are caged individually or in groups, also affect the measured toxicity.

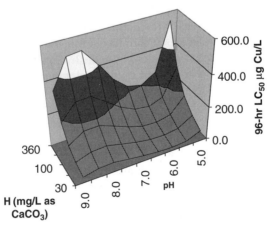

Figure 17.2 Hardness and pH interaction in copper toxicity. (From Rand and Petrocelli, 1985.)

17.3   INDIVIDUAL VARIABILITY

Even when given the same exposure to a toxin, individuals vary in their responses. The causes of variation include genetic, nutritional, age and sex, metabolic activity level, life stage, or exposure history leading to lesions, sensitization, or enzyme induction.

Several examples in humans will show that individuals of the same species can have large genetic differences in responses to toxins. Many Orientals are genetically predisposed to a more rapid metabolism of ethanol to acetalydehyde. Since it is the metabolite that is responsible for many of the symptoms of ethanol intoxication, these people react strongly to even small doses, becoming flushed and uncomfortable. Another example is in the inherited propensity that some people have toward specific cancers, such as retinoblastoma, or skin or colon cancer. These people have inherited one of the "hits" required to convert a normal cell into a cancerous one. In a third example, millions of persons throughout the world have red blood cells with defects in their respiration pathways, such as a deficiency in glucose-6-phosphate dehydrogenase. These cells are not efficient in maintaining glutathione, which protects against peroxide attacks. People with this type of problem are especially sensitive to red blood cell hemolysis (breakdown) from certain chemicals, such as aspirin and naphthalene, and from substances in some foods, such as fava beans. This is an X-linked trait, exhibited only in males, who inherit it from their mothers (Section 6.1.2).

The sex of an animal can have a great influence as well. Male mice are much more sensitive than females to chloroform, possibly because males have a much higher concentration of cytochrome P450. Female rats are more sensitive than males to certain organophosphate pesticides. However, castration and hormone treatment render the males more sensitive.

Young animals are typically 1.5 to 10 times as sensitive to toxins as adults, possibly due to underdeveloped immunity or detoxification mechanisms. Malathion is about 28 times more toxic to newborn rats than to adults. However, this is not always the case. DDT is about 20 times less toxic to newborns, and Dieldrin was about 4.5 times less toxic. The young may absorb differently, and their blood–brain barrier is less efficient.

Preexisting disease can affect person's response to a toxin in several ways. If the disease affects the kidney or liver, it may affect the half-life of a compound by changing the rate of biotransformation or excretion. Of course, disease can also have an indirect effect by rendering the person more vulnerable to any type of additional damage.

The relationship between nutrition and toxicity is now known to be a major factor affecting toxic response. The effect occurs through altered absorption and renal function and by affecting toxin distribution in tissues. Fasting or low-protein diets may reduce cytochrome P450 activity. This can either increase toxicity (e.g., DDT) or decrease it (e.g., chloroform). Lipids in the diet delay absorption of lipophobic substances and enhance it for lipophilic substances. Essential fatty acids, such as linoleic acid, are important to the cytochrome P450s. Fatty tissues can store lipophilic toxins away from receptors. Thus, obesity actually protects against chronic toxicity of these compounds. However, high-fat diets enhance absorption and retention of lead and fluoride.

Epidemiological evidence indicates that vitamin A is protective against lung cancer. Exposing rats to PCB, DDT, and dieldrin significantly reduced the stores of vitamin A in the liver. However, vitamin A can be toxic at high levels. On the other hand, β-carotene, which is a precursor of vitamin A, is fairly nontoxic. Vitamins E and C are both important antioxidants. Lipophilic vitamin E acts to protect the membranes from free radicals and other oxidizers; vitamin C does the same in the cytoplasm. In addition, in vitro experiments suggest that vitamin C competes for nitrites, which are often found in preserved foods. This blocks their reaction with amines that would otherwise form carcinogenic nitroso compounds.

Zinc reduces the toxicity of lead and cadmium, and selenium is protective against cadmium and mercury. Also, selenium is a coenzyme for glutathione peroxidase. Together they destroy H_2O_2, which otherwise disrupts membranes by oxidation. Many metals are enzyme cofactors (e.g., iron is essential for the cytochrome P450 system).

Enzymes may be induced by previous exposure to the toxin or to other substances. This may persist for some time after the original exposure, affecting the response to a toxin until the induction effect dissipates.

17.4   TOXIC EFFECTS

The toxic modes of action result in numerous consequences, which are the toxic effects observed. The effect of a toxic substance on an organism may be direct, induced, or indirect. **Indirect effects** are those in which damage to an organism is caused by intermediary effects, such as physical or chemical changes to the environment, or by loss of food or shelter. **Induced effects** are those in which the toxicant renders the organism vulnerable to other environmental forces, such as infectious disease or predation. **Direct effects** are those that do not require such intermediate effects, such as toxins that produce damage to tissues and organs, and may result in its mortality.

The difference between direct and induced effects may be a matter of level of exposure. At lower exposures an organism may be weakened and made vulnerable to predation. The same organism exposed to the same toxicant at a higher level may suffer directly. Direct effects are easier to study under laboratory conditions. Often, it is difficult to determine if an effect is direct or induced. For example, neoplasms (tumorlike growths) have been observed in bivalve mollusks exposed to oil spills. However, experimental difficulties have prevented laboratory verification of a direct relationship.

Another way to classify toxic effects is as acute or chronic, or as lethal or nonlethal. **Acute effects** are those that develop rapidly over a short time scale, usually within hours to days, and usually from single or few exposures. Acute effects are usually associated with mortality (death), but not necessarily so. **Chronic effects** are those that have a long **latency** (time between exposure and occurrence of effect). They are often the result of repeated exposures to lower doses than would produce acute effects. Because of the long latency, chronic effects are inherently more difficult to study in the laboratory. It may be necessary to maintain exposures for more than 10% of an organism's lifetime in order to observe them. **Lethal effects** are those that result in extinction of the organism. Lethality is usually taken to mean mortality but can also be prevention of reproduction. **Sublethal** effects include changes in biochemistry, physiology, histology (cell structure), reproduction, or behavior. The latter includes changes in movement or aggressiveness. Behavioral changes can affect an individual's chances of survival in the wild. Table 17.1 summarizes a variety of sublethal toxicological endpoints.

Effects can be either **reversible** or **nonreversible**, depending on whether the damage can be healed. A distinction is made between local effects and systemic effects. **Local effects** occur where the toxicant was first contacted or absorbed, whereas **systemic effects** are those that occur at other sites, therefore requiring absorption and transport. Systemic

TABLE 17.1   Factors Affected by Sublethal Toxic Exposures

Uptake, accumulation, and excretion
  Complexation and storage, distribution within tissues and organs, kinetics of uptake and release, bioconcentration, bioaccumulation.
Physiological effects
  Photosynthesis and respiration, osmoregulation, feeding and nutrition, heartbeat rate, blood circulation, body temperature, water balance.
Biochemical effects
  Metabolism of carbohydrates, lipids, and proteins; pigmentation; enzyme activities; blood chemistry; hormonal functions; oxygen uptake rate.
Behavioral effects (individual responses)
  Sensory capacity, rhythmic activities, motor activity, appetite, equilibrium, motivation and learning phenomena.
Behavioral effects (interindividual responses)
  Migration, intraspecific attraction, aggregation, aggression, predation, vulnerability, mating.
Reproduction
  Viability of eggs and sperm, breeding/mating behavior, fertilization and fertility, survival, life stages, development of young.
Genetic
  Chromosome damage, mutagenic and teratogenic effects.
Growth alterations and delays
  Cell production, body and organ weights, developmental stages (e.g., larval and juvenile stages).
Histopathological effects
  Abnormal growths, respiratory and sensory membrane changes, structural changes in tissue and organ (e.g., reproductive organs).

Source: Connell and Miller (1984).

effects may be localized in the sense that a particular organ is sensitive to a chemical. For example, the air pollutant ozone has the local effect of causing damage to the linings of the lungs, whereas the systemic toxin carbon monoxide does its damage by binding to the hemoglobin of the red blood cells, preventing them from carrying oxygen to the tissues.

Another complicating factor is that the impact of a toxin may vary significantly according to the life stage of the organism. For example, DDT is fairly nontoxic to adult vertebrates at levels found in the environment. However, it causes the shells of eggs produced by birds with high body burdens to be so thin that the egg is unlikely to survive long enough to hatch.

A toxin can also have effects at multiple levels of the biological hierarchy, from biochemical and cellular to population and ecological.

17.4.1 Biochemical and Physiological Effects

Here the discussion is not on biochemical causes of toxic effects, as discussed in Section 17.1, but on the *secondary* effects that are manifested biochemically. These are the changes in the operation of biochemical pathways that do not involve the toxin directly. They are the chemical response of the cell to toxin-induced damage to enzymes, reactants, or cell structures involved in reactions. As mentioned above, the distinction is sometimes arbitrary. For example, enzyme disruption by the complexation of metallic cofactors, or production of free radicals by oxidants, could both be considered secondary effects. Some effects are more distinctly removed from the original chemical lesion. Examples include mammalian effects such as allergic response, histamine production, hormone mimics, and blood chemistry changes.

In mammals, many of the clinical measurements developed for medical use can also be used as indicators of toxic stress. These include blood measurements, such as cell counts and clotting times, and plasma chemical analysis such as chloride, cholesterol, glucose, and protein. The advantage of using such measurements is that they could be more sensitive than gross responses such as reduced growth or mortality, they could predict long-term responses in short-term tests, and because they provide insight into the mechanism of toxic action.

Many of the measurements developed for mammals have been applied to other animals, such as fish. However, the correlation between this type of data and environmental impacts has not been sufficiently researched to be of broad use. Several biochemical measurements have been developed specifically for fish. These include metabolic activity, as indicated by oxygen consumption. For example, fish exposed to dieldrin required more oxygen to swim against a current and had a reduced cruising speed.

17.4.2   Genotoxicity

Other irreversible changes include various forms of genotoxicity, or genetic damage. These can cause three types of effects:

- **Mutagenesis** is the formation of hereditable genetic changes.
- **Teratogenesis** is the production of birth defects.
- **Carcinogenesis** is the production of cancer.

17.4.3   Mutagenesis

Mutations were described in Section 6.2.3. Agents that can cause mutations are called **mutagens**. Mutations can result in hereditable diseases, birth defects, or cancer. According to one study, 90% of carcinogens show mutagenic activity in a microbial assay (the Ames test, Section 20.1.12). The same study found that few noncarcinogens are mutagenic. The correlation between mutagenicity in laboratory tests and that in humans is highly uncertain. Nevertheless, prudence dictates that substances known to be mutagenic in appropriate laboratory models should be presumed to present a high risk to humans as well.

Other types of genetic damage besides mutations can occur, including chromosomal alterations such as broken chromosomes or change in number of chromosomes, or inexact copying of the DNA during normal cellular replication. All of these can be caused by toxic agents. Their results are typically more serious than mutations.

17.4.4   Teratogenesis

Birth defects are abnormal developments in embryos that are manifested either structurally or functionally. Teratogenic agents act in a narrow range of doses between no observable effect levels and levels that are lethal to the embryo. Furthermore, they are preferentially induced when the exposure occurs during a narrow time span during gestation, which corresponds to the time when organs are being formed. For humans this is

TABLE 17.2   Teratogenic Substances

Physical agents	Hypothermia, hyperthermia, hypoxia, radiation
Agents causing hypoxia	Carbon monoxide, carbon dioxide
Infections	Rubella viruses, syphilis
Dietary deficiency or excess	Vitamins A, D, and E, ascorbic acid, nicotinaminde, Zn, Mn, Mg, Co
Hormone deficiency or excess	Cortisone, insulin, androgens, estrogens
Natural toxins	Aflatoxin B_1, nicotine
Heavy metals	Methyl mercury, lead, thallium, strontium, selenium
Solvents	Benzene, carbon tetrachloride, 1,1-dichloroethane, dimethyl sulfoxide, propylene glycol, xylene
Pesticides	Insecticides, herbicides, fungicides
Other	Azo dyes, antibiotics, sulfonamides, drugs (caffeine)

Source: Lu (1991).

approximately from day 26 to day 56 of gestation. The same dosage can cause different birth defects, depending on when during this period the dose is administered. For example, a cleft palate can be caused by exposure on one day, whereas heart defects may be caused on another.

Although teratogenesis is classified here with genotoxic effects, its causes are not well understood, and a number of other mechanisms probably also act. It seems that anything that interferes with cell division can be a teratogen. This includes agents known to block DNA expression, heavy metals such as lead and cadmium that inhibit enzymes, and even substances that simply cause a delay in cell replication.

Other forms of toxicity do not correlate well with teratogenicity. Substances that are teratogenic may not be toxic to the mother, and vice versa. Surprisingly, many mutagens are not teratogens. The sensitivity of the embryo can be decreased if the toxic substance does not easily cross the placental barrier. On the other hand, the embryo may be more sensitive than the mother to some compounds if the compounds act on cell division, since embryos undergo this process at a high rate. Both deficiencies and excesses of some vitamins, such as vitamin A, are teratogenic. Other teratogens are listed in Table 17.2.

17.4.5   Carcinogenesis

Cancer is "possibly the most dreaded toxic event and probably the hardest for which to provide reassuring safety precautions" (Williams and Burson, 1985). Normal cells respond to the presence of adjacent cells by ceasing to replicate. In cancer, the mechanisms limiting the growth of cells become damaged. The damaged cells grow uncontrollably, forming tissue masses called **tumors** (except in some cases, such as leukemia, the cancer of the white blood cells). This growth consumes the resources of the organism; impinges on nearby tissues, causing them to atrophy; and ultimately results in mortality. A great amount is understood about the causes and progresson of cancer, yet we are still far from having a complete understanding, or even from knowing enough to treat the disease effectively.

The Stages of Cancer   Cancer progresses in distinct stages. The first step, which produces no symptoms, is a first mutation that predisposes the cell to cancer. This step is called **initiation**. In the second step, **promotion**, the first clinical manifestations begin with the formation of benign tumors. Finally, in **progression,** the tumors become malignant, which in turn can spread to other tissues to form secondary tumors. In the nomenclature of cancer, the suffix **-oma** is appended to a tissue name to denote a benign tumor; for example, hepatoma and osteoma are benign tumors of the liver and bone, respectively. Malignant tumors are described using either **carcinoma** or **sarcoma**, for mesothelial or epithelial tumors, respectively. Thus, names for the malignant tumors for liver and bone are hepatocellular carcinoma and osteosarcoma.

Benign tumors exhibit cellular differentiation and grow by expansion, causing adjacent tissues to atrophy. The tumor shows some differentiation and has a clear boundary. Benign tumor cells appear similar to normal cells under microscopic examination. Benign tumors are usually not fatal, except when they impinge on critical tissues such as in the brain. Brain tumors rarely become malignant because they are fatal before reaching that stage. Benign tumors do not inevitably progress to the next stage, although their clinical removal obviates that possibility.

Malignant tumors are undifferentiated. Instead of forming a discrete bounded structure as benign tumors do, they grow invasively into neighboring tissues. Malignant cells appear obviously deranged. They are capable of the process of **metastasis**, in which clumps of malignant cells migrate to other tissues through blood and lymph vessels, forming secondary tumors. This rapidly increases the growth of cancerous tissue and accelerates the progression of clinical symptoms, especially weakness, a large amount of weight loss, loss of various bodily functions, and pain.

What accounts for the observed stages of cancer? It is known that genetic damage is the root cause of cancer. The damage can be either mutations or aberrations. The latter include chromosome breakage, deletion of chromosome segments, or swapping of segments between chromosomes. Other evidence, however, suggested involvement of nongenotoxic agents. For example, when polynuclear aromatic hydrocarbons (PAHs) are applied to the skin of a mouse, cancer does not occur until followed by application of another chemical, such as phorbol esters from croton oil. It does not even matter if the second application is delayed for up to a year. Clearly, the PAH predisposes the skin cells to cancer, and the esters stimulate progression to other stages. Furthermore, the compounds that predispose were often found to be mutagenic, whereas the ones that only stimulate progression often were not.

Types of Carcinogens   The knowledge that carcinogens act by different mechanisms led to a distinction in two types of carcinogens. The first type are called **genotoxic carcinogens**, which act either themselves or via metabolites to either damage DNA directly or impair the processes of repair or transcription. This is initiation, as defined above, and the chemicals are called **initiators**. Examples include nitrosamines, epoxides, and metals such as cadmium, chromium, or nickel. The direct-acting genotoxins are often electrophilic compounds that bind to DNA, similar to the action of mutagens. Others must be biotransformed to be genotoxic and are called **precarcinogens**. Most genotoxic environmental pollutants are in this category, including chlorinated hydrocarbons, aromatics such as benzene, and PAHs. The mechanism for carcinogenic metals, such as arsenic, chromium, and nickel, is not understood. They are thought to impair DNA replication or transcription by complexing with the DNA or associated proteins. Several nonchemical carcinogens act by changing the cellular DNA and therefore may be classified as genotoxic. These include ionizing radiation and certain viruses.

The second type are called **epigenetic carcinogens** or **promoters**. They do not affect the DNA, but enhance the progression to cancer subsequent to initiation of genetic damage by the genotoxic carcinogen. Epigenetic carcinogens act by (1) encouraging cell division (promotion), (2) inhibiting intercellular communication, or (3) impeding mechanisms for destroying aberrant cells. Tobacco smoke contains both initiators and promoters.

Destruction of damaged cells is part of the function of the immune system, and immunosuppressant drugs used in organ transplants are known to cause cancer by inhibiting this function. One important group of pollutants that may act in this way is the dioxins. They seem not to be genotoxic, but are both strong promoters and highly immunosuppressive. This is especially true for the form known as 2,3,7,8-tetrachlorodibenzodioxin (2,3,7,8-TCDD).

Intercellular communication normally helps limit cell growth. Cells send each other chemical signals that stop their division process when they are in contact. Cancer cells do not respond to these signals.

Many environmental pollutants are promoters, not initiators. Cell division can be stimulated a number of ways. Anything that kills or damages cells stimulates growth as part of the healing process. This can include chemical toxins which act by other means, and physical trauma such as burns, freezing, or possibly even mechanical injury. Even implanted foreign solid materials, such as asbestos, plastics, metal, and glass, can promote cancer. These are called **solid-state carcinogens**. A possible mechanism for this is that their presence stimulates fibrosis, connective tissue cell growth, as the organism attempts to encapsulate the foreign material. The more cell growth is stimulated, the greater the chance that any cell, previously initiated by a genotoxic carcinogen, will be activated; and the greater the chance of a transcription error causing an initiation.

Hormones are known to act as promoters. For example, estrogen administered to menopausal females increases the risk of endometrial cancer. The synthetic hormone diethylstilbestrol (DES) used to be given to pregnant women with high miscarriage risk to improve their chances of carrying the pregnancy to full term. Tragically, it has been found that daughters produced by those pregnancies are at a high risk of contracting cervical cancer in their late teens or early 20s. Testosterone, or more precisely its metabolite (*dihydrotestosterone*), promotes prostate cancer in men.

More important in an environmental context, many pollutants have been found to either mimic or influence hormones. The herbicide Amitrole (aminotriazole) inhibits an enzyme that uses iodine to form thyroxine. The pituitary responds to the low level of thyroxine by stimulating thyroid growth. This in turn, can lead to cancer of the thyroid.

A variety of anthropogenic compounds found in nature have been shown to mimic the hormone estrogen: **xenoestrogens**, **endocrine disruptors**, or **environmental estrogens**. These include the chlorinated pesticides atrazine, chlordane, DDT, endosulfan, kepone, and methoxychlor, as well as dioxins and some polychlorinated biphenyl (PCB) congeners. Several compounds associated with plastics are xenoestrogens as well: Bisphenol A is released by polycarbonates when heated. Nonylphenol is a softener for plastics used in packaging and in flexible plastic tubing. Phthalates are also plastic softeners that are commonly used in food packaging and which have been found in laboratory experiments to cause reproductive disorders.The xenoestrogenic properties of some of these materials were discovered when laboratory investigations were confounded by their presense. In one case, cultures of breast cancer cells grew more rapidly than expected because of contamination from laboratory plasticware. Possible xenoestrogenic effects have been observed in the environment as well. Male fish living near municipal sewer outlets were found to be producing a protein associated with females. Alligators hatched in Lake Apopka, Florida, following a spill of organochlorine insecticides had altered hormone levels and abnormally small penises.

Another type of epigenetic carcinogen is the **cocarcinogens**: These increase the concentration of an initiator by affecting absorption, biotransformation, or detoxification. For example, they may decrease detoxification by inhibiting enzymes or depleting detoxification substrates such as glutathione. Ferric oxide and asbestos may facilitate cellular uptake of genotoxics.

The distinction between genotoxic and epigenetic carcinogenesis may have significance in the risk assessment process. Specifically, the presence or absence of a toxic threshold, a level of exposure below which essentially no effect is found, may depend on the mechanism. This is discussed in Section 19.4.1.

Genetic Basis of Cancer   A fairly detailed understanding of the causes of cancer progression is starting to emerge. Early in the twentieth century, the microscopic observation that tumors tended to have damaged chromosomes, plus the fact that susceptibility to some cancers could be inherited, led to the realization that genetic damage was the root cause of all cancers. In addition, the fact that daughters of a cancerous cell would continue to be cancerous if transplanted to another organism led to the **one-hit hypothesis**: Malignant cancer was caused by a single mutation to a critical gene.

Eventually, the one-hit hypothesis gave way to other facts. If a single gene caused cancer, and it was inherited, the cancer should develop immediately after birth. However, even in cases of inherited susceptibility, inception of cancer still takes some years. This implies that at least two changes are required, leading to the **two-hit hypothesis**. The fact that some cancers go through a series of stages with increasing virulence, and daughter cells from each stage maintain their characteristics when transplanted, imply that for such cancers even more than two mutations occur.

In recent years, the tools of genetic engineering have revealed many of the detailed steps involved for some cancers. The newer discoveries started with the examination of viruses that cause cancer. It was found that the viruses carried a mutated human gene that they inserted into the infected cell. The normal human gene was called a **proto-oncogene**, which is responsible for stimulating cell growth during gestation. Normally, this gene is turned off in adults. The proto-oncogene can be mutated into a more active form, called an **oncogene**. The oncogene sends growth signals despite the presence of signals that would tell other cells to stop. Thus, a single mutation is all that is necessary for the change to be expressed. This is called a *dominant mutation* or an *activated mutation*. Recall that a cell contains two copies of each gene. Only one mutation is necessary if the resulting gene produces a protein that causes the disease. However, although the presence of one or more oncogenes facilitates a cell's transformation into a cancer cell, by itself it is not potent enough to produce a malignancy.

Other genes, called **tumor suppressor genes**, code for proteins that inhibit cell replication. To produce cancer by damaging these genes, it would be necessary to mutate or delete both alleles, or else the undamaged copy would continue to produce the protein. This supported the two-hit hypothesis.

The further progression of changes was found by examining tumors at different stages and detecting genetic changes present. For example, colon tumors were studied because the malignant tumor could sometimes be found next to the benign tumor, or polyp, from which it had developed. The malignant tumor would always have all the mutations found in the benign tumor, plus additional ones. The results supported the theory of **clonal evolution** of cancer, which states that cancer progresses from a single cell that receives a single hit enabling it to replicate with less inhibition than normal cells; later, one of its daughter cells receives a second hit, which makes it replicate faster; and so on until enough hits occur to produce a malignant cell. Because each mutation causes the cells to replicate faster, the chance of further mutations is enhanced. The more replications, the greater the chance of an error.

In the case of colon cancer, the first hit was to a gene called *APC*, whose function is unknown. Then an oncogene called *ras* and two tumor suppressor genes, *p53* and *DCC*, are mutated. The order of the last three apparently does not matter, although the *ras* mutation seems to occur first most often.

The gene *p53* seems to be responsible for slowing cell division when DNA damage has occurred, presumably to allow time for DNA repair. Irradiation of a cell by ultraviolet light increases the amount of protein coded for by *p53*. It also may cause cell death when damage is too great. This can prevent mutations from being passed on. As a result, mutations to *p53* can reduce these protective actions.

It may not be evident from this discussion that mutations and transcription errors are rare events. Cancer is common in humans only because cell growth and occasions for DNA repair occur with high frequency. Even without genotoxic chemicals in the environment, natural or not, radiation from cosmic rays and natural radionuclides in the body is probably responsible for the great majority of human cancers. It is estimated that each cell in our bodies must make tens of thousands of DNA repairs per day as a result of this radiation. Each human has up to 10^{14} cells. About 25% of all humans contract cancer in their lifetimes. Even if we assume that all of these cancers are caused by radiation damage and a 70-year lifetime, this would indicate that the probability of a single DNA repair resulting in cancer is about 10^{-25}! This provides no consolation since the great number of repair events produces such a high risk at the individual level.

Several other factors at the genetic level reduce the risk of cancer. One is that the number of our 25,000 genes that are proto-oncogenes or tumor suppressor genes is very small. Also, many of the cellular mutations probably result in the death of the cell. Finally, a large portion of the DNA actually consists of introns, which are apparently not expressed.

Classification of Carcinogens   Several U.S. federal agencies have developed classification systems for carcinogens based on the strength of the supporting evidence. For example, the U.S. EPA adopted the following classification system in 1986:

Group A: *Human Carcinogen*: Chemicals are placed in this group only if there is sufficient evidence from epidemiological studies to support a causal relationship between exposure to the chemical and cancer in humans.

Group B: *Probable Human Carcinogen*: This includes two subgroups:

Group B1: Agents for which there is "limited" epidemiological evidence for carcinogenicity in humans.

Group B2: Agents for which there is 'sufficient' evidence of carcinogenicity in animals but "inadequate evidence" or "no data" from epidemiological studies.

Group C: *Possible Human Carcinogen*: Includes agents with "limited" evidence of carcinogenicity in animals. Such limited evidence would include marginally statistically significant findings or finding of benign tumors caused by substances that are not mutagenic.

Group D: *Not Classifiable as to Human Carcinogenicity*: Includes agents for which there is inadequate evidence for human or animal carcinogenicity.

Group E: *Evidence of Noncarcinogenicity for Humans*: No evidence of carcinogenicity was found in at least two adequate animal tests in difference species or in both one animal test and one epidemiological study.

The International Agency for Research on Cancer (IARC) has a similar classification scheme used by many countries. Appendix C lists some substances classified as carcinogens by the U.S. EPA, along with their classification and levels of evidence. Group A is often called *known human carcinogens*. Only about 50 chemicals or substances have sufficient evidence associated with them to be placed in this classification. About 13 of these are of significant environmental concern:

Arsenic	Hexavalent chromium	Plutonium-239
Asbestos	Diethylstilbestrol	Radium-226
Benzene	2-Naphthylamine	Radon-222
Benzidine	Nickel	Vinyl chloride
Bis(chloromethyl) ether		

In 1996 the U.S. EPA proposed a new set of guidelines for carcinogenic risk assessment in which they propose replacing the foregoing categories by three descriptors: *known/likely*, *cannot be determined*, and *not likely*.

17.4.6   Histological Effects

Histological effects are those observable in structures and functions at the cell and tissue level by microscopic examination. The gross effects of toxicity observable at the individual level are almost always preceded by histological or biochemical effects. Thus, these effects are more sensitive indicators of toxicity than mortality or behavioral changes. Table 17.3 lists some histological changes that can be caused by toxic substances.

Inflammation is a response of connective tissues to repair and isolate, or "wall off," damage. Special cells release *histamine*, *heparin*, potassium, and *prostaglandins*. These cause nearby blood vessles to dilate, making the area appear reddened and to feel warm. Phagocytes and macrophages are attracted to remove pathogens and cell debris. The outcome of inflamed tissues may be repair, chronic inflammation, or necrosis, depending on how successful the organism is in repairing the damage. Chronic inflammation results from the spread of whatever caused the original damage. Fibrosis often results as the organism attempts to isolate the irritant and replace lost cells. An organism can die from excessive inflammatory reponse. For example, damage to lungs or gills from fibrosis can reduce oxygen transport in those organs.

Necrosis results at a sufficient level of cellular damage. Degenerative changes leading to necrosis are often the result of the inability of the cell to maintain ionic balance across intracellular and extracellular membranes. This, in turn, can be caused by loss of ATP production. Direct damage to intracellular membranes can result in their rupture. Rupture

TABLE 17.3   Summary of Some Histological Effects

Anemia	Loss of oxygen transport capacity of the blood
Atrophy	Wasting of tissues
Dysplasia	Abnormal change in size or shape of cells or tissues
Edema	Excessive water collecting in body cavities
Erythema	Redness of the skin caused by inflammation
Fibrosis	Production of new fibrous connective tissue (scar tissue) as a result of inflammation or to replace necrotic cells
Hemorrhage	Presence of blood outside the vessels due to injury
Inflammation	Process in which cells exude substances as part of a repair process; exudate may include blood serum, fibrin, pus, erythrocytes, mucous, or lymphocytes
Lesion	Any abnormal change to a cell, tissue, or organ
Necrosis	Cell death within a living organism
Neoplasm	Tumor, normal cells that have been transformed into a form with uncontrolled growth that may or may not be malignant
Vasodilation	Enlargement of blood vessels

Source: Rand and Petrocelli (1985).

of lysosomes releases hydrolytic enzymes to the cytoplasm, resulting in further damage. Unless cell death occurs very rapidly, it is usually accompanied by some inflammatory response. Epithelial and connective tissues can **regenerate**, or repair damage, relatively well. Smooth and skeletal muscle can also regenerate, although not as well. Cardiac muscle and nervous tissue cannot regenerate at all.

Tumors or neoplasms result from changes in the genetic and nuclear mechanisms for cell growth and reproduction. Their cells usually resemble the cells from which they derived but lack the normal structure or function of the original tissues. Tumors may be either benign or malignant, as described in Section 17.4.5.

17.4.7   Effects on Particular Organs or Organ Systems

The discussion in this section concerns effects in organs or organ systems of vertebrate animals. Within that group the emphasis is, of course, on humans.

As mentioned above, particular toxins will tend to target particular organs or organ systems. The liver and the kidney are common targets of toxic activity because of their role in detoxification and their large blood flow. The skin and eyes, lungs, and digestive tract are vulnerable to the more reactive toxicants, as they are the sites of first entry to the organism. The nervous system has both unique protection and vulnerability.

Liver   The liver receives almost all the venous blood flow prior to its return to the heart and lungs. All of the blood from the stomach and intestines go directly to the liver. Thus, substances absorbed in the GI tract may be biotransformed before reaching other organs. Although the liver detoxifies most compounds, some are converted to more toxic forms. As the generator of most of the bioactivated compounds (see Section 18.5), the liver is also the site of first and most concentrated contact. The liver is the main site of toxic damage for a number of chlorinated organics, such as carbon tetrachloride, trichloroethylene, PCBs, and lindane.

Hepatic (pertaining to the liver) damage can occur in a number of ways. **Fatty liver** is an accumulation of lipid globules inside the liver cells and is caused by several mechanisms that impair the release of triglycerides to the blood. Carbon tetrachloride and ethanol are among the substances that can cause this. Necrosis is caused by carbon tetra-chloride, which forms free radicals in the liver, as well as by other halogenated hydrocar-bons. **Cirrhosis** is the formation of scar tissue in the liver. It is also caused by carbon tetrachloride, although ethanol is most commonly associated with this condition. Although there is evidence to the contrary, the effect of ethanol may be related to nutri-tional deficiency associated with alcoholism. **Cholestasis** is an inflammation of the ducts carrying bile or a decrease in bile flow by other mechanisms. There are many types of liver cancer, and many chemicals are known to cause cancer in laboratory animals. The role of chemicals in human liver cancer is less clear, except for the notable case of vinyl chloride, which is known as a potent cause of angiosarcoma.

Carbon tetrachloride may act in several ways. Besides forming radicals, it can also be converted to phosgene. It also has an indirect effect. It causes a large release of the hormone epinephrine by sympathetic nerves. As with other hormones, epinephrine is quickly broken down after performing its function. This takes place in the liver, and the high levels caused by carbon tetrachloride can result in liver damage.

Liver damage can be detected clinically by several functional tests, such as the rate at which it clears certain injected dyes or *bilirubin*, a chemical produced normally by the breakdown of heme from red blood cells. More sensitive than the liver function tests are tests for certain enzymes. Often when the excretory capability of the liver is impaired, the effects are visible as **jaundice**, the yellowing of the skin and the whites of the eyes caused by the accumulation of bilirubin.

Kidney   A toxin that affects the kidney is called a **nephrotoxin**. The kidney is very sensi-tive to toxins, both because of its high rate of **perfusion** (blood flow) and because it is an important site of biotransformation. This may be somewhat of a surprise because the kidney is usually thought of exclusively as an excretory organ. In fact, it has several other important metabolic and physiologic functions. The major functions are (1) removal of waste products from the blood; (2) regulation of red blood cell levels in response to oxygen levels; (3) regulation of blood pressure, volume, electrolytes, and pH; and (4) metabolism of calcium and vitamin D. Thus, kidney damage can be manifested in diverse ways. However, toxic effects on the excretion function are the most commonly observed.

Elimination in the kidney occurs by three processes: (1) glomerular filtration, (2) tubular resorption, and (3) tubular secretion. Most nephrotoxins seem first to attack the proximal tubule of the nephron. The glomerulus is also susceptible, and the entire nephron may be affected if exposure is high enough. For example, mercury reduces active resorption of sodium in the proximal tubule. This reduces the passive resorption of water, increasing urine production initially. Subsequently, glomerular damage shuts down urine production completely, allowing waste products to accumulate dangerously in the blood. If recovery occurs, another period of high urine production occurs. The entire cycle may take several months. Lead affects the upper part of the proximal tubule, where glucose and amino acid resorption occurs. Thus, this damage can be observed by detecting these compounds in the urine. Although an important effect of carbon tetrachloride is severe hepatic necrosis, if it causes mortality it is usually a result of kidney failure.

Kidney damage can also be detected by changes in blood chemistry. Blood urea nitrogen (BUN), a product of protein breakdown, and creatinine, a metabolite of creatine, are excreted by glomerular filtration. Damage to the glomerus reduces the filtration rate, causing these substances to accumulate in the blood, where they can be detected. The normal kidney allows only low-molar-mass proteins to pass into the urine; therefore, elevated high-molar-mass protein content in the urine indicates damage.

The kidney uses a lot of oxygen, making it sensitive to **ischemia** (lack of oxygen due to reduced blood flow). The kidney is the organ most sensitive to cadmium and is also harmed by inorganic divalent mercury, lead, and other metals, as well as chlorinated hydrocarbons. Glycol metabolizes to oxalic acid, which forms salt deposits in the tubules as calcium oxalate. Rhubarb leaves have sufficiently high oxalic acid levels to cause this problem when eaten. Some chemicals, including arsine gas, benzene, lead, methyl chloride, napththalene, trinitrotoluene, and analine dyes, cause hemolysis of red blood cells. This releases pigments such as hemoglobin, which in turn are associated with acute renal (kidney) failure.

Skin and Eye   The stratum corneum (see Section 9.1) is the main barrier to chemical toxins. Chemicals that penetrate it tend to be absorbed systemically. The skin does not just passively transport contaminants. It is capable of metabolizing them. One of the earliest known chemical causes of cancer in humans is that of the skin caused by the topical application of oils containing polynuclear aromatic hydrocarbons (PAHs). The skin biotransforms PAHs into compounds more carcinogenic than the original. Many petroleum-based or coal-tar materials are photoactive; that is, sunlight increases their toxicity. Biotransformation of toxins may take place in the epidermis or the dermis.

The skin suffers toxic effects itself, including cancer, primary irritation, allergic reactions, hair loss, pigment disturbances, ulceration, and chloracne. **Dermatitis** is an inflammation of the dermis. Irritant contact dermatitis and allergic dermatitis can both be caused by exposure to chemicals and produce similar symptoms, including hives, rashes, blistering, eczema, or skin thickening. The difference between them is that a true allergy takes time to develop, typically at least two weeks; whereas irritation does not require a previous exposure. For example, no one reacts to poison ivy when first exposed. Only after a second or subsequent exposure does the itchy rash develop.

Common causes of ulceration include acids and burns. In addition, contact with cement and chromium-containing materials are well known to cause skin ulcers. The latter includes leathers that have been tanned with chromium compounds. **Chloracne** is an disease characterized by acute formation of an acnelike skin rash. It is caused by halogenated hydrocarbons, especially polychlorinated polyaromatic hydrocarbons. Chloracne is a classic symptom of dioxin poisoning.

Ultraviolet light from the sun is a form of ionizing radiation that can cause skin cancer, or **melanoma**. The basal cells and melanocytes are vulnerable to UV. It is feared that the destruction of stratospheric ozone by substances such as chlorofluorocarbons will result in increased UV radiation at the ground level, leading to increased melanoma. Such an increase has already been observed in Australia, which has also experienced declines in stratospheric ozone.

The *eye* is vulnerable to irritants such as smog, solvents, detergents, and corrosive substances. Other pollutants act systemically and can damage the optic nerve. For example, methanol and carbon disulfide damage the central vision in this way, and pentavalent arsenic and carbon monoxide affect peripheral vision.

Lungs   The lungs are vulnerable to a wide variety of chemical irritants, such as ozone, chlorine, ammonia, and components of photochemical smog. Damage can be either acute or chronic, depending on the level of exposure. Acute effects include bronchial constriction and **pulmonary edema** (accumulation of fluid in the airways). Arsenical compounds can cause irritation, but chronic exposure can result in lung cancer.

Some inhaled solvents, such as perchloroethyene and xylene, are transported to the liver and biotransformed. The resulting metabolites then return to the lungs, where they can cause cell damage and edema.

Many types of particles also harm the lungs, including smoke from cigarettes or other combustion sources, or dusts from industrial operations producing particles of asbestos, silicates, coal or even cotton, flax, or hemp. In the disease called **silicosis**, particles of certain crystalline forms of silica are engulfed by macrophages in the lungs, which then attempt to sequester the particles in lysosomes. However, the particles rupture the lysosome membranes. This releases the lysosome enzymes into the cytoplasm and destroys the cell, as well as causing damage to the lung tissue. In addition, the particles are released to continue the cycle of damage. Ultimately, fibrosis results, making breathing more difficult. In late stages the heart is affected, leading to congestive heart failure.

Asbestos particles also cause fibrosis in the lungs, which can lead to lung cancer. Many other particles cause fibrosis, including coal dust, kaolin, talc, and a number of metal or metal oxide particles. Another result of chronic damage is **emphysema**, in which the walls separating one alveolus from another are destroyed.

The mucociliary escalator is responsible for removing many of the particles trapped in the bronchial tubes, including infectious microorganisms. However, some toxic substances, notably tobacco smoke, paralyze it for 20 to 40 minutes. Mucus stagnates, but the irritation actually increases mucus secretion. These effects can partially or totally block smaller bronchi and exposes the habitual smoker to the possible indirect effects of lower respiratory tract infections and chronic **bronchitis** (inflammation of the bronchi). The inhaled irritants described above can have a similar effect.

Occupational exposure to a variety of substances is known to be capable of causing **asthma**. This is an allergic reaction in which exposure causes histamine to be released. Histamine stimulates the bronchi to contract, greatly increasing breathing resistance. This is known to affect bakers exposed to flour and workers exposed to wood dust, as well as butchers exposed to fumes caused by heat-sealing PVC films for wrapping meat. Some people become sensitized to toluene diisocyanate, which is used in polyurethane products. Subsequent exposures to very small amounts can cause a severe asthma attack.

Other exposures cause a different allergic reaction, deeper in the lungs at the bronchiolar level leading into the alveoli. The symptoms, which include coughing, sputum production, fever, and fatigue, resemble pneumonia. A fraction of the people affected slowly develop shortness of breath without fever. Both of these often result in misdiagnosis. This condition has been found in farmers exposed to thermophilic actinomycete spores (farmer's lung), workers exposed to bird droppings, laboratory technicians who become sensitive to the urine of experimental rats, and strangely, archeologists who remove wrappings from Egyptian mummies.

Tests on the lungs can be used to detect damage to exposed persons or to form a baseline for workers who are at risk of exposure. The chest x-ray is one such test. A fairly simple, noninvasive test is the **spirogram**, in which the subject breathes through a device that measures volume and flow (see Figure 9.11). Measurements include the **forced vital capacity** (FVC), which is the maximum volume the person can exhale, and the flow rate over certain periods of the exhalation. **Respiratory frequency**, commonly called *breathing rate*, is sensitive to certain irritants and correlated to their concentration. Ozone and NO_2 increase the frequency, whereas SO_2 and formaldehyde decrease it.

Immune System   The immune system consists of various organs, including the bone marrow, spleen, thymus, and lymph nodes, plus specialized cells and plasma proteins produced by those organs which circulate in the blood and lymphatic system. The cells are the lymphocytes, or white blood cells, and include T cells and B cells. The proteins include the immunoglobulins, interleukins, and the complement system. Together they act to rid the body of chemical and biological contaminants. There are three types of immune system derangements. **Immunosuppressants** reduce immune response and render the body more vulnerable to foreign substances. **Immunostimulants** cause hypersensitivity reactions or allergies. **Autoimmune disease** is the condition where the immune system attacks its own substances as if they were foreign.

Many pollutants are immunosuppresive. For example, PCBs are thought to reduce production of immunoglobulins. Others include:

- Heavy metals: lead, cadmium, chromium, methyl mercury, sodium arsenite and arsenate, arsenic trioxide.
- Pesticides: DDT, dieldrin, carbaryl, carbofuran, methylparathion, maneb, chlordane, hexachlorobenzene.
- Halogenated hydrocarbons: PCB, polybrominated biphenyls, TCDD, TCE, chloroform, pentachlorophenol.
- Others: benzo[*a*]pyrene, DES, benzene.

Immunostimulants usually cause their reaction within 15 minutes of exposure. A first exposure does not cause a reaction since the immune system must generate immunoglobulin antibodies. Second and later exposures cause the release of histamine, heparin, serotonin, prostaglandins, and other compounds. These result in symptoms such as asthma and rhinitis. Examples of immunostimulant toxins include nickel, beryllium, and the pesticide pyrethrum. A delayed hypersensitivity reaction may also appear between 12 and 48 hours after exposure to substances such as nickel, chromium, and formaldehyde. Symptoms include contact dermatitis. Autoimmune diseases are reportedly caused by agents such as dieldrin, gold, and mercury.

Nervous System   The nervous system is exceedingly sensitive, partly because its role is so critical in higher animals. It also is relatively vulnerable to ischemia because it has a high metabolic rate and a low capacity for anaerobic metabolism. Thus, it is the first site of action for carbon monoxide. The central nervous system also has special protection in the blood–brain barrier (see Section 18.4). Peripheral nerves have a similar, although less effective barrier.

Here we will mention three main types of neurotoxic effects:

- Physical damage to neurons, including the myelin covering
- Disruption of signal transmission by individual neurons, such as by changing the permeability of the neuron membrane to ions, which includes stimulants and depressants
- Interference with signal transmission at the synapses between neurons or between neurons and muscles

TABLE 17.4 Toxins That Damage the Nervous System Physically

Demyelinating Toxins	Peripheral Motor Toxins	Brain Toxins
Acetyl tetramethyl tetralin	Acrylamide	Acetylpyridine
Bicycloheanone oxalydihydrazone	Arsenic	DDT
Chronic carbon monoxide	Azide	Mercury
Chronic cyanide	Bromophenylacetyluria	Manganese
Cyanate	Carbon disulfide	
Diphtheria toxin	Chlorodinitrobenzene	
Ethidium dibromide	Cyanoacetate	
Ethylnitrosourea	Diisopropyl fluorophosphate	
Hexachlorophene	Dinitrobenzene	
Isoniazid	Dinitrotoluene	
Lead	Disulfiram	
Lysolecithin	Doxorubicin	
Pyrithiamine	Ethambutol	
Salicylanilides	Ethylene glycol	
Tellurium	Formate	
Thallium	Hexane and 2,5-hexanedione	
Triethyltin	Iminodipropronitrile	
	Iodoform	
	Methanol	
	Methyl-N-butyl ketone and 2,5-hexanediol	
	Methyl mercury	
	Perhexilene	
	Phosphorus	
	Tetraethyllead	
	Triorthocresyl phosphate	
	Vincristine	

Source: Williams and Burson (1985).

Physical damage may target either the long axons by which neurons transmit signals over distance, or the cells of peripheral nerves or the brain (Table 17.4). Lead, especially organic lead such as tetraethyllead, mercury, organic arsenicals, chronic alcoholism, and hexachlorophene intoxication, are all associated with edema (extracellular fluid) in the brain.

A toxin that affects neural transmission is *Saxitoxin* (from dinoflagelates), which prevents sodium from entering the neuron as the signal passes. *Pyrethrins* (natural and synthetic insecticides) and DDT increase sodium permeability, causing repeated firing of the neuron, resulting in convulsant activity. Some compounds, such as lead, triethyltin, cyanate, hexachlorophene, chronic carbon monoxide, and isoniazid, decrease the myelin insulation on axons. If the brain is affected, symptoms include mental dullness, restlessness, muscle tremor, convulsions, memory loss, and epilepsy.

The category of interference with transmission includes stimulants and depressants. **Stimulants** increase the excitability of neurons. They include the rat poison strychnine as well as the popular *xanthines*. The latter include *caffeine*, found in coffee and tea, and *theophylline* and *theobromine* (both found in chocolate) (Table 17.5). Xanthines prevent the breakdown of cyclic adenosine monophosphate (cAMP), which is connected with the regulation of the active transport of sodium and potassium across the neuronal

TABLE 17.5   Caffeine Amounts in Popular Beverages

	Milligrams
Coffee	
5-oz cup, drip method	146
5-oz cup, percolator method	110
5-oz cup, instant	53
Tea	
5-oz cup, brewed 1 minute	9–33
5-oz cup, brewed 3–5 minutes	20–50
Cocoa and chocolate	
1 oz milk chocolate	6
1 oz baking chocolate	35
Soft drinks	
12 oz Mountain Dew	52
12 oz Pepsi Cola	37
12 oz Coca-Cola	34
Nonprescription drugs (standard dosage)	
NoDoz	200
Excedrin	132
Anacin	64

Source: Simpson and Ogorzaly (1995); original *Consumer Report*, 1981, Vol. 46, pp. 598–599.

membrane. Caffeine mimics adrenaline in its action, stimulating the heart, increasing stomach acidity and urine output, and increasing the metabolic rate by 10%.

Depressants act by decreasing the excitability of neurons. Ethanol is known to depress neural excitability by decreasing sodium and potassium conductance. However, the most important from an environmental point of view are the agents that cause **narcosis**, or central nervous system depression. This is one of the most common toxic effects and is caused by solvent partitioning of lipophilic solvents to plasma membranes. The mechanism is not well understood but may be related to a decrease in ion fluxes through the membrane. The symptoms include disorientation, giddiness, and euphoria; at higher levels the symptoms progress to paralysis, unconsciousness, and death.

This property of solvents accounts for their use as anaesthetics. Ethyl ether was one of the earliest applied this way. Chloroform came into common use because its vapors are not explosive. However, its use for this purpose has been discontinued because of other toxic effects, including its carcinogenicity. The potency of these solvents as an anaesthetic, along with their toxic effects, is well correlated with their polarity as measured by the octanol–water partition coefficient, K_{OW}. Octanol is a surrogate for biological lipids, so K_{OW} also measures the tendency of a substance to partition into the plasma membrane.

Synaptic transmission can be affected by a toxin binding with receptors (e.g., the toxin produced by *Clostridium tetani* in tetanus), by blocking neurotransmitter release (botulinum toxin produced by *Clostridium botulinum*) or by inhibiting cholinesterases (organophosphate or carbamate pesticides; Section 17.1).

Reproductive System   The reproductive system suffers effects independent of those that directly affect offspring. The testes are protected by a barrier similar to, although less

TABLE 17.6 Male Reproductive System Toxins

Physical agents	Organic chemicals
Microwaves	Alkylating agents
Ionizing radiation	Anaesthetic gases
High temperatures	Carbon disulfide
Social habits	Chloroprene
Alcohol ingestion	Vinyl chloride
Cigarette smoking	Chemotherapeutic agents
Marijuana smoking	Dibromochloropropane
Metals	Kepone
Lead	DDT
Cadmium	Female oral contraceptives

Source: Williams and Burson (1985).

effective than, the blood–brain barrier. There is an efficient DNA repair mechanism in the cells that generate sperm, but not in the sperm cells themselves. Blood lead levels in excess of 41 to 53 μg/dL were associated with reduced sperm counts and infertility. Other effects include reduced testosterone levels, testicular atrophy, loss of libido, and impotence. Table 17.6 lists some toxins known to affect the male reproductive system.

Toxic effects on the female reproductive system include reduced fertility due to ovum loss or prevention of implantation, ovary damage, and early menopause. Apparently, before puberty the oocytes are relatively resistant to toxic damage, other than that caused by ionizing radiation. Cigarette smoking is known to cause early menopause, possibly because of the PAHs present in the smoke.

Systemic Effects and Toxic Effects in the Blood Some toxins cause the body to release substances to the bloodstream in harmful amounts. Allergic responses cause cells to release stored histamine. Besides the bronchial constriction described above, histamine causes blood vessels to **dilate** (increase in diameter) and their permeability increases. Both of these have the effect of reducing blood pressure, which can result in vascular collapse. These are the events in **anaphylactic shock**. Another example of an indirect effect is the release of epinephrine in response to exposure to carbon tetrachloride.

Hydrogen sulfide (H_2S) and hydrogen cyanide (HCN) both can be absorbed through the lungs (although HCN can also be ingested), and both act by poisoning the cytochrome system, bringing aerobic respiration to a halt in every cell in the body. Both also have strong odors. HCN has a bitter almond odor, but 20 to 40% of the population is unable to detect it. H_2S produces a strong 'rotten egg' smell. H_2S may be generated by bacteria in lethal quantities whenever biodegradable sulfur-containing material is present in confined spaces, such as in sewage manholes or food transport containers. It is often responsible for particularly tragic accidents with multiple fatalities. A typical scenario is as follows: A worker enters a confined space such as a tank and collapses in moments from the exposure. One or more co-workers attempt a rescue and are also overcome. There have even been cases where professional rescuers wearing a self-contained breathing apparatus received lethal doses due to leakage around the mask.

Oxygen is carried in the blood by complexation with hemoglobin in the red blood cells. The ability to carry oxygen to the tissues can be reduced by several toxins. The best known is carbon monoxide, which competes with oxygen for the hemoglobin. The CO–hemoglobin complex is called **carboxyhemoglobin**. The ratio of carboxyhemoglobin to oxygenated hemoglobin depends on the partial pressures of the two compounds as described by the *Haldane equation*:

$$\frac{[HbCO]}{[HbO_2]} = \frac{MP_{CO}}{P_{O_2}} \tag{17.2}$$

where M has a value somewhere from 210 to 245. Thus, if the CO and O_2 were at equal concentrations, more than 200 times as much of the hemoglobin would be bound to CO as to O_2. As an application, this equation can be used to determine what partial pressure of CO would tie up half of the hemoglobin, a level that could result in coma or death. This is found by setting the left-hand side of equation [17.2] to 1.0, resulting in

$$P_{CO,1/2sat} = \frac{P_{O_2}}{M} \tag{17.3}$$

Thus, expressing the partial pressures as a percentage of 1 atm, we get $P_{CO,\frac{1}{2}sat} = P_{O_2}/210 = 21\%/210 = 0.1\%$, which is 1000 ppmv. The relation between percent carboxyhemoglobin and health effects is described in Section 21.5.

Some compounds can reduce the oxygen-carrying capacity of hemoglobin by oxiding its ferrous (+II) iron atom to the the ferric (+III) state. This form of hemoglobin, *methemoglobin*, does not bind oxygen or carbon monoxide. Nitrite is one of the substances that do this. It is usually ingested as nitrate either in cured meats or in drinking water, and then converted to nitrite by intestinal flora. The problem is usually restricted to infants and causes a disease called **methemoglobinemia**, or "blue baby" syndrome. The drinking water standard for nitrate of 10 mg/L is set with this sensitive population in mind.

Chemical toxins can cause other blood problems: low red blood cell counts (**anemia**), low white blood cell counts (**agranulocytosis**), low platelet counts (**thrombocytopenia**), or damage to bone marrow tissue that produces all of the above (**aplastic anemia**). As described in Section 17.4.7 on the kidney, some substances cause lysis of red blood cells, or hemolytic anemia.

Some toxins affect the heart and the blood vessels. High levels of cobalt causes calcium accumulation in the mitochondria of heart muscle. Lead, mercury, and a number of other substances damage the endothelium of capillaries in the brain. This causes edema (swelling) and harms the blood–brain barrier.

17.4.8 Effects at the Individual Level

Often, the underlying physiological mechanism of toxicity is not detected. Instead, gross organismal effects are observed. Symptoms of toxicity observable at the whole organism level are, of course, highly varied, depending on the mechanism of damage. The most significant is **mortality**, the death of the organism. This results, of course, when any of the previously described effects become serious enough to sufficiently impair a vital function. This can occur at the subcellular level, such as cyanide's role in halting the electron transport chain or organophosphates in causing paralysis of the respiratory system. Necrosis of selective tissues, such as by carbon tetrachloride in the liver, can eventually impair liver function and cause mortality indirectly. Sometimes the only effect observed will be weight loss or gain, or indications of gastrointestinal disturbances such as vomiting or diarrhea.

Behavioral effects may be much easier to detect than underlying physiological effects in living organisms. Behavioral effects can affect the survivability of an organism in various ways:

- Response to stimuli, such as temperature or salinity for fish or avoidance of noxious chemicals
- Mating behavior
- Feeding behavior
- Locomotor behavior
- Learning ability
- Social behavior, such as territoriality, courtship, care of young, or forming groups (schooling in fish)
- Other activities such as burrowing or nest building

Numerous apparatuses have been developed to expose fish to gradients of physical or chemical stimuli. The fish position themselves in the gradient, and their preference is recorded. For example, blue crabs have been shown in this way to be able to detect naphthalene at concentrations starting at 10^{-7} mg/L! A simple test of predatory behavior involves placing a predator and a prey species, one of which has previously been exposed to a toxicant, in a tank. After a period of time, the number of prey remaining are counted.

Motor activities include qualitative descriptions of movement such as repetitiveness or randomness of motion compared to control animals. Experimental psychology has developed a large database of behavior with rats, mice, pigeons, and primates. These data are useful for making comparisons with those animals.

17.4.9   Effects at the Ecological Level

Again, underlying physiological effects have consequences at higher levels of organization, beyond the individual, to populations and to relationships between species and even to effects on the physical and chemical environment. Population effects can be placed into the following categories:

Population reductions. These may be due to direct toxic effects, elimination of prey, or to ingestion of prey which have themselves taken up toxins from the environment.

Population increases. There are due to elimination of competitors or predators, resulting in increases of nontarget species and replacement of one species by another.

Sublethal effects. These include bioaccumulation, changes in behavior such as predator–prey relationships, and differential survival and reproduction.

Genetic changes. These include selection pressure for resistance, and the bottleneck effect, by which the distribution of traits may change significantly. Resistance is particularly troublesome, because it tends to cause users to increase their application rates, increasing risks to nontarget organisms.

Ecosystem effects can be summarized as follows:

Reduction in number of species, in species diversity, and changes in energy flows.

Removal of organisms with longer life-spans and increased predominance of those with shorter life spans.

Outbreaks of some nontarget species, especially in lower trophic levels. This results from a disturbance in normal ecological relationships, such as predator–prey, competition, and symbiosis.

Population instability of predators and parasites at higher trophic levels due to disturbances at lower trophic levels.

Data showing these effects have been collected from both experimental and accidental ecosystem exposures to xenobiotics. These show that it is necessary to address not just acute and chronic toxicity, but also fate and transport of contaminants and impacts on all the trophic levels in an ecosystem.

All of these effects are associated with either mortality of individual organisms or with reduced reproductive success. The classic case of the latter is the effect of DDT on birds of prey, including the American bald eagle. Even though the levels that accumulated in the birds was below those toxic to the adult, they were lethal in the sense that the birds became unable to produce young. The pesticide would become concentrated in the egg yolks at levels that prevented the chicks from surviving to hatch, or the egg shell produced by the parent would be too thin to survive the incubation period.

Organisms with short life spans tend to reproduce rapidly, and rush to fill environmental niches vacated by other species. This lengthens the time to recovery of an affected ecosystem. Examples of these fast-growing species are various types of "weed" plants and animals such as the Norway rat.

Physical changes to the environment can result from toxic effects. Changes in plant distribution can result in erosion, siltation of surface waters, and elimination of animal habitats such as nesting sites or specific types of shelter required by certain species.

17.4.10 Microbial Toxicity

Bacteria are among the organisms that are most insensitive to toxic substances. Several reasons for this are:

- They have much simpler metabolic processes than higher organisms, reducing the number of ways a toxicant can interfere. For example, organophosphate pesticides attack acetocholinesterase, which bacteria do not produce. However, toxins that affect all kinds of organisms, such as cyanide, can also affect bacteria.

- Bacteria can form resistant stages or slow their metabolism to survive adverse conditions. Thus, many toxicants may stop growth but not kill cells.

- Bacterial membranes are more selective in their transport mechanisms. This is in contrast to animals, which have specialized transport organs such as the digestive tract and kidneys.

Thus, if a substance is toxic to bacteria, there is a good chance that it will be even more toxic to higher organisms. This makes bacteria useful for toxicological screening purposes. An exception, of course, is the antibiotics, which are selected to inhibit specialized metabolic processes in bacteria.

Even if a toxic substance does kill or inhibit a type of bacteria, it may be difficult to measure the effect for several reasons:

- Because bacteria are capable of rapid growth, if only a part of the population is killed, the remainder can quickly reproduce to replace them.
- Bacteria are specialized in metabolic mechanisms, and related genera may have related capabilities. Thus, if one species is inhibited, a previously minor relative may grow rapidly to occupy the niche. An exception is the nitrifiers. These are typically slow growing and slow to take advantage.
- In any nonsterile environment, bacteria are ubiquitous and diverse. That is, there will be many different strains with overlapping metabolic capabilities and many with different capabilities. Thus, there will usually be one or more strains available that are resistant to an introduced toxicant and able to take advantage of available nutrients to grow.

These factors argue against the likelihood of the effectiveness of bioaugmentation. **Bioaugmentation** is the practice of inoculating biological treatment processes, such as activated sludge or septic tanks, with bacterial cultures to replace "killed" bacteria or to supply organisms capable of degrading refractory substances.

PROBLEMS

17.1. The LC_{50} of copper in mg/L to salmonid fish vs. hardness (H) in mg $CaCO_3$/L is given by equation (17.1). Use this equation to compute the LC_{50} value for a soft water, $H = 50$ mg $CaCO_3$/L, and for a hard water, $H = 350$ mg $CaCO_3$/L.

17.2. What is the receptor for a carcinogenic initiator? For a promoter?

17.3. By what mechanism could a free radical cause cancer as an intiator; by what mechanism as a promoter?

17.4. Name a carcinogen that would be considered a local toxin.

17.5. How many toxic effects of ethanol can you think of?

17.6. Name one chronic and one acute effect of alcohol on humans. Do the same for chloroform.

REFERENCES

Cavenee, W. K., and R. L. White, 1995. The genetic basis of cancer, *Scientific American*, March.

Connell, D. W., and G. J. Miller, 1984. *Chemistry and Ecotoxicology of Pollution*, Wiley, New York.

Davis, D. Lee, and H. L. Bradlow, 1995. Can environmental estrogens cause breast cancer? *Scientific American*, October.

Klaassen, C. D., M. O. Amdur, and J. Doull (Eds.), 1996. *Casarett and Doull's Toxicology: The Basic Science of Poisons*, 5th ed., McGraw-Hill, New York.

LaGrega, M. D., P. L. Buckingham, and J. C. Evans, 2001. *Hazardous Waste Management*, McGraw-Hill, New York, Chapters 5 and 14.

Landis, W. G., and Ming-Ho Lu, 1995. *Introduction to Environmental Toxicology: Impacts of Chemicals upon Ecological Systems*, Lewis Publishers, Chelsea, MI.

Lu, F. C., 1991. *Basic Toxicology: Fundamentals, Target Organs, and Risk Assessment*, 2nd ed., Hemisphere Publishing, Bristol, PA.

Masters, G. M., 1998. *Introduction to Environmental Engineering and Science*, Prentice Hall, Upper Saddle River, NJ.

Newman, M. C., 1998. *Fundamentals of Ecotoxicology*, Ann Arbor Press, Ann Arbor, MI.

Rand, G. M., and S. R. Petrocelli, 1985. *Fundamentals of Aquatic Toxicology: Methods and Applications*, Hemisphere Publishing, New York.

Simpson B. B., and M. C. Ogorzaly, 1995. *Economic Botany: Plants in Our World*, McGraw-Hill, New York.

Weast, R. C. (Ed.), 1979. *CRC Handbook of Chemistry and Physics*, 60th ed., CRC Press, Boca Raton, FL.

Williams, P. L., and J. L. Burson (Eds.), 1985. *Industrial Toxicology: Safety and Health Applications in the Workplace*, Van Nostrand Reinhold, New York.

18

FATE AND TRANSPORT OF TOXINS

The ability of a toxin to produce an effect depends on factors other than its ability to react with a receptor. These include factors that influence how much of the toxin actually arrives at the receptor. Physicochemical parameters describe the equilibrium distribution of substances among phases in the environment and within organisms. However, the environment and organisms are often far from equilibrium. Therefore, we need to take a kinetic approach, one that describes the rates at which substances move within organisms. In the next few sections we focus on the fate and transport of substances within organisms, or **pharmacokinetics**. In later sections we discuss transport among compartments of the environment and various parts of the ecosystem. Environmental transport is described using many of the same principles and methods as used here.

The way that a chemical is distributed in the environment and within an organism varies with physicochemical properties that are independent of its toxicity. Toxins may be absorbed by different mechanisms and by different routes, such as by inhalation or ingestion, each with its own efficiency at delivering the toxin. Once absorbed, the toxin finds its way to various parts of an organism, including storage sites. The organism can eliminate the toxin by biochemical reactions or excretion, or may convert it to a more toxic substance. These processes can be described mathematically, giving models that can be used for understanding and predicting the phenomena.

18.1 PHYSICOCHEMICAL PROPERTIES

Besides the obvious fact that the structure of a chemical determines how it reacts chemically with the receptor, structure affects its equilibrium distribution in the environment and within organisms. Here we describe the key physicochemical properties affecting

---

Environmental Biology for Engineers and Scientists, by David A. Vaccari, Peter F. Strom, and James E. Alleman
Copyright © 2006 John Wiley & Sons, Inc.

this distribution, which are used in Section 18.7 in our discussion of pharmacokinetics and environmental modeling.

The properties we are interested in here all relate to the tendency for a chemical to distribute itself between two connected phases. The phases involved may include air, water, soil, or sediment in the environment, and tissues, membranes, lipids, and so on, in organisms. The molecules of a chemical will be distributed according to the physico-chemical attractions they have for the other molecules surrounding them in each phase. The physicochemical attractive forces are the same as those described in Section 3.3: van der Waals, dipole moment, hydrogen bonds, and so on. The process by which a chemical distributes itself between two phases is called **partitioning**.

The equilibrium between two phases is described by exactly the same thermodynamics as was developed for biochemical reactions in Chapter 5. In phase transfer the "reaction" is comparatively simple:

$$\text{compound in phase A} \Longleftrightarrow \text{compound in phase B}$$

This leads to a definition of the phase equilibrium constant, which states that the compound simply distributes itself between the two phases at a constant ratio, called the **partition coefficient**, K_P:

$$K_P = \frac{C_A}{C_B} \tag{18.1}$$

where C_A and C_B are the concentrations of the compound in phases A and B, respectively. (For use in partition constants, the concentrations can be expressed either in molar or mass concentration units.)

One of the most basic partition parameters describing a chemical is its **vapor pressure**, P_v. This is the partial pressure at a given temperature that a chemical will have in equilibrium with its pure liquid or solid phase. For example, if a vial is filled partway with benzene liquid and sealed, the benzene will vaporize into the airspace until its partical pressure reaches an equilibrium value, which is the vapor pressure. The vapor pressure can be considered to be a measure of the attractive forces among a compound's molecules. The stronger the attraction, the lower the vapor pressure. A high vapor pressure can facilitate exposure to toxic substances by inhalation or atmospheric transport.

Because it is such a basic property, the vapor pressure can serve as a sort of reference concentration for comparison wtih concentrations in other phases. To be more precise, the concentrations in different phases can be converted to units of **fugacity**, which are equivalent to gas-phase partial pressure under ideal gas conditions. Put another way, the fugacity can be thought of as the partial pressure of a substance in equilibrium with whatever concentration and phase is under consideration. The vapor pressure, then, corresponds to a reference fugacity, or the concentration in any phase in equilibrium with the pure liquid. Some investigators use fugacity in place of concentration units to express the amounts of a contaminant in the environment.

At ambient pressures, the ideal gas law is very accurate for both pure and mixed gases and rarely needs correction. Expressed in terms that relate the partial pressure of compound i, P_i, to concentration in mass/volume units, C_i, and to temperature, T, and the gas law constant, R, the ideal gas law is

$$P_i \mathcal{M} = C_i RT \tag{18.2}$$

where $\mathcal{M}$ is the molar mass. Concentrations of chemicals in air may be expressed as mole/volume, mass/volume, or commonly as volume/volume. The latter is usually given as the parts per million by volume (ppmv). In ideal gas mixtures, the following expression relates ppmv with the partial pressure in equation (18.2):

$$\text{ppmv} = 10^6 \frac{P_i}{P_T} \tag{18.3}$$

where P_T is the total pressure. At 20°C and 1.0 atm total pressure, we can combine equations (18.2) and (18.3):

$$C_i(\text{mg/m}^3) = \text{ppmv} \cdot \frac{\mathcal{M}(\text{g/mol})}{24} \tag{18.4}$$

Example 18.1   What would be the concentration of benzene in air, C_b, in a poorly ventilated room containing an unsealed drum of liquid benzene?

Answer   In this case the partial pressure of the benzene in the air will be approximately equal to the vapor pressure. The vapor pressure of benzene is 95.2 mmHg at 25°C. From equation (18.2) (note that 1.0 atm = 760 mmHg):

$$C_b = \frac{P_v \mathcal{M}}{RT} = \frac{95.2\,\text{mmHg}}{760\,\text{mmHg/atm}} \left(78.11 \frac{\text{g}}{\text{mol}}\right) \left(\frac{\text{mol} \cdot \text{K}}{0.08205\,\text{L} \cdot \text{atm}}\right) \left(\frac{1}{298.15\,\text{K}}\right) = 0.40\,\text{g/L}$$

Example 18.2   The ambient air quality standard for ozone ($\mathcal{M} = 48\,\text{g/mol}$) is 0.120 ppmv. What is the concentration in mg/m^3?

Answer   From equation (18.3), we have $P_{O3} = P_t \cdot \text{ppmv}/10^6 = 1.2 \times 10^{-7}$ atm. From equation (18.2), $C_{O3} = P_{O3}\mathcal{M}/RT = [(1.2 \times 10^{-7}\,\text{atm})\,(48\,\text{g/mol})]/(0.08205\,\text{L/atm} \cdot \text{mol} \cdot \text{K} \cdot 293\text{K}) = 2.40 \times 10^{-7}\,\text{g/L} = 0.240\,\text{mg/m}^3$.

Another important partition coefficient is **Henry's constant**, H_c, which describes how a chemical distributes itself between vapor phase and a liquid mixture (usually water solution in the cases of interest here). The amounts in the vapor and water phases can be measured in numerous ways, including partical pressure, fugacity, mole fraction, or molar or mass concentration. When expressed as molar or mass concentration units, Henry's constant is

$$H_c = \frac{C_g}{C_l} \tag{18.5}$$

where C_g is the gas-phase concentration and C_l is the concentration in the liquid, both phases at equilibrium with each other. Whereas the vapor pressure indicates the tendency of a compound to distribute itself between its pure liquid and the vapor phase, Henry's constant relates to the distribution between a solvent and a vapor. The less the attraction between solute and solvent molecules, the higher the Henry's constant value. In fact, because vapor pressure represents the tendency to be in the gas phase, and aqueous solubility the tendency to be in the aqueous phase, Henry's constant can be estimated as the ratio of vapor pressure to aqueous solubility (with appropriate unit conversions).

Example 18.3   The OSHA standard for toluene states that the average air concentration that a worker may be exposed to during an 8-hour work shift may not exceed 200 ppmv.

If someone were working in a poorly ventilated enclosed space over a wet well, what is the highest permissible concentration of toluene in the water so that if the air were in equilibrium with the water, it would not exceed the OSHA standard? The molar mass of toluene is 92.13 g/mol, and $H_c = 0.265$.

Answer   Using equations (18.3) and (18.4), we have

$$
\begin{aligned}
C_g &= \frac{P_i \mathcal{M}}{RT} = \frac{P_T \cdot \text{ppmv}}{10^6} \frac{\mathcal{M}}{RT} \\
&= \left(\frac{1\,\text{atm}\,(200\,\text{ppmv})}{10^6\,\text{ppmv}} \right) \left(\frac{92.13\,\text{g/mol}}{0.08205\,\text{L} \cdot \text{atm/mol} \cdot \text{K}} \right) \left(\frac{1000\,\text{mg/g}}{298.15\,\text{K}} \right) = 0.753\,\text{mg/L}
\end{aligned}
$$

From equation (18.5),

$$
C_l = \frac{C_g}{H_c} = \frac{0.753\,\text{mg/L}}{0.265} = 2.84\,\text{mg/L}
$$

The most important solvent in biological systems is, of course, water. Many organics, such as ethanol, mix with it readily. Other substances are immiscible with the water. That is, two separate phases are formed at equilibrium, the aqueous phase consisting mostly of water and the organic phase consisting mostly of the organic. The concentration of the organic compound in the water in this situation is the **aqueous solubility** (c_s). For example, water in contact with liquid benzene dissolves 1750 mg per liter of water. Substances tend to be more soluble in water: As polarity increases, molar mass decreases if they ionize or if they are capable of forming hydrogen bonds. Aqueous solubility and vapor pressure are the key physicochemical parameters governing the fate and transport of many synthetic chemicals. In fact, Henry's constant is the ratio of vapor pressure to aqueous solubility.

As their name implies, hydrophobic compounds have very low aqueous solubility. These compounds distribute themselves between a hydrophobic phase and a hydrophilic phase. In biological systems this means lipids and water, respectively. Experimentally, this relationship is determined using a surrogate compound for the lipids: *octanol*. Octanol is fairly nonpolar, although not extremely so. Water has an appreciable solubility in it (21% as a mole fraction, or about 31.5 g/L), yet water and octanol do form distinct phases. In the laboratory a solute can be added to a vial containing both water and octanol and allowed to come to equilibrium. The solute will distribute itself between the two phases in a constant proportion similar to how it might be distributed between, say, blood plasma and plasma membrane lipids. The molar concentrations of solute in the octanol and water in this experiment, c_O and c_W, respectively, can be determined, and their ratio is a physicochemical property of the solute called the **octanol–water partition coefficient**, K_{OW}:

$$
K_{OW} = \frac{c_O}{c_W} \tag{18.6}
$$

A number of empirical correlations have been developed to estimate K_{OW} from aqueous solubility. One such relationship (Chiou et al., 1983) is (the aqueous solubility, c_s, must be given in units of molarity):

$$
\log K_{OW} = 0.710 - 0.862 \log c_s \tag{18.7}
$$

Appendix A lists the values of physicochemical parameters described above for several important toxic substances.

Besides describing how substances distribute themselves within an organism, physicochemical parameters are important in the distribution between organisms and their environment. Of particular significance is the K_{OW} value. This is a good predictor of how substances will distribute between the lipid tissues of aquatic organisms and the water they live in. Since K_{OW} for hydrophobic organics is very large, lipid tissues of these organisms often store concentrations of toxins far higher than is found in the environment. This phenomenon is called *bioaccumulation* or *bioconcentration*.

Example 18.4   The aqueous solubility of Freon ($CCl_2\, F_2$) is 280 mg/L. If its concentration in water were 0.25 mg/L, estimate the concentration in fatty tissues. Assume that there is no elimination.

Answer   The molar mass of Freon is 120.9 g/L. The solubility in molar units is

$$c_s = \frac{0.280\,g/L}{120.9\,g/mol} = 0.00232\,mol/L$$

From equation (18.7),

$$K_{OW} = 10^{0.710-0.862\,\log c_s} = 10^{0.710-0.862\,\log(0.00232)} = 10^{2.98} = 959$$

The Freon concentration in the lipid phase can then be estimated by assuming that it partitions according to the octanol–water partition coefficient. Then, using equation (18.6) [or, equivalently, equation (18.1)], we have

$$C_O = K_{OW}\, C_l = 959 \times 0.25\,mg/L = 240\,mg/L$$

Lipid storage prevents excretion or detoxification of a compound, and this results in another phenomenon, called *biomagnification*. In biomagnification, because animals that eat others which have already bioconcentrated the toxics will receive a higher dose of the toxin, the concentration of toxin increases toward the top of the food pyramid. Thus, predators at high trophic levels will be most sensitive to the presence of toxins that bioaccumulate. The K_{OW} is a good predictor of the tendency of a substance to bioaccumulate, bioconcentrate, or biomagnify. These concepts are discussed in more detail in Section 18.7.

Adsorption is another type of partitioning. It is the transport of a solute from gas or liquid to the surface of a solid, or to other interfaces, such as of membranes. Adsorbing materials in the environment include soil, sediment, and suspended solids in air or water. Adsorption can reduce the availability of compounds to organisms by tying them up, or it can increase availability by serving as a contact mechanism. An example of the latter is the role of atmospheric dust in carrying gaseous radon into lungs.

Two types of chemical equilibrium often need to be considered together with physicochemical equilibrium: acid–base reactions and complex formation. This is because their equilibrium occurs fairly rapidly, so that any change in concentration of one reactant causes rapid conversion to another form. The importance of this is that different forms may have different physicochemical and biological behaviors. For example, many organic acids exist in both undissociated (HA) and dissociated forms (H^+ and A^-). However, the

**TABLE 18.1   Acid and Base *p*-Dissociation Constants
for Several Toxic Pollutants**[a]

Compound	pK_a
Hydrogen sulfide	−6.99
Hypochlorous acid	−9.40
Hydrogen cyanide (1)	−9.21
Phenol	−9.96
p-Chlorophenol	0.80
Pentachlorophenol (2)	4.75
Trichloroacetic (1)	−0.52
Analine-H$^+$ (3)	−10.87
Ammonium	−9.26

[a]Note that aniline and ammonium are given in protonated form.
These are organic bases that tend to be ionized at pH levels below
the pK_a, as opposed to the organic acids, which are ionized
predominantly at pH levels above the pK_a.

Source: (1) Petrucci et al. (2001); (2) Christodoulatos and
Mohiuddin (1996); (3) Austin Community College (2004).

anion, A$^-$, tends to concentrate in the polar aqueous phase, whereas the undissociated
form, HA, is more likely to concentrate in lipid or gas phases.

The distribution of dissociated and undissociated forms as a function of pH was developed in equation (3.3), and Figure 3.3 shows how this distribution would appear for
acetic acid. You will recall that an acid is 50% dissociated when pH equals pK_a, whereas
it will be mostly undissociated at lower pH, and mostly in the dissociated form at higher
pH. Table 18.1 shows the pK_a values for several pollutants.

In the case of acetic acid, for example, the dissociated form will be essentially
insoluble in lipids and thus will not pass through plasma membranes. On the other
hand, the undissociated form will have greater solubility in lipids, and therefore a greater
ability to pass through them. As a result, the effective Henry's constant, aqueous solubility, and octanol–water partition coefficient will all vary with pH for organic acids and
bases. If the undissociated form has a phase equilibrium constant of K_u and for the
dissociated form it is K_d, equation (3.3) can be used to derive the effective equilibrium
constant, K_{eff}:

$$K_{eff} = \frac{K_u - K_d}{1 + 10^{pH-pK_a}} + K_d \tag{18.8}$$

In each equilibrium constant of equation (18.8), the aqueous concentration is in the denominator. Each equilibrium constant can represent a Henry's constant, an octanol–water
partition coefficient, or a linear adsorption partition coefficient.

Equation (18.8) can describe the variation of toxicity of many substances with pH. An
example is hydrogen cyanide, which has a pK_a of 9.37. Both the undissociated and dissociated forms are toxic to fish, but the former is about twice as toxic. Thus, the toxicity
decreases somewhat as pH increases in the area of the pK_a.

Example 18.5   Henry's constant for undissociated hydrogen sulfide (H_2S) is 0.0042 and the pK_a (for the first dissociation) is 7.1. The dissociated form is nonvolatile. What is the effective Henry's constant of H_2S at pH 7.5?

Answer   Use equation (18.3) with $K_u = 0.0042$ and $K_d = 0.00$. Then

$$H_{c,\text{eff}} = \frac{H_{c,H_2S} - 0.0}{1 + 10^{pH-pK_a}} + 0.0 = \frac{0.0042}{1 + 10^{7.5-7.1}} = 0.0012$$

An important example is ammonia. Being basic, it ionizes as pH goes below a pK_a value of 9.26. In ionized form it is relatively impermeable to cell membranes and is much less toxic to fish. However, although only a small fraction is undissociated at low pH, that fraction increases by a factor of about 10 for each pH unit of increase. For example, at pH 6, 0.056% of the ammonia is undissociated, but at pH 7 it is 0.55%, and at pH 8 it is 5.3%. From pH 7 to 8 the 96-h LC_{50} to rainbow trout goes from 80 to 20 mg/L NH_3-N, a fourfold increase in toxicity.

Another kind of chemical interaction with physicochemical processes is the formation of soluble complexes. A **complex** is a combination of two or more species in solution held together by physicochemical or ionic forces. A notable example is the metal ions, many of which form complexes with hydroxide and carbonate species present in natural water. Often, a variety of different complexes can form from a particular metal, resulting in species of varying composition and ionic charge. For example, in computing the solubility of copper in natural waters, it is necessary to account for the formation of the following species:

$$Cu^{2+}, CuOH^+, Cu(OH)_2, Cu(OH)_3^-, Cu(OH)_4^{2-}, CuCO_3, Cu(CO_3)^{2-}$$

All of these species tend to be in equilibrium with each other. At a given total copper concentration, the relative amounts of these species will depend mostly on pH and alkalinity. Each of these copper species may have different toxicities to an organism. Therefore, it follows that the toxicity of copper to fish, for example, will depend on pH and alkalinity. Many other ions complex with metals, including sulfide and phosphate. Chloride and sulfate are significant ligands at seawater concentrations. Metals also form complexes with organic compounds, such as to electronegative atoms; π-bonds (which make up the $C=C$ bond), as in unsaturated fats or aromatics; or by forming salts of carboxylic acids.

As a result of processes such as dissociation or complexation, a chemical may be present in a phase in more than one form. The partition coefficient, however, only describes the ratio of *like species* in the two phases. For example, in the case of acetic acid in water, the partition coefficient is only the ratio of undissociated acid concentrations. However, when a phase is analyzed, it is typically the total concentration of all species, including complexes and dissociated forms, which is measured. Therefore, the measured ratio of total concentrations for the two phases is given a different name, the **distribution coefficient**. The partition coefficient does not change with factors such as pH, but the distribution coefficient does, in the same manner as equation (18.8).

Since structure has such an important influence on toxic behavior of chemicals, efforts have been made to empirically correlate structural characteristics, as well as the physicochemical properties, to toxic effects. The ability to predict the toxicity of substances is useful because given the variety of chemicals and effects, it is not feasible to perform all the possible tests. Furthermore, the ability to predict effects could even be applied to chemicals that have not yet been synthesized or purified.

Such correlations are called **quantitative structure–activity relationships** (QSARs). The QSAR method uses statistical techniques, especially multiple regression or discriminant analysis, to find relationships between a variety of predictors and a single effect. Regression is used to predict continuous variables, such as the dose lethal to 50% of organisms (the LD_{50} value, described below). Discriminant analysis is more useful for categorical (either/or) endpoints, such as carcinogenicity or skin irritation. Examples of QSARs are the correlation between narcosis and polarity, or the correlation between toxicity and the side-chain length of organophosphate pesticides, as described in Section 18.3.

Various types of predictors can be used. The most common are (1) physicochemical properties, such as solubility or K_{OW}; (2) functional group descriptors; and (3) molecular indices. Prediction of narcosis from polarity is an example of the first type. Organophosphate toxicity is an example of the second type.

The use of functional groups is preferable to using properties, since it can be applied even to never-synthesized chemicals in the proposal stage of a research project. On the other hand, since the physicochemical properties in some cases are directly related to the toxic effect, they can provide better predictions. The functional group descriptors typically go well beyond the kinds of groups listed in Table 3.4. Also included are molecular descriptors such as number of double bonds, number of rings indexed by type and number of atoms forming the ring, and length of hydrocarbon chains.

Use of the third type of data, molecular indices, can be more powerful predictors of toxic activity than functional groups. These include the molecular connectivity indices, electronic charge distributions, and kappa environmental descriptors. Molecular connectivity indices (MCIs) are a group of parameters that describe the topological structure of a chemical. It contains information on molecular size, shape, branching structure, cyclization, unsaturation, and heteroatom content (presence of atoms other than carbon in the molecule's "backbone"). Physicochemical and biological parameters are correlated to the MCIs in the same way as they might be correlated with the K_{OW}.

Despite the apparent promise of such predictors, they typically work well only when limited to a single type of compound and a single toxic effect. For example, one might be able to develop a model to predict the mutagenicity of polychlorinated biphenyls, but it would be much more difficult to develop a single model to predict the mutagenicity of all chlorinated hydrocarbons. A single effect, such as mortality, may be a consequence of different modes of action. Thus, a model to predict LD_{50} might be limited to a particular group of compounds. Jørgensen et al. (1997) describe the use of a variety of correlation and QSAR techniques in the prediction of a variety of physicochemical and biological parameters, including:

Aqueous solubility

Partition coefficient (including K_{OW})

Henry's constant

Soil adsorption and exchange coefficient

Rate constants for hydrolysis and photolysis

Enzyme inhibition

Microbial and mammalian toxicity (LC_{50}, EC_{50}, etc.)

Uptake of organics by algae

Biosorption

Bioconcentration, bioaccumulation, and biomagnification factors

Biodegradation rate

No-effect concentration

Growth rate

Uptake rate, uptake efficiency, and excretion rate

18.2   UPTAKE MECHANISMS

Uptake is the transfer of a chemical from the environment into an organism. A toxic substance must pass through a cell membrane to enter the organism. This can occur naturally by several mechanisms, which were described in Section 4.1. Passive transport occurs whenever there is a concentration gradient across the membrane. The situation is illustrated in Figure 18.1.

The passive transport flux, F_{pt} [$M \cdot t^{-1} \cdot L^{-2}$], can be modeled in a simplified way as being proportional to the concentration difference across the membrane by the diffusion coefficient, D, and inversely proportional to the membrane thickness, h:

$$F_{pt} = \frac{D}{h}(C_2 - C_3) \tag{18.9}$$

The diffusion coefficient depends on the properties of the solute and of the membrane, but in general, it is inversely related to the square root of the molar mass of the diffusing species. Thus, small molecules diffuse faster. The concentrations at the faces of the plasma lipid membrane can be assumed to be in equilibrium with the aqueous phases in which they are in contact. Thus, $C_2 = K_M C_1$ and $C_3 = K_M C_4$, where K_M is the partition coefficient between the membrane and the adjacent phases. Substituting these relations into

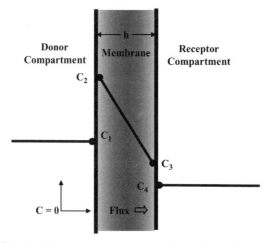

Figure 18.1   Passive transport across a membrane by molecular diffusion.

equation (18.9), we can express the flux in terms of the concentrations in the aqueous phases:

$$F_{pt} = \frac{DK_M}{h}(C_1 - C_4) \qquad (18.10)$$

Thus, it can be seen that the partition coefficient has a direct impact on the rate of transport. Since the octanol–water partition coefficient is a model for the lipid–water system, compounds with a high K_{OW} will be absorbed easily by passive transport. The constants in equation (18.10) can be lumped into a single parameter, the **mass transfer coefficient**, k:

$$F_{pt} = k(C_1 - C_4) \qquad (18.11)$$

The mass transfer coefficient is an inverse measure of the resistance posed by the membrane to movement of the substance. A similar equation can be used to describe the transfer between two phases in the absence of a membrane, such as from air to water or from blood plasma to a lipid phase. This is the **film theory model** of interphase mass transfer. The presence of two immiscible phases in contact with each other implies that they will be physicochemically different and have different affinities for the solute as described by the partition coefficient. The flux of solute can be described similar to equation (18.11), but modifying the concentration of one of the phases by the partition coefficient, so that the flux will be zero when the two phases are in equilibrium. For example, the model for transfer from an organic phase with concentration C_O to an aqueous phase with concentration C_W will be

$$F = k(K_P C_W - C_O) \qquad (18.12)$$

In this case the mass transfer coefficient, k, will depend on the diffusion coefficients of the solute in the two phases, as well as other factors, such as the amount of mixing between the interface and the bulk fluid.

Other transport mechanisms were described in Section 4.1. These include filtration, facilitated diffusion, active transport, and endocytosis. Facilitated diffusion and active transport, which depend on a carrier in the membrane, have a maximum flux associated with saturation of the carrier. As a result, their dynamics can be described by a form of Michaelis–Menten kinetics.

18.3   ABSORPTION AND ROUTES OF EXPOSURE

Absorption refers to uptake into the systemic circulation of an organism. Absorption occurs through three main **routes of exposure**:

- **Ingestion**: absorption through the lining of the gastrointestinal tract from food or other particles
- **Inhalation**: absorption through the lungs
- **Dermal contact**: absorption directly through the skin

Ingestion and inhalation are forms of **intake**. All three are mechanisms of uptake. The route of exposure determines how much of a toxin enters and how different organs are exposed. The toxicity of a substance depends on the route of exposure. For example, arsenic is more carcinogenic by inhalation than by ingestion, whereas vinyl chloride is the reverse. Asbestos danger from inhalation is well documented, but little evidence exists yet for its toxicity by ingestion.

Uptake by ingestion occurs via absorption through cell membranes lining the gastro-intestinal tract. Most of the surface area of the gastrointestinal (GI) tract is associated with the villi of the small intestines. Thus, most neutral molecules will be absorbed there. Organic acids and bases, on the other hand, depend on pH relationships for their absorption. The pH of the stomach ranges from 1 to 3, and the intestines from 5 to 8. Thus, weak acids will be undissociated in the stomach and tend to be absorbed there. Conversely, weak bases will be ionized in the stomach and therefore unable to pass membrane barriers by passive transport. However, they can be absorbed in the intestines at the higher pH.

The pH relationships in the GI tract form a special mechanism for facilitating absorption of acids and bases. Because blood plasma has a pH close to neutral, weak acids will be mostly dissociated in it, and only a small fraction will be in the undissociated form. However, in the low pH environment of the stomach, weak acids will be mostly undissociated. The undissociated form, being uncharged, can pass through the membranes of the cells lining the stomach. After it crosses, it enters the bloodstream, where it become ionized due to the high pH. This keeps the concentration gradient high, maximizing the passive diffusion flux.

Figure 18.2 illustrates what is occurring in the case of benzoic acid. The concentration gradient of non-ionized acid is $100 - 1 = 99$. If the driving force were based on the total concentration instead of the non-ionized concentration, the gradient would change sign ($101 - 2513 = -2412$). Benzoic acid would move from the blood into the stomach instead of the reverse. A similar situation occurs in the intestines, except in that case the pH of the lumen is higher than that of the blood. This facilitates absorption of weak bases by a similar mechanism.

Ingestion does not only involve food. Risk assessment for contaminated land some-times must take into account the ingestion of soil or other solids by children. A typical assumption is that children eat 200 mg of soil per day.

The respiratory tract is the site of absorption by inhalation for gases, such as carbon monoxide, vapors of mercury, and high-vapor pressure liquids such as benzene

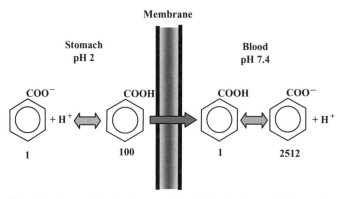

Figure 18.2   How pH relationships in the stomach and blood plasma facilitate absorption of weak acids such as benzoic acid.

or perchloroethylene, as well as liquid aerosols and solid particles. Gases and vapors are absorbed by the large surface area of the alveoli. The absorptive flux will be largely controlled by the aqueous solubility of the solute. Very soluble gases and vapors, such as chloroform, are quickly absorbed, and the rate of absorption is effectively controlled by the respiration rate. Such compounds are said to be **ventilation limited**. The absorption rate for less soluble gases and vapors is controlled by blood flow to the lung. These substances are thus said to be **flow limited**. The absorption of flow-limited toxins is also affected by how completely the blood is cleared of the toxin as it passes through the body before returning to the lungs. Toxins are removed from the blood in various ways, including the storage of lipophilic substances in fatty tissues and complexing of the toxin with various receptors.

Particles larger than $10 \, \mu m$ tend to be trapped in the nose and expelled with mucus. Particles smaller than about $0.01 \, \mu m$ remain effectively suspended in the airflow and tend to be exhaled instead of affecting lung surfaces. Particles in the range between these values tend to be deposited in various parts of the respiratory tract. The larger ones are deposited in the upper respiratory tract, and then may be swallowed along with mucus, entering the GI tract. Smaller particles go deeper, first to the trachea or bronchi. Some of these may be transported to the upper respiratory tract by the mucociliary escalator, and either be coughed up or swallowed. Others are engulfed by phagocytes and absorbed by the lymphatic system. Particles smaller than $1 \, \mu m$ may reach the alveoli, where they may be absorbed into the bloodstream. Particles of cigarette smoke range between 0.1 and 1.0 μm, and most are between 0.2 to 0.25 μm. Thus, they tend to be deposited in the tracheobronchial airways and alveoli.

Particles may be a source of solid toxins such as the heavy metals or polynuclear aromatic hydrocarbons present in smoke. For example, cigarette smoke is a source of cadmium as well as PAHs, not to mention nicotine. Furthermore, regular smoking paralyzes the mucociliary escalator, which otherwise would help clear some of the particles from the lungs. Particles may also help transport gases or vapors into the lungs by adsorbing them. An example of this is radioactive radon, which otherwise is not very easily absorbed. It has even been suggested that one reason that cigarette smoke is such a potent cause of lung cancer is that it carries radon, naturally present in the air, into the lungs.

The human lungs contain about $70 \, m^2$ of surface area for absorption. Adults breathe about 8 L of air per minute at rest. For risk assessment purposes it is common to assume a breathing rate of $20 \, m^3/day$ (a bit less than 14 L/min). In aquatic organisms the gills play a role similar to that of the lungs. Gills have been shown to absorb the pesticide dieldrin, and cadmium was observed to be taken up by passive diffusion.

The stratum corneum is the main barrier to uptake of toxic substances through the skin. Once past it, the epidermis offers little further resistance. The toxin then can enter the dermis, from which it can be transported to other organs by the blood. Hair follicles offer a diffusion shunt through the epidermis. Thus, the scalp and male face absorb toxins more readily than other areas do. The palms have a thicker stratum corneum, but it is relatively porous and thus is not a good barrier for toxins. Abrasion removes the stratum corneum, greatly increasing absorption. Some solvents, such as dimethyl sulfoxide (DMSO), can help carry toxins through the skin. This is taken advantage of in drug delivery by means of transdermal patches. Toxins that are absorbed but do not penetrate the epidermis become washed away, since the epidermis replenishes itself about once a month. The stratum corneum is replaced about every 14 days. It seems that all toxins that can be absorbed by the skin do so by passive diffusion.

The skin can biotransform substances, often increasing their toxicity. Repetitive exposure to coal tar induces the enzyme arylhydrocarbon hydroxylase (AHH) in the epidermis. This and other enzymes convert PAHs into more carcinogenic forms. It is interesting that people with psoriasis cannot induce AHH and do not seem to develop skin cancer upon exposure to coal tar.

Examples of toxins absorbed through the skin include hexane, carbon tetrachloride, and some insecticides. Toxins can also be absorbed by the eyes, although in that case it is the eye as a target organ that is usually of concern. Various invertebrates have been shown to take up toxins through their body surface, including arthropods and crustaceans such as *Daphnia*. Molting has been shown to be a way to shed heavy metals. In addition, other routes of exposure are used experimentally, such as injection subcutaneously, intramuscularly, or into body cavities.

Often, the amount of uptake is reduced because some of the toxic substance is bonded, complexed, or limited by diffusion in the environmental source. The fraction that is available for uptake is called the **bioavailability**. For example, a fraction of some pesticides can become so strongly bound to soil particles that they are practically unavailable. It may be necessary to take multiple routes of exposure into account for some toxins. A worker applying pesticides by spray may absorb by both inhalation and dermal contact.

18.4   DISTRIBUTION AND STORAGE

Once a toxin enters the bloodstream, it is distributed rapidly throughout the body. However, organs do not all receive equal distributions, nor do they store them with equal efficiency. Moreover, the location where most of a toxin is stored is not necessarily the primary site of toxic effect. For example, 90% of the lead in adult humans is stored in the bones, but its effects are on the kidney, nervous system, and blood cell production.

Individual organs and tissues will then take toxins according to the blood flow through the organ (organ **perfusion**), their affinity for the toxin, and the presence of any transport barriers. Most capillaries have large pores between the cells that form their wall. In some tissues, however, there are few or no pores. The most notable case is the **blood–brain barrier**. This prevents passage of polar compounds of medium molar mass. Its behavior is similar to that of an intact plasma membrane in that it is permeable to nonpolar compounds. Thus, mercuric chloride, which is mainly in ionic form, does not penetrate, whereas methyl mercury does. Other tissues have barriers as well, including the peripheral nerves, the placenta, the eyes, and the testes.

The major sites of storage for toxins are (1) bound to plasma proteins, (2) the liver and kidneys, and (3) adipose tissue. Plasma proteins form complexes with many toxicants, serving to solubilize and transport them. The effect of protein binding depends on how the proteins compete with processes that detoxify or excrete them. If they give up the toxicants readily, they may help to transport them to the detoxification site. If, on the other hand, protein binding is relatively strong, it may sequester the toxins away from detoxification.

Toxins are often concentrated in the liver or kidneys, possibly due to their role in detoxifying and excreting them. They contain their own binding proteins. An example is *metallothionein*, which figures in cadmium storage and in the transfer of cadmium from the liver to the kidney. Toxins absorbed in the stomach and intestines must pass through the liver before reaching other parts of the body, thus giving that organ a chance to biotransform them. Toxins that have been inhaled or absorbed dermally may reach other tissues before the liver.

Adipose tissue (also called **depot fat**) is the tissue where energy is stored in lipid form. The cells in these tissues contain droplets of triglycerides occupying more than 90% of the cytoplasm. Naturally, they tend to concentrate hydrophobic toxins. Even nonhydrophobic compounds can be stored in lipids after being conjugated with fatty acids. Toxins can be stored in adipose tissue without causing harm there. However, lipid storage of toxins has resulted in acute toxicity in humans who had stored subacute doses in their fatty tissues, later liberating an acute dose to the blood by depleting their fats in a weight-loss program.

The bones also store some toxins, notably lead, radium, and fluoride.

18.5   BIOTRANSFORMATION

Once inside an organism, many toxic compounds undergo chemical changes by biochemical reactions to form **metabolites**. Such biochemical changes are called **biotransformation**. Yet another distinction in types of toxicity is between those compounds that react directly to cause damage and those in which metabolites are responsible for the toxicity. In the former, biotransformation produces a reduction in toxicity; in the latter it increases it. When biotransformation increases the toxicity of a chemical, the process is called **bioactivation**. Ultimately, biotransformation serves to produce more water-soluble compounds that are more easily eliminated by the organism, or even to eliminate them internally by mineralization to CO_2 and water. Chemicals that are not biotransformed tend to be persistent in the environment.

An example of biotransformation increasing toxicity is sodium fluoroacetate (rat poison 1080), which is not extremely toxic itself. However, by the mechanism *lethal synthesis* fluoroacetate is converted to fluorocitrate by the same Krebs cycle pathway that converts acetate to citrate. The enzyme for the next step combines with the fluorocitrate, but the reaction does not occur, and the fluorocitrate does not easily desorb from the active site of the enzyme. This amounts to reversible inhibition of the enzyme, which slows the Krebs cycle to a crawl.

In vertebrates, much of the biotransformation is done by the liver. For example, methanol is oxidized in the liver. The oxidation by-products cause the symptoms of methanol poisoning, which include headache, back and abdominal pain, blurry vision, and potentially blindness. A latency period of 8 to 36 hours may precede the onset of symptoms as oxidation products accumulate. Because ethanol is metabolized by the same pathway, methanol poisoning can actually be treated with ethanol. The presence of ethanol slows the oxidation of methanol by competing for the pathway.

Of course, many biotransformation reactions occur in other tissues; these are called **extrahepatic**. Whether in the liver or not, biotransformations can be either catabolic (breakdown reactions), such as oxidation, reduction, or hydrolysis, or they may be anabolic (synthesis) reactions, which are called **conjugations**. The catabolic reactions are also called **phase I** reactions. They usually expose or add a functional group such as $-OH$, $-NH_2$, $-SH$, or $-COOH$. Hydrophilicity is increased, but only slightly.

The anabolic reactions are called **phase II**, to indicate that they often follow phase I. The phase II reactions involve conjugating (combining) the substrate with another molecule, such as an amino acid, or by adding sulfate, methyl, or acetyl groups. These groups are added to the functional groups added in phase I. Phase II reactions greatly increase hydrophilicity, speeding excretion of the compound.

18.5.1   Phase I Reactions

Phase I reactions include oxidation, reduction, and hydrolysis. These reactions may prepare xenobiotics for the phase II reactions.

Oxidation   Toxins can be oxidized by numerous pathways. However, the most important system is the group of enzymes called **monooxygenases**, containing the cytochrome P450 enzyme system (also called **mixed-function monooxygenases**). These are nonspecific membrane-bound enzymes found mainly in the endoplasmic reticulum of the liver, but also in other tissues. Their function is to take one of the oxygen atoms from an O_2 molecule and insert it into the toxin molecule while the other is combined with two protons to produce H_2O:

$$RH + O_2 + NADPH_2 \Rightarrow ROH + H_2O + NADP$$

Specific examples are shown in Figure 18.3. Although not shown in the figure, all of these reactions involve $NADPH_2$ and O_2 as reactants and monooxygenase as a catalyst.

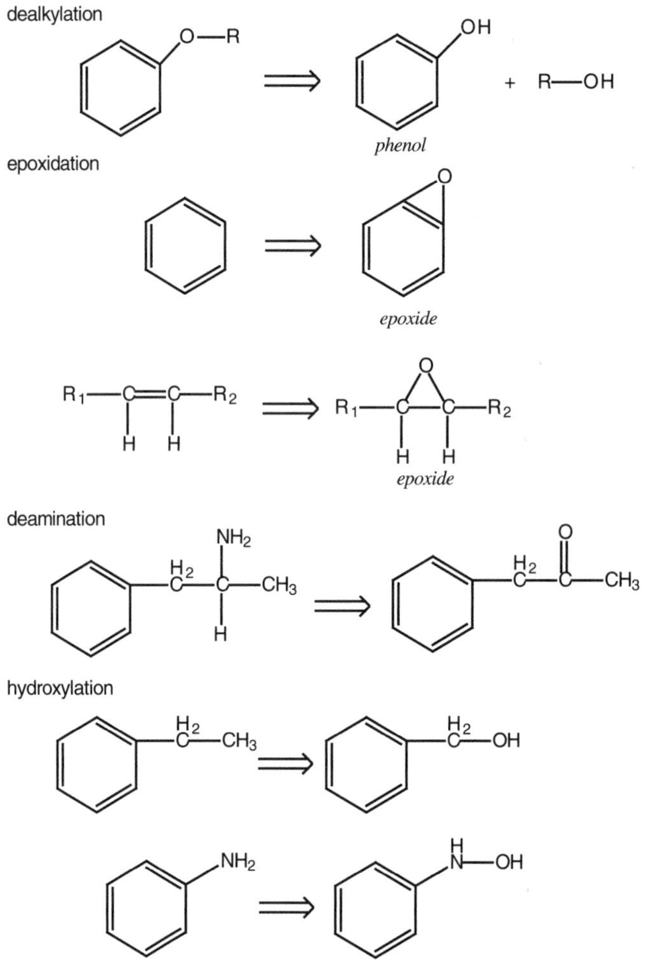

Figure 18.3   Microsomal oxidation biotransformation reactions.

TABLE 18.2 Pollutants and Other Substances That Induce Cytochrome P450

Drugs	Industrial Chemicals		PAHs
Barbiturates	Aldrin/dieldrin	Lindane	Benzo[a]pyrene
Ethanol	Chlordane	PCBs	Dibenzanthracene
Nicotine	Chloroform	Piperonyl butoxide	3-Methylcholanthrene
Steroids	DDT, DDD	Pyrethrum	
Heptachlor	Urethane	Ketones	
Toxaphene			

Source: Williams and Burson (1985).

TABLE 18.3 Substances That Inhibit Cytochrome P450

Piperonyl butoxide	Aminotriazole	Carbon tetrachloride
Chloramphenical	α-Napthyl isocyanate	Bromobenzene
Cobaltous chloride	Carbon disulfide	

Source: Williams and Burson (1985).

One important effect of cytochrome P450 monooxygenases is to convert aromatics into epoxides. It is thought that this metabolite is responsible for the cancer-causing behavior of aromatics such as benzene and benzo[a]pyrene. The epoxides are unstable and convert readily to phenols. Phenol is found in the urine of persons exposed to benzene. Other enzymes can convert epoxides into dihydrodiols, which are nontoxic and easily excreted.

Some chemicals, called **inducers**, increase the rate of biotransformation of other compounds by stimulating the synthesis of more cytochrome P450. Thus, these substances are likely to interact with other substances in a more than additive manner (Section 19.5). Table 18.2 lists some of these compounds.

Other compounds inhibit cytochrome P450 and thus may interact in a less than additive manner with other substances. Some of these are listed in Table 18.3. The effect of inducers and inhibitors can be used to show whether a substance or its metabolite is causing the toxic effect. For example, bromobenzene causes necrosis in the liver, which is increased by phenobarbital and decreased by other inhibitors. Therefore, the damage must be caused mostly by a metabolite.

The liver enzyme *alcohol dehydrogenase* converts ethanol into acetaldehyde, which is even more toxic than ethanol. This is followed by the action of aldehyde dehydrogenase, which forms easily excreted acids.

Reduction Under low-oxygen tension conditions, the cytochrome P450 monooxygenase system is capable of catalyzing reduction of azo and nitro compounds to amines (Figure 18.4). Other types of compounds that are reacted reductively are those containing an aldehyde, ketone, disulfide, sulfoxide, quinone, N-oxide, or alkene group. Aldehydes are converted to alcohols and ketones to secondary alcohols.

Another reduction mechanism is the replacement of a halogen bonded to a carbon by a hydrogen (**reductive dehalogenation**). This is responsible for the hepatotoxicity of carbon tetrachloride and similar halogenated alkanes. (Dehalogenation can also be

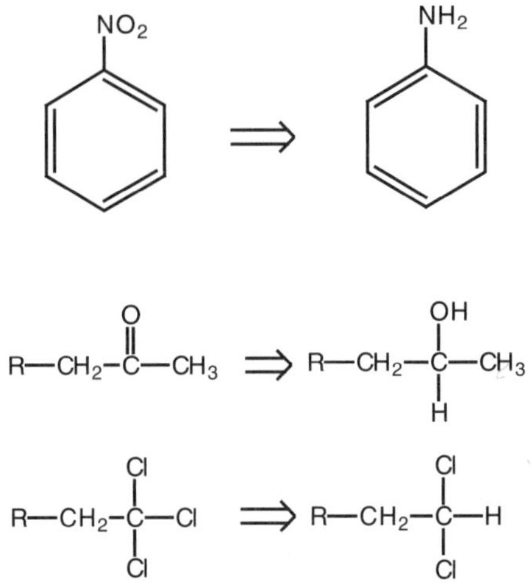

Figure 18.4 Reduction biotransformation reactions.

performed oxidatively by replacing a hydrogen and a halogen with oxygen, or by eliminating two adjacent halogens and forming a carbon–carbon double bond.)

Hydrolysis Esters, epoxides, and amides are subject to hydrolysis catalyzed by a variety of enzymes (Figure 18.5). One important group is the cholinesterases, which include acetylcholinesterase. This was described above, although in the context of the toxic effect of pesticides. Carboxylesterases bind or hydrolyze organophosphorus pesticides. Epoxide hydrolase converts epoxides, such as those formed by cytochrome P450 enzymes, into diols. This protects against the genotoxic effect of the epoxides.

18.5.2 Phase II Reactions

Conjugations These phase II reactions involve combining the toxin with another molecule or functional group. These usually increase polarity, and therefore solubility, to facilitate excretion. Phase II reactions usually take place in the cytoplasm rather than in microsomes. They tend to be faster than phase I reactions, making the latter a rate-limiting step.

The most important detoxification mechanism is the conjugation with glucuronic acid (Figure 18.6). The enzyme responsible for it is found in the endoplasmic reticulum of many tissues, but especially in the liver. Many classes of compounds can be glucuronidated, including aliphatic and aromatic alcohols and amines.

Other conjugations involve addition of amino acids such as glutamine or glycine, or sulfate, methyl, or acetyl groups. The enzymes for these processes are found in the endoplasmic reticulum and the mitochondria, but mostly in the cytoplasm of hepatic cells. Methylation differs from the other phase II conjugations in that it tends to reduce aqueous solubility rather than increase it. The methyl group is usually added to a heteroatom (O, N, or S). A rare exception is the methylation of a carbon in benzo[*a*]pyrene to form 6-methylbenzo[*a*]pyrene.

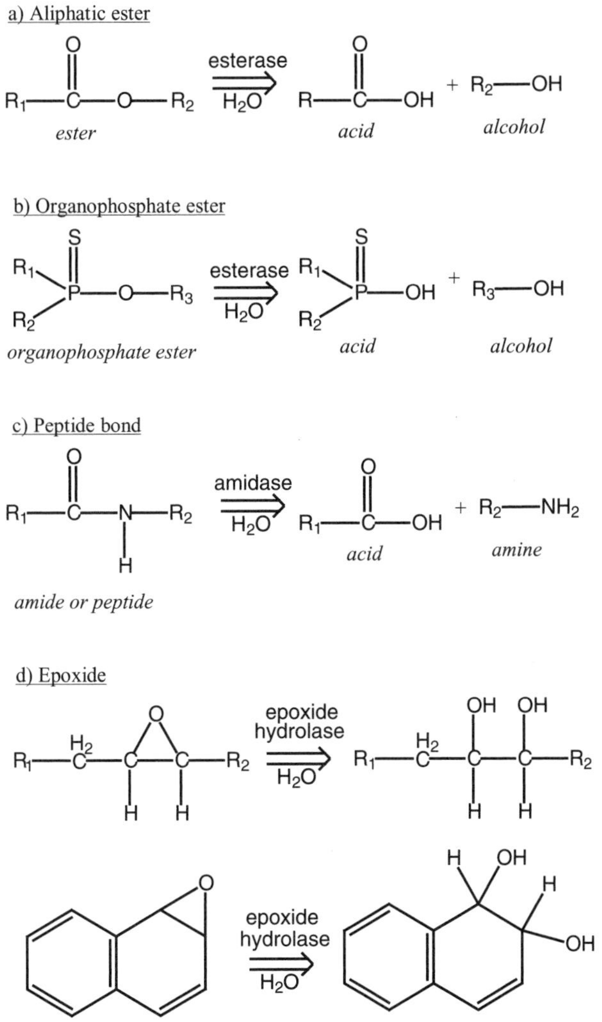

a) Aliphatic ester

b) Organophosphate ester

c) Peptide bond

d) Epoxide

Figure 18.5 Hydrolysis biotransformation reactions.

One of the most important detoxification mechanisms is the formation of mercapturic acid derivatives. This mechanism acts on a wide variety of xenobiotics. The reaction sequence starts with conjugation of the toxin with glutathione. Glutathione is a tripeptide:

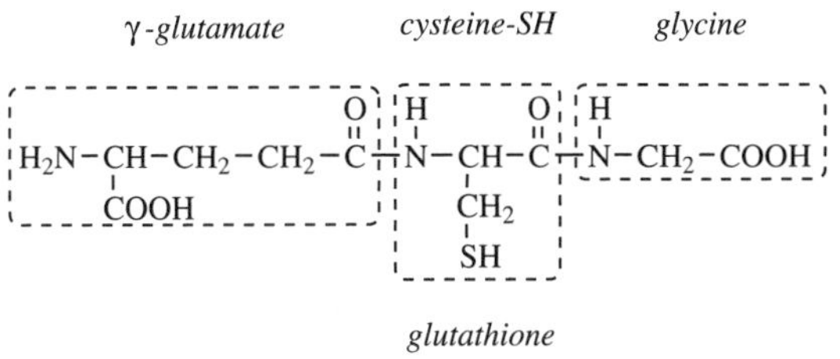

glutathione

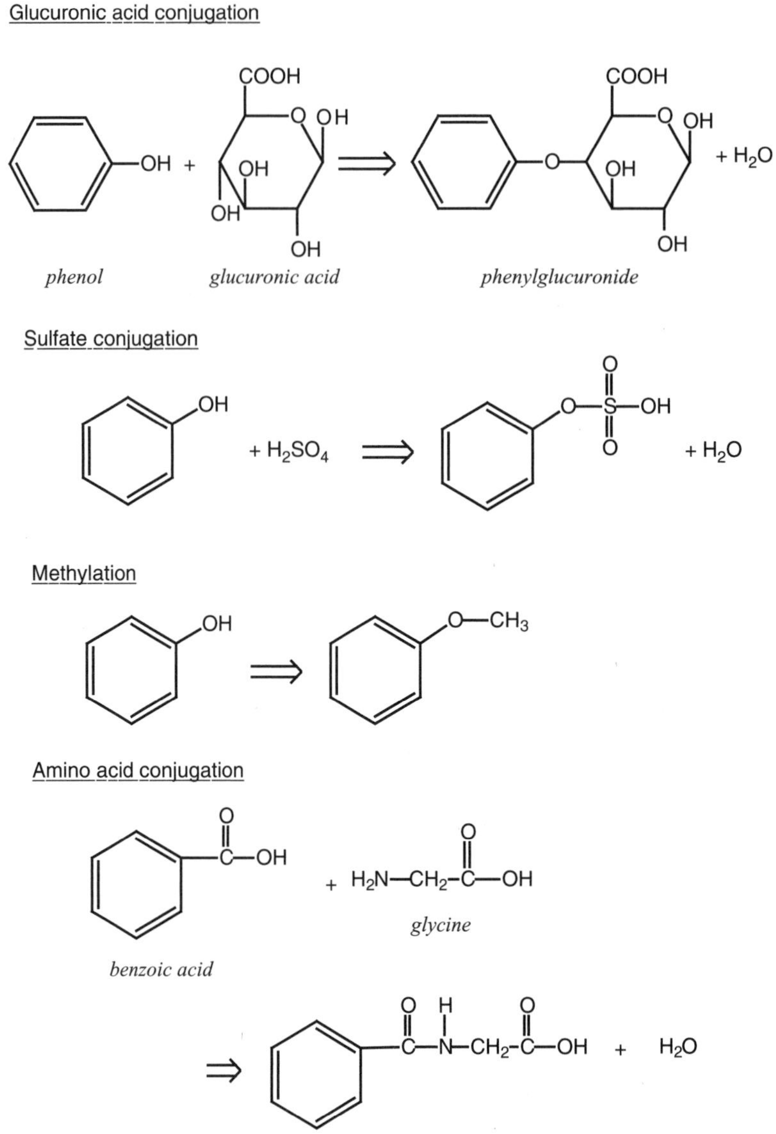

Glucuronic acid conjugation

phenol   *glucuronic acid*   *phenylglucuronide*

Sulfate conjugation

Methylation

Amino acid conjugation

glycine

benzoic acid

Figure 18.6   Conjugation (phase II) biotransformation reactions.

Notice the sulfhydryl (—SH) group on the cysteine. It is this that is bonded to the foreign molecule in the first step. For example, consider the reaction of glutathione with 1-chloro-2,4-dinitrobenzene shown in Figure 18.7a. The second step is removal of the two free amino acids leaving a thiol ester with cysteine. This is followed by the third and final step in Figure 18.7b, a reaction with acetyl-CoA to produce 2,4-dinitrophenyl mercapturic acid. This mechanism is effective with electrophilic compounds. It is an important detoxification mechanism for elimination of carcinogenic epoxides produced by the monooxygenases.

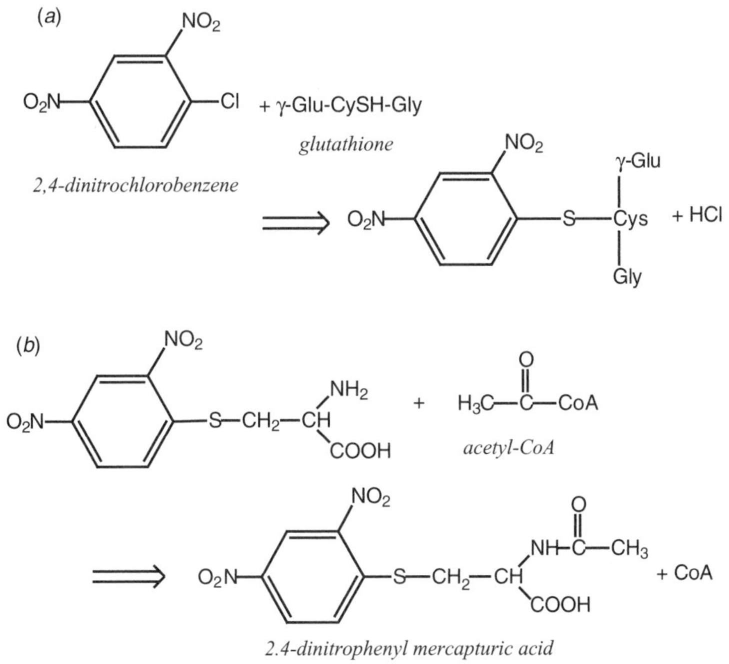

Figure 18.7 Glutathione conjugation (phase II) biotransformation reaction.

Bioactivation may involve a complicated series of reactions in several tissues. For example, 2,6-dinitrotoluene is first oxidized by cytochrome P450 and conjugated with glucuronic acid in the liver. It is then excreted in the bile, where bacteria reduce one or both nitro groups to amines and split off the glucuronide. These are reabsorbed by the intestines and return to the liver. The amine group is then hydroxylated by P450 enzymes. The resulting compound can react with DNA, causing mutations that can lead to liver cancer. Table 18.4 lists some compounds that are bioactivated.

18.6   EXCRETION

Compounds that are not biotransformed or stored are excreted to the exterior of the organism. It is remarkable that the body can do this for many xenobiotic compounds as well as natural ones. The kidney is the most important organ for excretion of toxins or their metabolites. The liver is next in importance, followed by the lungs. Lesser roles are played by the hair, skin, sweat, nails, and milk.

As you may recall from Chapter 9, the kidney works by first passively filtering about 125 mL/min of plasma from the 600 mL/min of blood perfusing it. This passes solutes and macromolecules up to molar mass 60,000, thus retaining cells and most plasma proteins. Then, by tubular reabsorption involving both active and passive transport, the nephron conserves water, several salts, and other nutrients, such as sugars and amino acids, and forms concentrated urine. Finally, tubular secretion actively transports wastes, including salts, drugs, and toxins. The final urine flow is about 1.0 mL/min.

TABLE 18.4   Bioactivated Compounds

Parent Compound	Toxic Metabolite	Toxic Effect
Allyl formate	Acrolein	Hepatic necrosis
Benzene	Benzene epoxide	Bone marrow depression
Bromobenzene	Bromobenzene epoxide	Hepatic, renal, bronchiolar necrosis
Carbon tetrachloride	Trichloromethane free radical	Hepatic necrosis, hepatic cancer
Carcinogenic alkylnitrosamines	α-Hydroxylation	Hepatic cancer
Carcinogenic aminoazo dyes	N-Hydroxy derivatives	Hepatic cancer
Chloroform	Phosgene	Hepatic, renal necrosis
Ethanol	Acetaldehyde	Varied
Fluoroacetate	Fluorocitrate	General toxicity
Halothane	Free radical	Hepatic necrosis
Hemolysis-producing aromatic amines	N-Hydroxy metabolites	Hemolysis
Methanol	Formaldehyde	Retinal and general toxicity
Methemoglobin-producing aromatic amines and nitro compounds	N-Hydroxy metabolites	Methemoglobinemia
Naphthylamine	N-Hydroxy naphthlamine	Bladder cancer
Nitrates	Nitrites	Methemoglobinemia
Nitrites plus secondary or tertiary amines	Nitrosamines	Hepatic, pulmonary cancers
Olefins	Epoxides	Skin cancer
Parathion	Paraoxon	Neuromuscular paralysis
Urethane	N-Hydroxyurethane	Cancers, cytotoxicity

Source: Lu (1991).

The rate of excretion (mass per unit time) divided by the concentration in the blood is called the **renal clearance**. If a toxin is filtered, but neither reabsorbed nor secreted, the full 125 mL/min are "cleared" of toxin. If the toxin is partially absorbed and not secreted, the renal clearance will be less than 125 mL/min. Toxins that are actively secreted can have a renal clearance as high as the perfusion rate of 600 mL/min.

Lipophilic toxins tend to be protein bound in the blood, and thus are retained at the filtration step. Those that are dissolved in the plasma are fairly easily reabsorbed. Active transport can excrete even those compounds that are bound. For example, it was found that DDA, a polar metabolite of DDT, was excreted 250 times as rapidly as DDT in mosquito fish, although both were protein bound to a similar extent. The fact that DDA was being actively transported was established by the observation that metabolic inhibitors decreased its excretion rate. The DDT, on the other hand, was probably being reabsorbed by passive diffusion.

Ionic molecules have less tendency to be reabsorbed. Therefore, excretion of weak acids and bases will depend on their dissociation constant and the pH of the urine. Urine is usually acidic, making passive diffusion more important for organic bases than for organic acids. Excretion of acids has been manipulated therapeutically by increasing the pH of the urine by ingestion of bicarbonate. This increases the ionization of the acids, preventing their reabsorption. In addition, the nephrons have separate mechanisms for active secretion of organic acids and bases. Biotransformation by the liver and other organs serves to increase excretion by the kidney by increasing the polarity of the toxins. They also tend to lower the pK_a of organic acids, increasing their ionization.

Besides its role in biotransformation, the liver excretes directly via bile. The liver tends to be more effective than the kidney in excretion of protein-bound compounds and compounds with molar mass greater than about 300 to 400. The liver has separate active transport mechanisms for organic acids, bases, and neutrals, and possibly a separate system for metals.

Although most toxins excreted in the bile are eliminated with the feces, many can be reabsorbed in the intestines. All the blood from the intestines passes through the liver before going elsewhere in the body; thus, compounds can be reexcreted. This results in a cycle called the **enterohepatic circulation**. This is especially true of lipophilic compounds. In such cases, elimination can be facilitated by ingesting an adsorbent such as activated carbon. This will compete with the intestines to absorb the toxin and then carry it out with the feces. Sometimes, lipophilic compounds which have been rendered more soluble by conjugation reactions in the liver can have the bonds hydrolyzed by intestinal microorganisms, increasing their reabsorption.

The importance of excretion by the bile has been demonstrated by the observation that toxicity can increase severalfold in animals with experimentally blocked bile ducts. For example, this increases the toxicity of diethylstilbestrol (DES) to rats by 130 times.

The lungs are important in excretion by the passive diffusion of compounds with high vapor pressure or Henry's constants. This means smaller, relatively nonpolar molecules such as hydrocarbons and chlorinated hydrocarbons. Soluble compounds such as ethanol are found if the blood concentration is high enough. Thus, the concentration of chloroform in exhaled breath is proportional to the blood concentration. Since storage in depot fats acts as a reservoir for these compounds, they may be detected in the breath long after exposure has ceased. In fish, excretion through the gills also seems to be most important for lipophilic compounds that are not biotransformed rapidly.

Metals are concentrated on hair. Although in humans this is not an important excretion mechanism, it provides an easy way to test for exposure. In fact, since hair grows about 1 cm per month, it is possible to test segments of hair to obtain an indication of the time history of exposure. Excretion by hair is important in furred animals, and this must be taken into account when attempting to extrapolate to humans from laboratory tests on rats or other small mammals.

Due to its fat content of 3 to 5%, milk can be important for excretion of lipophilic compounds. These include chlorinated hydrocarbons such as DDT and PCBs. The processes that transport calcium into the milk are thought to do the same for lead. Unfortunately, this route of excretion is more important as a source of these contaminants for nursing young than as a way of clearing the toxicant from the mother.

18.7 PHARMAKOKINETIC MODELS

A model is a kind of analogy. It is a surrogate for a real process that is easier to analyze and experiment with. A **mathematical model** is thus a set of equations that behave similar to a real system. Often, the process of developing a mathematical model helps to develop our mental model, leading to improved understanding. Furthermore, the model may be useful in making predictions, and for designing engineered systems to obtain desired results.

Ultimately, we would like to predict actual toxic effects. The models described here, however, will only predict the exposure of an organism to substances in the environment and the concentration of various tissues that results. Our interest will first be in **pharmacokinetic models**, models of the behavior of toxic substances within organisms: how they are taken up, distributed and stored, biotransformed, and excreted. The same principles can then be used for **environmental models**, models of the fate and transport of toxic substances between and among compartments of the environment (air, water, soil, etc., and different organisms in the environment), which control exposure.

A common type of model is the **compartment model**, which represents organisms by one or more volumes characterized by a single concentration for each substance being modeled and which is physicochemically and biologically uniform. For example, an organism could be represented by three compartments: the gut, systemic fluids (blood), and target organs. The compartments are connected to each other and to the environment by transport mechanisms. Biochemical reactions create or consume the modeled substance within the compartments. Mathematical relations describe the rate at which each of these processes adds or removes material from each compartment. The process of model development can be broken down into the following steps:

1. Define the *compartments* that will make up the model.
2. Identify the *transport mechanisms* that move substances between compartments.
3. Identify the biochemical *reactions* producing or destroying the modeled substance in each compartment.
4. Write the *mathematical description* of the individual transport fluxes and reaction processes.
5. Link the processes with *material balance* considerations to produce the final mathematical form of the model.
6. Mathematically *solve* the model using the appropriate initial and boundary conditions.
7. *Validate* the model by comparing with experimental results.

Material balance expresses the law of conservation of matter. The balance of the matter that enters or leaves a compartment, or is created or destroyed within it, must result in an accumulation. This can be expressed in this way:

$$\text{accumulation} = \Sigma \, \text{inputs} + \Sigma \, \text{outputs} + \Sigma \, \text{reactions} \tag{18.13}$$

The accumulation term is the change in mass, M, within the compartment, which in turn is the product of volume, V, and concentration, C. All the terms in equation (18.13) should have the units $[M \cdot t^{-1}]$. It is common to assume that the volume of the compartment is constant (no growth). In this case, the accumulation can be expressed in terms of the change in concentration in the compartment:

$$\text{accumulation} = \frac{dM}{dt} = \frac{d(VC)}{dt} = V \frac{dC}{dt} \tag{18.14}$$

The *fluxes* are the mass transport flows moving matter from one compartment to another, or to or from the environment. These include the various forms of absorption, such as passive diffusion [equation (18.11) or film theory equation (18.12)]. An important type of flux is **advection**, the result of a substance being carried into a compartment by fluid flow. Examples include transport of a toxin to the liver by blood flow, or water flowing through the gills of a fish. Advective flux, F, can be represented by the product of flow, Q, and concentration:

$$F = QC \qquad (18.15)$$

Advective flux can also describe uptake by ingestion, wherein the flow is the mass of food ingested per unit time, and the concentration is the average toxin concentration in the food. Fluxes must have positive signs for transport into a compartment, and negative for movement out.

The biochemical reactions are described using the methods from chemical kinetics. The general form of the reaction term is

$$R = Vr \qquad (18.16)$$

where r is the **reaction rate**, that is, the mass produced or consumed per unit volume per unit time. The rate is described, in turn, by a rate law, which gives the dependence of the rate on other factors, especially concentrations of reactants. Often, simple elementary rate laws can be assumed. In some cases more complex kinetics may be assumed, such as a rate based on Michaelis–Menten kinetics [equation (5.36)], or one of the other rate equations from Section 5.3.1. Here we describe two of the simpler kinetic rate laws, which are often sufficiently accurate to describe many reactions. The first is the **zero-order reaction**, in which the rate is constant independent of the concentration of reactants:

$$r = \pm k \qquad (18.17)$$

The sign will be positive for production reactions and negative for destruction. An example of a zero-order reaction is the metabolism of ethanol in humans (Problem 18.1).

The other simple elementary rate law is the **first-order reaction**, in which the rate of reaction is proportional to the amount of reactant present:

$$r = \pm kC \qquad (18.18)$$

First-order reaction is probably the most common rate law. The negative form is called **first-order decay**. Its use will lead us to several important parameters for describing toxicant behavior which are developed in the next two sections.

Models developed as just described are differential equations which, when solved with appropriate initial conditions, predict the concentration of toxin in a compartment vs. time. Simpler models can be solved analytically giving the result as a function. More complex models must often be solved numerically using a computer, giving the concentration vs. time as a table of values. Models that predict how results change with time are called **dynamic models**.

Often, if enough time is allowed to elapse, dynamic models show that concentrations approach a constant value. In the limit as time goes to infinity, all changes cease, and a **steady-state** condition exists. Mathematically, the steady-state condition could be found by taking the limit of an analytical solution as time goes to infinity. Even simpler, and applicable to models that require numerical solution as well, is to set all derivatives (accumulation terms) to zero. The differential equations then become algebraic and are solved as such. Do not confuse steady state with equilibrium. As long as a net reaction is occurring in a compartment, it may still be balanced by flux terms. For example, an organism could be ingesting benzene on a daily basis and biotransforming it to phenol for excretion. As long as the rate of ingestion equals the rate of reaction, there will be no accumulation of benzene, and the system will be at steady state.

Equilibrium formally refers to a situation when the chemical potential of reactants and products are equal for all reactions. In practical terms, equilibrium means that instead of using the rate laws and mass transfer flux equations to describe the reactions, one substitutes equilibrium relationships such as equation (5.9) for reaction equilibrium or equation (18.1) for mass transfer equilibrium. Just as steady state does not mean "no reaction," equilibrium does not mean "no reaction." For example, chloroform in respired air may be assumed to be in mass transfer equilibrium with its concentration in blood plasma, yet a continuous transfer of the solute will continue as long as the air is changed continually.

In the next several sections some simple compartment models are developed, both to illustrate the modeling process and because they have several important features that are used to describe the fate and transport of toxins in biological systems.

18.7.1   Dynamic Model and the Half-Life

The most basic model, of course, is the **one-compartment model**, in which the compartment represents a whole organism. As a hypothetical case, consider how to model a fish that ingests zooplankton contaminated with a hydrocarbon. Having decided on the one-compartment model, we have finished the first step of the model development. We postulate only two processes: absorption by ingestion and elimination by kidney excretion. Let us suppose that the hydrocarbon is biotransformed completely. We treat the hydrocarbon and its metabolite as a single compound. Thus, it is not eliminated until the metabolite is excreted. Finally, let us assume that the metabolite is removed by the kidney by glomerular filtration only and is not reabsorbed. Thus, the rate of excretion, r_e, will be negatively proportional to the concentration in the blood plasma:

$$r_e = -k_e C \qquad (18.19)$$

where C is the concentration of hydrocarbon plus metabolite and k_e is a coefficient related to the renal clearance rate. Note that this mass transfer process can be formulated as a rate instead of a flux. The same will be true for ingestion. The rate of absorption, r_a, is the product of the assimilation efficiency, α (the fraction of ingested toxicant that is absorbed), the mass of food ingested per unit time, W, and the average concentration of toxicant in the food, C_f (in units of mass of solute per unit mass of food):

$$r_a = \alpha W C_f \qquad (18.20)$$

This situation is shown schematically in Figure 18.8. Equations (18.19) and (18.20), having units of $[M \cdot t^{-1}]$, must be multiplied by the volume before substitution into equation (18.13) along with equation (18.14). Canceling the volume yields

$$\frac{dC}{dt} = \alpha W C_f - k_e C \qquad (18.21)$$

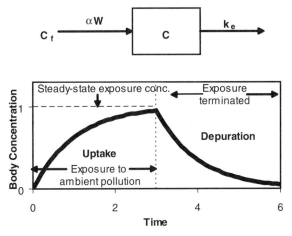

Figure 18.8   Schematic of a one-compartment model, and a plot of uptake and depuration of a contaminant in the case of a fixed environmental concentration C_f.

For the initial condition that $C_i = 0.0$ at $t = 0.0$, equation (18.21) can be solved analytically:

$$C = \frac{\alpha W}{k_e} C_f (1 - e^{-k_e t}) \tag{18.22}$$

This represents a period of increasing **body burden** (the mass of contaminant contained by an organism), as shown in Figure 18.8. If the contamination in the food supply were suddenly eliminated, uptake would cease, and the only process would be elimination. In aquatic biology, this situation is called **depuration**. Depuration is used commercially to eliminate contaminants from shellfish grown in polluted waters by transferring them to clean waters for a period prior to marketing. The model of equation (18.21) applies, with C_f set to zero. This leaves the following:

$$\frac{dC}{dt} = -k_e C \tag{18.23}$$

This is a first-order decay model. The solution, starting from an initial concentration of C_i, is

$$C = C_i e^{-k_e t} \tag{18.24}$$

Thus, first-order decay results in an exponential decrease in concentration. This has the important property that the concentration decreases by equal fractions in equal time intervals. For example, the time it takes for the concentration to drop from 100 mg/L to 50 mg/L is the same as the time to go from 2.0 mg/L to 1.0 mg/L. The time it takes for the concentration to drop by half is called the **half-life** ($t_{1/2}$). It is used as a convenient index of how rapidly a toxin is eliminated. The rate coefficient gives the same information but does not have such a tangible interpretation. In fact, the half-life and the rate coefficient can be computed from one another. From equation (18.24), using $C/C_i = 0.5$ yields

$$t_{1/2} = \frac{-\ln(0.5)}{k_e} = \frac{0.693}{k_e} \tag{18.25}$$

Other characteristics times, such as t_{90}, the time required for a 90% reduction, could be computed in a similar way. Keep in mind that the half-life and similar parameters can be used only if the elimination can be approximated as first order. (See Problem 18.3 for an example where this is not the case.) Conveniently, this is true quite often, and half-lives have been measured for a wide variety of compounds.

Example 18.6   Predatory animals living in a contaminated ecosystem are found to have a tissue concentration of a toxin equal to 80 mg toxin per kilogram of tissue. Samples of their usual prey were found to have a body burden of 10 mg/kg. Several animals are taken into captivity and fed food without the toxin. After 100 days of this treatment, the body burden is found to have decreased by 40%. Assuming that the animals were at a steady-state body burden when collected and that the assimilation efficiency equals 1.0, estimate the elimination rate coefficient, the half-life, and the mass loading rate in the wild.

Answer   Rearranging equation (18.24) gives us $k_e = -\ln(C/C_i)/t = 0.511/(40\,\text{days}) = 0.0128\,\text{day}^{-1}$. From equation (18.25), $t_{1/2} = 0.693/k_e = 0.693/(0.013\,\text{day}^{-1}) = 54.3$ days. Assuming steady state and that $\alpha = 1.0$, equation (18.21) becomes $0 = WC_f - k_e C_{ss}$. Therefore,

$$W = k_e \frac{C_{ss}}{C_f} = (0.0128\,\text{day}^{-1}) \left(\frac{80\,\text{mg toxin}}{\text{kg body weight}} \right) \left(\frac{\text{kg food}}{10\,\text{mg toxin}} \right)$$

$$= 0.102 \frac{\text{kg food/day}}{\text{kg body weight}}$$

The animal must eat one-tenth its body weight each day to maintain the stated body burden.

18.7.2   Steady-State Model and Bioaccumulation

The uptake and retention of contaminants in tissues in excess of concentrations in the source of the contaminant (such as food or water) is called **bioaccumulation**. In Figure 18.8, uptake is being held constant, but elimination increases as concentration increases. A situation is approached asymptotically where elimination balances uptake, and the system will be at steady state. The concentration at which this will occur can be determined in two equivalent ways, as mentioned above. One would be to take the limit of equation (18.22) as $t \to \infty$. The other is to set equation (18.21) to zero and solve for C. Either way gives the same result for the steady-state concentration, C_{ss}. The ratio of C_{ss} to C_f is the **bioaccumulation factor**, K_B (also designated BAF):

$$K_B = \frac{C_{ss}}{C_f} = \frac{\alpha W_f}{k_e} \tag{18.26}$$

Bioaccumulation can occur with other routes of exposure. One would develop different models for K_B if this were the case. In this case, since C_{ss} is expressed per unit body weight and C_f is per weight of food, K_B will depend on the mass of food ingested per day per body weight. The caloric requirement of organisms is related to their surface area; therefore, smaller animals tend to have higher bioaccumulation by ingestion (although other routes of exposure may be more important). It can also depend on other things that influence metabolic rate, such as stress or temperature. If elimination occurs predominantly by interphase mass transfer, as is often the case with xenobiotic compounds, k_e will be inversely related to the partition coefficient, making K_B directly related. This accounts for the observation that bioaccumulation is strongly correlated with K_{OW}, especially if the concentration in fatty tissues is used in place of whole body concentration.

Example 18.7 In Example 18.6, what is the bioaccumulation factor?
   Answer $K_B = 80 \, \mathrm{mg/kg}/(10 \, \mathrm{mg/kg}) = 8$; also, $K_B = \alpha W_f/k_e = (1.0) \, (0.102 \, \mathrm{day}^{-1})/ 0.0128 \, \mathrm{day}^{-1} = 8$.

18.7.3 Equilibrium Model and Bioconcentration

Other mechanisms can give proportional bioaccumulation relationships. For example, consider a one-compartment model in which both uptake and elimination are by mass transfer between the environmental concentration, C_{env}, and the concentration within the organism, C_{org}. For example, this might apply to the uptake of hydrocarbons from the air by terrestrial animals or from the water for aquatic animals. In such cases, uptake and elimination could be described by equation (18.12), with mass transfer coefficients k_u and k_e for the respective rates r_u and r_e:

$$r_u = k_u(K_P C_{\mathrm{env}} - C_{\mathrm{org}})$$
$$r_e = -k_e(K_P C_{\mathrm{env}} - C_{\mathrm{org}}) \tag{18.27}$$

Assuming steady state, the mass balance leads to the relationship that the rate of uptake equals the rate of elimination:

$$r_u = r_e \tag{18.28}$$

The process in which bioaccumulation occurs by direct uptake from the environment is called **bioconcentration**. It is distinguished from bioaccumulation in that bioaccumulation includes all routes of exposure, whereas bioconcentration only considers uptake directly from the environmental medium in which the organism lives. The term *bioconcentration* is usually reserved for aquatic systems.

Equations (18.27) and (18.28) can be combined and solved for the ratio of organism concentration to environmental concentration, called the **bioconcentration factor** (K_C, also designated **BCF**). In this case,

$$K_C = \frac{C_{\mathrm{org}}}{C_{\mathrm{env}}} = \frac{k_u K_P + k_e K_P}{k_u + k_e} = K_P \tag{18.29}$$

Thus, for this model the bioconcentration factor is equal to the partition coefficient. Again, it must be emphasized that both the bioaccumulation factor and the bioconcentration factor can be derived from different models using different assumptions. For example, if we assume simple first-order rate processes for uptake and elimination:

$$\frac{dC_{\mathrm{org}}}{dt} = k_u C_a - k_e C_{\mathrm{org}} \tag{18.30}$$

where in this case k_u and k_e are the uptake and elimination rate coefficients, respectively, then the bioconcentration factor would be written as

$$K_C = \frac{k_u}{k_e} \tag{18.31}$$

Thus, the bioconcentration and bioaccumulation factors are not defined uniquely in terms of a particular model. More generally, they are defined as the ratio of organism concentration to the concentration in the environment and/or the food supply.

Whatever the theoretical mechanisms are to explain the bioconcentration factor, experimental measurements have shown that it correlates with K_{OW}. For example, measurements with 64 organics in fish have been used to develop the following empirical relationship:

$$\log_{10} K_C = 0.76 \log_{10} K_{OW} - 0.23 \qquad r^2 = 0.82 \tag{18.32}$$

Relationships such as this should be used cautiously, however. Although they are useful over a wide range of K_{OW}, at a particular values the upper and lower 95% confidence levels can differ by a factor of 63. Furthermore, numerous correlations of this form have been developed for different groups of chemicals and different species. Some studies have correlated K_C with aqueous solubility instead of K_{OW}. Several of these physicochemical parameters are given in Table 18.5.

A more complete model would separately take into account the processes of environmental uptake, ingestion, biotransformation, and excretion:

$$\frac{dC_{\text{org}}}{dt} = \alpha W C_f + k_u C_{\text{env}} - k_e C_{\text{org}} \tag{18.33}$$

Assuming steady state, rearranging, and substituting for K_C and K_B from equations (18.26) and (18.31) shows the contribution to body burden by environmental uptake and ingestion:

$$C_{\text{org}} = K_C C_{\text{env}} + K_B C_f \tag{18.34}$$

TABLE 18.5   Bioconcentration Factors and Physicochemical Parameters for Some Compounds That Bioaccumulate

Compound	Log K_{OW}	$\text{Log}_{10} K_C$	C_S (mg/L)
Chlordane	5.51	4.05	0.056
DDT	5.98	4.78	0.0017
Dieldrin	5.48	3.76	0.022
Lindane	4.82	2.51	0.150
Chlorpyrifos	4.99	2.65	0.3
Cycexatin	5.38	2.79	<1.0
2,4-D	1.57	1.30	900
2,4,5-T	0.6	1.63	2.38
TCDD	6.15	4.73	0.0002

Source: LaGrega et al. (2001); original source D. W. Connell (1990), *Bioaccumulation of Xenobiotic Compounds*, CRC Press, Boca Raton, FL.

Note that K_B in this case should be interpreted as an accumulation factor for ingestion alone. More generally, bioaccumulation includes all sources of contaminants.

18.7.4 Food Chain Transfer and Biomagnification

A bioaccumulation model such as equation (18.34) can be applied separately to each trophic level in an ecosystem, with the food at one level being the organism concentration at the lower level:

$$C_i = K_{C,i}C_{\text{env}} + K_{B,i}C_{i-1} \qquad (18.35)$$

Here C_i is the concentration of toxin in the organisms of trophic level i, C_{i-1} is the concentration of the next-lower trophic level, K_{Ci} is the bioconcentration factor for trophic level i, and K_{Bi} is the bioaccumulation factor for trophic level i.

For example, suppose that the environmental concentration of a toxic substance is 1 mg/kg and $K_C = K_B = 10$ for all trophic levels. The primary producers will have a concentration of 10 mg/kg due to bioconcentration alone. The herbivores will have $C = 10 \cdot 1 + 10 \cdot 10 = 110$ mg/kg. The concentration in carnivores will be $C = 10 \cdot 1 + 10 \cdot 110 = 1110$ mg/kg. Secondary carnivores will have a concentration 11,110 mg/kg.

An increase of tissue concentration of a toxic substance with increasing trophic level is called **biomagnification**. Figure 18.9 illustrates the relationship for two food chains. Substances that bioaccumulate also tend to biomagnify. Thus, the tendency to biomagnify is usually also related to the K_{OW}. However, a low rate of elimination, especially by biotransformation, is also necessary. These conditions are particularly true for chlorinated organics. Biomagnification can produce extremely high body burdens in organisms at

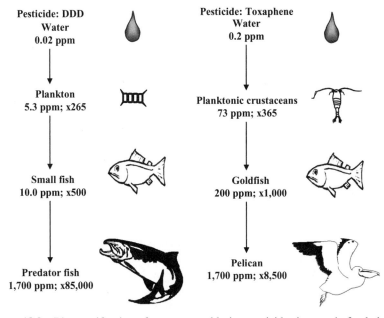

Figure 18.9 Biomagnification of two organochlorine pesticides in aquatic food chains.

TABLE 18.6   Thoman's Food Chain Multipliers

Log K_{OW}	Trophic Level		
	2	3	4
3.5	1.0	1.0	1.0
4.0	1.1	1.0	1.0
4.5	1.2	1.2	1.2
5.0	1.6	2.1	2.6
5.5	2.8	5.9	11.0
6.0	6.8	21	67.0
6.5	19	45	100.0

Source: U.S. EPA (1994).

the top of the food chain, the secondary carnivores or predators. The bioaccumulation factor (and therefore, biomagnification) can be related to the bioconcentration factor by a **food chain multiplier (FM)**:

$$BAF = FM \cdot BCF \tag{18.36}$$

The U.S. EPA recommends using food chain multipliers based on K_{OW} and trophic level as shown in Table 18.6. At log K_{OW} values higher than those in the table, the FM becomes uncertain and may even decrease due to slow transport kinetics and bioavailability. Thus, if log K_{OW} is 5.0, the bioaccumulation factor for carnivores (trophic level 3) will be about twice the bioconcentration factor.

Because organochlorine pesticides are not only lipophilic and persistent but also toxic by their nature, they are the substances most capable of causing harm by biomagnification. Furthermore, biomagnification limits the effectiveness of pesticides. Since in agriculture pesticides are often used to control plant pests, they are applied at levels toxic to the base of the food pyramid. If the pesticide biomagnifies, this will necessarily result in more harm to predatory insects, which are the ones needed to control the pests. Thus, ironically, it was sometimes found that applications of these substances resulted in increased pest problems. For these reasons, organochlorine pesticides have been replaced in many applications by other pesticides that do not biomagnify.

18.7.5   Multicompartment Models

If sufficient information is available, or if contaminant behaviors that are more complex are to be studied, multiple-compartment models can be used. For example, Figure 18.10 shows a model that could be used to predict narcosis by inhalation of a hydrocarbon. A separate mass-balance equation must be written for each compartment. Each arrow represents a flux, and a term must be added to the liver compartment for elimination by biotransformation. Although it may not be possible to define a half-life or a bioaccumulation factor for some of the more complex multicomponent systems, their behavior often may be approximated by one of those parameters.

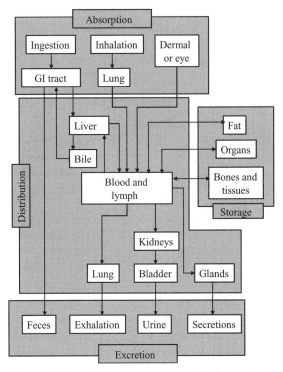

Figure 18.10 Multicompartment model of an animal.

18.8 EFFECT OF EXPOSURE TIME AND MODE OF EXPOSURE

Toxic effects can be affected by the time distribution of exposure. Consider three cases: (1) a single, instantaneous exposure; (2) a continuous, chronic exposure; and (3) periodic or intermittent exposures. The difference between these modes will depend on the type of effect, such as whether it is a local or a systemic toxin, whether or not there is a threshold, and whether the damage can be repaired or accumulates. If there is a threshold, or if the effect depends on the concentration of the toxicant or a metabolite in a compartment, the effect would be correlated with the concentration in the compartment containing the receptor. If the damage accumulates, an effect could be seen even at a low dose if sufficient prior doses were applied. Here we might expect the effect to correlate with the time integral of the concentration in the compartment.

For the first case, Figure 18.11 shows how concentration will vary for an instantaneous dose in a two-compartment model. The first, or systemic, compartment, peaks immediately. This is what would be found in the bloodstream if a substance were to be injected intravenously by hypodermic syringe, for example. The systemic compartment then clears by some process, such as the first-order process illustrated. The second compartment represents one containing a receptor, such as neurons in the brain. It exchanges toxicant only with the systemic compartment. Its concentration increases until it exceeds that in the first compartment. Then it decreases slowly by exchange with the first compartment. Here there would not be a great difference with type of effect, except that integrated

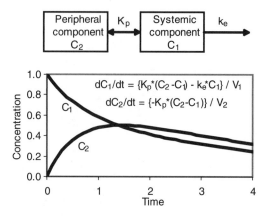

Figure 18.11 Typical concentration–time plot for two-compartment model. The systemic compartment typically represents blood concentration. The peripheral compartment can represent another tissue or organ, such as the brain or muscle. C and V are the concentration and volume, respectively, of the compartments. In this simulation k_e, K_p, and $V_2 = 1.0$, and $V_1 = 3.0$.

effects would take longer to appear. Biotransformation reactions are neglected in these examples. If the system behaved more like a one-compartment model, the toxicant would be cleared in a depuration process similar to that shown in the second part of Figure 18.8.

Figure 18.12 shows a similar plot for the second two cases. Intermittent exposure results in higher peak concentrations, which may cause a greater effect either if the effect is concentration associated or if there is a threshold. On the other hand, the periodic decrease in concentration may enable repair mechanisms to eliminate previous damage. This has been observed in exposing rats to ozone. A concentration that caused pulmonary edema when applied continuously had no effect when applied intermittently. However, if damage is cumulative, the integral of both curves will be similar, as will be their effect. An example is the application of organophosphate pesticides, which cumulatively deplete acetocholinesterase. There is a repair mechanism, in the form of hydrolysis of the enzyme-OP compound and replacement of the enzyme, but it can be overwhelmed at moderate doses.

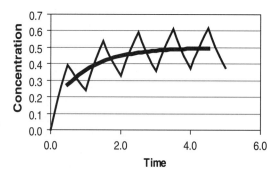

Figure 18.12 Concentration–time plot for continuous and intermittent exposure.

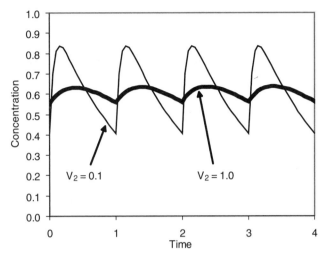

Figure 18.13 Effect of differing pharmacokinetics on the time course of concentration in an organism. k_e, K_p, and $V_1 = 1.0$, and $V_2 = 1.0$ or 0.1 as indicated.

Finally, the same dosing regimen could result in very different concentration–time relationships, depending on the pharmacokinetics involved. Slow adsorption and elimination or the presence of a large storage compartment could smooth out changes in concentration. In Figure 18.13 the concentration in the peripheral compartment does not vary greatly when the two compartments have the same volume ($V_1 = V_2 = 1.0$). But when the peripheral compartment is much smaller ($V_2 = 0.1$), large concentration swings occur.

PROBLEMS

18.1. The OSHA eight-hour inhalation standard for phenol is approximately $19\,mg/m^3$. What is this in ppmv?

18.2. The OSHA eight-hour inhalation standard for chloroform is 50 ppmv, and its Henry's constant is 0.148 (conc./conc.). If wastewater flowing through a sewer was contaminated with 0.50 mg of chloroform per liter, the air space in the sewer and in the manholes would eventually achieve equilibrium with the chloroform in the water. Would the air concentration of the TCE in the manhole exceed the OSHA limit?

18.3. Use equations (18.7) and (18.32) to estimate the octanol–water partition coefficient and the bioconcentration factor of chlordane. Compare to the values given in Table 18.5.

18.4. Ethyl alcohol is eliminated from the body by a zero-order process. That is, it is eliminated by a constant mass per unit time. Assume that the removal rate is

9.0 g/h. If a can of beer contains 12 g of alcohol, how long does it take to eliminate all of the alcohol from the body? (Assume a one-compartment model.) If a person drinks one can of beer per hour, how much alcohol will be present 1 hour after drinking the sixth can? Why would ethanol be eliminated by a zero-order process instead of a first-order process?

18.5. For the case of Example 18.6, use equation (18.21) (solved for t) to compute how long it would take for the body burden of an individual predator to increase from zero to 40.0 mg/kg after it is placed in the contaminated environment. Note that 40.0 mg/kg is halfway from the initial to the steady-state condition. How does the time required compare to the half-life for elimination? Why?

18.6. What do narcosis-causing compounds have in common with compounds that tend to bioaccumulate? What properties would cause a compound to cause narcosis but not bioaccumulate?

18.7. Why would narcosis-causing compounds tend to be less amenable to biological wastewater treatment by the activated sludge process?

18.8. Convert equation (18.7) to the nonlogarithmic form. What does the result say about the qualitative relationship between K_{OW} and C_s? (Specifically, what is the shape of the curve of K_{OW} vs. C_s?)

18.9. The pK_a value of pentachlorophenol is 4.75. What will be the concentration of undissociated pentachlorophenol in a solution with a total pentachlorophenol concentration of 100 mg/L at pH 6.0? At pH well below the pK_a, the adsorption partition coefficient for pentachlorophenol to a soil is 840; at pH well above the pK_a, it is 6.10. What will be the effective adsorption coefficient at pH 4.75? At pH 6.0?

18.10. If the environmental half-life of an organophosphorus pesticide were 100 days, compute the rate coefficient, k, and t_{99}.

18.11. What do bioconcentration, bioaccumulation, and biomagnification have in common? What distinguishes them from each other?

REFERENCES

Austin Community College, Chemistry Department, 2004. www.austin.cc.tx.us/chemlab/weakbase.htm.

Chiou, C. T., P. E. Porter, and D. W. Schmedding, 1983. Partition equilibria of nonionic organic compounds between soil organic matter and water. *Environmental Science and Technology*, Vol. 17, No. 4, pp. 227–231.

Christodoulatos, C., and M. Mohiuddin, 1996. Generalized models for prediction of pentachlorophenol adsorption by natural soils, *Water Environment Research*, Vol. 68, No. 3.

Jørgensen, S. E., B. Halling-Sørensen, and H. Mahler, 1998. *Handbook of Estimation Methods in Ecotoxicology and Environmental Chemistry*, CRC Press, Boca Raton, FL.

LaGrega, M. D., P. L. Buckingham, and J. C. Evans, 2001. *Hazardous Waste Management*, McGraw-Hill, New York.

Lu, F. C., 1991. *Basic Toxicology: Fundamentals, Target Organs, and Risk Assessment*, 2nd ed., Hemisphere Publishing, Bristol, PA.

Petrucci, R. H., W. S. Harwood, and F. G. Herring, 2001. *General Chemistry*, Prentice Hall, Upper Saddle River, NJ.

U.S. EPA, 1994. *Water Quality Standards Handbook*, 2nd ed., EPA/823-B94/005a, U.S. Environmental Protection Agency, Washington, DC, August.

Williams, P. L., and J. L. Burson (Eds.), 1985. *Industrial Toxicology: Safety and Health Applications in the Workplace*, Van Nostrand Reinhold, New York.

19

DOSE–RESPONSE RELATIONSHIPS

Toxicity is a relative measure of the ability of an agent to cause a harmful effect on a living organism. All substances have toxic properties. Even water is an irritant to the skin, and oxygen is toxic to humans at a high enough partial pressure and duration of exposure. On the other hand, some substances that are common industrial toxins are beneficial or even necessary to life at lower doses. This is particularly true with some of the metals, such as chromium or nickel. Even ionizing radiation can fit this category. Ultraviolet radiation from the sun converts 7-dehydrocholesterol in the skin into a form of vitamin D, a necessary nutrient. Figure 19.1 compares the type of response of necessary compounds with those that are not. Curve (a) represents the case in which the substance is a required nutrient; curve (b) is the case for a substance that is not required. For case (a) there is a deficiency of the compound below the concentration C_1, whereas it is toxic above C_2. Concentration C_1 would be related to the **minimum daily dietary requirement** for the substance, and C_2 is related to the experimental value, called the **no observed adverse effect concentration** (NOAEC).

The situation of curve (a) in Figure 19.1 brings to mind a famous statement by the sixteenth-century physician Paracelsus, often paraphrased as "the dose makes the poison": "All substances are poisons; there is none which is not a poison. The right dose differentiates a poison and a remedy."

Table 19.1 lists a number of substances that can be both toxic and essential. In the spirit of Paracelsus, then, we must be interested in determining the **dose–response relationship**: what dose of a toxin produces what level of undesirable effect. Leaving until later the discussion of selecting the particular effect of concern, and the experimental details, let us assume situations similar to the following examples:

1. Two hundred white rats are divided into 10 groups of 20; each rat receives a one-time subcutaneous (below the skin) injection of a quantity of a toxic substance

Environmental Biology for Engineers and Scientists, by David A. Vaccari, Peter F. Strom, and James E. Alleman
Copyright © 2006 John Wiley & Sons, Inc.

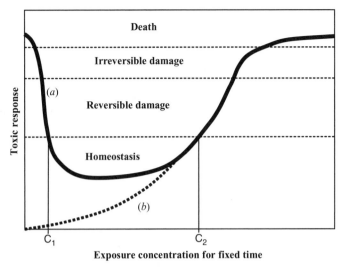

Figure 19.1   Hypothetical dose–response relationship for (*a*) a substance required at low dosages and (*b*) a nonessential substance.

adjusted to the weight of the individual rat. The number of rats with tumors is counted at 80 days.

2. Ten test tubes are prepared, each containing 20 *Daphnia*, a small crustacean. The concentration of a toxicant in the water of each tube is adjusted to a different concentration. Percent mortality is measured at 96 hours.

In both cases, the individuals in the 10 groups each receive different dosages among the following: 0 (control), 1, 2, 5, 10, 25, 50, 100, 250, and 500. The units in the former case may be milligrams toxicant per kilogram of body weight; for the latter it could be mg/L.

TABLE 19.1   Several Substances That Are Required for Growth But Are Toxic at Higher Levels

Substance	Recommended Dietary Allowances (mg/day)	Toxic Level
Iodine	0.15	
Zinc	15	60 mg/day (LOAEL)
Selenium	0.05–0.2	0.8 – 1.0 mg/day
Copper	2.0–3.0	>3.0 mg/day
Chromium	0.05–0.2	
Nickel		3500 mg/day
Iron	10 in males, 18 in females	0.8 mg/kg/day
Potassium	1875–5625	
Sodium	2400	3500 mg/day
Vitamin D	10 μg	100 μg
Vitamin A	800 male, 1000 female (μg RE[a])	500,000 μg RE
Vitamin E	8 male, 10 female (mg TE[a])	800–3200 mg/day

[a]RE, retinol equivalents; TE, α-tocopherol equivalents.

Dose can have different meanings, as the examples show. The use of mass per unit body weight is ideal, as it ensures that the exposure of each individual is accurately known. However, concentration units are often appropriate, especially with reference to environmental pollution (mass/volume in air or water, or mass/mass in solids such as food or soil). The concentration can refer to that of ambient air or water, of ingested food or water, or of a preparation to be applied to the skin or elsewhere. Or, in the case where the toxicant is a complex mixture such as wastewater treatment plant effluent, one uses dilutions measured in percent (volume/volume). In these cases it is more proper to talk about the **dose–concentration relationship**.

Dose–response relationships can be divided into two types, based on the kind of information used to describe them. One is the simpler empirical model, which begins with assumptions about the shape of the dose–response relationship. These are called **tolerance distribution models**. The other type is fundamental, based on knowledge or assumptions about how a toxin acts. These are called **mechanistic models**.

19.1   TOLERANCE DISTRIBUTION AND DOSE–RESPONSE RELATIONSHIPS

Due to individual variability, we do not expect a sharp threshold for an effect; that is, a single value with no mortality at lower dose, and 100% mortality at a dose above that level. If the range of concentrations is chosen properly, the usual situation will exhibit a sigmoidal response vs. either the dose or the logarithm of the dose (Figure 19.2a). That is, as dosage increases, the percent mortality will increase in a smooth, increasing curve. The logarithm of the dose is often used, especially when the mean of the frequency distribution is no more than two or three standard deviations above zero. This prevents the theoretical problem of the distribution predicting toxic effects at negative dosages.

Each point on Figure 19.2a is the percent response, P_i, at one dosage, d_i. In the *Daphnia* experiment example, this would be the result from one of the test tubes. If a very large number of test tubes were tested covering a much larger number of different dosages (d_i), a histogram could be prepared showing the frequency distribution for mortality, called a **tolerance distribution**, $T(d)$. In typical toxicity tests there aren't enough dosage levels to do this. Instead, $T(d)$ can be estimated by numerically differentiating the dose–response curve, $P(d)$, using successive pairs of points (P_i, d_i), (P_{i+1}, d_{i+1}), plotted at the midpoint of each pair:

$$T \frac{d_i + d_{i+1}}{2} \simeq \frac{P_{i+1} - P_i}{d_{i+1} - d_i} \tag{19.1}$$

A sigmoidal dose–response curve would result in a bell-shaped tolerance distribution as a function of dose. This leads naturally to an assumption of the normal distribution for the shape of $T(d)$ (Figure 19.2b). If this is a valid assumption, the dose–response curve will be given by the integral of the normal distribution:

$$P(d) = \int_{-\infty}^{Y} \frac{1}{\sqrt{2\pi}} \exp\left(\frac{-u^2}{2}\right) du \qquad Y = \alpha + \beta x \tag{19.2}$$

where x is either the dose, d, or the logarithm of the dose, as described above. The distribution has two parameters, α and β, related to the mean, d_{mean}, and standard deviation,

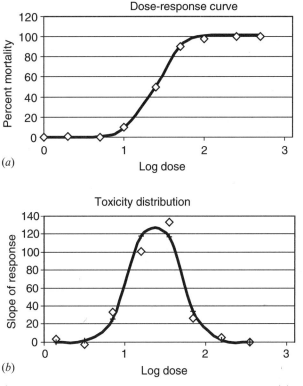

Figure 19.2   (*a*) Sigmoidal logarithmic dose–response relationship; (*b*) normal tolerance distribution of mortality frequency.

s, of the tolerance distribution. Several methods are available for estimating them. They could be computed directly from the points of $T(d)$ computed using the **method of moments**. The procedure is as follows. First, equation (19.1) is used to compute $T(d)$. Then the following equations are used to compute the mean and standard deviation:

$$d_{\text{mean}} = \frac{\sum_i d_i T(d_i)}{\sum_i T(d_i)} \tag{19.3}$$

$$s = \sqrt{\frac{\sum_i d_i^2 T(d_i)}{\sum_i T(d_i)} - d_{\text{mean}}^2} \tag{19.4}$$

Several regression methods are also available, including nonlinear regression and maximum likelihood estimation. These have the advantage of producing not only the parameters but also a confidence interval for the dose–response curve predicted. The disadvantage of all the foregoing methods is that they may produce poor results if the data are not normally distributed or if the data do not cover the range of toxic effects well. That is, the data must include at least one dosage with essentially zero response and one with about 100% response. For example, if a bioassay on a wastewater produces 30% mortality at 100% concentration, these methods will not work.

TABLE 19.2 Qualitative Classification of Toxic Compounds

Classification	LD$_{50}$ Range (mg/kg)	Example	LD$_{50}$ of Example
Supertoxic	5 or less	Dioxin in guinea pig	0.002
Extremely toxic	5–50	Parathion in goats	42
Highly toxic	50–500	DDT in rat	100
Moderately toxic	500–5,000	Strychnine in rat	2,000
Slightly toxic	5,000–15,000	Ethanol in mouse	10,000
Practically nontoxic	> 15,000		

There are several procedures, called **nonparametric methods**, that do not make distributional assumptions. One of these is the *trimmed Spearman–Karber method*, which is a numerical procedure for estimating the centroid of the tolerance distribution. Another is the *binomial confidence interval*, which is used when the data do not include partial kills. The nonparametric methods, however, often cannot estimate the standard deviation of the tolerance distribution.

The d_{mean} corresponds to the dose at which half of an exposed population would be expected to suffer the effect. This is called the **median lethal dose** (LD$_{50}$) or **median effective dose** (ED$_{50}$). Of course, LD, refers only to results of tests where the measured response is mortality; ED, on the other hand, can be based on any toxicological effect. If the dose is measured in concentration units, the median may be refered to as *lethal concentration* (LC$_{50}$) or *effective concentration* (EC$_{50}$) instead. If the effect is reduction in growth rate, as commonly applied to microorganisms, a parameter known as the **median inhibitory concentration** (IC$_{50}$) may be used. This is the concentration that would reduce the growth rate by 50%. These parameters are commonly used as a basis to compare the toxicity of various toxic agents. A common qualitative classification of LD$_{50}$ is given in Table 19.2. Keep in mind that although only 50% may experience mortality at a given dosage, it is likely that all the organisms will suffer deleterious toxicological effects but that some may recover or at least survive. In the following discussion of toxicity, we will refer to LD$_{50}$, although the same principles apply to the other measures of median effect.

Of course, it must be noted that the LD$_{50}$ is a property not only of the toxic agent but also of experimental variables such as the organism, the time scale of the test, the method of dosing, and so on. One can also refer to other percentiles, such as the lethal dose that causes 1% or 10% mortality, the LD$_{01}$ or LD$_{10}$, respectively.

The standard deviation, s, is related to the "steepness" of the dose–response relationship. A steep slope indicates that the organism responds strongly to increases in dosage. A flatter response curve can be caused by slow absorption, rapid excretion or detoxification, or delayed bioactivation. In a normal distribution, 68.3% of the population will be within one standard deviation of the mean, 95.5% within two, and so on. Table 19.3 summarizes the relationship. Precise standard deviations are given for several of the percentiles of interest (e.g., 1% and 10%). These can be used to compute LD$_{01}$ and LD$_{10}$ from s and LD$_{50}$, keeping in mind that the standard deviation is computed in log-transformed units:

$$LD_{01} = LD_{50} \times 10^{-2.326s} \tag{19.5}$$

$$LD_{10} = LD_{50} \times 10^{-1.282s} \tag{19.6}$$

TABLE 19.3   Normal Distribution and the Relation to Probits

Standard Deviations from the Mean	Probits	Cumulative Percentage
−3	2	0.14
−2	3	2.28
−1	4	15.9
0	5	50.0
1	6	84.1
2	7	97.7
3	8	99.87
−3.090	1.910	0.10
−2.326	2.674	1.00
−1.282	3.718	10.0
−0.674	4.326	25.0

Thus, you should be able to see that two compounds could have the same LD_{50}, but if one has a larger value of s (a less steep dose–response curve), its LD_{10} and LD_{01} will be lower. This has important consequences for the use of toxicological data in environmental applications. A mortality of 50% would be far too high to be tolerated in real life. However, handbooks often report only the LD_{50}. Thus, the potential exists to greatly overestimate, or worse, underestimate the relative toxicity of a compound. Thus, it would be useful for slope information or lower percentile LD values to be given also. However, these data are not commonly available. Note, on the other hand, that because the dose–response relationship is flatter near the LD_{01} or LD_{10}, the uncertainty in their values will be much greater than for LD_{50}.

Example 19.1   Table 19.4 contains hypothetical results from an aquatic bioassay on the toxicity of wastewater to *Daphnia*. The first column, d, is the dosage as the percentage of wastewater in the solution to which the *Daphnia* were exposed. The second column, n, is the total number of organisms tested at each dosage; the third column, m, is the number killed after 48 hours of exposure; and M is the percent mortality, m/n. Since the control ($d = 0\%$) shows 4.7% mortality, we should first decrease both n and m by 4.7% of m. The ratio of the resulting numbers is the adjusted percent mortality, M'. A plot of M' versus d gives the dose–response curve shown in Figure 19.2a.

TABLE 19.4   Hypothetical Aquatic Bioassay Results and Tolerance Distribution Analysis

d	n	m	M	M	d	$T(d)$	$d \times T$	$d^2 \times T$
0%	43	2	4.7%	0.0%	20%	0.103	0.021	0.004
40%	35	3	8.6%	4.1%	45%	1.309	0.589	0.265
50%	38	8	21.1%	17.2%	55%	2.877	1.582	0.870
60%	33	16	48.5%	46.0%	65%	1.994	1.296	0.843
70%	40	27	67.5%	65.9%	75%	2.305	1.728	1.296
80%	38	34	89.5%	89.0%	85%	0.521	0.443	0.377
90%	36	34	94.4%	94.2%	95%	0.265	0.252	0.239
100%	33	32	97.0%	96.8%		9.374	5.911	3.894

From the data we can see that 50% mortality falls between dosages of 60 and 70%. By interpolating linearly between these two points, we can estimate LC_{50} as 62.0%. Similarly, LC_{10} falls between 40 and 50% concentration, and by interpolation we find LC_{10} to be 44.5%. However, we would expect the latter value to be too high, because the response curve in this region will be curved upward. Better accuracy can be obtained using the method of moments. First, it is necessary to compute the tolerance distribution using equation (19.1). This gives the results shown in the sixth and seventh columns of Table 19.4 [d and $T(d)$] and plotted in Figure 19.2b. Note that the dosages correspond to the midpoints between those in the first column.

Equations (19.2) and (19.3) require the sums of T, dT, and d^2T. These values are tabulated in the last three columns of Table 19.4, and the sums are the numbers in the bottom row. Thus, the mean of the distribution is $d_{mean} = 5.911/9.374 = 63.1\%$, and $s = (3.894/9.374 - 0.631^2)^{1/2} = 13.3\%$. Thus, $LC_{50} = 63.1\%$, and from equations (19.5) and (19.6), $LC_{10} = 42.6$ and $LC_{01} = 30.9\%$. You can see that in this case, the interpolation estimates of LC_{50} and LC_{10} were not far off and that the LC_{10} was overestimated, as expected. The curves in Figure 19.2a and b are the normal distribution with the mean and standard deviation from this example.

Toxicology retains a curious legacy from precomputer days. To simplify hand calculations involving the normal distribution, toxicologists avoided the use of negative numbers by the expedient of adding an arbitrary value of 5 to the standard deviations. The resulting units are called **probits**. Thus, -2 standard deviations from the mean is the same as $+3$ probits (see Table 19.3). The median, or LD_{50} (0 standard deviations) becomes 5 probits. This has become standard practice and remains in use today.

The assumption that the tolerance distribution is normal is not based on fundamental principles, and a number of other distributions have been proposed. The best known is the **logit**, in which the following equation replaces equation (19.2):

$$P = \frac{e^Y}{1 + e^Y} \tag{19.7}$$

Another is the **logistic function**, which is based on *hit theory* (described in the next section):

$$P = \frac{1}{1 + e^{-(\alpha + \beta x)}} \tag{19.8}$$

These models have advantages over the probit model in that the parameters of the logit model can be estimated by linear least-squares methods, and the logistic equation has a mechanistic basis. However, the probit or log-probit model is more established. The practical differences between these models are not very great, and their selection is often a matter of personal preference and custom.

19.2   MECHANISTIC DOSE–RESPONSE MODELS

These models developed from hit theory explanations for carcinogenesis, and cancer remains their most important application. The **one-hit model** (originally developed for radiation effects) fits the assumption that a single reaction event can transform a cell to produce the toxic effect, and that the probability of a hit occurring is proportional to the dosage:

$$P(d) = 1 - \exp(-\beta d) \qquad \beta > 0 \tag{19.9}$$

However, the one-hit model did not fit data well. A multihit model improved this situation. It assumed that several hits were necessary for a toxic effect, all of which had the same probability. This led to the **multistage model**, to fit the theory of carcinogenesis, which requires several hits to the same cell, each having different probabilities, to produce a malignancy:

$$P(d) = 1 - \exp\left(-\sum_{i=1}^{k} \beta_i d^i\right) \qquad \beta_i \geq 0 \tag{19.10}$$

The parameter k is the number of stages. Although evidence may indicate many stages (e.g., five stages in the case of colon cancer described above), for statistical validity at least $k + 1$ different dosage levels must be tested to determine k coefficients. It is often cost prohibitive to have so many levels. Note that dosage in these models is not log-transformed, as is often done in the probit model.

Both of these models approach linearity as dose goes to zero. However, when equation (19.10) is extrapolated to doses far below those used in the calculation, the uncertainty becomes large. This is handled by replacing the linear term in equation (19.10) (the term with $i = 1$) with its upper 95% confidence limit. This results in the **linearized multistage model**, which is the preferred model in use by the U.S. EPA for carcinogenesis.

All of these models are linear in the low-concentration range. However, the one-hit model cannot describe a sigmoidal curve. The multistage model can, and has the further advantage over the probit or normal distribution model that it can describe asymmetric tolerance distributions.

19.3   BACKGROUND RESPONSE

For the probit model any toxic response in zero dosage control is usually subtracted from all the data. Thus, what is used in the calculations is the *excess* mortality or toxic effect. For stochastic biological models, where a hit can be caused by natural background causes, the response can be adjusted for the probability of that occurring. In this case, $P(d)$ is interpreted as the additional risk over background, which can be attributed to the dose d:

$$P(d) = \frac{P^*(d) - p}{1 - p} \tag{19.11}$$

where $P^*(d)$ is the observed response and p is the probability of a spontaneous hit. The value of p may be so small that it cannot be measured directly in laboratory animals. In such a case, it could be estimated by extrapolating dose–response data to zero dose, as described in the next section.

19.3.1   Low-Dose Extrapolation

For many toxic effects, especially carcinogenesis, the goal is to control environmental exposures to keep the risk at an extremely low level. A typical regulatory goal for a

TABLE 19.5   Data and Model Results for Dieldrin

Dose (ppm)	Fraction Having Tumors	Excess Risk	Predicted P	Predicted Excess Risk
0.0	$17/156 = 0.109$	0.000	0.107	0.000
1.25	$11/60 = 0.183$	0.074	0.213	0.106
2.5	$25/58 = 0.431$	0.322	0.389	0.282
5.0	$44/60 = 0.733$	0.624	0.746	0.640

Source: Crump et al. (1977); original source Walder et al., 1972, *Food Cosmet. Toxicol.* Vol. 11, pp. 415–432.

carcinogen is to keep lifetime risk below 10^{-6} (0.0001%). However, for practical and economic reasons, the dosages in laboratory experiments are selected to produce effects with probability no lower than 10^{-1}. The biological uncertainties with such an extrapolation are discussed below; here we discuss the mathematical aspects of the problem.

Table 19.5 and Figure 19.3 show a typical situation for the carcinogen dieldrin. The data are shown in the first and second columns of the table. The excess risk (the risk due to the toxin) is found by subtracting out the spontaneous risk (0.109 in this case). Cancer bioassays are commonly run at only two or three dosages, plus a control. In the example, the dosages at the three levels produced tumors in 18%, 43%, and 73% of the test animals. How, then, should one extrapolate to estimate the dosage that would produce tumors in 0.0001%? One of the simplest ways would be a linear extrapolation of the two lowest data points, (P_1, d_1) and (P_2, d_2). Thus, to compute the dose, d, resulting in risk P:

$$d = d_2 - \frac{P_2 - P}{P_2 - P_1}(d_2 - d_1) \tag{19.12}$$

Or, simple linear regression could be used on all the data. Both of these approaches, however, often produce a threshold below which the probability of harm would be

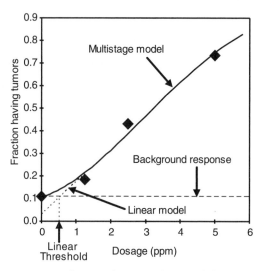

Figure 19.3   Dose–response curve for a carcinogen and extrapolation to low dosages by linear and multistage models.

predicted to be zero. For biological reasons described below, this model is ruled out for carcinogens. The next simplest model is linear extrapolation of the lowest data point to the origin:

$$d = \frac{P}{P_1} d_1 \qquad (19.13)$$

This produces a very conservative result, one producing a low predicted dose for a given risk. For example, using the data from Table 19.5, the lowest excess risk is 0.074 at a dosage of 1.25 ppm. Thus, the dosage that would produce a risk of 10^{-6} is

$$d = \frac{10^{-6}}{0.074} (1.25\,\text{ppm}) = 1.69 \times 10^{-5}\,\text{ppm}$$

Even more conservative is to use the upper confidence limit for P_1, in place of P_1 in equation (19.13). It is thought that every likely response would be below this line, making this the most conservative model. Ideally, one of the mechanistic models would apply instead of these empirical approachs, although two data or three points may be insufficient to estimate their coefficients.

The major models used by regulatory agencies are the multistage, one-hit, multihit, logit, probit, and Weibull models. The U.S. Occupational Health and Safety Administration (OSHA) has rejected the use of any models for setting standards. The U.S. EPA has selected the linearized multistage (LMS) model. A maximum-likelihood method was developed by Crump et al. (1977) to estimate the values of the β's. Crump et al. fitted the multistage model to the data in Table 19.5, obtaining

$$P = 1 - \exp[-(0.11296 + 0.05148\,d + 0.04007\,d^2)] \qquad (19.14)$$

The probability of an animal having tumors computed by this model is shown in the fourth column of Table 19.5. The rate of spontaneous tumors can be estimated from this model by setting d equal to zero, yielding 0.107 in the example.

At extremely low concentrations the quadratic term of the multistage model is insignificant, and the model is approximately linear with d. The maximum likelihood procedure also yields a 95% confidence interval around the prediction of the model. The upper confidence limit of the model also is linear with d at low doses. The slope of the upper confidence interval thus provides a conservative estimate of the sensitivity of excess risk to dose:

$$P = \text{CPF} \cdot d \qquad (19.15)$$

where CPF, the slope of the upper confidence interval, is called the **carcinogenic potency factor** or **slope factor**. For example, the CPF for chloroform is 0.13 kg·day/mg. Thus, when CPF is found from the slope of the upper 95% confidence interval of the multistage model, equation (19.15), constitutes the linearized multistage model.

In 1996, the U.S. EPA proposed changes in its approach to low-dose extrapolation. The proposed guideline recommends biologically based extrapolation when sufficient data are available. One method that is part of the proposal is a linear extrapolation based on the lower 95% confidence level of ED_{10}, which is called LED_{10}.

19.3.2   Thresholds

If a straight line is fitted to the three nonzero dosages and corresponding observed fractional incidences from Table 19.5 by least-squares linear regression, the resulting model is

$$P = 0.0322 + 0.143\, d \tag{19.16}$$

Extrapolation of this linear model at low doses is shown in Figure 19.3. The dosage at which the line crosses the background risk is the threshold. The threshold can be computed for this model from equation (19.16) by setting P equal to the background level of 0.109 and solving for d. The resulting threshold is

$$d_{\text{threshold}} = \frac{0.109 - 0.322}{0.143} = 0.537\,\text{ppm}$$

Below the threshold, the linear model predicts that there is no effect. In fact, for regulatory purposes the U.S. EPA assumes that a threshold exists for all *noncarcinogenic* toxins. Put another way, it is assumed that for every noncarcinogenic toxin, there exists a "safe" dose at or below which one would expect a zero probability of having an adverse reaction. At low doses damage becomes progressively small, until it is overwhelmed by repair mechanisms or compensating factors. However, a tumor can be formed from a single cell that has one or very few mutations. The tremendous amplification that occurs with genetic effects makes the threshold assumption unreasonable in the case of genotoxicity.

Many factors could produce inaccuracy in dose–response extrapolation. The high doses used in animal studies can overload bioactivation mechanisms, resulting in underestimation of risk. One may also overload repair or elimination mechanisms, overestimating risks. There may be lower tumor incidence in a high-dose group, due to poor survival among them. However, even if there is overestimation of risk, the notion of a threshold is not supported.

Even if a threshold exists theoretically, as it may for some epigenetic carcinogens, the fact that we are exposed not to a single toxin at a time, but to many, may rule out the argument for assuming a threshold exists in actuality. In general, regulatory agencies assume that there is no threshold for carcinogenesis.

19.4   INTERACTIONS

Since the environment contains numerous toxins, we are interested in their combined effect. We worry that two pesticides taken up by the same organism may have a much greater effect than either one alone. To detect such an effect, we must first define the situation expected when two toxins act independently.

First, we need to establish the following fact: If d is the dosage of toxin A producing a given response, $1/d$ is a direct measure of the toxicity of A. By direct measure we mean a quantity that is proportional to toxicity. For example, the LD_{50} is the dosage that results in 50% mortality. However, LD_{50} is inversely related to toxicity. We would consider a toxin that is twice as toxic as A to have half the LD_{50}, and one 10 times as toxic to have one-tenth the LD_{50}. Thus, $1/LD_{50}$ is a satisfactory measure of toxicity.

Next, let us define a parameter to describe the mixed dosage of two toxins, A and B, with $LD_{50,A}$ and $LD_{50,B}$, respectively. The mixed dosage, d_t, will simply be the sum of the individual dosages, d_A and d_B, respectively. Then the fraction of the total dosage, λ_A, will be

$$\lambda_A = \frac{d_A}{d_t} \tag{19.17}$$

A similar definition could be made for λ_B. Note that $\lambda_B = 1 - \lambda_A$ for a binary mixture. The range of values for λ_A is 0.0 (pure B) to 1.0 (pure A).

Now consider an experiment in which the toxicity of a mixture is determined. The fraction λ_A is held constant but the total dosage d_t is varied to determine the LD_{50} of the mixture $(LD_{50,M})$. In general, $LD_{50,M}$ will depend on the fraction λ_A, and will equal $LD_{50,B}$ when λ_A is zero and will equal $LD_{50,A}$ when λ_A is 1.0. The question is: What happens in between?

We can define the situation expected when the two toxins act independently: Two toxins are said to show **additive toxicity** when the toxicity of a mixture of them (as measured by inverse dosage) varies linearly with the mixture fraction. Figure 19.4a shows this relationship. In the case where the dosage is the LD_{50}, the equation of the straight line is

$$\frac{1}{LD_{50,M}} = \frac{1}{LD_{50,B}} + \lambda_A \left(\frac{1}{LD_{50,A}} - \frac{1}{LD_{50,B}} \right) \tag{19.18}$$

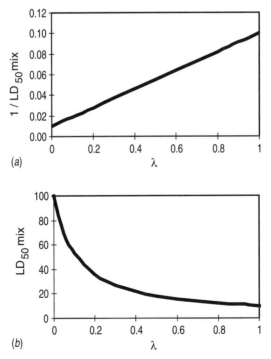

(a)

(b)

Figure 19.4 (a) Linear relationship for toxicity (the inverse of LD_{50}); (b) effect of mixture fraction (λ) directly on LD_{50}.

This can be rearranged to the following equation, which can be used to predict the additive $LD_{50,M}$:

$$\frac{1}{LD_{50,M}} = \frac{\lambda_A}{LD_{50,A}} + \frac{\lambda_B}{LD_{50,B}} \tag{19.19}$$

Figure 19.4b shows an example of how $LD_{50,M}$ will vary with λ_A according to equation (19.18).

It is common to multiply equation (19.19) by $LD_{50,M}$ to obtain Marking and Dawson's (1975) equation for additive toxicity. The result is defined as S, the sum of toxic action, and will equal 1.0 for additive toxicity:

$$S = \frac{d_A}{LD_{50,A}} + \frac{d_B}{LD_{50,B}} \tag{19.20}$$

If the $LD_{50,M}$ measured equals that computed by equation (19.19) (or $S = 1.0$), one can conclude that the toxicity is additive.

Example 19.2   If compound A has an LC_{50} of 100 mg/L, and compound B has an LC_{50} of 10 mg/L, what would the additive model predict for the $LC_{50,M}$ of a 75 : 25 mixture of the two?

Answer

$$\lambda_A = 0.75 \qquad LD_{50,M} = \frac{1}{0.75/100 + 0.25/10} = 30.8$$

That is, a mixture of 23.1 mg/L A and 7.7 mg/L B will be expected to cause 50% mortality.

More than two toxic substances: Equations (19.19) and (19.20) can be applied to any number of toxins simply by adding similar terms. The sum of the λ's must be 1.0. Thus, a more general form of equation (19.19), which applies to n different toxic substances, is

$$\frac{1}{LD_{50,M}} = \sum_{i=1}^{n} \left(\frac{\lambda_i}{LD_{50,i}} \right) \tag{19.21}$$

where $\lambda_i = d_i/d_t$.

Example 19.3   What would be the expected EC_{10} and associated individual concentrations, according to the additive model, of a mixture consisting of 10% A, 20% B, and 70% C? (All percentages are by weight.) The EC_{10} values for the three compounds individually are 10, 15, and 80 mg/L, respectively.

Answer

$$EC_{10,M} = \frac{1}{0.1/10 + 0.2/15 + 0.7/80} = 31.17\,mg/L$$
$$d_A = (0.10)(31.17) = 3.12\,mg/L$$
$$d_B = (0.20)(31.17) = 6.23\,mg/L$$
$$d_C = (0.70)(31.17) = 21.82\,mg/L$$

Many organophosphate pesticides are additive toxins. This is because they have similar properties and act by a very similar mechanism—covalent bonding with acetylcholinesterase.

19.4.1 Nonadditive Interactions

If $LD_{50,M}$ measured in the laboratory is less than that computed by equation (19.19) (or $S < 1.0$), the two toxins interact in a positive way. This is called **synergistic** interaction, also called "more than additive." Another type of positive interaction is when A is essentially nontoxic by itself but causes an increase in the toxicity of B when present. This is called **potentiation** (although this term is sometimes also used in the literature to indicate synergistic interaction).

Unfortunately for the additive model, synergistic interactions are common, as is to be expected, especially for an effect such as lethality. This is because, even if the two toxins have completely independent effects, one may sicken the organisms, making them more vulnerable to the effect of the other. The models of equations (19.19) or (19.21) could be extended to nonlinear interactions by adding terms involving products of toxicities and fitting the coefficients by regression. However, data on mixture toxicity are hard to come by. Furthermore, they may not be consistent at different effect levels. That is, if there is positive interaction between the LD_{50}s of two compounds, it is not necessarily true that there will be positive interaction to a similar extent, or at all, in their LD_{01} values, let alone at the 10^{-6} risk level.

In an exception to what was noted about organophosphate interactions above, malathion interacts synergistically with some other organophosphates, as much as 50 times as strong in combination. As Rachel Carson put it in *Silent Spring*: "1/100 of the lethal dose of each compound may be fatal when the two are combined."

Ethanol and carbon tetrachloride have a synergistic effect on the liver, and tobacco smoke and asbestos interact in lung cancer. In the latter case the interaction was defined differently from that above. It was observed that asbestos workers who did not smoke experienced a fivefold increase in their risk of lung cancer, and cigarette smokers who did not work with asbestos had an 11-fold increase. However, asbestos workers who smoked experienced a 55-fold increase in lung cancer rate. Thus, the interaction here is defined in terms of additivity of *risk* instead of additivity of *toxicity*. An example of potentiation is isopropanol, which does not harm the liver by itself but greatly increases the toxicity of carbon tetrachloride.

Positive interactions can be caused by a number of mechanisms, including:

- By interfering with an enzyme that detoxifies other compound
- By affecting absorption, blood transport, or excretion
- By reacting to form a more toxic substance (e.g., nitrites and some amines react in the stomach to form carcinogenic nitrosomines)
- Induction of biotransformation enzymes

If two toxins interact, resulting in a measured toxic effect less than that predicted by the additive model, this is a negative, or **antagonistic**, interaction, also called "less than additive." Antagonistic interactions are detected when the LD_{50} value of a mixture is greater than predicted by equations (19.19) or (19.21), or $S > 1.0$. Some antidotes take advantage of antagonistic interactions. Mechanisms causing negative interactions include:

- Chemical reactions, such as complexation between EDTA and heavy metals
- Inhibition of bioactivation (e.g., toluene competitively inhibits the enzymes that biotransform benzene)
- Two toxins produce opposite effects, such as central nervous system depressants and stimulants
- Competition for the same receptor
- Induction of detoxification enzymes

Enzyme induction was mentioned above as a mechanism of both positive and negative interactions. The cytochrome P450 enzymes are a common target of induction. Benzene can interact positively with other solvents by inducing cytochrome P450.

Example 19.4   Doudoroff (1952) measured the 8-hour LC_{50} of zinc and copper to fathead minnows (*Pimephales promelas*), obtaining $LC_{50,zinc} = 8\,mg/L$ and $LC_{50,copper} = 0.2\,mg/L$. He then tested a mixture with a fixed 1.0 mg/L zinc and varied the copper concentration to obtain the $LC_{50,M}$. This was found to occur at a copper concentration of 0.025 mg/L. What kind of interaction is this?

Answer

$$\lambda_{Zn} = \frac{1.0}{1.025} = 0.9756 \quad \lambda_{Cu} = \frac{0.025}{1.025} = 0.02439$$

$$LC_{50,M} \text{ expected} = \frac{1}{0.9756/8.0 + 0.02439/0.2} = 4.1$$

$$\text{ratio of observed } LC_{50,M} \text{ to expected } LC_{50,M} = \frac{1.025}{4.1} = 0.25$$

Therefore, the toxicity of the mixture was four times that predicted by the additive model. Using Marking and Dawson's equation (19.20), we obtain $S = 0.250$, which also indicates *more than additive* toxicity.

It is easy to show that Marking and Dawson's (1975) parameter corresponds to the ratio

$$S = \frac{LD_{50,M} \text{ observed}}{\text{additive } LD_{50,M}} \tag{19.22}$$

Therefore, $S < 1.0$ indicates positive interaction or synergism, and $S > 1.0$ indicates less than additive or antagonistic interaction. Despite this simple interpretation for S, toxicologists prefer to transform it into an *additive toxicity index*, I, which eliminates the physical interpretation:

$$I = \frac{1}{S} - 1 \qquad \text{for } S < 1.0 \text{ (greater than additive interaction)} \tag{19.23a}$$

$$I = 1 - S \qquad \text{for } S > 1.0 \text{ (less than additive interaction)} \tag{19.23b}$$

The resulting index produces a *I* value of 0.0 for additive toxicity, greater than 0.0 for positive interaction, and less than 0.0 for negative interaction.

19.5   TIME–RESPONSE RELATIONSHIP

The picture of dose–response relationships we have been describing so far is a static one. It assumes that all measurements are taken at a fixed interval after the first exposure. Inclusion of time adds another dimension to the response and is an important qualifier in any toxicity data. A 96-hour LD_{50} may seriously underestimate the risk if the situation being studied involves chronic, ongoing exposure.

The progress of a toxicity test can be followed by examining the **toxicity curve,** made by plotting the logarithm of time vs. the logarithm of the LD_{50}. The toxicity curve indicates when acute lethality has ceased by leveling off parallel to the time axis (Figure 19.5). The LD_{50} associated with the plateau region is called the **threshold LD_{50}** (denoted as d_t on Figure 19.5), also called **asymptotic** or **incipient LD_{50}**. If a plateau is not reached, the LD_{50} computed will be sensitive to the experimental interval selected, and it would be advisable to extend the test. The toxicity curve may assume other shapes as well. This can provide the experimenter with information about how the compound acts or may indicate the presence of multiple toxic agents in a tested mixture. In Figure 19.5 the threshold LD_{50} is 5 mg/L. The figure also indicates that the LD_{50} would be 13.6 if the organisms were exposed for just 24 hours.

19.6   OTHER MEASURES OF TOXIC EFFECT

Bioassay results are used to compute a variety of parameters related to the toxicity of an agent. Table 19.6 summarizes some of the more common ones, along with acronyms used

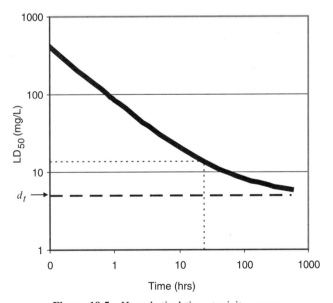

Figure 19.5   Hypothetical time–toxicity curve.

TABLE 19.6   Acronyms Describing Toxic Doses or Concentrations

LD_x, LC_x	Lethal dose (or concentration)	Dose or concentration resulting in mortality to x percent of a test population.
ED_x, EC_x	Effective dose (or concentration)	Dose or concentration resulting in some specified effect in x percent of a population.
ID_x, IC_x	Inhibitory dose (or concentration)	Dose or concentration resulting in a reduction of the normal response of an organism (such as growth) by x percent.
NOEL, NOEC	No observed effect level (or concentration)	Level or concentration below which no effect has been measured experimentally.
NOAEL, NOAEC	No observed adverse effect level (or concentration)	Level or concentration below which no adverse toxic effect has been measured.
LOEL, LOEC	Lowest observed effect level (or concentration)	Lowest level or concentration at which an effect could be observed.
MTC	Minimum threshold concentration	
MATC	Maximum allowable toxicant concentration	Greater than NOEC and less than LOEC as found from chronic toxicity test.

Source: Based on Landis and Yu (1995).

for them. The MATC is usually given as a range: NOEC < MATC < LOEC. These parameters imply the existence of a toxic threshold. Often, only acute toxicity data, such as LD_{50} values, are available. One would like to estimate the MATC from acute data. Although an estimated MATC would be of limited validity for application to the environment, it would be useful for determining the dosages to use in chronic toxicity tests. The **application factor** (AP) is the ratio of acute to chronic toxicity. For example, the application factor could be defined as LD_{50}/MATC and would thus be given as a range. Application factors for toxicity of xenobiotics to fish have been found to vary from 0.01 to 0.50. More extreme cases exist, such as 0.003 for lead and chromium, and even lower for certain pesticides.

All of the levels referred to in Table 19.16 are implicitly linked to an experimental and statistical procedure for their measurement. For example, unless a compound exhibits a true threshold effect, it should always be possible to detect adverse effects at a lower dosage, by increasing the number of organisms in the experiment.

PROBLEMS

19.1. For the following simplified artificial toxicity data, plot the response curve and the estimated tolerance distribution.

Dosage	0%	40%	50%	60%	100%
Mortality	0%	0%	50%	100%	100%

(a) Estimate LC_{50} and LC_{10} by interpolating the response curve. (b) Compute the mean and standard deviation of the tolerance distribution using the method of moments, and use these to compute LC_{50} and LC_{10}. (c) Compare the results and explain any discrepancies. Which do you think is more reliable?

19.2. Many medicinal drugs are effective above a threshold dosage but toxic above a higher threshold. This range is called the **therapeutic window**. Consider an antibiotic being used to treat a potentially fatal infection. How would Figure 19.1 appear for such an antibiotic? What implications does the range between the two thresholds have for how the drug is administered (dosage, frequency)? Which nutrient in Table 19.1 has the smallest ratio between the toxicity and deficiency thresholds?

19.3. In a bioassay experiment, the lowest dose was 20 mg/kg and produced tumors in three out of 13 animals. By linear extrapolation, what dosage will be expected to produce a 10^{-6} risk?

19.4. The LD_{50} of substance A is 10 mg/kg, and for substance B it is 1.5 mg/kg. Assuming that there is no interaction, what would be the expected LD_{50} of a mixture that is 70% A and 30% B?

19.5. In the definition of additive toxicity of mixtures, why not say that compound B is twice as toxic as A if it results in double the mortality at the same dosage? (*Hint*: Consider what would happen with respect to the dose–response curve if the two substances, A and B, were the same thing.)

19.6. Two substances, A and B, have LC_{50} values of 80 and 25 mg/L, respectively. A mixture of the two containing A and B in a $3:1$ ratio by mass has an LC_{50} of 40 mg/L. Are the toxins additive, more than additive, or less than additive? Compute the $LC_{50,M}$ predicted and Marking and Dawson's S.

REFERENCES

Crump, K. S., H. A. Guess, and K. L. Deal, 1977. Confidence intervals and tests of hypothesis concerning dose response relations inferred from animal carcinogenicity data, *Biometrics*, Vol. 33, p. 437.

Doudoroff, P., B. G. Anderson, G. E. Burdick, P. S. Galtsoff, W. B. Hart, R. Patrick, E. R. Strong, E. W. Surber, and W. M. Van Horn, 1951. Bioassay methods for the evaluation of acute toxicity of industrial wastes to fish, *Sewage and Industrial Wastes*, Vol. 23, p. 1381.

Freidman, G. D., 1987. *Primer of Epidemiology*, 3rd ed., McGraw-Hill, New York.

Landis, W. G., and Ming-Ho Yu, 1995. *Introduction to Environmental Toxicology: Impacts of Chemicals Upon Ecological Systems*, Lewis Publishers, Boca Raton, FL.

Marking, L. L., and V. K. Dawson, 1975. Method for assessment for toxicity or efficacy of mixtures of chemicals. *U.S. Fish Wildl. Serv. Fish Control*, Vol. 67, pp. 1–8.

Merck, 2004. *Merck Manual*, http://www.merck.com/pubs/mmanual/section1/chapter3/3g.htm.

National Research Council, 1980. *Recommended Dietary Allowances*, Food and Nutrition Board, NRC, National Academy of Sciences, Washington, DC.

National Toxicology Program. http://ntp-server.niehs.nih.gov/Main_Pages/Chem-HS.html.

World Health Organization, 2004. http://www.who.int/water_sanitation_health/GDWQ/Chemicals/coppersum.htm.

20

FIELD AND LABORATORY TOXICOLOGY

In the first part of the chapter, we discuss toxicity testing in detail, including descriptions of the types of tests commonly used, the variety of organisms used in testing, and extrapolation of results from animals to humans. Then we describe epidemiology and how it is used in the study of human disease.

20.1 TOXICITY TESTING

The laboratory measurement of toxicity is one of the largest activities in toxicology. We can distinguish three ways to detect toxic effects: in vivo, in vitro, and in situ. An **in vivo** test means a laboratory experiment with whole living organisms. Most of the discussion here concerns this type of testing. **In vitro** tests (literally "in-glass") are performed with living cells extracted from plants or, more often, from animals. These can be faster and less expensive, but less precise. Therefore, they are best used for screening purposes. **In situ** measurements are performed on organisms in their natural habitat. These can be the most revealing, but in practice are limited by economic and logistical constraints.

Here we are concerned with toxicity testing to determine the probability of particular adverse effects, as distinguished from laboratory research experiments that seek to understand toxic mechanisms. Toxicity tests are often called **bioassays**. This term is more generally used to describe experiments in which organisms are used as a kind of chemical detector. For example, the growth rate of microorganisms is used to determine the potency of vitamins.

Environmental Biology for Engineers and Scientists, by David A. Vaccari, Peter F. Strom, and James E. Alleman
Copyright © 2006 John Wiley & Sons, Inc.

Consider the task of understanding the toxicity of even a single chemical substance. All of the following parameters must be selected:

- Species and strain
- Age or life stage of organisms
- Sex
- Toxic endpoint (mortality, tumor formation, etc.)
- Route of administration
- Duration of test
- Number of organisms per test
- Dosage levels

In addition, there are numerous decisions to be made about experimental protocol and data handling. Clearly, the reporting of the toxicity of a substance in terms of a single number is very incomplete. There is little assurance that the effect of ingestion of a chemical on the growth of adult rats of both sexes has any relation to the 96-hour LD_{50} by inhalation in juvenile hamsters, let alone to the NOAEL for human males exposed dermally in an occupational setting.

Although the possible variety may seem daunting, toxicologists have created a variety of standardized approaches and test protocols that make it easier to compare results obtained by different experimenters. However, one should keep the following in mind: The inherent variability of living things in their response to toxins, and their sensitivity to numerous environmental factors, can produce large differences in results from different laboratories, or even between tests conducted at the same lab at different times. For example, duplicate tests of endosulfan toxicity to fish conducted at four labs produced LC_{50} values with a geometric mean of 1.34, but which ranged from 0.68 to 3.30, a factor of almost 5.

20.1.1   Design of Conventional Toxicity Tests

As in any project, the first decision to be made in selecting or developing a toxicity test is to carefully state the objectives. Furthermore, given the limitations of extrapolating from laboratory to field, the objectives often must be circumscribed in light of what can be discovered by laboratory tests. Examples of uses for toxicity data include support for decisions on product development or use, permitting activities for waste discharges to the environment, risk assessment, or in support of environmental litigation. Considerations in the design of **conventional toxicity tests** are described here. These exclude specialized tests, such as for carcinogenicity, mutagenicity, or teratogenicity. Additional considerations that apply to those are described in the following sections.

20.1.2   Test Duration

The time scale of the effect commonly falls into three categories. *Acute* toxicity tests are usually complete within 24 to 96 hours. *Short-term* (also called *subacute* or *subchronic*) toxicity tests take longer, up to about 10% of the life span of the organism. *Long-term* or chronic toxicity tests cover most of the life span of the organisms. The life span of mice is about 18 months, 24 months for rats, and 7 to 10 years for dogs and monkeys.

20.1.3 Selecting Organisms

The goal of toxicity testing may be generalized into two categories according to the target of interest: humans or all other living things. Humans themselves may be used as subjects in toxicological studies under carefully proscribed conditions. For example, in human clinical trials, subjects may be administered dosages below the expected NOAEL and then have their blood and urine sampled periodically to determine rates and mechanisms of biotransformation and elimination.

However, for the most part, other mammals are used as surrogates for humans in the determination of adverse effects. The mouse or rat is generally used for determination of the LD_{50} because they are fairly economical and easy to handle. They are also preferred because a large number of toxicological data have been accumulated using them, which makes it possible to compare results between different toxic substances. The pig is better for purposes of extrapolation to humans because it is phylogenetically closer and more similar to humans in diet as well. Of course, nonhuman primates such as monkeys are closest of all. However, expense often precludes their use.

For environmental toxicity testing, a wider variety of organisms may be used. It may be necessary to consider toxic effects on vertebrates other than mammals (fish, birds, amphibians), on invertebrates, and on plants. Test organisms may be collected from the wild. Whenever possible, organisms should be selected for environmental toxicity testing that:

- Represent a broad range of sensitivities
- Are abundant and easily available species
- Are indigenous to any area that will be affected
- Include species that are important recreationally, commercially, or ecologically
- Are easy to culture and maintain in the laboratory
- Have adequate background information available, such as physiology, genetics, and behavior

Table 20.1 lists some organisms commonly used for acute aquatic toxicity tests. Of these, the most commonly used are the water flea (*Daphnia* sp.), fathead minnow, bluegill, rainbow trout, mysid shrimp, and sheepshead minnow. *Daphnia magna* and mysid shrimp have been found useful for predicting the MATC for fish.

Among terrestrial animals, the mammals used include several species of rats and mice, as well as hamsters, guinea pigs, rabbits, and dogs. Common avian species are Northern bobwhite (*Colinus virginianus*), Japanese quail (*Coturnix japonica*), mallard (*Anas platyrhynchos*), ring-necked pheasant (*Phasianus colchicus*), American kestrel, and screech owl (*Otus asio*). A teratogenicity test has been developed using the South African clawed frog, *Xenopus laevis*. Earthworms have been used to assess the toxicity of soil samples contaminated with hazardous wastes.

Finally, some toxicity tests use a combination of species in a single test. A small-scale ecosystem with at least two interacting organisms is called a **microcosm**. At a medium scale, they are called **mesocosms**. These are useful for evaluating ecosystem effects such as biomagnification or predation. They may range from two organisms cultured together in a test-tube microcosm, to a farm pond mesocosm used to evaluate agricultural pesticide effects. Terrestrial microcosms may consist of soil cores with associated microorganisms and invertebrates.

TABLE 20.1   Common Organisms Used in Aquatic Acute Toxicity Tests

Freshwater
  Vertebrates
    Rainbow trout, *Salmo gairdneri*
    Brook trout, *Salvelinus fontinalis*
    Fathead minnow, *Pimephales promelas*
    Channel catfish, *Ictalurus punctatus*
    Bluegill, *Lepomis macrochirus*
    Frog, *Rana* sp.; toad, *Bufo* sp.
  Invertebrates
    Daphnids: *D. magna, D. pulex, D. pulicaria, Ceriodaphnia dubia*
    Amphipods: *Gammarus lacustris, G. fasciatus, G. pseudolimnaeus*
    Crayfish: *Orconectes* sp., *Cambarus* sp., *Procambarus* sp., or *Pacifastacus leniusculus*
    Midges: *Chironomus* sp., Stoneflies, *Pteronarcys* sp.
    Mayflies: *Baetis* sp., *Ephemerella* sp., *Hexagenia limbata, H. bilineata*
    Snails: *Physa integra*; planaria, *Dugesia tigrina*
  Algae
    Chlorphyta (green algae), *Selenastrum capricornutum*
    Cyanophyta (blue-green bacteria), *Anabaena flos-aquae, Microcystis aeruginosa*
    Chrysophyta (brown algae and diatoms), *Navicula pelliculosa, Cyclotella* sp., *Synura petersenii*
Saltwater
  Vertebrates
    Sheepshead minnow, *Cyprinodon variegatus*
    Mummichog, *Fundulus heteroclitus*
    Longnose killifish, *Fundulus similis*
    Silverside, *Menidia* sp.
    Threespine stickleback, *Gasterosteus aculeatus*
    Pinfish, *Lagodon rhomboides*
    Spot, *Leiostomus xanthurus*
    Sand dab, *Citharichthys stigmaeus*
  Invertebrates
    Copepods: *Acartia tonsa, A. clausi*
    Shrimp: *Penaeus setiferus, P. duorarum, P. aztecus*
    Grass shrimp: *Palaemonetes pugio, P. vulgaris*
    Sand shrimp: *Crangon septemspinosa*
    Mysid shrimp: *Mysidopsis bahia*
    Blue crab: *Callinextes sapidus*
    Fiddler crab: *Uca* sp.
    Oyster: *Crassostrea virginica, C. gigas*
    Polychaetes: *Capitella capitata, Neanthes* sp.
  Algae
    Chlorophyta (green algae): *Chlorella* sp., *Chlorococcum* sp., *Dunaliella tertiolecta*
    Chrysophyta (brown algae and diatoms): *Isochrysis galbana, Nitzschia closterium, Pyrmnesium parvum, Skeletonema costatum, Thalassiosira pseudonana*
    Rhodophyta (red algae): *Pophyridium cruentum*

Source: Based on Rand and Petrucelli (1985); Landis and Yu (1995).

Several standard microcosms have been developed. For example, the **standardized aquatic microcosm** (SAM) is conducted in 4-L jars and provides a highly controlled environment. Four species of invertebrates, 10 of algae, and one of bacteria are added to a sterilized aquatic medium. Temperature, illumination, and chemical and biological sampling schedule are specified over the 63-day test duration. Another example is the **soil core microcosm** (SCM). It consists of 60-cm-long by 17-cm-diameter soil columns obtained from the field, containing naturally occurring organisms, and is tested over a period of at least 12 weeks.

Testing should preferably be done with animals of both sexes, and both adults and juveniles. Sometimes tests are done with equal numbers of both sexes, but they are lumped together for measurement purposes.

20.1.4   Toxic Endpoint and Other Observations

All of the effects described in Table 17.1 are candidate endpoints for toxicity testing. These include biochemical, histological, and individual effects. Population effects can be revealed by measurement of growth, death, and fertility rates. Even ecological interactions may be measured by using microcosms and mesocosms. Although the test objective may require a single endpoint, good use of an experiment calls for the recording of numerous observations. Mortality is commonly the primary effect to be observed. However, other changes can reveal toxic effects that may result in mortality beyond the test period. These may include body weight and food consumption, which may be needed for computating dosage. Autopsies should be performed on all the dead animals as well as some of the survivors. This practice can identify the organs or tissues that have been affected. General observations include changes in color or consistency of stool or urine, and appearance of eyes, skin, and hair coat.

In aquatic toxicity tests, a variety of chemical and physical measurements should be performed regularly during the test. These may include pH, alkalinity, hardness, temperature, oxygen concentration, and salinity or conductivity of the medium.

In plant toxicity tests, endpoints may include growth rates, rate of seed germination, and root elongation. The response of algae is typically measured in terms of growth. This, in turn, is measured in cell mass, by extracting chlorophyll and measuring spectrophotometrically, or by cell counts (microscopic or electronic).

20.1.5   Route of Administration and Dosage

All of the natural routes of exposure (oral, dermal, or by inhalation) may be used experimentally. The oral route may be administered with the food or by tube directly into the gastrointestinal tract. The latter method is called **gavage**. In addition, chemicals may be administered by injection **intravenously** (into the circulatory system), **intraperitoneally** (into the abdominal body cavity), **intramuscularly** (into the muscle), or **subcutaneously** (below the skin). These injection methods, however, run the risk of causing local effects.

It is preferable to apply dosages of toxin normalized to the body weight of the animal. However, it is often more practical to dose by a fixed concentration in the food, air, or water. Because in this case some animals will obtain a relatively higher or lower dose than others, the standard deviation of the dose–response tolerance distribution will be greater. The toxicant in an aquatic toxicity test may be an aqueous mixture such as wastewater treatment plant effluent. In this case, dilutions are used in place of concentration, expressed as a percentage of the whole effluent.

Concentration levels are chosen to bracket the ED_{50} expected value, based on experience with comparable substances. Initially, it may be necessary to perform a *range-finding test* with dosage levels spaced out approximately by factors of 10. For example, three levels might be chosen with concentrations of 0.1, 1.0, and 10 mg/L. Effluents will be diluted successively by factors of 10; thus, one might use 100%, 10%, and 1%. Subsequently, more closely spaced dosage levels may be tested to obtain definitive information. The ratio between successive dosage levels will typically be between 1.2 and 2.0. For example, one might use nominal dosages of 10.0, 15.0, 25.0, 35.0, and 50.0 (the ratio of one value to the next need not be constant but should be in a narrow range). The goal is to have several dosage levels show some toxic effect, with at least one dosage with less than 35% of the organisms showing toxic effect, and at least one showing more than 65%.

When the number of treatment levels is small, duplicates of each concentration may be called for. The purpose of duplicates is to help determine confidence intervals. Since this determination can also be made with regression techniques in fitting the dose–response curve, similar information can be obtained by doubling the number of treatment levels within the same range.

One test group, with at least as many organisms as receive each treatment level, is left untreated as a control. If a significant number of control organisms suffer toxic effects (say, more than 10%), the entire test may be invalidated. If a chemical toxin is applied by mixing with a solvent such as methanol or DMSO, another control group should be given a treatment containing the solvent only.

20.1.6   Number of Organisms Per Test

The number of organisms exposed to each treatment level varies greatly with the requirements of the test and other practical factors such as cost. Fewer than 10 organisms limits the resolution of the response. For example, if there were only five organisms per test, mortality could only have values in multiples of 20%. Nevertheless, for large organisms such as dogs, small numbers are commonly used because of cost. In aquatic toxicity tests with fish, 10 organisms per treatment are recommended for static tests, and 20 for flow-through.

20.1.7   Other Experimental Variables

There are numerous seemingly minor issues, such as the design of organism containment structures, which nevertheless can significantly affect the response of organisms. Of great importance is the amount of space allotted to each organism. Too small a space is a cause of stress, which can interact with the toxicity. Sometimes the design is specified completely as to materials and dimensions. For example, one test specifies that cages for rats have a floor area from 110 cm^2 for a 100-g animal to 450 cm^2 for one that is 500 g. Handling, lighting, and temperature may be specified. A test may require an acclimation period before initiating the toxic exposure. In acute and subchronic tests, feeding is sometimes withheld.

Aquatic tests may be static, static renewal, or flow-through. In **static tests**, the water is not changed after the beginning of the test. In **static renewal tests**, it is changed periodically, whereas **flow-through tests** have a continuous change. Aeration is sometimes required and sometimes not, but in either case the oxygen levels may be specified as a percentage of the saturation value. The volume of solution needed is based on the mass of organisms in the test. Recommended values are 0.5 to 0.8 g of biomass per liter for static tests, 1 to 10 g/L·day for flow-through tests. The temperature for many species is 20 to 22°C, but some organisms must be tested in colder water. For example, some crayfish and mayflies, threespine stickleback, sand shrimp and bay shrimp are tested at 17°C. Salmon, trout, stoneflies, flounder, herring, and marine copepods are tested at 12°C. On the other hand, mysid shrimp should be cultured at 27°C.

If the test involves plants, lighting must be specified not only as to intensity, but also as to the photoperiod.

Concern about the care and well-being of terrestrial vertebrates (e.g., rats and mice) has led to strict guidelines for their husbandry and humane treatment. Research organizations have animal use committees that review protocols for proposed tests. The guidelines seek to ensure that any pain and suffering experienced by these animals is justified by the potential gain in information. They have forced investigators to design their studies more carefully in order to maximize the information gained, reducing the number of animals used. It has also spurred the development of alternative tests that do not use live animals, such as the in vitro tests described below.

20.1.8 Conventional Toxicity Tests

Table 20.2 give references to a number of toxicity tests, to give an idea of the variety available. *Daphnia* species, water fleas, are one of the most important organisms for aquatic toxicity testing because of their small size and ease of cultivation. They are crustaceans that measure about 0.2 to 3 mm, appearing to the unaided eye as a swimming speck. *D. magna* requires relatively hard water, about 80 to 100 mg/L as $CaCO_3$, which is about double the hardness requirement for *D. pulex*. Testing is usually done with organisms less than 24 hours old. A single test with 10 organisms can be conducted in a 125-mL container. They can be fed algae or fishmeal. An acute toxicity test conducted with *Daphnia* has 48 hours' duration. Mortality is difficult to determine, so the endpoint chosen is usually immobilization, defined as not swimming even after gentle prodding with a glass rod. The chronic toxicity or partial life cycle test runs 21 days and measures growth in terms of length or mass, and numbers of offspring produced per organism. *Ceriodaphnia nubia* is much smaller than the other species, and it reproduces faster. A faster, less expensive chronic toxicity test has been developed with it, called the *three-brood renewal toxicity test*. Ten organisms are placed in 15 mL of test solution, one to a container. Reproduction is measured after 7 days.

The most definitive chronic toxicity test, called an *embryo-to-embryo test*, would encompass the entire life-cycle of an organism. However, this can be costly and time consuming. As an alternative, research has found that the MATC for fish can be established by testing embryos, larvae, and juveniles. This is called **early life stage** (ELS) **toxicity testing**. Numerous endpoints have been used, such as mortality, motility reduction, reproduction reduction, and so on.

Toxicity to bacteria has been assayed by various means, such as the effect on plate counts, cell growth rates, or oxygen uptake rate. Particular reactions such as nitrification rate may also be tested, as this tends to be particularly sensitive.

TABLE 20.2   References on Standard Toxicity Tests and Protocols

Method Number	Date	Title
D 4229-84	1993	Conducting static acute toxicity tests on wastewaters with *Daphnia*.
E 724-89	1993	Standard guides for conducting static acute toxicity tests starting with embryos of four species of bivalve molluscs.
E 729-88a	1993	Standard guide for conducting acute toxicity tests with fishes, macroinvertebrates and amphibians.
E 1191-90	1993	Standard guide for conducting the renewal life-cycle toxicity tests with saltwater mysids.
E 1197-89	1993	Standard uide for conducting renewal life-cycle toxicity tests with *Daphnia magna*.
E 1197-87	1993	Standard guide for conducting a terrestrial soil-core microcosm test.
E 1218-90	1993	Standard guide for conducting static 96-h toxicity tests with microalgae.
E 1241-88	1993	Standard guide for conducting early life stage toxicity tests with fishes.
E 1295-89	1993	Standard guide for conducting three-brood, renewal toxicity tests with *Ceriodaphnia dubia*.
E 1366-91	1993	Standard practice for standardized aquatic microcosms: freshwater.
E 1367-90	1993	Standard guide for conducting 10-day static sediment toxicity tests with marine and estuarine amphipods.
E 1383-90	1993	Standard guide for conducting sediment toxicity tests with freshwater invertebrates.
E 1391-90	1993	Standard guide for collection, storage, characterization, and manipulation of sediments for toxicological testing.
E 1415-91	1993	Standard guide for conducting the static toxicity tests with *Lemna gibba* G3.
E 1439-91	1993	Standard guide for conducting the frog embryo teratogenesis assay—*Xenopus* (FETAX).
E 1463-92	1993	Standard guide for conducting static and flow-through acute toxicity tests with mysids from the west coast of the United States.
D 4229-84	1984	Conducting static acute toxicity tests on wastewaters with *Daphnia*.
E 729-88a	1991	Standard guide for conducting acute toxicity tests with fishes, macroinvertebrates, and amphibians.
E 1193-87	1987	Standard guide for conducting renewal life-cycle toxicity tests with *Daphnia magna*.
E 1197-87	1987	Standard guide for conducting a terrestrial soil-core microcosm test.
E 1218-90	1990	Standard guide for conducting static 96-h toxicity tests with microalgae.
E 1241-88	1991	Standard guide for conducting early life stage toxicity tests with fishes.
E 1367-90	1990	Standard guide for conducting 10-day static sediment toxicity tests with marine and estuarine amphipods.
1383-90	1990	Standard guide for conducting sediment toxicity tests with freshwater invertebrates.

U.S. Environmental Protection Agency references

EPA/600/ 2-83/054	1983	Protocols for bioassessment of hazardous waste sites
EPA/600/ 4-78/012	1987	Methods for measuring the acute toxicity of effluents to aquatic organisms
EPA/600/2	1988	Protocols for short term toxicity screening of hazardous waste sites

Source: Based on Landis and Yu (1995). Methods from the *Annual Book of ASTM Standards,* American Society of Testing and Materials, Philadelphia.

The **frog embryo teratogenesis assay** (FETAX) is a 96-hour in vitro test using embryos of the South African clawed frog, *Xenopus laevis*. It has wide acceptance as an alternative to mammalian teratogenesis testing for several reasons. The cost is reasonable because the frog is easy to raise and produces large amounts of eggs and sperm. A large database of results is available. The test is very repeatable and the results correlate well with known mammalian teratogens. The concentration of toxin that produces 50% mortality (the LC_{50}) and that produces abnormalities in half the embryos [50% **teratogenic concentration** (TC_{50})] are recorded. The ratio LC_{50}/TC_{50}, called the **teratogenic index** (TI), is an indication of the developmental hazard posed by the toxin. Values of TI below 1.5 are considered to pose little hazard.

Mammalian tests are conducted primarily for extrapolation to humans. Rats are popular test subjects. They should be selected to be less than six weeks old and to have a variation in weight of less than 20% for each sex. A typical test duration is 90 days. Subchronic inhalation toxicity is tested by introducing the toxin directly into the chamber air. Oral toxicity may be administered in the food or water or by gavage (injection into the stomach by a tube). Mortality is the most common endpoint, but examinations are also made by urinalysis, blood chemistry, and so on.

20.1.9   In Situ Measurement of Conventional Toxicity

A criticism of laboratory toxicity tests is that the artificial environment, especially for vertebrates, can create stresses different from what the organisms experience in their habitat. Furthermore, the modes of exposure in the field will be mixed, and this may be difficult to simulate experimentally. In situ measurements overcome these problems, although introducing others. It is harder to get reproducible results since there is no control over the culturing and handling of the organisms. A negative control cannot be assured to differ only in its exposure to a toxicant. Therefore, a greater number of samples may be needed to establish statistical significance than for laboratory experiments. On the other hand, whatever unknown variables there may be, at least one can be fairly certain that they are representative of a natural setting. Finally, although in situ measurements may indicate that a disturbance has occurred, the cause cannot always be attributed to toxic exposure.

Biochemical tests on organisms in their habitats that can indicate toxic effects are called **biomarkers**. Biomarkers are of particular interest for compounds that do not bioaccumulate, and that are thus difficult to measure in tissues directly. Any of the biochemical effects described above could be used. Examples include activity of enzymes such as cytochrome P450 (a positive indicator of toxicity) or acetylcholinesterase (a negative indication). Increased metallothionein is a sensitive biomarker for metal exposure. Despite their potential, they have not been widely found useful in environmental applications. Some have proven useful in clinical application for cancer detection.

Indicator species are species that are naturally present and that are sensitive to the toxin. They may be used in one of two ways: (1) to indicate the health of an ecosystem by their presence or absence; or (2) to bioaccumulate toxins, making the toxins more easily detectible than in the environment. For example, bivalves are often used as indicator species for bacterial pollution in coastal and estuarine ecosystems since, as filter feeders, they tend to accumulate them.

Sentinel organisms are organisms deliberately introduced for later recovery and testing. The classic case is the canary in the coal mine. Mussels have been used as sentinels by suspending them in the water in plastic trays. Organisms that are sessile or otherwise easy to recover are required for this type of study. Sentinels can provide early warning to take necessary action in time. For example, sentinels may be used in wells at the boundary of contaminated sites.

20.1.10   Occupational Monitoring

Several tests can detect the effect of exposure to organophosphate or carbamate pesticides on acetylcholinesterase. Red blood cell cholinesterase can be measured but is thought to be overly sensitive. Surface electromyography (EMG) indicates more directly the presence of any dysfunction in the neuromuscular junction.

20.1.11   Population and Community Parameters

Above the individual level, toxic effects can be detected by changes within a population or in the structure or function of a community of populations. Since toxic effects often target the young, or reduce reproduction, changes in age distribution may result. In particular, an exposed population may shift to older organisms.

Community structure can be described with more or less detail using food webs or population diversity indices. A diversity index is meaningful only by comparison with similar ecosystems with different exposure histories or by tracking changes in an index with time or distance from exposure. Other things being equal, increased diversity is thought to represent more stability in an ecosystem. However, similar normal ecosystems can vary greatly in their index values due to variations in developmental stage, the proximity to boundaries or ecotomes, and other effects.

A number of different measures of community function exist. One is the ratio of respiration to photosynthesis. However, this indicates nutrient enrichment pollution more directly than does the presence of toxic substances.

20.1.12   Testing for Carcinogenicity and Teratogenicity

Carcinogenic potential can be detected by three types of tests: long-term carcinogenicity studies, rapid screening tests, and biomarkers. Long-term tests are the most definitive. These generally use mice or rats and last the lifetime of the animals (18 and 24 months, respectively). Two or three dose levels are usually used, the highest being the "maximum tolerated dose" (MTD). The MTD is estimated from 90-day studies and is chosen so as not to produce severe noncarcinogenic toxicological effects or to reduce significantly the life of any organisms that do not develop tumors. The other doses are typically one-fourth to one-half of the MTD. The number of tumors is recorded by animal (location, whether benign or malignant, the presence of any unusual tumors), the number of animals with tumors, the number of tumors per animal, the number of tumor sites, and the time to onset.

In the United States, a set of minimum test standards for carcinogenicity has been developed by the National Toxicology Program. These standards require that at least two species of rodents be used, at least two doses plus a control must be administered, and at least 50 males and 50 females of each species must be tested at each dose. Thus, the minimum number of animals in a study is 600. Such testing costs from $500,000 to $1.5 million for each chemical, yet can only detect excess tumor rates above 5 or 10%. The largest bioassay ever performed used over 24,000 mice, yet still was not sensitive enough to measure excess risk of 1%. Despite these difficulties, the results of animal bioassays are routinely extrapolated to risk levels of 10^{-6}, as described in Section 19.3.1.

Rapid screening tests have been developed using especially sensitive strains of animals. Mouse skin is sensitive to tumor formation from topical application, apparently because it is active in biotransformation. Strain A mice spontaneously develop lung tumors in 100% of animals by 24 months of age. They can be used to determine carcinogenicity in tests as short as 12 to 24 weeks. Several rat strains have been developed with naturally high rates of breast cancer.

Biomarkers are detectable biochemical changes. A variety has been developed that can discriminate between cancer initiation, promotion, and progression. The binding of chemicals such as benzo[a]pyrene to DNA can be detected by use of radiolabeled compounds or by radioimmunoassay. Promotion has been correlated with increased prostaglandin synthesis and with changes in the levels of some enzymes, such as ATPase. A variety of chemicals is produced by malignant cancer cells. Fetal liver and hepatocarcinoma, and sometimes other organs, produce α-fetoprotein (AFP). Mammary carcinomas are associated with a group of chemicals, and prostate cancer is marked by elevated levels of prostate-specific antigen (PSA), which can be detected by a blood test.

A **positive control** is a test group that receives a known carcinogen at a dose that is expected to produce a significant response. This provides a check on the sensitivity of the organisms used in the test and on the overall testing procedure.

An important problem in the testing of carcinogens is that some may only be initiators and others only promotors. Thus, it may be necessary to test a chemical with a known initiator, and separately with a known promoter, to detect carcinogenicity. Since initiators are genotoxic, compounds that test positively for mutagenicity should also be suspected of carcinogenicity.

Teratogen tests are commonly done with pregnant rats, rabbits, mice, and hamsters, but also with pigs, dogs, and primates. An assay using the developing chick embryo was once widely used but was found to produce too many false positives. Because teratogenesis is sensitive to time of administration, in tests a chemical is usually dosed continuously over the period of organ formation. Fetuses are removed surgically about one day before their expected birth so that dead fetuses can be counted. The fetuses are examined for abnormalities. Some of the pregnant females may be allowed to deliver so that delayed effects can be discovered.

Extrapolation of teratogenesis in animal models to humans is particularly susceptible to false negatives. The most potent human teratogen known is thalidomide, which acts at a dose of 0.5 to 1.0 mg/kg. However, it has no teratogenic effect in rats and mice at 4000 mg/kg. On the other hand, acetylsalicylic acid (aspirin) is known to be safe for use by pregnant humans but is a strong teratogen in rats, mice, and hamsters.

20.1.13   Mutagenicity Testing and In Vitro Tests

A variety of mutagenicity tests has been developed. They can be categorized as tests that detect gene mutations, chromosome effects, DNA repair and recombination, or others. The others include tests for transformation of mammalian cells into malignant cancer cells in vitro.

Gene mutation tests detect mutations directly. They may be conducted in vitro or in vivo, and with prokaryotic or eukaryotic organisms. One of the best known is the **Ames test**. This test uses a strain of *Salmonella typhimurium* that has been mutated artificially so that it cannot synthesize the amino acid histidine. In the test, this organism is cultured in a histidine-deficient medium with the chemical being tested. Mutagens cause a reverse mutation, again allowing the organism to produce its own histidine. The presence of growth in the histidine-deficient medium thus serves to indicate the occurrence of a mutation. Since many mutagens require bioactivation, microsomes extracted from animal livers are added to the medium. These supply active cytochrome enzyme systems. Similar tests have been developed using eukaryotic yeast cell cultures and mammalian cell cultures.

(The developer of this test, Bruce Ames, is an outspoken critic of the way that tests are used to identify carcinogens. He states: "The effort to eliminate synthetic pesticides because of unsubstantiated fears about residues in food will make fruits and vegetables more expensive, decrease consumption, and thus increase cancer rates. The levels of synthetic pesticide residues are trivial in comparison to natural chemicals, and thus their potential for cancer causation is extremely low.")

A type of in vivo gene mutation test is the **host-mediated assay**. In this test, cells that are susceptible to reverse mutation (whether bacterial or eukaryotic) are injected with the toxicant into a live animal, such as a mouse. After several hours the animal is sacrificed and the injected cells are cultured to detect mutations. This has the advantage of using all the biotransformation mechanisms of the living animal.

Mutations can also be detected directly in test animals. The most popular is the fruit fly *Drosphila melanogaster*, but tests have also been developed to detect mutations in mice.

Chromosomal effects include structural aberrations such as deletion or duplication of chromosomes, or **translocation**, involving switching of sections between different chromosomes. These are conducted either in vitro using mammalian cell cultures or in vitro with *Drosophila* sp. or mammals. In either case the effect is detected by microscopic examination a period of time after dosing the system with the toxicant.

The third type of mutagenicity test involve the DNA repair mechanisms to detect mutations. Some strains of *Escherichia coli* and *Bacillus subtilis* are deficient in repair enzymes. Their growth will be inhibited relative to normal strains when mutations occur, and this inhibition can be detected by an assay. In another test of this type, DNA synthesis stimulated by mutations can be detected in human cell culture by the uptake of radioactive thymidine.

20.1.14 Extrapolation from Animals to Humans

Differences in the toxicity of specific substances between humans and laboratory animals cause considerable uncertainty in the application of animal data to humans. The differences can sometimes be isolated to differences in absorption, distribution, biotransformation, and excretion. For example, 2-naphthalamine is converted to the carcinogen 2-naphthyl hydroxylamine by dogs and humans. This is excreted by the kidneys and causes bladder cancer. However rats, rabbits, and guinea pigs do not excrete this metabolite and do not get bladder cancer from the original compound. Ethylene glycol is metabolized by two pathways with the separate end products oxalic acid or CO_2. Cats metabolize more of the ethylene glycol to oxalic acid, whereas rabbits produce more CO_2. Consequently, ethylene glycol is more toxic to cats than to rabbits. The LD_{50} for dioxin (TCDD) is 5 mg/kg in hamsters but only 0.001 mg/kg in guinea pigs. This large range in toxicity between species has produced considerable controversy over the setting of dioxin standards for humans.

Because of this uncertainty, the use of animal toxicity data for humans is often attacked as not being "scientifically valid." However, there are degrees of scientific validity, and there is certainly correlation between the toxicities to two different mammalian species. We must ask ourselves whether it would be prudent to expose large populations of humans to substances that harm rats if we don't have specific information that it would be safe to do so.

The U.S. EPA suggests selecting data from tests on animals that respond most like humans. If this type of data is not available, all acceptable data should be used, with emphasis placed on the most sensitive species, strain, and sex.

Animal data are scaled to human dosages by the ratio of either body mass or body surface area. Body surface area is preferred by the U.S. EPA. It is more conservative, producing risks that are higher by a factor of 5 when extrapolating from rat data, and 12 times higher when using mouse data. Furthermore, safety factors are usually applied to extrapolate from animals to humans to provide additional conservatism.

20.2   EPIDEMIOLOGY

The exposure of people to toxins in the workplace or in the environment can result in an unintentional, uncontrolled "experiment." This kind of information can be exploited to find information that could not be found in the laboratory. Laboratory experiments often cannot reproduce the large number of exposures, and experiments with humans are closely constrained by ethical considerations. Epidemiology includes techniques to analyze "natural experiments." These kinds of studies are termed **observational**, in contrast with experimental. For example, the effect of fluoride on humans can be studied by comparing the effect on populations with concentrations in their drinking water.

Epidemiology is the study of the aggregate occurrence of disease in human populations. It is not concerned specifically with occurrence in individuals. Thus, it often uses the language and tools of probability and statistics to answer questions such as: Is a group of workers exposed to pesticides in a manufacturing plant more likely to get a certain type of cancer? If some people at a picnic came down with food poisoning, which food is most likely to have caused it? If a certain community has more cases of childhood leukemia than the state average, what is the probability that the "cluster" occurred by chance?

Epidemiological studies may find correlations between one or more factors and disease. Such correlations are suggestive of causation but are not proof. Correlation may occur by chance or because the factor and the result are both caused by another factor. For example, a correlation was found between stork populations in communities near Hamburg, Germany, and the human birth rate. Such a correlation might suggest that storks bring babies! The real reason for the correlation was because there were more storks in rural areas than in developed ones, and people in rural communities tended to have larger families.

A correlation turns to proof when additional information supports causation, such as discovery of a fundamental mechanism that predicts such causation, or evidence from a controlled experiment. **Hill's criteria** (Table 20.3) is a list of factors that provides support for causation. Not all of the criteria need to be positive, although the more that are satisfied, the stronger is the support. Also, negative results do not rule out a causal relationship but may indicate an incomplete understanding of the relationship.

TABLE 20.3   Hill's Criteria for Causal Relationships

1. **Strength**: A high magnitude of effect is associated with exposure to the stressor.
2. **Consistency**: The association is observed repeatedly under different circumstances.
3. **Specificity**: The effect is diagnostic of a stressor.
4. **Temporality**: The stressor precedes the effect in time.
5. **Presence of a biological gradient**: Greater amount of stressor produces a stronger effect.
6. **Mechanism**: A plausible mechanism of action exists.
7. **Coherence**: The hypothesis does not conflict with knowledge of natural history and biology.
8. **Experimental evidence**: A controlled experiment supports the relationship.
9. **Analogy**: Similar stressors cause similar responses.

One of the simplest types of epidemiological analysis performed is a comparison of occurrence rates. The number of occurrences can be displayed in tabular form:

	With Disease	Without Disease
Exposed	a	b
Not exposed	c	d

The **relative risk**, r, is the ratio of the probability of exposed persons contracting the disease to the probability for unexposed persons. It is computed from the table as

$$r = \frac{a/(a+b)}{c/(c+d)} \tag{20.1}$$

Example 20.1   Consider as an example the cancer risk associated with cigarette smoking. Approximately one-fourth of the adult U.S. population smokes. About 20% of the U.S. population dies of cancer. What is the rate for smokers, and what is the relative risk for a smoker getting cancer compared to a nonsmoker?

Answer   We will select as a basis $a + b = 1000$ smokers. If this is one-fourth of the total population, there are $c + d = 3000$ nonsmokers, and 4000 is the total basis population. The 20% of the total population that dies from cancer is represented by 800 persons, leaving 3200 without cancer. The risk for nonsmokers is 10%, or $c = 300$ persons, and therefore there are 2700 nonsmokers without cancer. By difference, $a = 500$ and $b = 500$. The rate comparison table will be:

	With Cancer	Without Cancer	Total
Smoker	500	500	1000
Nonsmoker	300	2700	3000
Total	800	3200	4000

Now the probability of a smoker getting cancer can be calculated to be $a/(a+b) = 0.50$, and the relative risk is thus $0.50/0.10 = 5.0$. In other words, smokers are five times more likely to die from (any kind of) cancer in their lifetime than are nonsmokers. (For lung cancer in particular, the relative risk is about 10 to 15!)

Statistical techniques can refine relative risk calculations. Suppose that the relative risk ratio was only 1.1. Does that represent a significant increase in prevalence? If the total number of people in each group were large enough, such a ratio could be statistically significant. In a small experiment, such a ratio is more likely to occur by chance. Statistical hypothesis-testing methods can estimate whether that chance is significant. Other statistical techniques used in epidemiology include analysis of variance (ANOVA), multilinear regression, and logistic regression.

Because the data used by epidemiologists are usually not generated in controlled experiments, special techniques have been developed to establish the kind of information produced by controls. The example given above is a **cross-sectional study**, in which disease in a single population is studied at one particular time. The prevalence of disease in subgroups possessing different factors is determined and analyzed. A **case–control** (or **retrospective**) **study** is similar, but begins with one population that has a disease and then locates another control or comparison group that is without it. A **cohort** (or **prospective**) **study** starts with a group that is free of disease and follows it over a period of time, measuring which individuals contract the disease. Measurements are made of various potential risk factors, and at the end of the period the prevalence of disease is correlated with those factors. Retrospective and prospective are also used in another sense: The former is a study that is done on data collected previously, whereas the latter is a study done on data after the beginning of, and explicitly for, the study.

PROBLEMS

20.1. Typical carcinogenic toxicity tests can only detect substances that will cause at least 10% response in the test animals. In terms of carcinogenic toxicity (ED_{10}) and noncarcinogenic toxicity (MTD), what circumstances would make the carcinogenicity of a substance undetectable by a bioassay? What could be done about this situation?

20.2. What are the advantages and disadvantages of using fish as an aquatic testing organism in place of *Daphnia*?

20.3. The relative risk for cancer for a smoker compared to a nonsmoker increases with the number of years of smoking. The estimated risk of dying from lung cancer in a single year for a 70-year-old who has been smoking for 50 years is 649 per 100,000 people. This is 27 times the risk for a 70-year-old nonsmoker. What is the risk for the population of 70-year-olds as a whole, assuming that one-third of the population smokes?

REFERENCES

Ames, B. N. *The Causes and Prevention of Cancer*, http://www.ultranet.com/~jkimball/Biology-Pages/A/Ames_Causes.html.

Landis, W. G., and Ming-Ho Yu, 1995. *Introduction to Environmental Toxicology: Impacts of Chemicals upon Ecological Systems,* Lewis Publishers, Boca Raton, FL.

Lu, F. C., 1991. *Basic Toxicology: Fundamentals, Target Organs, and Risk Assessment*, 2nd ed., Hemisphere Publishing, Bristol, PA.

Newman, M. C., 1998. *Fundamentals of Ecotoxicology*, Ann Arbor Press, Ann Arbor, MI.

Rand, G. M., and S. R. Petrocelli, 1985. *Fundamentals of Aquatic Toxicology: Methods and Applications*, Hemisphere Publishing, New York.

21

TOXICITY OF SPECIFIC SUBSTANCES

Here we will summarize information relevant to toxicity for substances classed together in various ways, such as by chemical or physical properties (metals, hydrocarbons, radionuclides), by usage (pesticides or solvents), or by medium (air or water pollution). For each, we discuss briefly some of the more important anthropogenic sources of exposure and their physical and chemical properties, fate and transport, and toxic mechanisms and effects.

21.1 METALS

Metals are unique toxins. They are neither created nor destroyed (neglecting nuclear reactions), yet they readily change form and activity. A subgroup in which we are most interested is the heavy metals. **Heavy metals** are those with atomic number 22 to 34 and the elements below them in the periodic table, plus the lanthanides and actinides. Metals are widely and unevenly dispersed in the environment. Human activities that result in exposure to significant amounts of metals include mining and smelting operations, metal plating, fuel combustion, leather tanning, and their use in products from pigments to pipes.

Some metals are essential nutrients but are toxic at higher levels. Examples include chromium, cobalt, copper, iron, selenium, and zinc. Even arsenic may be required. There are three main mechanisms of toxicity for metals:

1. Complexation with proteins, especially with sulfhydryl groups of cysteine, resulting in enzyme disruption
2. The same complexation mechanism as (1), but affecting other proteins, such as those embedded in membranes
3. Competition with essential metals, such as enzyme cofactors

Environmental Biology for Engineers and Scientists, by David A. Vaccari, Peter F. Strom, and James E. Alleman
Copyright © 2006 John Wiley & Sons, Inc.

Metals are excreted by the kidney, by reversing absorption in the intestines, and by enterohepatic circulation. Because methylmercury concentrates in the latter pathway, the mercury can be removed by oral administration of adsorbents such as activated carbon.

Some aquatic plant and animal species growing in contaminated areas have been observed to develop tolerance for copper and zinc. Animals respond to cadmium, copper, and mercury by producing the low-molar-mass protein metallothionein in the liver. The metallothionein binds to these metals and to zinc, and transports them to the kidney for storage. Additional dosages of cadmium or mercury can displace the less toxic zinc or copper, resulting in less harm than the cadmium or mercury would otherwise cause.

Copper has many sources, including its use in drinking water pipes and its deliberate addition to surface water as an algicide. Copper is essential for many enzymes. However, higher animals have mechanisms to conserve it and other essential trace metals during periods of deficiency, and increase excretion at higher dosages. Copper is also critical to plant life, playing a role in electron transport in photosynthesis. It is also very toxic to aquatic plants, accounting for its use as an algicide in recreational lakes and pools.

Cadmium is a particularly toxic metal. Sources of exposure include wastewater produced in metal plating, sewage sludge applied to plants, smelting, and other occupational exposures. Cigarette smoking is an important source and may double the average body burden. Cadmium becomes complexed with metallothionein and stored in the kidney, where it accumulates over a lifetime. It is excreted very slowly, having a half-life of about 20 years. Damage occurs when a "critical concentration" of about 200 to 300 µg/g occurs in the kidney. It affects proximal tubules, reducing resorption of glucose and amino acids, and causing hypertension (high blood pressure). Inhalation can cause fibrosis, chronic bronchitis, and emphysema. Prostate cancer has been reported in occupational exposures. Calcium is lost, causing bone disorders, including pain, osteoporosis, and deformities. In the over 200 enzymes that require zinc as a cofactor, cadmium can displace the zinc competitively. Zinc is also a metabolic antagonist of cadmium. Therefore, high zinc intakes give some protection against cadmium toxicity.

Mercury is released to the environment by the burning of fossil fuel and refuse. It was formerly a major pollutant generated by the chloro-alkali industry. Food has about 5 to 20 µg/kg, except fish: Tuna and swordfish range from 200 to 1000 µg/kg. In the environment, mercury (Hg) is oxidized to inorganic ions. Anaerobic bacteria convert it to organic mercury, especially methylmercury or dimethylmercury. Organic mercury is much more toxic, since it is more readily absorbed. Dimethylmercury is volatile and can escape to the atmosphere from contaminated water and sediment. In the body, methylmercury is converted to the divalent inorganic form. Mercury has a very high affinity for sulfhydryl groups and thus can affect almost any enzyme and membrane-bound proteins. The effect on membranes is to increase their permeability to sodium and potassium, affecting cell osmolarity. Inorganic mercury is toxic to the kidney. Organic forms (methyl- or ethylmercury) target the nervous system. Divalent mercury is corrosive. If ingested, it causes abdominal cramps and bloody diarrhea, followed by renal damage. Calomel (monovalent or mercurous chloride) is less toxic but is associated with skin reactions. Elemental mercury vapor is an occupational hazard affecting the central nervous system. Neurological symptoms appear at exposure to concentrations as low as 0.05 mg Hg/m^3. The half-life for elimination of ingested mercury in seals, fish, and crabs has been measured to range from 267 to 700 days.

Lead is found in batteries, gasoline additives, paint, lead pipe, and brass fixtures. Many uses have been phased out recently, particularly indoor paint, gasoline, and pipe solder. Even without the lead pipe and pipe solder sources, brass remains an important source in plumbing systems. This is because brass fixtures are made with about 8% lead to improve machinability. Inorganic lead affects heme; organic lead affects the nervous system. Anemia is found at a blood lead level of 50 µg/dL; enzyme effects are found at as low as 10. Above 80 µg/dL (70 in children) encephalophathy can occur with symptoms including vomiting, lethargy, irritability, and so on, and possibly proceeding to coma and death. In children, a moderate exposure can produce measurable mental defects; 40 to 50 µg/dL can result in hyperactivity, a decrease in attention span, and a slight reduction in IQ score. In occupational exposures, 40 µg/dL can cause peripheral **neuropathy**, nerve damage causing numbness and the classic symptoms "footdrop" and "wristdrop." A decreased calcium intake causes lead to redistribute from storage in the bone, where it is harmless, to the more vulnerable kidney. The organic lead in auto exhaust from burning leaded fuel degrades rapidly but becomes a source of inorganic lead.

Organotin compounds are used in metal coatings to prevent biological growth on ship hulls and in cooling towers. Tributyltin is one of the common compounds used in this way. However, it has been found to exert significant toxic effects in estuaries after leaching from ship hulls.

Chromium is used in leather tanning, paints, and pigments. It is found in valences from +II to +VI, but the trivalent and hexavalent are the most important biologically. Trivalent chromium is the most common form. It is an essential nutrient, being needed as a cofactor for insulin. Hexavalent chromium is corrosive and causes allergic skin reactions on dermal exposure and cancer by inhalation. The toxicity of hexavalent chromium may be caused by reactions involving the reduction of the hexavalent to the trivalent form within cells.

Zinc is commonly used in galvanic protection of iron. The mechanism of protection involves the oxidation of the zinc to soluble ionic forms. Zinc toxicity causes anemia in mammals by interfering with absorption and utilization of copper and iron. The zinc level that causes toxic effects depends strongly on the ratio of zinc to copper.

Arsenic is found as high as 5 mg/kg in seafood. Trivalent arsenite is the most toxic form, followed by pentavalent. It combines reversibly with thiol groups in tissues and enzymes and can substitute for phosphorus in biochemical reactions such as oxidative phosphorylation. Occupational exposure has been associated with lung cancer but has not been shown to cause cancer in lab animals. Ingestion has been related to cancer of the skin, liver, bladder, and lung. Noncarcinogenic effects range from loss of appetite to irritation of epithelial tissues to paralysis of the hands and feet. The U.S. drinking water standard is 10 µg/L as of January 2006. Natural levels of arsenic as high as 280 µg/L are found in drinking waters in New Hampshire. Some 20 to 60 million people in Bangladesh use water above 50 µg/L, some as high as 2000 µg/L. Another effect commonly found in Bangladesh is thickening and discoloration of the skin. As a result, victims there have been socially ostracized. Twenty to thirty years after exposure to 500 µg/L, about 10% of people develop cancer; 60 mg/L in drinking water is rapidly fatal.

Aluminum has been found to be concentrated in the brains of people with Alzheimer's disease (AD). However, it is not known if this is a cause or a result of the disease. There are a number of epidemiological studies linking dementia to aluminum in drinking water. However, other evidence contradicts this, such as the fact that studies of humans and animals exposed to large amounts of aluminum do not show development of AD. Aluminum has also been found to cause skeletal problems in adults and infants who received large doses as part of a medical therapy.

Beryllium causes pulmonary fibrosis when inhaled, hypersensitivity reactions on dermal contact, and is a probable human carcinogen. Inhaled *nickel* is an occupational

TABLE 21.1   Acute Toxicity: 48- to 96-h LC_{50} or EC_{50} of Several Aquatic and Marine Phyla[a]

Organism	Cadmium	Chromium	Copper	Mercury	Lead	Zinc
Arthropods	0.005–0.55	0.008–10.1	0.05–3.0	0.00002–0.040	NA	0.030–9.0
(crustaceans)	0.015–47	10	0.05–100	0.004–0.4	0.7–3.0	0.2–5.0
Freshwater insects	<0.003–18.	2.0–64.0	0.007–1.2		0.28–5.5	0.04–5.5
Annelids	NA	NA	0.06–0.90	NA	NA	NA
			0.1–0.5	0.01–0.09	NA	0.8–50.
Mollusks	NA	NA	0.04–9.0	0.09–2.0	NA	0.5–20.
	2.2–3.5	14–105	0.2–8.0	0.004–30.	0.8–30.	0.1–40.
Vertebrates	NA	NA	0.01–0.9	0.003–20.	1.0–500.	0.05–7.
Salmonid fish			0.03–0.5	NA	NA	20.–70.
Centrachidae	12.6–126.	36.2–40.	0.7–10.0	3.0–10.	20.–400.	1.0–20.
or other fish	22–55	91	NA	NA	NA	NA
Algae	NA	NA	0.001–8.0	<0.0008–2.0	0.5–1.0	0.03–8.0
Chlorophyta			NA	<0.005–0.4	NA	0.05–7.0
Diatoms	NA	NA	0.005–0.8	NA	NA	NA
			0.005–0.05	0.0001–0.010	NA	0.2–0.5

[a]Results are given in mg/L. The first set of numbers is for aquatic organisms, the second is for marine. NA, not available.
Source: Based on Connell and Miller (1984) and Rand and Petrucelli (1984).

carcinogen. Nasal cancer predominates but also causes cancer in the lung, larynx, stomach, and possibly the kidney.

Metals toxicity testing in aquatic environments can be difficult, due to the sensitivity of metals to organic complexation. The test organisms themselves can produce such organics, possibly confounding the test. Table 21.1 shows acute toxicity data to aquatic and marine organisms for several metals. The ranges for aquatic and marine organisms do not differ greatly, overlapping substantially in almost every case. Acute toxicity of metals to fish is usually associated with damage to the gills. Chronic toxicity is usually measured using the more sensitive embryonic and larval life stages of aquatic animals. Most metals bioaccumulate but do not biomagnify. Bioaccumulation factors range from 100 to 10^6. One metal that does biomagnify is mercury in the organic form. Sediments tend to have much higher metal concentrations than the water column; thus, animals that feed in the sediment tend to have higher body burdens. Another way to summarize the toxicity of metals is to rank them by toxicity for various types of organisms, as in Table 21.2.

TABLE 21.2   Ranking of Metal Toxicity for Several Groups of Organisms

Organism	Ranked Sequence of Toxicity
Algae (*Chlorella vulgaris*)	Hg > Cu > Cd > Fe > Cr > Zn > Ni > Co > Mn
Fungi	Ag > Hg > Cu > Cd > Cr > Ni > Pd > Co > Zn > Fe > Ca
Plants (barley)	Hg > Pb > Cu > Cd > Cr > Ni > Zn
Protozoa (*Paramecium*)	Hg, Pb > Ag > Cu, Cd > Ni, Co > Mn > Zn
Annelida (polychaete)	Hg > Cu > Zn > Pb > Cd
Vertebrates (stickleback fish)	Ag > Hg > Cu > Pb > Cd > Au > Al > Zn > H > Ni > Cr > Co > Mn > K > Ba > Mg > Sr > Ca > Na
Mammals (rat, mouse, rabbit)	Ag, Hg, Tl, Cd > Cu, Pb, Co Sn, Be, In, Ba, Mn, Zn, Ni, Fe, Cr > Y, La > Sr, Sc > Cs, Li, Al

Source: Connell and Miller (1984).

21.2   PESTICIDES

Pesticides can be classified by function and divided into subclasses by structure (Table 21.3). However, some of the structural types are used for several functions. For example, carbamates are used as both insecticides and herbicides, as is the organochlorine hexachlorobenzene. Before the development of synthetic pesticides during World War II, other compounds were used, such as arsenicals and nicotine. What they all have in common is that they are intended to control or eliminate undesirable organisms. The emphasis here will be on the insecticides and herbicides because of their wider distribution in the environment. The structures of some of these compounds are shown in Figures 21.1 and 21.2.

Organochlorine (OC) pesticides are persistent, bioaccumulate, and biomagnify. Thus, their use is avoided now. The organophosphorus (OP) compounds do not have these characteristics but are extremely toxic to mammals. The carbamates are similar to OP compounds but less acutely toxic to mammals. The botanically derived compounds include pyrethrin, which is isolated from the chrysanthemum, and permethrin, a synthetic compound. They are effective and relatively safe. They readily degrade on exposure to sunlight. Half-lives in soil for OCs range from 1 to 12 years; OPs from 0.2 to 0.5 year; carbamates from 0.05 to 1.0 year.

The atmosphere is a major route of dispersion. Many pesticides are applied by spraying. The less soluble substances tend to be distributed in the environment adsorbed to particles. As a result, some are transported in significant quantities over long distances attached to dust particles in the atmosphere or as suspended solids in aquatic systems.

Many pesticides are degraded by photoxidation and by microbes in the environment into other toxic substances. DDT is converted to DDE, aldrin to dieldrin, and heptachlor to heptachlor epoxide.

TABLE 21.3   Types of Pesticides and Some Examples

Insecticides	
Organophosphorus	Diazinon, dichlorvos, dimethoate, malathion, parathion
Carbamate	Carbaryl (Sevin), propoxur (Baygon), Aldicarb (Temik) methiocarb
Organochlorine	DDT, methoxychlor, toxaphene, mirex, Kepone (chlordecone)
Cyclodienes	Aldrin, chlordane, dieldrin, endrin, endosulfan, heptachlor
Botanicals	Pyrethrin, permethrin
Herbicides	Chlorophenoxy acids (2,4-D, 2,4,5-T), hexachlorobenzene (HCB)
	Nitrogen-based: Picloram, Atrazine, diquat, paraquat
	Organophosphates: glyphosate (Roundup)
Antimicrobial	Chlorine, quaternary alcohols
Fungicides	Nitrogen-containing: triazines, dicarboximides, phthalimide
	Wood preservatives: creosote and pentachlorophenol, hexachlorobenzene
Rodenticides	Warfarin, sodium fluoroacetate, strychnine
Fumigants	Acrylonitrile, chloropicrin, ethylene dibromide, 1,3-dichloropropane
Phytotoxins	Copper and copper compounds
Antifoulants	Organic tins (low phytotoxicity): tributyltin

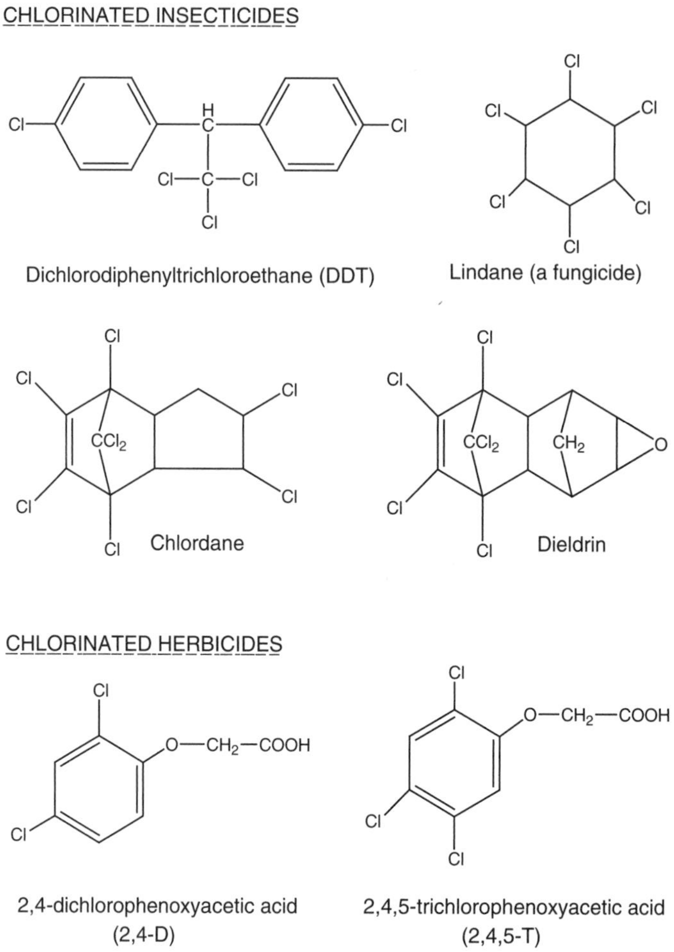

CHLORINATED INSECTICIDES

Dichlorodiphenyltrichloroethane (DDT)   Lindane (a fungicide)

Chlordane   Dieldrin

CHLORINATED HERBICIDES

2,4-dichlorophenoxyacetic acid   2,4,5-trichlorophenoxyacetic acid
(2,4-D)   (2,4,5-T)

Figure 21.1   Structures of some organochlorine pesticides.

21.2.1   Toxic Effects

Tables 21.4 and 21.5 contains toxicity data on some pesticides. Most insecticides act by interfering with the nervous system. For example, DDT binds to lipoproteins in axon membrane, holding a gate open for sodium. Organochlorines in general stimulate the nervous system, causing irritability, tremor, and convulsions.

OP and carbamate inhibit acetylcholinesterase. In the Central nervous system (CNS) they cause tremor and convulsion. In the autonomic nervous system they cause diarrhea, involuntary urination, and bronchoconstriction. At the neuromuscular junction they cause muscle contraction, weakness, and loss of reflexes, followed by paralysis. Chronic exposure causes **neuropathy** (destruction of nerve fibers). Motor effects predominate, including muscle wasting, difficulty with movement, and inability to locate one's joints in space, making walking difficult. Exposure to OPs were observed in both human and animal tests to result in **myopathy** (damage to muscle tissue), including necrosis of skeletal muscles. This is thought to be caused by the high levels of acetylcholine resulting from inhibition

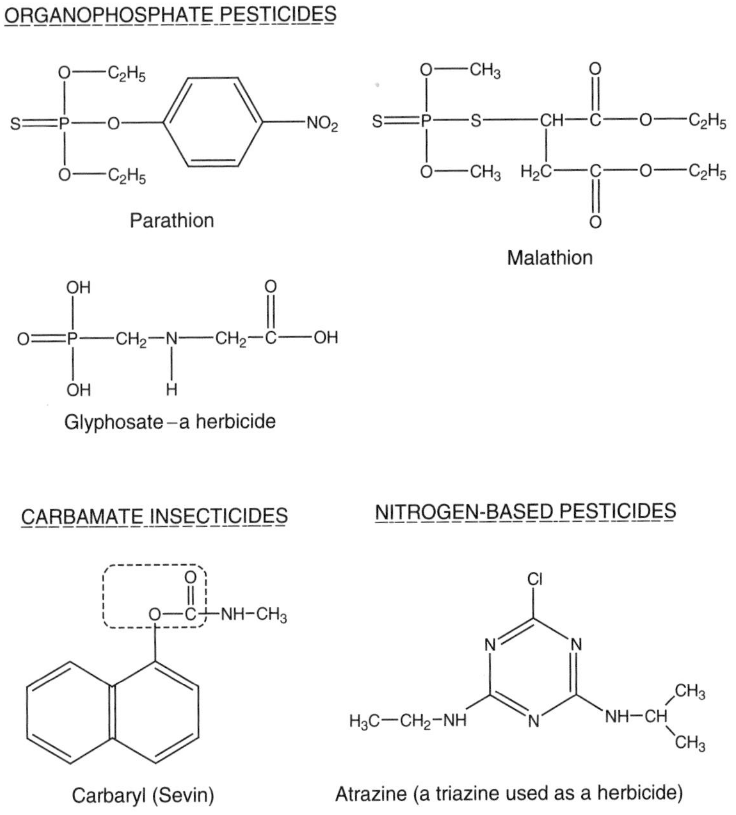

ORGANOPHOSPHATE PESTICIDES

Parathion

Malathion

Glyphosate – a herbicide

CARBAMATE INSECTICIDES

NITROGEN-BASED PESTICIDES

Carbaryl (Sevin)

Atrazine (a triazine used as a herbicide)

Figure 21.2 Structures of some unchlorinated pesticides.

TABLE 21.4 Mammalian Toxicity of Some Insecticides[a]

Pesticide	LD_{50} (mg/kg)	NOEL (mg/kg)			ADI (mg/kg)
		Rat	Dog	Human	
Diazinon	108	0.1	0.02	0.02	0.002
Dichlorvos	80	NA	NA	0.033	0.004
Malathion	1375	5.0	0.2	NA	0.02
Parathion	13	NA	NA	0.05	0.005
Trichlorfon	630	2.5	1.2	NA	0.005
Aldicarb	0.8	0.125	0.25		0.005
Carbaryl	850	10	NA	0.06	0.01
DDT	113	0.05	NA	NA	0.01
Aldrin/dieldrin	40	0.025	0.025	NA	0.001
Chlordane	335	0.25	0.075	NA	0.001
Endrin	18	0.05	0.02	NA	0.0002
Heptachlor	100	0.25	0.0625	NA	0.0005
Lindane	88	1.25	1.6	NA	0.01
Methoxychlor	6000	10	NA	NA	0.1

[a]NA: Not available.
Source: Lu (1991).

TABLE 21.5   Acute Aquatic Toxicity of Example Pesticides[a]

Pesticide	Photosyn. % Red.[b]	*Daphnia* 48-h EC_{50}	Amphipod	Stonefly Naiad	Fish	Tadpole
Organochlorine						
Aldrin	85	28	9800	1.3	NA	150
Chlordane	94	29	26	15	69	NA
DDT	77	0.36	1.0	7.0	34	800–1000
Dieldrin	85	250	460	0.5	16	100–150
Endrin	46	20	3.0	0.25	1.3	120–180
Heptachlor	94	42	29	1.1	56	440
Lindane	28	460	48	4.5	56	2700–4400
Methoxychlor	81	0.78	0.8	1.4	35	330
Toxaphene	42	15	26	2.3	5.1	140–500
Organophosphorus						
Diazinon	7	0.9	200	25	NA	NA
Dibrom	56	0.35	110	8.0	NA	1700
Dichlorvos	NA	0.066	0.5	0.1	NA	NA
Malathion	7	1.8	1.8	10	12,500	200–420
Methyl parathion	NA	0.14	NA	NA	7,500	NA
Parathion	NA	0.6	0.6	5.4	1,600	1000
Phosdrin	NA	0.16	0.16	5.0	NA	NA
Herbicides						
Dinitrocresol	NA	0.014[c]	NA	0.32	NA	NA
Diquat	NA	NA	NA	NA	91[d]	NA
2,4-D acid	NA	1.4	NA	15	NA	NA
2,4-D butoxyethanol ester	NA	>100[c]	0.44	1.6	2.1[d]	NA
Naphtha	NA	3.7	0.84	2.8	NA	NA
Paraquat	NA	3.7	11	>100	400[d]	28
Picloram	NA	NA	27	48	26.5[d]	NA
Silvex	NA	2	NA	0.34	83[e]	10
Trifluralin	NA	0.24	2.2	3.0	8.4[d]	0.1[e]
2,4,5-T	NA	NA	NA	NA	0.5[d]	NA

[a]The fish are freshwater fathead minnow for OC and OP pesticides, and bluegill for the herbicides; 96-h LC50 except where noted. Concentrations in μg/L for organochlorine and organophosphorus pesticides and mg/L for herbicides. NA, not available.
[b]Reduction of estuarine phytoplankton photosynthetic productivity during 4 h exposed to a concentration of 1.0 mg/L.
[c]26-h EC_{50}.
[d]48-h LC_{50}.
[e]24-h LC_{50}.
Source: Connell and Miller (1984).

of acetylcholinesterase. Symptoms include muscle tenderness. Adverse psychological disturbances have been identified which lasted from six months to a year. These included depression, nightmares, and emotional instability.

There is positive interaction between many OPs. One of the strongest is a 100-fold increase in the combined toxicity of malathion and tri-*o*-cresyl phosphate (TOCP). Some pesticide formulations include potentiators to increase their effect, although this increases mammalian toxicity. Antagonistic effects occur as well. Substituted urea herbicides reduce the toxicity of parathion, probably by inducing MFO detoxification.

OP and carbamates are not carcinogens, in general, except some that are chlorinated, or some like carbaryl, which can be nitrosated to a carcinogenic form. All organochlorines cause hepatoma in mice. DDT is controversial because it does not cause cancer in rats, hamsters, and a few other animals. Only for hexachlorocyclohexane is there epidemiological evidence of cancer in humans. The fumigants ethylene dibromide (EDB) and 1,2-dibromo-3-chloropropane (DBCP) cause stomach cancer in rats and mice and have been restricted. The herbicide amitrole (aminotriazole) and the ethylenebisdithiocarbamates (such as maneb) produce thyroid tumors. Carbaryl (sevin) is a teratogen in dogs but not in other animals. The dithiocarbamate fungicides, captan, paraquat, and others, are reportedly teratogenic. The herbicide 2,4,5-T is known to cause teratological effects in rats.

Carbaryl produced effects attributed to decreased renal resorption in human experiments. Paraquat causes edema, hemorrhage, and fibrosis in the lungs when inhaled. Pyrethrum is associated with hypersensitivity, contact dermatitis, and asthma. OCs are hepatotoxic and induce microsomal monooxygenases. Some pesticides from all the major groups have an effect on the immune system. These include malathion, methylparathion, carbaryl, DDT, paraquat, and diquat.

Herbicides act on plants by affecting either growth or photosynthesis. The phenoxyacids simulate natural auxins (plant growth hormones). Urea herbicides inhibit photosystem II, preventing ATP and NADPH formation. Diquat and paraquat inhibit photosynthesis. Some are selective for either monocots (grassy plants) or dicots (broadleaf).

The manufacture of the pesticide 2,4,5-T produces 2,3,7,8-tetrachlorodibenzo-*p*-dioxin (TCDD) as a trace impurity. This is one of the most toxic substances known. The manufacture and use of 2,4,5-T and related herbicides has resulted in some of the most notorious environmental contamination incidents known, which involve TCDD (see Section 21.4).

Dinitrophenols are used for weed control. They uncouple oxidative phosphorylation in respiration, causing metabolic activity to increase out of control. Effects of acute exposure in humans include the sensation of heat, and rapid breathing and heart rate. Chronic symptoms include anxiety, sweating, thirst, and fatigue. Paraquat causes lung, kidney, and liver damage and can cause pulmonary fibrosis even from routes other than inhalation.

Fate and Transport: Dieldrin depuration half-life ranges from 148 days for the guppy *Lebistes reticulatus* to 12.2 days for tubificid worms (*Tubifex* sp.). Depuration half-lives are correlated with lipophilicity. Plants cannot excrete toxins, but instead, segregate them in vacuoles within the cells. Table 21.6 summarizes some of the environmental fate and transport properties for a number of pesticides.

21.2.2 Ecosystem Effects

Biomagnification is a problem mostly with organochlorine pesticides. The classic case is DDT and its metabolite DDD. Figure 18.9 illustrates two examples of organochlorine biomagnification.

By their very nature, pesticides can produce pronounced effects on populations and ecosystems. These include many that were dramatically publicized by Rachel Carson in her book *Silent Spring*, including fish and bird kills, development of insect resistance, and actually causing an increase in some pests by destroying their predators. In the latter

TABLE 21.6 Fate and Transport Properties of Some Pesticides

Pesticide	Aqueous Solubility (mg/L)	$Log_{10} K_{OW}$	Log_{10} BCF (Fish) (L/kg)	Typical Environmental Half-Life
Organochlorines				1–12 years
2,3,7,8-TCDD	0.0002	6.15	4.73	
DDT	0.0012	5.98	4.78	
Aldrin	0.01			
Chlordane	0.056	5.15	4.05	>100 days
Heptachlor	0.056			
Dieldrin	0.18	5.48	3.76	
Lindane	7.0	4.82	2.51	>100 days
Organophosphates				72–180 days
Chlorpyrophos	0.3	4.99	2.65	
Parathion	24			
Diazinon	40			
Malathion	145			<30 days
Dimethoate	2,500			
Carbamates				20–365 days
Carbaryl	40			
Carbofuran	700			
Pyrethrin				<1day
Herbicides				
Simazine	5.0			30–100 days
Atrazine	34			30–100 days
2,4,5-T	280	0.6	1.63	<30 days
2,4-D	890	1.57	1.30	<30 days
Glyphosate	12,000			30–100 days
Diquat	(70%)			
Dalapon	(80%)			

Source: Rand and Petrucelli (1985); LaGrega et al. (2001); Rao et al. (2004); Ferruzzi and Gan (2004).

case, the target species is sometimes the beneficiary of a pesticide application, if its predators and competitors are more strongly affected. An example of this counterproductive situation is the application of pesticide to control boll weevils and leafworms infesting cotton farms in Central America. From 1950 to 1955 the number of pesticide applications had to increase from few or none to 8 to 10 per season. By 1960, 28 applications per season were needed. Over the same period, the number of pest species increased from the original two in 1950, to five in 1955, and then eight in 1960. Many similar examples exist.

Many pesticides, such as insecticides, fungicides, and fumigants, target soil organisms and have their greatest environmental effect on nontarget organisms in the soil. Susceptible invertebrates include insects, earthworms, slugs, and gastropods. Some of these organisms contribute to topsoil formation. Insecticides applied to natural systems, such as for the control of forest pests, can reduce the species diversity within the invertebrates.

Among the terrestrial vertebrates, avian species are especially susceptible to the insecticides. LD_{50} values are often less than 100 mg/kg. However, the sublethal effects are of greatest concern. In particular, bird reproduction is very sensitive to organochlorine insecticides. The best known effect is the thinning of eggshells due to DDT, its metabolite DDE, and dieldrin, among others. These prevent the embryos from surviving to hatch. However, many other reproductive effects are also found, including direct embryo toxicity and aberrant parental behavior such as destruction of eggs. The greatest concern with these effects is with birds of prey, because the organochlorine pesticides are biomagnified. Species that have experienced serious population declines include the bald eagle (*Haliacetus leucophalus*) and the osprey (*Pondion haliatetus*).

DDT has not been found to seriously affect wild mammals. However, other organochlorine pesticides, especially the cyclodienes such as heptachlor, dieldrin, and endrin, have resulted in mortalities. Although the cholinesterase inhibitors are very toxic to mammals, no significant effects have been observed in the wild. Development of resistance has been observed in a variety of aquatic invertebrates.

Aquatic ecosystems have been affected both by direct application of pesticides, such as for control of mosquitoes, and by runoff from agricultural, forest spraying, and so on. Arthropods such as crustraceans are very susceptible to insecticides. Heavy killings of crabs and shrimp have occurred in marshes sprayed with organochlorines. The microcrustaceans are an important component of the food chain and are also affected heavily. Mollusks and annelids, on the other hand, are relatively tolerant. Fish tend to be very susceptible to the organochlorines, especially the cyclodienes. Some warm-water, rapidly reproducing species of fish in heavily sprayed areas of Mississippi have become resistant to levels of some of the organochlorines 100 times higher than that which is normally toxic.

Chronic toxicity is reported as the MATC in Section 19.7. Application factors (LC_{50}/ MATC) for chronic toxicity vary greatly. For Kepone the application factor for sheepshead minnows ranges from 0.001 to 0.002. However, for endrin the AP is 0.35 to 0.91. In other words, Kepone causes chronic effects at concentrations as much as 1000 times lower than the LC_{50}, whereas the chronic effects of endrin do not appear until the concentration reaches one-third of the LC_{50}.

Pesticides interact with each other, often positively. The toxicity of pesticides to aquatic organisms also interacts with the abiotic factors temperature and pH, as described in Section 17.3. Both positive and negative effects of temperature have been observed. Endrin was found to be several hundred times as toxic to carp (*Cyprinus carpio*) at 27°C as it was at 7°C. DDT and methoxychlor, on the other hand, were more toxic at lower temperatures. Hardness does not seem to produce an effect independent of pH, and many of the pH effects are associated with ionizable pesticides or those that react to form other products at certain pH ranges.

21.3 HYDROCARBONS, SOLVENTS, PAHs, AND SIMILAR COMPOUNDS

Here we consider other anthropogenic organic chemicals and their by-products. Solvents include many types of liquid hydrocarbons, which are used as carriers for other materials, such as pigments in paints, or for rinsing oil and grease from manufactured items. Some, such as benzene, may also be components of fuel or raw materials for synthesis of other compounds. Some of the larger molecules we consider include the polynuclear aromatic hydrocarbons (PAHs), phthalate esters, and surfactants. Chlorinated organics are considered separately in the next section.

Many hydrocarbons are biotransformed by cytochrome P450 enzymes. Thus, they can interact with other toxins either positively by increasing cytochrome activity (as benzene does) or negatively by competing for the enzyme (as toluene does with benzene).

The solvents are hydrocarbons up to five or six carbons in size. They have appreciable vapor pressures, which may result in significant inhalation exposure. Solvents are also absorbed significantly through the skin in occupational settings. The larger compounds are more likely to be adsorbed to particles. Exposure can result from inhalation or ingestion of the particles. Some of them enter the food chain, where they can be bioaccumulated and ingested.

The major toxic effects common to these compounds are CNS depression, including narcosis, and irritation. These are of special concern with acute exposure to the solvents. However, these types of compounds can cause other effects, ranging from liver or kidney damage to carcinogenesis. CNS depression is essentially the action of a general anaesthetic. The potency of organics to produce CNS depression increases with the chain length of the compound. Halogenation, addition of an alcohol group, or unsaturation (removal of hydrogen to form a double carbon–carbon bond) increase potency as well.

One mechanism of irritation by hydrocarbons and surfactants is by the extraction of fats from the skin, lungs, eyes, or other cell membranes contacted. Addition of functional groups to an organic molecule tends to increase irritant properties. Amines and acidic groups make the compound corrosive, and alcohol, aldehyde, and ketone groups can precipitate and denature proteins associated with the membranes.

The *aliphatic* (saturated) hydrocarbons have relatively less toxicity than others do in this group. Ingestion of more than 1 mL/kg can produce systemic effects. For lower amounts, aspiration to the lungs is the principal concern. Chronic exposure, such as to hexane or heptane, produces neuropathy, probably due to metabolism to alcoholic and ketone forms. Olefinic (unsaturated) forms are stronger CNS depressants. Interestingly, the presence of double bonds eliminates the neurotoxicity of hexane and heptane. The cyclic hydrocarbons, such as cyclohexane, are similar to the open-chain forms, except that they are higher in irritancy, and do not seem to produce chronic effects.

Alcohols, including glycols, are much stronger CNS depressants than aliphatics are and slightly more irritating. As carbon chain length increases, irritation decreases but lipophilicity increases, as does systemic toxicity. Methanol is less inebriating than ethanol but has the unusual property of destroying the optic nerve. Fifteen milliliters can cause blindness. As with ethanol, it is metabolized by a zero-order rate mechanism, but at one-seventh the rate. Ethanol acts as an irritant by dehydrating protoplasm. An initial stimulant effect is caused by depression of control mechanisms in the brain. Pain sensitivity is greatly reduced. Cutaneous (skin) blood vessels become dilated. The resulting increased heat loss can be dangerous in cold weather. It increases gastric secretion, which can aggravate stomach ulcers. It causes fat accumulation and cirrhosis in the liver. The latter can be fatal itself or can cause progression to cancer. Ethanol increases urine flow through a mechanism involving pituitary and adrenal hormones. The resulting water loss, along with the acetaldehyde by-product, may be a cause of the headache in a hangover. On the other hand, there is evidence that consumption of small amounts of alcoholic beverages with meals may have some benefits for the cardiovascular system.

Isopropyl alcohol is less toxic than *n*-propanol, but both these and butanol are more toxic than ethanol. Allyl alcohol is highly irritating. It can be absorbed through the skin, resulting in deep pain and possible burns of the eye. Glycols are compounds that have two hydroxyls on adjacent carbons. They are less toxic than the monohydroxy alcohols. Ethylene glycol can be fatal to humans with a single dose of 100 mL. It is biotransformed to oxalic acid, which blocks kidney function. Ethanol can inhibit this transformation, making it an antidote.

Aldehydes tend to be more irritating than they are CNS depressants. One unique toxicity of the aldehydes, especially formaldehyde, is sensitization. That is, it can increase a person's response to other chemicals. Formaldehyde is a common industrial chemical used in plastics and resins. The LD_{50} in humans for formalin (37 to 50% formaldehyde solution) is about 45 g, although deaths have been reported at as low as 30 g. Ingestion can produce headaches, GI tract corrosion, pulmonary edema, fatty liver, kidney necrosis, unconsciousness, and vascular collapse. Formaldehyde has been associated with mutagenicity and carcinogenicity in laboratory tests, but a steep dose–response relationship, the lack of epidemiological evidence for carcinogenicity in humans, plus the following facts, suggest the presence of a carcinogenicity threshold. Formaldehyde is a metabolic by-product, normally present at several ppm in tissues. Thus, it appears that carcinogenicity is associated with exposures high enough to cause irritation and tissue injury. Acetaldehyde is less irritating and toxic. Acrolein is an unsaturated propionaldehyde. The double bond greatly increases its reactivity as well as its irritant and toxic effect.

Ketones produce fewer occupational health problems, possibly because their irritant properties serve as a warning before other effects occur. Acetone causes skin irritation only after repeated lengthy contact. Eye irritation occurs with unacclimated persons at air concentrations of 500 ppmv, and repeated exposure can produce tolerance up to 2500 ppmv. *Carboxylic acids* are mostly irritants. Acidity, and irritation, decreases with chain length and increases with halogenation. *Esters* are stronger anesthetics than alcohols. The lower-carbon esters are stronger irritants than alcohols and can cause **lacrimation** (eye tearing). Phosphate esters are used as plasticizers and can cause CNS damage. *Ethers* are effective anesthetics and slightly irritating.

Amines have many uses, including as disinfectants in consumer products. They include some of the most toxic common industrial chemicals. Most simple amines have pK_a values between 10.5 and 11.0, making them ionized at physiological pH and corrosive to tissues.

primary amine   secondary amine   tertiary amine

An important property of amines is that they are easily absorbed by all routes of exposure, including the skin. Thus, the dermal toxicity is often similar to the ingested toxicity (Table 21.7). Systemic effects include pulmonary edema and hemorrhage, liver and

TABLE 21.7   Toxicity of Some Amines in Animal Tests (mg/kg)

Amine	Oral LD_{50}	Skin LD_{50}
Methylamine	0.02	0.04
Ethylamine	0.4	0.4
Propylamine	0.4	0.4
Butylamine	0.5	0.5
Hexylamine	0.7	0.5

Source: Williams and Burson (1985).

kidney necrosis, and coronary degeneration. They react to form methemoglobin in the blood. Repeated exposure can produce strong allergic reactions. Aromatic amines or diphenylamine compounds are carcinogens, such as benzidine, 2-naphthylamine, or 4-aminodiphenyl, which cause "aniline tumors" in dye industry workers. Any alkyl amine (such as those that can be produced in digestion of food) can react with nitrites in food and acid in the stomach to form nitrosamines, which can potentially cause liver cancer.

Analine itself is an important industrial compound that is also used in household products such as polishes, paints, and inks. It is considered very toxic and is easily absorbed through the skin. In the blood, aniline oxidizes Fe(II) to Fe(III), turning the blood to a brown-black color and eliminating its ability to transport oxygen.

Nitro compounds are mild irritants. Unsaturated compounds are easily absorbed dermally. Nitrated aromatics produce a variety of effects, including methemoglobinemia. Some produce bladder tumors. Trinitrotoluene and dinitrobenzene are easily absorbed by all routes of exposure. Besides the other toxic effects, they can uncouple oxidative phosphorylation and cause liver damage. *Nitriles* (compounds with structure $R-C\equiv N$) are easily absorbed, and many slowly degrade to cyanide. Acetonitrile is considered moderately to very toxic. Acrylonitrile is classified as highly toxic and is a suspected human carcinogen.

$$H_2C=CH-C\equiv N$$
acrylonitrile

Single-ring aromatics include benzene and its derivatives. Aromatics tend to be more toxic than aliphatics with a similar number of carbons. CNS depression caused by aromatics differs qualitatively from that produced by aliphatics. Coma induced by aliphatics, such as by gasoline, exhibit inhibited reflexes. On the other hand, aromatics produce hyperactive reflexes, sometimes with convulsions.

Benzene is reportedly fatal at about 10 to 15 mL for an adult, and possibly as low as 2 to 5 mL. Inhalation at 250 ppmv produces headache, nausea, vertigo, and drowsiness; 3000 ppmv in air is irritating to the eyes and respiratory tract. Acute symptoms appear after 30 to 60 minutes at 7500 ppmv; 20,000 ppmv for 5 to 10 minutes has been reported to be fatal. Acute symptoms are respiratory inflammation, pulmonary edema and hemorrhage, renal congestion, and cerebral edema. Chronic dermal exposure can produce damage resembling first- or second-degree burns. Blood disorders progress through three stages: First is a reversible reduction in blood clotting and a mild anemia. Continued exposure produces increased leukocyte production. Finally, a more severe anemia and hemorrhage occurs. Less commonly, benzene also causes a cancer, leukemia. It is a class A carcinogen. Chomosome damage has also been observed in humans. Toluene is a stronger CNS depressant than benzene but is not known to cause permanent blood disorders. The xylenes are even stronger CNS depressants than toluene.

Phenols are cytotoxic, acting as a local anesthetic and CNS depressant. They are easily absorbed but produce strong irritation and burns. Phenol easily penetrates the skin, where it causes deep burns. Burns from ingestion can lead to extreme shock and death. Phenol can also cause necrosis systemically in the liver and kidneys as well as the heart and urinary tract. Death from systemic poisoning is usually from respiratory depression as a result of its CNS depression activity. Ingestion of as little as 1 or 2 g reportedly can be fatal. The dihydroxyphenols catechol, resorcinol and hydroquinone, and cresol, behave similarly to phenol, except that they are more toxic.

benzene   *phenol*   *hydroquinone*   *p-cresol*

Polynuclear aromatic hydrocarbons (PAHs, also referred to as PNAs or polycyclic aromatic hydrocarbons) are fused benzene rings and their derivatives. Many are prevalent in coal and crude oil. They also are significant products of partial combustion; that is, they are found in soot and smoke particles, and in pyrolytically produced tars and oils. The simplest PAH is naphthalene.

naphthalene   *anthracene*   *phenanthrene*

benzo[a]pyrene   *pyrene*

fluoranthene   *fluorene*

PAHs are strongly absorbed to soil and sediment particles in the environment and can even be transported long distances by the wind. Aqueous solubility decreases with molar mass and is lower for linear ring arrangements than more condensed forms (compare anthracene and phenanthrene in Table 21.8). In aquatic environments, PAHs tend to be most concentrated in sediments, least in the water column, and intermediate in organisms.

Some bacteria and fungi can mineralize some PAHs, but others can only cometabolize them. Fungi and some animals utilize the cytochrome P450 (MFO) system to hydroxylate PAHs to diols. Certain intermediates of this conversion in mammals are thought to be responsible for carcinogenicity in this group. Injection of PAHs in fish can induce MFO activity severalfold within several days.

PAHs cause acute toxic effects in humans only at high doses that are not common in environmental or occupational settings. Napthalene inhalation causes headache, confusion, nausea, and profuse perspiration in humans. For a variety of marine and

TABLE 21.8.   Transport Properties of Some PAHs and Bioaccumulation Factors for *Daphnia pulex*

Compound	No. of Aromatic Rings	Molar Mass	Solubility (μg/kg)	Bioaccumulation Factor ± SE
Naphthalene	2	128.2	22,000–31,700	131±10
Fluorene	2	166.2	1685–1980	
Anthracene	3	178.2	30–73	917±48
Phenanthrene	3	178.2	1002–1290	325±56
Fluoranthene	3	202.3	206–260	NA
Pyrene	4	202.3	132–171	2702±245
Benzo[a]pyrene	5	228.3	9.4–14	NA

[a]NA, not available.
Source: Rand and Petrucelli (1985).

aquatic vertebrates and invertebrates, LC_{50} ranged from 2.0 to 3.8 mg/L. For phenanthrene, LC_{50} was 0.3 mg/L to the grass shrimp *Palaemonetes pugio* and 0.6 mg/L to the marine copepod *Neanthes arenaceodentata*. The concentration of PAH that causes a 50% reduction in the photosynthetic activity of the freshwater algae *Chlamydomonas angulosa* was 9.6 mg/L for naphthalene and 0.9 mg/L for phenanthrene.

Of greater importance are the mutagenic and carcinogenic properties of this class of compounds. Benzo[a]pyrene (BaP) is the first known human carcinogen. In London, Pott noticed the association between scrotal cancer in men working as chimney sweeps in 1775. They are exposed to high dosages of BaP and other PAHs in creosote. MFO activation is required for genotoxicity. Intermediates in biotransformation have been shown to bind to cellular DNA.

Crude and fuel oils are mixtures of aliphatic, olefinic, aromatic, and polyaromatic hydrocarbons. Because of their complex composition, fate and transport in the environment result in large changes in toxic potency. The processes of volatilization, dissolution, and degradation produce changes in composition known as **weathering**. Despite their hydrophobicity, petroleum hydrcarbons do not significantly bioaccumulate. Apparently, this is because of their ability to be metabolized to soluble forms for excretion. The 96-hour LC_{50} of unweathered crude oil ranges from 1.34 to more than 19.8 mg/L in adult fish and from 0.81 to above 19.8 mg/L in aquatic invertebrates. No. 2 fuel oil ranged from 0.7 to 11.6 times as toxic, with a median of about 3.

Acute exposures, such as those due to oil spills, can cause considerable mortality, especially of intertidal organisms and seabirds. This leads to successional changes in populations. Substituted aromatics represent the relatively soluble fraction of these oils. These can affect organisms throughout the water column. Chronic exposure to oil pollution, such as in highly industrialized estuaries, have been found to reduce the number of species and individuals, to increase the proportion of opportunistic species, and to decrease ecological energy flow. Benthic amphipods, such as the genus *Ampelisca*, have been found to be a sensitive indicator organism for petroleum pollution.

Phthalate esters are added to plastics to control their mechanical properties, especially to reduce brittleness. In the structural formula shown below, the R groups may be a variety of long-chain hydrocarbons, such as isooctyl phthalate, 2-ethylhexyl phthalate, or isodecyl phthalate. Although they have a very low vapor pressure, they can diffuse from plastic materials into the environment, especially at elevated temperatures. Bioconcentration factors as high as 720 and 3600 have been measured. Their acute toxicity is fairly low. The oral LD_{50} in lab animals range from 8 to 64 g/kg. Short-chain phthalates tend to be more toxic. The 96-hour LC_{50} value for dibutyl phthalate ranged from 0.7 to 6.5 mg/L, whereas di-2-ethylhexyl phthalate was usually greater than 10 mg/L. Terrestrial organisms are not as likely to suffer exposures, whereas aquatic organisms receive exposure from water pollution, and humans through various routes, such as food packaging. Phthalates have also been implicated as possible xenoestrogens.

phthalate ester

Surfactants, surface-active agents, are substances that lower the surface tension of liquids, especially water. They include soaps and synthetic detergents, of which the latter are of greatest environmental interest. Structurally, their molecules are designed to have a lipophilic (often hydrocarbon) portion and a hydrophilic portion. They may be anionic, cationic, zwitterionic (capable of forming cations or anions), or nonionic. Above a characteristic concentration, the **critical micelle concentration** (CMC) surfactants form enclosed particles called **micelles** (see Figure 3.9). These can concentrate lipophilic substances from the water, decreasing their aqueous concentration but increasing the capacity of the overall solution (water plus micelles) to dissolve those substances. They may increase the uptake of lipophilic substances, increasing their toxicity. Biodegradability of surfactants decreases with branching of the hydrocarbon portion and increases with its length.

Toxicity occurs by various mechanisms, including the formation of complexes with membranes and enzymes. Toxicity tends to increase with chain length. Some results have shown greater correlation with surface tension than with surfactant concentration. For example, the lethal concentration for hydra occurred at a surface tension of 49 ± 4 dyn/cm. This corresponded with the level that would disrupt the cell membrane. Some specific acute toxicity results include a 24-hour EC_{50} for *Daphnia* ranging from 12 to 13 mg/L, and a 96-hour LC_{50} for fathead minnow ranging from 3.5 to 23 mg/L. A number of investigators have noted damage to gills and external sensory organs at sublethal concentrations.

21.4 HALOGENATED ORGANICS

Halogenated organic solvents displaced industrial uses for many hydrocarbon solvents because their vapors have a flash point at much higher temperatures, making them non-explosive. Others were developed for their special properties, such as the high dielectric constant of polychlorinated biphenyls. Yet another group, such as chlorinated dioxins and trihalomethanes in drinking water, are unintended by-products of other processes. Compounding their toxic properties are their lipophilicity and their low rate of biodegradation, both of which contribute to tendency to persist and bioaccumulate in the environment.

The *halogenated alkanes* refer mostly to chlorinated one- and two-carbon compounds, various brominated compounds included among the trihalomethanes (THMs) formed during chlorination of drinking water, and fluorinated alkanes such as a variety of Freons.

The properties of the halogenated alkanes are similar to their nonhalogenated counterparts. CNS depressant action and liver and the potential for kidney damage both increase with degree of chlorination. Unsaturated compounds are more toxic than saturated. Replacing chlorine with bromine increases toxicity, whereas replacing it with fluorine decreases it. Many halogenated compounds induce liver cancer in laboratory tests with rodents.

Methylene chloride may be the least toxic of the chlorinated methanes. It has the lowest irritancy and potential for CNS depression. It is commonly used as a laboratory solvent. Chloroform and carbon tetrachloride are potent CNS depressants and liver and kidney toxins. A summary of the inhalation effects of chloroform in humans is:

200–300 ppmv	Odor threshold
1000 ppmv	Fatigue and headache
1500 ppmv	Dizziness and salivation within minutes
4000 ppmv	Fainting and nausea
14,000–16,000 ppmv	Rapid narcosis

Vinyl chloride, a class A carcinogen, is one of the most dangerous compounds in this group. CNS depression and death have been reported upon acute exposure. Chronic exposure results in *arthroosteolysis* (destruction of bone at the joints), *Raynaud's phenomenon* (a vascular disease that can lead to gangrene), and *sclerodermatous skin changes* (hardening and thickening of the skin). It is mutagenic and is a well-established cause of liver cancer at levels formerly thought to be acceptable for occupational exposure.

Polychlorinated biphenyls (PCBs) have the general structure shown below, with chlorines substituted at various positions, forming a great variety of possible congeners. Their empirical formula is $C_{12}H_{12-n}Cl_n$, where n is the number of chlorine substitutions. Thus, a congener with five chlorines will have a molar mass of 328.5 and consists of 54% chlorine by weight. PCBs were produced as mixtures of various congeners plus impurities and sold by the Monsanto Company under the trade name Aroclor 12*xx*, where *xx* indicates the average percent chlorine. Aroclor 1254, one of the common mixtures sold, thus has an average of five chlorines per molecule. Their vapor pressure and aqueous solubility is very low, and they adsorb strongly to soil and sediment. The solubility of Aroclor 1248 ranges from 0.034 to 0.175 mg/L; that of Aroclor 1268 is less than 0.007 mg/L. They both bioconcentrate and biomagnify in aquatic systems. They have the chemical property of being highly stable, even at high temperatures, thus are very persistent in the environment. They are resistant to biochemical degradation. However, some breakdown has been observed with PCBs having fewer than four chlorine atoms.

biphenyl

2,2',3,4'- tetrachlorobiphenyl,
a polychlorinated biphenyl (PCB)

Their use has been discontinued because of the environmental problems they cause. However, because of their persistance and of previous releases, they remain a problem in a number of aquatic and marine environments. For example, an advisory has been issued recommending that consumption of striped bass taken from the Hudson River estuary be limited to not more than one per month because of PCB contamination of that river.

PCBs are stored in fatty tissues and excreted in milk. They are strong inducers of microsomal enzymes. This results in enhancement of metabolism of some of the sex hormones and produces interaction with other toxins. They can also act as a carcinogenic promoter. In the liver they cause tumor formation and fibrosis. Their reproductive effects include reduced fertility in mink and rhesus monkey. They also produce immunosuppression. Other effects included immunosuppression, numbness, coughing and expectoration, fatigue, and eye discharge.

Chloracne is one of the major symptoms of exposure to halogenated PAHs, including chlorinated phenols, PCBs, and chlorinated dioxins. **Chloracne** is an acute rash of acne that may resemble a severe form of ordinary acne, but characterized by hundreds and hundreds of *comedomes*, very persistent blackheads that develop into inflamed papules.

In 1968, 1000 people in Japan were poisoned by PCB-contaminated rice oil. Chloracne was the major symptom. Polybrominated biphenyls (PBBs) have toxic properties similar to those of PCBs. They are less common in the environment, although a major exposure occurred in Michigan in 1973 when cattle feed was contaminated.

PCBs have low acute toxicity to most mammals tested (Table 21.9). Mink is an exception, as are rhesus monkeys. The latter show toxic effects after one year of eating food with 2.5 to 5.0 mg/day of PCB 1254; 100 to 300 mg/kg in food per day results in mortality in two to three months. Toxicity to birds is lower than that of DDT and seems to increase with degree of chlorination. In contrast to birds, toxicity of PCBs to fish decreases with degree of chlorination, and the overall toxicity is much higher.

Because of the typically low exposures, sublethal effects are more important in mammals than are lethal effects. Effects include liver enlargement and damage, reduced growth, immunosuppresion, and liver enzyme induction that increases with degree of

TABLE 21.9 Acute Toxicity of PCBs[a]

Organism	PCB Grade	Conditions	Test	Toxicity
Mammals				
Rats	1254	NA	LD_{50}	1.3–2.5 g/kg
Rats, mice, rabbits	1242	NA	LD_{50}	4.2–29 g/kg
Rats	1254	100% m. in 53 days	PCB in food	1000 ppm
Rats	1254	50% m., 8 months	PCB in food	1000 ppm
Mink	1254	100% m., 105 days	PCB in food	3.6 ppm
Birds				
Mallard, pheasant, bobwhite, quail	Six Aroclors (1232–1264)	50% in 5 days of toxic diet	LC_{50}	0.745–5.0 g/kg
Fish				
Gammarus oceanicus	1254	PCB colloidal and solubilized in emulsion	LC_{50}	0.001–0.1 mg/L
Goldfish	Clophen A50	5, 21 days	LC_{50}	4, 0.5 mg/L
Pinfish, spotfish	1254	12–18 days	LC_{50}	0.005 mg/L
Cutthrout trout	1221–1260	96 h	LC_{50}	1.2–61.0 mg/L
Fathead minnows	1242–1254	Newly hatched, 96 h	LC_{50}	0.008, 0.015 mg/L
Crustaceans				
Pink shrimp	PCBs	48 h	LC_{50}	0.1 mg/L

[a]NA, not available; m., mortality.
Source: Connell and Miller (1984).

chlorination. PCBs cause liver tumors in rats but are not mutagenic or teratogenic in mammals. Birds, on the other hand, have exhibited teratogenesis from PCBs as well as eggshell thinning and embryotoxicity. The carcinogenic potency of PCBs also varies with number of chlorines. Molecules with five chlorines have the highest potency; those with six are about half as potent; those with four are one-tenth as carcinogenic; molecules with fewer than four or more than six are less than one-tenth as potent.

A number of reproductive effects have been noted in mammals. A single injection of 20 mg PCB per kilogram of body weight lengthens the estrus cycle of mice; 100 mg/kg in the diet of Aroclor 1242 or 1254 decreased survival of rat offspring, and 20 mg/kg of Aroclor 1254 reduced the number of pups per litter. However, 100 mg of Aroclor 1260 per kilogram showed no effect in rats. Monkeys showed lower fertility and reduced birthweights at a dosage of 2.5 ppm of Aroclor 1248 in the diet.

PCBs are of concern in ecosystems because of their persistence and tendency to bioconcentrate and biomagnify. In addition, they show chronic toxicity at exceedingly low concentrations to certain species, which can cause significant population effects. As low as 4.4 ppb decreases hatching, survival, and growth of coho salmon; 19 ppb produces a 50% reduction in cell population growth of *Daphnia magna*. In addition, in an effect that strikes at the heart of a natural aquatic ecosystem's productivity, Aroclor 1242 inhibited net phytoplankton production by 50% at exposures of 10 to 25 ppb. Changes also occur in phytoplankton species distribution, with smaller species favored. This causes changes in zooplankton species distribution, which may in turn affect community structure at higher trophic levels.

Dioxins refer to chlorinated dibenzo-*p*-dioxins. They are produced as a by-product of various manufacturing processes, such as the production of the herbicide 2,4,5-T or the wood preservative pentachlorophenol, and are impurities in those products. They are also produced in all combustion processes in which the feed material contains chlorine in any form, such as chlorides. Thus, they may be found in fumes from incineration of refuse, and even of plant material. However, they are not produced significantly in exhausts from fossil-fuel combustion. The different congeners have different toxic and physicochemical properties. Of greatest concern is 2,3,7,8-tetrachlorodibenzo-*p*-dioxin (2,3,7,8-TCDD). This substance has the reputation of being the most toxic human-made substance known.

dioxin   *2,3,7,8-tetrachlorodibenzodioxin (TCDD)*

Like PCBs, chlorinated dioxins tend to be very stable thermally and environmentally. They photodegrade in the presence of organic material. However, they adsorb strongly to particulate matter such as soil and sediment, protecting them from exposure to the necessary ultraviolet radiation. They have very low solubility in both polar and nonpolar solvents. The solubility in *o*-dichlorobenzene is 1.4 g/L, in *n*-octanol it is 0.048 g/L, and in water it is 2×10^{-7} g/L. Thus, the K_{OW} is 2.4×10^5.

Table 21.10 shows the ranges of TCDD bioconcentration factors that have been observed in model ecosystems. Data from the accident in Seveso, Italy (see below) and elsewhere have shown that terrestrial organisms can bioconcentrate dioxin from the soil.

TABLE 21.10   TCDD Bioconcentration Factors

Organism	BCF
Plants	
Algae	6–2083
Duck weed	4000
Invertebrates	
Daphnia	2198–48,000
Snails	735–24,000
Fish	
Silverside	54
Mosquito fish	676–24,000
Catfish	2000

Source: Connell and Miller (1984).

TCDD is absorbed in the liver by first-order kinetics and stored. It is slowly converted to polar metabolites and excreted. Elimination in rats is also first order and can be described by a one-compartment model. Half-lives of 17 and 31 days have been measured. However, biomagnification has not been proven for dioxin.

The toxic effects of chlorinated dioxins are similar to PCBs. They are strong inducers of microsomal enzymes. They are toxic to the liver, immune system, and to fetuses. Teratogenic effects have been observed at dosages as low as 1/400 of the LD_{50} for the maternal rat. Dioxins cause cancer in laboratory animals, including rats at 2 ppb. Tumors appear in the liver, respiratory tract, and oral cavity, among other places. In humans, chlorinated dioxins are known to produce chloracne, but no evidence of carcinogenicity has been found. The extrapolation to humans is complicated by the extreme variability in toxicity between different animals. The oral LD_{50} for guinea pigs is 0.6 to 2 μg per kilogram of body weight in a single oral dose, for mouse it is 114 to 284 μg/kg, and in hamsters it is 5 mg/kg.

Dioxin has been responsible for a number of notorious pollution episodes. It is an ingredient of Agent Orange, the military defoliant used in Vietnam to deprive the enemy of cover. In that application it was blamed for numerous toxic effects to exposed American soldiers. Catfish and carp taken from the Dong Nai River in South Vietnam were found to have from 0.320 to 1.02 μg/kg body burdens. Agent Orange manufacturing plants in the United States, including in Cleveland, Ohio and Newark, New Jersey, were found to be heavily contaminated as well.

In 1971, a company that manufactured 2,4,5-trichlorophenol hired a used oil dealer to dispose of manufacturing wastes. The oil dealer mixed the material with waste oil and was subsequently hired to spray oil for dust control at several riding stables in Times Beach, Missouri. Shortly after the spraying, it was noticed that birds and insects nearby had begun to die. The horses also became ill, and many died, including 65 of 125 at one stable. The incident came to the attention of authorities when a 6-year-old girl became ill with hemorrhagic cystitis. Others affected included a 10-year-old girl with nosebleeds and headaches and two 3-year-old boys with chloracne. It took three years before the problem was traced to dioxin and the oil dealer. Up to 37 sites had been sprayed. Many had soil contamination levels higher than 100 μg/kg, and some more than 1.0 mg/kg. A flood subsequently spread the contamination over a large area, which was evacuated by the U.S. EPA and subsequently remediated.

Near Seveso, Italy (near Milan), in 1976, a trichlorophenol reaction vessel at a 2,4,5-T manufacturing plant released an unknown amount of dioxin over an area of 2000 ha (about 5000 acres). Soil TCDD dosages ranged from <0.75 to 5000 $\mu g/m^2$, resulting in concentrations averaging 3.5 $\mu g/kg$ in the top 7 cm of the soil. Extensive plant and animal mortality occurred, including hundreds of small pets. Body burden measured in small wild animals include 4.5 $\mu g/kg$ in field mice and 12.0 $\mu g/kg$ in earthworms. In humans 135 cases of chloracne were confirmed, mostly in children, but no deaths; 14 required hospitalization within 8 days and some required treatment for some years. An area of 108 ha (267 acres) was evacuated of all residents and remains so today.

21.5   AIR POLLUTANTS

Air pollutants include species discharged directly to the atmosphere (such as carbon monoxide, sulfur dioxide, and nitrogen oxide) and those formed from other pollutants by reactions in the atmosphere (ozone and nitrogen dioxide). Some of the sulfur and nitrogen species contribute to acid deposition (acid rain and dry acid deposition).

Nitrogen oxides are formed whenever oxygen and nitrogen are heated together to flame temperatures. Thus, all combustion processes using air produce nitric oxide (NO). The product formed remains when the mixture is cooled. The NO then reacts with O_2 in the atmosphere to form nitrogen dioxide (NO_2). Other species include nitrous oxide (N_2O), nitrogen trioxide (N_2O_3), nitrogen tetroxide (N_2O_4), and nitrogen pentoxide (N_2O_5). These are often collectively called NO_x. Only NO and NO_2 are environmentally significant. NO_2 is the most abundant and most toxic form. It is also involved in reactions that generate ozone and photochemical smog (see below). NO_2 is absorbed in water, but NO is very insoluble. A seventh species, nitric acid (HNO_3), is significant both toxicologically and as a component of acid deposition. Nitric acid is formed primarily by reaction of NO_2 and hydroxide radicals, but also by a series of reactions involving NO_2, ozone, and water.

Plants take up and respond to NO_2 similarly to SO_2. Plants can use atmospheric NO_2 nutritionally, assimilating them through the pathway

$$NO_x \rightarrow NO_3^- \rightarrow NO_2^- \rightarrow NH_3 \rightarrow aminoacids \rightarrow proteins$$

This pathway can compete with carbon assimilation for $NADPH_2$. Toxic effects can be caused by accumulation of ammonia as well as other mechanisms, such as oxidation of unsaturated fatty acids. However, it is believed that none of these effects are significant at ambient NO_x concentrations.

In humans and animals, most studies have been done at concentrations much higher than ambient. Various cellular and structural changes have been observed in lungs exposed to 10 to 25 ppmv of NO_2 for 24 hours. Brief inhalation of 200 to 700 ppmv can be fatal. Systemically, NO_2 is converted to nitrite and nitrate in the lungs and can be detected in the blood and urine. Some blood changes have been observed, although methemoglobin is only formed at high concentrations. Lipids from the lungs of rats exposed to NO_2 are oxidized, and more so if the rats are deficient in vitamin E. Asthmatics have been shown to experience significant increase in hyperreactivity of the airway at 0.1 ppmv of NO_2. NO_x exposure from ambient air is less likely to be a problem than the indoor air pollution produced by unvented use of combustion for cooking or heating in homes.

Ozone (O_3) is formed in various ways, including direct photolysis of oxygen (O_2) by ultraviolet light from sunlight or from equipment, or by spark discharges from motors, static discharges, electrostatic air purifiers, and by lightning. UV from sunlight produces the stratospheric *ozone layer* that protects the surface from the same ultraviolet radiation. However, in the troposphere (the lower atmosphere) an environmentally more significant mechanism is the complex sequence of reactions that form *photochemical smog*. In these reactions, NO_2 and oxygen react to form ozone and NO. Normally, the equilibrium would favor a low ozone concentration. However, atmospheric hydrocarbons, such as from incomplete products of combustion, react to form free radicals. The organic radicals scavenge the NO, shifting the equilibrium to the right. The entire sequence is a chain reaction regenerating the NO_2, resulting in many ozone molecules produced for each NO_2 initially available. Another important product of photochemical smog reactions is *peroxyacyl nitrate* (PAN), formed by the reaction of the free radical ROO•, NO_2, and oxygen:

$$ROO^{\cdot} + NO_2 + O_2 \implies R-\overset{\displaystyle O}{\overset{\displaystyle \|}{C}}-O-O-NO_2$$

peroxyacyl nitrate

In plants, ozone produces numerous growth and metabolic changes. Both ozone and PAN can oxidize protein sulfhydryl groups, affecting enzyme function. For example, exposure to 0.05 ppmv ozone inhibited the hydrolysis of starch for 2 to 6 hours in cucumber, bean, and monkey flower.

Ozone and other oxidants, such as chlorine gas, can produce eye and respiratory tract irritation. The effect has been likened to a "sunburn inside the lungs." The ambient air quality standard for ozone is 0.1 ppmv. Levels of 0.6 to 0.8 ppmv O_3 for 1 hour caused headaches, nausea, anorexia, and increased airway resistance. Exposure has been shown to produce hyperreactivity (asthma) in dogs and humans. Pulmonary edema has been observed at concentrations only slightly higher than levels that occur ambiently in Los Angeles, California. Humans and other animals are known to develop tolerance to ozone exposure. Exposing rodents to 0.3 ppmv ozone makes them tolerant of later exposures of several ppmv, which if applied to the same animals at the outset would cause massive pulmonary edema. Inhalation of 10 to 20 ppmv of Cl_2 causes immediate irritation, and 1000 ppmv can be fatal after only brief exposure.

One incident of the release of an extremely hazardous air pollutant has had great influence on the regulation of these materials. On December 2, 1984, at an Indian subsidiary of the Union Carbide Corporation in Bhopal, India, water was added to a storage tank containing the pesticide precursor *methyl isocyanate* (MIC). The addition may have been deliberate sabotage. In any case, the water reacted with the MIC to create a large amount of heat, vaporizing the MIC. Several tons of vapors escaped into a residential area surrounding the plant, waking the residents to choking, blinding fumes. Some 3800 people died as an immediate result, and 11,000 were left permanently disabled. (Unofficial estimates put the total number of deaths as high as 16,000.) Chronic effects may be affecting tens or hundreds of thousands more. Regardless of the exact number, there is little doubt that this was the greatest single industrial disaster in history. The incident has led to strict control over the storage of extremely hazardous air pollutants, including the chlorine gas used at many water and wastewater treatment plants for disinfection.

$$H_3C-N=C=O$$

methyl isocyanate

TABLE 21.11   Relation Between Percent Carboxyhemoglobin and Equilibrium Exposure Concentration with Health Effects in Humans

Health Effect	Percent Carboxyhemoglobin	Exposure Concentration (ppmv)
No symptom to reduced acuity	0–10	0–200
Headache and reduced acuity	10–20	200–400
Throbbing headache	20–30	400–600
Vomit and collapse	30–40	600–800
Coma	40–60	800–1200
Death	Over 60	Over 1200

Source: Based on Seinfeld (1975).

Carbon monoxide (CO) is produced by automobile engines and other combustion processes. It accumulates to toxicologically significant levels in high-traffic areas and areas where vehicles must accelerate, such as at intersections and toll booths. Typical levels in air range from 5 to 100 ppmv. The toxic effect of CO is related to the percent carboxyhemoglobin in the blood. Equation (17.2) and Table 21.11 show the equilibrium relation between this and the CO concentration in the air. It takes 8 to 10 hours for the blood to reach equilibrium. Thus an exposure to 50 ppmv for 90 minutes has been found to be equivalent to 10 to 15 ppmv for 8 hours or more.

Sulfur dioxide in the atmosphere originates primarily from combustion of fossil fuels, especially coal and some petroleum oils, and from metals smelting. Emissions are in the form of SO_2. The solubility of sulfur dioxide is 11.3 g per 100 mL, facilitating its absorption to aqueous aerosols. In solution it forms sulfite (SO_3^{2-}) and bisulfite (HSO_3^-). It reacts both in the gas phase and in the aerosols to form H_2SO_4. The sulfuric acid form is removed from the atmosphere by dry and wet deposition.

Sulfur dioxide enters plant leaves by molecular diffusion through the stomata. Photosynthesis and transpiration are increased by short-term low-concentration exposures, but decreased by longer-term or higher-concentration levels. The resulting sulfite and bisulfite are toxic to the plant. They can be oxidized with or without enzymes to the less toxic sulfate, but damaging free radicals can be formed. Effects include **chlorosis** (loss of chlorophyll) and necrosis as well as changes in phosphorus metabolism and photosynthesis. The effects are highly species dependent. Atmospheric sulfate is also a contributor to acid deposition.

In animals, sulfur dioxide is irritating to the eyes and upper respiratory tract. However, these effects appear only at levels 50 or 100 times ambient levels. Necrosis of nasal epithelium occurred in mice exposed to 10 ppmv for 72 hours. However, at high humidity or atmospheric particulate levels, increased irritation occurs. It is likely that it is the conversion to sulfuric acid that is responsible. At 5 ppmv, humans exhibit increased respiratory frequency and decreased tidal volume (faster and shallower breathing). Asthmatics are more sensitive than the rest of the population, but less so when breathing through the nose than through the mouth. SO_2 interacts positively with ozone and smoke particles. Concentrations of SO_2 and smoke both exceeding 500 $\mu g/m^3$ are thought to produce excess mortality among the elderly and the chronically ill; 130 $\mu g/m^3$ of SO_2 (0.046 ppmv) together with 130 $\mu g/m^3$ of particulates resulted in increased number and severity of respiratory disease in schoolchildren.

Particulates include dust and smoke. Smoke is made up of carbon and products of incomplete combustion, including polynuclear aromatic compounds. The toxic effects of smoke are thought to be primarily due to compounds that are adsorbed onto its surface. The solids serve to carry the compounds into the lungs.

Ambient particulate concentrations as low as 80 to 100 $\mu g/m^3$ (annual geometric mean) have been reported to result in increased death rates for people over 50 years old. A 24-hour level of 300 $\mu g/m^3$ together with 630 $\mu g/m^3$ of SO_2 (0.22 ppmv) produced acute worsening of symptoms of patients suffering from chronic bronchitis.

Cigarette smoke contains more than 2000 identified compounds. It is a significant source of radionuclides and carbon monoxide. Local toxic effects include chronic bronchitis, emphysema, and cancer of the lung and other parts of the respiratory tract. Systemic effects include arteriosclerosis of coronary arteries and cancer of the bladder, pancreas, and other organs. Nicotine in tobacco smoke causes addiction. Smoking paralyzes the mucociliary escalator that clears particles from the respiratory tract. It also inhibits macrophages in the alveoli, making the lungs more susceptible to disease. Pregnant women who smoke have more miscarriages and their babies have lower birthweights than the infants of nonsmokers. Users of snuff or cigars who do not inhale smoke are susceptible to cancer of the lips and oral cavity. Nonsmokers subjected to "secondhand smoke" also are at increased risk. Children of smokers have more respiratory illnesses than do children of nonsmokers.

Epidemiological data have been collected on cigarette smoking and cancer only since the 1940s. As mentioned above, the relative risk between smokers and nonsmokers for all cancers is about 5.0, and for lung cancer is 10 to 15. The data on lung cancer can be summarized in more detail by the following models. The rate of death by lung cancer increases with the first to second power of the number of cigarettes smoked per day, and with the fourth to fifth power of the number of years the person smoked. This means that doubling the number of cigarettes smoked per day triples the risk of lung cancer, but doubling the number of years that one smokes increases the risk by a factor of 20. Another model that takes into account the smoker's age in years (a) and years of exposure (y) is

$$\text{probability of death by lung cancer per year} = 10^{-11}a^4 + 10^{-9}y^4 \qquad (21.2)$$

According to this model, the annual lung cancer death rate for 70-year-old nonsmokers is 24 per 100,000. For 70-year-olds that have been smoking since the age of 20, the rate is 649 per 100,000. This is a relative risk of 27.

Smoking is also known to increase the cancer rates of the oral cavity, the bladder, the kidney, the pancreas, and the esophagus, as well noncancer diseases such as cerebrovascular disease (e.g., strokes), coronary heart disease, and pulmonary disease (e.g., emphysema).

21.6 WATER POLLUTANTS

The nitrogen species of interest in water are *ammonia, nitrite,* and *nitrate.* The transformations and other relationships among these species are described in detail elsewhere. Here we focus only on their toxicological properties. Ammonia distributes between the nonionized form and the ammonium ion. The distribution, and thus the toxicity, depend on the pH (Section 3.4). The nonionized form seems to be much more toxic, possibly because of a greater ability to pass through membranes. Ammonia also increases in toxicity at lower dissolved oxygen levels.

At high concentrations, ammonia seems to affect the central nervous system of fish, resulting in hyperventilation, convulsions, and death. At chronic levels of exposure, the toxicity seems to involve histological and reproductive effects. The LC_{50} falls in the range 0.49 to 4.6 mg of nonionized ammonia per liter for a variety of aquatic animals. These include *Daphnia*, crayfish, and the fish bluegills, red shiners, channel catfish, and large-mouth bass. One group that seems more sensitive is the salmonids, which include pink salmon (0.08 to 0.1 mg/L) and rainbow trout (0.2 to 1.1 mg/L). Chronic effects include reduced uptake of food and reduced growth at levels as low as 0.002 to 0.15 mg/L NH_3.

Ammonia is oxidized by bacteria such as *Nitrosomonas* to nitrite (NO_2^-), which is then converted to nitrate (NO_3^-) by, among others, *Nitrobacter*. The first step is usually rate limiting. Thus, nitrite does not normally accumulate to appreciable levels in aquatic systems. Nonionized ammonia inhibits *Nitrobacter* at 0.1 to 1.0 mg NH_3/L and inhibits *Nitrosomonas* at 10 to 150 mg/L. Thus, in between these two levels, the second step is inhibited and not the first, and nitrite can accumulate. Thus, lowering the pH can cause nitrite accumulation by its effect on ammonia. Furthermore, nitrite is in equilibrium with nitrous acid (HNO_2), with a K_a of about 3.39 at 10°C. Nitrous acid inhibits *Nitrobacter* and *Nitrosomonas* at concentrations starting at about 0.22 to 2.8 mg/L. This inhibition also favors nitrite formation. Some organic chemicals have similar effects. Concentrations of nitrite up to 73 mg NO_2^--N /L have been found in lakes and streams.

Nitrite is toxic to fish and mammals by forming methemoglobin, limiting the oxygen-carrying capacity of the blood (see Section) LC_{50} for fish has been measured at 0.1 to 0.4 mg NO_2^--N/L for rainbow trout, 1.6 mg NO_2^--N/L for mosquitofish, 7.5 to 13 mg NO_2^--N /L for catfish, and greater than 67 mg NO_2^--N/L for the mottled sculpin. Toxicity to fish decreases above pH 6.4, apparently due to decreasing HNO_2 levels.

Nitrate toxicity is very low. One of the most sensitive aquatic species is the guppy (*Poecilia reticulata*), with a 96-hour LC_{50} of 180 to 200 mg NO_3^--N/L. Other species range from 420 to 2000 mg NO_3^--N/L. Some of this toxicity may be related to a salinity effect rather than a specific toxicity of nitrate.

In mammals, high concentrations of nitrites can react with some amines under the acid conditions of the stomach to form **N-nitroso compounds** (nitrosamines and nitrosamides). Many of these are highly carcinogenic. Nitrate is used to cure meats and to preserve other foods and is found in water contaminated by agricultural fertilizers or domestic wastewater. It is also found naturally in some vegetables such as turnip greens, beets, and celery. Nitrate in food or water can be converted to nitrite by microorganisms in the mouth and stomach. Nitrites and nitrates also cause ''aging'' of coronary arteries. N-nitroso compounds can also be formed by cooking or drying foods over open flames when the NO_x compounds formed in combustion react with amines in the food.

$$
\begin{array}{c}
N=O \\
| \\
R_1-N-R_2
\end{array}
$$

nitrosamine

Suspended solids may come from inert materials such as soil particles or from pollution sources such as sewage. In the case of sewage, other harmful effects, such as oxygen limitation, predominate over the physical effect of the solids. Forest streams typically have very low suspended solids: less than 20 to 50 mg/L. Some rivers, such as the Mississippi River in Louisiana, can reach almost 300 mg/L. The higher levels can be abrasive for aquatic plants, and can coat the bottom, reducing survival of benthic organisms and fish eggs.

Fish production is thought to be reduced at levels from 25 to 80 mg/L. From 80 to 400 mg/L, fish production is likely to be affected severely. Lethal effects in fish have been measured at suspended solids levels ranging from 40,000 to 160,000 mg/L. However, a single value cannot be determined because it will depend on shape and size distribution of the particles. Presumably the lethality is due to clogging or abrasion of the gills.

21.7   TOXICITY TO MICROBES

Bacteria tend to be less sensitive than eukaryotic organisms to toxic substances. This may be due to the relative simplicity of their metabolic processes and the lower degree of specialization. For example, organophosphate pesticides act on acetylcholinesterase in the nervous system, which has no counterpart among bacteria. Thus, bacteria are fairly insensitive to these substances. General metabolic inhibitors, however, can affect bacteria in a manner similar to that of higher organisms. Bacterial membranes, cell walls, and slime layers offer greater resistance to many toxins than animal structures such as gills, lungs, and intestinal surfaces. Furthermore, bacteria can become dormant when conditions are not favorable for growth or survival. Cyst formation can be another barrier to a toxicant entering the cell.

Besides reduced sensitivity, three factors relating to microbial ecology serve to mitigate the impact of toxins on microbial communities:

1. Bacteria *grow rapidly* and can quickly regain their numbers if only part of the population is eliminated.
2. Bacterial groups have *redundant capabilities*. That is, although bacteria may be metabolically specialized individually, many different types share similar specialization. Consequently, when one type is poisoned, there will often be others ready to take their place. There are some exceptions, such as the nitrifying bacteria. These grow slowly, and are replaced slowly, yet are important to the ecosystem. Thus, toxicity assessments should consider possible effects on this group.
3. Bacteria are *ubiquitous* in natural systems. Even in laboratory or industrial systems it is very difficult to sterilize an open system chemically and at the same time to isolate it from external resupply. In the natural environment, it is all but impossible.

Nevertheless, inhibition of bacteria has been measured by various means, including plate counts, biochemical measurements, respirometry, and biomass growth rates. Table 21.12

TABLE 21.12   Effect of Herbicides on Microbial Respiration

Herbicide	*Erwinia carotovora*		*Pseudomonas fluorescens*		*Bacillus sp.*	
	Conc. (µg/L)	% Inhibition	Conc. (µg/L)	% Inhibition	Conc. (µg/L)	% Inhibition
Diquat	25	40.8	50	0	25	100
Loxynil	25	71.0	50	0	10	29.0
Paraquat	50	25.5	25	43.7	5	100
PCP	10	29.6	25	65.8	5	100
Picloram	50	0	50	28.8	50	0
Nitralin	50	0	25	22.0	50	0

Source: Rand and Petrucelli (1985).

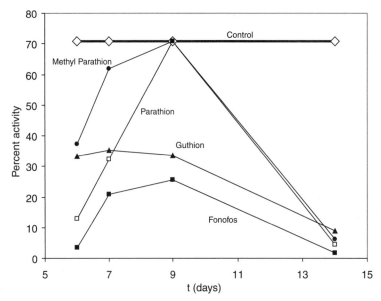

Figure 21.3   Effect of thiophosphorus pesticides on ammonia oxidation by estuarine nitrifiers. (From Rand and Petrucelli, 1985; original source: Jones and Hood, 1980, *Can. J. Microbiol.*, Vol. 26, pp. 1296–1299.)

shows the reduction in respiration measured with several herbicides. It is interesting that *Bacillus* spp. were completely inhibited by diquat at levels that had no effect on *P. fluorescens* respiration. Note that inhibition of respiration does not necessarily mean that the organisms were killed.

Toxic effects on bacteria can be studied by determination of effects on specific functions, such as saprophytic activity, methanogenesis, nitrification, or other biogeochemical cycle activities. Figure 21.3 shows an example of this. The amount of inhibition was observed to vary with incubation time. For example, methylparathion's inhibition decreases up to 9 days, then increases again. This may be due to the degradation of the pesticide, followed by an accumulation of an even more toxic by-product, aminophenol.

Nitrogen fixation can be studied by measuring the conversion of acetylene to ethylene in the absence of nitrogen. This is because the enzyme nitrogenase catalyzes both reactions. Stimulation is often found with pesticides, as well as inhibition. For example, malathion causes an initial period of inhibition, followed by more than double the activity after 8 days. Other combinations of pesticides and species produced up to five times the activity.

Soil fungi and actinomycetes were found to bioaccumulate organochlorine pesticides. Bioconcentration factors ranged from 10 for lindane to 59,000 for chlordane. Bacteria are also involved in biotransformation reactions that can enhance or reduce toxicity of substances to other organisms. One of the most important is the conversion of inorganic mercury to methylmercury. Other examples include the conversion of crude oil components to genotoxic agents, the conversion of P=S groups in organophosphate pesticides to the more active P=O groups, and the formation of nitrosamines from secondary amines and nitrites.

21.8 IONIZING RADIATION

An important physical cause of toxic effects is a variety of forms of radiation. **Ionizing radiation** possesses enough energy to strip an electron from an atom. This can result in the formation of damaging free radicals or directly damage bonds in biochemical substances. The most sensitive system in living things is the DNA, since damage to a single molecule can transform a cell to malignancy. It is not necessary for a radioactive emission to damage a DNA molecule directly. The most abundant molecule in living things is water. Water can form free radicals when irradiated, and these in turn can produce toxic effects, including genotoxicity.

Ionizing radiation may be electromagnetic (γ-rays, x-rays, ultraviolet rays) or particulate. The major particulate forms of radiation are α and β, but they may include neutrons, other subatomic particles, or larger particles; such as various atomic nuclei found in cosmic rays. Of primary interest here the α, β, and γ emissions from the decay of radioactive atoms, or **radionuclides**. Each particle of radiation possesses energy, commonly measured in millions of electron volts (MeV). A change to SI units is under way that will replace the MeV with the joule: $1 \text{MeV} = 1.6 \times 10^{-13} \text{J}$.

The energy of a particle relates to the amount of damage a particle can do to a tissue. As radiation passes through material such as tissue, it is absorbed, depositing its energy in the material. Much of this energy goes into stripping electrons from atoms. This requires an average of 33.85 eV per electron. This ability to eject electrons confers a capacity to disrupt chemical bonds with potentially damaging results.

Gamma (γ) radiation is electromagnetic radiation (photons) of very high energy, usually less than 1 MeV. They are produced by radioactive decay of unstable atomic nuclei. **X-rays** are somewhat less energetic, up to 0.25 MeV for radiation from medical x-ray machines. **Ultraviolet** (UV) are even less energetic, ranging from about 4.13 to 155 eV. X-ray and UV photons are produced by excitation of an atom's electrons. **Alpha** (α) particles consist of two protons and two neutrons (helium nuclei). They have energies that range from 3 to 9 MeV. **Beta** (β) particles are high-energy electrons, with energies ranging up to 3.5 MeV. Alpha, beta, and gamma particles are all emitted by spontaneous reactions in the nuclei of unstable (radioactive) isotopes.

Although alpha particles are high in energy, they have the least penetrating power, due to their mass and charge. They are strongly absorbed and deposit their energy rapidly, producing a large amount of damage. For example, alpha particles with 5 MeV penetrate about 40 μm of soft tissue or about 5 cm in air. Thus, they do not penetrate the skin. However, if they are inhaled or ingested, they do not need to penetrate very far to reach cell nuclei, potentially causing genetic damage. Beta particles are charged, but have little mass. They penetrate about 3 m in air, but they are stopped by thin sheets of many solids. Thus, alpha and beta radiation is a problem mostly if their emitters are ingested. Gamma rays are the most penetrating. Their range in air is about 4 m, and they will penetrate several centimeters of lead. Consequently, γ rays can cause their effects on living things either when emitters are ingested or by external radiation sources.

Many of the environmentally important radionuclides are products of the sequence of nuclear decay reactions that start with ^{238}U. Each beta emission increases the atomic number without significantly changing the atomic mass by converting a neutron to a proton. An alpha emission reduces the atomic number by two and the atomic mass by four. Figure 21.4 shows the members of the series, with their half-lives and the energy carried

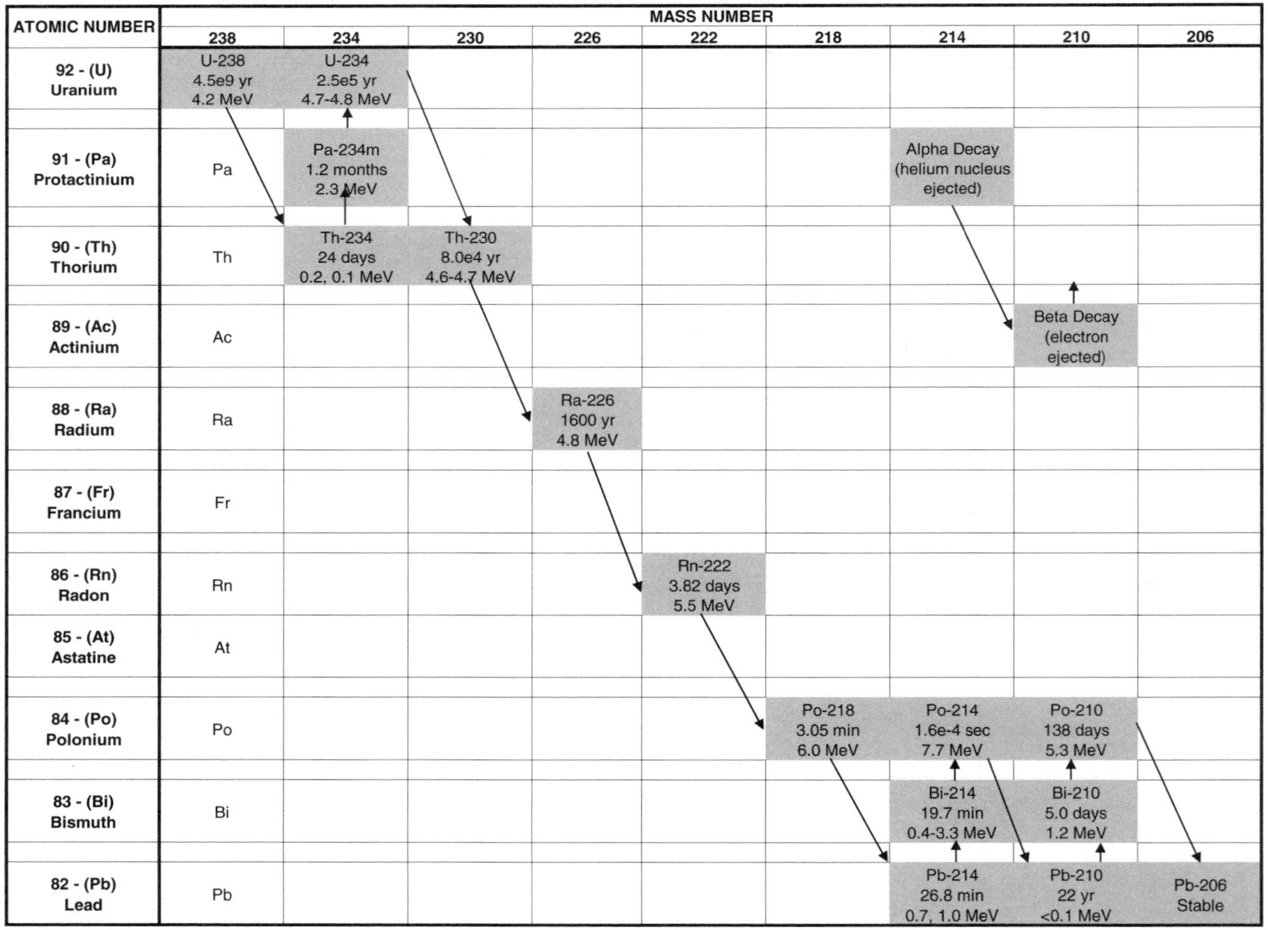

Figure 21.4 Uranium-238 decay series.

TABLE 21.13   Some Important Radionuclides and Energies of Dominant Radiation Particle[a]

Isotope	Half-Life	Dominant Particle Emission	Avg. Particle Energy (MeV) (% intensity)	Nutrient Analogs and Critical Tissue	Relative Food Chain Transport (BAF)	Source
^{3}H	12.26 yr	β	0.01861	H, whole body	High (~1)	Nuclear weapons, nuclear power
^{14}C	5730 yr	β	0.156	C	NA	Natural, nuclear power
^{40}K	1.28×10^9 yr	β	1.35 (1.35%)	K, whole body	High (~1)	Natural
^{32}P	14.3 days	β	1.71	P	NA	Natural
^{85}Kr	10.76 yr	β	0.67	NA	NA	Nuclear weapons, nuclear power
^{87}Rb	4.8×10^{11} yr	β	0.274	NA	NA	Natural
^{90}Sr	28.1 yr	β	0.546	Ca, bones	High (~1)	Nuclear weapons
^{106}Ru	367 days	β	0.0392	NA	NA	Nuclear weapons
^{129}I	1.6×10^7 yr	β	0.189	NA	NA	Nuclear power
^{131}I	8.07 days	β	0.582	I, thyroid	High (~10)	Nuclear weapons, power, medical
^{134}Cs	2.05 yr	β	0.499	NA	NA	Nuclear power
^{135}Cs	3×10^6 yr	β	0.210	NA	NA	Nuclear weapon tests
^{137}Cs	30.23 yr	β	0.551	K, whole body	High (~3.0)	Nuclear power
^{222}Rn	3.82 days	α	NA	None, lungs, S, Se?	Negligible	Natural
^{238}U	3.5×10^9 yr	α	4.2	GI tract, kidney, lung	Low–moderate (<1.0)	Natural, contaminant in depleted uranium projectiles
^{239}Pu	24,400 yr	α	5.149	None, bone, lung	Very low (<0.01)	Nuclear weapons

[a]Energy of dominant radiation particle given, or weighted average for type of particles. NA, not available.
Source: Based on Connell and Miller (1984); Weast (1979); Newman (1998).

by their alpha and beta particles. Some of these reactions may also produce γ radiation, usually at lower energies.

Table 21.13 summarizes the emissions from some other important radionuclides. Note that the smaller radionuclides tend not to emit alpha radiation.

Each type of decay is characterized by its half-life, the type of particle(s) emitted, and the energy carried by the particles. The half-life defines the rate of disintegrations. The half-life is related to the decay rate coefficient, k, by equation (18.25). The rate coefficient can then be used in equation (18.23) to compute the rate of decay.

Example 21.1   One gram of ^{226}Ra has a half-life of 1600 years. Thus, $k = 0.693/1600$ years $= 4.33 \times 10^{-4}$ yr^{-1} or 1.374×10^{-11} s^{-1}. The number of atoms in 1 g of radium is

$$\frac{1\,\text{g}}{226\,\text{g/mol}}(6.022 \times 10^{23}) = 2.665 \times 10^{21}\ \text{atoms}$$

Then the rate of decay can be computed as

$$r = kc = (1.374 \times 10^{-11}\,\text{s}^{-1})(2.665 \times 10^{21}\,\text{atoms}) = 3.66 \times 10^{10}\,\text{atoms/s}$$

That is, 1 g of radium disintegrates at the rate of 3.66×10^{10} atoms per second.

21.8.1 Dosimetry

In measuring radiation, it is necessary to discriminate among *emission, exposure,* and *dose.* **Emission** is the rate at which particles are produced or energy is released. One **disintegration per second** is termed a **becquerel** (Bq). The **curie** (Ci) is defined as $1\,\text{Ci} = 3.70 \times 10^{10}$ Bq. Note that this is approximately the number of becquerels emitted by 1 g of radium. Historically, the Curie was defined in terms of radium decay, although now it is given the fixed value. Also, note that 1 curie emitted by different radionuclides can have different amounts of energy. Thus, the number of curies by itself does not indicate the amount of potential harm that might be caused by a radionuclide. Environmental concentrations of radionuclides are often expressed in *picocuries*: $1\,\text{pCi} = 10^{-12}\,\text{Ci} = 0.037$ Bq. For example, radon in drinking water is regulated in terms of pCi/L. The U.S. EPA has proposed a maximum contaminant level for radon in drinking water of 300 pCi/L and has an action level in ambient air of 4 pCi/L.

Exposure is defined only for electromagnetic radiation such as γ-rays. The SI unit is based on the number of charges of one sign produced by complete absorption in air. The unit is **coulombs per kilogram** of air. The older unit of exposure is the **roentgen** (R):

$$1\,\text{roentgen} = 2.58 \times 10^{-4}\,\text{C/kg}$$

Exposure can occur due to **external** sources, such as γ-, β-, or x-ray emissions from outside the body. Alternatively, they may be from **internal** sources, such as any radionuclide that is ingested or absorbed and decays within the body.

The **dose** refers to the amount of energy actually absorbed. The SI unit is the **gray** (Gy), which is 1 J/kg. The older unit is the **rad**. The conversion is $100\,\text{rad} = 1\,\text{Gy}$. Because of the previously mentioned average ionization energy, exposure can be converted into dosage in either system of units assuming 100% absorption of the energy:

$$1\,\text{C/kg} = 33.85\,\text{Gy}$$
$$1\,\text{roentgen} = 0.87\,\text{rad(in air)}$$

For tissues this conversion is approximate and may vary with type of tissue and γ energy. Exposure and dose have frequently been confused with each other, partly because in the units of roentgens and rads they are numerically similar. Exposure refers only to ionization that would occur in air due to electromagnetic radiation such as γ- or x-rays. It is the radiation that would be measured by an external dosimeter, such as a film badge clipped to the shirt of a worker. However, it only indirectly indicates how much radiation has been absorbed by the worker. Dose applies to energy absorbed by any material and due to any type of radiation.

The differing ability of various emissions to damage living tissue is accounted for by what is called the **relative biological effectiveness** (RBE) in biology. It is defined as the ratio of the toxicological effect of the radiation in question to the effect produced by γ

radiation of the same dose. The value depends not only on the radiation but on the type of endpoint (e.g., cancer, cell necrosis). Because the RBE is difficult to determine for human health effects, the relative rate at which energy is deposited in water by the emission, compared to γ radiation, is used in place of RBE. This ratio is called the **radiation weighting factor** (W_r; formerly the **quality factor**). Thus; γ radiation has a W_r value of 1. The W_r of α radiation is about 20. The W_r for β is also about 1. The product of dose and W_r gives the **dose equivalent** (H), which is measured in sieverts (Sv); $Gy \times W_r = Sv$. (In the older measurement units, $rad \times W_r = \textbf{rem}$. Then $100\,rem = 1\,Sv$.) An interpretation of dose equivalent is that it is the dose of γ radiation that would produce the same amount of tissue damage as the radiation under consideration.

Dose equivalent, however, still does not provide enough information to estimate the risk of cancer. The reason is that all parts of the body rarely receive the same exposure, and in the case of ingested radionuclides, they often concentrate in various organs or tissues with varying sensitivities. For example, dental x-rays are concentrated on the head and neck area, and ingested radioactive iodine tends to concentrate in the thyroid gland. To account for factors such as these, dose equivalent is multiplied by a **tissue weighting factor**, w_t, which is the ratio of the cancer risk to the organ per sievert divided by the cancer risk per sievert to the whole body. The product of dose equivalent and w_t is called the **effective dose equivalent** (H_E). The w_t for the gonads is 0.20; for bone marrow, colon, lung, and stomach it is 0.12; for the bladder, breast, liver, esophagus, and thyroid it is 0.05; for the skin and bone surface it is 0.01; for the rest of the body it is 0.05. The effective dose for parts of the body are computed separately because the dose received by the different parts are rarely the same. The sum of the effective dose equivalents for all the parts of the body is used to compute overall effective dose equivalent. This value is then used to compute a person's risk of cancer. Table 21.14 summarizes the units used in radiation dosimetry.

The dosage values described so far are in units of energy deposited per unit mass of tissue. However, the dosage may decrease with time because of the elimination of the radionuclide by radioactive decay, by biological elimination, or by both. The decrease

TABLE 21.14 Units and Conversion Factors Used in Radiation Health Measurements[a]

Fundamental Units	Traditional Units	SI Units
Disintegration Rate		
1 disintegration/s	1 becquerel (Bq)	27.0 picocuries (pCi)
	3.70×10^{10} Bq	1 curie (Ci)
Exposure (γ- or X-Rays): Number of Charges of One Sign Produced by Complete Absorption in Air		
2.58×10^{-4} C/kg	1 roentgen (R)	2.58×10^{-4} C/kg
	3876 R	1 C/kg
Dose: Amount of Energy Actually Absorbed		
1 J/kg	100 rad	1 gray (Gy)
Dose Equivalent: Dose $\times$ RBE or Dose $\times$ W_r: Capability to Damage Living Tissue		
	100 rem	1 sievert (Sv)
Effective Dose Equivalent: Dose Equivalent $\times$ Tissue Weighting Factor		
	100 rem	1 sievert (Sv)

[a]Units on the same row are equal to each other.

is a first-order process, with an effective rate constant, k_E, equal to the sum of the rate constants for radioactive decay, k_R, and for biological elimination, k_B. The initial rate at which energy is being deposited in a tissue is D_0. Then the amount that will be deposited in the first t time units will be

$$D = \frac{D_0}{k_E}[1 - \exp(-k_E t)] \tag{21.3}$$

The dose that is deposited over a lifetime (assumed to be 50 years for adults and 70 years for exposure to children) is called the **committed effective dose** (CED). For an infinite time period, or for practical purposes when the lifetime is four or more times the effective half-life $(1/k_E)$, the committed effective dose is equal to the **total dose**:

$$D = \frac{D_0}{k_E} \tag{21.4}$$

Internal nuclides with long effective half-lives do not change appreciably in concentration in the body over a person's lifetime. In such cases it is appropriate to expressed the dosage as a rate of energy delivered per mass of tissue per unit time, D_0.

Example 21.2 ^{131}I is a β emitter with a half-life of 8 days and a 180-day biological half-life in the thyroid. Thus, $k_R = 0.693/8.07\,\text{days} = 0.0859\text{day}^{-1}$ and $k_B = 0.693/180\,\text{days} = 0.00385\,\text{day}^{-1}$. Then $k_E = 0.0897$ and the effective half-life is 7.72 days. About 60% of the iodine in the body concentrates in the thyroid gland. If a person receives 3 mCi of ^{131}I, what will be the effective dose rate to the thyroid? What will be the effective dose delivered over the first 15 days, and what will be the total dose?

Answer   From Table 21.13, ^{131}I produces 0.582 MeV per disintegration; 60% of the total dose results in $(0.6)(3)\,\text{mCi} = 1.8\,\text{mCi}$ dose to the thyroid. Thus, the dose rate will be

$$(1.8\,\text{mCi})\left(\frac{3.7 \times 10^4\,\text{Bq}}{\text{mCi}}\right)\left(\frac{0.582\,\text{MeV}}{\text{Bq}\cdot\text{s}}\right)\left(\frac{1.6 \times 10^{-13}\,\text{J}}{\text{MeV}}\right)\left(\frac{86,400\,\text{s}}{\text{day}}\right) = 5.36 \times 10^{-4}\,\text{J/day}$$

This is also the dose equivalent, H, since the radiation weighting factor can be taken as 1.0 for β emissions. Also, since we are dealing with a known dose to an organ rather than a whole-body dose, a tissue weighting factor is not needed and the dosage we are computing is the effective dose equivalent, H_E. Assume that the mass of the thyroid is 25 g. Since the penetration depth of β emissions is much smaller than the thyroid, one can assume that all the energy is deposited in that organ. Thus, the initial dose rate, D_0, will be

$$D_0 = \frac{5.36 \times 10^{-4}\,\text{J/day}}{0.025\,\text{kg}} = 0.0214\,\text{Gy/day} = 0.0214\,\text{Sv/day}$$

The dose delivered over the first 15 days will be

$$D = \frac{0.0214\,\text{Sv/day}}{0.0897\,\text{day}^{-1}}[1 - \exp(-0.0897\,\text{day}^{-1})(15\,\text{day})] = 0.177\,\text{Sv}$$

Over the long term, the total dose will be

$$D = \frac{0.0214 \, \text{Sv/day}}{0.0897 \, \text{day}^{-1}} = 0.239 \, \text{Sv}$$

Dosimetry calculations can easily become complicated by additional factors not considered here. The distribution of radionuclides can be described by pharmacokinetic models to more accurately describe their spatial and temporal distributions in the body. Conversion of exposure to absorbed dose can be complicated by external and internal shielding, which can be difficult to quantify. Often, it is necessary to take into account not only a particular radionuclide, but its daughters as well. Radon is one example where this is critical (see below). In such cases the daughters can have greatly different emission characteristics and pharmacokinetics. Once the radionuclide loading to a particular tissue is determined, it may be necessary to consider secondary forms of radiation as the energy interacts with the matter (e.g., neutrons or positrons). It may be necessary to examine radiation to tissues adjacent to the organ containing the radionuclide. An example of the latter is that some of the radioactive isotopes of strontium may deposit at the inner surface of hard bone tissue, where their emissions can irradiate the sensitive bone marrow tissue. Many radioisotopes produce more than one type of emission. For example, some 8% of the emission from ^{131}I is γ radiation. In the case of γ-ray dosimetry from internal emitters, the spatial distribution of the nuclides affects the absorbed dose. This can be taken into account either by detailed integration or by use of predetermined *geometry factors*.

21.8.2   Radiation Exposure and Risks

The major sources of radionuclides in the environment are natural sources, fallout from nuclear detonations in the atmosphere, and sources associated with nuclear power. Natural sources include a number of minerals, such as ^{238}U and its derivatives (Figure 21.4), ^{40}K and the gas tritium (^{3}H), and ^{14}C, which is continually produced by cosmic ray bombardment of the atmosphere. One of the more important by-products of ^{238}U is radon, ^{222}Rn. The natural background radiation to which the public is exposed is about 3 mSv per year, as shown in Table 21.15.

Inhalation of radon is the largest source of natural exposure. Outdoor radiation in the United States has been measured at an average of 7 Bq/m^3 (0.19 pCi/L); indoor levels tend

TABLE 21.15   Estimated Total Effective Dose from Several Forms of Natural Radiation for the U.S. and Canadian Populations

Source	Dose Rate (mSv/yr)
Cosmic	0.27
Cosmogenic	0.01
Terrestrial	0.28
Inhaled	2.0
In body	0.40
Total	3.0

Source: Harley (1996).

to be much higher, averaging 40 to 80 Bq/m^3 (1 to 2 pCi/L). This is because the decay of each gram of radium naturally present in the ground produces about 0.10 mL of radon gas per day. As an ideal gas, radon moves readily through the soil, entering buildings through the foundation. About 1.18% of potassium consists of the radioactive isotope ^{40}K. The ^{40}K in our bodies contributes about 0.17 mSv/yr. Cosmic rays produce an exposure of 35 mR/yr at sea level, and 100 mR/yr at an altitude of 10,000 ft. Exposure to granitic bedrock produces about 90 mR/yr from natural isotopes, and sedimentary rock produces 23 mR/yr.

Nuclear power results in emissions at a number of points in the fuel cycle, such as mining, fuel processing, and power generation. A wide variety of radionuclides are produced in the power reactor. Besides the direct reaction products, neutrons striking other materials, such as air and impurities in cooling water, produce new radionuclides by **neutron activation**. This creates radioactive forms of carbon, nitrogen, argon, oxygen, copper, sodium, and many others. Some of these products are released to the environment. Atmospheric nuclear weapons testing up to 1973 released far larger quantities of radionuclides to the environment than nuclear power has.

The primary risk of concern with radiation exposure is cancer and hereditable genetic disorders. The International Commission on Radiation Protection (ICRP, 1990) estimates the probability of fatal cancer of all types for the general population to be 5% per sievert dosage and the probability of severe hereditary disorders to be 1% per sievert. Table 21.16 details these risks for individual tissues. Also given is the aggregated detriment, which

TABLE 21.16   Estimated Radiation Risk Probabilities for Specific Tissues and Organs

Tissue or Organ	Probability of Fatal Cancer (% per Sv)		Aggregated Detriment (% per Sv)	
	Whole Population	Workers	Whole Population	Workers
Bladder	0.30	0.24	0.29	0.24
Bone marrow	0.50	0.40	1.04	0.83
Bone surface	0.05	0.04	0.07	0.06
Breast	0.20	0.16	0.36	0.29
Colon	0.85	0.68	1.03	0.82
Liver	0.15	0.12	0.16	0.13
Lung	0.85	0.68	0.80	0.64
Esophagus	0.30	0.24	0.24	0.19
Ovary	0.10	0.08	0.15	0.12
Skin	0.02	0.02	0.04	0.03
Stomach	1.10	0.88	1.00	0.80
Thyroid	0.08	0.06	0.15	0.12
Remainder	0.50	0.40	0.59	0.47
Total	5.00	4.00	5.92	4.74
Probability of Severe Hereditary Disorders				
Gonads	1.00	0.60	1.33	0.80
Grand Total			7.3	5.6

Source: Harley (1996).

takes into account fatal and nonfatal cancers, severe hereditary effects, and loss of relative length of life.

The ICRP occupational guideline for radiation protection (1990) limit the dosage received to an average of 20 mSv (2 rem) per year over five years, with a maximum of 100 mSv in any one year. A dosage of 20 mSv with a fatal cancer occupational risk factor of 4% per sievert corresponds to an 8×10^{-4} risk. The NCRP (1993) has defined a **negligible individual dose** (NID) below which efforts to reduce radiation exposure further are unwarranted. The NID was selected to be 0.01 mSv (1 mrem) per year. This corresponds to 5×10^{-7} added risk of fatal cancer per year.

Acute effects in humans are found at higher dosages. The 30-day LD_{50} for whole-body radiation depends on individual factors and the level of medical support provided, but is considered to be about 4 to 7 Gy (400 to 700 rad). At such levels the first symptoms include nausea and vomiting within about a week. After two weeks sore throat and loss of appetite occurs, followed by diarrhea, emaciation, and death. Suppression of the immune system can result in infections that contribute to death rates. Exposure can cause skin lesions, damage to leukocytes and lymphatic tissue, and ultimately the complete elimination of antibody production. Bone marrow depression is observed at about 2 Gy, and changes in blood counts result at exposures from 0.14 to 0.50 Gy.

In organisms other than humans, the genotoxicity has been shown to be strongly related to the amount of haploid DNA in the organism (Figure 21.5). This relationship seems to be reflected in the lethal toxicity of radiation to various groups shown in Table 21.17.

The fate and transport of radionuclides is as varied as their chemical composition. Radon, an ideal gas, is fairly soluble in water. Some of its daughters, however, are solids that attach to particles in the air that in turn may be deposited in the lungs (see Figure 21.4). Many of the metal radionuclides released by nuclear power plants have been found to be strongly adsorbed by sediments. The concentration of plutonium in aquatic organisms has been found not to exceed the concentration found in associated sediments. It has also been found to decrease with trophic level.

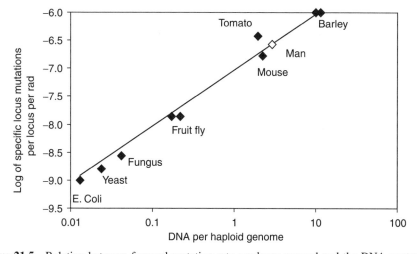

Figure 21.5 Relation between forward mutation rate per locus per rad and the DNA content per haploid genome. The line is a regression line. The point for humans is estimated from the DNA content.

**TABLE 21.17   Lethal Dosage of Radiation for Adult
Organisms of Several Groups**

Group	Lethal Dose (krad)	Toxic Measurement[a]
Bacteria	4.5–735	LD_{90}
Cyanobacter	<400–>1200	LD_{90}
Other algae	3–120	LD_{50}
Protozoans	?–600	LD_{50}
Mollusks	20–109	$LD_{50/30}$
Crustaceans	1.5–56.6	$LD_{50/30}$
Fish	1.1–5.6	$LD_{50/30}$

[a]$LD_{50/30}$ is the 30-day LD_{50}.
Source: Connell and Miller (1984); original source I.L. Ophel, 1976, Tech.
Rep. Ser. 172, International Atomic Energy Agency, Vienna.

PROBLEMS

Write a research paper on one of the following topics:

21.1. Update Rachel Carson's *Silent Spring*: Which of her fears were proved right; which wrong? How did the pesticide industry respond to her book? What is the stance of the pesticide industry today?

21.2. Look into the medical literature to find and summarize case histories of toxic effects on humans of a chosen environmental toxin or class of toxins, such as a type of pesticide or solvent.

21.3. Select several chemicals in each of two or three different categories, such as metals, pesticides, and solvents, for which you can find information on the slope of the toxicity curve. See if you can generalize about what you find. Do particular groups have similar slopes? Does a particular chemical tend to have similar slopes in different organisms? What does the slope indicate about the toxic mechanisms involved?

21.4. Find journal articles discussing the extrapolation of animal toxicity test data to humans, and summarize the information given and arguments made.

21.5. Research the stimulation of chemical sensitivity by formaldehyde.

21.6. Suppose that a woman has absorbed 1 mCi of ^{90}Sr into her bones. Compute the dose rate, the initial dose rate, the dose delivered over 50 years, and the resulting added risk of cancer. Assume that the total mass of bones is 5 kg.

REFERENCES

Anon., 1998. Arsenic crisis in Bangladesh, *Chemical and Engineering News*, November 16, pp. 27–29.

Brown, A. W. A., 1978. *Ecology of Pesticides*, Wiley, New York.

Connell, D. W., and G. J. Miller, 1984. *Chemistry and Ecotoxicology of Pollution*, Wiley, New York.

Esposito, M. P., T. O. Tiernan, and F. E. Dryden, 1980. *Dioxin*, EPA/600/2/80/197, U.S. Environmental Protection Agency, Washington, DC.

Ferruzzi, G., and J. Gan, 2004. *Pesticide Selection to Reduce Impacts on Water Quality*, Publication 8119, University of California Division of Agriculture and Natural Resources, Berkeley, CA.

Gofman, J. W., 1981. *Radiation and Human Health*, Sierra Club Books, San Francisco, New York.

Grosch, D. S., and L. E. Hopwood, 1979. *Biological Effects of Radiations*, Academic Press, New York.

Harley, N., 1996. Toxic effects of radiation and radioactive materials, Chapter 25 in *Casarett and Doull's Toxicology*, 5th ed., C. D. Klaassen (Ed.), McGraw-Hill, New York.

ICRP, 1990. 1990 Recommendations of the International Commission on Radiological Protection. ICRP Publication 60. International Commission on Radiation Protection, Pergamon Press, Oxford, England.

IRIS database, other databases.

LaGrega, M. D., P. L. Buckingham, and J. C. Evans, 2001. *Hazardous Waste Management*, McGraw-Hill, New York.

Lu, F. C., 1991. *Basic Toxicology: Fundamentals, Target Organs, and Risk Assessment*, 2nd ed., Hemisphere Publishing, Bristol, PA.

Manahan, S. E., 1989. *Toxicological Chemistry*, Lewis Publishers, Boca Raton, FL.

Newman, M. C., 1998. *Fundamentals of Ecotoxicology*, Ann Arbor Press, Ann Arbor, MI.

Rand, G. M., and S. R. Petrocelli, 1985. *Fundamentals of Aquatic Toxicology: Methods and Applications*, Hemisphere Publishing, New York.

Rao, P. S. C., R. S. Mansell, L. B. Baldwin, and M. F. Laurent, 2004. *Pesticides and Their Behavior in Soil and Water*, University of Florida, http://pmep.cce.cornell.edu/facts-slides-self/facts/gen-pubre-soil-water.html.

Seinfeld, J., 1975. *Air Pollution: Physical and Chemical Fundamentals*, McGraw-Hill, New York.

U.S. EPA, 1985. *Draft Superfund Public Health Evaluation Manual,* U.S. Environmental Protection Agency, Washington, DC.

Weast, R. C. (Ed.) 1979. *CRC Handbook of Chemistry and Physics*, 60th ed., CRC Press, Boca Raton, FL.

Williams, P. L., and J. L. Burson (Eds.), 1985. *Industrial Toxicology: Safety and Health Applications in the Workplace*, Van Nostrand Reinhold, New York.

22

APPLICATIONS OF TOXICOLOGY

The results of toxicological studies are used by a number of government agencies to protect the health of individuals or of ecosystems. Risk assessment is used to provide a rational basis for regulatory decision making. It may even be used in cost–benefit analyses for regulations. In what is called a toxicity reduction evaluation, bioassays may be used to find and eliminate the source of toxicity in wastewaters.

22.1   RISK ASSESSMENT

Risk is defined as the probability and magnitude of an adverse effect. **Quantitative risk assessment** is the process of determining that probability and magnitude. Examples of risk assessment include *financial risk analysis*, the determination of exposure to property or other economic loss, and *safety hazard analysis*, the determination of the risk of low-probability high-consequence effects. Here we are concerned with the following kinds of risk assessment: **Health risk assessment** is the determination of the likelihood of adverse effects in humans, generally from chronic low-level exposure to toxic agents. **Ecological risk assessment** (also called **hazard assessment**) examines the risk of harm to species other than humans. Risk assessment is usually considered to include the following four steps:

1. **Hazard identification** is the selection of environmental agents that may be responsible for a cause–effect relationship between an environmental agent and some harmful effect. The identification may be made by the use of epidemiological studies or laboratory animal studies compared with levels of the environmental agent present.

Environmental Biology for Engineers and Scientists,   by David A. Vaccari, Peter F. Strom, and James E. Alleman
Copyright © 2006 John Wiley & Sons, Inc.

2. **Toxicity assessment** is the determination of a quantitative relationship between observed effects and environmental causes. These relationships can include the determination of dose–response relationships, LC_{50} or other measurements, from laboratory toxicity tests.

3. **Exposure assessment** is the measurement or prediction of the dosage to which particular organisms may be subjected. This often involves tools such as mathematical modeling of the fate and transport of pollutants in the environment.

4. **Risk characterization** (or **risk estimation**) is the determination of the risk, based on integration of the toxicity and exposure information found in the preceding steps. Furthermore, risk characterization should include a discussion of the uncertainty in the computed risk and qualitative factors. This discussion could include comparisons between different toxicity and exposure assessment approaches, and comparisons between different target populations.

These steps may be taken in different orders. For example, exposure assessment precedes toxicity assessment in the Superfund risk assessment process. If the exposure is found to exceed established standards, such as drinking water MCLs or air NAAQSs (see Appendixes D and E, respectively), remediation is considered. In environmental risk assessment, toxicity assessment and exposure assessment are considered to be parallel steps rather than in sequence.

Whereas the first two steps above are the province of biologists, the last two steps contain roles for the environmental engineer. Furthermore, risk assessment is typically followed by risk management, which may also involve engineers. **Risk management** is the selection and implementation of actions to control the risk. This may require design, construction, and operation of facilities to remediate or treat sources of environmental pollution.

Virtually this entire textbook might be thought of as providing an understanding of normal and perturbed functioning of living things so as to be able to attribute effects observed to particular environmental causes. These topics covers the hazard identification portion of risk assessment. In Chapter 19 and Section 20.1 we described the techniques of toxicity assessment. In the following sections we describe the remaining steps: exposure assessment and risk characterization.

22.1.1 Human Health Risk Assessment

Human health risk assessments are used for many regulatory purposes. These include the setting of standards and guidelines for drinking water and air quality and the determination of cleanup goals under the Resource Conservation and Recovery Act (RCRA) or the Comprehensive Environmental Responsibility and Liability Act (CERCLA, the Superfund Act).

Exposure Assessment   Exposure assessment has three parts. First is the determination of exposure pathways, such as breathing of air or ingestion of food, soil, or water. Second, mathematical models of pollutant fate and transport are often used to predict exposure based on release rates from the sources coupled with the physicochemical properties of the pollutant. **Exposure** is defined as the amount of pollutant contacting body boundaries (skin, lungs, gastrointestinal tract, etc.). Exposure may also be measured directly by personal monitoring devices, such as air samplers that are carried by people for a period of time. Prediction of fate and transport is a part of the risk assessment process that has the major role for engineers.

Once the concentration of a pollutant at the point of exposure is determined, the third and final step of the exposure assessment in human risk assessment is the calculation of **receptor dosage** or **chemical intake rate**. This is usually expressed in units of milligrams of intake per kilogram of body weight per day. The calculation depends on the route of exposure and other factors. A typical form of the relationship could be

$$I = C \frac{\text{CR} \cdot \text{EF} \cdot \text{ED} \cdot \text{ABS}}{\text{BW}} \frac{1}{\text{AT}} \tag{22.1}$$

where

I = intake rate $(\text{mg/kg} \cdot \text{day}^{-1})$

C = exposure concentration (mg/L for water; mg/m^3 for air; mg/g for soil)

CR = contact rate $(\text{L/day, m}^3/\text{day, or g/day})$

EF = exposure frequency (days/yr)

ED = exposure duration (yr)

ABS = absorption efficiency (fraction)

BW = body weight (kg)

AT = averaging time (yr)

Formulations that are more complex may be used. For example, if the exposure concentration is time varying, an expression that integrates the exposure could be used. The absorption efficiency reflects the fact that not all contaminants inhaled or ingested are actually taken into the bloodstream. Inhaled gases may not be absorbed completely because of the limited time for mass transfer. Contaminants associated with particulates, such as dust and soil, may be sequestered in the interior of particles or in highly insoluble forms. The fraction that is not sequestered is called the **bioavailability**.

The exposure concentration may be based on measurements or best estimates based on modeling results. If the risk assessment is retrospective (i.e, it is for the purpose of computing risk based on historical exposure), the best estimate of concentration is used. However, if the risk assessment is prospective (i.e., it is a prediction of risks for the future), it is common practice to use an upper 95% confidence limit for the exposure concentration. This could be done by adding 1.96 times the standard deviation to the average value. For example, if historical data indicate that a population has been exposed to an average concentration of 8.0 mg/m^3 of a toxin, with a standard deviation of 0.5 mg/m^3, the exposure concentration should be taken to be 8.0 + (1.96) (0.5) = 8.98 mg/m^3.

The averaging time, AT, is 70 years for carcinogens and is equal to the exposure duration (ED) for noncarcinogens. The contact rate, exposure frequency, and duration are usually based on average or reasonable maximum (e.g., 90th percentile) values, as shown in Table 22.1.

The intake rate may be based on long-term exposures as shown in Table 22.1, or up to a lifetime of 70 years. This produces a **chronic daily intake** (CDI). Risk assessment may also be based on short-term concentrations averaged over a 10- to 90-day period. The

Table 22.1 Exposure Factors Used by the EPA for Risk Assessment

Land Use	Exposure Pathway	Daily Intake	Exposure Frequency (days/yr)	Exposure Duration (yr)	Body Weight (kg)
Residential	Ingestion of potable water	2 L (adult) 1 L (child)	350	30	70 (adult) 15 (child)
	Ingestion of soil and dust	100 mg (adult) 200 mg (child)	350	24 as adult +6 as child	70 (adult) 15 (child)
	Food intake (wet weight)	1500 g			
	Inhalation of contaminants	20 m^3 (adult) 10 m^3 (child)	350	30	70
Industrial and commercial	Ingestion of potable water	1 L	250	25	70
	Ingestion of soil and dust	50 mg	250	25	70
	Inhalation of contaminants	20 m^3 (workday)	250	25	70
Agricultural	Consumption of homegrown produce	42 g (fruit) 80 g (veg.)	350	30	70
Recreational	Consumption of locally caught fish	54 g	350	30	70

Source: Kolluru et al. (1996); Doull et al. (1980).

resulting intake is called the **subchronic daily intake** (SDI). In some cases, instead of estimating intake of a contaminant, the body burden will be measured directly, such as by blood or tissue sampling.

Risk Characterization   Toxic effects in humans are broadly classified for risk assessment purposes as either carcinogenic or noncarcinogenic. **Carcinogenic effects** are the formation of tumors. **Noncarcinogenic effects** include all toxic responses other than tumor formation, which may include, for example, respiratory, neurological, or reproductive effects. The U.S. EPA makes the following important distinction between these two categories: Carcinogenic effects are always assumed to have no threshold; any finite dosage is capable of producing cancer. Noncarcinogenic effects, on the other hand, are always assumed to have a threshold below which no effect will occur. The characterization of risk is different for the two types. Some chemicals may exhibit both types of effect, so both need to be checked when doing a risk assessment for any chemical.

Noncarcinogenic Risk Characterization   Toxicological effects other than carcinogenesis are assumed to produce their effect only above a threshold. The threshold is determined experimentally from bioassay and epidemiological data. The highest level at which no toxic effect is found is termed the **no observed adverse effect level** (NOAEL). The NOAEL is often divided by a safety factor of 10 to take into account differences between the typical members of a population and its more sensitive members.

The resulting safe threshold is called the **reference dose** (RfD) [formerly called the **acceptable daily intake** (ADI)]. Appendix C gives RfDs for several important environmental toxins. Unless information to the contrary is available, noncarcinogenic risks are assumed to be additive.

The RfD is used to compute a **hazard index** (HI), which represents a summary of all exposures over all pathways:

$$\text{HI} = \sum_i \frac{E_i}{\text{RfD}_i} \tag{22.2}$$

where E_i is the exposure or intake in mg/kg·day. If the hazard index exceeds 1.0, a more detailed study on the risk and control measures would be required. If several chemicals are known to produce their noncarcinogenic effect on different organs, the summation may be carried out separately for each organ.

Carcinogenic Risk Characterization   The different models for extrapolating carcinogenic risk, including the linearized multistage model, result in a linear relationship between risk and dose at low dose. The slope of that relationship is the carcinogenic potency factor (CPF) or slope factor (SF) and relates the probability of contracting cancer, p, to the exposure or intake:

$$p = \text{CDI} \cdot \text{CPF} \tag{22.3}$$

where CDI is the chronic daily intake in mg/kg·day (equivalent to exposure E). CPFs for some pollutants are given in Appendix C. The potency factor includes a safety factor to take into account the uncertainty in "mouse-to-human" extrapolation. The value of the safety factor varies from 10 to 1000, and depends on the quality and quantity of the evidence of toxicity. If toxicity has been tested in only a single species, a larger safety factor is called for. If multiple species have been tested, a smaller safety factor, together with using the most sensitive species, would give reasonable protection. Other considerations affecting the safety factor include whether there is specific knowledge of toxic mechanisms such as specific biotransformations that occur, or of pharmacokinetic behavior. There may be different potency factors for different routes of absorption.

Both the RfD and the potency factor include safety factors and conservative assumptions. To summarize, the risk assessment is conservative because of the emphasis on data from the most sensitive species, strain, and sex; extrapolation by body surface area instead of body weight, the use of safety factors for extrapolation from animals to humans; use of the upper confidence limit of the dose–response model; and the assumption of no threshold for carcinogens.

In performing human health risk assessments for exposures involving the public, an excess lifetime risk less than 10^{-6} is generally considered acceptable for regulatory purposes. In some cases, such as occupational exposures or other populations that have a stake in the costs, higher risks may be acceptable. It must be noted that the public perception of risk is quite different from the probabilistic definition given here. Although engineers and scientists recognize that "zero risk" is an impossibility, the public rightfully demands that the goal should be zero risk. That is, although the goal may not be achievable, all reasonable exertions should be made to attempt it.

Example Risk Characterization CPF and RfD values for some pollutants are given in Appendix C. More complete and up-to-date information can obtained online from U.S. EPA's Integrated Risk Information Systems (IRIS) database at http://www.epa.gov/iris.

Example 22.1 Estimate and characterize the risk to an adult from a lifetime of drinking water containing 1 μg/L of perchloroethylene (PCE) and 10 μg/L of carbon tetrachloride (CT).

	RfD (mg/kg·day)	Oral CPF (mg/kg·day)
PCE	0.010	0.13
CT	$7.0e \times 10^{-4}$	0.051

Solution Calculation of intake:

$$CDI_{PCE} = \frac{(0.001\,\text{mg/L})(2.0\,\text{L/day})}{70\,\text{kg}} = 2.86 \times 10^{-5}\,\text{mg/kg} \cdot \text{day}$$

$$CDI_{CT} = \frac{(0.01\,\text{mg/L})(2.0\,\text{L/day})}{70\,\text{kg}} = 2.86 \times 10^{-4}\,\text{mg/kg} \cdot \text{day}$$

Calculation of hazard index:

$$HI = \frac{2.86 \times 10^{-5}}{0.010} + \frac{2.86 \times 10^{-4}}{0.0007} = 0.00286 + 0.408 = 0.411$$

Although most of the risk comes from carbon tetrachloride, the hazard index is well below 1.0, indicating that the person is below the threshold for noncarcinogenic effects. The carcinogenic risk is calculated as follows:

$$p_{PCE} = (0.13)(2.86 \times 10^{-5}) = 3.72 \times 10^{-6}$$
$$p_{CF} = (0.051)(2.86 \times 10^{-4}) = 1.46 \times 10^{-5}$$

Both of these risks exceed the acceptable level of 10^{-6}. The water supply may need treatment or remediation, or a substitute water supply may have to be provided. Keep in mind all the sources of uncertainty in these calculations, notably measurement error in the bioassay and error associated with low-dose extrapolation and extrapolation between species.

Having computed a numeric risk, it is tempting to believe in it too much. Remember all of the sources of uncertainty: error in laboratory experiments, low-dose extrapolation, animal-to-human extrapolation, variation in human populations, and so on. Nevertheless, the computed risk is conservative, is based on what is known, and may often be the best information available for planning purposes.

22.1.2 Ecological Risk Assessment

Like human health risk assessments, ecological risk assessments (ERAs) are called for under provisions of numerous laws and regulations, including RCRA, CERCLA, and the Clean Water Act. Their purpose is to assess injury (adverse effects) and damages (economic costs) to natural resources caused by the release of harmful materials. Several approaches to ERAs have been developed. Here we focus on the U.S. EPA framework. Under this framework, the ERA "evaluates the likelihood that adverse ecological effects may occur or are occurring as a result of exposure to one or more stressors." Under this definition, *likelihood* can be expressed in deterministic, probabilistic, or even qualitative terms.

The U.S. EPA framework has the same risk assessment steps as described above for human health risk assessment, with slightly different terminology:

- *Problem formulation* (hazard identification) determines stressors, populations at risk, ecological effects of concern, and measurement endpoints.
- *Analysis*: includes two concurrent steps:
  Characterization of exposure (exposure assessment)
  Characterization of ecological effects (toxicity assessment)
- *Risk characterization*

The exposure and ecological effects characterization are considered parallel steps as part of an *analysis* phase. The U.S. EPA framework also makes explicit the interactions between risk assessment and risk management on one hand and data acquisition on the other. A dialogue between the risk assessor and the risk manager is expected to ensure that the risk assessment will satisfy both scientific and societal goals.

Ecological risk assessment has many more sources of uncertainty than human health risk assessment. The release of hazardous chemicals to the environment can affect numerous species. Even if a particular sensitive species were selected, often less is known about its sensitivity. The response of cold-blooded species may depend on environmental temperature, whereas laboratory toxicity studies are typically made at a single temperature.

Problem Formulation Phase In the problem formulation step, an identification and preliminary characterization are made of the stressor, the population affected, and the ecological effects. This information is used to select assessment and measurement endpoints, and a conceptual site model predicting how the stressor affects the ecosystem and the endpoints.

A **stressor** is a physical, chemical, or biological factor that can cause adverse effects on individuals, populations, communities, or ecosystems. Examples of physical stressors are erosion, flooding, or thermal pollution. Chemical stressors may include pesticides, heavy metals, and so on. Biological stressors include exotic species, pathogens, or pests. Characteristics of stressors that should be determined include their intensity, duration, frequency, timing relative to biological cycles, and spatial distribution.

The *ecosystem or ecological components potentially at risk* may include the broadest definition of ecosystem to include all living and nonliving things in an area, or may focus selectively on a particular species, the indicator species. The indicator species may be chosen because of their sensitivity to the stressor, or because they are endangered or threatened, or because they represent significant economic or recreational populations.

The *ecological effects* may be selected based on the observations that prompted the risk assessment in the first place, such as fish kills or population depletion. Alternatively, they may be selected based on known relationships between stressors and their effects on the ecological components at risk. Examples of ecological effects are changes in species abundance and diversity or habitat and ecosystem function and capacity, natural resource damage, endangered species impacts, or toxic effects on individual populations measured in laboratory bioassays.

The information on stressors, populations at risk, and ecological effects is used to select two types of endpoints. **Assessment endpoints** are the ultimate values that are in need of protection. Assessment endpoints should have ecological relevance, they should reflect policy goals and societal values, and they should be both affected by the stressor and related to the ecological effects. **Measurement endpoints** are qualitative or quantitative outcomes that can be measured directly. For example, the survival of a sport fishery may be an assessment endpoint. A corresponding measurement endpoint might be mortality in a different fish species used in a laboratory bioassay. In some cases, the assessment endpoint may be quantifiable, in which case it is also a measurement endpoint. When the assessment endpoint cannot be measured directly, measurement endpoints should be selected that correlate or imply changes in the assessment endpoint. Other examples of measurement endpoints are:

- Population abundance
- Age structure
- Reproductive potential and fecundity
- Species diversity
- Food web diversity
- Nutrient retention or loss
- Standing crop
- Productivity

Finally, a working hypothesis is formed linking the stressor to ecological components at risk and linking assessment endpoints to measurement endpoints. This hypothesis, called the **conceptual model**, includes qualitative scenarios describing how the stressor affects the ecological components. For example, the fate and transport of chemical and potential routes of exposure will be postulated.

Analysis Phase   The analysis phase is the technical evaluation of potential effects of the stressor on the population exposed. It is based on the conceptual model. As in human health risk assessment, there is a *characterization of exposure* (exposure assessment) and a *characterization of ecological effects* (toxicity assessment). The outcomes of the analysis phase are an **exposure profile**, which quantifies the magnitude and spatial and temporal patterns of exposure, and a **stressor-response profile**, which quantifies and summarizes the relationship of the stressor to the assessment endpoint. These profiles are inputs to the next phase, risk characterization.

The exposure profile may be expressed in various units, depending on the receptor. Individual exposure may be given as a dosage (mg/kg body weight per day). Exposure of an entire ecosystem may be expressed as concentration per unit area per time unit.

The stressor–response profile details the relationship between stressors and effects, relationships between assessment and measurement endpoints, and information-supporting cause–effect relationships. In many practical cases, the stressor–response profile consists simply of a parameter, such as the LC_{50} from bioassay tests.

Risk Characterization Phase   With the exposure profile and the stressor–response profile in hand, the final phase of the ecological risk assessment may be performed. **Risk characterization** relates the two profiles and develops a qualitative or quantitative description of the likelihood of adverse ecological effects. It consists of two steps: risk estimation and risk description.

Risk estimation is the integration of the exposure profile and the stressor–response profile. Several approaches are used. The simplest is simply to compare a single exposure value to a single toxicity parameter. If distributions are known for either the exposure or the toxicity, a range of ratios can be developed. This yields a distribution that can be used to estimate the probability of an effect occurring. The same distributional information can be used in **Monte Carlo simulation**, in which a large number of predictions are made using values randomly drawn from the exposure and/or toxicity distributions. The distribution of the results is used both to give a more accurate estimate of the probability of the effect occurring and to estimate the probable range of outcomes.

A part of risk estimation is the characterization of the amount of uncertainty in the result. The latter two approaches produce some estimate of the uncertainty due to the variability in the amount of stressor present and the variability in the response. Other sources of uncertainty include error in sampling or data analysis, incorrect assumptions in the formulation of the conceptual model, or insufficient data quality. Errors in the conceptual model are of the most difficult type to detect and correct. Problems with insufficient data can be rectified, in principle, by collecting additional data.

The second step of risk characterization, and the last of the risk assessment, is **risk description**. This includes a summary of the results of the risk estimation. Ideally, the results could be presented in a quantitative fashion, such as "there is a 5% chance of 50% mortality in the population." The sources of uncertainty should be described, along with the weight of evidence information, including a description of the sufficiency and quality of the data, any corroborative information from the literature, and evidence of causality. The results of the risk estimation may suggest secondary risks to other populations. For example, if high zooplankton mortality is predicted, effects on fish populations may be expected. In such cases, additional analysis steps would be called for.

The final part of the risk description is **interpretation of ecological significance**. This discussion places the results in context as a step toward the communication of the results. Considerations include a description of the nature and magnitude of the effects, their spatial and temporal patterns, and the potential for the ecosystem to recover if the stressor is removed.

22.2   TOXICITY REDUCTION EVALUATION

Under the National Pollutant Discharge Elimination System (NPDES) provisions of the Clean Water Act, facilities that discharge directly into the navigable waterways of the United States must have a NPDES permit. The NPDES permit specifies the quantity and quality of waste that may be discharged. Many permits specify limits on the toxicity of the waste as measured by an acute aquatic bioassay test. Because the exact chemical composition of many wastes cannot be determined easily, the test measures toxicity as a function of the dilution of the waste rather than in terms of the concentration of any particular chemical. Toxicity measured in this way is called **whole effluent toxicity**. Permit requirements that include toxicity testing are called **water quality–based standards**. This distinguishes them from **technology-based standards**, which specify the type of treatment that must be applied to a waste, or from **effluent standards**, which simply specify the concentrations or amounts of particular contaminants that may be discharged.

If a facility fails to meet its NPDES requirement for whole effluent toxicity, it must find a way to reduce that toxicity. The procedure for doing this is called a **toxicity reduction evaluation** (TRE). The details of the TRE procedure depend on the type of facility. In particular, a different protocol has been promulgated for domestic wastewater treatment plants than for industrial facilities. However, both can be summarized as follows:

- Data acquisition
- Evaluation and optimization of process performance
- Toxicity identification evaluation (TIE)
- Source identification evaluation and/or process modification for control
- Implementation
- Follow-up and confirmation

We focus here on just one of these steps, the **toxicity identification evaluation** (TIE). The TIE is a procedure to identify the specific substance or substances responsible for the toxicity of the whole effluent, or at the minimum, the physicochemical characteristics of those substances so that process modification could be implemented. The TIE is composed of three phases that characterize, specifically identify, then confirm the cause of the toxicity.

Phase I. **Toxicity characterization** is the determination of the physicochemical characteristics of the cause of toxicity. The procedure consists of a suite of lab-scale treatments, each of which is selective for a type of chemical. Bioassays are performed on each treated wastewater. The treatments used are:

- *Baseline.* Untreated samples are tested for initial toxicity.
- *Oxidant reduction.* Addition of a reducing agent (sodium thiosulfate) can implicate toxic oxidants such as chlorine.
- *Chelation.* Chelating agents such as EDTA can sequester cationic toxicants such as heavy metals.
- *pH adjustment and graduated pH test.* The effect of pH manipulation on toxicity is determined. Can be used to suggest whether the toxicant is an acid or a base. The tests described below are combined with pH adjustment.
- *Degradation.* testing of several samples over a period of time determines if the toxicant degrades with time.
- *Aeration.* Information is provided to suggest whether the toxicant is volatile, oxidizable, or participates in acid–base reactions. Volatile acids and bases (e.g., H_2S or NH_3) could be suggested by the results.
- *Filtration.* Whether toxicity is caused by insoluble materials or substances that form precipitates upon pH change is determined.
- C_{18} *solid-phase extraction.* Passing the sample through a C_{18} high-performance liquid chromatography (HPLC) column removes high-molar-mass nonpolar compounds.

- *Others*. If the procedures above are insufficient to characterize the toxicant, other tests may be performed, such as treatment with activated carbon (for a variety of organic or inorganic compounds), ion-exchange resins (for cations or anions), zeolite resins (for ammonia), or molecular sieves (which remove compounds based on molar mass).

The acute toxicity bioassay for phase I should be performed using a variety of species. The most common species used are *Ceriodaphnia*, *Daphnia magna*, and *Pimephales promelas* (fathead minnows). Only freshwater species are recommended, even if the receiving water is saline. At least six dilutions should be tested, including 0% (pure dilution water) and 100% (pure effluent). LC_{50} is computed using probit analysis.

The phase I tests may suggest a way to eliminate the problem. If not, the procedure moves on to phase II.

Phase II. **Identification of specific toxicants** is a chemical analysis of the wastewater which, guided by the results of the phase I tests, may suggest the specific compounds responsible for the toxicity. This identification is often performed with instrumental methods of chemical analysis, such as gas chromatography/mass spectrometry for volatile or semivolatile organics, HPLC for nonvolatile compounds, atomic absorption spectrophotometry (AAS) or inductively coupled plasma emission spectrophotometry (ICP) for heavy metals, as well as various wet chemistry techniques.

Phase III. **Confirmation of identifications** from phase I or II is made by a variety of methods, depending on the type of identification. Some methods are:

- Compare the dose–response curve of the wastewater with data from the literature.
- Use treated wastewater spiked with similar concentrations of the toxicant to reproduce the NOAEL or LC_{50}.
- Detect correlations between wastewater toxicity and the concentration of toxicant identified.
- Compare the relative sensitivity of various species to the wastewater with that for the toxicant identified.
- Observe specific effects in aquatic organisms, such as nonlethal effects, and compare with those produced by spiked controls.

Following identification of the cause of the toxicity, the source is identified and removed or the process should be modified to eliminate the toxicity. If the wastewater is from a wastewater treatment plant, source identification must deal with the added complexity of the effect of the treatment on the influent toxicant. This is handled by performing batch treatments on all samples taken for source identification purposes. This approach is called a **refractory toxicity assessment**.

PROBLEMS

22.1. Pick several chlorinated organic solvents from Appendix C, such as TCE, PCE, and chloroform. Compute the concentrations that would produce at 10^{-6} cancer risk in drinking water. Compare these levels with the federal primary drinking water standards in Appendix D.

22.2. Identify one compound that would be removed by each of the treatments used in the toxicity characterization phase of the TIE procedure.

REFERENCES

Canter, L. W., 1996. *Environmental Impact Assessment*, 2nd ed., McGraw-Hill, New York.

Doull, J., C. D. Klaassen, and M. O. Amdur (Eds.), 1980. *Casarett and Doull's Toxicology: The Basic Science of Poisons*, 2nd ed., McGraw-Hill, New York.

http://www.epa.gov/Region9/waste/sfund/prg/index.htm.

http://www.epa.gov/Region9/waste/sfund/prg/files/background.pdf.

Kolluru, R., S. Bartell, R. Pitblado, and S. Stricoff, 1996. *Risk Assessment and Management Handbook for Environmental, Health and Safety Professionals*, McGraw-Hill, New York.

Seinfeld, J. H., 1975. *Air Pollution: Physical and Chemical Fundamentals*, McGraw-Hill, New York.

U.S. Congress, Office of Technology Assessment, 1987. *Identifying and Regulating Carcinogens*, OTA-BP-H-42, U.S. Government Printing Office, Washington, DC. November.

U.S. EPA, 1985a. *Principles of Risk Assessment: A Nontechnical Review*, U.S. Environmental Protection Agency, Washington, DC.

U.S. EPA, 1985b. *Draft Superfund Exposure Assessment Manual*, U.S. Environmental Protection Agency, Washington, DC.

U.S. EPA, 1992. *A Framework for Ecological Risk Assessment*, EPA/630/R-92/001, U.S. Environmental Protection Agency, Washington, DC. February.

U.S. EPA, 1994. *Water Quality Standards Handbook*; 2nd ed., EPA/823/B-94/005a, U.S. Environmental Protection Agency, Washington, DC. August.

U.S. EPA, 1996. *Proposed Guidelines for Carcinogen Risk Assessment*, EPA/600/P-92/003c U.S. Environmental Protection Agency, Washington, DC. April.

APPENDIXES

Environmental Biology for Engineers and Scientists, by David A. Vaccari, Peter F. Strom, and James E. Alleman
Copyright © 2006 John Wiley & Sons, Inc.

APPENDIX A   Physicochemical Properties of Common Pollutants

Also see Table 21.6 for properties of additional pesticides and Table 21.8 for additional PAHs.

	Chemical Name	CASRN[a]	Empirical Formula[b]	Molecular Weight[b]	Water Solubility[c] mg/L	°C	Vapor Pressure mm Hg	°C	Ref.	Henry's Constant at 298.15 K[b] (–)	Ref.	Log KoW[d]	Fish BCF[c] L/kg
1	acenaphthene	83-32-9	$C_{12}H_{10}$	154.21	3.42	25	1.55E-03	25	c	0.0063	b	3.92	242
2	acrolein	107-02-8	C_3H_4O	56.064			210	20	d	0.0050	b	–0.09	
3	acrylonitrile	107-13-1	C_3H_3N	53.063	73,500	20	100	23	d	0.0037	b	0.25	48
4	benzene	71-43-2	C_6H_6	78.113	1,780	25	95.2	25	d	0.233	c	2.13	5
5	benzidine	92-87-5	$C_{12}H_{12}N_2$	184.24	400	12	10	176	c			1.34	88
6	carbon tetrachloride	56-23-5	CCl_4	153.82	800	20	91.3	20	d	1.25	c	2.62–2.83	19
	Chlorinated benzenes												
7	chlorobenzene	108-90-7	C_6H_5Cl	112.56	500	20	11.8	25	d	0.159	c	2.18–2.84	10
8	1,2,4-trichlorobenzene	120-82-1	$C_6H_3Cl_3$	181.45	30	25	0.29	25	d	0.0866	c	4.3[c]	2,800
9	hexachlorobenzene	118-74-1	C_6Cl_6	284.78	0.006	25	1.90E-05	20	d	0.0696	b	5.31	8,690
	Chlorinated ethanes												
10	1,2-dichloroethane	107-06-2	$C_2H_4Cl_2$	98.96	8,690	20	61	20	d	0.0634	c	1.48	1
11	1,1,1-trichloroethane	71-55-6	$C_2H_3Cl_3$	133.4	4,400	20	127	25	d	0.697	c	2.49	6
12	hexachloroethane	67-72-1	C_2Cl_6	236.74	42,000	20	0.4	20	d	0.332	c	3.34	87
13	1,1-dichloroethane	75-34-3	$C_2H_4Cl_2$	98.96	5,500	20	180	20	c	0.223	c	1.9	
14	1,1,2-trichloroethane	79-00-5	$C_2H_3Cl_3$	133.4	4,500	20	23	25	d	0.0367	b	2.17	5
15	1,1,2,2-tetrachloroethane	79-34-5	$C_2H_2Cl_4$	167.85						0.0168	b		
16	chloroethane	75-00-3	C_2H_5Cl	64.515			1000	20	d	0.480	b	1.43	
	Chloroalkyl ethers												
17	bis(chloromethyl) ether	542-88-1	$C_2H_4Cl_2O$	114.96	22,000	25	30	22	d			–0.38	1
18	bis(2-chloroethyl) ether	111-44-4	$C_4H_8Cl_2O$	143.01	12,000	25	0.71	20	d	0.00086	b	1.58	7
19	2-chloroethyl vinyl ether (mixed)	110-75-8	C_4H_7ClO	106.55									
	Chlorinated naphthalene												
20	2-chloronaphthalene	91-58-7	$C_{10}H_7Cl$	162.62						0.0130	b		

#	Compound	CAS	Formula	MW		T		T					
	Chlorinated phenols												
21	2,4,6-trichlorophenol	88-06-2	$C_6H_3Cl_3O$	197.45	800	25	400	25	c		c	3.69	110
22	p-chloro-m-cresol	59-50-7	C_7H_7ClO	142.58								3.1	
23	chloroform	67-66-3	$CHCl_3$	119.38	8,000	20	160	20	d	0.148	c	1.97	4
24	2-chlorophenol	95-57-8	C_6H_5ClO	128.56		20	2.2	20	d	0.00034	b	2.15	
	Dichlorobenzenes												
25	1,2-dichlorobenzene	95-50-1	$C_6H_4Cl_2$	147	100	20	1	20	c	0.0747	b	3.38	56
26	1,3-dichlorobenzene	541-73-1	$C_6H_4Cl_2$	147	123	25	2.28	25	d	0.136	c	3.6[c]	56
27	1,4-dichlorobenzene	106-46-7	$C_6H_4Cl_2$	147	79	25	1.18	25	d	0.130	c	3.6[c]	56
	Dichlorobenzidine												
28	3,3-dichlorobenzidine	91-94-1	$C_{12}H_{10}Cl_2N_2$	253.13	3.99	22	1.00E-05	22	d			3.02	312
	Dichloroethylenes												
29	1,1-dichloroethylene	75-35-4	$C_2H_2Cl_2$	96.944	2250	20	591	25	d	1.041	c	1.32	6
30	1,2-trans-dichloroethylene	156-60-5	$C_2H_2Cl_2$	96.944	600	20	326	20	c	0.379	c	2.06	2
31	2,4-dichlorophenol	120-83-2	$C_6H_4Cl_2O$	163	4,500	25	5.90E-02	20	c			2.9[c]	41
	Dichloropropylene												
32	1,2-dichloropropane	78-87-5	$C_3H_6Cl_2$	112.99	2,700	20	50	25	d	0.106	c	2.28	
33	1,3-dichloropropene	542-75-6	$C_3H_4Cl_2$	110.97	2,800	25	38	25	c	0.0791	b	1.39	2
34	2,4-dimethylphenol	105-67-9	$C_8H_{10}O$	122.17			10	92.3	d	0.212	b	2.3	
	Dinitrotoluene												
35	2,4-dinitrotoluene	121-14-2	$C_7H_6N_2O_4$	182.14	270	22	5.10E-03	20	c	0.0019	b	1.98	4
36	2,6-dinitrotoluene	606-20-2	$C_7H_6N_2O_4$	182.14	180	20	1.80E-02	20	c			2[c]	4
37	1,2-diphenylhydrazine	122-66-7	$C_{12}H_{12}N_2$	184.24	1,840	20	2.60E-05	25	c			2.94	25
38	ethylbenzene	100-41-4	C_8H_{10}	106.17	152	20	10	25.9	d	0.348	c	3.15	38
39	fluoranthene	206-44-0	$C_{16}H_{10}$	202.26	0.206	25	5.00E-06	25	c	0.00037	b	4.9	1,150
	Haloethers												
40	4-chlorophenyl phenyl ether	7005-72-3	$C_{12}H_9ClO$	204.66									
41	4-bromophenyl phenyl ether	101-55-3	$C_{12}H_9BrO$	249.11									
42	bis(2-chloroisopropyl) ether	108-60-1	$C_6H_{12}Cl_2O$	171.07	1,700	25	0.85	20	d	0.0062	b	2.58	0
43	bis(2-chloroethoxy) methane	111-91-1	$C_5H_{10}Cl_2O_2$	173.04		25	1.40E-04	25	d	0.000016	b	0.75	

APPENDIX A (Continued)

Chemical Name	CASRN[a]	Empirical Formula[b]	Molecular Weight[b]	Water Solubility[c] mg/L	°C	Vapor Pressure mm Hg	°C	Ref.	Henry's Constant at 298.15 K[b] (—)	Ref.	Log KoW[d]	Fish BCF[c] L/kg
Halomethanes												
44 dichloromethane	75-09-2	CH_2Cl_2	84.933	20,000	20	400	24.1	d	0.119	c	1.3	5
45 chloromethane	74-87-3	CH_3Cl	50.488	5,150	20	3800	20	d	0.248	c	0.91	
46 bromomethane	74-83-9	CH_3Br	94.939	3,190	30	1420	20	d	0.00011	b	1.19	
47 bromoform	75-25-2	$CHBr_3$	252.73			5.6	25	d	0.0237	b	2.38	
48 dichlorobromomethane	75-27-4	$CHBrCl_2$	163.83			50	20	d	0.101	b	1.88	
49 trichlorofluoromethane	75-69-4	CCl_3F	137.37	1,100	20	796	25	d	4.03	b	2.53	
50 dicholorodifluoromethane	75-71-8	CCl_2F_2	120.91	280	25	4.87	25	c	13.0	c	2.16	
51 dibromochloromethane	124-48-1	$CHBr_2Cl$	208.28	4,750	20	15	20	d	0.0367	b	2.09	
52 hexachlorobutadiene	87-68-3	C_4Cl_6	260.76	2	20	0.15	20	d	1.026	b	4.9	3
53 hexachlorocyclopentadiene	77-47-4	C_5Cl_6	272.77	1.8	25	8.00E-02	25	d	0.672	b	3.99	4
54 isophorone	78-59-1	$C_9H_{14}O$	138.21	10,400	20	0.38	20	c	0.00024	b	1.7	
55 naphthalene	91-20-3	$C_{10}H_8$	128.17						0.0202	b	3.01–3.59	
56 nitrobenzene	98-95-3	$C_6H_5NO_2$	123.11	1900	20	0.15	20	d	0.00086	b	1.85	
Nitrophenols												
57 2-nitrophenol	88-75-5	$C_6H_5NO_3$	139.11						0.00058	b	1.79	
58 p-nitrophenol	100-02-7	$C_6H_5NO_3$	139.11						0.000041	b	1.91	
59 2,4-dinitrophenol	51-28-5	$C_6H_4N_2O_5$	184.11	5,600	18	1.49E-05	18	d			1.5[c]	0
60 4,6-dinitro-o-cresol	534-52-1	$C_7H_6N_2O_5$	198.13	290	25	0.05	20	c			2.564	0
Nitrosamines												
61 N-nitrosodimethylamine	62-75-9	$C_2H_6N_2O$	74.082	1,000,000		8.1	25	d			−0.57	0
62 N-nitrosodiphenylamine	86-30-6	$C_{12}H_{10}N_2O$	198.22			0.1	25	d			3.13	
63 M-nitrosodi-n-propylamine	621-64-7	$C_6H_{14}N_2O$	130.19	9,900	25	0.1	25	c			1.36	
64 pentachlorophenol	87-86-5	C_6HCl_5O	266.34	14	20	1.10E-04	25	d	0.00367	b	5.12	770
Phthalate Esters												
65 di(2-ethylhexyl) phthalate	117-81-7	$C_{24}H_{38}O_4$	390.56								4.89	
66 butyl benzyl phthalate	85-68-7	$C_{19}H_{20}O_4$	312.36			8.60E-06	20	d			4.77	
67 dibutyl phthalate	84-74-2	$C_{16}H_{22}O_4$	278.35	13	25	7.28E-05	20	d			4.9	
68 n-octyl phthalate	117-84-0	$C_{24}H_{38}O_4$	390.56								5.22	

No.	Compound	CAS	Formula	MW	Sol.	T	Vapor pressure	T		Henry		log Kow	
69	diethyl phthalate	84-66-2	$C_{12}H_{14}O_4$	222.24	896	25	3.50E-03	25	c	0.000034	b	2.47	117
70	dimethyl phthalate	131-11-3	$C_{10}H_{10}O_4$	194.19						0.000013	b	2.12	
	Polynuclear aromatic hydrocarbons												
71	benz[a]anthracene	56-55-3	$C_{18}H_{12}$	228.29	0.0038	20	5.00E-09	20	d			5.61	
72	benzo[a]pyrene	50-32-8	$C_{20}H_{12}$	252.31	0.014	25	5.60E-09	25	c	0.000018	b	6.04	
73	3,4-benzofluoranthene	205-99-2	$C_{20}H_{12}$	252.31	0.0043	20	5.00E-07	20	c			6.06[c]	
74	benzo[k]fluoranthane	207-08-9	$C_{20}H_{12}$	252.31	0.0018	25	5.10E-07	25	c			6.84	
75	chrysene	218-01-9	$C_{18}H_{12}$	228.29		25	6.30E-09	25	c			5.61–5.91	
76	acenaphthylene	208-96-8	$C_{12}H_8$	152.2								4.07	
77	anthracene	120-12-7	$C_{14}H_{10}$	178.23	0.045	25	1.70E-05	25	c	0.00269	b	4.45	
78	benzo[g, h, l]perylene	191-24-2	$C_{22}H_{12}$	276.34	0.00026	25	1.00E-10	25	d	0.000013	b	6.58	
79	fluorene	86-73-7	$C_{13}H_{10}$	166.22	1.69	25	7.10E-04	20	c	0.00408	b		1,300
80	phenanthrene	85-01-8	$C_{14}H_{10}$	178.23	1	15	9.60E-04	25	c	0.00429	b	4.57	2,630
81	dibenzo[a, h]anthracene	53-70-3	$C_{22}H_{14}$	278.35	0.0005	25	0.1	20	c			6.5	
82	indeno[1, 2, 3, -cd] pyrene	193-39-5	$C_{22}H_{12}$	276.34	0.00053	25	0.1	20	c			6.584	
83	pyrene	129-00-0	$C_{16}H_{10}$	202.26	0.13	25	2.50E-06	25	c	0.00049	b	4.88[c]	
84	tetrachloroethylene	127-18-4	C_2Cl_4	165.83	150	25	18.47	25	d	0.6762	c	3.4	31
85	toluene	108-88-3	C_7H_8	92.14	515	20	36.7	30	d	0.2756	c	2.69	11
86	trichloroethylene	79-01-6	C_2HCl_3	131.39	1.1	25	57.8	20	d	0.3978	c	2.29	11
87	vinyl chloride	75-01-4	C_2H_3Cl	62.499	4,270	20	2660	20	d	1.0680	c	0.6	1
	Pesticides and metabolites												
88	aldrin	309-00-2	$C_{12}H_8Cl_6$	364.91						0.00112	b		
89	dieldrin	60-57-1	$C_{12}H_8Cl_6O$	380.91	0.195	25	1.78E-07	20		0.00044	b	3.5	
90	chlordane (technical mixture and metabolites)	12789-03-6	$C_{10}H_6Cl_8$	409.78					c				4,760

APPENDIX A (*Continued*)

Chemical Name	CASRN[a]	Empirical Formula[b]	Molecular Weight[b]	Water Solubility[c]		Vapor Pressure			Henry's Constant at 298.15 K[b]		Log KoW[d]	Fish BCF[c]
				mg/L	°C	mm Hg	°C	Ref.	(–)	Ref.		L/kg
91 4,4'-DDT	50-29-3	$C_{14}H_9Cl_5$	354.49	0.0055	25	1.90E-07	25	c	0.00212	b	6.19	54,000
92 4,4'-DDE	72-55-9	$C_{14}H_8Cl_4$	318.03	0.04	20	6.50E-06	20	d			6.51	51,000
93 4,4'-DDD	72-54-8	$C_{14}H_{10}Cl_4$	320.05	0.02–0.09	25	1.89E-06	30	c			6.2	
Endosulfan and metabolites												
94 endosulfan	115-29-7	$C_9H_6Cl_6O_3S$	406.92									
95 endosulfan sulfate	1031-07-8	$C_9H_6Cl_6O_4S$	422.92									
Endrin and metabolites												
96 endrin	72-20-8	$C_{12}H_8Cl_6O$	380.91									
97 endrin aldehyde	7421-93-4	$C_{12}H_{10}Cl_6O$	382.93			2.00E-07	25	d			5.6	
Heptachlor and metabolites												
98 heptachlor	76-44-8	$C_{10}H_5Cl_7$	373.32	0.18	25	3.00E-04	25	d			3.87–5.44	15,700
99 heptachlor epoxide	1024-57-3	$C_{10}H_5Cl_7O$	389.32	0.35	25	3.00E-04	25	c			5.4	14,400
Hexachlorocyclohexane												
100 α-hexachlorocyclohexane	319-84-6	$C_6H_6Cl_6$	290.83	1.63	25	2.50E-05	20	c	0.00031	b	3.8	130
101 β-hexachlorocyclohexane	319-85-7	$C_6H_6Cl_6$	290.83	0.24	25	2.80E-07	20	c			3.78	130
102 γ-hexachlorocyclohexane	58-89-9	$C_6H_6Cl_6$	290.83	7.8	25	9.40E-06	20	c	0.00014	b	3.72	130
103 δ-hexachlorocyclohexane	319-86-8	$C_6H_6Cl_6$	290.83	31.4	25	2.00E-02	20	d			4.14	130
Polychlorinated biphenyls (PCBs)												
104 PCB-1242 (Arochlor 1242)	53469-21-9	Mixture				50	25	d			4.11	
105 PCB-1254 (Arochlor 1254)	11097-69-1	Mixture				7.70E-05	25	d			6.3	
106 PCB-1221 (Arochlor 1221)	11104-28-2	Mixture										
107 PCB-1232 (Arochlor 1232)	11141-16-5	Mixture										
108 PCB-1248 (Arochlor 1248)	12672-29-6	Mixture										

No.	Compound	CAS Number	Formula							
109	PCB-1260 (Arochlor 1260)	11096-82-5	Mixture						6.11	
110	PCB-1016 (Arochlor 1016)	12674-11-2	Mixture							
111	toxaphene	8001-35-2	$C_{10}H_{10}Cl_8$	413.8	0.5				3.3	13100
112	antimony (total)	7440-36-0	Sb	121.76		4.05E-05	25	d		
113	arsenic (total)	7440-38-2	As	74.9216	25	0.4	25	d		
114	asbestos (fibrous)	1332-21-4	Mixture							
115	beryllium (total)	7440-41-7	Be	9.01218		10	1860	d		
116	cadmium (total)	7440-43-9	Cd	112.41		1	394	d		81
117	chromium III, insoluble salts	16065-83-1	Cr^{3+}	51.996		1	1840	c		
118	chromium VI	18540-29-9	Cr^{6+}	51.996						
119	copper (total)	7440-50-8	Cu	63.546		1	1628	d		200
120	cyanide (free)	57-12-5	CN^-	26.018						
121	lead and compounds (inorganic)	7439-92-1	Pb	207.2		1.77	1000	d		49
122	mercury (total)	7439-97-6	Hg	200.59	0.002	2.00E-03	25	d	0.454 [b]	5,500
123	nickel (total)	7440-02-0	Ni	58.6934		1	1810	d		47
124	selenium and compounds	7782-49-2	Se	78.96	356	1.00E-03	20	d		16
125	silver (total)	7440-22-4	Ag	107.868	825					
126	thallium (total)	7440-28-0	Tl	204.383	1	1	825	d		
127	zinc and compounds	7440-66-6	Zn	65.39		1	487	d		47
128	2,3,7,8-tetrachlorodibenzo-p-dioxin	1746-01-6	$C_{12}H_4Cl_4O_2$	321.97	0.0002	1.00E-06	25	c	6.72	5,000
129	methyl tert-butyl ether	1634-04-4	$C_5H_{12}O$	88.17	51000	245	25	d	0.0261 [b]	

References:

[a] U.S. EPA Integrated Risk Information System, www.epa.gov/iris/subst/index.html.

[b] Gen. Ed. Mallard, W. G. 2000. NIST Chemistry WebBook, http://webbook.nist.gov/chemistry. National Institute of Standards, Washington, DC.; Henry's law constants compiled by R. Sander.

[c] LaGrega, M. D., P. L. Buckingham, and J. C. Evans, *Hazardous Waste Management*, McGraw-Hill, New York.

[d] Prager, Jan C., 1995. *Environmental Contaminant Reference Data Book*, Vols. 1 and 2, Van Nostrand Reinhold, A Division of International Thomas Publishing, New York.

APPENDIX B Biodegradability of Common Pollutants

Also see Table 21.6 for properties of some pesticides

| | Chemical Name | CASRN | Aqueous Biodegradation (Unacclimated) | | | |
| | | | Aerobic Half-Life (days) | | Anaerobic Half-Live (days) | |
			Maximum	Minimum	Maximum	Minimum
1	acenaphthene	83-32-9	102	12.3	408	49.2
2	acrolein	107-02-8	28	7	120	28
3	acrylonitrile	107-13-1	23	1.25	92	5
4	benzene	71-43-2	16	5	720	112
5	benzidine	92-87-5	8	2	32	8
6	carbon tetrachloride	56-23-5	365	180	28	7
	Chlorinated benzenes					
7	chlorobenzene	108-90-7	150	68	600	272
8	1,2,4-trichlorobenzene	120-82-1	180	28	720	112
9	hexachlorobenzene	118-74-1	2080.5	985.5	8358.5	3869
	Chlorinated ethanes					
10	1,2-dichloroethane	107-06-2	180	100	720	400
11	1,1,1-trichloroethane	71-55-6	273	140	1092	560
12	hexachloroethane	67-72-1	180	28	720	112
13	1,1-dichloroethane	75-34-3	154	32	616	128
14	1,1,2-trichloroethane	79-00-5	365	180	1460	720
15	1,1,2,2-tetrachloroethane	79-34-5	180	28	28	7
16	chloroethane	75-00-3	28	7	112	28
	Chloroalkyl ethers					
17	bis(chloromethyl) ether	542-88-1	28	7	112	28
18	bis(2-chloroethyl) ether	111-44-4	180	28	720	112
19	2-chloroethyl vinyl ether (mixed)	110-75-8				
	Chlorinated naphthalene					
20	2-chloronaphthalene	91-58-7				

No.	Compound	CAS				
	Chlorinated phenols					
21	2,4,6-trichlorophenol	88-06-2	7	70	1825	169
22	p-chloro-m-cresol	59-50-7				7
23	chloroform	67-66-3	28	180	28	
24	2-chlorophenol	95-57-8				
	Dichlorobenzenes					
25	1,2-dichlorobenzene	95-50-1	28	180	720	112
26	1,3-dichlorobenzene	541-73-1	28	180	720	112
27	1,4-dichlorobenzene	106-46-7	28	180	720	112
	Dichlorobenzidine					
28	3,3-dichlorobenzidine	91-94-1	28	180	720	112
	Dichloroethylenes					
29	1,1-dichloroethylene	75-35-4	28	180	173	81
30	1,2-trans-dichloroethylene	156-60-5				
31	2,4-dichlorophenol	120-83-2	2.78	8.3	43	13.5
	Dichloropropylene					
32	1,2-dichloropropane	78-87-5	167	1277.5	5146.5	668
33	1,3-dichloropropene	542-75-6	7	28	112	28
34	2,4-dimethylphenol	105-67-9	1	7	28	4
	Dinitrotoluene					
35	2,4-dinitrotoluene	121-14-2	28	180	10	2
36	2,6-dinitrotoluene	606-20-2	28	180	13	2
37	1,2-diphenylhydrazine	122-66-7	28	180	720	112
38	ethylbenzene	100-41-4	3	10	228	176
39	fluoranthene	206-44-0	140	440	1759.3	558.45
	Haloethers					
40	4-chlorophenyl phenyl ether	7005-72-3				
41	4-bromophenyl phenyl ether	101-55-3				
42	bis(2-chloroisopropyl) ether	108-60-1				
43	bis(2-chloroethoxy) methane	111-91-1	18	180	720	72

863

APPENDIX B *(Continued)*

Chemical Name	CASRN	Aqueous Biodegradation (Unacclimated)			
		Aerobic Half-Life (days)		Anaerobic Half-Live (days)	
		Maximum	Minimum	Maximum	Minimum
Halomethanes					
44 dichloromethane	75-09-2	28	7	112	196
45 chloromethane	74-87-3	28	7	112	28
46 bromomethane	74-83-9	28	7	112	28
47 bromoform	75-25-2	180	28	720	112
48 dichlorobromomethane	75-27-4				
49 trichlorofluoromethane	75-69-4	365	180	1460	720
50 dicholorodifluoromethane	75-71-8	180	28	720	112
51 dibromochloromethane	124-48-1	180	28	180	28
52 hexachlorobutadiene	87-68-3	180	28	720	112
53 hexachlorocyclopentadiene	77-47-4	28	7	112	28
54 isophorone	78-59-1	28	7	112	28
55 naphthalene	91-20-3	20	0.5	258	25
56 nitrobenzene	98-95-3	198	13.4	13	2
Nitrophenols					
57 2-nitrophenol	88-75-5	28	7	28	7
58 p-nitrophenol	100-02-7	7	0.758333	9.8	6.8
59 2,4-dinitrophenol	51-28-5	263.1	67.5	7.1	2.8
60 4,6-dinitro-o-cresol	534-52-1	21	7	7.1	2.8
Nitrosamines					
61 N-nitrosodimethylamine	62-75-9	180	21	720	84
62 N-nitrosodiphenylamine	86-30-6	34	10	136	40
63 M-nitrosodi-n-propylamine	621-64-7	180	21	720	84
64 pentachlorophenol	87-86-5	178	23	1533	42

	Phthalate esters					
65	phenol phthalate esters					
66	di(2-ethylhexyl) phthalate	117-81-7	23	5	389	42
67	butyl benzyl phthalate	85-68-7	7	1	180	28
68	dibutyl phthalate	84-74-2	23	1	23	2
69	n-octyl phthalate	117-84-0	28	7	365	180
70	diethyl phthalate	84-66-2	56	3	224	28
71	dimethyl phthalate	131-11-3	7	1	28	4
	Polynuclear aromatic hydrocarbons					
72	benz[a]anthracene	56-55-3	678.9	102	2719.25	408.8
73	benzo[a]pyrene	50-32-8	529.25	57	2117	228
74	3,4-benzofluoranthene	205-99-2	609.55	360	2438.2	1442
75	benzo[k]fluoranthane	207-08-9	2138.9	908.85	8577.5	3639
76	chrysene	218-01-9	992.8	372.3	4015	1482
77	acenaphthylene	208-96-8	60	42.5	240	170
78	anthracene	120-12-7	459.9	50	1839.6	200
79	benzo[g,h,I]perylene	191-24-2	650	590	2591.5	2373
80	fluorene	86-73-7	60	32	240	128
81	phenanthrene	85-01-8	200	16	799.35	2.67
82	dibenzo[a,h]anthracene	53-70-3	941.7	361	3759.5	1445
83	indeno[1,2,3-cd]pyrene	193-39-5	730	598.6	2920	2402
84	pyrene	129-00-0	1898	210	7592	839.5
85	tetrachloroethylene	127-18-4	365	180	1642.5	98
86	toluene	108-88-3	22	4	210	56
87	trichloroethylene	79-01-6	365	180	1642.5	98
88	vinyl chloride	75-01-4	180	28	720	112
	Pesticides and metabolites					
89	aldrin	309-00-2	584	21	7	1
90	dieldrin	60-57-1	1095	175	7	1
91	chlordane (technical mixture and metabolites)	12789-03-6				
92	4,4'-DDT	50-29-3	5694	730	100	16

APPENDIX B (*Continued*)

Chemical Name	CASRN	Aqueous Biodegradation (Unacclimated)			
		Aerobic Half-Life (days)		Anaerobic Half-Live (days)	
		Maximum	Minimum	Maximum	Minimum
93	4,4'-DDE	72-55-9	5694	730	100	16
94	4,4'-DDD	72-54-8	5694	730	294	70
	Endosulfan and metabolites					
95	endosulfan	115-29-7	14	2	56	8
	Heptachlor and metabolites					
99	heptachlor	76-44-8	65	15	260	60
100	heptachlor epoxide	1024-57-3	552	33	7	1
	Hexachlorocyclohexane					
101	α-hexachlorocyclohexane	319-84-6	135	80	40	7
102	β-hexachlorocyclohexane	319-85-7	124	60	94	30
103	γ-hexachlorocyclohexane	58-89-9	413	31	30	5.9
104	δ-hexachlorocyclohexane	319-86-8	100	40	100	30
	Polychlorinated biphenyls (PCBs)					
129	2,3,7,8- tetrachlorodibenzo-p-dioxin	1746-01-6	591.3	419.75	2354.25	1672

Reference:
Howard, P. H., R. S. Boethling, W. F. Jarvis, W. M. Meylan, and E. M. Michalenko (Eds.), 1991. Handbook of Environmental Degradation Rates, Heather Taub Printup, Lewis Publishers, Chelsea, MI.

Appendix C Toxicological Properties of Common Pollutants (Ref. 'a' except where noted)

			Oral Noncarcinogenic Toxicity						Carcinogenicity		
			RfD	LD_{50}^b		RfC	LC_{50}^b at $T = 298K$, $P = 1$ atm		Oral CPF	Inhalation Unit Risk	Carc. Class.
	Chemical Name	CASRN	mg/kg·day	mg/kg	Species	mg/m³	ppmv (h)	Species	kg/day·mg	m³/μg	
1	acenaphthene	83-32-9	6E-2	600, 1,700	mice	NA			NA		
2	acrolein	107-02-8	NA	46 7	rat rabbit	2.00E-05	26 (1) 8.3 (4)	rat rat	NA		C
3	acrylonitrile	107-13-1	NA	78 90	rat guinea pig	2.00E-03			5.40E-01	6.80E-05	B1
4	benzene	71-43-2	NA			NA	Inadequate results		2.90E-02	1.00E+00	A
5	benzidine	92-87-5	3.00E-03								
6	carbon tetrachloride	56-23-5	7.00E-04			NA			1.30E-01	1.50E-05	B2
	Chlorinated benzenes										
7	chlorobenzene	108-90-7	2.00E-02	2,290 1,440 2,250 5,060	rat mouse rabbit guinea pig	NA			None		D
8	1,2,4-trichlorobenzene	120-82-1	1.00E-02								
9	hexachlorobenzene	118-74-1	8.00E-04			NA	Inadequate results		None		D
	Chlorinated ethanes										
10	1,2-dichloroethane	107-06-2	NA	870-950 860-970 670-890	mouse rabbit rat	NA	12,000 (31.8) 3,000 (165) 1,000 (432)	rat rat rat	9.10E-02	2.60E-05	B2
11	1,1,1-trichloroethane	71-55-6		Pending further review 5,660 5,660	rabbit guinea pig		18,000 (3) 14,000 (7) 13,500 (10)	rat rat rat mouse	None		D

Appendix C (*Continued*)

	Chemical Name	CASRN	Oral Noncarcinogenic Toxicity						Carcinogenicity		
			RfD	LD_{50}^{d}		RfC	LC_{50}^{d} at $T = 298K$, $P = 1$atm		Oral CPF	Inhalation Unit Risk	Carc. Class.
			mg/kg·day	mg/kg	Species	mg/m³	ppmv (h)	Species	kg/day·mg	m³/µg	
12	hexachloroethane	67-72-1	1.00E-03			NA			1.40E-02	4.00E-06	C
13	1,1-dichloroethane	75-34-3	NA	14,100	rat	NA	17,300 (2) 16,000 (8)	mouse rat	None		C
14	1,1,2-trichloroethane	79-00-5	4.00E-03	100–200	rat	NA	2,000 (4)	rat	5.70E-02	1.60E-05	C
15	1,1,2,2-tetrachloroethane	79-34-5	NA			NA			2.00E-01	5.80E-05	C
16	chloroethane	75-00-3	NA			1.00E+01	57,614 (2)	rat	NA		
	Chloroalkyl ethers										
17	bis(chloromethyl) ether	542-88-1	NA				Inadequate results				A
18	bis(2-chloroethyl) ether	111-44-4	NA	136 126 750	mouse rabbit rat		Inadequate results				B2
19	2-chloroethyl vinyl ether (mixed)	110-75-8	NA								
	Chlorinated naphthalene										
20	2-chloronaphthalene	91-58-7	8.00E-02			NA	Inadequate results		NA		
	Chlorinated phenols										
21	2,4,6-trichlorophenol	88-06-2	NA	2,800	rat						
22	parachlorometa cresol	59-50-7	NA	710	mouse						
23	chloroform	67-66-3	1.00E-02	908/1,117 9,827 2,250	male/ female rat rabbit dog	NA			6.10E-03	2.30E-05	B2
24	2-chlorophenol	95-57-8	5.00E-03			NA			NA		

No.	Compound	CAS									
	Dichlorobenzenes										
25	1,2-dichlorobenzene	95-50-1	9.00E-02	0.8E-3–2E-3	guinea pig	NA	1,135 (6)	mouse	None		D
26	1,3-dichlorobenzene	541-73-1	NA	3,863	rat	NA			None		D
27	1,4-dichlorobenzene	106-46-7	NA			8.00E-01			NA		
	Dichlorobenzidine										
28	3,3-dichlorobenzidine	91-94-1	NA	386/352	male/female mouse	NA		4.50E-01	None		B2
	Dichloroethylenes										
29	1,1-dichloroethylene	75-35-4	9.00E-03	200 / 1,500	mouse / rat	NA	10,000–15,000 (4)	nonfasted rat	6.00E-01		C
							500–2500 (4)	fasted rat		5.00E-05	
30	1,2-trans-dichloroethylene	156-60-5	2.00E-02			NA			NA		
31	2,4-dichlorophenol	120-83-2	3.00E-03			NA			NA		
	Dichloropropylene										
32	1,2-dichloropropane	78-87-5	NA	2,000–4,000	guinea pig	4.00E-03	720 (10)	mouse	NA		
33	1,3-dichloropropene	542-75-6	3.00E-04	140 / 300 / 504	rat / mouse / rabbit	2.00E-02	998 (1)	rat/mouse	NA		B2
34	2,4-dimethylphenol	105-67-9	2.00E-02			NA			NA		
	Dinitrotoluene										35
36	2,4-dinitrotoluene	121-14-2	2.00E-03			NA	Inadequate results			6.80E-01	NA
	2,4-/2,6-dinitrotoluene		NA								B2
37	1,2-diphenylhydrazine	122-66-7	NA				Inadequate results			NA	
38	ethylbenzene	100-41-4	1.00E-01	5,460	rat	1.00E+00			NA		D
39	fluoranthene	206-44-0	4.00E-02			NA			None		D

869

Appendix C (*Continued*)

	Chemical Name	CASRN	Oral Noncarcinogenic Toxicity RfD mg/kg·day	LD50 mg/kg	LD50 Species	RfC mg/m³	LC50 at T = 298K, P = 1atm ppmv (h)	LC50 Species	Carcinogenicity Oral CPF kg/day·mg	Inhalation Unit Risk m³/μg	Carc. Class.
	Haloethers										
40	4-chlorophenyl phenyl ether	7005-72-3	NA								
41	4-bromophenyl phenyl ether	101-55-3	NA			NA			None		D
42	bis(2-chloroisopropyl) ether	108-60-1	4.00E-02	220–270	rat	NA	16,000 (7)	mouse	NA		
43	bis(2-chloroethoxy) methane	111-91-1	NA			NA			None		D
	Halomethanes										
44	dichloromethane	75-09-2	6.00E-02	1,600	rat	NA	16,000 (7)	mouse	7.50E-03	4.00E-07	B2
45	chloromethane	74-87-3	NA	1,800	rat		2,228 (6)	male mouse			
							8,476 (6)	female mouse			
46	bromomethane	74-83-9	1.40E-03			5.00E-03	1,200 (1)	mouse	NA		D
							780 (4)	rat			
47	bromoform	75-25-2	2.00E-02	1,388/ 1,147	male/ female rat		Inadequate results				
			1,500			female mouse					
48	dichlorobromomethane	75-27-4	2.00E-02	916/969	male/ female rat	NA			6.20E-02	NA	B2
49	trichlorofluoromethane	75-69-4	3.00E-01			NA	250,000 (0.5)	guinea pig, rabbit		NA	
							100,000 (0.5)	rat			

870

No.	Compound	CAS No.									
50	dicholorodifluoromethane	75-71-8	2.00E-01			NA		mouse	NA	NA	
51	dibromochloromethane	124-48-1	2.00E-02	1,186/848; 800/1,200	male/female rat; male/female rat	NA	760,000 (0.5); 800,000 (0.5)	mouse	8.00E-04	NA	C
52	hexachlorobutadiene	87-68-3	Withdrawn on 5/1/93 for further review	90; 87–116	guinea pig mouse; rat	NA			7.80E-02	2.20E-05	C
53	hexachlorocyclopentadiene	77-47-4	7.00E-03	300–630	rat	NA			None		D
54	isophorone	78-59-1	2.00E-01	1,000–3,450	rat	Inadequate results					C
55	naphthalene	91-20-3	2.00E-02	2,600	rat	3.00E-03			NA		C
56	nitrobenzene	98-95-3	5.00E-04	640	rat	NA			None		D
	Nitrophenols										
57	2-nitrophenol	88-75-5	NA				Inadequate results				
58	*p*-nitrophenol	100-02-7	NA	620	rat		Inadequate results				
59	2,4-dinitrophenol	51-28-5	2.00E-03								
60	4,6-dinitro-*o*-cresol	534-52-1	NA	26–65; 21	rat; mouse						
	Nitrosamines										
61	*N*-nitrosodimethylamine	62-75-9	NA			NA	78 (4); 57 (4)	rat; mouse	5.00E+01	1.40E-02	B2
62	*N*-nitrosodiphenylamine	86-30-6	NA			NA			4.90E-03	NA	B2
63	*N*-nitrosodi-*n*-propylamine	621-64-7	NA			NA			7.00E+00	NA	B2
64	pentachlorophenol	87-86-5	3.00E-02	146/175	male/female rat	NA			1.20E-01	NA	B2

Appendix C *(Continued)*

| | | | Oral Noncarcinogenic Toxicity | | | | | | Carcinogenicity | | |
| | | | RfD | LD$_{50}^{d}$ | | RfC | LC$_{50}^{d}$ at $T = 298K$, $P = 1$atm | | Oral CPF | Inhalation Unit Risk | Carc. Class. |
	Chemical Name	CASRN	mg/kg·day	mg/kg	Species	mg/m^3	ppmv (h)	Species	kg/day·mg	m^3/µg	
	Phthalate esters										
65	di(2-ethylhexyl) phthalate	117-81-7	2.00E-02			NA			1.40E-02	NA	B2
66	butyl benzyl phthalate	85-68-7	2.00E-01			NA			NA		C
67	dibutyl phthalate	84-74-2	1.00E-01	9,000	mouse		Inadequate results				
68	n-octyl phthalate	117-84-0	NA	30,000	rat						
				13,000	mouse						
69	diethyl phthalate	84-66-2	8.00E-01	9,000	rat	NA	538 (1)	mouse	NA		D
							826 (1)	rat			
70	dimethyl phthalate	131-11-3	NA	7,200	mouse		Inadequate results				
				2,400	rat, guinea pig						
	Polynuclear aromatic hydrocarbons										
71	benz[a]anthracene	56-55-3	NA			NA			NA		B2
72	benzo[a]pyrene	50-32-8	NA			NA			7.30E+00	NA	B2
73	3,4-benzofluoranthene	205-99-2	NA			NA			NA		B2
74	benzo[k]fluoranthane	207-08-9	NA			NA			NA		B2
75	chrysene	218-01-9	NA			NA			NA		B2
76	acenaphthylene	208-96-8	NA			NA			None		D
77	anthracene	120-12-7	3.00E-01			NA			None		D
78	benzo[g,h,i]perylene	191-24-2	NA			NA			None		D
79	fluorene	86-73-7	4.00E-02			NA			None		D
80	phenanthrene	85-01-8	NA			NA			None		D
81	dibenzo[a,h]anthracene	53-70-3	NA			NA			NA		B2
82	indeno[1,2,3-cd]pyrene	193-39-5	NA			NA			NA		B2
83	pyrene	129-00-0	3.00E-02			NA			None		D
84	tetrachloroethylene	127-18-4	1.00E-02			NA			NA		
85	toluene	108-88-3	2.00E-01	2,600–7500	rat	4.00E-01			NA		D

No.	Chemical	CAS No.		5,680	dog	NA		Withdrawn for further review		
86	trichloroethylene	79-01-6	NA	5,680	dog	NA				
87	vinyl chloride	75-01-4	NA				2,6000 (1) rat 12,000 (4) rat 8,450 (4) mouse 49,000 (0.5) mouse 5,500 (10) mouse			
	Pesticides and metabolites									
88	aldrin	309-00-2	3.00E-05			NA		1.70E+01	4.90E-03	B2
89	dieldrin	60-57-1	5.00E-05			NA		1.60E+01	4.60E-03	B2
90	chlordane (technical mixture and metabolites)	12789-03-6	5.00E-04			7.00E-04		3.50E-01	1.00E-04	B2
91	4,4'-DDT	50-29-3	5.00E-04			NA		3.40E-01	9.70E-05	B2
92	4,4'-DDE	72-55-9	NA			NA		3.40E-01	NA	B2
93	4,4'-DDD	72-54-8	NA			NA		2.40E-01	NA	B2
	Endosulfan and metabolites									
94	endosulfan	115-29-7								
95	endosulfan sulfate	1031-07-8	NA							
	Endrin and metabolites									
96	endrin	72-20-8	3.00E-04			NA		NA		D
97	endrin aldehyde	7421-93-4	NA							
	Heptachlor and metabolites									
98	heptachlor	76-44-8	5.00E-04	40–188	rat	NA		4.50E+00	1.30E-03	B2
99	heptachlor epoxide	1024-57-3	1.30E-05			NA		9.10E+00	2.60E-03	B2
	Hexachlorocyclohexane									
100	α-hexachlorocyclohexane	319-84-6	NA			NA		6.30E+00	1.80E-03	B2
101	β-hexachlorocyclohexane	319-85-7	NA			NA		1.80E+00	5.30E-04	C
102	γ-hexachlorocyclohexane	58-89-9	3.00E-04			NA		NA	NA	
103	δ-hexachlorocyclohexane	319-86-8	NA			NA		NA	NA	D
	Polychlorinated biphenyls (PCBs)									
104	PCB-1242 (Arochlor 1242)	53469-21-9	NA	794–1,269	rat	NA				

Appendix C *(Continued)*

| | | | Oral Noncarcinogenic Toxicity | | | | | Carcinogenicity | | |
| | | RfD | LD$_{50}^{d}$ | | RfC | LC$_{50}^{d}$ at $T = 298K$, $P = 1$atm | | Oral CPF | Inhalation Unit Risk | Carc. Class. |
Chemical Name	CASRN	mg/kg·day	mg/kg	Species	mg/m^3	ppmv (h)	Species	kg/day·mg	m^3/µg	
105 PCB-1254 (Arochlor 1254)	11097-69-1	2.00E-05	1,300	30-day-old rat	NA			NA		
			1,400	60-day-old rat						
			2,000	120-day-old rat						
106 PCB-1221 (Arochlor 1221)	11104-28-2	NA								
107 PCB-1232 (Arochlor 1232)	11141-16-5	NA								
108 PCB-1248 (Arochlor 1248)	12672-29-6	Inadequate results						NA		
109 PCB-1260 (Arochlor 1260)	11096-82-5	NA	4,000–10,000	adult rat						
			1,300	weanling rat						
110 PCB-1016 (Arochlor 1016)	12674-11-2	7.00E-05			NA			NA		
111 toxaphene	8001-35-2	NA	80-90	rat	NA			1.10E+00	3.40E-04	B2
Heavy metals and minerals										
112 antimony (total)	7440-36-0	4.00E-04			NA			NA		
113 arsenic (total)	7440-38-2	3.00E-04			NA			1.50E+00	4.30E-03	A
114 asbestos (fibrous)	1332-21-4	NA			NA			NA	0.23 (fibers per mL)	A
115 beryllium (total)	7440-41-7	2.00E-03			2.00E-02			Inadequate results		B1
cadmium (total)	7440-43-9	5E-04 (water) 1E-03 (food)			NA			NA	1.80E-03	B1
116										

No.		CAS number				Inadequate results				D/A
117	chromium III, insoluble salts	16065-83-1	1.50E+00							
118	chromium VI	18540-29-9	3.00E-03			8E-06-1E-04			Cannot be determined	D
119	copper (total)	7440-50-8	NA			NA			NA	D
120	cyanide (free)	57-12-5	2.00E-02			NA			NA	D
121	lead and compounds (inorganic)	7439-92-1	NA			NA			NA	B2
122	mercury (total)	7439-97-6	NA			3.00E-04			None	D
123	nickel (soluble salts)	7440-02-0	NA			NA			NA	
124	selenium and compounds	7782-49-2	5.00E-03			NA			NA	D
125	silver (total)	7440-22-4	5.00E-03			NA			NA	D
126	thallium (total)	7440-28-0	NA							
127	zinc and compounds	7440-66-6	3.00E-01			NA			None	D
128	2,3,7,8-tetrachloro dibenzo-p-dioxin	1746-01-6	NA							
129	methyl tert-butyl ether	1634-04-4	NA	2962	rat	3	23 (4)	rat	NA	

NA, not available.

References:

[a] U.S. EPA, Integrated Risk Information System, www.epa.gov/iris/subst/index.html.

[b] Prager, Jan C., 1995. Environmental Contaminant Reference Data Book, Vols. 1 and 2, Van Nostrand Reinhold, A Division of International Thomas Publishing, New York.

Chemical Name	CASRN[a]	Empirical Formula[b]	Maximum Contaminant Level (mg/L)	OSHA 8-h Standard (mg/m^3)
2 acrolein	107-02-8	C_3H_4O		0.25
4 benzene	71-43-2	C_6H_6	0.005	
6 carbon tetrachloride	56-23-5	CCl_4	0.005	
Chlorinated benzenes				
7 chlorobenzene	108-90-7	C_6H_5Cl	0.1	350
8 1,2,4-trichlorobenzene	120-82-1	$C_6H_3Cl_3$	0.07	
9 hexachlorobenzene	118-74-1	C_6Cl_6	0.001	
Chlorinated ethanes				
10 1,2-dichloroethane	107-06-2	$C_2H_4Cl_2$	0.005	
11 1,1,1-trichloroethane	71-55-6	$C_2H_3Cl_3$	0.2	
12 hexachloroethane	67-72-1	C_2Cl_6		10
13 1,1-dichloroethane	75-34-3	$C_2H_4Cl_2$		40
14 1,1,2-trichloroethane	79-00-5	$C_2H_3Cl_3$	0.2	45
15 1,1,2,2-tetrachloroethane	79-34-5	$C_2H_2Cl_4$		35
23 chloroform	67-66-3	$CHCl_3$		240
Dichlorobenzens				
25 1,2-dichlorobenzene	95-50-1	$C_6H_4Cl_2$	0.6	300
27 1,4-dichlorobenzene	106-46-7	$C_6H_4Cl_2$	0.075	450
Dichlorobenzidine				
28 3,3-dichlorobenzidine	91-94-1	$C_{12}H_{10}Cl_2N_2$		
Dichloroethylenes				
29 1,1-dichloroethylene	75-35-4	$C_2H_2Cl_2$	0.007	
30 1,2-trans-dichloroethylene	156-60-5	$C_2H_2Cl_2$	0.1	
Dichloropropylene				
32 1,2-dichloropropane	78-87-5	$C_3H_6Cl_2$	0.005	350
38 ethylbenzene	100-41-4	C_8H_{10}		435
Halomethanes				
44 dichloromethane	75-09-2	CH_2Cl_2	0.005	
47 bromoform	75-25-2	$CHBr_3$		5
50 dicholorodifluoromethane	75-71-8	CCl_2F_2		495
54 isophorone	78-59-1	$C_9H_{14}O$		140
55 naphthalene	91-20-3	$C_{10}H_8$		50
56 nitrobenzene	98-95-3	$C_6H_5NO_2$		5
Nitrophenols				
60 4,6-dinitro-o-cresol	534-52-1	$C_7H_6N_2O_5$		0.2
Nitrosamines				
64 pentachlorophenol	87-86-5	C_6HCl_5O		0.5
Phthalate esters				
65 di(2-ethylhexyl) phthalate	117-81-7	$C_{24}H_{38}O_4$	0.006	5
67 dibutyl phthalate	84-74-2	$C_{16}H_{22}O_4$		5
70 dimethyl phthalate	131-11-3	$C_{10}H_{10}O_4$		5
84 tetrachloroethylene	127-18-4	C_2Cl_4	0.005	
85 toluene	108-88-3	C_7H_8	1	
Pesticides and metabolites				
88 aldrin	309-00-2	$C_{12}H_8Cl_6$		0.25
89 dieldrin	60-57-1	$C_{12}H_8Cl_6O$		0.25
90 chlordane (technical mixture and metabolites)	12789-03-6	$C_{10}H_6Cl_8$		0.5

Chemical Name	CASRN[a]	Empirical Formula[b]	Maximum Contaminant Level (mg/L)	OSHA 8-h Standard (mg/m^3)
Endrin and metabolites				
96 endrin	72-20-8	$C_{12}H_8Cl_6O$		0.1
Heptachlor and metabolites				
98 heptachlor	76-44-8	$C_{10}H_5Cl_7$	0.0004	0.5
99 heptachlor epoxide	1024-57-3	$C_{10}H_5Cl_7O$	0.0002	
Polychlorinated biphenyls (PCBs)				
104 PCB-1242 (Arochlor 1242)	53469-21-9	Mixture	0.0005	1
105 PCB-1254 (Arochlor 1254)	11097-69-1	Mixture	0.0005	0.5
106 PCB-1221 (Arochlor 1221)	11104-28-2	Mixture	0.0005	
107 PCB-1232 (Arochlor 1232)	11141-16-5	Mixture	0.0005	
108 PCB-1248 (Arochlor 1248)	12672-29-6	Mixture	0.0005	
109 PCB-1260 (Arochlor 1260)	11096-82-5	Mixture	0.0005	
110 PCB-1016 (Arochlor 1016)	12674-11-2	Mixture	0.0005	
111 toxaphene	8001-35-2	$C_{10}H_{10}Cl_8$	0.003	0.5
112 antimony (total)	7440-36-0	Sb	0.006	0.5
113 arsenic (total)	7440-38-2	As	0.010	0.5
114 asbestos (fibrous)	1332-21-4	Mixture	7	
115 beryllium (total)	7440-41-7	Be	0.004	
116 cadmium (total)	7440-43-9	Cd	0.005	
119 copper (total)	7440-50-8	Cu	1.3	
120 cyanide (free)	57-12-5	CN^-	0.20	
121 lead and compounds (inorganic)	7439-92-1	Pb	0.015	
122 mercury (total)	7439-97-6	Hg	0.0020	
123 nickel (total)	7440-02-0	Ni		1
124 selenium and compounds	7782-49-2	Se	0.050	0.2
125 silver (total)	7440-22-4	Ag	0.10	0.01
126 thallium (total)	7440-28-0	Tl	0.002	0.1
127 zinc and compounds	7440-66-6	Zn	5	

APPENDIX E   Ambient Air Quality Standards

Pollutant	Primary Standards	Averaging Times	Secondary Standards
Carbon monoxide	9 ppm (10 mg/m^3)	8-h[1]	None
	35 ppm (40 mg/m^3)	1-h[1]	None
Lead	1.5 μg/m^3	Quarterly average	Same as primary
Nitrogen dioxide	0.053 ppm (100 μg/m^3)	Annual (arith. mean)	Same as primary
Particulate matter	50 μg/m^3	Annual[2] (arith. mean)	Same as primary
PM$_{10}$	150 μg/m^3	24-h[1]	
PM$_{2.5}$	15.0 μg/m^3	Annual[3] (arith. mean)	Same as primary
	65 μg/m^3	24-h[4]	
Ozone	0.08 ppm	8-h[5]	Same as primary
	0.12 ppm	1-h[6]	Same as primary
Sulfur oxides	0.03 ppm	Annual (arith. mean)	
	0.14 ppm	24-h[1]	
		3-h[1]	0.5 ppm (1300 μg/m^3)

[1] Not to be exceeded more than once per year.

[2] To attain this standard, the expected annual arithmetic mean PM$_{10}$ concentration at each monitor within an area must not exceed 50 μg/m^3.

[3] To attain this standard, the 3-yr average of the annual arithmetic mean PM$_{2.5}$ concentrations from single or multiple community-oriented monitors must not exceed 15.0 μg/m^3.

[4] To attain this standard, the 3-yr average of the 98th percentile of 24-h concentrations at each population-oriented monitor within an area must not exceed 65 μg/m^3.

[5] To attain this standard, the 3-yr average of the fourth-highest daily maximum 8-h average ozone concentrations measured at each monitor within an area over each year must not exceed 0.08 ppm.

[6] (a) The standard is attained when the expected number of days per calendar year with maximum hourly average concentrations above 0.12 ppm is ≤ 1.

(b) The 1-h NAAQS will no longer apply to an area one year after the effective date of the designation of that area for the 8-h ozone NAAQS. The effective designation date for most areas is June 15, 2004 [40 CFR 50.9; see *Federal Register*, of April 30, 2004 (69 FR 23996)].

Physical constants

Avogadro's no., $N = 6.022 \times 10^{23}$ mol^{-1}

Gas law constant:

$$R = 0.0820562 \, \text{L} \cdot \text{atm}/\text{mol} \cdot \text{K}$$
$$= 8.3144 \, \text{J}/\text{mol} \cdot \text{K}$$
$$= 1.987 \, \text{cal}/\text{mol} \cdot \text{K}$$

Planck's constant, $h = 6.6255 \times 10^{-34}$ J $\cdot$ s

Metric prefixes and abbreviations

tera $= 10^{12}$	T
giga $= 10^{9}$	G
mega $= 10^{6}$	M
kilo $= 10^{3}$	k
hecto $= 10^{2}$	h
deka $= 10$	da
deci $= 10^{-1}$	d
centi $= 10^{-2}$	c
milli $= 10^{-3}$	m
micro $= 10^{-6}$	μ
nano $= 10^{-9}$	n
pico $= 10^{-12}$	p
femto $= 10^{-15}$	f
atto $= 10^{-18}$	a

Unit abbreviations and definitions (conversions given are for exact conversions):

atm	atmosphere
Btu	British thermal unit
°C	Celsius
cal	calorie
Cal	Calorie (kcal)
eV	electron-volt
°F	Fahrenheit
ft	foot (0.3048 m)
ft^3	cubic foot
g	gram
gal	gallon
hp	horsepower (550 ft-lbf/s)
h	hour
in.	inch
J	joule (kg $\cdot$ m^2/s^2, N $\cdot$ m, or W $\cdot$ s)
K	kelvin
L	liter (0.001 m^3)
lbf	pound force
lbm	pound mass
mi	mile (5280 ft)

N	newton $(kg \cdot m/s^2)$
Pa	pascal (kN/m^2)
psi	pound per square inch
s	second
W	watt (J/s)

Deriving and using conversion factors

The units on one side of the following table are equal to those opposite them in the same row. The equation formed by equating them is called an **identity**. The ratio of one side of an identity to the other is equal to **unity** (i.e., the number 1 with no units).

You can multiply or divide any quantity by unity without actually changing the value, only its units. Thus, unity written as a ratio of two sides of an identity forms a **conversion factor**. For example: From the identity 0.3048 m = 1 ft, we obtain the conversion factor $0.3048\,m/1\,ft = 1$, which is unity.

or since 800 gal/day = 4.45 ft/h,

$$\frac{800\,gal/day}{4.45\,ft/h} = 179.8 \frac{gal/day}{ft/h} = 1$$

For example, to compute the number of meters in a mile:

$$1\,mi \times \frac{5280\,ft}{mi} \times \frac{0.3048\,m}{ft} = 1609.334\,m$$

(Note that because in this case both conversion factors are exact values, it is not necessary to round off.) As this example shows, we can use a series of conversion factors. For another example, to convert 1 lbf into newtons, we could use the following identities:

$$1\,lbf = 32.174\,ft\text{-}lbm/s^2$$
$$1\,ft = 0.3048\,m$$
$$1\,lbm = 0.45359\,kg$$
$$1\,N = 1\,kg \cdot m/s^2$$

Then, combining these in unity form as conversion factors yields

$$1\,lbf \times \frac{32.174\,ft\text{-}lbm}{lbf\text{-}s^2} \times \frac{0.3048\,m}{ft} \times \frac{0.45359\,kg}{lbm} \times \frac{1\,N \cdot s^2}{kg \cdot m} = 4.448\,N$$

Unit conversion identities and data

SI units are in bold. "STP" indicates standard temperature and pressure: 25°C and 1 atm. Note that some physical properties are given as identities (e.g., the density of water or of dry air). These can still be used to derive conversion factors for other physical quantities.

LENGTH (L)

0.3048	meter	=	1	foot
2.54	centimeters	=	1	inch
5280	feet	=	1	mile

AREA (L^2)

10.76	square feet	=	1	square meter
1	acre	=	43,560	square feet

1	million acres	=	39.53	miles square
1	hectare	=	100	meters square
1	hectare	=	10,000	square meters
1	hectare	=	2.471	acres

VOLUME (L³)

1000	liters	=	1	cubic meter
3.78543	liters	=	1	gallon
28.32	liters	=	1	cubic foot
1	cubic meter	=	35.31	cubic feet
1		=	264.2	gallons
7.48052	gallons	=	1	cubic foot
1	million gallons	=	3785.4	cubic meters
1	fluid ounce	=	29.6	millimeters
1	mole ideal gas	=	22.41	liters (STP)
1	lb-mol ideal gas	=	359	cubic feet (STP)

FLOW (L³·T⁻¹)

1	million gallons per day	=	1.547	cubic feet/ second
1	million gallons per day	=	43.8	liters per second

MASS (M)

453.59	grams	=	1	pound mass
32.2	pounds	=	1	slug
1	ounce	=	28.35	grams
1	pound-mole	=	453.59	gram-moles

VELOCITY OR SURFACE LOADING (L · T⁻¹)

0.508	cm/ s	=	1	foot per minute
60	mi/h	=	88.00	feet per second
800	gal/day-ft²	=	4.45	feet per hour
800		=	32.60	meters per day

ACCELERATION (L · T⁻²)

Standard acceleration of gravity:

32.2	ft/s-s	=	9.815	m/s·s

DENSITY OR CONCENTRATION (M · L⁻³)

Density of water (4°C)

1	g/ cm³	=	8.3455	lb / gal
1	g/ cm³	=	62.43	lb / ft³

Density of dry air at 20°C and 1 atm

1.290	g/L	=	80.56	lb per 1000 ft³

Parts per million in water (20°C)

1	mg/L	=	8.34	lb per million gallons
1	mg/L	=	62.38	lb per million cubic feet

Density of mercury (0°C)

		=	13.6	g/cm³

FORCE (M · T · L⁻²)

1	newton	=	1	kg·m/s·s
1	pound force	=	32.174	ft-lbm/s-s
1	pound force	=	1	ft-slug/s-s

PRESSURE (M · T · L⁻²)

1	atmosphere	=	76	cm Hg (0°C)
1	atmosphere	=	29.921	in. Hg
1	atmosphere	=	1.01325	bar
1	atmosphere	=	14.696	psi
1	atmosphere	=	101,326	Pa (N/m²)
1	bar	=	100,000	Pa

Hydraulic Head (L)

1	psi	=	2.31	ft water head
1	atmosphere	=	33.9	ft water head

ENERGY (M · T^{-2} · L)

1	Btu	=	252	calories
1	Btu	=	1054.4	joules (N·m or W·s)
1	Btu	=	778.15	ft-lbf
1	calorie	=	4.184	joules
1	Calorie (dietary)	=	1	kilocalorie
1	kWh	=	3412	Btu
1	eV	=	1.60E-19	joules
1	MeV	=	1.60E-13	joules

Intrinsic Energy

1	kcal/g	=	1801.53	Btu/lbm

Latent Heat of Water

40.6	kJ/mol	=	9704	cal/mol
970.34	Btu/lb	=	539	cal/g

POWER (M · T^{-3} · L)

1	Watt (J/s)	=	3.414	Btu/h
1	horsepower	=	745.7	watts
1	horsepower	=	0.7068	Btu/s
1	horsepower	=	550	ft-lbf/s

HEAT CAPACITY

Water: =

1	cal/g·K	=	1	Btu/lbm- F

Radiological conversion identities. See Table 21.14.

APPENDIX G The Elements
The elements that are essential to at least some living things are shaded.

Name	Sym	AN	Name	Sym	AN	Name	Sym	AN	Name	Sym	AN	Name	Sym	AN
Actinium	Ac	89	Copper	Cu	29	Iron	Fe	26	Phosphorus	P	15	Strontium	Sr	38
Aluminium	Al	13	Curium	Cm	96	Krypton	Kr	36	Platinum	Pt	78	Sulfur	S	16
Americium	Am	95	Darmstadtium	Ds	110	Lanthanum	La	57	Plutonium	Pu	94	Tantalum	Ta	73
Antimony	Sb	51	Dubnium	Db	105	Lawrencium	Lr	103	Polonium	Po	84	Technetium	Tc	43
Argon	Ar	18	Dysprosium	Dy	66	Lead	Pb	82	Potassium	K	19	Tellurium	Te	52
Arsenic	As	33	Einsteinium	Es	99	Lithium	Li	3	Praseodymium	Pr	59	Terbium	Tb	65
Astatine	At	85	Erbium	Er	68	Lutetium	Lu	71	Promethium	Pm	61	Thallium	Tl	81
Barium	Ba	56	Europium	Eu	63	Magnesium	Mg	12	Protactinium	Pa	91	Thorium	Th	90
Berkelium	Bk	97	Fermium	Fm	100	Manganese	Mn	25	Radium	Ra	88	Thulium	Tm	69
Beryllium	Be	4	Fluorine	F	9	Meitnerium	Mt	109	Radon	Rn	86	Tin	Sn	50
Bismuth	Bi	83	Francium	Fr	87	Mendelevium	Md	101	Rhenium	Re	75	Titanium	Ti	22
Bohrium	Bh	107	Gadolinium	Gd	64	Mercury	Hg	80	Rhodium	Rh	45	Tungsten	W	74
Boron	B	5	Gallium	Ga	31	Molybdenum	Mo	42	Roentgenium	Rg	111	Ununbium	Uub	112
Bromine	Br	35	Germanium	Ge	32	Neodymium	Nd	60	Rubidium	Rb	37	Ununhexium	Uuh	116
Cadmium	Cd	48	Gold	Au	79	Neon	Ne	10	Ruthenium	Ru	44	Ununoctium	Uuo	118
Cesium	Cs	55	Hafnium	Hf	72	Neptunium	Np	93	Rutherfordium	Rf	104	Ununquadium	Uuq	114
Calcium	Ca	20	Hassium	Hs	108	Nickel	Ni	28	Samarium	Sm	62	Uranium	U	92
Californium	Cf	98	Helium	He	2	Niobium	Nb	41	Scandium	Sc	21	Vanadium	V	23
Carbon	C	6	Holmium	Ho	67	Nitrogen	N	7	Seaborgium	Sg	106	Xenon	Xe	54
Cerium	Ce	58	Hydrogen	H	1	Nobelium	No	102	Selenium	Se	34	Ytterbium	Yb	70
Chlorine	Cl	17	Indium	In	49	Osmium	Os	76	Silicon	Si	14	Yttrium	Y	39
Chromium	Cr	24	Iodine	I	53	Oxygen	O	8	Silver	Ag	47	Zinc	Zn	30
Cobalt	Co	27	Iridium	Ir	77	Palladium	Pd	46	Sodium	Na	11	Zirconium	Zr	40

APPENDIX H   Periodic Table of the Elements

Key:

Symbol
Atomic Number
Atomic Mass (g/mol)

^{12}C = 12.0000 amu

[Atomic mass in brackets is for the longest-lived or best-known isotope]

The elements that are essential to at least some living things are shaded

1	2	3	4	5	6	7	8	9	10	11	12	13	14	15	16	17	18
H 1 1.008																	He 2 4.003
Li 3 6.941	Be 4 9.012											B 5 10.81	C 6 12.01	N 7 14.01	O 8 16	F 9 19	Ne 10 20.18
Na 11 22.99	Mg 12 24.31											Al 13 26.98	Si 14 28.09	P 15 30.97	S 16 32.07	Cl 17 35.45	Ar 18 39.95
K 19 39.10	Ca 20 40.08	Sc 21 44.96	Ti 22 47.87	V 23 50.94	Cr 24 52.00	Mn 25 54.94	Fe 26 55.85	Co 27 58.93	Ni 28 58.69	Cu 29 63.55	Zn 30 65.41	Ga 31 69.72	Ge 32 72.64	As 33 74.92	Se 34 78.96	Br 35 79.90	Kr 36 83.80
Rb 37 85.47	Sr 38 87.62	Y 39 88.91	Zr 40 91.22	Nb 41 92.91	Mo 42 95.94	Tc 43 [98]	Ru 44 101.1	Rh 45 102.9	Pd 46 106.4	Ag 47 107.9	Cd 48 112.4	In 49 114.8	Sn 50 118.7	Sb 51 121.8	Te 52 127.6	I 53 126.9	Xe 54 131.3
Cs 55 132.9	Ba 56 137.3	*	Hf 72 178.5	Ta 73 180.9	W 74 183.8	Re 75 186.2	Os 76 190.2	Ir 77 192.2	Pt 78 195.1	Au 79 197	Hg 80 200.6	Tl 81 204.4	Pb 82 207.2	Bi 83 209	Po 84 [209]	At 85 [210]	Rn 86 [222]
Fr 87 [223]	Ra 88 [226]	#	Rf 104 [261]	Db 105 [262]	Sg 106 [266]	Bh 107 [264]	Hs 108 [277]	Mt 109 [268]	Ds 110 [271]	Uuu 111 (272)	Uub 112 (277)		Uuq 114 (296)		Uuh 116 (298)		Uuo 118 (?)

* Lanthanides	La 57 138.9	Ce 58 140.1	Pr 59 140.9	Nd 60 144.2	Pm 61 [145]	Sm 62 150.4	Eu 63 152.0	Gd 64 157.3	Tb 65 158.9	Dy 66 162.5	Ho 67 164.9	Er 68 167.3	Tm 69 168.9	Yb 70 173.0	Lu 71 175.0
# Actinides	Ac 89 [227]	Th 90 232.0	Pa 91 231.0	U 92 238.0	Np 93 [237]	Pu 94 [244]	Am 95 [243]	Cm 96 [247]	Bk 97 [247]	Cf 98 [251]	Es 99 [252]	Fm 100 [257]	Md 101 [258]	No 102 [259]	Lr 103 [262]

INDEX

Pages on which definitions or main discussions are found appear in **boldface**. Pages referring to tables or figures are indicated by "tab" or "fig," respectively, after the page number.

Environmental Biology for Engineers and Scientists, by David A. Vaccari, Peter F. Strom, and James E. Alleman
Copyright © 2006 John Wiley & Sons, Inc.